Robert B. Heimann

Classic and Advanced Ceramics

Related Titles

Aldinger, F., Weberruss, V. A.

Advanced Ceramics and Future Materials

An Introduction to Structures, Properties and Technologies

2010
Hardcover
ISBN: 978-3-527-32157-5

Riedel, R., Chen, I-W. (eds.)

Ceramics Science and Technology

4 Volume Set

Hardcover
ISBN: 978-3-527-31149-1

Volume 1: Structures

2008
Hardcover
ISBN: 978-3-527-31155-2

Volume 2: Properties

2010
Hardcover
ISBN: 978-3-527-31156-9

Volume 3: Synthesis and Processing

2010
Hardcover
ISBN: 978-3-527-31157-6

Volume 4: Applications

2012
Hardcover
ISBN: 978-3-527-31158-3

Öchsner, A., Ahmed, W. (eds.)

Biomechanics of Hard Tissues

Modeling, Testing, and Materials

2010
Hardcover
ISBN: 978-3-527-32431-6

Krenkel, W. (ed.)

Ceramic Matrix Composites

Fiber Reinforced Ceramics and their Applications

2008
Hardcover
ISBN: 978-3-527-31361-7

Öchsner, A., Murch, G. E., de Lemos, M. J. S. (eds.)

Cellular and Porous Materials

Thermal Properties Simulation and Prediction

2008
Hardcover
ISBN: 978-3-527-31938-1

Robert B. Heimann

Classic and Advanced Ceramics

From Fundamentals to Applications

WILEY-
VCH

WILEY-VCH Verlag GmbH & Co. KGaA

The Author

Prof. Dr. Robert B. Heimann
Am Stadtpark 2A
02826 Görlitz

Library of Congress Card No.: applied for

British Library Cataloguing-in-Publication Data
A catalogue record for this book is available from the British Library.

Bibliographic information published by the Deutsche Nationalbibliothek
The Deutsche Nationalbibliothek lists this publication in the Deutsche Nationalbibliografie; detailed bibliographic data are available on the Internet at http://dnb.d-nb.de.

Cover Design Formgeber, Eppelheim
Typesetting Toppan Best-set Premedia Limited
Printing and Binding Betz-Druck GmbH, Darmstadt

Printed in the Federal Republic of Germany
Printed on acid-free paper

ISBN: 978-3-527-32517-7

To Gabriele whose love, support, and patience were indispensable for creating this text.

Classic and Advanced Ceramics: From Fundamentals to Applications. Robert B. Heimann

ISBN: 978-3-527-32517-7

Contents

Classic and Advanced Ceramics: From Fundamentals to Applications. Robert B. Heimann

ISBN: 978-3-527-32517-7

Preface

Modern materials science and engineering technology rely on the three principal classes of material, distinguished by their nature of chemical bonding: metals; ceramics and polymers; and the alloys and composites of these materials. The aim of this treatise is to educate not only graduate and doctoral students but also professionals in mineralogy, chemistry, materials science and related disciplines on the subject of ceramics, both traditional and advanced. Hopefully, it will also serve as a primer for more involved studies in ceramic engineering *proper*, and thus lay the foundation for a more detailed knowledge acquisition.

Ceramics, by definition, are inorganic, nonmetallic and predominantly polycrystalline materials that may be shaped at room temperature from a variety of raw materials. They obtain their typical properties by sintering at high temperatures. Unlike the German custom of distinguishing between inorganic (poly)crystalline (ceramics *sensu strictu*) and noncrystalline (glasses) materials, the English usage includes glasses in the generic term "ceramics." However, in this treatise the author will follow the German tradition, and consequently glasses and other amorphous materials will be excluded from the discussions. Nonetheless, silicate-based chemically bonded ceramics (CBCs) such as cements/concrete will be included as, with time, they undergo crystallization processes.

Ceramics are the oldest man-made materials, dating back to the dawn of human civilization. They possess an overwhelmingly wide variability in terms of their origin, history, utilization, and mechanical, thermal, optical, biological and electronic properties. Traditional ceramics are based almost exclusively on naturally occurring raw materials, most commonly silicaceous minerals such as clays, micas, quartz and feldspars, although for special applications synthetically produced clay minerals may also be utilized. A smattering of other nonsilicate minerals may also be included, such as gibbsite, magnesite, calcite, and dolomite. In contrast to this, advanced ceramics are produced predominantly from chemically synthesized micro- or nanoscaled pure alumina, titania, zirconia, magnesia and other oxides and their compounds, as well as from the carbides and nitrides of silicon, boron and aluminum, and a host of transitional elements. The processing technologies used include the high-temperature transformation of raw materials into desired ceramic bodies, with highly controlled mechanical, thermal, electrical, tribological and optical properties, in addition to the low-temperature hydrolysis

Classic and Advanced Ceramics: From Fundamentals to Applications. Robert B. Heimann

ISBN: 978-3-527-32517-7

of calcium silicates and aluminates to synthesize CBCs such as concrete. Hence, the application of ceramics spans the chasm between traditional silicate-based structural materials such as bricks, earthenware, stoneware, porcelain and concrete, and "high-tech" functionally advanced ceramics such as thermal barrier coatings for aerospace gas turbine blades, electrolyte layers for high-temperature solid oxide fuel cells, ferroic ceramics for sensor and actuator applications, diamond single crystals for future carbon-based integrated circuits, and bioconductive monolithic parts and coatings for bone reconstruction and dental and endoprosthetic hip implants.

The aim of this book is to cover pertinent aspects of the processing, structure, technology and properties of classic and advanced ceramic materials, but without claiming to exhaust the topic even remotely in an encyclopedic fashion. Instead, typical examples will be described that stand *pars pro toto* for the totality of ceramic materials in existence today. Consequently, special emphasis is placed on the mineralogy of the materials described, the basic crystallographic aspects of the thermal transformation processes during the firing of natural ceramic raw materials to arrive at traditional structural ceramics, as well as on the general physical principles of functionally advanced ceramics such as zirconia or silicon nitride, the technically important class of ferroic and superconducting ceramic materials on which many modern "high-tech" applications such as sensors and actuators are based, and last – but not least – bioceramics to replace diseased bone and restore lost functions of the human body.

The plethora of excellent books on ceramics produced during the past thirty years have been devoted predominantly to either the fundamentals, to the process technology, or to the engineering applications of their subject matter, while paying considerably less attention to other subjects. Likewise, books on advanced ceramics are generally replete with highly complex solid-state physics that do not always match the level of interest, let alone the comprehension of their intended audience, from areas of mineralogy, chemistry, and materials engineering. Instead, the present book attempts to take a "middle road" between process engineering and solid-state physics approaches by providing a technical (applied) mineralogy approach. The intention is, therefore, to bridge the perceived abyss between the more deductively oriented realm of physics, chemistry and materials science, and the more inductively and empirically oriented realm of the geosciences. As inherent in the role of technical mineralogy, this different approach will combine – in a synergistic manner – the viewpoints and expertise of geosciences and materials science, and will therefore find its main audience among graduate and doctoral students and professionals of mineralogy that, in this context, can be defined as "the materials science of the solid earth."

The text is largely based on a series of lectures given to graduate students of geosciences at Technische Universität Bergakademie Freiberg between 1993 and 2004, to undergraduate and graduate students of physics and chemistry at Chiang Mai University, Chiang Mai, Thailand, and to graduate students of materials science at Chulalongkorn University, Bangkok, Thailand between 1998 and 2001. The subject matter of ceramics – and in particular of advanced ceramics – is a lively area of research and development endeavor, with several thousands of reports made annually among a host of scientific and trade journals. It would be futile to attempt to cover even a small fraction of this trove of information in a single book; hence, what is provided in the following pages is a mere "snapshot" of past and ongoing developments – no more, no less.

Since in the previous paragraphs the viewpoints of technical (applied) mineralogy have been invoked, a general paradigmatic positioning of this specific discipline should be appended here. Research, development and teaching in the field of technical mineralogy provide a modern, tractable bridge between the classical geosciences and modern materials science. Technical (applied) mineralogy can be defined as that discipline of "mineralogical sciences" that studies the mineralogical structure and properties, the technological fundamentals, and the characterization of raw materials, technical products and processes that include the mineralogy of residual and waste product streams, and pertinent environmental issues. Hence, it is positioned at all crossroads of the "modern materials cycle" (Figure P.1). In particular, it assists in the enhancement of traditional materials, and in the development of novel advanced materials.

The arena of activities of technical mineralogists in academia, government, and industry is extremely diverse and includes, but is not limited to:

- The beneficiation of raw materials (ore, industrial minerals, coal, salts, stone, clay).
- The design, development, synthesis, processing, testing and quality management of technical products (ceramics *per se*, glass, cement, construction materials, pigments), including single crystal growth and mass crystallization as well as their characterization with polarization microscopy and X-rays, but also increasingly modern high-resolution analytical surface techniques.
- The control, remediation, and risk analysis of historical and modern tailings of mining, and ore dressing and smelting activities, as well as the development and validation of environmentally safe materials for sound disposal concepts of domestic and industrial wastes, including radioactive matter.
- Environmental activities to foster a sustainable raw materials and energy economy, including the management of minerals that occur as secondary products of industrial processes, such as gypsum derived from flue gas desulfurization, and other residual and waste materials.
- Damage analysis and the restoration of ancient monuments, as well as the determination of provenance, age, type of material, and manufacturing technologies of historical objects of art (archaeometry).

This wide professional range attesting to the heterogeneity of the discipline creates lively interdisciplinary collaboration among neighboring fields of scientific and engineering endeavors. These fields include solid-state chemistry and physics, materials technology and engineering, process engineering, mining, geology and geophysics, geoecology, biology, medicine, environmental sciences, as well

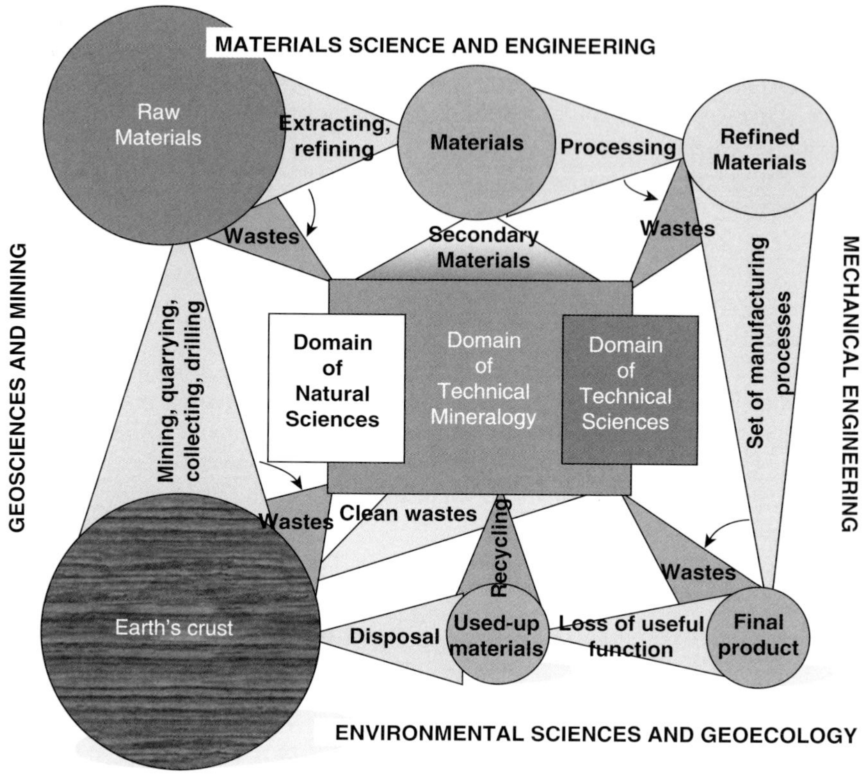

Figure P.1 The domain of technical mineralogy within the materials cycle. During all operations, from mining to the production of raw and refined materials to the manufacture of end products and to their eventual disposal and/or recycling, several waste material streams are created that challenge R&D in technical (applied) mineralogy. The sizes of the circles symbolize the different relative volumina of the mass streams, whereby the contribution of the Earth's crust is grossly underrepresented.

as archeology and social and cultural sciences. Hence, the curriculum of technical (applied) mineralogy is both versatile and involved. As opposed to chemistry or mechanical engineering, the lack of an industry that directly mirrors the scope of academic research within technical mineralogy somewhat impedes any fruitful research interaction with colleagues in industry. Yet, whilst the variability of the fields of endeavor of technical mineralogy and increasing cross-pollination among neighboring disciplines preclude a clear distinction of responsibilities, the old adage still applies: "Technical mineralogy is what technical mineralogists do."

I am highly indebted to Prof. Dr Dr.h.c. Walter Heywang (München), Prof. Horst J. Pentinghaus (Karlsruhe), Prof. Herbert Pöllmann (Halle) and Dipl.-Phys. Wolfram Wersing (Bergen, Chiemgau) for providing advice and valuable critical comments, and, in particular, to Prof. Hans Hermann Otto (Clausthal-Zellerfeld) for contributing Chapter 9 and Appendix E. The publishing house Wiley-VCH Weinheim, represented by Dr Heike Nöthe, lent important editorial and logistic support.

Robert B. Heimann
Görlitz

1
Introduction to Classic Ceramics

1.1
Ceramics through the Ages, and Technological Progress

Throughout the ages of humankind, materials have been the overwhelmingly crucial determinant of the competitiveness of individuals and societies. Today, a better understanding of the atomic and molecular structure of materials is becoming indispensable for the development of new materials, and the improvement of existing materials. As a result, materials are being tailored to meet specific applications to address pressing industrial and societal challenges in the highly competitive contemporary world. In this process, ceramics technology plays a particularly important role, and hence has emerged as a driver of technological progress in many industrial sectors.

It is a widely accepted paradigm that such technological progress takes place in a highly competitive environment where only a limited amount of the required resources exist. Hunger for raw materials has always been a strong driving force in world history. Throughout the history of humankind, the information contained within each newly developed or significantly improved material or technology has increased exponentially. Figure 1.1 suggests that the knowledge required to make pottery – that is, the mining/collecting, processing, forming, and firing of clay, including the knowledge and skill to construct and operate kilns and flues – were orders of magnitude higher than those needed to fashion rather simple tools and implements from bone or stone. The quantification of the "technology information content," plotted logarithmically on the ordinate of Figure 1.1, is – of course – highly subjective. Nevertheless, it suggests that the knowledge acquired in pottery making has later been put to use to mine, dress, and smelt ore, and to purify and alloy metals. As is evident from the figure, technological development stagnated in the Western societies during the Dark and Middle ages, but eventually took off dramatically during the Renaissance and the emerging Age of Science. Since the rate of change in materials technology is ever-accelerating, the increase in information content – that is, entropy – leads to an ever-decreasing technological half-life of newly invented materials and technologies. The consequences of this effect have been estimated and projected onto future economical and societal trends of developed and developing nations

Classic and Advanced Ceramics: From Fundamentals to Applications. Robert B. Heimann

ISBN: 978-3-527-32517-7

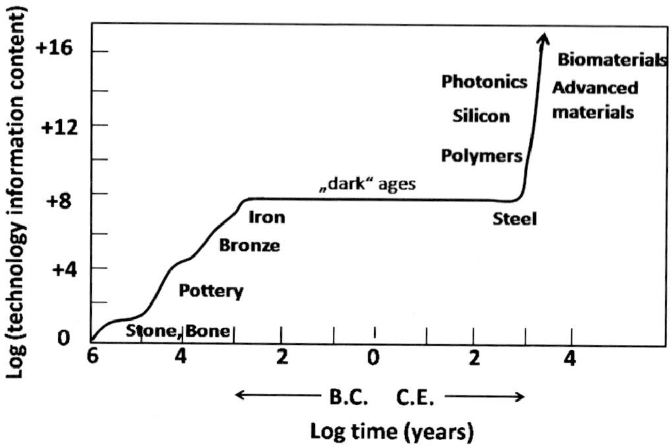

Figure 1.1 Materials development over time: increase of technology information content. Adapted from Hench (1988).

(see, for example, Hench, 1988; Franklin, 1990; Heimann, 1991; Marchetti, 1997; Heimann, 2004).

The rate of change in the information content of advanced materials duplicates the equally fast rate of information and technology transfer within societies of the developed world (Heimann, 1991). As pointed out by Hench (1988), a positive feedback mode connects the two rates, leading to an autocatalytic relationship between materials and technology. This relationship thrives in technological niches that compete with each other for survival and growth, and is controlled by complicated mechanisms involving small random effects which, however, can accumulate and become magnified by positive feedbacks (Arthur, 1990).

Ceramics *sensu strictu* are the oldest man-made materials. By definition, they are inorganic, nonmetallic, silicate-based materials, insoluble in water and many acids and alkalis, and contain at least 30% crystalline compounds. In general, ceramics are shaped at ambient temperature from a specific raw materials mix by a large variety of forming techniques and tools (see, for example, Brownell, 1976), and obtain their typical properties by firing beyond 800 °C (Hennicke, 1967).

While at the dawn of civilization naturally available "ceramics" such as hard rock and flint were utilized for tools (Figure 1.2), with the advent of fire it became apparent that soft and pliable clay and loam raw materials could eventually be changed into hard, durable shapes that were capable of holding liquids, and consequently these were used as storage containers and cooking pots. This development is thought to have been triggered by the transition from hunter–gatherer to agrarian societies. Through the firing process, clay minerals generated by the weathering of granitic rocks could be transformed back into something resembling an artificial "stone" (Heimann and Franklin, 1979). Later, construction materials such as bricks, tiles, and pipes were produced from fired clay. As early as 1600 B.C., the technology of glazing of bricks was known and exploited by the Babylonians.

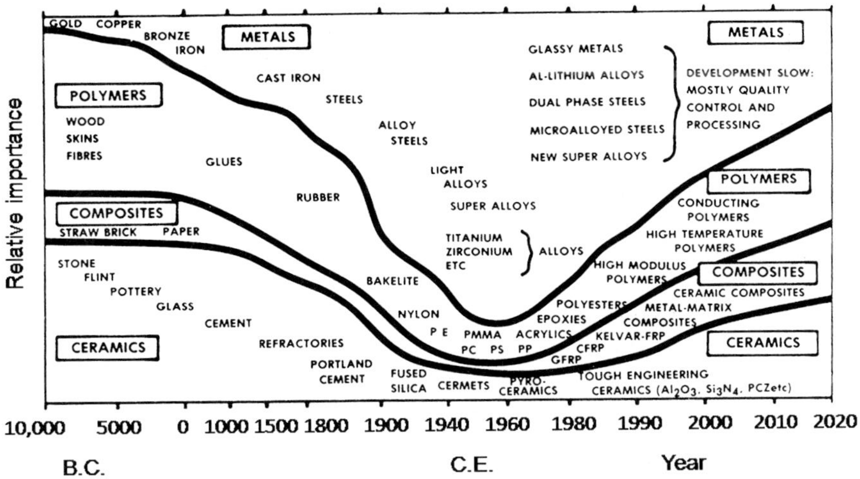

Figure 1.2 Historical timeline of development of materials (Froes, 1990).

The early history of the ceramic technology is difficult to assess in both geographical and temporal context. Arguably, among the first objects fashioned from clay were maternal goddess images such as the famous Upper Paleolithic "Venus of Dolni Věstonice," Moravia, and fragments of animal and human figurines dating from between 25 000 and 29 000 years ago (Klima, 1962). Near the end of the Mesolithic (13 000–12 000 B.P.), hunter–gatherers living in Japan independently rediscovered ceramic technology, but this time applied it to manufacture the world's oldest known ceramic vessels of the Jōmon culture (Chard, 1974; Sherratt, 1980). Very recently, still earlier remnants of ceramic technology were found in a cave in southern China and dated to between 18 300 and 15 430 cal B.P (Boaretto *et al.*, 2009). Since ceramic shards are well preserved in most soils, they are of overriding importance in archeology to date, and distinguish prehistoric cultures by the unique and enduring physical and stylistic features of their pottery. Highlights in ceramic art and technology are the Greek Attic red-on-black and black-on-red vases of the sixth and fifth centuries B.C., the Roman Terra sigillata ware (first century B.C. to third century C.E.), Chinese Song (960–1279 C.E.) and Ming wares (1368–1644 C.E.), as well as the European developments surrounding the inventions of Faience and Majolica (late fifteenth to early sixteenth century C.E.), soft-paste (Sèvres, France) and triaxial hard-paste (Meissen, Saxony) porcelains of the eighteenth century C.E., and soapstone porcelain and bone china in eighteenth-century England. The art, structure and technology of these ceramics have been magnificently researched and displayed in the seminal work "Ceramic Masterpieces" by Kingery and Vandiver (1986). The British development lines in particular were described by Freestone (1999) and Norton (1978).

In parallel, a second line of development emerged concerned with technical refractory ceramics for applications in ancient metal-working activities, including

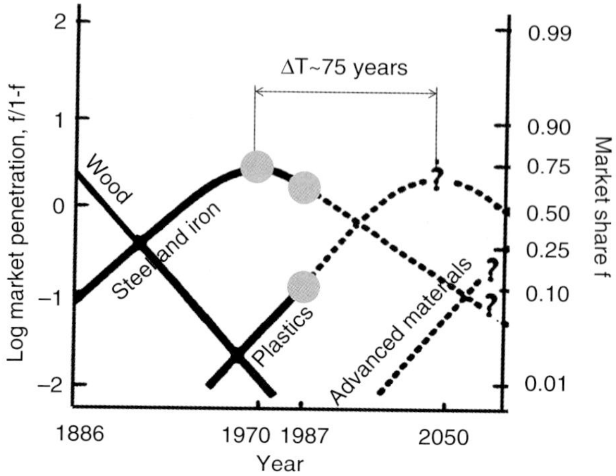

Figure 1.3 Logistic substitution of structural engineering materials between 1886 and 2050 plotted according to the Marchetti–Nakicenovic model (Marchetti and Nakicenovic 1979; Marchetti, 1997). The maxima of the evolutionary curves are spaced about 75 years apart (i.e., 1.5 times the Kondratieff cycle).

tuyères, kilns, furnace linings, smelting and casting crucibles (Rehren, 1997), glass smelting pots, and saggars for firing delicate – and hence high-priced – pottery (Freestone and Tite, 1986).

As indicated in Figure 1.2, ceramics and ceramics-based composite materials played a very important role during the early technological development period of mankind until about 1500 C.E., when metals technology took over. This lasted until the 1970s, when the ubiquitous application of engineering polymers and their composites reduced the impact of metals (Figure 1.3). However, in parallel a second "ceramic age" emerged, highlighted by the development and practical use of tough engineering, functional, and other advanced ceramics. Today, the production volume of classic ceramics such as bricks, tiles and cement/concrete still drastically outperforms that of advanced ceramics. For example, the present world tonnage of cement produced is in excess of a staggering 2×10^9 tons annually (see Section 5.2.1). In contrast, the volume of advanced ceramic materials produced is ridiculously small, although owing to their high value-added nature their sales figures approach those of classic ceramics (see Section 6.2).

Around 1970, metal technology – exemplified by the most common construction materials of steel and iron – reached its maximum market penetration of approximately 75%, and then began to decline. Today, these materials are gradually being replaced by engineering plastics, the use of which is predicted to peak around the year 2050. Simultaneously, the use of advanced materials, including advanced ceramics, is on the rise and will presumably reach a market share of about 10% by the year 2050. This model is based on the logistic Volterra–Lotka equation (Prigogine and Stengers, 1984), that is a measure of the continuous competition

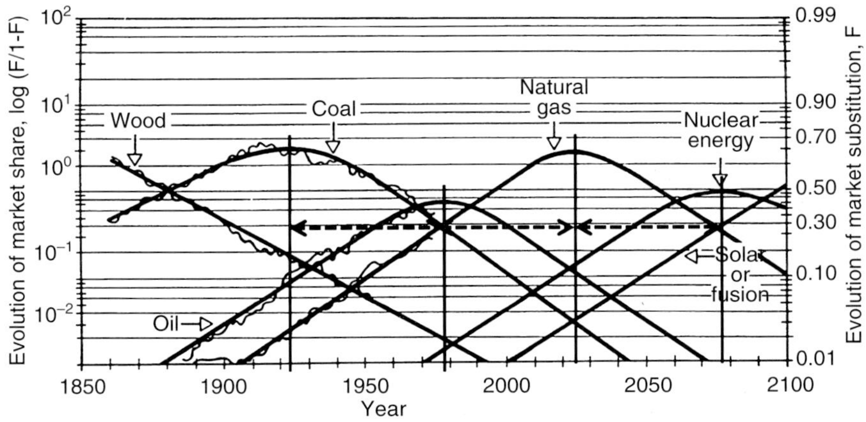

Figure 1.4 Global use of primary energy sources since 1850 (Marchetti, 1989, 1997). The maxima of the Verhulst logistic curves are spaced 50–55 years apart (Kondratieff cycles). Data beyond 1970 are extrapolated.

of materials and technologies, and the fight for technological niches (Heimann, 1991). The maxima of the overlapping logistic equations (Verhulst equations) are shown to be spaced approximately 75 years apart. This offset, however, does not match the well-known Kondratieff cycle of 50–55 years, which arguably is a series of recurring long-range economic cycles that have been shown to govern numerous evolutionary developments, including discoveries (inventions), innovations,[1] industrial production figures, and primary energy uses (Figure 1.4) (Marchetti, 1981, 1997; see also Heimann, 1991, 2004).

In order to underscore the overriding role that raw materials play in society, two additional scenarios will be juxtaposed: (i) the worldwide industry production; and (ii) the individual use of raw materials per capita and lifetime in present-day Germany. The major growth industries are considered to be energy production and distribution, the chemical industry, and microelectronics. The proportions of these industrial sectors of the total industry production worldwide for 1960 and 1990, and extrapolated to 2025, are shown in Table 1.1. While the energy-producing and chemical industries are assumed to remain constant, microelectronics are predicted to double between 1990 and 2025, whereas the metal-based industries (including processing and machining industries) will show a remarkable decline.

Figure 1.5 lists the tonnage of raw materials used per capita within a person's average lifetime of 70 years in contemporary Germany, representative of the raw materials "hunger" of a developed nation with a high technological and societal efficiency (Millendorfer and Gaspari, 1971; Marchetti, 1981).

1) *Innovations* start new industries; *inventions* are discoveries that are at the base of innovations (Marchetti, 1981).

Table 1.1 Proportion (%) of growth for industries of the total industry production worldwide. Data from United Nations Yearbook (1998).

Industry	**Year**		
	1960	**1990**	**2025**
Energy	4.8	6.5	7.0
Chemistry	8.5	12.9	14.0
Microelectronics	1.9	10.3	**25.0**
Metal-based industries	26.5	23.0	**9.4**

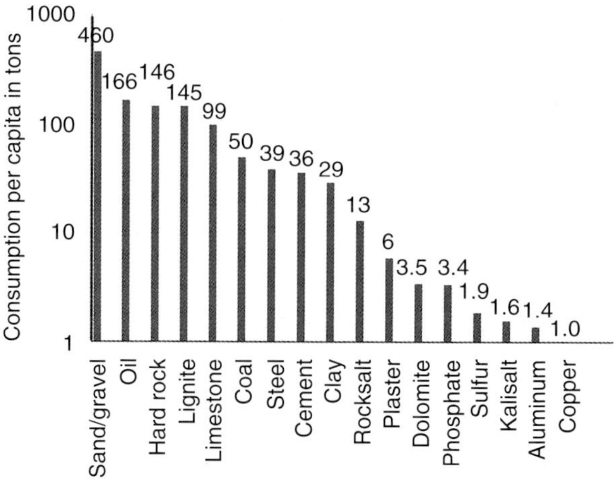

Figure 1.5 Per capita consumption of material resources in an average lifetime in Germany. Data from Bundesanstalt für Geowissenschaften und Rohstoffe (BGR), Hannover, Germany, Global-Report 2859, 1995).

1.2 Classification of Ceramics

A systematic treatment of inorganic–nonmetallic materials is best accomplished by considering a hierarchical approach, as shown in Figure 1.6. The first triangle of level 1 contains the three materials supergroups – metals, polymers, and ceramics – *sensu lato* that are distinguished by their differing chemical bonding relations. The second level of triangles shows at its apices the inorganic–nonmetallic materials classes n that is, ceramics *sensu strictu*, glasses, and hydraulic adhesive materials. These classes can further be subdivided into silicatic, oxidic, and nonoxidic materials (the third hierarchical triangle). Eventually, the chemical components characterize the individual properties (fourth hierarchical triangle). Figure 1.6 is

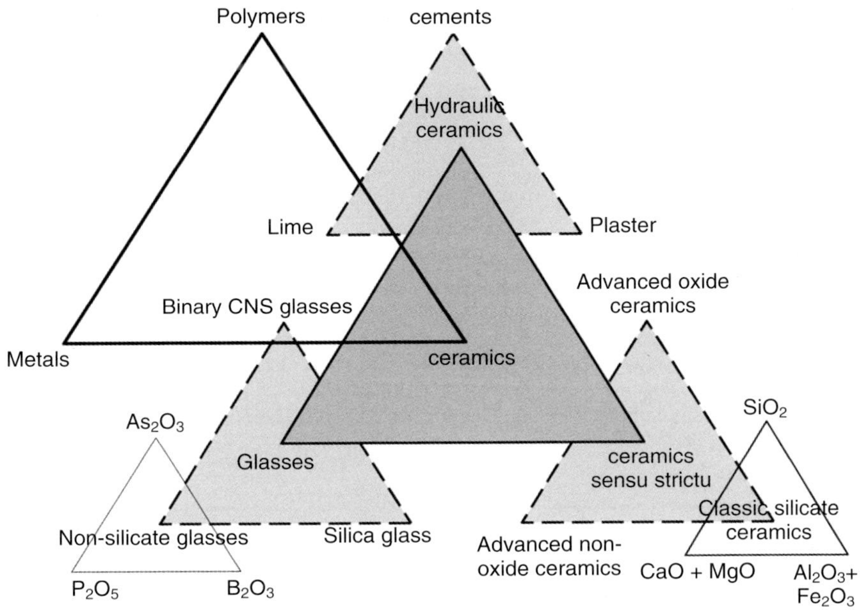

Figure 1.6 Four levels of hierarchical triangles relating different groups of materials. Level 1 (materials supergroups): metals, polymers, **ceramics**; level 2 (ceramics *sensu lato*): glasses, hydraulic ceramics, **ceramics *sensu strictu***; level 3 (ceramic subgroups): advanced oxide ceramics, advanced non-oxide ceramics, classic silicate ceramics; level 4 (phase diagrams): SiO_2, CaO + MgO, $Al_2O_3 + Fe_2O_3$.

Table 1.2 The three main groups of silicatic ceramic materials (level 2 of Figure 1.6).

Material	Processing steps[a)]			T_{max} (°C)	Time of invention
Ceramics *sensu strictu*	P	F	H	<1450	<6000 B.C.
Glasses	P	H	F	1500	<3000 B.C.
Cements (CBCs)[b)]	H	P	F	>1500	Around 1850

a) P = powder production; H = heating; F = forming.
b) CBC = chemically bonded ceramic.

intended to show only the principle of the approach; in reality, such a succession of hierarchical triangles would be more complex. For example, the huge variation of chemical compositions inherent in silicate ceramics would require replacing the triangles by higher-dimensional shapes.

The three main groups of ceramics of level 2 are distinguished by their processing temperatures, the succession of processing steps (F = forming, H = heating, P = powder production), and the time of invention (Table 1.2).

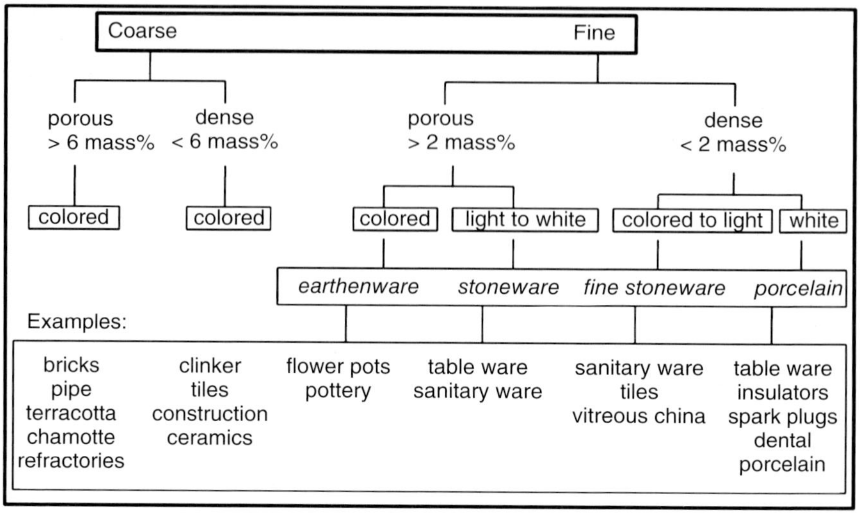

Figure 1.7 Classification of silicate-based ceramics (after Hennicke, 1967).

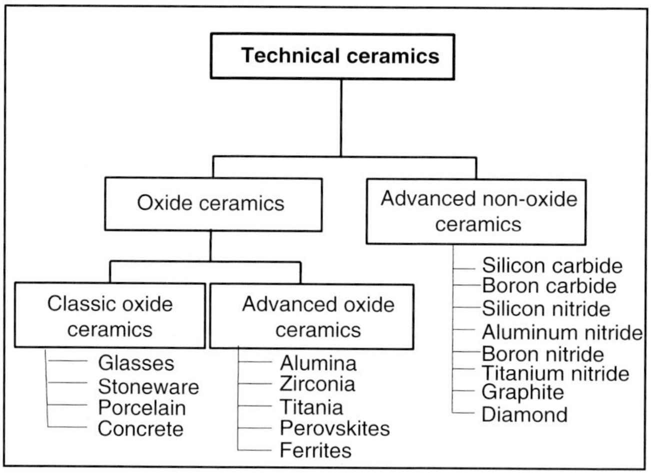

Figure 1.8 Classification of technical ceramics (level 3 of Figure 1.6).

Historically, silicate-based ceramics have been classified in various ways. One of the most useful schemes (Hennicke, 1967) divides different classic ceramic wares according to their starting powder grain sizes (coarse: >0.1 ... 0.2 mm; fine: <0.1 ... 0.2 mm), porosity of the fired product, water absorption capacity (<2 ... >6 mass%), and color of the fired ceramic body (Figure 1.7).

A classification of the field of technical ceramics is shown in Figure 1.8.

In the chapters following this introduction, the path will be traced from natural silicate-based ceramic raw materials, rheological principles of clay–water interaction, and important ceramic phase diagrams to the mineralogy and chemistry of the ceramic firing process. A basic approach to cement and concrete will conclude the first part of the volume dealing with classic ceramics. Glasses, however, have deliberately been excluded from discussion as they do not comply with the definition of ceramics according to Hennicke (1967), as detailed above.

References

Arthur, W.B. (1990) Positive feedbacks in the economy. *Sci. Am.*, **262** (2), 92–99.

Boaretto, E., Wu, X., Yuan, J., Bar-Yosef, O., Chu, V., Pan, Y., Liu, K., Cohen, D., Jiao, T., Li, S., Gu, H., Goldberg, P., and Weiner, S. (2009) Radiocarbon dating of charcoal and bone collagen associated with early pottery at Yuchanyan Cave, Hunan Province, China. *Proc. Natl Acad. Sci. USA*, **106** (24), 9595–9600.

Brownell, W.E. (1976) *Structural Clay Products*, Applied Mineralogy, vol. **9** (eds V.D. Fréchette, H. Kirsch, L.B. Sand, and F. Trojer), Springer, Wien, New York, 231 pp.

Chard, C.S. (1974) *Northeast Asia in Prehistory*, University of Wisconsin Press, Madison, London.

Franklin, U.M. (1990) *The Real World of Technology*, CBC Massey Lectures Series, CBC Enterprises, Montreal, Québec, Canada.

Freestone, I. (1999) The science of early British porcelain. *Br. Ceram. Proc.*, **60**, 11–17.

Freestone, I.C. and Tite, M.S. (1986) Refractories in the ancient and pre-industrial world, in *Ceramics and Civilisation*, vol. **3**, *High-Technology Ceramics Past, Present and Future* (eds W.D. Kingery and E. Lense), American Ceramics Society, Westerville, OH, pp. 35–65.

Froes, F.H. (1990) Aerospace materials for the twenty-first century. *Swiss Mater.*, **2** (2), 23–36.

Heimann, R.B. (1991) Technological progress and market penetration of advanced ceramics in Canada. *Bull. Am. Ceram. Soc.*, **70** (7), 1120–1127.

Heimann, R.B. (2004) Applied mineralogy –an important driving force towards a sustained development of future technologies, in *Applied Mineralogy. Developments in Science and Technology*, vol. **1** (eds M. Pecchio, *et al.*), International Council for Applied Mineralogy (ICAM), Sao Paulo, Brazil, pp. 3–11.

Heimann, R. and Franklin, U.M. (1979) Archaeo-thermometry: the assessment of firing temperatures of ancient ceramics. *J. Int. Inst. Conserv., Can. Group*, **4** (2), 23–45.

Hench, L.L. (1988) Ceramics and the challenge of change. *Adv. Ceram. Mater.*, **3** (3), 203–206.

Hennicke, H.W. (1967) Zum Begriff Keramik und zur Einteilung keramischer. *Werkstoffe Ber. Dtsch. Keram. Ges.*, **44**, 209–211.

Kingery, W.D. and Vandiver, P.B. (1986) *Ceramic Masterpieces. Art, Structure, and Technology*, The Free Press, New York.

Klima, B. (1962) The first ground plan of an Upper Paleolithic loess settlement in Middle Europe and its meaning, in *Courses Towards Urban Life* (eds R.J. Braidswood and G.R. Willey), Chicago Press, Chicago, pp 193–210.

Marchetti, C. (1980) Society as a learning system: discovery, invention, and innovation cycles revisited. *Technol. Forecast. Social Change*, **18**, 267–282.

Marchetti, C. (1989) Lebenszyklen und Energiesysteme. *Energiewirtschaftliche Tagesfragen*, **39**, 8–12.

Marchetti, C. (1997) Energy supply: the long-time horizon. Conference on Energy and the Greenhouse Issue, ABB/IEA GHG Forum, 3 March 1997, Baden-Dättwil, Switzerland.

Marchetti, C. and Nakicenovic, N. (1979) The dynamics of energy systems and the logistic substitution model. Report RR-79-13. International Institute of Applied Systems Analysis, Laxenburg, Austria.

Millendorfer, H. and Gaspari, C. (1971) Immaterielle und materielle Faktoren der Entwicklung. Ansätze zu einer allgemeinen Produktionsfunktion. *Z. Nationalökon.*, **31**, 81–120.

Norton, F.H. (1978) *Fine Ceramics: Technology and Applications*, Krieger Publ. Co, Malabar (FL), ISBN: 0-88-275-582X.

Prigogine, I. and Stengers, I. (1984) *Order out of Chaos. Man's New Dialogue with Nature*, Bantam, Toronto.

Rehren, Th. (1997) Tiegelmetallurgie. Tiegelprozesse und ihre Stellung in der Archäo-metallurgie. Habilitation Thesis, Technische Universität Bergakademie Freiberg, Freiberg, Germany.

Sherratt, A. (ed.) (1980) *The Cambridge Encyclopedia of Archaeology*, Prentice Hall of Canada Ltd., Scarborough, Ontario, pp. 90–91.

Yearbook of the United Nations (1998) Economic and social questions, Part Three. United Nations, New York, 10017. Available at: http://unyearbook.un.org/1998YUN/1998 (accessed date 7 December 2009).

2
Mineralogy, Structure, and Green Processing of Clay Minerals

This chapter deals with the formation, structure, and properties of important clay minerals, the nomenclature of clay minerals, the green processing of clays (forming, drying), and the interaction of clay particles with water. Particular emphasis is devoted to modeling the rheological behavior of clay–water systems, the influence of pH, size and charge of electrolyte ions on clay particle dispersion and aggregation, and the origin and significance of the zeta potential, ζ.

The natural raw materials utilized to produce silicate-based ceramics can be divided into: (i) *highly plastic materials,* such as clays comprising the minerals kaolinite, illite, or montmorillonite; (ii) *sparingly plastic minerals* for special (electro)ceramic applications, such as pyrophyllite and talc; and (iii) *nonplastic materials,* such as tempering additives (quartz, chamotte) and fluxes (feldspar, apatite, nepheline, calcite, dolomite, etc.) that are added to clays to alter the chemistry, workability and sintering behavior of the ceramic masses. Synthetic raw materials include precursors of glazes (lead oxide, barium carbonate, tin oxide) and special ceramic masses (alumina, zirconia, magnesia), as well as hydrothermally synthesized wollastonite and diopside, and synthetic kaolinite with a very narrow grain size distribution and high plasticity for high-performance electroceramics. Wet chemical techniques such as coprecipitation, freeze- and spray-drying, and sol–gel synthesis are also applied to produce raw materials for special applications.

Plastic natural ceramic raw materials, consisting predominately of kaolinite, illite and/or montmorillonite, are accompanied by residual quartz, feldspar, mica, and calcite as well as organic residues. In particular, the limestone content has been used to distinguish between clay (<4 mass% lime), marly clay (4–10 mass% lime), clayey marl (10–40 mass% lime), marl (40–75 mass% lime), calcareous marl (75–90 mass% lime), marly limestone (90–96 mass% line), and limestone (>96 mass% lime). Kaolinitic raw materials formed *in situ* (autochthoneous) are called *kaoline,* while kaolinitic raw materials found in secondary deposits (allochthoneous) are called *clays.* Marl and marly limestones are important raw materials for Portland cement production (see Section 5.2.1).

Classic and Advanced Ceramics: From Fundamentals to Applications. Robert B. Heimann

ISBN: 978-3-527-32517-7

2.1 Natural Clay Minerals

The basis of understanding the complex processes occurring during the oxidizing or reducing firing of clay minerals is a clear understanding of the mineralogical nature of clays and their interaction with weathering and soil solutions. Clays are weathering products of feldspars, micas and other rock-forming minerals, and as such mechanical mixtures of very different components, each with a different and characteristic grain size distribution. This mechanical mixture generally consists of four main components: (i) fine-grained weathering relics such as quartz, feldspar, sericite, and interlayer-deficient micas (formerly called hydromicas; Rieder *et al.*, 1998); (ii) newly formed clay minerals such as kaolinite, halloysite, illite, and montmorillonite; (iii) remainders of organisms consisting of calcite or aragonite (shells), silica (chert) or graphitic carbon; and (iv) neoformations occurring after deposition, such as pyrite, dolomite, or glauconite. The clay minerals have grain sizes below 2 μm, whereas other weathering relics such as quartz, feldspar and micas may have grain sizes up to 20 μm (see Figure 4.4).

The genetic heritage and origin of clay minerals have been described by many research groups, including Esquevin (1958), Millot (1970), Velde (1977), and Eberl (1984) and, most recently, Velde and Meunier (2008). Three basic genetic mechanisms were found to be operational: inheritance; neoformation; and transformation:

- *Inheritance* means that clay minerals can originate from reactions occurring in a different environment during a previous stage of the rock cycle.
- Origin by *neoformation* means that clay minerals were formed by precipitation from dilute soil solutions or by reactions of amorphous materials.
- *Transformation* finally consists of reaction sequences involving alteration of the inherited structure by chemical reactions, either by ion-exchange or layer transformation during diagenesis (Eberl, 1984). As the results of layer transformation are preserved in the geological records, investigations of their nature provides valuable information on the environmental conditions in the sediment source area as a function of time.

While inheritance dominates in the sedimentary environment at generally ambient conditions characterized by slow reaction rates, layer transformation requires a considerable input of activation energy, and thus is found preferentially in the diagenetic and hydrothermal realms, where higher temperatures prevail. In between these two environments, the weathering environment exists in which all three mechanisms discussed above can be operational. Hence, when these three mechanisms occur in three different geologic environments, it leads to nine possibilities of clay mineral formation in nature, attesting to the exceptional variability and complexity of clay mineral chemistries.

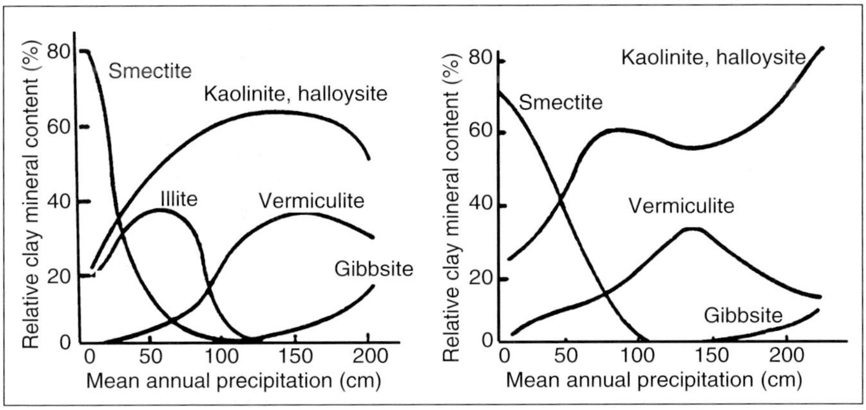

Figure 2.1 General scheme of the relationship between the frequency distribution of clay minerals and the amount of precipitation in residual soils of acid (left) and basic (right) igneous rocks (after Barshad, 1966; Eberl, 1984).

2.1.1 Formation of Clay Minerals

All clay minerals are products of the interaction of rocks with aqueous solutions of the weathering environment (Velde and Meunier, 2008). These interactions are essentially a series of leaching and precipitation processes whereby the pH of the solutions is of great importance. In addition to the type of rock and its mineral constituents (Figure 2.1), one of the most important variables is the *climate*. There is a basic distinction to be made between leaching reactions that occur in temperate or tropical climates, and precipitation reactions that occur in humid or arid environments. Thus, the compositional and structural variability of clay minerals can be understood on the basis of the different modes of environmental interaction schemes (see Table 2.1).

Rain water contains considerable amounts of carbon dioxide, and also a certain amount of nitric acid, which renders the overall pH value slightly acidic at between 5 and 6. In the temperate–humid climate zone, this acidity is increased to below 5 by humic acids of the soil overlying the bedrock. On the other hand, in the tropical–humid climate zone the groundwater may be close to neutral, owing to the interaction with soil solutions produced by intensified bacterial decomposition of plant material.

While rain and soil solutions dissolve easily soluble ions such as K^+, Na^+, Ca^{2+}, and Mg^{2+} from the minerals of the parent rocks, the degree to which Fe^{2+}, Al^{3+}, and Si^{4+} are attacked and solubilized depends to a large extent on the pH and Eh values of the leaching solutions. The environmental conditions determine also the degree of reprecipitation of hydroxides into the surrounding soil, in particular iron and aluminum hydroxides and oxihydroxides. Freely moving water and

Table 2.1 Origin, formation, and transformation relations of clay minerals (Heimann and Franklin, 1979).

Mineral	Origin	Formation	Transformation	State
Muscovite	Igneous and metamorphic rocks	Crystallization mostly under higher pressure–temperature conditions	To illite, montmorillonite and glauconite	Residual
Biotite	As above	As above	To montmorillonite, vermiculite and chlorite	Residual in clays with short transportation path
Illite	Sedimentary rocks	a) From muscovite or biotite by leaching of potassium ions b) From montmorillonite by adsorption of potassium ions c) As neoformation from weathering solutions	To muscovite/biotite, to chlorite in marine environments (by addition of magnesium ions) or brines	Either residual or neoformation
Glauconite	Sedimentary rocks, low hydrothermal in igneous rocks (?)	a) From illite syn-sedimentary or by diagenesis (dissolution–reprecipitation) b) From colloidal solutions in pore spaces of marine sediments	To illite by leaching	Neoformation in sediments or synsedimentary
Smectite group (montmorillonite, beidellite, nontronite, etc.)	Silicatic rocks of either origin	a) Incomplete leaching of silicates (feldspars, micas) due to restricted water circulation b) After deposition either by removal of potassium from micas or neoformation from solutions	To kaolinite by subsequent leaching, to illite or glauconite by addition of potassium and iron, to chlorite	Either residual or neoformation
Chlorite group	Igneous, metamorphic and sedimentary rocks	a) Leaching of biotite, hornblende b) Neoformation after deposition c) Very low- to low-grade metamorphosis	To smectites and vermiculite	Residual or neoformation in sediments
Kaolinite	Mostly silica-rich igneous rocks	Complete leaching of silicates (mostly feldspars) by free water circulation	Seldom observed	Neoformation in autochthone environment

high precipitation rates favor the formation of gibbsite, whereas smectites form predominantly under dry conditions. Kaolinite formation occurs in the intermediate region (Figure 2.1).

The reaction of water with feldspars leaches soluble ions such as K^+, Na^+, Ca^{2+}, and Mg^{2+} from the lattice that are transported in true solution to the sea. The residual Fe–Al–Si–O lattice will initially be stabilized by the adsorption of protons and $(H_3O)^+$ ions, respectively. Further hydrolysis depends on the chemical properties as well as the amount and mobility of the solute. The individual ions of the residual feldspar lattice behave as follows:

- Fe^{2+} is relatively soluble in weakly acidic water but will be quickly oxidized under neutral or weakly alkaline conditions, and precipitated as iron oxihydroxides (goethite, α-FeOOH) that cause soils in temperate–humid climates to be yellow-brown (siallitic type of weathering). Under tropical–humid climatic conditions, silicon will be removed completely and iron ions fixed in hematite, thus coloring tropical soils (laterites) intensively red (allitic type of weathering).
- Si^{4+} is weakly, but noticeably, soluble in water (2–4 mmol l^{-1}) within a large range of pH (4–9), presumably as $[SiO_4]^{4-}$ ion. Under dry conditions (arid climates) it can be precipitated as opal (desert varnish).
- The remaining Al–Si–O lattice is profoundly distorted and largely amorphous. Al^{3+} is easily soluble in strongly acidic waters (pH < 4) but almost insoluble in surface waters (pH = 5–7). Under humid climate conditions, clay minerals will be formed, the type of which depends on the parameters of the weathering system (as discussed above). Freely circulating water causes the removal of soluble ions and the formation of kaolinite (siallitic weathering). In contrast, under the regime of restricted circulation, less-soluble Mg^{2+}- and Fe^{2+} ions form together with the Al–Si–O residual lattice minerals of the montmorillonite family. Under arid climate conditions (allitic weathering) with a neutral or weakly alkaline pH, aluminum is much less soluble than silicon, and remains as gibbsite (γ-$Al(OH)_3$). These principles can be represented qualitatively in the stability diagram shown in Figure 2.2, which is constructed from thermochemical data (Garrels and Christ, 1965; Drever, 1982).

If the original mineral was mica rather than feldspar, then the weathering process would follow a somewhat different reaction path. The weakly bonded K^+ ions are very quickly dissolved, forming a residual lattice that is stabilized by $(H_3O)^+$ ions. Thus, so-called "interlayer-deficient micas" are formed, of which the most commonly occurring is illite (see Section 2.1.2.2). These minerals are broken up into small fragments by mechanical strain between geometrically incompatible silicate layers caused by the ion exchange. The majority of natural clays consist predominantly of illites that are, to a large extent, responsible for the plastic behavior of typical ceramic clays. Clays with high illite content are sometimes referred to as "immature," compared to "mature" clays with high contents of kaolinite and/or montmorillonite.

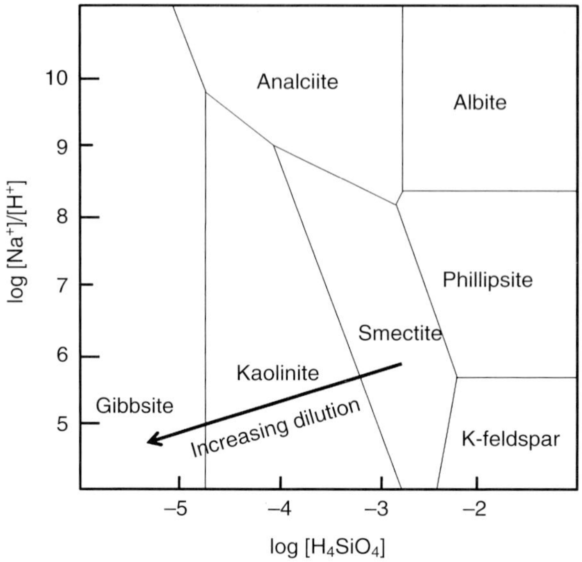

Figure 2.2 Calculated stability diagram of the system Na_2O–K_2O–Al_2O_3–SiO_2–H_2O for log $[K^+]/[H^+] = 4$. Adapted from Garrels and Christ, 1965; Drever, 1982.

The question of the formation and transformation of illite into other clays minerals (Table 2.1) remains a matter of intensive discussion and research (Velde, 1977; Eberl, 1984; Lindgren *et al.*, 2002). This topic is complicated not only by the complexity of the reactions involved but also by the exceptional small size and often amorphous nature of the reaction products (Heimann and Franklin, 1979). Some current ideas on this issue will be presented below. Common intermediate structures of transformation of clay minerals are mixed-layer or interstratified clays. For example, during the diagenetic transformation of trioctahedral micas such as biotite, potassium will be removed so that expandable smectite-like layers may alternate with nonexpandable illite-type layers, resulting in illite–smectite (I/S) interstratified structures. If the illite and smectite layers alternate regularly, the resulting mineral is termed *rectorite* (allevardite). On the other hand, residual illite can be vermiculitized in response to acid leaching of potassium, following a sequence illite → I/S (ordered) → I/S (random) with increasing intensity (Eberl, 1984). These relationships are highly complex and, in many details, still rather mysterious. For further detail, the reader is referred to more specialized literature (e.g., Reynolds, 1980; Srodon, 1980).

2.1.2
Structure of Important Clay Minerals

Clay minerals consist of hexagonal networks of SiO_4 tetrahedra. The basal planes of the tetrahedra are in the plane of the sheet silicate network, and their tips point

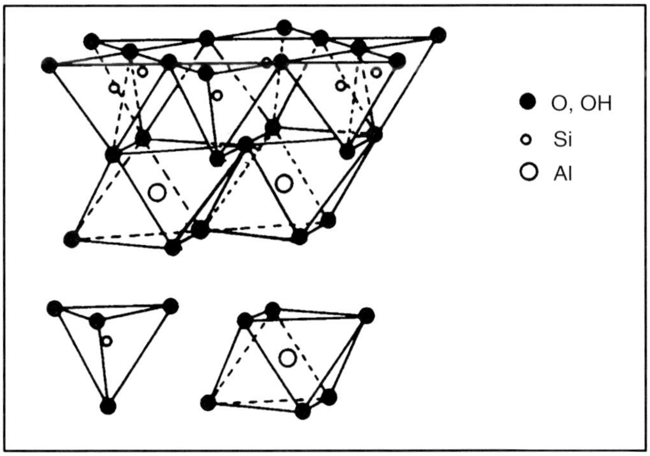

Figure 2.3 TO-structure of two-layer sheet silicates (kaolinite) (after Millot, 1979).

all in the same direction. The oxygen atoms at these free apices are bound to either Al or Mg atoms, and the residual valencies are saturated by OH^- ions. Consequently, the Al^{3+} and Mg^{2+} cations are in a sixfold coordinated (octahedral) position such that these sheet silicate minerals consist essentially of two layers – a tetrahedral SiO_4 layer (T), and an octahedral $AlO_2(OH)_4$ gibbsite-type or $MgO_2(OH)_4$ brucite-type layer (O) (Figure 2.3). In the gibbsite layer there are always two aluminum atoms bound to each group of $6(OH)^-$ ions (dioctahedral two-layer minerals), while in the brucite layer three magnesium atoms combine with $6(OH)^-$ ions (trioctahedral two-layer minerals). Members of these groups are kaolinite with gibbsite-like and serpentine with brucite-like layers, respectively (Table 2.2).

In the same structural category, three-layer minerals are found in which the octahedral Al or Mg layer is sandwiched between two SiO_4 tetrahedral layers, the free apices of which point towards each other (Figure 2.4). Talc, $Mg_3[(OH)_2/Si_4O_{10}]$ is one common example of a three-layer trioctahedral sheet silicate, whereas pyrophyllite, $Al_2[(OH)_2/Si_4O_{10}]$ is the three-layer equivalent of kaolinite with dioctahedral nature (Table 2.2).

If, in three-layer sheet silicates, part of the Si in the tetrahedral layer is substituted by Al, then negative surface charges will occur that are compensated for by alkali cations, rather weakly bound between the three-layer stacks. In that way the large group of mica is formed. On the other hand, the formal partial substitution of Al by Mg in the pyrophyllite lattice produces a charge deficiency that will be compensated for by monovalent or divalent atoms such as Na^+ or Ca^{2+} to create the minerals of the smectite group.

The group of chlorite shown in Table 2.2 is sometimes referred to as a "four-layer structure," as they are composed of modified three-layered talc units connected by a brucite-type layer. However, it may be more appropriate to call them "mixed-layer structures with ordered intermediate layers."

Table 2.2 Classification of sheet silicates.

Sheet silicates	
Dioctahedral (with gibbsite layers)	**Trioctahedral (with brucite layers)**
Two-layer structures	
Kaolinite $Al_4[(OH)_8/Si_4O_{10}]$	Serpentine $Mg_3[(OH)_4/Si_2O_5]$
Halloysite $Al_4[(OH)_8/Si_4O_{10}]\cdot 4H_2O$	
Three-layer structures	
Pyrophyllite $Al_2[(OH)_2/Si_4O_{10}]$	Talc $Mg_3[(OH)_2/Si_4O_{10}]$
Montmorillonite $(Al_{2-x}Mg_x)[(OH)_2/Si_4O_{10}]^{x-}\cdot nH_2O + xNa^+$	Vermiculite $Mg_3[(OH)_2/Si_4O_{10}]\cdot nH_2O$
Beidellite $Al_2[(OH)_2/Al_{0.5}Si_{3.5}O_{10}]^{0.5-}\cdot nH_2O$	Saponite $Mg_3[(OH)_2/Al_{0.5}Si_{3.5}O_{10}]^{0.5-}\cdot nH_2O$
Nontronite $Fe_2^{3+}[(OH)_2/Al_{0.5}Si_{3.5}O_{10}]^{0.5-}\cdot nH_2O$	Hectorite $Mg_{3-x}Li_x[(OH)_2/Si_4O_{10}]^{x-}\cdot nH_2O$
Muscovite $KAl_2[(OH)_2/AlSi_3O_{10}]$	Phlogopite $KMg_3[(OH)_2/AlSi_3O_{10}]$
Margarite $CaAl_2[(OH)_2/Al_2Si_2O_{10}]$	Clintonite $CaMg_3[(OH)_2/Al_2Si_2O_{10}]$
Sudoite $Al_3Fe[(OH)_8/Si_4O_{10}]$	"Chlorite" $Mg_5Al[(OH)_8/AlSi_3O_{10}]$

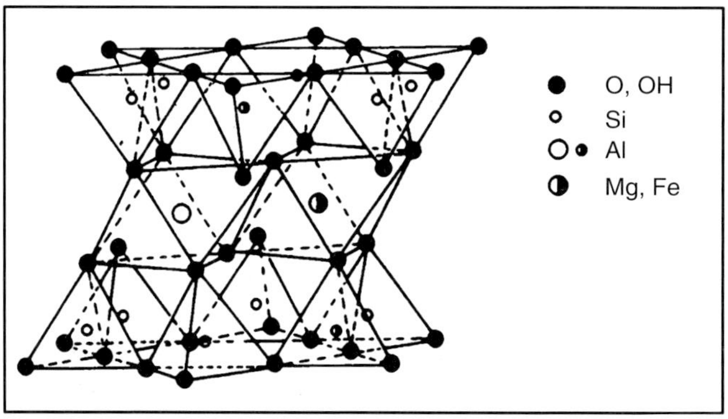

Figure 2.4 TOT-structure of three-layer sheet silicates (montmorillonite) (after Millot, 1979).

2.1.2.1 **Kaolinite**

Kaolinite, $Al_4[(OH)_8/Si_4O_{10}]$, is the typical weathering product of feldspar under temperate–humid climate conditions, and in the presence of surplus water of slightly acidic pH that is sufficient to remove completely the alkali and alkali earth ions of the parent feldspars. The mineral exists in four main structural variants: triclinic kaolinite [space group (S.G.) *P1*, a = 515 pm, b = 894 pm, c = 739 pm, α = 91.9°, β = 105.0°, γ = 89.8°, Z = 1); monoclinic dickite (S.G. *Cc*, a = 515 pm, b = 894 pm, c = 1442 pm, β = 96.7°, Z = 2); monoclinic nacrite (S.G. *Cc*, a = 891 pm, b = 515 pm, c = 1570 pm, β = 113.7°, Z = 2); as well as a b-axis distorted fireclay. The structures of kaolinite, dickite, and nacrite are shown in Figure 2.5.

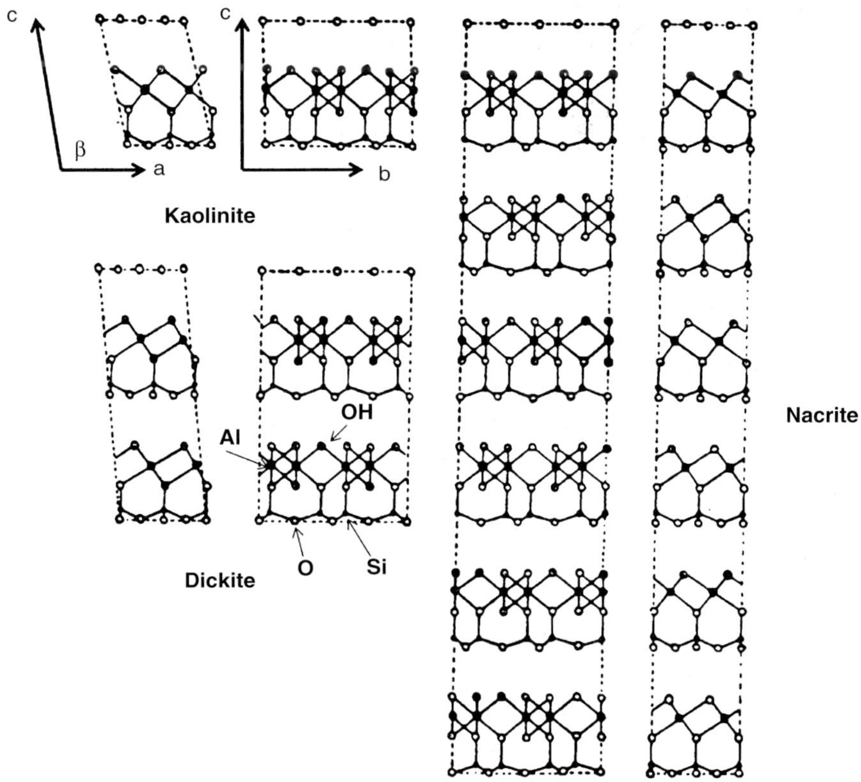

Figure 2.5 Structures of kaolinite, dickite, and nacrite.

The structure of nacrite has been reinvestigated by Zheng and Bailey (1994), and recently refined by Mercier and Le Page (2008). The latter authors found a complicated polytypic pattern of 36 transformations between adjacent kaolinite layers, 20 of which are energetically distinguishable transformations (EDTs) and 16 enantiomorphic transitions (EDT*). Kaolinite and dickite are lowest-energy models at zero temperature and pressure, whereas nacrite is the lowest-enthalpy model under moderate pressure.

2.1.2.2 **Illite**

Despite the fact that members of the illite family appear to be the most abundant clay minerals next to kaolinite, their state has not yet been defined unequivocally by the Association Internationale pour L'Etude des Argiles (AIPEA) (Jambor *et al.*, 1998). Stubborn mysteries still surround the formation and transformation of this most abundant clay mineral, mostly related to its widely varying chemical composition, small crystal size, degree of crystallinity or lack thereof, as well as the complexity of transformation sequences in the geologic environment over time. It has been a longstanding agreement that illite *sensu lato* can form basically by

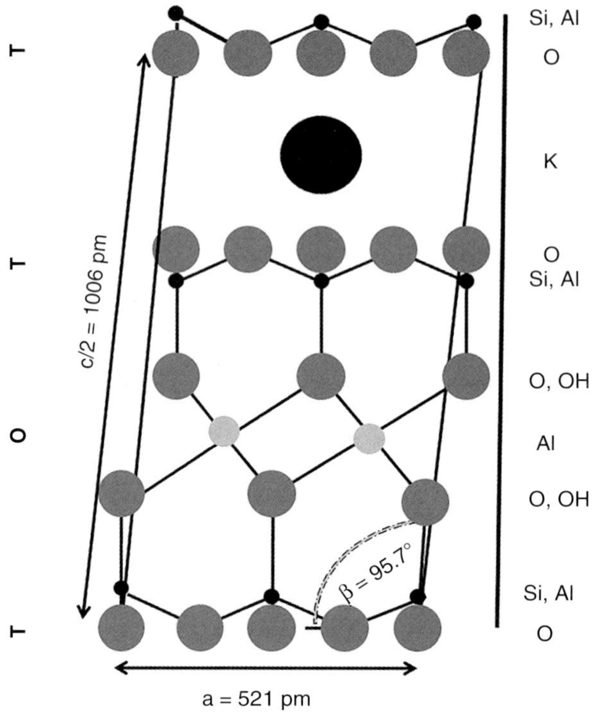

Figure 2.6 Projection of the structure of muscovite -2M$_1$ onto the a–c plane.

inheritance; that is, through a loss of potassium (degradation) during leaching of muscovite (dioctahedral illites) or biotite (trioctahedral illites); by transformation through the addition of potassium (aggradation) to montmorillonite; and possibly also by neoformation involving precipitation from dilute colloidal weathering solutions (Table 2.1). However, the potassium/argon age dates of oceanic illites suggest that they have been derived by inheritance from the continents rather than by neoformation or transformation reactions on the ocean floor (Lisitzin, 1972; Eberl, 1984).

The loss of easily soluble potassium from the trioctahedral mica biotite will be compensated for by ion exchange with H_3O^+ ions, by oxidation of Fe^{2+} ions, and by the replacement of Al in the octahedral layer by Si. On the other hand, the dioctahedral mica muscovite (Figure 2.6) undergoes similar potassium loss by degradation and associated charge deficiency (Table 2.3). In the resulting dioctahedral illites, the ideal Si/Al ratio of 3 in the tetrahedral layer of muscovite is changed to between 5 and 40 (Hower and Mowatt, 1966). In addition, water is intercalated between the layer stacks.

Starting from an ideal muscovite lattice, three possible reaction paths have been suggested, as shown in Table 2.3: (i) K^+ ions in the interlayer space will be replaced by H_3O^+ ions; (ii) one K^+ ion together with one OH^- group of the octahedral layer will be replaced by two H_2O molecules; and (iii) one Si^{4+} ion in the tetrahedral

Table 2.3 Three feasible ways of illite formation from dioctahedral muscovites.

	Interlayer	Octahedral layer	Tetrahedral layer
Muscovite	K_2	$Al_4(OH)_4$	$Al_2Si_6O_{20}$
Path 1	$K(H_3O)$	$Al_4(OH)_4$	$Al_2Si_6O_{20}$
Path 2	$K(H_2O)$	$Al_4(H_2O)(OH)_3$	$Al_2Si_6O_{20}$
Path 3	K_2	$Al_4(OH)_4$	$Al_2Si_4H_8O_{20}$

layer will be replaced by four protons. It should be emphasized that illite with increasing degradation approaches the chemical composition of kaolinite, but the structural state of montmorillonite; hence, numerous interstratified ordered and disordered structure variants exist. However, the higher proportion of aluminum in the tetrahedral layer compared to montmorillonite requires more potassium as the interlayer cation. As a result, very little intracrystalline expansion occurs in illite, fixing even in the presence of water and electrolytes the c-axis dimensions at 1018 pm in illite-1M or 2012 pm in illite-$2M_1$.

Illite-1M (S.G. *C2/m*, a = 520 pm, b = 895 pm, c = 1018 pm, β = 101.7°, Z = 2) and illite-$2M_1$ (S.G. *C2/c*, a = 519 pm, b = 900 pm, c = 2012 pm, β = 95.2°, Z = 4) have been considered polytypes of a common structure that can be related to the parent muscovite structure. Indeed, the lattice dimensions are very close and the space group *C2/c* (15) is being maintained for the $2M_1$ structure variants of muscovite and illite. However, in contrast to the hitherto accepted opinion, both illite-1M and illite-$2M_1$ have been found to be individual and well-defined minerals separated by a miscibility gap (Zöller and Brockamp, 1997). The tetrahedral layers of illite-$2M_1$ are slightly undulating, whereas in illite-M1 they are coplanar. The angles between Si atoms are 120° in illite-M1, but alternate between 118° and 124° in illite-$2M_1$; consequently, the former material has a more regular structure.

Since the degree of crystallinity of illite in sediments increases with temperature, this can be used as a marker to estimate the diagenetic–metamorphic zones (grades) of metasedimentary rocks of marine fine-clastic origin that are widespread in sedimentary basins and in the outer fold-and-thrust zones of the orogenic belts. Systematic changes in sharpness of the X-ray powder diffraction (XRPD) 1.0 nm-basal reflection of illite upon burial (increasing temperature and pressure) were first recognized by Weaver (1960), who introduced the so-called "sharpness ratio" for petrogenetic purposes. The term illite crystallinity (IC) in its current form – that is, the full width at half maximum (FWHM) of the 1.0 nm XRD peak of illite and/or muscovite, measured on the <2 μm size fraction – was first introduced by Kübler (1967, 1984) and is now being increasingly used worldwide, being referred to as the Kübler Index (KI). Warr and Rice (1994) introduced standards referred to as the crystallinity index standard (CSI), while a recent contribution attempted to correlate the KI and the CSI (Kisch *et al.*, 2004).

The physical interpretation of KI is based on the mean number of layers in coherent scattering domains (i.e., illite particle size) and the smectite-layer percentage in illite/smectite minerals for a given stacking order, since the KI is sensitive to low percentages of expandable layers (Jaboyedoff *et al.*, 2001). Strain-induced defect migration during diagenetic and low-metamorphosis transformation contributes to the polytypic transformation from the illite-1M_d polytype to smectite-rich I/S to dominant 2M_1 polytypism in illite-rich I/S (Dong and Peacor, 1996). The increased dominance of illite-2M_1 directly influences crystal thickening (Dong *et al.*, 1997), and hence the KI. In summary, strain decreases with increasing metamorphic grade, which implies an increasing perfection of the crystal lattice of sheet silicates.

2.1.2.3 Montmorillonite: Structure and Application

Clay minerals of the smectite–vermiculite group are formed within the rock cycle by several mechanisms including:

- weathering of feldspar and other rock-forming minerals in humid climates in soils with restricted water access;
- marine weathering of palagonite lavas (vitreous basaltic lavas) to yield secondary minerals such as nontronite, $Fe_2^{3+}[(OH)_2/Al_{0.5}Si_{3.5}O_{10}]^{0.5-}Na_{0.5}\cdot 4H_2O$ and phillipsite, $KCa[Al_3Si_5O_{16}]\cdot 6H_2O$;
- terrestrial weathering of rhyolitic volcanic tuffs to form bentonites (Wyoming, Saskatchewan, Turkey);
- neoformation in soils and at the ocean floor; and
- layer transformation of chlorites.

From a structural point of view, smectites can be thought of as being formed by the partial substitution of Al by Mg in the three-layer sheet silicate pyrophyllite, $Al_2[(OH)_2/Si_4O_{10}]$, producing a charge deficiency that will be compensated for by monovalent or divalent atoms such as Na^+ or Ca^{2+}. More specifically, the replacement of Al by Mg in the octahedral layer leads to *montmorillonite* with an approximate composition $(Al_{2-x}Mg_x)[(OH)_2/Si_4O_{10}]^{x-}\cdot nH_2O$, whereas the replacement of Si by Al in the tetrahedral layer produces *beidellite* $Al_2[(OH)_2/Al_{0.5}Si_{3.5}O_{10}]^{0.5-}\cdot nH_2O$ (Figure 2.7). In *nontronite*, $Fe_2^{3+}[(OH)_2/Al_{0.5}Si_{3.5}O_{10}]^{0.5-}\cdot Na_{0.5}\cdot 4H_2O$, aluminum in the octahedral layer has been more or less completely replaced by trivalent iron, and silicon in the tetrahedral layer partially by aluminum.

As the charge-compensating alkali or alkali earth ions located between layer stacks are easily hydrated, minerals of the smectite group such as montmorillonite are able to absorb not only large quantities of water but also bulky organic molecules, including fatty acids, nucleic acids, and proteins. This absorption causes the mineral to expand parallel to the c-axis direction; such expansion can be followed using XRPD by measuring the shift of 00*l* interplanar spacings. The d(001) basis distance of air-dried montmorillonite is about 1.4 to 1.5 nm, depending on the nature of the interlayer cations, the humidity, and the temperature. As the hydrate layer surrounding the interlayer cations is only loosely bound by van der Waals-type intermolecular forces, water can be reversibly removed by gentle heating

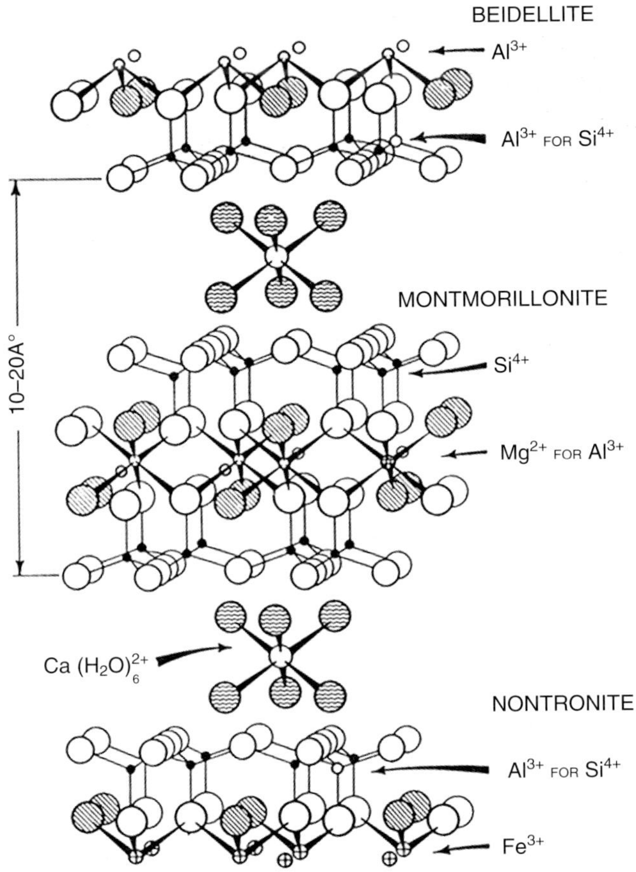

Figure 2.7 Structures of beidellite (top), montmorillonite (center), and nontronite (bottom).

to 120 °C. This causes the d(001) distance to shrink to about 0.9 nm, which corresponds to d(001) of pyrophyllite. The subsequent intercalation of organic molecules generates a lattice expansion in the c-axis direction to about 1.7 nm (diethylene glycol or glycerol) or even 2.7 nm (hexane, heptane). The intercalation of a bimolecular layer of acrylamide yields a c-axis distance or 2.0–2.1 nm (Ogawa *et al.*, 1989; Gao and Heimann, 1993; see Figure 2.8). Montmorillonites with Na^+ as exchangeable interlayer cation show, on treatment with distilled water, an exceptionally high expansion such that in many cases a d(001) spacing can no longer be observed.

The fact that not only water but also organic molecules such as amphiphilic fatty acids can be intercalated by smectites is of great significance. In the past, this unique property has been exploited to remove oil and grease from animal hides during the tanning process, or from woolen cloth during fulling. Hence bentonites – that is, sedimentary rocks with a high proportion of smectites – were referred to as "Fuller's Earth" (Robertson, 1986).

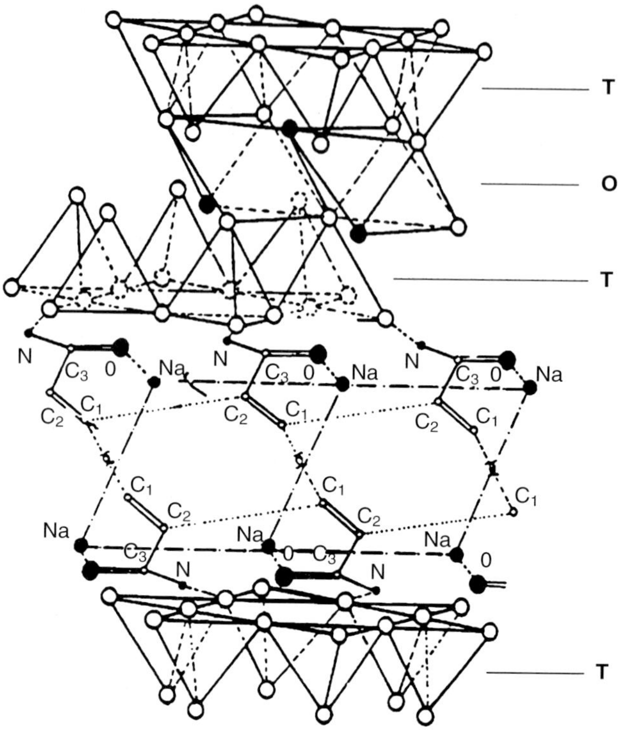

Figure 2.8 Schematic conformation of a bimolecular layer of acrylamide intercalated into the interlamellar space of an expanding three-layer (TOT) clay mineral parallel <001> (Ogawa, Kuroda, and Kato, 1989; Gao and Heimann, 1993). In the TOT lattice only oxygen atoms are shown. Reprinted with permission from Elsevier.

Today, there is also the intriguing notion that montmorillonite may have acted as a catalyst and template for the origin of life on earth (Cairns-Smith, 1986; Ertem and Ferris, 2000; see also Güven, 2009). Montmorillonites assist fatty acid micelles to self-organize into membrane vesicles that are then able to growth and divide in response to changes of the pH, bringing together the key components of early life. Montmorillonite particles often become encapsulated in these vesicles, thus providing a pathway for the prebiotic formation of catalytically active surfaces within membrane vesicles. In addition, montmorillonite is known to catalyze the polymerization of ribonucleic acid (RNA) from activated ribonucleotides. RNA, when adsorbed onto clay, can also be encapsulated within vesicles (Srivatsan, 2004) which, once formed, are able to grow by incorporating additional fatty acids supplied as micelles; the vesicles can then divide without any dilution of their contents by extrusion through small pores. As these processes mediate vesicle replication through cycles of growth and division, they may serve as models for the origin of the earliest living cells in a proposed "RNA world" (Ertem and Ferris, 2000; Hanczyc *et al.*, 2003). The geographically widespread custom of eating clay (geophagy; Carretero and Lagaly, 2007) for gastrointestinal ailments, and also applying clays externally to remedy skin infections (Williams *et al.*, 2009), may represent a faint echo of the crucial importance of clays for the origin of life.

Today, bentonites (i.e., montmorillonite-based rocks) have an extraordinarily wide range of industrial applications. Typical uses of bentonite mined in the USA (ca. 5 million metric tons in 2006) includes application as a bonding agent of sand in foundry molds (16%), in drilling fluids (25%), pet litters (25%), as bonding agents for iron ore pelletizing (16%), and various uses as bleaching agents for vegetable and mineral oils, as fining additives to remove proteins from wine and fruit juices, as desiccants, in paper-making (including carbonless copy paper), as environmental sealants for waste disposal sites, as slope-stabilizing additives when digging trenches and boring tunnels, as animal feeds, and as fillers and extenders for polymers (18%) (USGS, 2008).

Expanding three-layer clay minerals intercalated with polymers such as (poly) acrylamide or alkali acrylates results in so-called superabsorbing polymer-clay nanocomposites (SAPC) (Theng, 1974; Gao and Heimann, 1993), which have remarkable rheologic (Gao *et al.*, 1999) and swelling (Li *et al.*, 2004) properties. Currently, these composites are being developed for widespread applications in areas as diverse as environmental protection (dehydration of sludges, sorbing agent for toxic organic chemicals and heavy-metal contaminants), agriculture and silviculture (as a soil amelioration agent), civil engineering (friction-reducing material for placing pipes, additives to special cements/concretes; Gao *et al.*, 1997; corrosion protection of steel; Hoang, 2007), energy management (water retention in enhanced oil recovery operations; Rong *et al.*, 1985), dewatering of adulterated diesel fuel and gasoline; Gao *et al.* 1996), and health and domestic uses (sanitary napkins and towels, paint, adhesive, food packaging). Modern polymer–clay nanocomposites (PCNs) containing only a few percent of montmorillonite are being increasingly used to provide structural reinforcement and mechanical strength (e.g., in automotive body panels), as gas-permeation barriers and flame retardants, as surface erosion protection for heat shields in aerospace modules, as well as conducting nanocomposites and bionanocomposites for tissue engineering (Carrado and Komadel, 2009).

Figure 2.8 shows, schematically, the mode of intercalation of poly(acrylamide) (PAM) into the interlayer space of montmorillonite. The oxygen atoms of the carbonyl group form van der Waals-type bonds with the exchangeable cation such as Na^+, while the NH_2^- group forms hydrogen bonds with the terminal oxygen atoms of the tetrahedral silicate sheet. As further shown in Figure 2.8, the orientation relations are $(010)_{PAM} \parallel (010)_{clay}$ and $\langle c^* \rangle_{PAM} \parallel \langle 001 \rangle_{clay}$. In the desolvated state, the intramolecular distance between equivalent $CH_2 =$ groups (C_1) of neighboring molecules is 0.38 nm (dashed lines). The dotted lines between C_1 and C_2 groups of adjacent molecules indicate the formation of a polymer chain by "head-to-tail" arrangement with a distance of 0.45 nm (Gao and Heimann, 1993).

PCNs, acid-activated bentonites, organophilic montmorillonites, and pillared clays with designed micro- and mesoporosity (Güven, 2009) find specific and selective applications as catalysts and catalyst carriers in the petrochemical industry, and also in environmental applications. The capability of smectites to absorb not only water but also toxic organic contaminants, heavy-metal pollutants and radionuclides, has led to many investigations for applying smectite-bearing clays as buffer and sealing materials with high absorption selectivities for the fission products in nuclear waste disposal vaults. In fact, this role has been proposed by several national nuclear waste management programs in countries, including Canada (e.g., Bird and Cameron, 1982; Rummery and Rosinger, 1984; Heimann, 1987), the United States (Komarneni and Roy, 1980), Sweden (Pusch and Kamland, 1996), Switzerland (Van Loon and Glaus, 2008), and others. The increasing use of swelling clays such as bentonites to seal immobilized/stabilized industrial and domestic waste repositories also requires the retention of swelling and adsorption capabilities over the lifetime of the waste-disposal facility (Gates *et al.*, 2009). Hence, a control of the ion-exchange mechanisms is required that may compromise the desirable properties of the expandable smectite clays (see, e.g., Heimann, 1993; Heling, 1994). In contrast to this concept, a safe disposal philosophy has been proposed that involves constructing final disposal sites below the groundwater level, without technical barriers (Pentinghaus and Kienzler, 1993).

Further industrial uses of montmorillonitic minerals are based on their thixotropic nature (Kelessidis, 2008). Thixotropy is a property exhibited by certain gels (semisolid, jellylike colloids) that appear to be solid and maintain a shape of their own until subjected to a shearing force or some other disturbance, such as shaking; they then act as a sol (a semifluid colloid) and flow freely. Thixotropic behavior is reversible, and when allowed to stand undisturbed the sol slowly reverts to a gel. Common thixotropic gels include oil well drilling mud, certain paints, and printing inks (Eisenhour and Brown, 2009). (For details on thixotropy and its physical underpinning, see Section 2.4.2.5 on the rheology of clay–water suspensions.) A special issue of the international magazine *Elements* has been devoted to the origin, mineralogy, geochemistry and modern applications of bentonite (Bain, 2009)

2.1.3
Nomenclature of Clays

The importance of clays as raw materials for classic ceramics, their widespread occurrence, chemical and structural variability, and the dependence of processing and firing properties on the phase composition of the precursor materials of ceramic products has led to several attempts to develop a comprehensive system of clay nomenclature. Brindley (1951) reported the earliest efforts to obtain international collaboration on nomenclature and classification of clay minerals. Since then various national clay groups have been formed proposing a variety of changes of nomenclature at group meetings of the International Clay Conferences. Most national clay groups have representation on the Nomenclature Committee of the Association Internationale pour l'Etude des Argiles (AIPEA; International Association for the Study of Clays), which was established in 1966. This Association has worked closely with other international groups, including the Commission on New Minerals and Mineral Names (CNMMN) of the International Mineralogical Association (IMA), which is responsible for the formal recognition of new minerals and mineral names, and the International Union of Crystallography (IUCr). In contrast to the other national clay groups, however, the Clay Minerals Society (CMS) Nomenclature Committee, established in 1963, remains in existence and occasionally produces recommendations. The precursor to this committee was the Nomenclature SubCommittee, which was organized in 1961 by the (US) National Research Council. From time to time, the AIPEA issues recommendations in close contact with the national organizations (Guggenheim *et al.*, 2006).

In Germany, the much simplified system introduced by Ernst *et al.* (1959) has been widely used by the ceramic industry as a simple and versatile tool to describe, specify, and standardize clay mixtures (Table 2.4).

For example, a clay denominated qualitatively KIQ contains kaolin > illite > quartz. A more quantitative description would be $K(70)I(\overline{10})Q(\overline{20})$; that is, 70 mass% minerals of the kaolin group, and approximately 10 mass% illite and 20 mass% quartz, whereby the overstrike indicates an approximate value. The main group minerals with <8 mass% will be separated by a hyphen; that is, KIQ-Y characterizes a clay with predominately kaolin + illite + quartz, and less than 8 mass% of accessory minerals. The system sensibly also considers special features such as composition, cation-exchange capacity, or grain size. For example, $K_{ka+K}(65+x)$ describes a kaolin containing 65 mass% kaolinite and an unknown amount of another mineral of the kaolin group K. The identifier $M_{mo}(55;\ Na^+;\ 95\,mval$ per $100\,g)$ describes a clay of the montmorin group M containing 55 mass% Na-montmorillonite with a cation exchange capacity of 95 mval per 100 g dry mass. Finally, a 0.5/40 KQI clay is a clay with kaolin > quartz > illite having a median grain size diameter of 0.5 μm and a specific surface of $40\,m^2\,g^{-1}$.

2.2
Synthetic Raw Materials

Synthetic raw materials include precursors of glazes (lead oxide, barium carbonate, tin oxide) and special ceramic masses (alumina, zirconia, magnesia), as well as hydrothermally synthesized wollastonite and diopside, and synthetic clays such as kaolinite with very narrow grain size distribution and high plasticity for high-performance ceramics. Wet chemical techniques such as coprecipitation, freeze- and spray-drying, and sol–gel synthesis are also applied to produce raw materials for special applications.

Wollastonite and diopside are utilized as abrasion-resistant materials for road surfaces and fillers for rubber and plastics (Jacob, 1976), and also as thermally insulating materials for a variety of applications (Demidenko *et al.*, 2001). Currently, these calcium and calcium magnesium silicates are synthesized starting with mixtures of moist lime (CaO) or burned dolomite (CaO + MgO) and finely ground quartz flour. The mixtures are treated hydrothermally between 200 and 250 °C for 7–16 h to yield xonotlite or mixtures of xonotlite and serpentine according to:[1)]

1) Cement chemical notation: C = CaO, S = SiO_2, H = H_2O, M = MgO

Table 2.4 Comprehensive nomenclature system of clays. After Ernst *et al.* (1959).

Main group		Subgroup		Mineral	
K	Kaolin group			ka	kaolinite
				dc	dickite
				fc	fireclay mineral
		HL	Halloysite		
M	Montmorin group (expanding)			mo	montmorillonite
				bd	beidellite
				nt	nontronite
				sa	saponite
I	Mica-like minerals (nonexpanding)	IL	Clay minerals of the mica group	hm	hydromuscovite
				il	illite
				ld	ledicite
				mu	muscovite
				bt	biotite
				ph	phlogopite
				gl	glauconite
X	Other silicatic clay minerals			tc	talc
				pg	palygorskite
		ML	Mixed-layer structures	co	corrensite
				hb	hydrobiotite
		VM	Vermiculite group	vm	vermiculite
		CH	Chlorite group		
		SP	Serpentine group		
A	Aluminum oxides			gi	gibbsite
				di	diaspor
				bo	böhmite
Q	Quartz minerals			qz	quartz, chalcedon
				cr	cristobalite
				op	opal
F	Feldspar group			or	orthoclase
				pl	plagioclase
S	Soluble salts	CL	Chlorides		
		SF	Sulfates	gy	gypsum
C	Carbonates			cc	calcite
				do	dolomite
				ak	ankerite
				sd	siderite
Y	Other accessory minerals			fe	goethite
				hm	hematite
				mt	magnetite
				il	ilmenite
				ru	rutile
				zr	zircon
O	Organic materials				

$$6Ca(OH)_2 + 6SiO_2 \rightarrow C_6S_6H + 5H_2O \text{ or} \tag{2.1a}$$

$$6Ca(OH)_2 + 6Mg(OH)_2 + 12SiO_2 \rightarrow C_6S_6H + M_6S_4H_4 + 2S + 7H_2O \tag{2.1b}$$

Calcination at 1150–1350 °C yields wollastonite or diopside according to:

$$C_6S_6H \rightarrow 6CS + H_2O \tag{2.2a}$$

$$C_6S_6H + M_6S_4H_4 + 2S \rightarrow 6CMS_2 + 5H_2O. \tag{2.2b}$$

The treatment of mixtures of silica and calcium carbonate at 1000–1100 °C in the dry state yields wollastonite directly by a solid-state reaction (Jacob, 1976).

The hydrothermal synthesis of kaolinite from high-purity silica and alumina yields materials with excellent purity, high crystallinity, very small grain size, and high plasticity. The median size of the crystallites is about 0.1 μm, while the thickness of the hexagonal platelets varies between 1 and 100 nm. Figure 2.9 shows the grain size distribution of the artificial kaolinite in comparison with two natural materials from Georgia, USA and Japan.

The wet chemical synthesis of precursor materials for ferrites with spinel structure by coprecipitation of metal chlorides and subsequent calcinations yields fine-grained powders with a homogeneous distribution of the constituent oxides, for example:

$$NiCl_2 + FeCl_3 + Na_2CO_3 \rightarrow (Ni,Fe)CO_3 \tag{2.3a}$$

$$3(Ni,Fe)CO_3 + \frac{1}{2}O_2 \rightarrow (Ni,Fe)_3O_4 + 3CO_2 \tag{2.3b}$$

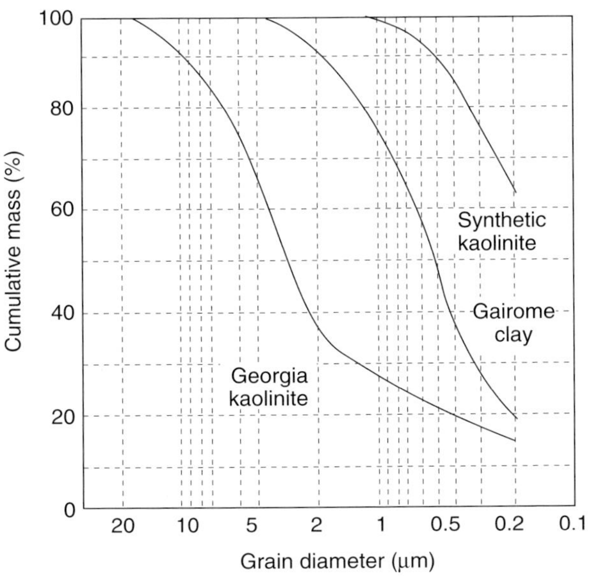

Figure 2.9 Grain size distribution of hydrothermally synthesized kaolinite compared to natural kaolinitic clays from Georgia, USA and Gairome, Japan. Reprinted with permission of Japan Fine Ceramics Center, Nagoya, Japan.

In advanced ceramic processing, a chemical precursor to synthesized powder is frequently a solution of a metal salt in an appropriate solvent. These metal carriers can either be metal alkoxides or metal salts such as nitrides or chlorides in true solution, and in both cases the metal precursor will be hydrolyzed, followed by condensation at elevated temperature. The reaction steps for metal alkoxides are as follows:

1) Hydrolysis of alkoxides $M(OR)_x$ with $R = CH_3–$, $CH_3CH_2–$, etc., to yield:

$$M(OR)_x + yH_2O \rightarrow M(OH)_x(OR)_{x-y} + yROH \tag{2.4a}$$

2) Polymerization–condensation of alkoxides to yield:

$$–M–OH + HO–M– \rightarrow –M–O–M– + H_2O \tag{2.4b}$$

The reaction steps for metal salts M^{x+} involve:

1) Slow hydrolysis of metal salts at elevated temperature:

$$M^{x+} + yH_2O \rightarrow M(OH)_x + yH^+ \text{ or } M^{x+} + OH^- \rightarrow M(OH)_x \tag{2.4c}$$

2) Polymerization–condensation:

$$–M–OH + HO–M– \rightarrow –M–O–M– + H_2O. \tag{2.4d}$$

Frequently, the hydrolysis process is followed by a wash with excess electrolyte, and dispersion (peptization) at elevated temperature with a strong acid. The resultant sol is stabilized by the positively charged H_3O^+-exchanged surface at pH 2 ... 3. The condensation process proceeds through particle nucleation, the Gibbs free enthalpy of which is governed by an increase in surface free energy proportional to r^2 ($\Delta G_0 = 4\pi r^2 \cdot \delta'$), and a decrease proportional to the volume of the newly formed particle, that is, r^3 ($\Delta G_V = -4/3\pi \cdot r^3 \cdot \Delta g_V$, where $\Delta g_V = S \cdot \Delta T$). (For details, see textbooks on general thermodynamics of classical nucleation theory.)

2.3
Processing and Forming of Clay Powders

In contrast to metals and some engineering polymers, neither casting nor hot or cold working represent viable forming routes for ceramics. As most properties of any ceramic composition have already been decided at the forming stage, there is a minimal possibility for any in-process or post-process modifications of sintered ceramics. As an example, defects such as cracks that might be introduced during the forming stage have little chance of being healed at the sintering stage, as solid-state diffusion in ceramics, even at high temperatures, is very slow. Hence, close attention must be paid to the effective shaping and consolidation of ceramics in the green stage – that is, prior to sintering. The formation processes of ceramic green bodies can be classified as either *dry forming*, or *wet forming*, which requires that a certain amount of water is added to the ceramic precursor powder.

2.3.1
Dry Forming Methods

These include uniaxial dry pressing and cold isostatic pressing, that are governed by the laws of powder mechanics. Powder mechanics, when applied to the dry forming of ceramics, focus on powder compaction being achieved without the disintegration of individual powder particles. As a maximum packing density is desired so as to produce mechanically and thermally stable end products, methods to increase the packing density have been developed that include:

- Grading powders by blending fractions of different particle size, so as to fill any gaps between coarse particles with fine particles.
- Fluid-assisted packing of dispersed powders.
- Reducing particle interaction by spheroidization, an increase in the average size, or the addition of lubricants such as poly(vinyl alcohol) (PVA) or poly(ethylene glycol) (PEG).
- The application of an external force, although this may involve risks related to the elastic spring-back and/or cracking of the particles.

Powder grading is frequently used in industrial practice. The maximum packing factor can be calculated according to the Furnas model (Messing and Onoda, 2006), using the additivity of the individual powder fractions. For example, three fractions with different grain size and hence packing factors Φ_c (coarse), Φ_m (medium) and Φ_f (fine), yield a theoretical maximum packing factor:

$$\Phi_{max} = \Phi_c + (1-\Phi_c)\Phi_m + (1-\Phi_c)(1-\Phi_m)\Phi_f .$$

During uniaxial dry pressing, a considerable pressure gradient develops along the pressure axis, as well as across the diameter of the pressed shape. These gradients create gradients in both density and elasticity, and a loss of tolerance after densification by sintering. Upon ejection from the mold, a relaxation of the differential strain created by the elasticity gradient can lead to fracturing, delamination, and ring capping. The addition of binders and/or lubricants with only a small elastic recovery on stress relief will counteract such failures. Both, dry (0–5% water) and semi-dry (5–8% water) pressing are used preferentially to form high-throughput advanced ceramic parts with simple geometries.

2.3.2
Wet Forming Methods

These include extrusion, injection-molding, tape-casting, slip-casting, and wet colloidal processing, including sol–gel techniques. These methods are dominated by the rheological behavior of the clay particle suspensions, that in turn control the flow of the suspension and, therefore, the ease of shape formation. For details of forming clay green bodies, the reader is referred to specialized literature.

However, the basic principles of the interaction of clay with water, and the rheology of clay particle suspensions, are discussed in the following section.

2.4
Rheology of Clay Particle Suspensions

2.4.1
Modeling of Rheological Behavior

The study of the rheological nature and behavior involves phenomena that occur during the deformation and flow of materials. The responses of solids and liquids to mechanical deformation are governed by two fundamental functional relationships: (i) the stress (σ)–strain (ε) curve; and (ii) the flow curve – that is, the relationship between shear stress (τ) and velocity gradient ($D = \mathrm{d}v/\mathrm{d}y$). The two limiting equations are:

- Hooke's law for ideally elastic solids: here, $\tau = G \cdot \gamma$ (where τ is the shear stress, G is the shear modulus, and γ the shear deformation), or $\sigma = E \cdot \varepsilon$ (where σ is the stress, E is Young's modulus of elasticity, and ε is the strain).
- Newton's law for ideally viscous liquids: here, $\tau = \eta \cdot D = \eta \cdot (\mathrm{d}\gamma/\mathrm{d}t)$ (where η is the dynamic viscosity, D is the $\mathrm{d}v/\mathrm{d}y$ velocity gradient, and γ is the shear deformation).

In ideally elastic solids, the deformation energy will be stored within the solid without any losses, whereas in ideally viscous liquids the deformation energy will be completely transformed into frictional heat. In the "real world" however, the elastic constants connecting the tensorial components of stress and strain, including the modulus of elasticity E, the shear modulus G, the Poisson number ν, the compressibility κ, and other parameters, are not time-invariant. Consequently, even after releasing very small stresses, both residual and time-dependent deformations remain in engineering solids that require a certain relaxation time to restore the equilibrium state of the system (e.g., see Meyers and Chawla, 2009).

To account for the nonideal nature of real solids and liquids, the theory of linear viscoelasticity provides a generalization of the two classical approaches to the mechanics of the continuum – that is, the theory of elasticity and the theory of hydromechanics of viscous liquids. Simulation of the ideal boundary properties "elastic" and "viscous" requires mechanical models that contain a combination of the ideal element "spring" to describe the elastic behavior as expressed by Hooke's law, and the ideal element "dash pot" (damper) to simulate the viscosity of an ideal Newton liquid, as expressed by the law of internal friction of a liquid. The former follows the equation $F = D^* \cdot x$ (where F = force, x = extension, and D^* = directional force or spring constant). As D^* is time-invariant, the spring element stores mechanical energy without losses. The force F then corresponds to the stress σ, while the extension x corresponds to the strain ε to yield $\sigma = E \cdot \varepsilon$.

On the other hand, the "dash pot" is a damping element and is thought to express the viscous behavior of an ideal liquid by the friction to which a piston is subjected when gliding in a cylinder filled with a viscous liquid (Figure 2.10). Owing to this internal friction, the deformation energy will be transformed completely into heat. The equation $\sigma = \eta \cdot \dot{\varepsilon} = \eta \cdot (d\varepsilon/dt)$ means that the stress is proportional to the deformation (strain) rate, with the dynamic viscosity η as proportionality constant. The nonlinear spring ($E = E(\varepsilon)$) and dash pot ($\eta = \eta(\dot{\varepsilon})$) elements account for nonlinear relations between stress and strain, and between stress and deformation rates, respectively (Figure 2.10).

The slide block (Saint–Venant body) simulates ideal plastic behavior with no strain at all below a critical yield stress θ (Figure 2.10), although at and above the critical yield stress the strain increases without limit. This behavior is frequently invoked in problems of slope stability for wet fine-grained materials such as silt and clay, as the slide block will not move due to friction until sufficient stress is applied.

Coupling of the model elements can be achieved either in series or in parallel, this being analogous to capacitors in an electrical circuit (Figure 2.11).

Coupling in series. The partial shear stresses τ_S of the spring elements and τ_D of the damping element equal the total shear stress $\tau = \tau_S = \tau_D$, analogous to the total amount of electricity transported through a series of capacitors that is $Q = Q_1 = Q_2 = \ldots$ In contrast, the total shear deformation is the sum of the partial shear deformation γ_S and γ_D, $\gamma = \gamma_S + \gamma_D$. In the same way, the total voltage of a series of capacitors is $U = U_1 + U_2 + \ldots$ (Figure 2.11a). These relations pertain to the presence of shear forces. In the case of (pure) tensile forces, the shear stress τ will be replaced by the tensile stress σ, and the shear deformation γ by the strain ε.

Coupling in parallel. The partial shear stresses add up to the total shear stress, $\tau = \tau_S + \tau_D$, as do the capacities $C = C_1 + C_2 + \ldots$ of an electrical circuit consisting

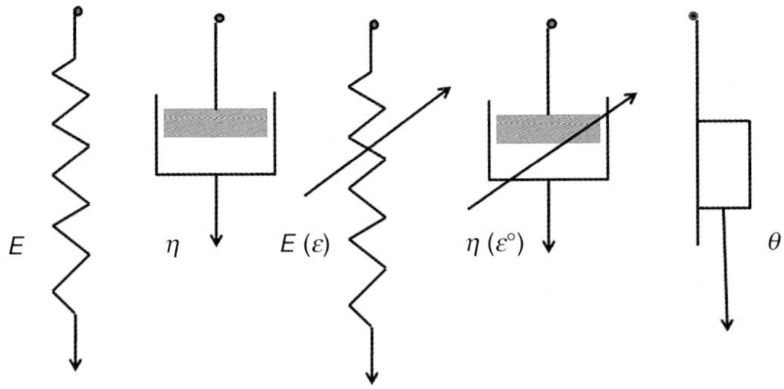

Figure 2.10 Model elements for viscoelastic simulations (from left: linear spring element, linear dash pot element, nonlinear spring element, nonlinear dash pot element, sliding block (Saint–Venant body) with yield stress θ (after Hennicke, 1978).

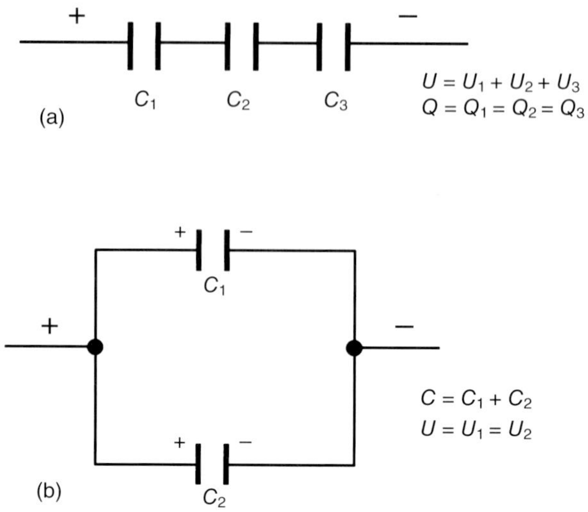

Figure 2.11 Electrical equivalent circuits for arranging rheological model elements. (a) Capacities in series; (b) Capacities in parallel.

of capacitors in parallel. The partial shear deformations γ_S of the spring elements and γ_D of the damping element equal the total shear deformation $\gamma = \gamma_S = \gamma_D$, analogous to the total amount of voltage transported through capacitors in parallel that is $U = U_1 = U_2 = \ldots$ (Figure 2.11b). These relations pertain to the presence of shear forces. In the case of (pure) tensile forces, the shear stress τ will be replaced by the tensile stress σ, and the shear deformation γ by the strain ε.

To account for the fact that neither ideal solids nor ideal liquids exist in the real word, the rheological behavior of real materials can be approximated by a combination of the individual model elements, either in linear two-element models of Maxwell and Voigt–Kelvin types, or in linear three-element models. In many real cases, nonlinear models have to be invoked (see Section 2.4.1.3).

2.4.1.1 **Linear Two-Element Models**

The behavior of linear spring and dash pot elements coupled in series (Maxwell model) can be characterized by the following functional relations, that are depicted graphically in Figure 2.12a. From the analogy to the in-series arrangement of capacities in electrical circuits (Figure 2.11a) it follows that:

$$\tau = \tau_S = \tau_D;\ \gamma = \gamma_S + \gamma_D \text{ and} \tag{2.5a}$$

$$d\gamma/dt = d\gamma_S/dt + d\gamma_D/dt \tag{2.5b}$$

For the spring element:

$$d\tau/dt = G \cdot d\gamma_S/dt,\ d\gamma_S/dt = (1/G) \cdot d\tau/dt \tag{2.6a}$$

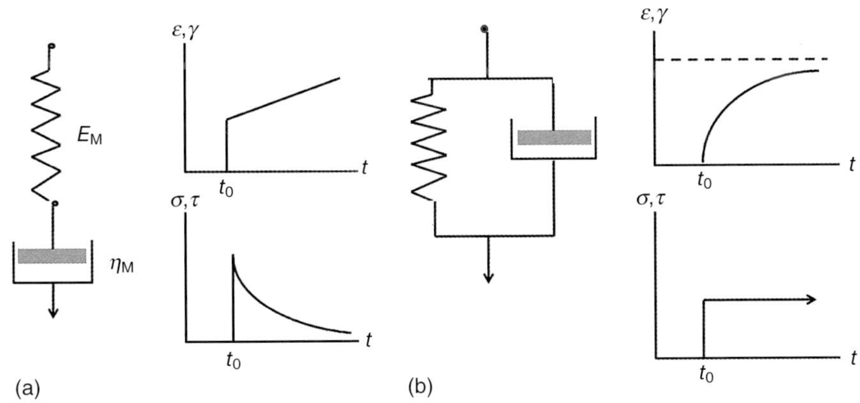

Figure 2.12 Behavior of two-element models. (a) Maxwell model (spring and dash pot elements in series); (b) Voigt–Kelvin model (spring and dash pot elements in parallel). Time-dependent responses of tensile strain ε or shear deformation γ at constant stress σ or shear stress τ (top), and tensile stress σ or shear stress τ at constant strain ε or shear deformation γ (bottom) (after Hennicke, 1978).

For the dash pot element:

$$\tau = \eta \cdot d\gamma_D/dt,\ d\gamma_D/dt = (1/\eta)\cdot\tau. \tag{2.6b}$$

Substitution of Eq. (2.6a) and Eq. (2.6b) into Eq. (2.5b) yields:

$$d\gamma/dt = (1/G)\cdot d\tau/dt + (1/\eta)\cdot\tau. \tag{2.7}$$

For $\gamma =$ const. ($d\gamma/dt = 0$), the stress τ decreases with time in an exponential fashion according to $\tau(t) = \tau_0 \cdot \exp(-t/t^*)$ (Figure 2.12a, bottom right) with a *relaxation time* $t^* = \eta/G$.

Accordingly, the behavior of parallel coupling of spring and dash pot elements (Voigt–Kelvin model, Figure 2.12b) is governed by:

$$\gamma = \gamma_S = \gamma_D;\ \tau = \tau_S + \tau_D. \tag{2.8}$$

Substitution of $\tau_S = G \cdot \gamma_S$ (spring element) and $\tau_D = \eta \cdot (d\gamma_D/dt)$ (dash pot element) into Eq. (2.8) yields

$$\tau = G\cdot\gamma_S + \eta\cdot(d\gamma_D/dt), \tag{2.9a}$$

and, since $\gamma_S = \gamma_D = \gamma$, the shear stress is $\tau = G\cdot\gamma + \eta\cdot(d\gamma/dt)$. (2.9b)

The ratio $\eta/G = t^{**}$, in this case called the *retardation time*, is defined as the decrease of the time derivative of the (shear) deformation, ($d\gamma/dt$) at constant (shear) stress τ. Equation (2.9b) can thus be written as:

$$\tau = G[\gamma + t^{**}(d\gamma/dt)]. \tag{2.9c}$$

For $\tau =$ const. ($d\tau/dt = 0$), the shear deformation γ approaches with increasing time t a constant value according to the exponential law $\gamma(t) = (\tau_0/G)[1 - \exp(-t/t^{**})]$ (Figure 2.12b, top right).

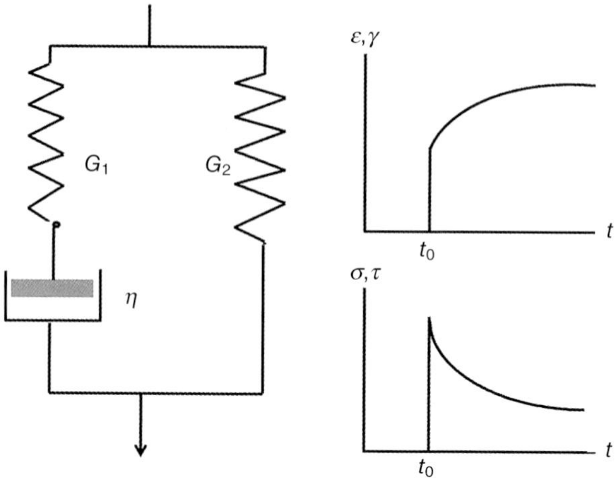

Figure 2.13 Three-element model for viscoelastic behavior of liquids (general relaxation model) as a combination of Maxwell and Voigt–Kelvin models.

2.4.1.2 Linear Three-Element Models

The functional relations follow a rather complex linear differential equation of second order:

$$a_0\tau + a_1\dot{\tau} + a_2\ddot{\tau} = b_0\gamma + b_1\dot{\gamma} + b_2\ddot{\gamma}. \quad (2.10)$$

The coefficients a_i and b_i of Eq. (2.10) can be adjusted to account for deviation from reality.

As shown in Figure 2.13, the general relaxation model can be thought of as being composed of a spring element characterized by the shear modulus G_2 and a Maxwell two-element combination characterized by the shear modulus G_1. Treatment analogous to the linear two-element models yields the equation

$$\tau + (\eta/G_1)(d\tau/dt) = G_2\cdot\gamma + \eta((G_1 + G_2)/G_1]\cdot(d\gamma/dt). \quad (2.11)$$

Note that the form of the $\varepsilon(t)$ and $\gamma(t)$, and $\sigma(t)$ and $\tau(t)$ curves, respectively, of the three-element model can be derived by superposition of the individual curves shown in Figure 2.12. This superposition of Maxwell and Voigt–Kelvin models in the form shown in Figure 2.13 is particularly suitable to describe the rheological behavior of viscoelastic clay–water suspensions under large shear loads. In particular, they have profound consequences for the formation of ceramic powders in the presence of large amounts of binders, as encountered during the extrusion, injection-molding and tape-casting techniques that are used increasingly in the economic fashioning of mass-produced ceramic parts for a variety of industries. As the strain rates in these forming methods can vary within five orders of magnitude, from 1% per second to 100 000% per second, they may have a major impact on the flow properties of the clay and on advanced ceramic precursor powders created using these techniques.

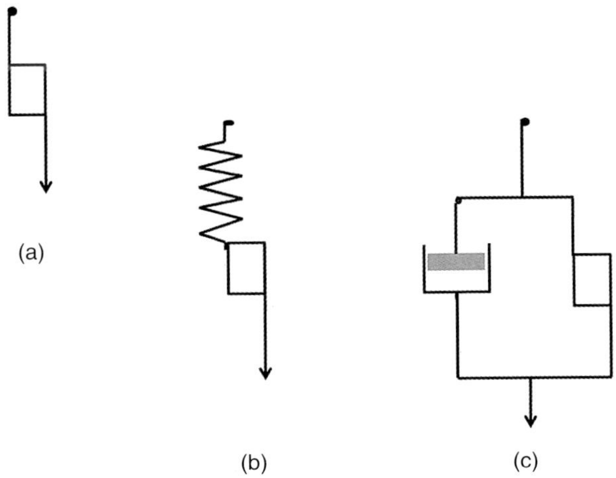

Figure 2.14 Nonlinear rheological models. (a) Saint–Venant model; (b) Prandtl–Reuss model; (c) Bingham model.

2.4.1.3 Nonlinear Models

Nonlinear models of rheological behavior can be approximated by step functions, whereby the existence of a finite yield stress θ plays a dominant role. Three typical nonlinear models include the Saint–Venant model of ideal plastic behavior, the Prandtl–Reuss model of an elastoplastic material, and the Bingham model of viscoelastic behavior. The first model can be mechanically approximated by a sliding block, the second by a Maxwell element and a sliding block in series, and the third by a dash pot damping element and a sliding block in parallel (Figure 2.14).

Below the yield stress θ the rate of shear deformation (strain) ($d\gamma/dt$) is zero in all three nonlinear models. This is also true of the shear strain γ for the Saint–Venant and Bingham models, whereas in the Prandtl–Reuss model γ increases slowly with the shear stress τ from zero to the yield stress θ. At this point, the value of γ increases limitless as a step function, as it does in the two other models (Figure 2.15).

2.4.2 Colloidal Processing of Clay

2.4.2.1 Structural Viscosity

Like all small particles, clay particles have the tendency to strongly agglomerate under the influence of van der Waals forces, that for submicron-sized particles, override the gravitational force. The small size of clay particles (typically <2 µm) and local surface charges are the result of the exposure of broken bonds and ion exchange with surrounding liquids. The local surface charges can be attributed to unsaturated bonds at cleavage planes, from distorted surface layers, and from

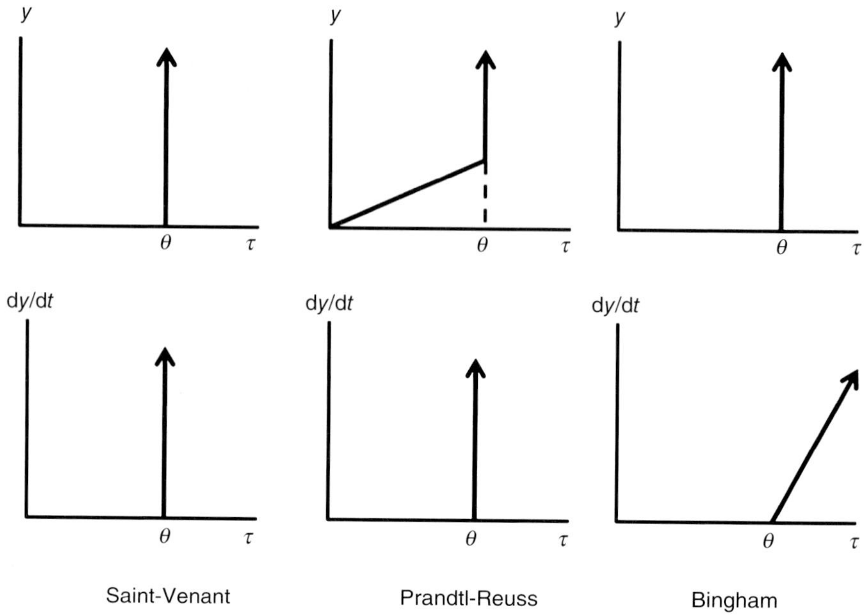

Figure 2.15 Responses of shear strain γ and rate of shear strain ($d\gamma/dt$) to increasing shear stress for three nonlinear rheological models.

polarized ions adsorbed at the surface. Typically, the edges of the particles are positively charged, and the faces negatively charged. The attraction of H_3O^+ ions to within 1 to 6 nm of the surface results in formation of a dense, rather stiff water layer surrounding the clay particles. This strongly adhering layer promotes the dispersion of individual clay particles at high concentrations of water, and the workability at lower concentration. Hence, the addition of surplus water to a clay results in a decrease of the yield stress (flow limit) θ and, consequently, an increase of the plastic deformation expressed as the velocity gradient $D = dv/dt = d\gamma/dt$ (see Section 2.4.1), and also a decrease of the apparent dynamic viscosity η^*. The functional relationship between these parameters can be expressed by:

$$D = (1/\eta^*)\cdot\tau^n. \quad (2.12)$$

Here, the exponent n characterizes the dependency of the apparent viscosity η^* on the shear stress τ. For $n = 1$, the system follows the ideal Newton equation, and $\eta^* = \eta$. However, for $n < 1$ the viscosity increases with increasing shear stress (dilatancy), while for $n > 1$ the viscosity decreases with increasing shear stress (structural viscosity); both are of great significance for many clay–water suspensions (Hawlader *et al.*, 2003). The dependence of apparent viscosity η^* and rate of shear deformation $d\gamma/dt$ on shear stress τ are shown in Figure 2.16a and b, respectively.

Structural viscosity – the decrease of apparent viscosity with increasing shear stress – is related to the solid–solid aggregation of clay particles as a response to

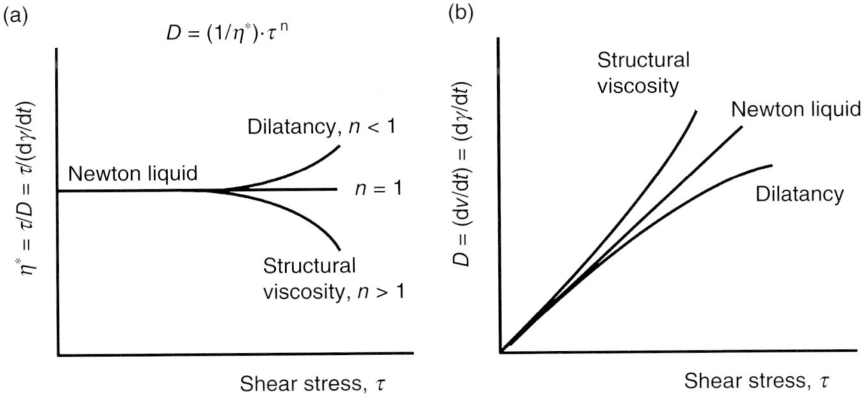

Figure 2.16 Dependency of apparent viscosity η^* (a) and rate of shear deformation $D = (\mathrm{d}\nu/\mathrm{d}t) = (\mathrm{d}\gamma/\mathrm{d}t)$ (b) on shear stress τ for viscoelastic liquids and suspensions, respectively.

electric interaction forces, and its decrease with increasing stress can be explained phenomenologically by shearing off the dense and stiff water layer surrounding the individual clay particles. On the other hand, dilatancy behavior occurs owing to density changes, so-called "shear thickening." Positive dilatancy–a decrease in density–can be observed in suspensions of spherical particles, such as quartz sand, that exhibit a maximum dense packing during rest but a less dense packing during mechanical shear loading. Negative dilatancy–an increase in density–occurs during the deformation of loam that prefers a loose packing, akin to a "house of cards" structure (see Figure 2.22) during rest that will be destroyed during mechanical loading, resulting in a denser structure.

If the structure of the viscoelastic suspension changes during flow (as discussed above), the shear stress τ is a nonlinear function of the shear deformation rate $(\mathrm{d}\gamma/\mathrm{d}t)$, expressed by the general relationship:

$$\tau = \alpha \cdot (\mathrm{d}\gamma/\mathrm{d}t)^{\beta}. \tag{2.13}$$

In this case, the apparent viscosity η^* is defined as the slope of the τ versus $\mathrm{d}\gamma/\mathrm{d}t$ line and yields

$$\eta^* = \mathrm{d}\tau/(\mathrm{d}^2\gamma/\mathrm{d}t^2) = \alpha' \cdot (\mathrm{d}\gamma/\mathrm{d}t)^{\beta-1}. \tag{2.14}$$

If $\beta < 1$, the materials are termed *pseudoplastic*, and their viscosity decreases with the rate of shear deformation as the structure of the flowing material gradually becomes more ordered. If $\beta > 1$, the material displays so-called *Bingham behavior* (see Section 2.4.1.3), and the viscosity increases with increasing shear deformation rate. Such systems are also characterized by a yield stress θ below which there is no deformation (shear deformation rate $D = 0$). If $\beta \equiv 1$, ideal Newtonian behavior is observed, with $\alpha = \eta$.

Figure 2.17 shows how the addition of varying concentrations of NaOH to a suspension of H-loaded kaolinite changes the rheological behavior, as expressed

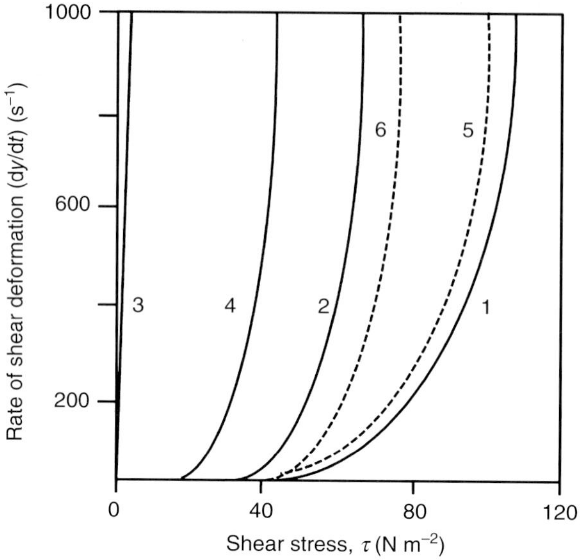

Figure 2.17 Flow curves of H-loaded kaolinite (200 g dry kaolinite per 500 ml solution). Modified after Salmang *et al.*, 2007. For an explanation, see the text.

by the dependency of the velocity gradient (dv/dt) and the rate of shear deformation ($d\gamma/dt$), respectively, on the shear stress τ. Here, a suspension of 200 g kaolinite in 500 ml distilled water displays Bingham structural viscosity behavior; that is, a yield stress of about 40 N m^{-2} is required to induce flow (curve 1), and the rate of shear deformation increases with increasing shear stress τ. Adding as little as 1 mval NaOH per 100 g dry kaolinite shifts the flow curve to a lower value of yield stress; that is, less stress is required to establish a steep flow velocity gradient (curve 2). Increasing the amount of monovalent electrolyte ions to 10 mval NaOH per 100 g dry kaolinite causes a dramatic change in the flow curve, to display quasi-Newtonian behavior; that is, a linear dependency of ($d\gamma/dt$) on τ (curve 3). Even close to an infinitesimal increase of shear stress induces an easy flow. However, large amounts of NaOH (i.e., a high pH, as encountered by the addition of 1000 mval NaOH per 100 g dry kaolinite) reverses this trend, as the high density of negative OH^- surface charges allows the particles to come into close contact again, with an ensuing increase in viscosity and also a required shear stress (curve 4). Cations with higher charge, such as Ca^{2+}, cause easy flocculation and hence lead to high shear stresses [curve 5 = 1 mval $Ca(OH)_2$ per 100 g dry kaolinite; curve 6 = 10 mval $Ca(OH)_2$ per 100 g dry kaolinite].

2.4.2.2 Hofmeister Series

As shown in Figure 2.17, the flow behavior of clay–water suspensions changes in the presence of soluble electrolyte ions that can be exchanged with the ions adsorbed at the surface of the clay particles. Natural clays yield predominantly Ca^{2+}

and Mg^{2+} ions in exchange sites and, in an acidic environment, also H_3O^+ ions. The kinetics – that is, the ease with which cations and (more pronounced) anions will be exchanged – follows the so-called *Hofmeister* (lyotropic) series of preferential adsorption of ions on clay particles. The ions on the left will be more easily exchanged by ions positioned further right, than *vice versa* (Kunz *et al.*, 2004):

$$Li^+ < Na^+ < K^+ < NH_4^+ < Mg^{2+} < Ca^{2+} < Sr^{2+} < Ba^{2+} < Al^{3+} < H^+. \quad (2.15)$$

This series is an expression of the increase of adsorption energies and a decrease of the thickness of the diffuse electrical double layer (Gouy–Chapman layer) surrounding the clay particles. This effect offers the principal possibility to influence the thickness of the double layer by adding to the clay–water suspension ions with variable thicknesses of hydration shells, and this will in turn influence the workability (the "plasticity") of the clays. For example, an exchange of Ca^{2+} and Mg^{2+} ions by Na^+ ions through the addition of NaCl, Na_2SiO_3 or $Na_6(PO_3)_6$ (sodium hexametaphosphate; "Calgon") results in a build-up of the double layer and thus a decreased interaction of individual clay particles. In Na-exchanged clays, each negative surface charge is exactly compensated by one Na^+ ion within the Stern layer (see below); hence, there is a maximum screening that reduces effectively the particle–particle interaction. The viscosity of the clay slurry and the degree of aggregation decrease in turn. In contrast, clay suspensions can be flocculated by adding small and more highly charged ions such as Ca^{2+}, Mg^{2+}, or Al^{3+}. These ions tend to reduce the thickness of the electrical double layer and cause the agglomeration (flocculation) of clay particles, owing to increased van der Waals attractions.

The Hofmeister series ranks the relative influence of ions on the physical behavior of a wide variety of aqueous processes, ranging from the colloidal assembly of clay minerals to protein folding. Originally, it was thought that an ion's influence on macromolecular properties was caused, at least in part, by "making" or "breaking" the bulk water structure. However, recent time-resolved and thermodynamic studies of water molecules in salt solutions have shown that a bulk water structure is not central to the Hofmeister effect. Instead, models are being developed that depend upon direct ion–macromolecule interactions as well as interactions with water molecules in the first hydration shell of the macromolecule (Zhang and Cremer, 2006).

2.4.2.3 **Effect of pH on Clay–Water Suspensions**

The interaction of protons with charged clay particles leads to several effects that are important for the rheological behavior of clay–water suspensions, and hence their workability.[2] At *low* pH (< 5), the clay particles maintain strongly positive edge charges and negative surface charges owing to the adsorption of protons at unsaturated O^{2-} bonds as well as OH^- bonds at edge sites. As a consequence, the positive edges and the negative surfaces of neighboring particles attract each other, so as to attain a mechanically rather stable configuration of a "house of cards"

2) The term "workability" (in German: Bildsamkeit) is preferred here over the term, "plasticity", since in crystal physics plasticity has a different connotation.

structure (see Figure 2.22). This structure imparts a high viscosity to the highly flocculated clay suspension. At extremely low pH (< 2), however, the negatively charged surfaces can become neutralized, with the consequence that the "house of cards" structure will collapse, causing a decrease in viscosity. Such exceptionally low pH values have been observed by the hydrolysis of exchangeable surface Al^{3+} ions involving the release of protons and the subsequent intercalation of gibbsite-like hydroxyl–Al complexes formed by this Brønsted acidification into the smectite lattice to form a nonexpandable Al–montmorillonite (Heimann, 1993).

At *intermediate* pH (5–8), protons will be gradually released from their edge positions. As a consequence, the positive charge at the clay particle edges is decreased and the mutual attraction of the particles and the slurry viscosity are likewise decreased. If the pH is further increased, the edge charge will be decreased to zero, and deflocculation will occur so as to form a sol, as the still negatively charged clay particle surfaces are sufficient to compensate the van der Waals attraction forces and thus to disperse the particles.

At *high* pH (8–12), significant proton release render both the edges and the surface negatively charged, and the strong repulsion of the particles causes highly dispersed, low-viscosity slurries to appear. At this high pH the slurry can be tailored to contain a very substantial proportion of solids (up to 50%), without any significant increase in viscosity.

At *very high* pH (> 12), flocculation occurs again since the high density of the negative OH^- surface charges allows the particles to come into close contact with each other, with an ensuing increase in viscosity (see Figure 2.17, curve 4).

It should be noted that a pH of exactly 7 is quite difficult to maintain for clay–water suspensions, since any addition of water to clay immediately causes a drop in pH to about 5 (this occurs because protons are rapidly released from the particle edges, as described above).

Ion adsorption at clay particles is a dynamic process, so that an exchange of ions can take place readily in response to the changing pH. These changes in pH influence the thickness of the Helmholtz–Gouy–Chapman electrical double layer, and in turn the value of the so-called *zeta potential* (ζ) that behaves inversely to the viscosity (see Figure 2.20).

2.4.2.4 **Zeta (Electrokinetic) Potential**

The concept of the zeta potential, ζ, is one of the most significant concepts in the science of colloidal processing of clays and ceramics. The functional dependency of this property on the type and concentration of ions, including pH, in the suspension serves to characterize the suitability of clays for a variety of ceramic wares.

The potential distribution around a clay particle with a negatively charged surface is shown in Figure 2.18. The surface charge $-Q_0$ will be compensated by an equal charge $+Q_0$ in the surrounding solution. This countercharge is distributed between two shells. Immediately adjacent to the surface a tightly adhering layer of positive counterions (Stern layer) exists at the distance r_1 with charge Q_1. Further outward, a second charge Q_2 extends as a diffuse layer that finally attains equilibrium with the polar liquid shell. Hence, there are three potentials: (i) the Nernst

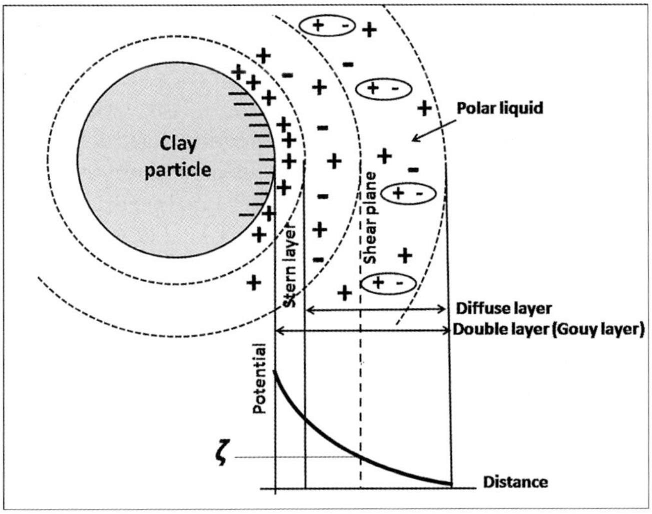

Figure 2.18 Potential distribution around a clay particle suspended in a polar liquid (water). The thickness of the Gouy electrical double layer is the sum of the Stern and the diffuse layers. ζ is the zeta potential.

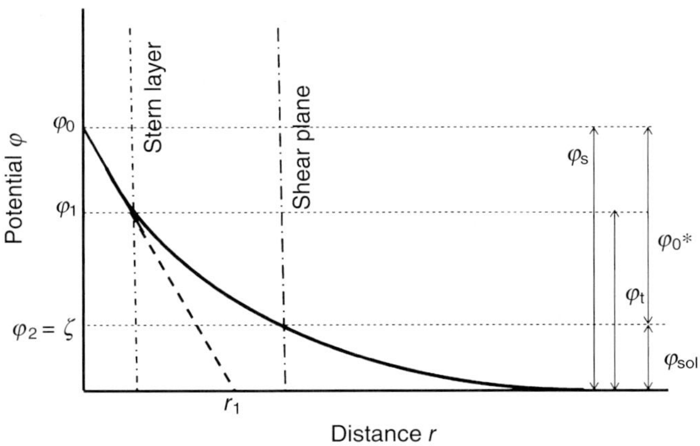

Figure 2.19 Details of the potential distribution shown in Figure 2.18.

surface potential φ_0 immediately at the clay particle surface; (ii) the Stern potential φ_1 at the transition between Stern layer and diffuse layer; and (iii) the ζ-potential φ_2, that may be defined as the potential at the intersection between the hydrodynamic shear (slipping) plane and the φ–r curve (Figure 2.19).

Historically, the electrical double layer concept has been proposed by Helmholtz (1879). According to this theory, two rigid layers of positive and negative ions exist close to the particle surface with a linear decay in potential, akin to an electric capacitor [$U = (4\pi \cdot Q \cdot d)/A$, where U is the potential, Q the electric charge, d the distance, and A the surface area]. However, this description is correct only to the first approximation. Later, the presence of a diffuse double layer was proposed by Gouy (1910) and Chapman (1913) with an exponentially decaying potential, and which appears to be much closer to reality.

Following the Gouy–Chapman theory, the quantitative treatment of the potential relations starts with the expression of the exponential change of the electrical potential φ in an electrolyte solution at the distance r from the charged clay particle (Figure 2.19):

$$\varphi = \varphi_0 \cdot \exp(-k \cdot r) \tag{2.16}$$

with φ_0 = (Nernst) surface potential at $r = 0$, and $1/k = L$ = Debye length (thickness of the electrical double layer). The Nernst potential φ_0 is a function of the surface charge Q_0 and the Debye length, and for surface potentials lower than ~25 mV can be approximated by $\varphi_0 = Q_0/(k \cdot \varepsilon_s)$, where ε_s is the dielectric permittivity of the solvent. For clay particles suspended in water, typical values are $Q_0 = 0.01 \ldots 0.1\,\mathrm{C\,m^{-2}}$, $\varepsilon_s = 80 \cdot \varepsilon_0$, $\varepsilon_0 = 8.85 \times 10^{-12}\,\mathrm{C\,V\,m^{-1}}$, $1/k = L = 1 \ldots 10\,\mathrm{nm}$.

The potential at the distance $r = L$ from the particle surface is defined as the zeta potential, ζ, and is equivalent to the electrokinetic potential. More specifically, it is the electrical potential at the location of the hydrodynamic shear (slipping) plane against a point in the bulk fluid far removed from the particle's surface (Figure 2.18). Hence, the zeta potential is the potential difference between the dispersion medium and the stationary layer of fluid attached to the dispersed particle (Figure 2.19). Quantitatively, this can be calculated from the equation:

$$\zeta = \varphi_{r=L} = \varphi_0 \cdot \exp(-1). \tag{2.17}$$

Since L is a function of the concentration of all ions c_j in solution and their valencies z_j, it follows that:

$$L \sim \sqrt{(1/\textstyle\sum c_j \cdot z_j^2)}. \tag{2.18}$$

Maximizing L involves a low concentration of large monovalent ions. In this case, the zeta potential will also be large, and the colloidal solution will be stabilized against collapse and flocculation. Typical values of L for a NaCl solution are 30 nm at $10^{-4}\,\mathrm{mol\,l^{-1}}$, and 1 nm at $0.1\,\mathrm{mol\,l^{-1}}$. Similarly for an $MgSO_4$ solution, L = 15 nm at $10^{-4}\,\mathrm{mol\,l^{-1}}$ and 0.5 nm at $0.1\,\mathrm{mol\,l^{-1}}$.

Figure 2.19 shows the potential distribution around a clay particle in greater detail. Here, the potential φ_s corresponds to the Nernst potential φ_0; that is, the potential of the surface against ground. It can be thought of as being composed of the potential φ_0^* and φ_{sol}, the potential of the aqueous solution; then, φ_t is the potential of the effective surface that equals φ_1 at the distance of the Stern layer. The zeta potential φ_2 is the difference of φ_t and φ_{sol}, and depends on the number of interacting ions – that is, the concentration of the exchangeable ions in solution, their effective charge, and the pH – and hence defines the thickness of the electrical double layer, L [see Eq. (2.18)].

Another quantitative expression of the zeta potential is based on the assumption of the validity of the Helmholtz approach of the electrical double layer behaving like a simple capacitor consisting of two parallel plates. Here, the capacity is $C = \varepsilon \cdot \varepsilon_0 \cdot A/d$, where ε is the dielectric permittivity of the suspension medium such as water, ε_0 is the dielectric permittivity of the vacuum, A is the particle surface, and d is the distance between two parallel, oppositely charged platelets. The effective charge is $Q = \sigma \cdot A$, where σ is the charge density per platelet. Since the potential is defined as the ratio of charge and capacity it follows that:

$$\varphi = \zeta = \sigma \cdot A \cdot d / \varepsilon \cdot \varepsilon_0 \cdot A = \sigma \cdot d / \varepsilon \cdot \varepsilon_0 . \tag{2.19}$$

Measurements of the zeta potential of a clay–water suspension in the presence of one or more electrolytes can be performed using a U-shaped tube (Burton tube) in which the suspension is subjected to an electric field applied between two electrodes. The electrophoretic velocity v of the electrolyte ions is related to the zeta potential. Then, the electric force $\sigma \cdot E$ (where E = electrostatic field strength) acting on the charged particles is balanced by the frictional force $\eta \cdot v/d$ (η = dynamic viscosity); that is:

$$\sigma \cdot E = \eta \cdot v/d. \tag{2.20}$$

A combination of Eqs (2.19) and (2.20) yields the Helmholtz–Smoluchowski equation:

$$\zeta = \eta \cdot v / \varepsilon \cdot \varepsilon_0 \cdot E. \tag{2.21}$$

This equation is at the basis of the measurement of the zeta potential from the electrophoretic mobility in a convenient instrument, the *Zeta meter*. However, the equation is invalid in the region where the mobility is affected by the surface conductivity of the particles and the polarization of the electrical double layer (Egorova, 2005).

The higher the (negative) zeta potential, the better is the dispersion of clay particles in clay–water suspensions. Some typical values are: Ca-loaded clay −10 mV; H-loaded clay −20 mV; Mg-loaded clay −40 mV; and Na-loaded clay −80 mV. The ranking of the exchangeable ions corresponds to the Hofmeister series discussed above. As clays treated with Na hexametaphosphate ("Calgon") reach zeta potential values as high as −135 mV, this addition is used ubiquitously in the processing of clay to achieve the maximum particle separation as required for slurry preparation for the slip-casting of pottery. The zeta potential and apparent viscosity behave in opposing fashion, as shown schematically in Figure 2.20.

It should be mentioned in passing that a similar dispersion of kaolinitic clays was achieved by ancient Chinese and medieval European potters through a process called "aging," whereby urine and feces were added to the raw clay. In this case, the urea serves as a peptization agent, penetrating the clay structure and being adsorbed preferentially between the layer stacks, and in turn allowing the clay mineral platelets to slide very easily over each other. It was suggested that bacterial activity might assist in this process (Oberlies and Pohlmann, 1958).

Knowledge of the zeta potential of a clay slurry, and the ability to manipulate this by adding particular monovalent or polyvalent ions, by modifying the pH, and/or by altering the viscosity by altering the proportions of solids, are at the

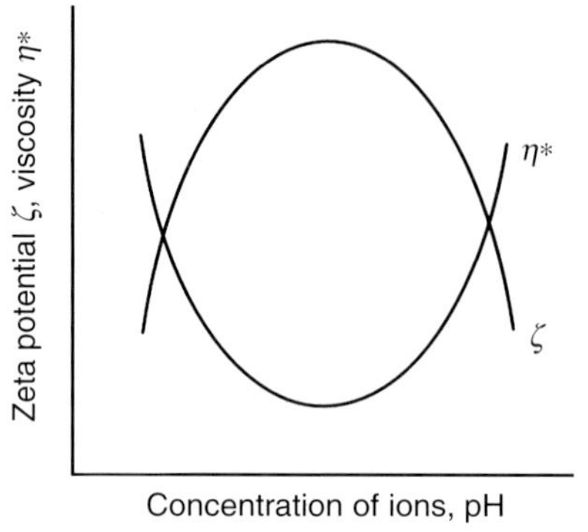

Figure 2.20 Schematics of the inverse behavior of the zeta potential (ζ) and apparent viscosity, η^*.

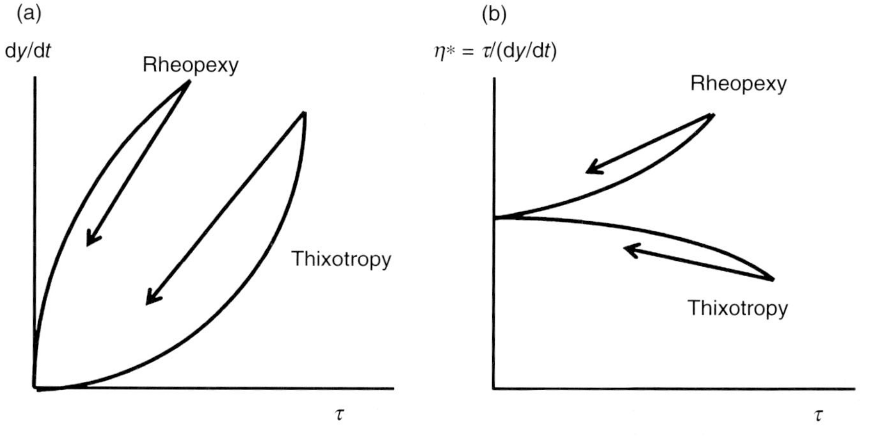

Figure 2.21 Dependence of shear deformation rate $d\gamma/dt$ (a) and apparent viscosity η^* (b) on shear stress τ for thixotropic and rheopectic materials.

heart of achieving high-quality ceramic products such as chinaware, fine porcelain, and some items of fine stoneware.

2.4.2.5 **Thixotropy and Rheopexy**

Thixotropy is the property of a gel to become a liquefied sol by mechanical loading such as shaking, but then to stiffen again after resting so as to regain the viscosity of a gel. In terms of rheological parameters, thixotropy is defined as the decrease of viscosity under constant shear stress or shear rate, followed by a time-dependent recovery after removal of the shear load (Figure 2.21). The opposite phenomenon, termed *rheopexy* (flow strengthening), involves an increase of (shear) viscosity with increasing shear stress. Rheopectic behavior is generally observed in systems with a high proportion of macromolecules that become thoroughly entangled during shear loading, with a resultant increase in viscosity. One simple system with a

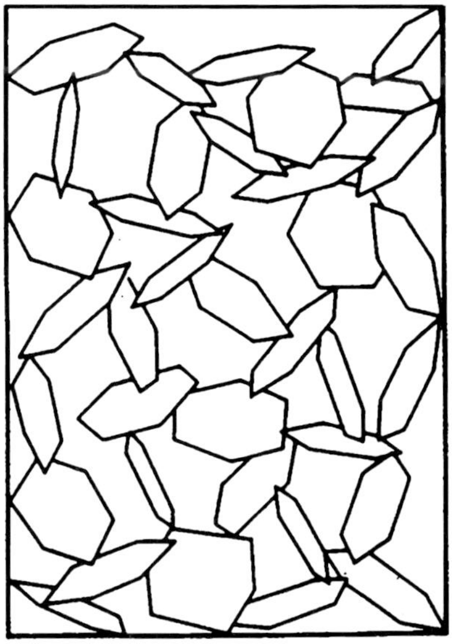

Figure 2.22 The "house-of-cards" structure of a thixotropically solidified kaolinite suspension.

rheopectic nature is that of a starch–water suspension, where hysteresis loops are formed as shown in Figure 2.21.

Classical studies have included those of Freundlich (1935), who introduced the term "thixotropy" ("change by touching") and Cashen (1963); more recent reports were made by Cheng (2007), Li *et al.* (2008), and Kelessidis (2008). This reversible sol–gel transition is caused by particle interactions induced by van der Waals forces that lead to aggregation under the occlusion of liquid so as to form a "house-of-cards" structure. The structure of a thixotropically solidified gel of kaolinite is shown schematically in Figure 2.22.

Although the $d\gamma/dt(\tau)$ and $\eta^*(\tau)$ dependencies of suspensions with structural viscosity/dilatancy (Figure 2.16) and thixotropy/rheopexy (Figure 2.21) behaviors show similar tendencies, there is a fundamental difference grounded in the different temporal characteristics of these phenomena. While structurally viscous systems with Newtonian or Bingham characteristics respond spontaneously to a change in shear stress, thixotropic and rheopectic systems require some time to recover–that is, to reconstruct the "house-of-cards" structure after collapse.

2.5 Drying of Green Clay Bodies

Before a permanent densification of formed clay bodies by sintering at high temperature can be attained, thorough drying is required in order to avoid cracking of the ceramic during firing; such cracking is caused by occluded water building

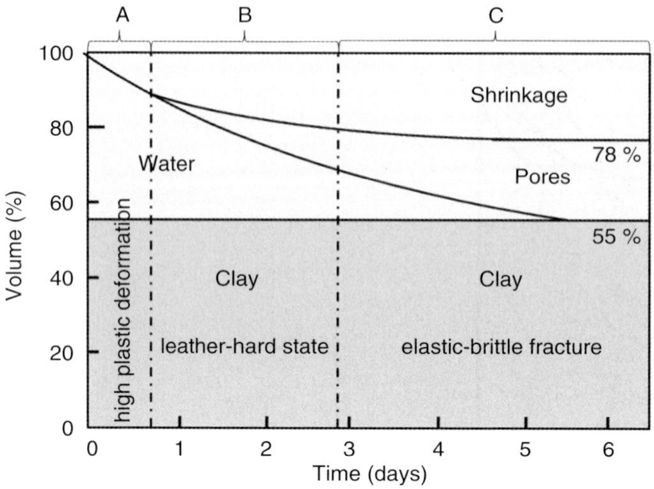

Figure 2.23 Typical drying (Bourry) diagram for clay masses.

up a considerable pressure inside the ceramic shard. Hence, great care must be expended so as to remove water, but without jeopardizing the microstructure of the green body. Since, depending on the forming process, up to 30 mass% water can be contained in the green body, this is not an easy task. The graphical representation of the volume–time relationship is the *Bourry diagram* (Figure 2.23) (Ackermann *et al.*, 1964, von Chiari and Hennicke, 1986).

According to this diagram, the drying process for moist model clays with a solid content of 55 vol% and a water content of 45 vol% can be divided into three stages:

- Stage A, which lasts for about 12–18 h, is characterized by the expulsion of water. The water loss corresponds to the volume loss; that is, the shrinkage is linearly proportional to the water loss. During this stage the clay is still highly plastic.
- In stage B, which lasts for up to three days, drying is accompanied by shrinkage, the volume percentage of which is smaller than that of the evaporated water, owing to the increasing formation of air-filled pores. The mechanical state during stage B corresponding to a water content of 85% to about 70% is called "leather-hard"; that is, the clay body is very firm and only slightly pliable so that trimming, the attachment of handles, and decoration by incising is still possible.
- In stage C, shrinkage ceases and stabilizes at about 22%. The volume of the expelled water is completely filled with air, and the clay body is said to be "air-dry," with a water content between 3% and 1%. Elevated drying temperatures up to 250 °C can reduce the water content to zero.

At the end of the drying process, the clay body attains elastic-brittle fracture characteristics, and has shrunk by about 22 vol%. The final dry clay body consists of about 70 vol% solid clay and 30 vol% pores.

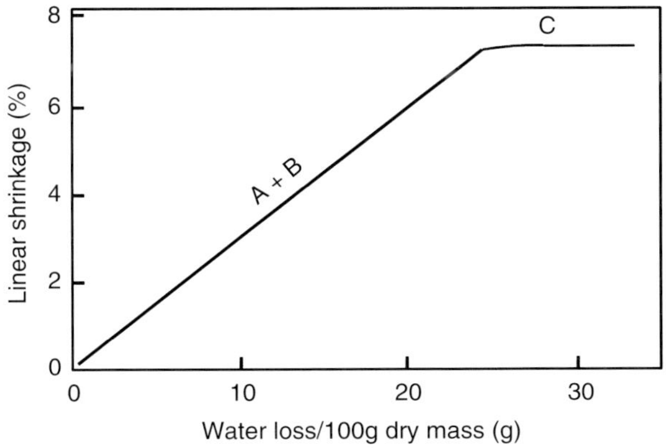

Figure 2.24 Bigot diagram of the drying of the ceramic mass, the Bourry diagram of which is shown in Figure 2.23.

Another way to monitor quantitatively the drying sensitivity of clay masses is through observing the linear shrinkage (expressed as a percentage) of a rod or bar of wet clay, and to plot such shrinkage against the water loss in g per 100 g dry clay. The resultant *Bigot curve* is composed of two lines with different slopes (Figure 2.24), that correspond to stages A + B, and C, respectively of the Bourry diagram, assuming that the volume shrinkage is equal to three times the linear shrinkage.

The Drying Sensitivity Index (DSI) after Bigot can by correlated with a different DSI expressed by the Ratzenberger Index (Ratzenberger and Vogt, 1993; Aungatichart and Wada, 2009).

One very important property of dry ceramic bodies is their *dry flexural strength*; that is, the fracture strength of the green body at the end of the shrinkage phase. This is dependent in a complex way on many parameters, as follows:

- Drying temperature: increasing the drying temperature increases the flexural strength.
- Raw density: that is, the packing density as a function of the forming process. The flexural strength increases from slip-casting to turning to pressing.
- Particle size: clays with a high proportion of particles <2 μm have a higher flexural strength than kaolinitic clays.
- Cation-exchange capacity: the flexural strength of Na^+-loaded clays is higher than that of Ca^{2+}-loaded clays, since the former are easier to arrange in a parallel fashion than the latter.
- Charge of cations: the flexural strength increases from La^{3+} to Ca^{2+} to Na^+.
- Drying rate: the slower the drying, the higher is the flexural strength.
- Porosity of ceramic body: a decrease of the porosity of the dried body through lowering the moisture content increases the flexural strength.

References

Ackermann, Ch., Gauglitz, R., and Hennicke, H.W. (1964) Das Bourry-Diagramm und die Trockenschwindung trocken verpresster Massen. Proceedings IX. International Ceramics Conference, Brussels, pp. 115–123.

Aungatichart, P. and Wada, S. (2009) Correlation between Bigot and Ratzenberger drying sensitivity indices of red clay from Ratchaburi province (Thailand). *Appl. Clay Sci.*, **43** (2), 182–185.

Bain, D.C. (ed.) (2009) Bentonite – versatile clays. *Elements*, **5** (2), 83–116.

Barshad, I. (1966) The effect of variation in precipitation on the nature of clay mineral formation in soils from acid and basic igneous rocks. Proceedings, International Clay Conference, Jerusalem, Vol. **1**, pp. 167–173.

Bird, G.W. and Cameron, D.J. (1982) Vault Sealing Research for the Canadian Nuclear Fuel Waste Management Program. Atomic Energy of Canada Limited Technical Record, TR-145.

Brindley, G.W. (1951) *X-Ray Identification and Crystal Structure of Clay Minerals*, The Mineralogical Society, London, UK.

Cairns-Smith, A.G. (1986) Introducing clays, in *Clay Minerals and the Origin of Life* (eds A.G. Cairns-Smith and H. Hartman), Cambridge University Press, Cambridge, pp. 13–22.

Carrado, K.A. and Komadel, P. (2009) Acid activation of bentonites and polymer-clay nanocomposites. *Elements*, **5** (2), 111–116.

Carretero, M.I. and Lagaly, G. (2007) Clays and health: an introduction. *Appl. Clay Sci.*, **36**, 1–3.

Cashen, G.H. (1963) Electric charges and thixotropy of clays. *Nature*, **197**, 349–350.

Chapman, D.L. (1913) A contribution to the theory of electro-capillarity. *Philos. Mag.*, **25**, 475–481.

Cheng, D.C.-H. (2007) Thixotropy. *Int. J. Cosmet. Sci.*, **9** (4), 151–191.

Demidenko, N.I., Podzorova, L.I., Rozanova, V.S., Skhorokhodov, V.A., and Shevchenko, V.Yu (2001) Wollastonite as a new kind of natural material (a review). *Glass Ceram.*, **58** (9-10), 308–310.

Dong, H. and Peacor, D.R. (1996) TEM observations of coherent stacking relations in smectite, I/S and illite of shales: evidence for MacEwan crystallites. *Clays Clay Miner.*, **44**, 257–275.

Dong, H., Peacor, D.R., and Freed, R.L. (1997) Phase relations among smectite, R1 I/S and illite. *Am. Mineral.*, **82**, 379–391.

Drever, J.I. (1982) *The Geochemistry of Natural Waters*, Prentice-Hall, Englewood Cliffs, New Jersey.

Eberl, D.D. (1984) Clay mineral formation and transformation in rocks and soils. *Philos. Trans. R. Soc. Lond.*, **A311**, 241–257.

Egorova, E.M. (2005) The validity of the Smoluchowski equation in electrophoretic studies of lipid membranes. *Electrophoresis*, **15** (1), 1125–1131.

Eisenhour, D.D. and Brown, R.K. (2009) Bentonite and its impact on modern life. *Elements*, **5** (2), 83–88.

Ernst, Th., Forkel, W., and von Gehlen, K. (1959) Vollständiges Nomenklatursystem der Tone. *Ber. Dtsch. Keram. Ges.*, **36**, 11–18.

Ertem, G. and Ferris, J.P. (2000) Sequence and region selectivity in the montmorillonite-catalyzed synthesis of RNA. *Origins Life Evol. Biosphere*, **30**, 411–422.

Esquevin, J. (1958) Les silicates de zinc. Etude de produits de synthèse et des minèraux naturels. Thèse Sci., Département de Science du Sol, Versailles, 345 pp.

Freundlich, H.M.F. (1935) *Thixotropy*, Hermann et Cie., Paris, France.

Gao, D. and Heimann, R.B. (1993) Structure and mechanical properties of superabsorbent poly(acrylamide)-montmorillonite composite hydrogels. *Polym. Gels Networks*, **1**, 225–246.

Gao, D., Liu, Y., Hou, J., and Heimann, R.B. (1996) Selective dewatering of gasoline and paraffin using SAP. *Chem. Eng.*, **3** (54), 7–10.

Gao, D., Heimann, R.B., and Alexander, S.D.B. (1997) Box-Behnken design applied to study the strengthening of aluminate concrete modified by a superabsorbent polymer/clay composite. *Adv. Cement Res.*, **9** (35), 93–97.

Gao, D., Heimann, R.B., Williams, M.C., Wardhaugh, L.T., and Muhammad, M. (1999) Rheological properties of poly(acrylamide)-bentonite composite hydrogels. *J. Mater. Sci.*, **34**, 1543–1552.

Garrels, R.M. and Christ, C.L. (1965) *Solutions, Minerals and Equilibria*, Freeman, Cooper & Cie., San Francisco, CA.

Gates, W.P., Bouazza, A., and Churchman, G.J. (2009) Bentonite clay keeps pollutants at bay. *Elements*, **5** (2), 105–110.

Gouy, G. (1910) Sur la constitution de la charge électrique à la surface d'un electrolyte. *J. Phys. Theor. Appl.*, **9**, 457–468.

Guggenheim, S., Adams, J.M., Bain, D.C., Bergaya, F., Brigatti, M.F., Drits, V.A., Formosa, M.L.L., Galán, E., Kogure, T., and Stanjek, H. (2006) Summary of the recommendations of nomenclature committees relevant to clay mineralogy: report of the Association Internationale pour L'Etude des Argiles (AIPEA) Nomenclature Committee for 2006. *Clays Clay Miner.*, **54** (6), 761–772.

Güven, N. (2009) Bentonites-clays for molecular engineering. *Elements*, **5** (2), 89–92.

Hanczyc, M.M., Fujikawa, S.M., and Szostack, J.W. (2003) Experimental models of primitive cellular compartments: encapsulation, growth, and division. *Science*, **302**, 618–622.

Hawlader, B.C., Muhunthan, B., and Imai, G. (2003) Viscosity effects on one-dimensional consolidation of clay. *Int. J. Geomech.*, **3** (1), 99–110.

Heimann, R.B. (1987) A statistical approach to evaluating durability of a simulated nuclear waste glass, in *The Geological Disposal of High-Level Radioactive Wastes* (ed. D.G. Brookins), Theophrastus Publ., S.A., Athens, Greece, pp. 181–206.

Heimann, R.B. (1993) Brønsted acidification observed during hydrothermal treatment of a calcium montmorillonite. *Clays Clay Miner.*, **41** (6), 718–725.

Heimann, R. and Franklin, U.M. (1979) Archaeo-Thermometry: the assessment of firing temperatures of ancient ceramics. *J. Int. Inst. Conserv., Can. Group*, **4** (2), 23–45.

Heling, D. (1994) Clay liners for waste dumps, in *Mineral Matter in Space, Mantle, Ocean Floor, Biosphere, Environmental Management, and Jewelry, Applied Mineralogy*, vol. **3** (ed. A.S. Marfunin), Springer, International, ch. 5, pp. 352–357.

Hennicke, H.W. (1978) Anelastizität und innere Reibung keramischer Werkstoffe, in *Handbuch Der Keramik*, Gruppe III K 3, Verlag Schmidt GmbH, Freiburg, Germany, pp. 1–13.

Hoang, H.V. (2007) Electrochemical synthesis of novel polyaniline-montmorillonite nanocomposites and corrosion protection of steel. Unpublished Ph.D. dissertation, TU Chemnitz, Germany.

Hower, J. and Mowatt, T.C. (1966) The mineralogy of illites and mixed layer illites/montmorillonites. *Am. Miner.*, **51**, 825–854.

Jaboyedoff, M., Bussy, F., Kübler, B., and Thelin, P. (2001) Illite "crystallinity" revisited. *Clays Clay Miner.*, **49**, 156–167.

Jacob, C.J. (1976) Synthesis of wollastonite from natural materials without fusion. US Patent 3,966,884.

Jambor, J.L., Kovalenker, V.A., and Robert, A.C. (1998) New mineral names. *Am. Miner.*, **83**, 1117–1121.

Kelessidis, V.C. (2008) Investigations on the thixotropy of bentonite suspensions. *Energy Sources Part A*, **30** (18), 1729–1746.

Kisch, H.J., Árkai, P., and Brime, C. (2004) On the calibration of the illite Kübler index (illite "crystallinity"). *Schweiz. Mineral. Petrogr. Mitt.*, **84**, 323–331.

Komarneni, S. and Roy, D.M. (1980) Hydrothermal effects on cesium sorption and fixation by clay minerals and shales. *Clays Clay Miner.*, **28**, 142–148.

Kübler, B. (1967) La cristallinité de l'illite et les zones tout a fait supérieures de métamorphisme, in *Colloque Sur Les Étages Tectoniques* (ed. J.P. Schaer), Baconniere, Neuchâtel, pp. 105–112.

Kübler, B. (1984) Les indicateurs des transformations physiques et chimiques dans la diagenese, temperature et calorimétrie, in *Thermométrie Et Barométrie Géologiques* (ed. M. Lagache),

Societe Française Minéralogie Cristallographie, Paris, pp. 489–596.

Kunz, W., Henle, J., and Ninham, B.W. (2004) 'Zur Lehre von der Wirkung der Salze' (On the Science of the Effect of Salts): Franz Hofmeister's historical papers. *Curr. Opin. Colloid Interface Sci.*, **9**, 19–37.

Li, H., Gao, D., Heimann, R.B., and Thomas, B. (2004) Studies on the mechanical properties of superabsorbent poly(acrylamide)/montmorillonite composite. *J. Funct. Polym.*, **17** (3), 496–498 (in Chinese).

Li, S.P., An, X.Q., and Zhu, Y.Y. (2008) Studies on the thixotropy and viscoelastic properties of HTlc/MT suspensions. *Colloids Surf., A*, **317** (1-3), 201–210.

Lindgren, H., Drits, V.A., Sakharov, B.A., Jakobsen, H.J., Salyn, A.L., Dainyak, L.G., and Krøyer, H. (2002) The structure and diagenetic transformation of illite-smectite and chlorite-smectite from North Sea Cretaceous-Tertiary chalk. *Clay Miner.*, **37** (3), 429–450.

Lisitzin, A.P. (1972) *Sedimentation in the World Ocean* (ed. K.S. Rodolfo), Society of Economic Palaeontologists and Mineralogists, Specialist Publication 17, Tulsa, USA, pp. 1–128.

Mercier, P.H.J. and Le Page, Y. (2008) Kaolin polytypes revisited *ab initio*. *Acta Crystallogr.*, **B64**, 131–143.

Messing, G.L. and Onoda, G.Y. (2006) Inhomogeneity-packing density relations in binary powders – experimental studies. *J. Am. Ceram. Soc.*, **61** (7–8), 363–366.

Meyers, M. and Chawla, K. (2009) *Mechanical Behavior of Materials*, 2nd edn, Cambridge University Press, Cambridge, New York, Melbourne, Madrid, Cape Town, Singapore, Sao Paulo, Delhi.

Millot, G. (1970) *Geology of Clays*, Springer, New York.

Millot, G. (1979) *Clays Sci. Am.*, **240** (4), 109–118.

Oberlies, F. and Pohlmann, G. (1958) Über die Einwirkung von Mikroorganismen auf Ton, Feldspat und Kaolin. Proceedings, VI International Ceramics Congress, Wiesbaden, pp. 149–168.

Ogawa, M., Kuroda, K., and Kato, C. (1989) Preparation of montmorillonite-poly(acrylamide) intercalation compounds and the water absorbing property. *Clay Sci.*, **7**, 243–251.

Pentinghaus, H. and Kienzler, B. (1993) Safe management of mega amounts of inorganic residues from energy production and waste treatment, in *Contaminated Soil '93* (eds F. Arendt, *et al.*), Kluwer Academic Publishers, Dordrecht, pp. 1537–1541.

Pusch, R. and Kamland, O. (1996) Physicochemical stability of smectite clays. *Eng. Geol.*, **41** (1-4), 73–85.

Ratzenberger, H. and Vogt, S. (1993) Möglichkeiten zur Vorausberechnung des Formgebungs-, Trocknungs- und Brennverhaltens Grobkeramischer Tone und Massen. *ZI-Jahrbuch*, Bauverlag, Wiesbaden, Berlin, pp. 70–111.

Reynolds, R.C. (1980) Interstratified clay minerals, in *Crystal Structures of Clay Minerals and Their X-Ray Identification* (eds G.W. Brindley and G. Brown), Mineralogical Society, London, pp. 249–303.

Rieder, M., Cavazzini, G., D'yakonov, Y.S., *et al.* (1998) Nomenclature of the micas. *Can. Mineral.*, **36**, 41–48.

Robertson, R.H.S. (1986) *Fuller's Earth – A History of Calcium Montmorillonite*, Volturna Press, Hythe, Kent, UK. 421 pp.

Rong, Y., Jiang, Y., Li, Y., Shen, L., and Sun, G. (1985) Chinese Patent CN 85 1 01256A.

Rummery, T.E. and Rosinger, E.L.J. (1984) The Canadian nuclear fuel waste management program, in *Fuel Reprocessing and Waste Management*, American Nuclear Society, vol. **1**, pp. 14–32.

Salmang, H., Scholze, H., and Telle, R. (2007) *Keramik*, 7th edn, Springer, Berlin.

Srivatsan, S.G. (2004) Modeling prebiotic catalysis with nucleic acid-like polymers and its implication for the proposed RNA world. *Pure Appl. Chem.*, **76** (12), 2085–2099.

Srodon, J. (1980) Precise identification of illite/smectites interstratifications by x-ray powder diffraction. *Clay Clay Miner.*, **28**, 401–411.

Theng, B.K.G. (1974) *Formation and Properties of Clay-Polymer Complexes*, Elsevier, Amsterdam.

US Geological Survey (2008) *Mineral Commodity Summaries 2008*, US Government Printing Office, Washington, DC, 199 pp.

Van Loon, L.R. and Glaus, M.A. (2008) Mechanical compaction of smectite clays increases ion exchange selectivity for cesium. *Environ. Sci. Technol.*, **42** (5), 1600–1604.

Velde, B. (1977) *Clays and Clay Minerals in Natural and Synthetic Systems. Developments in Sedimentology*, vol. **21**, Elsevier, Amsterdam, Oxford, New York.

Velde, B. and Meunier, A. (2008) *The Origin of Clay Minerals in Soils and Weathered Rocks*, Springer, Heidelberg, New York, ISBN 978-3-540-75633-0. 406 pp.

von Chiari, B. and Hennicke, H.W. (1986) Das Trockenverhalten plastischer keramischer Massen, Teil 1. *Keram. Z.*, **38**, 119–122.von Chiari, B. and Hennicke, H.W. (1986) Das Trockenverhalten plastischer keramischer Massen, Teil 2. *Keram. Z.*, **38**, 310–313.

von Helmholtz, H.L.F. (1879) Studien über elektrische Grenzschichten. *Ann. Phys. Chem.*, **7**, 337–382.

Warr, L.N., and Rice, A.H.N. (1994) Interlaboratory standardization and calibration of clay mineral crystallinity and crystallite size data. *J. Metamorph. Geol.*, **12**, 141–152.

Weaver, C.E. (1960) Possible uses of clay minerals in search for oil. *Am. Assoc. Petrol. Geol. Bull.*, **44**, 1505–1518.

Williams, L.B., Haydel, S.E., and Ferrell, R.E. (2009) Bentonite, bandaids, and borborygmi. *Elements*, **5** (2), 99–104.

Zhang, Y. and Cremer, P.S. (2006) Interaction between macromolecules and ions: the Hofmeister series. *Curr. Opin. Chem. Biol.*, **10**, 658–663.

Zheng, H. and Bailey, S.W. (1994) Refinement of the nacrite structure. *Clays Clay Miner.*, **42**, 46–52.

Zöller, M.H. and Brockamp, O. (1997) 1M- and $2M_1$-illites: different minerals and not polytypes. *Eur. J. Miner.*, **9**, 821–827.

3
Important Ceramic Phase Systems

In this chapter, the phase relations of those oxide components will be discussed that most commonly occur in ceramics, namely silica, alumina, calcia, magnesia, potassia, and iron oxide. The chapter is introduced by a cursory analysis of the anatomy and construction of phase diagrams based on Gibbs' phase rule, progressing from one component to two components, and from three components to multicomponent systems. The thermodynamics of these phase assemblies will be described, some important phase diagrams discussed, and conclusions drawn.

3.1
Fundamentals of Phase Diagrams

3.1.1
Gibbs' Phase Rule

Phase diagrams are chemographic representations of the famous Gibbs' phase rule (Gibbs, 1876; Gibbs, 1878) that relates the number of phases P, the number of components C, and the number of degrees of freedom F of a (closed) system by the simple equation

$$P + F = C + 2. \tag{3.1}$$

The components C are defined as those simple oxides that combine to constitute the phases P present at equilibrium;[1] that is, those portions of a system that are physically homogeneous and mechanically separable. The number of degrees of freedom, F, is the number of variables that can be altered without changing the number of phases present. The number "2" in Eq. (3.1) is applicable when

1) "Equilibrium" is defined as that state of any reversible system when no useful energy passes from or into the system. Then, the driving force is zero for a transformation of one phase into a different phase. However, it is sometimes difficult to determine experimentally whether equilibrium has been attained. Most, if not all, ceramic phase systems retain a metastable equilibrium that is characterized by very sluggish transformation rates. This slow kinetics often gives the system the appearance of being at true "thermodynamic" equilibrium.

Classic and Advanced Ceramics: From Fundamentals to Applications. Robert B. Heimann

ISBN: 978-3-527-32517-7

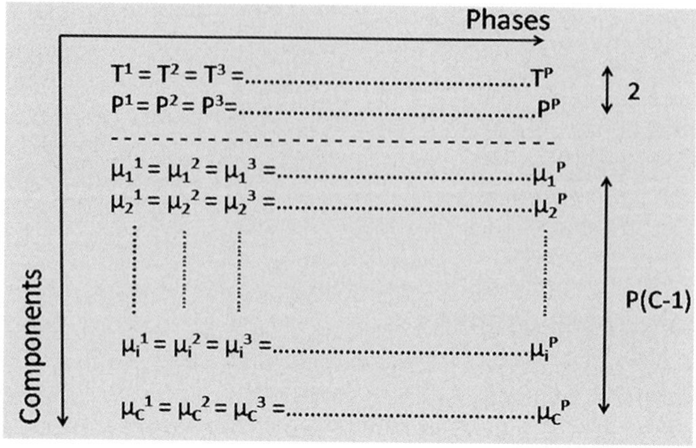

Figure 3.1 Matrix of independent variables.

both temperature and pressure have to be considered. However, in systems of importance for ceramics, the vapor pressure is very low over a wide range of temperatures; thus, the pressure variable can be safely neglected. Such systems with a reduced degree of freedom are termed "condensed" systems, and Eq. (3.1) reduces to

$$P + F = C + 1, \tag{3.2}$$

also called the "mineralogical phase rule".

The matrix of the number of independent variables such as temperature, pressure, and chemical potentials of the components is shown in Figure 3.1. The columns depict the phases indicated by superscripts, while the lines depict the components indicated by subscripts.

The basis for each phase transition is given by the general conditions of the law of mass action, $\mu_i^1 = \mu_i^2$; that is, two phases (1) and (2) can exist in a stable equilibrium only if the chemical potentials μ_i of each component in every phase are equal. This corresponds to the minimization of the Gibbs function $(\delta G/\delta n)_{P,T} = \Sigma_i \nu_i \cdot \mu_i = 0$, where ν_i refers to the stoichiometric coefficient. In the matrix shown in Figure 3.1 there are (C − 1) concentrations and P·(C − 1) + 2 independent concentration variables.

Hence, the information content of the matrix can be expressed by statements such as:

- In equilibrium the temperatures of all phases of a system must be equal; or
- if surface effects, membranes, shear forces, and so on, are absent, then under equilibrium condition the pressure of all phases of a system must be equal; or
- if each phase is open for a component so that this component is free to move between phases, the chemical potential of this component must be equal in all phases under equilibrium condition.

The total composition is completely defined by (C − 1) concentrations, since the concentration of the last component equals the difference of the sum of all other concentrations to 100%. To define the composition of all P phases, P·(C − 1) independent concentration variables are required. Hence, together with temperature and pressure, there are [P·(C − 1) + 2] parameters required to describe the system at each moment.

Under equilibrium conditions, (P − 1)·C variables are fixed, and hence:[2)]

$$F = [P \cdot (C - 1) + 2] - (P - 1) \cdot C = C - P + 2 \tag{3.3}$$

variables are not determined–that is, free. Hence, F describes the number of intensive (independent of the extension of the system) variables such as temperature, pressure, chemical potential, magnetic field strength, or electric field strength that can be altered independently and arbitrarily without causing the disappearance or formation of a phase. It is not always easy to unequivocally define the number of components C, since chemical reactions can influence the selection. The following definition may be helpful:

> "The number of components equals the total number of independent chemical compounds of a system minus the number of unequivocal, i.e. true chemical reactions that can occur among these compounds. The 'number of independent compounds' mean the total number minus the number of all constraints such as mass equilibria or charge neutrality. An 'unequivocal chemical reaction' means a reaction that cannot simply by formulated as an extension of other reactions within the system" (Moore, 1972).

For example, in a system containing calcium carbonate, calcium oxide, and carbon dioxide, there are three independent chemical compounds, namely $CaCO_3$, CaO, and CO_2. These are connected by a chemical reaction equation such as $CaCO_3 \rightarrow CaO + CO_2$. Consequently, the number of components is reduced by 1 to yield:

$$C = 3 - 1 = 2.$$

It should be emphasized that reaction equilibria can only be accounted for if they actually occur with measurable reaction rates.

3.1.2
One-Component Phase Diagrams

The number of independent variables in one-component systems is only two, namely temperature and pressure, since the composition is fixed (C = 1). Consequently, the system has two degrees of freedom (F = 2, bivariant) if only one phase

2) In a formal sense, Gibbs' phase rule is mathematically equivalent to Euler's polyhedron condition, V = E−F + 2.

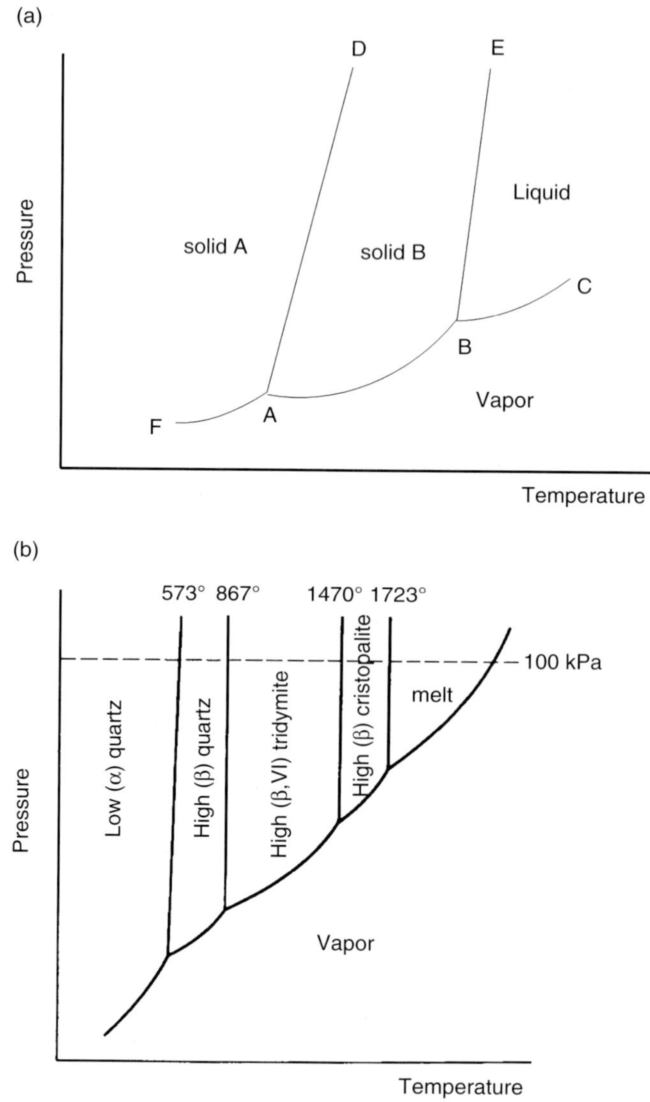

Figure 3.2 (a) General phase relations in a one-component system; (b) Hypothetical equilibrium phase diagram of silicon dioxide (after Fenner, 1913; Sosman, 1965).

is present, since T and P can be varied widely without moving the system into another *phase field*. As shown in Figure 3.2a, this is the case for P,T- conditions within the stability fields of the solid phases A and B, and the liquid and vapor phases. If two phases are present, the system is univariant (F = 1), as represented by the five *phase boundary lines*–that is, the melting curve B–E of phase B, the transition curve A–D between the solid phases A and B, the sublimation curves

F–A (phase A) and A–B (phase B), and the vapor pressure curve B–C of the liquid phase. Finally, if three phases are present, the system is invariant (F = 0), as represented by the two *triple points* A (equilibrium between solid A, solid B, and vapor) and B (equilibrium between solid B, vapor, and liquid).

It should be pointed out that the pressure indicated on the ordinates of Figure 3.2 is purely fictitious, since the vapor pressure of solid phases is too small to be measured accurately. Figure 3.2b shows the (hypothetical) equilibrium phase diagram of silica, that is, the phase diagram without metastable phases[3] (Fenner, 1913; Sosman, 1955, 1965; see also Kingery *et al.*, 1976; Heimann, 1977). The transformation temperatures between solid silica phases are given at 100 kPa (1 bar) pressure.

3.1.3 Two-Component (Binary) Phase Diagrams

The detailed explanation of the phase relationships of binary and ternary (see Section 3.1.4) systems has been adapted from earlier published work of the author (Heimann, 1989).

3.1.3.1 Simple Binary Phase Diagram without Intermediate Compound or Solid Solution

Figure 3.3 shows a schematic binary system of two components A and B without the formation of a compound A_mB_n. The abscissa shows the composition. At the point labeled A, 100% A (= 0% B) is present; the percentage of B increases towards the right. The ordinate depicts the temperature, and T_A and T_B refer to the melting points of the pure components A and B, respectively. Thus, the vertical boundaries represent one-component systems. With increasing admixture of B to A, and A to B, respectively the melting point of the mixture decreases according to the Raoult–van t' Hoff law. The diagram is actually a chemographical representation of the condensed Gibb's phase rule: P = (C + 1) − F (Rankin and Wright, 1915).

At any point along the liquidus curves T_A–e_{AB} and T_B–e_{AB}, there exist two phases – either solid A + melt, or solid B + melt. Consequently, the degree of freedom F is 1; that is, either the temperature or the composition can be altered without changing the number of phases present (univariant equilibrium). At the eutectic point e_{AB} the two solid phases A and B are in equilibrium with the melt. Thus, the number of phases is P = 3, and F = 0, since any variation of the temperature or the composition will invariably displace the system from point e_{AB} at which two solubility (liquidus) curves intersect (invariant equilibrium).

The mechanism of solidification of a homogeneous melt of composition X (75 mass% A, 25 mass% B) is represented by the isopleths, the vertical dashed line X_L–X. If a melt of composition X_L cools it reaches the liquidus curve at X_3 and a temperature T_3, and small amounts of A start to solidify. Upon further cooling,

3) Strictly speaking, tridymite does not truly belong to the pure one-component system SiO_2 as it is considered a "stuffed" derivative requiring foreign ions for its stabilization (see below).

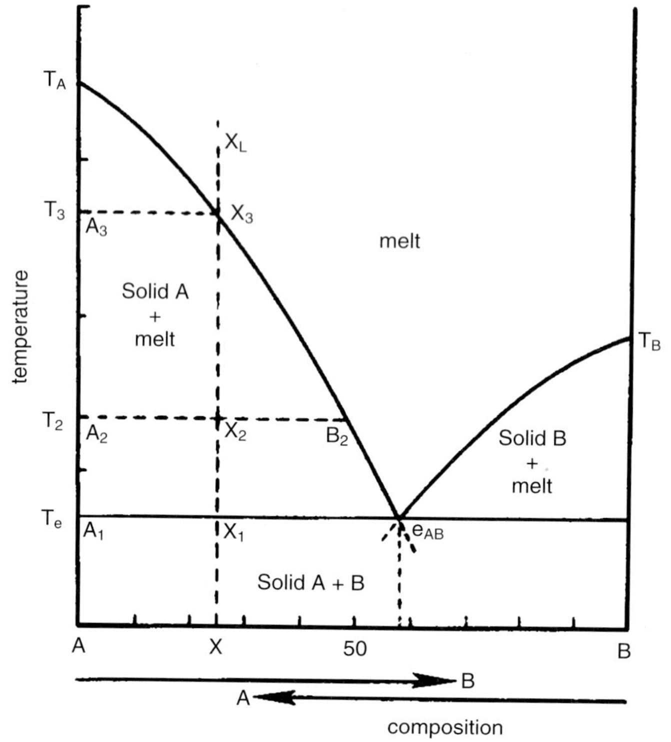

Figure 3.3 Simple binary phase diagram without intermediate compound or solid solution. A = anorthite, B = diopside. The composition is given in mass%. The eutectic composition x_{AB} is 42% anorthite, 58% diopside; the eutectic temperature T_e is 1270 °C.

more A solidifies and the composition of the melt in equilibrium with solid A follows the curve X_3–e_{AB}, since A is continuously removed and the melt becomes enriched in B. The amount of melt coexisting with solid A at each temperature can be determined applying the so-called "lever rule." For example, at T_2 solid A is in equilibrium with a melt of composition B_2 (52 mass% A, 48 mass% B) in the ratio of the line lengths $X_2B_2/A_2X_2 = 0.92$. Further cooling leads to more solidification of A until, at the eutectic temperature (T_e), the eutectic (invariant) point e_{AB} is reached below with no melt can further exist. At this point (42 mass% A, 58 mass% B), the temperature will remain momentarily constant until all traces of melt have solidified into a mixture of A and B. Their ratio can again be obtained by the lever rule to be A/B = XB/AX = 3, corresponding to the A/B ratio of the starting melt at X_L.

The binary phase diagram shown in Figure 3.3 actually relates to the system anorthite ($CaAl_2Si_2O_8$)–diopside ($CaMgSi_2O_6$), with melting points of the pure components of $T_A = 1550$ °C and $T_B = 1390$ °C, respectively. The eutectic temperature T_e is 1270 °C, with a composition e_{AB} of 42 mass% anorthite + 58 mass%

diopside. This system plays an important role in igneous petrology to describe the evolution of basaltic rocks, and also in highly fired calcareous clays used to produce alumina cement clinker (see Chapter 5).

3.1.3.2 **Complete Solid Solution of Two Components**

The principles described in Section 3.1.3.1 apply only to the situation when both solid components A and B are completely insoluble in each other. Strictly speaking, this is almost never the case in the real world. If there are close chemical and structural relationships between A and B, including comparable ionic radii of the constituting elements, then they can substitute for each other to form homogeneous solid solutions that are considered as only one phase. Consequently, the two solids A and B discussed above reduce to one phase (A, B) and the system can at no point become invariant; that is, the liquidus curves must become continuous. Also, there is no true eutectic point, and the equilibrium curves split into a liquidus curve at higher temperature (geometric locus of solidification of melt upon cooling) and a solidus curve at lower temperature (geometric locus of melting upon heating). These two curves meet at the melting points at the pure components and enclose a spindle-shaped region (Figure 3.4). Hence, the liquidus and solidus curves divide the phase diagram into three regions: (i) the single-phase region of the melt at high temperature; (ii) the single-phase region of the solid solution (A, B) at low temperature; and (iii) the two-phase region where solid and melt coexist.

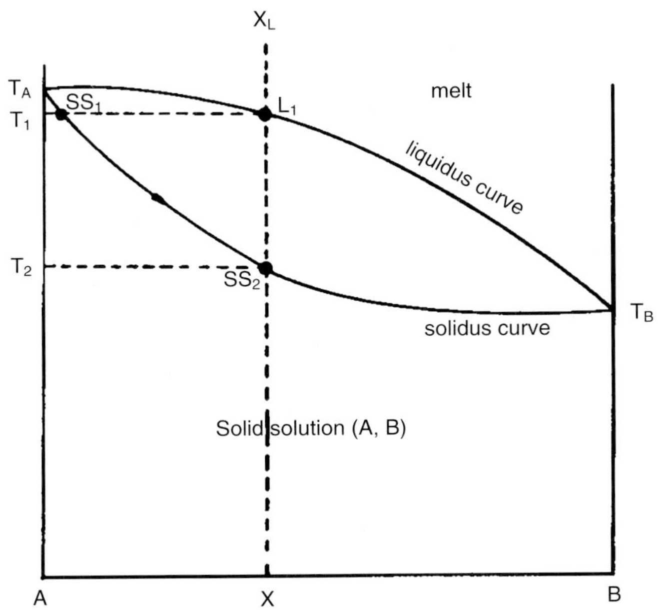

Figure 3.4 Complete solid solution of two components A and B. The tie line (conode) SS_1–L_1 shows the conjugate pair solid–liquid coexisting at the temperature T_1 (see also Figure A.3 in Appendix A).

However, only the first two possess physical relevance. The spindle-shaped region is only an "information storage room" (Rosenberger, 1979), since equilibria between solid and liquid solutions of identical composition are excluded by this space. Instead, as shown in Figure 3.4, at a temperature T_1 a solid solution SS_1 and a melt L_1 coexist as conjugate phases in thermodynamic equilibrium and are connected by a tie line or conode SS_1–L_1 parallel to the abscissa. During solidification, the composition of the solid solution changes from SS_1 to SS_2, at which composition the last trace of melt disappears and the composition equals that of the starting melt, X_L.

This type of binary phase diagram occurs in the system albite ($NaAlSi_3O_8$)–anorthite ($CaAl_2Si_2O_8$) that forms the plagioclase solid solution series $Na_{1-x}Ca_x[Al_{1+x}Si_{3-x}O_8]$. This system has important implications for the crystallization of granitic rocks from a melt. Details of the construction of the phase diagram from the temperature dependency of the Gibbs free energy can be found in Appendix A.

3.1.4 Three-Component (Ternary) Phase Diagrams

3.1.4.1 Composition of a Ternary Compound

If a third component is added to a binary diagram, a ternary triangular diagram is obtained. In this case, the temperature axis cannot be plotted directly but can be described only by isotherms that are projections down the temperature axis perpendicular to the triangle (see Figure 3.8). Figure 3.5a shows two methods to express the compositions of a phase assembly X in terms of the proportion of the components A, B, and C, and A, C, and D, respectively. These components can

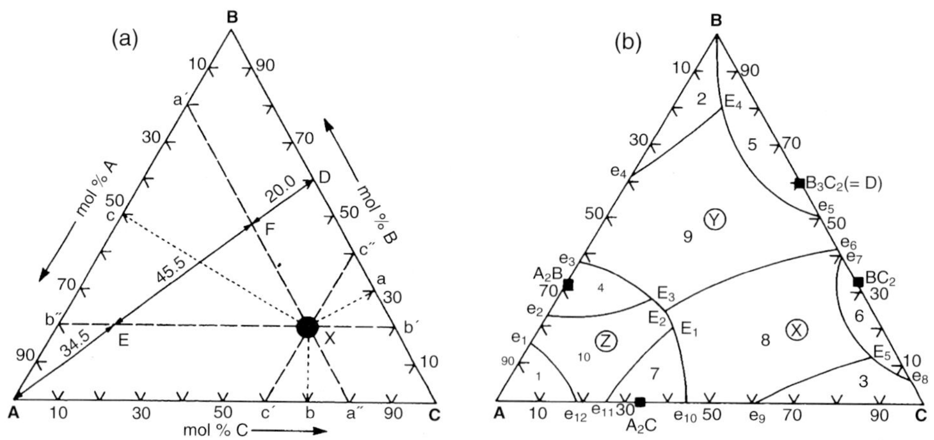

Figure 3.5 (a) Two methods of expressing the composition of compound X in the ternary system A–B–C; (b) Stability fields of components A, B, C (phase areas 1 to 3), binary components A_2B, B_3C_2, BC_2, and A_2C (phase areas 4 to 7), and ternary components X, Y, Z (phase areas 8 to 10).

be simple oxides such as CaO, Al_2O_3, and SiO_2. To represent larger families of ceramics manufactured from raw materials of widely variable compositions (see Chapter 2), the corners of the triangle are sometimes labeled "sand," "feldspar," and "clay." One simple method to obtain the composition of point X in Figure 3.5a consists of drawing lines (long dashes) parallel to the sides of the triangle through point X. These lines intersect the sides of the triangle at the points a′–a″, b′–b″, and c′–c″. The relative distance of X from the corners A, B, and C can then be expressed in percentages as read from the scales along the sides of the triangle, yielding 20 mol% A, 20 mol% B, and 60 mol% C. A second method consists of letting normals Xa, Xb, and Xc fall onto the sides of the triangle (short dashes). The lengths of the normals are proportional to the percentages of the component designated at the opposite end.

Occasionally, it is required to express the composition of a ternary phase X in terms of the components, located at the apices, and a binary phase, located on a side of the triangle. Then point X might be expressed in term of A, C, and D (binary phase B_3C_2) on the line (compatibility join, conode) AD. This construction yields A = DF·100/AD = 20%, D = AE·100/AD = 34.5%, and C = EF·100/AD = 45.5%. Since the binary phase D consists of 60 mol% B and 40 mol% C, one obtains approximately A = 20 mol%, B = 20 mol%, and C = 60 mol%, as above.

3.1.4.2 **Phase Boundary Lines, Eutectic Points, and Degrees of Freedom**

The point X in Figure 3.5a represents the composition of a hypothetical ternary phase, ABC_3. Exploration of the ternary system A–B–C by x-ray diffraction (XRD), optical methods, and differential thermal analysis (DTA) may have revealed two other ternary phases: Y, with the composition AB_2C; and Z, with a composition A_4BC, as well as four binary phases A_2B, BC_2, A_2C, and B_3C_2 (=D). These seven phases and their stability fields 4 to 10 are shown in Figure 3.5b. While points X, Y, and Z represent the exact stoichiometric compositions, the areas 8, 9, and 10 show those portions of the phase diagram in which the ternary phases are stable according to Gibbs' phase rule. In addition, the stability fields 1 to 3 of the pure components A, B, and C have been outlined. All of these phases are separated by phase boundary lines that indicate the fields of thermodynamic stability of the respective phases. It happens that three phase boundary lines intersect at five locations, E_1 to E_5, termed the ternary eutectic (invariant) points. Likewise, the intersections of phase boundary lines with the sides (binary joins AB, AC, BC) of the triangle yield 12 binary eutectic points, e_1 to e_{12}.

According to Gibbs' condensed phase rule, at the ternary eutectic points four phases coexist (C = 3, P = 4, F = 0): three solid phases and one liquid melt phase. For example, at the ternary eutectic point E_1 the two ternary phases X and Z, the binary phase A_2C, and the melt (not shown in projection) coexist. By definition, this point is invariant (F = 0) to changes in temperature and/or composition, as each departure from its values will displace the system from point E_1. A composition anywhere along a phase boundary lines is subject to a univariant equilibrium (F = 1), since it is possible to alter either the temperature or the concentration of one of the two components without leaving the boundary line between two

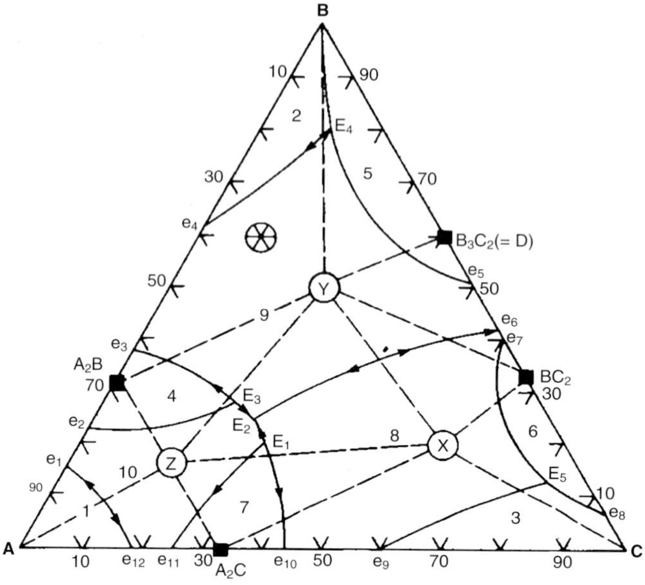

Figure 3.6 Compatibility triangles (dashed lines) superimposed on the stability fields 1 to 10, and directions of falling temperatures on the phase boundary curves (arrowheads), determined according to the Alkemade theorem.

solubility surfaces. Finally, any composition between phase boundary lines – that is, within a phase stability field – is subject to a divariant equilibrium (F = 2). The temperature and compositions of two components can be manipulated without forcing the system into a different phase stability field; that is, without a new phase appearing.

3.1.4.3 Compatibility Joins (Conodes) and Compatibility Triangles

Straight lines connecting the composition points of two primary phases[4] are termed "compatibility joins," "tie lines," or "conodes." A ternary conode, sometimes referred to as an *Alkemade line*, must intersect the phase boundary lines between adjacent (conjugate) primary phases (see Figure 3.6), and it determines the direction of falling temperature on the phase boundary curve. Since the direction of falling temperature is always away from the Alkemade line,[5] the intersec-

4) A primary phase is the only crystalline phase that can exist in equilibrium with a melt of a given composition. It is the first crystalline phase to appear on cooling of a composition from the molten state (Levin *et al.*, 1964).

5) Alkemade theorem: "The direction of falling temperature on the boundary curve of two intersecting primary phase areas is always away from the Alkemade line. If the Alkemade line intersects the boundary curve, the point of intersection represents a temperature maximum on the boundary curve. If the Alkemade line does not intersect the boundary curve, then the maximum on the boundary curve is represented by that end which, if prolonged, would intersect the Alkemade line" (Levin *et al.*, 1964).

tion of boundary curve and Alkemade line constitutes a temperature maximum or a saddle (col) point. Figure 3.6 shows the compatibility joins (dashed lines) connecting the primary phase composition points X, Y, and Z with each other, as well as the four binary compositions A_2C, B_3C_2, A_2B, and BC_2, and the pure components A, B, and C.

The arrows indicate the directions of falling temperatures as determined by the Alkemade theorem (see footnote 5). For clarity, not all possible arrows are shown. It is easy to see that the ternary eutectic point E_2 constitutes the composition with the lowest possible melting temperature, since the direction of the three arrows on the phase boundary curves e_3–E_2, e_{10}–E_2, and e_6–E_2 point to E_2 as the absolute temperature minimum within the ternary system A–B–C. The straight compatibility joins (conodes) dissect the triangle ABC into a collection of smaller triangles referred to as "compatibility triangles" or "composition triangles." These are extremely important in the determination of stable coexisting mineral (phase) assemblages, both in petrology and ceramic science. For example, a hypothetical clay with a composition of 30 mol% A, 60 mol% B, and 10 mol% C (hexagon in Figure 3.6) will develop on prolonged heating toward an equilibrium composition composed of the three stable phases B, Y, and A_2B that constitute the apices of the compatibility triangle in which the starting composition is located.

3.1.4.4 The Complete (3-D) Ternary Phase Diagram

So far, no explicit consideration has been given to the temperature. As the stability of a phase implies that it ought to possess the lowest Gibbs free energy $(\delta G/\delta n)_{P,T} = \Sigma_i \nu_i \cdot \mu_i$, under equilibrium conditions for a given temperature and pressure the state functions of the system must be manipulated in such a way that the Gibbs function of the stable phase is lower than that of any other component in the system. This requirement also implies that the chemical potential of the phase in question must be minimized according to the Gibbs–Duhem equation

$$d\mu = -s dT + v dP - m dH - k dF - \ldots \quad (3.4)$$

This equation shows that μ can be altered either by a change in temperature dT, pressure dP, magnetic field strength dH, electric field strength dF, or any other intensive variable. Technically most relevant, but easy to accomplish, is a change in temperature. Figure 3.7a shows schematically the $\mu(T)$ curves for a liquid and a solid in the vicinity of the melting point T_m. Since the two curves have different slopes, lowering the temperature will cause the chemical potential of the liquid melt, μ_L to increase over that of the solid until, at the melting point T_m, the chemical potential of the solid, μ_s becomes lower than that of the liquid.

The slopes of the chemical potential curves $\mu(T)$ correspond to the entropy, s; thus, the melt at high temperature has higher entropy, as indicated by the steeper slope of the μ_L curve. The phase transformation liquid–solid at the melting point is accompanied by an abrupt entropy decrease, signifying a higher ordered (crystalline) state of the system (Figure 3.7b). The heat capacity $C_p = T\,(ds/dT)_p \rightarrow \infty$ indicates a phase transformation of first order.

To obtain a feeling for the relationship between the phase composition and the temperature, it is necessary to consider a 3-D model, as shown in Figure 3.8. This

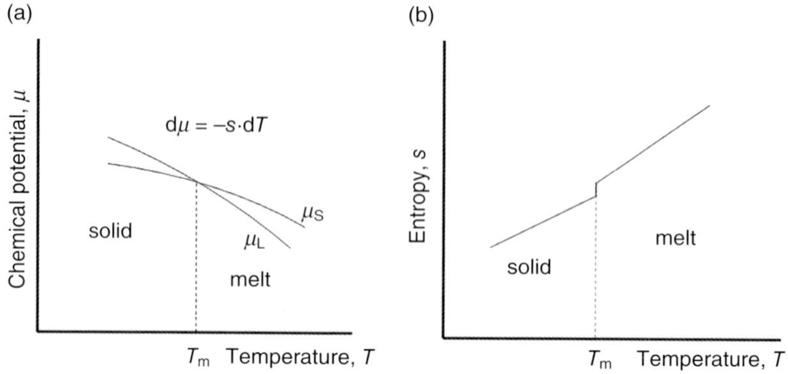

Figure 3.7 (a) Chemical potential versus temperature in the vicinity of a phase transformation (melting point T_m); (b) Entropy versus temperature (isobaric).

displays a simple ternary phase diagram without a stable ternary phase or solid solution. The temperature is represented by distances above the basal plane of the composition A–B–C. Three different surfaces are shown; the first surface occurring at the highest temperatures is the complex *liquidus surface* $T_A-e''_{AB}-T_B-e''_{BC}-T_C-e''_{AC}-E'$ that determines the maximum solubility of a solid phase in the melt at any one composition.

A curved liquidus surface more complex than that shown in Figure 3.8 can be conveniently visualized with the help of a (hypothetical) infinitely flexible membrane stretched between points T_A–T_B–T_C and weighted down by small point-like spheres of different density at the positions of the eutectic points. The downward deflection of the membrane caused by these balls is proportional to their density, so that larger deflections correspond to lower binary and ternary eutectic temperatures. Here, the physical model ends as the membrane must have a peculiar property of being able to "bulge" between points of deflection to account for the convex curvature of the liquidus surface and also the occurrence of saddles (cols), ridges, and dome-shaped surface areas, the positions of which are defined by the Alkemade lines (see above).

The liquidus surface determines the melting points of all given compositions, whereby the points T_A, T_B, and T_C are the melting points of the pure components A, B, and C. If, for example, component B is added to pure A, the melting temperature of this mixture decreases along the liquidus line $T_A-e''_{AB}$. At the binary eutectic point e''_{AB}, the lowest melting temperature of all compositions in the binary subsystem A–B has been reached. Adding component C leads eventually to the lowest possible melting temperature, T_E, of the ternary system, at the ternary eutectic point E′ close to the center of the triangle A–B–C.

The second surface shown in Figure 3.8 is the isothermal *solidus surface* tangent to the ternary eutectic point E′. This surface determines the compositions of the solid phases in equilibrium with the melt, and below this surface no further liquid phase can occur. The third surface shown is the projection of the 3-D phase

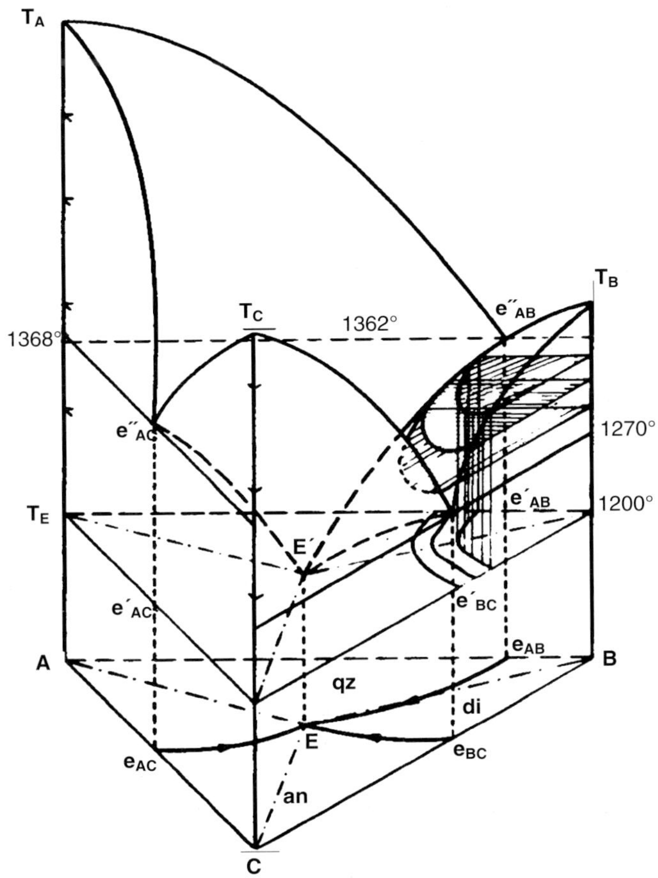

Figure 3.8 Three-dimensional phase diagram A–B–C with no ternary compound showing: (i) the complex curved liquidus surface (top); (ii) the isothermal solidus plane (center); and (iii) the planar subsolidus surface (bottom). At the upper right, isotherms have been projected from the liquidus surface $T_B-e''_{BC}-E'-e''_{AB}$ downward onto the solidus surface (Heimann, 1989). The diagram shows the ternary phase diagram quartz (A)–diopside (B)–anorthite (C), as investigated by Clark *et al.* (1962) and Osborn and Tait (1952).

diagram down to ambient temperature, the isothermal *subsolidus surface* A–B–C. Indicated on this surface are the stability fields of the phases A, B, and C, separated by the curved phase boundary lines e_{AC}–E, e_{BC}–E, and e_{AB}–E, as well as the conodes (chain lines) A–E, B–E, and C–E. In actual phase diagrams the isotherms shown in Figure 3.8 as (curved) planes parallel to the solidus surface and intersecting the liquidus surface $T_B-e''_{BC}-E'-e''_{AB}$ are projected downwards onto the solidus surface, in much the same way as elevation contours on a topographical map. The 3-D-ternary diagram shown in Figure 3.8 depicts the actual system quartz (A)–diopside (B)–anorthite (C) with the melting points T_{qz} = 1723 °C, T_{di} = 1390 °C,

T_{an} = 1550 °C; the binary eutectic temperatures 1362 °C (quartz–diopside), 1270 °C (diopside–anorthite), and 1368 °C (quartz–anorthite); and the ternary eutectic temperature of 1200 °C. The composition of the ternary eutectic is 30 mass% quartz, 33 mass% diopside, and 37 mass% anorthite (Clark *et al.*, 1962; Osborn and Tait, 1952). The ternary phase diagram discussed above refers to high-fired calcareous illitic clays with the four components SiO_2, Al_2O_3, CaO, and MgO, but fails to incorporate an important fifth component such as Fe_2O_3. Hence, even more complex phase relations must be considered in appropriate multicomponent (quaternary, quinary, hexanary, ... multinary) systems.

3.1.5 Four-Component (Quaternary) Phase Diagrams

The chemical compositions of clays are very complex, and frequently more than three components must be considered. Four-component (quaternary) phase diagrams can be represented by a tetrahedron. As an example, Figure 3.9 shows the quaternary system CaO–MgO–Al_2O_3–SiO_2 and several of its subsystems that divide the tetrahedra volume into four-component phase assemblies.

The binary compositions (F = 2) spinel (Sp), forsterite (Fo), wollastonite (Wo) and larnite (La) are situated on the binary joins (edges of the tetrahedron), while the ternary compositions (F = 1) anorthite (An) and diopside (Di) are situated on the ternary planes (areas within triangular planes). The only quaternary compound is melilite (Mel, $Ca_2(Mg,Al)[Si_2O_7]$), a solid solution between the ternary compounds gehlenite ($Ca_2Al[AlSiO_7]$) and åkermanite ($Ca_2Mg[Si_2O_7]$) that is situated within the volume of the tetrahedron. The quaternary systems (F = 0) Di-Fo-Sp-

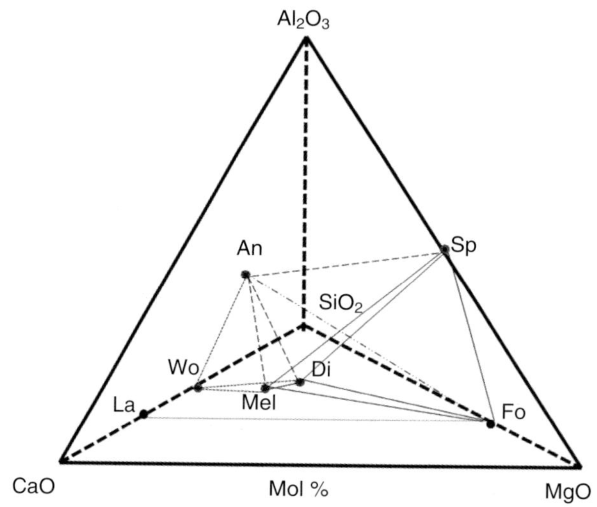

Figure 3.9 Quaternary system CaO–MgO–Al_2O_3–SiO_2, with some tetrahedral quaternary subsystems indicated.

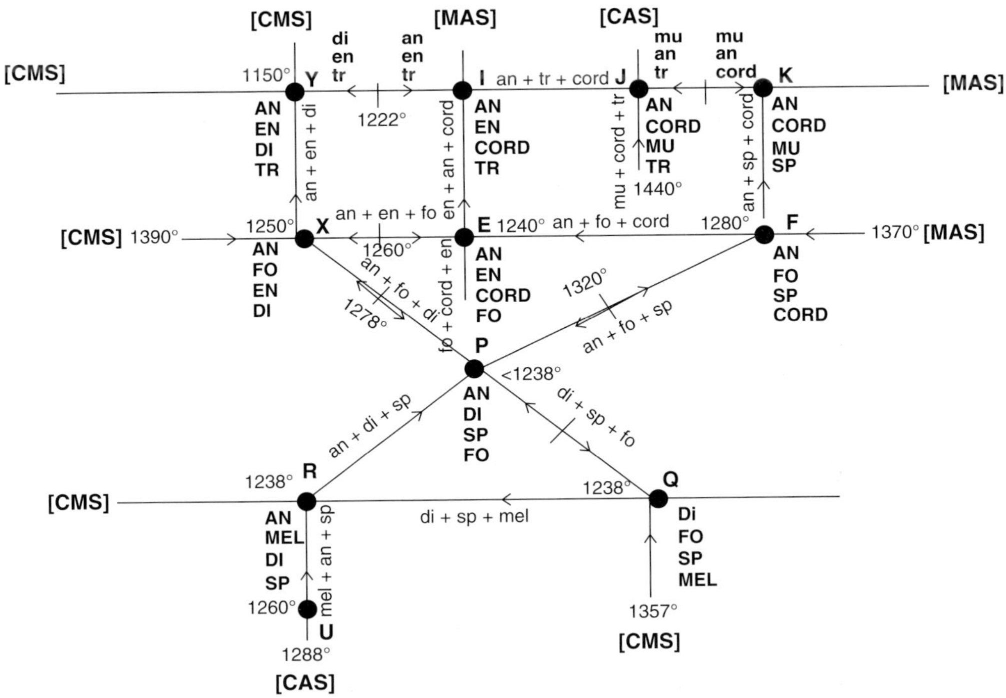

Figure 3.10 Network description of the four-component (quaternary) system CaO (C)–Al_2O_3 (A)–MgO (M)–SiO_2 (S) in the vicinity of the global quaternary eutectic P, in which the phase assembly anorthite (AN)–diopside (DI)–spinel (SP)–forsterite (FO) is stable. The nodes labeled Y, I, J, ... Q, R, U refer to quaternary compositions which are connected by phase boundary lines along which the ternary subsystems (labeled with lower-case letters) are stable. The eutectic melting temperatures are also given, and the directions of falling temperatures according to the Alkemade theorem are shown by arrows. Abbreviations: an = anorthite; cord = cordierite; di = diopside; en = enstatite; fo = forsterite; mel = melilite; mu = mullite; sp = spinel; tr = tridymite. Diagram adapted from Levin, Robbins, and McMurdie (1964).

Mel (+melt) (denoted "Q" in Figure 3.10) and the adjoining systems Di-Sp-An-Mel (+melt) (denoted "R"), An-Di-Sp-Fo (+melt) (denoted "P") as well as Di-Wo-An-Mel (+melt) are outlined.

Quaternary compounds will be located inside the phase volume of the tetrahedron; compositions representing the four associated ternary subsystems are positioned on the triangular surface planes of the tetrahedron, and the six binary marginal systems constitute the edges of the tetrahedron. For example, quaternary compositions are located within the subtetrahedron ("P" in Figure 3.10) An-Di-Fo-Sp bounded by the four ternary systems An-Fo-Di, An-Fo-Sp, An-Di-Sp, and Di-Sp-Fo. The six binary systems are An-Di, An-Sp, Sp-Fo, Fo-Di, Fo-An and Di-Fo.

On the line connecting the points representing the compositions of larnite (La, β-Ca_2SiO_4) and forsterite (Fo, Mg_2SiO_4) the compounds merwinite ($Ca_3Mg[SiO_4]_2$)

and monticellite ($CaMg[SiO_4]$) (not shown in Figure 3.9) are also positioned since the cation : Si ratios of these important cement minerals are 2 : 1 in all cases. Merwinite is monoclinic with space group (S.G.) $P2_1/a$, forsterite, monticellite and the low-temperature modification of larnite, calcio-olivine (α-Ca_2SiO_4) are orthorhombic with S.G. *Pnma*.

Since it is awkward to draw and interpret 3-D-tetrahedra, it is necessary to lay flat onto the paper plane one of the four triangular ternary diagrams and collapse the other three onto this plane. In many cases, however, it is preferred to apply a topologically equivalent situation by plotting the phase compositions as nodes in an orthogonal net connected by the appropriate phase boundaries as straight lines (Figure 3.10). Hence, the coexisting four phases are the nodes of this network and constitute invariant (quaternary eutectic) points. Along the connecting phase boundary lines, the three coexisting phases are indicated as projections of univariant planes. Arrows on these lines give the directions of falling temperatures according to the Alkemade theorem.

3.1.5.1 **Conclusion**

Figure 3.11 shows the correlation between the number of components (C), the maximum number of compounds in binary, ternary and quaternary phase systems, as well as the location of the respective phases in the phase diagrams. The construction, evaluation and interpretation of ternary and quaternary phase diagrams given above considered only very simple cases. More complex situations involving formation of ternary and quaternary compounds from the melt, occurrence of miscibility gaps, spinodal demixing, peritectic systems with incongruent melting, solid solutions, redox equilibria, construction of crystallization paths during freezing of a melt (see Section 5.2.3), liquid immiscibility conditions, and more peculiarities have not been considered. For this, the reader is referred to the pertinent

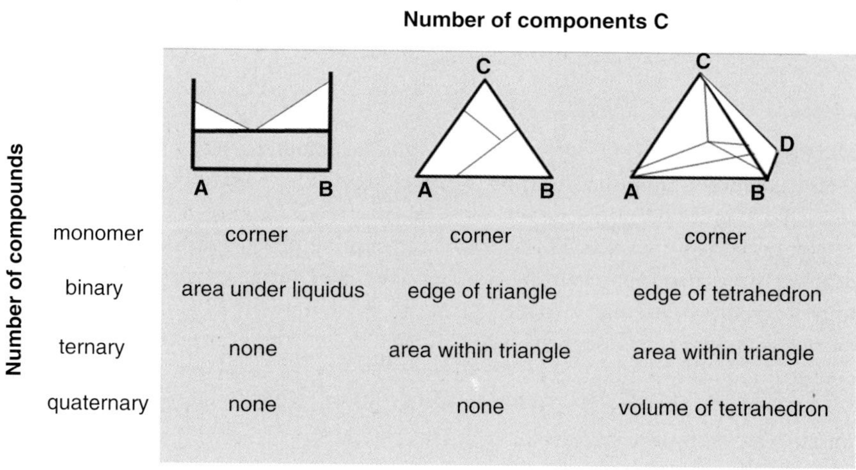

Figure 3.11 Summary of phase diagrams.

literature (see Kingery *et al.*, 1976; Levin *et al.*, 1964–1990; Rosenberger, 1979; West, 1982; The American Ceramic Society, 1992–1996).

With the advent of modern, high-speed computers, a vast number of software programs and thermodynamic modeling approaches are available, sometimes as freeware distributed via the Internet (Schultz and Chang, 1985; Andersen *et al.*, 1993; Chatterjee *et al.*, 1998). A useful guide to these programs and their capabilities has been published by Kattner (1997). *Ab initio* computer simulations of solid solutions, using both empirical representation and quantum mechanical modeling of interatomic distances and cation ordering interactions, were described and applied as tools to model the behavior of an ensemble of ordering cations as a function of temperature by Monte Carlo simulation techniques (Dove, 2001; see also Dove, 1999; Dove *et al.*, 2000a).

Older, but pedagogically excellent, information on the laws of stability and coexistence of minerals, including the application of Schreinemaker analysis to multicomponent systems of geological relevance, can be obtained from Niggli (1954). More detailed examples of the thermodynamic-based calculation of (simple) phase diagrams can be found in Appendix A.

3.2
Phase Systems with Ceramic Relevance

3.2.1
One-Component Systems

Arguably the most important one-component systems with relevance to ceramics are SiO_2 (silica) and Al_2O_3 (alumina). While these two oxides in their pure form are starting materials for several advanced ceramic applications, binary and ternary compounds of silica and alumina, together with CaO, MgO and FeO as well as alkalis, are formed during the firing of clays and thus are of relevance for classic ceramics. Alumina will be discussed in a different context in Chapter 7.

3.2.1.1 Silica

The first comprehensive phase diagram of SiO_2 was formulated by Fenner (Fenner, 1913), and was thought to be correct, with negligible variations, for about forty years (see Figure 3.2b). However, controversy gradually grew more general, and in particular was related to the stability and transformation of the high-temperature silica polymorphs tridymite and cristobalite (e.g., see Eitel, 1957). The most troublesome problem was that the displacive high–low transition of both phases did not occur at a constant temperature but rather varied over a wide range–a behavior that clearly was not consistent with Gibbs' phase rule. Eventually, X-ray crystallography assisted in solving this inconsistency by an unconstrained explanation of this uncommon behavior as related to the mode of the stacking of SiO_4 tetrahedra and the ensuing disorder (see below). Subsequent transmission electron microscopy (TEM) studies on purified synthetic tridymites showed that a series of low-temperature superstructures existed, the transformation temperatures and ranges of stability of which were influenced by many extrinsic (heat treatment, grinding, flux or fluid composition) and intrinsic (impurity content, defect density) factors (Carpenter and Wennemer, 1985). This not only allowed the formation and transformation of the silica phase "tridymite" complex, but also explained the earlier controversies that today are still not completely resolved.

More recent studies have included the investigation of local structures of silica polymorphs, and in particular their displacive transitions from low- to high-temperature forms by studying the total and partial radial electron density distribution functions obtained from X-ray and neutron-scattering experiment and by applying the rigid unit mode (RUM) model as well as Landau theory to their interpretation (Dove *et al.*, 1993; Dove *et al.*, 2000b; see also Dove *et al.*, 2000c). More general details on the RUM approach can be found in Appendix B, while the basics of the determination of radial electron density distributions are reviewed in Appendix C.

Many novel SiO_2 polymorphs have been added to the list of known phases during the past few decades, some of which are stable only under very high static or dynamic pressures, such as coesite (Coes, 1953; Chao *et al.*, 1960) and stishovite (Stishov and Popova, 1961), while some (such as keatite or silica K; Keat, 1954) are stable only under hydrothermal conditions, or are formed by disproportionation from the gas phase (e.g., silica W; Weiss and Weiss, 1954). More recently, moganite was described as a microcrystalline silica phase characterized by alternate stacking of layers of right- and left-handed α-quartz ("low-quartz") with periodic Brazilian law twinning along $\{11\bar{2}0\}$ at a unit cell scale (Flörke *et al.*, 1976; Miehe and Graetsch, 1992; Götze *et al.*, 1998). An entertaining (although highly debatable) "gedankenexperiment" conducted by Sosman (1955) identified some 23 more or less-well-defined silica phases and their transformation relations (see also Heimann, 1977).

Figure 3.12 shows the now generally accepted p,T-diagram of silica. At ambient pressure, α-quartz (low-temperature quartz) transforms by a displacive mechanism to β-quartz (high-temperature quartz) at 573 °C. This phase is stable up to 867 °C, at which point it theoretically converts by a reconstructive transformation mechanism to hexagonal primitive (HP)–tridymite (Kihara, 1978). However, since this transformation has never been observed in pure silica, it was assumed that the formation of tridymite would require stabilization by foreign ions (Flörke, 1956; Holmquist, 1968), despite there being experimental evidence to the contrary (for example, Hill and Roy, 1958; Rockett and Foster, 1967). HP–tridymite transforms at 1470 °C to β-cristobalite before melting at 1727 °C.

The low-temperature α-forms of quartz, tridymite and cristobalite represent reconstructive transformations[6] (Buerger, 1951) of the basic framework of corner-

6) Following a proposal by Heuer (1975), the terms "reconstructive" and "displacive," as introduced earlier by Buerger (1951) to characterize the types of bond (primary or secondary) altered during transition, should be replaced by the more general terms "heterogeneous polymorphic phase transition of first order" and "homogenous polymorphic phase transition of second or higher order." This is with respect to the place of formation and the type of shift of the boundary between parent and product phases.

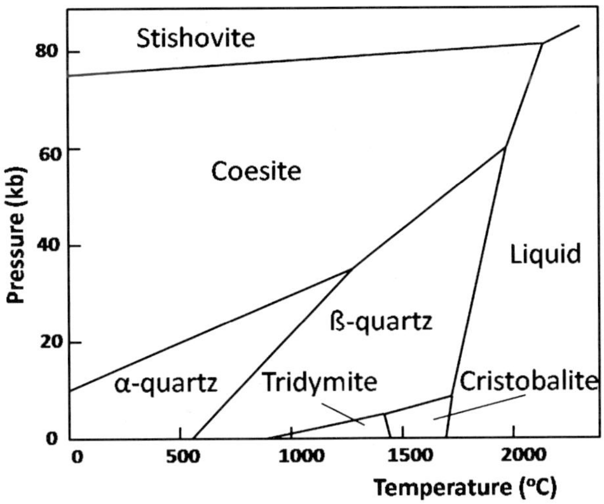

Figure 3.12 Pressure–temperature phase diagram of silica.

sharing SiO_4 tetrahedra, whereas the high-temperature β-forms of each of these polymorphic forms represent displacive transformations with respect to the α-forms (Strunz and Nickel, 2001) (these transitions are shown in Figure 3.19).

The high-pressure polymorphs coesite and stishovite are important indicators of the physical conditions in a deep-seated rock at the time these minerals are crystallized. Hence, their phase stability ranges, and their transformation temperatures and pressures have been studied copiously (e.g., see Weaver *et al.*, 1979; Chopin, 1984). Recently, theoretical evidence has been provided of one or more "post-stishovite" phases that are stable beyond the stability range of stishovite (S.G. $P4_2/mnm$) and which may be candidate structures to represent the silica component in the Earth's mantle close to the core-mantle boundary at 2600–2899 km depth (Belonoshko *et al.*, 1996). A Landau free energy expansion has been used by Carpenter *et al.* (2000) to calculate the elastic constant variations of stishovite under pressure, and transformation into silica with a $CaCl_2$-structure (S.G. *Pnnm*) has been suggested. This phase will transform to silica with an orthorhombic α-PbO_2 structure (S.G. *Pbcn*) between 112 and 118 GPa at 2500 K, and hence may be related to the $MgSiO_3$ perovskite to "post-perovskite" transition at the D″-discontinuity (Murakami *et al.*, 2004; Hirose and Lay, 2008).

α- and β-Quartz In quartz, the most common of the silica phases in nature and industry, the SiO_4 tetrahedra, are connected in such a way that they form helical chains along [00.1] (Figure 3.13). The symmetry group of α-quartz is *32*, and that of β-quartz *622*. Hence, the c-axis directions conform to threefold screw axes 3_1 or 3_2 in α-quartz, and sixfold screw axes 6_2 and 6_4 in β-quartz. These left- or right-handed helical screw axes introduce morphological chirality (handedness) that is responsible for a variety of physical properties. For example, owing to the absence of a center of symmetry, crystals of the symmetry group *32* are optically active,

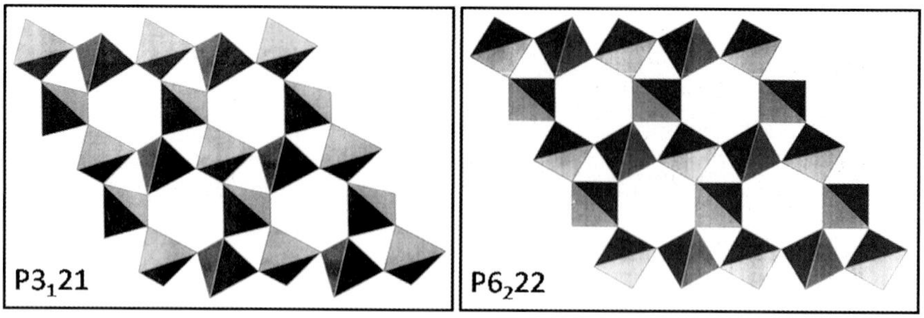

Figure 3.13 Helical conformation of SiO_4 tetrahedra in α-quartz (left) and β-quartz (right), as projected onto the (00.1) plane.

pyroelectric, and piezoelectric (see Chapter 8), and also show electro-optic (Pockels effect) and magneto-optic (Faraday effect) properties. The right- and left-handed quartz crystals are mirror images of each other (enantiomorphism). Quartz plates cut perpendicular to [00.1] rotate the plane of plane polarized light to the right and to the left, respectively. Triangular etch pits generated on (000.1) with hydrofluoric acid frequently show curved, beak-like extensions, the direction of rotation of which conform to the handedness of the threefold screw axis parallel [00.1] (Figure 3.14).[7)]

As shown on the right-hand side of Figure 3.14, the "beaks" are traces formed by the intersection of the edges of the trigonal dipyramid s {11.1} and the trigonal trapezohedron x {51.1} with the plane {00.1} (Heimann, 1979). Recently, a technique has been invented to determine the handedness directly by XRD (Tanaka *et al.*, 2008). While conventional Bragg diffraction is unable to distinguish between right- and left-handed quartz, resonant Bragg diffraction using circularly polarized X-rays reveals the handedness by coupling X-ray helicity to the sense of rotation of the trigonal screw axis in the crystal structure.

The polyhedral representation of Figure 3.13 shows that, during the transition from α- to β-quartz, the individual rigid SiO_4 tetrahedra rotate around their flexible links at shared oxygen sites, in accordance with the interpretation of the displacive transition by the RMC (Reverse Monte Carlo; McGreevy and Pusztai, 1988) and the RUM methods, respectively (e.g., Dove *et al.*, 2000a). As a result, the Si–O–Si bonding angle of 143.6° in α-quartz increases to 153.1° in the ideal β-quartz structure (Dorner *et al.*, 1981). Likewise, the Si–O distance decreases gradually from 161.0(2) pm at 0 K to 158.5(7) pm at 846 K (Dove *et al.*, 2000c). The increase of the bond angle with temperature may be related to a change of the electron density on the Si atom, such that of the basic sp^3-hybridization some density becomes d-type with π-bond character (Cruickshank, 1961).

7) This is only coincidental. Ironically, optically left-handed quartz (S.G. $P3_121$) contains a right-winding trigonal screw axis 3_1, while optically right-handed quartz (S.G. $P3_221$) contains a left-winding screw axis 3_2. This has led to much confusion in the literature (see Donnay and Le Page, 1978).

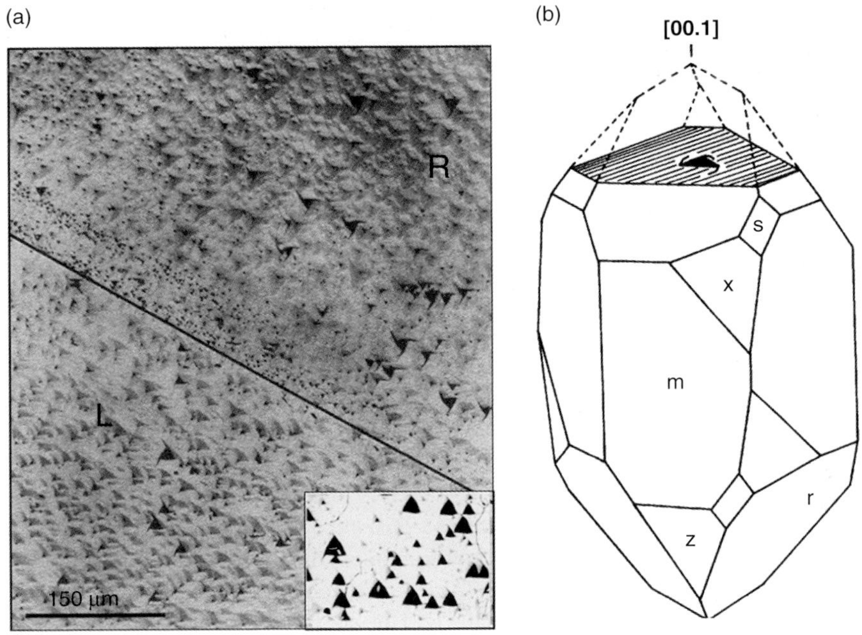

Figure 3.14 (a) Basal plane (00.1) of a Brazilian twin of α-quartz (twin plane {11.0}) etched in 20% HF for 24 h (Heimann, 1975). The direction of rotation of the "etch beaks" (Honess, 1927) indicates the orientation of the twin domains (R = right-handed, L = left-handed). On approaching the domain boundary the size of the etch pits decreases; (b) Right-handed α-quartz crystal, showing a triangular etch pit with clockwise-rotated "beaks" on the (00.1) plane (Heimann, 1979).

It is assumed that, during the low–high transition, domains with the new β-quartz symmetry and orientation will be formed that already are considerably below the transition temperature of 573 °C, and that the domain walls will shift with increasing temperature until the last trace of the α-phase disappears. These domains are twins after the Dauphiné law. (Dauphiné twins are penetration twins with a twinning axis [00.1] and a compositional plane {10.0}, whereby the handedness of the subindividuals are either R + R or L + L.) The driving force for twinning is the desire of the parent phase to attain the structural symmetry of the hexagonal HT product phase. As the HT form is approached, the frequency of the reversal of the structure increases and the scale of twinning becomes progressively smaller (Putnis, 1992). Eventually, the domains disappear into an incommensurate intermediate structure, modulated through micro-twinning by a first-order mechanism that converts by a second-order transformation to the HT β-quartz structure (Dolino, 1990). This behavior can be correlated with the soft phonon mode description of the displacive phase transition (see Section 8.2.1 and Appendix B). On approaching the transition temperature, the frequencies of the phonon modes rapidly decrease (Hammonds *et al.*, 1996), leading to an increase in their amplitudes. The increasing phonon amplitudes introduce orientational disorder, as seen in the occurrence of the Dauphiné twin domains. However, there is a twist to this otherwise consistent explanation of the low–high transformation of quartz. While the phase transition occurs as a result of the classical soft mode instability, in addition a large number of low-frequency modes will be excited that reinforce the orientational disorder (Dove *et al.*, 2000c).

The narrow channels along [00.1] in the quartz structure (Figure 3.13) are able to accommodate small Li atoms by the substitution of LiAl for Si. The β-eucryptite ($LiAlSiO_4$) substructures thus formed are isostructural with β-quartz, and can engage in solid solution with it (Heaney, 2000). This so-called "stuffed" derivative of β-quartz has an unusually low coefficient of thermal expansion, and is consequently used in applications that require a high thermal shock resistance, such as cookware and jet engine components. The substitution $Al^{3+} + Li^{+} \leftrightarrow Si^{4+}$ decreases the α–β transformation temperature of quartz by as much as 120 °C (Keith and Tuttle, 1952).

Tridymite and Cristobalite The idealized structure of HP-tridymite consists of layers of SiO_4 tetrahedra with the stacking sequence ABABA... (Figure 3.15a) that is compatible with the packing rhythm of the hexagonal close-packed (h.c.p.) structure to form a wurtzite-type lattice. The six-membered rings form sheets parallel (00.1) with apices of tetrahedra pointing alternatively in opposite directions (Figure 3.16a).

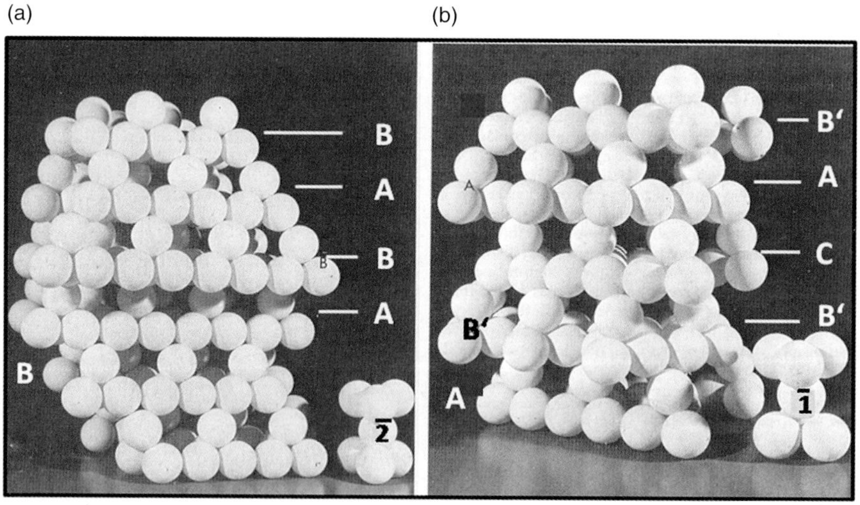

Figure 3.15 Two-layer arrangement of SiO_4 tetrahedra in the idealized tridymite structure with stacking sequence ABAB...(a) and three-layer arrangement in the idealized cristobalite structure with stacking sequence AB′CA...(b). The double Si_2O_7 tetrahedra show the symmetry $\bar{2}$ (*cis* form) in tridymite, but $\bar{1}$ (*trans* form) in cristobalite (Heimann, 1977).

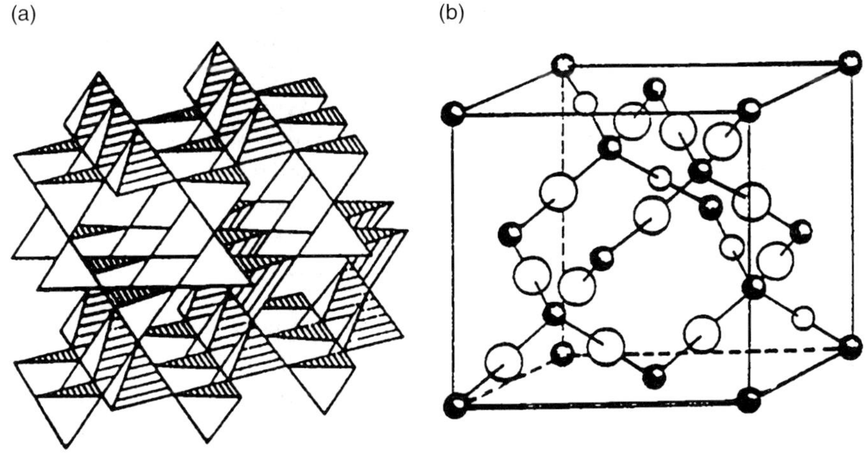

Figure 3.16 Structures of (a) HP-tridymite (high tridymite) and (b) β (high-)-cristobalite.

The two tetrahedra forming a Si_2O_7 unit are arranged with a mirror plane $\bar{2}$ through the connecting oxygen atom in *cis*-configuration. This symmetry element is responsible for an expansion of the layers as a result of the repulsion of the second-nearest oxygen neighbors.

In contrast to this, in the β-cristobalite structure (S.G. $Fd\bar{3}m$) the double tetrahedra are connected in *trans*-configuration corresponding to the symmetry element $\bar{1}$. This arrangement leads to the layer stacking sequence AB′CAB′... akin to the cubic close-packed (c.c.p.) structure. Figure 3.16b shows the zincblende-type lattice of β-cristobalite with a Si–O distance of 159 pm, an O–O distance of 260 pm, a Si–Si distance of 308 pm, and an Si–O–Si angle of 151°.

In the real structure, h.c.p.- and c.c.p.-stacking sequences alternate in random succession to form silica phases of prevailing three-layer-type ("cristobalite" *sensu lato*) or prevailing two-layer-type ("tridymite" *sensu lato*) (Flörke, 1967). These stacking sequences lead to higher periodicities – that is, superstructures causing the effect of polytypism in cristobalite and, in particular, tridymite (Carpenter and Wennemer, 1985). In this view, any discrimination between cristobalite and tridymite is practically meaningless if the h- and c-elements balance (Heimann, 1977). During the 1970s, the validity of this assumption was vigorously debated in the context of the structure of microcrystalline silica such as opal, chert, and chalcedony. By conducting XRD, electron diffraction and infra-red (IR) studies, Wilson *et al.* (1974) confirmed that the structure of natural opals and deep-sea cherts was consistent with that of low-temperature superstructures of tridymite. However, Jones and Segnit (1975) pleaded for the assumption of a significant cristobalite-type stacking in what they called "opal-CT." This phase was thought by Flörke *et al.* (1975) to be built up from a disordered interstratification of cristobalite-type and tridymite-type layers on an atomic scale.

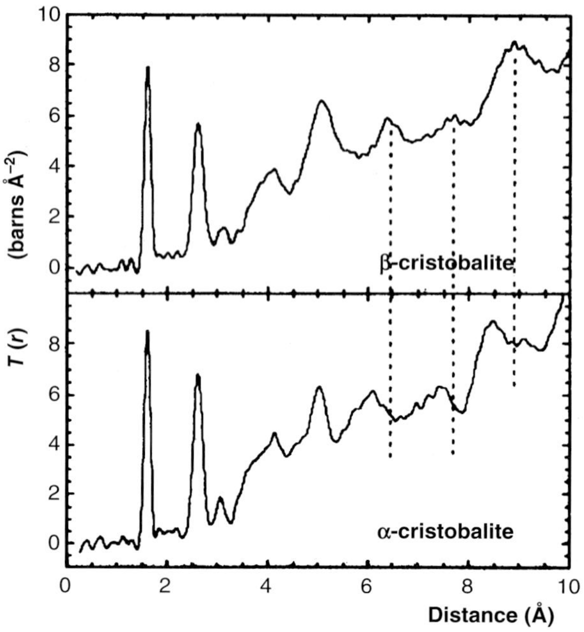

Figure 3.17 Pair distribution function $T(r)$ obtained by diffuse total neutron scattering on low (α)- and high (β)-cristobalite. The dashed lines highlight features in $T(r)$ unique for β-cristobalite, attesting to the different short-range ordered structures in both polymorphic forms (after Dove *et al.*, 2000c). Reprinted with permission from The Mineralogical Society of America.

Today, not only molecular dynamic calculations but also X-ray- and neutron-scattering experiments are invoked to provide a more complex, and hence sophisticated, view on the issue. In contrast to the α–β phase transition of quartz, with its distinct domain structure, the displacive transformation of cristobalite appears to proceed by random distortion of the SiO_4 tetrahedra six-rings. This has been confirmed by direct measurements of Si–O bond lengths of α-cristobalite and β-cristobalite (Figure 3.16) using neutron diffuse scattering data. Figure 3.17 shows the pair distribution function $T(r)$ of the total scattering function $S(Q)$ (see Appendix C) as a function of the distance r (Å) in α-cristobalite at 473 K and β-cristobalite at 573 K (Dove *et al.*, 1997). The first and second sharp peak corresponds to the Si–O bond length (1.62 Å) and O–O bond length (2.65 Å), respectively, of the SiO_4 tetrahedron. The third peak at around 3.1 Å corresponds to the nearest-neighbor Si–Si distance within two tetrahedra connected by a shared oxygen atom. While the positions and intensities of these first three peaks are similar in both cristobalite polymorphs, clear deviations occur for distances larger than 5 Å. This is evidence against the presence of domains of α-cristobalite in the structure of β-cristobalite (Dove *et al.*, 2000c) and points to differently disordered short-range structures, a view also supported by calculations using RMC and RUM approaches, and measurements of the Si–O bond angles. A similar situation applies to HP-tridymite.

In contrast, a Rietveld refinement of the Bragg peaks of β-cristobalite obtained using X-ray or neutron diffraction has resulted in a crystal structure with energetically unfavorable linear Si–O–Si bonds. On the other hand, total neutron-scattering experiments provided information on the instantaneous structure, such as bond lengths and bond angles through pair correlation functions (see Appendix C). With this technique, no tetrahedral contraction was found during the displacive α–β transition of cristobalite and tridymite, and also bond angles around 151°. Since, in most silicates, the Si–O–Si bond is close to 145° there is presumably considerable dynamic disorder that allows the bonds to bend in order to avoid local configurations with linear Si–O–Si bonds. This picture is consistent with the large thermal ellipsoids determined in crystal structure refinements, and the fact that the Si–O bond in the average structure is slightly shorter than normally found.

The displacive high–low transitions of both cristobalite and tridymite are associated with remarkable volume effects that, in the case of cristobalite, can be as high as 3 vol% between 200 and about 270 °C. This has important implications for a variety of technological processes. For example, the cooling of kilns lined with silica bricks to below the high–low transition temperature of cristobalite induces tensile stresses that cause cracking and spalling of the high-silica refractories. Because of the remarkable high-temperature corrosion stability of these refractories against acid fluxes and iron oxide compounds, refractories containing more than 90% SiO_2 are used as lining for glass melt furnaces, Siemens–Martin converters, electro-steel kilns, and coke ovens. The bricks formed and pressed from pulverized quartzite mixed with lime wash and fired at 1500 °C are almost completely transformed into tridymite via a metastable cristobalite phase (Schulle, 1972; Madden and van Vlack, 1967). Since even small additions of alumina will dramatically lower the heat resistance of silica bricks and their mechanical stability under pressure, due to formation of liquid phase at the SiO_2–Al_2O_3 (~5 mol% Al_2O_3) eutectic around 1600 °C (see Figure 3.20), *supersilica bricks* were developed starting from very pure quartzite or hydrothermal quartz raw materials. The transformation sequence among different silica phases in response to temperature and chemical interaction with iron and magnesium oxides has been expressed by the so-called DINAS[8] rule that describes the (reconstructive) transformations from α-quartz to β-tridymite, and finally to β-cristobalite that is accompanied by a distinct color change, from the yellow of the pristine silica brick through brown and black hues in bricks containing high proportions of tridymite together with fayalite and iron monticellite, to gray bricks with high amounts of cristobalite associated with magnetite (Figure 3.18). Since β-cristobalite inverts to α-cristobalite below 270 °C (with the large volume effect mentioned above), in normal operation steel kilns, coke ovens and open-hearth furnaces will not be cooled below this critical temperature, even during batch-type recharging.

8) The term DINAS refers to a locality in South Wales, UK, where the silica bricks were invented around 1820 by W.W. Young.

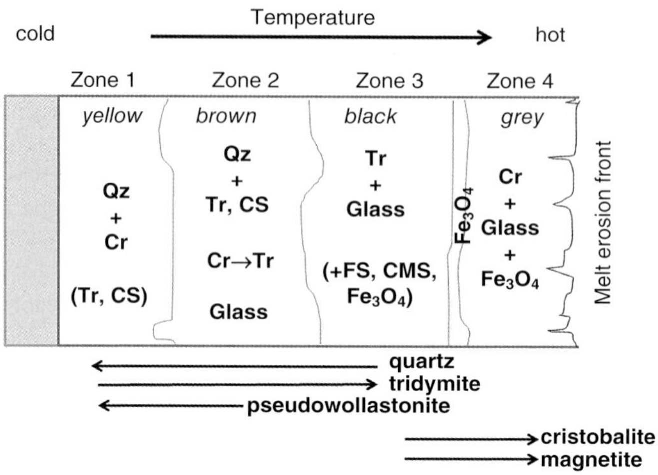

Figure 3.18 Schematic cross-section of a silica brick with transformations of the silica phases according to the DINAS rule.

Zone 1 shown in Figure 3.18 is the virgin silica brick consisting mostly of quartz, some (metastable) cristobalite, and traces of tridymite and pseudowollastonite, $Ca_3[Si_3O_9]$. In transition zone 2, which is closer to the heat source, the amount of tridymite and pseudowollastonite increases, the cristobalite has transformed to tridymite, and a brown-yellow glass has formed by the reaction of fluxing iron, calcium, and alkali oxides with silica. The adjacent tridymite zone (zone 3) comprises predominately HP-tridymite, a black glass with fayalite, Fe_2SiO_4 and isostructural iron monticellite, $CaFe[SiO_4]$ exsolutions, and magnetite. At the hot side of zone 3 a dense net of tridymite crystal forms interspersed with magnetite and some complex silicates. This zone is frequently separated from the adjoining cristobalite zone (zone 4) by a dense layer of magnetite crystals. In zone 4, complete transformation to twinned β-cristobalite has occurred, embedded in a glass phase colored gray by tiny magnetite exsolutions.

Figure 3.19 summarizes the polymorphic phase transitions of silica. The labeling of the tridymite superstructures corresponds to that introduced by Kitchin *et al.* (1996) as obtained by ^{29}Si magic angle spinning nuclear magnetic resonance (MAS-NMR) studies. Here, MC refers to low-temperature monoclinic tridymite transforming at 381 ± 1 K to orthorhombic OP tridymite ("middle" tridymite according to Sosman). At 438 ± 5 K, this phase transforms to a second orthorhombic OS phase which presumably is akin to HP-tridymite (orthorhombic–pseudohexagonal?) that finally transforms to a high-temperature orthorhombic OC phase at 479 ± 4 K (Kitchin *et al.*, 1996).

The transformation path from very pure β-quartz directly (i.e., avoiding the formation of tridymite) to cubic β-cristobalite at 1298 K has been proposed by Flörke (1956, 1967), based on experimental evidence.

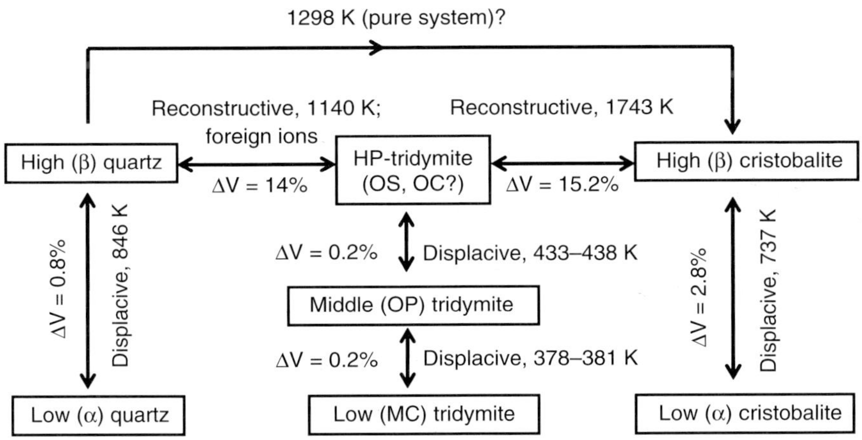

Figure 3.19 Summary of reconstructive and displacive phase transformations of silica at ambient pressure.

There exist close structural relationships between HP-tridymite and the stuffed derivative structures hexagonal nepheline, (Na,K)[$AlSiO_4$] and hexagonal kalsilite, K[$AlSiO_4$], which is stable at low temperature, and also hexagonal kaliophilite, K[$AlSiO_4$], stable at high temperature. The large cavities among the six-membered rings of SiO_4 tetrahedra in tridymite can accommodate Na atoms by substitution of NaAl for Si to form structures with planar wurtzite-type nets parallel (00.1). The space groups of nepheline ($P6_3$), kalsilite ($P6_3$ for kalsilite-1*H* and *P*31*c* for kalsilite-1*T*) and kaliophilite ($P6_3/m$) are all subgroups of the space group of HP-tridymite ($P6_3/mmc$) (Abbott, 1984). These stuffed derivatives display displacive transitions among low and high polymorphs (Kubo *et al.*, 1966), similar to their parent structures.

In cristobalite, even larger cavities can accommodate Na (and K) atoms to form carnegieite with zincblende-type nets (Richet and Mysen, 1999; Marcovic *et al.*, 2003).

3.2.2 Two-Component Systems

The three two-component systems Al_2O_3–SiO_2, MgO–SiO_2 and CaO–SiO_2 described in this section are of importance for several high-refractory and cement mineral compositions with high-temperature stability.

3.2.2.1 Al_2O_3–SiO_2

Phase Diagram Despite the wide occurrence of mullite in refractories and fine ceramics such as porcelain, and copious research having been carried out in this system, there remains considerable debate regarding the nature of the binary

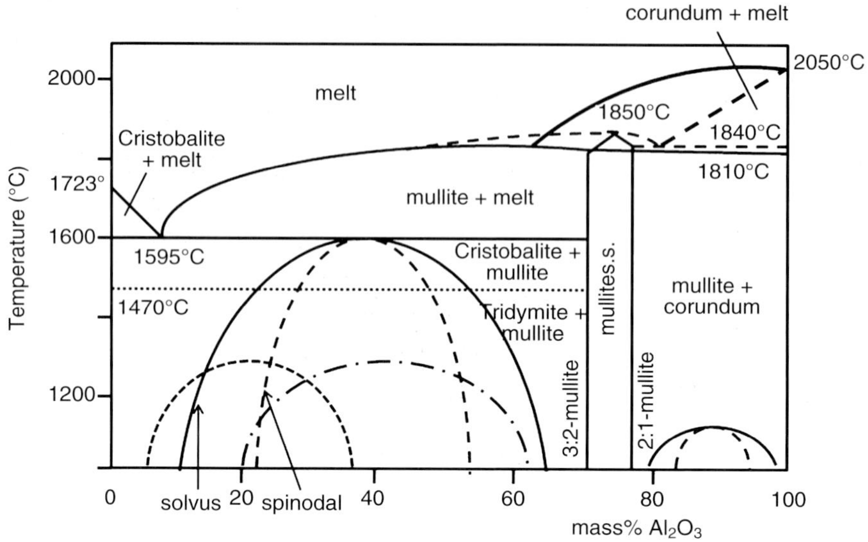

Figure 3.20 Binary system Al_2O_3–SiO_2 showing congruent (dashed liquidus; melting temperature 1850 °C, eutectic mullite–corundum at 1840 °C) and incongruent (solid line, peritectic at 1810 °C) melting curves of mullite. Several subsolidus immiscibility regions are also indicated. The limiting compositions of the mullite solid solution series are $3Al_2O_3{\cdot}2SiO_2$ (3 : 2-mullite) and $2Al_2O_3{\cdot}SiO_2$ (2 : 1-mullite).

phase diagram. This debate centers on the question of whether the only thermodynamically stable binary phase, mullite, melts congruently or incongruently. The discussion was triggered by two landmark reports that were made forty years apart by Bowen and Greig (1924) and by Aramaki and Roy (1962) (see also Toropov and Galakhov, 1953). The former authors found mullite to melt incongruently at 1828 °C, while the latter assumed congruent melting at about 1850 °C (Figure 3.20).

The problem is compounded by the closeness of the temperatures that would separate the incongruent from the congruent melting characteristics. As discussed in the most authoritative book on mullite in existence today (Schneider and Komarneni, 2005), the controlling factor for either behavior appears to be the position of the α-Al_2O_3 liquidus line that was found to depend on the starting material and the experimental method used. This means that the bond strength, free energies and crystal perfection of the precursor materials and participating phases will influence the position of the equilibrium. Today there is strong evidence that, under equilibrium conditions and in the presence of corundum, mullite melts incongruently, whereas in the absence of corundum it displays a nonequilibrium (congruent) melting behavior. This view has been supported by thorough experimental studies, for example, by Aksay and Pask (1975), Klug *et al.* (1987), and Schneider *et al.* (1994).

As shown in Figure 3.20, the phase diagram of alumina–silica is complicated even more owing to the existence of metastable regions characterized by subsolidus exsolutions with wide miscibility gaps. According to Risbud and Pask (1977), a region of immiscibility occurs between 10 and 60 mass% Al_2O_3 below 1300 and about 1600 °C, respectively. Another miscibility gap found between 80 and 98 mass% Al_2O_3 below 1100 °C is so far removed from the liquidus curve that it would be noticeable only in strongly quenched samples as present, for example, in plasma-sprayed coatings (McPherson, 1980; Heimann, 2008). The flat liquidus in the Al_2O_3-rich part of the binary system (Figure 3.20) already hints at the metastable subsolidus miscibility gap. As a consequence, slow nucleation kinetics in this range allows quenching of melts to obtain glassy products with high transformation temperatures (900–1000 °C). Indeed, glassy fibers drawn from such melts with approximate mullite compositions are used as efficient insulating materials for high-temperature applications. However, there is a strong tendency towards devitrification once the transformation temperature has been reached.

The outer contours of the subsolidus curves delineating the metastable immiscibility regions in Figure 3.20 correspond to the strain-free solvus or binodal (solid lines), the inner contours (dashed lines) to the spinodal. A brief introduction to the theory of spinodal demixing of crystals is provided in Appendix D.

Composition and Structure of Mullite Mullite forms a solid solution series with the limiting compositions $3Al_2O_3 \cdot 2SiO_2$ (3 : 2-mullite) with 72 mass% alumina, and $2Al_2O_3 \cdot SiO_2$ (2 : 1-mullite) with 78 mass% alumina. In high-fired ceramics such as porcelain or refractories, these two compositions occur in different morphologies, with primary 2 : 1-mullite forming flakes or scales, and secondary 3 : 2-mullit forming needle-like aggregates (see Figure 3.31). The former is stabilized at lower temperature, but converts to the latter by reacting with excess silica (cristobalite) at higher temperature (this point is discussed in greater detail in Chapter 4).

The crystal structure of mullite (S.G. *Pbam*) is very close to that of sillimanite, $Al^{VI}Al^{IV}[O/\ SiO_4]$ (Figure 3.21). In sillimanite (S.G. *Pbnm*), there exist chains of

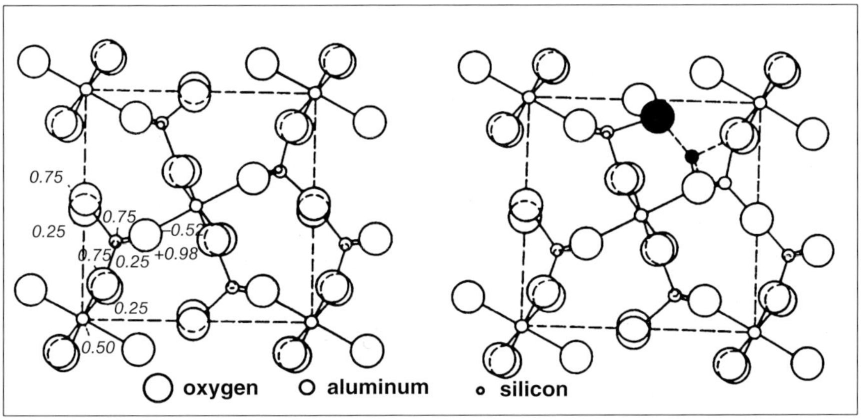

Figure 3.21 Unit cell of sillimanite (left) and mullite (right), projected onto (001).

edge-sharing [AlO_6] octahedra parallel [001], crosslinked by continuous chains of four-membered rings of corner-sharing AlO_4 and SiO_4 tetrahedra (Burnham, 1963; Angel and Prewitt, 1986). In mullite, some of the Si ions are replaced by Al according to $2Si^{4+} + O^{2-} \leftrightarrow 2Al^{3+} + \square$. The oxygen vacancy $\square$ causes the neighboring (Al,Si)O_4 tetrahedra chain to contract and to form an Al–O bridge, as indicated by the ions shown in black in Figure 3.21 (right).

The formal relationship between sillimanite and mullite is:

$$Al_2[Al_2Si_2O_{10}] \rightarrow Al_2[Al_{2+2x}Si_{2-2x}]O_{10-x}.$$

Pure sillimanite exists for $x = 0$, whereas $x = 1/4$ yields 3:2-mullite, $Al_2[Al_{2.5}Si_{1.5}]O_{9.75}$ (= $4/3\cdot(3Al_2O_3\cdot 2SiO_2)$), and $x = 2/5$ yields 2:1-mullite, $Al_2[Al_{2.8}Si_{1.2}]O_{9.6}$ (= $6/5\cdot(2Al_2O_3\cdot SiO_2)$). Since the Si–Al distribution in the [$AlSiO_4$] tetrahedra of sillimanite is ordered but statistically distributed in mullite, the lattice periodicity (c-axis length) of mullite (289 pm) is reduced to half of that of sillimanite (577 pm).

Fireclay Ceramics Technical ceramics composed of mullite and corundum are highly refractory, and are known as *fireclay* or *chamotte*. Fireclay bricks are produced from alumina-rich clays, whereby compositions with a predominant alumina content (>72 mass%) constitute a basic chamotte consisting of mullite + corundum, and compositions with higher silica content ($Al_2O_3 + TiO_2 \leq 32\%$) are known as *siliceous fireclay* or *Dinas chamotte*. The firing temperature of the green bodies varies between 600–900 °C and 1500–1600 °C. The former still contains reactive metakaolinite that interacts with added bonding clay during service in high-temperature furnaces, and thus attains its refractory properties. The latter consists of mullite, quartz, cristobalite and up to 50% glass, and excels with high density and volume stability. The technology of formation of bricks from chamotte and bonding clay varies widely, and the compositions of the masses vary accordingly: plastic forming requires masses consisting of 50–65% chamotte, 17–19% water, balance-bonding clay; dry pressing requires up to 95% chamotte; and slip casting requires 50–70% chamotte and >20% water. [For details of this technology, see Routschka and Wuthnow (2007).]

3.2.2.2 MgO–SiO_2

Phase Diagram In this binary system, two thermodynamically stable compounds occur, namely forsterite (Mg_2SiO_4) and enstatite ($MgSiO_3$) (Figure 3.22). Forsterite melts congruently at 1890 °C, crystallizes in the orthorhombic space group *Pbnm*, and consists structurally of an (almost) hexagonal close-packed arrangement of oxygen atoms. One half of the octahedrally coordinated sites are filled with Mg atoms, and one-eighth of the tetrahedrally coordinated sites are occupied by Si atoms. The $[MgO]_6$ octahedra share edges to form zigzag chains parallel [001]. Forsterite forms with isostructural fayalite, Fe_2SiO_4 the solid solution series of (α)-olivine. At high pressure (~14 GPa), α-olivine transforms to β-olivine (wadsleyite), and at ~17.5 GPa to γ-phase (ringwoodite) with a silicate spinel structure. These minerals are thought to be present in the transition zone of the Earth's

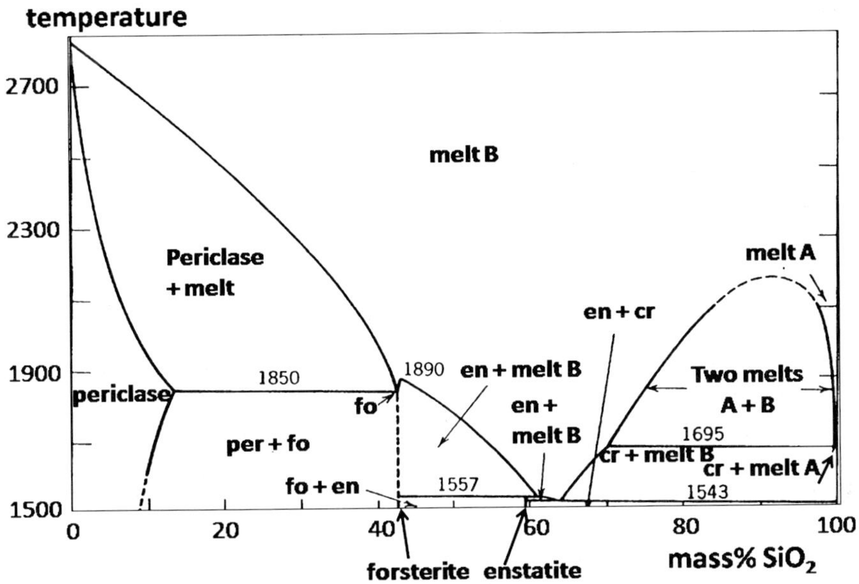

Figure 3.22 Binary system MgO–SiO_2 (per = periclase, MgO; fo = forsterite, Mg_2SiO_4; en = (proto)enstatite, $MgSiO_3$; cr = β-cristobalite, SiO_2).

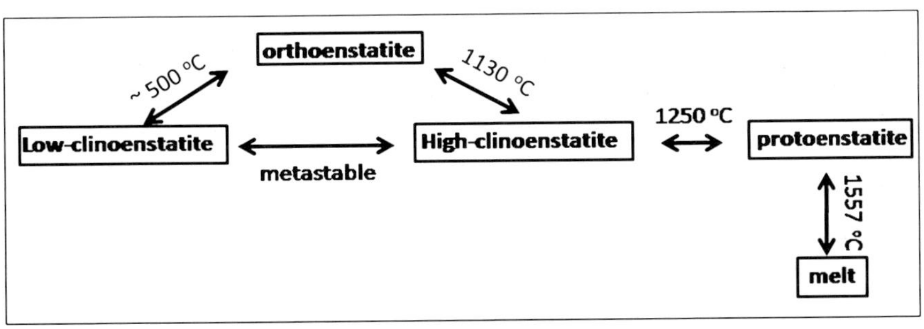

Figure 3.23 Transformation sequence of $MgSiO_3$ polymorphs.

mantle between about 410 and 660 km depth (Frost, 2008). At 24 GPa, corresponding to a depth of 660 km, the ringwoodite structure breaks down to an assemblage of (Mg,Fe)SiO_3 with perovskite structure and ferropericlase (Mg,Fe)O, a reaction associated with a seismic discontinuity.

On the other hand, orthorhombic enstatite (S.G. *Pbca*) belong to the orthopyroxene family, forming single chains parallel [001] of corner-sharing SiO_4 tetrahedra that are linked by double chains of edge-sharing $[MgO]_6$ octahedra. While (ortho)enstatite is stable at room temperature, it transforms via high-clinoenstatite (1130 °C) to protoenstatite (1250 °C) (Figure 3.23). At 1557 °C, protoenstatite melts incongruently under decomposition into forsterite and a melt of composition B (Figure 3.22).

The monoclinic clinoenstatite (S.G. *C2/m*) was found to be stable at high pressures. Recently, a phase diagram was calculated by (Choudhury and Chaplot, 2000) that is in overall qualitative agreement with experimental data. These studies have enabled a microscopic understanding of the factors contributing to the relative stability and indicate that, while the orthoenstatite to protoenstatite transition is temperature-driven, the orthoenstatite to clinoenstatite transition is pressure-driven. Experimentally, the sequence of transformation shown in Figure 3.23 was found.

$MgO–SiO_2$ Refractories In order to achieve a sufficient and manageable range of firing temperatures in $MgO–SiO_2$ refractories (so-called steatite ceramics), it is necessary that the amount of liquid phase does not change much with temperature (Kingery *et al.*, 1976). Hence, the starting composition of the green body prior to firing must be judiciously selected in such a way that it is located in the forsterite phase field, but is only slightly away from the exact forsterite composition at 42 mass% SiO_2 towards the enstatite phase field (Figure 3.22). In this case, a small amount of melt is formed at 1557 °C (melting point of enstatite) that, however, does not change very much with temperature as the liquidus curve is steep. In contrast, if the starting composition is <42 mass% SiO_2, no melt is formed below 1850 °C.

During the production of chromite refractories, the most common impurity of the raw materials used is serpentine, $Mg_3[Si_2/(OH)_4O_5]$, with about 50 mass% SiO_2. In this case, MgO must be added to shift the composition of the refractory product into the forsterite–periclase phase field in order to prevent the early formation of melt at low temperatures. Likewise, in steatite ceramics with >50 mass% SiO_2, the protoenstatite contracts by about 2.6 vol% during cooling, since (high)clinoenstatite is formed with a higher density. This leads to the introduction of porosity, such that the bricks tend to crumble during storage at ambient conditions. However, the selection of talc as a precursor material results in ceramics in which the protoenstatite will be stabilized by SiO_2 formed by the decomposition reaction:

$$Mg_3[Si_4(OH)_2O_{10}] \rightarrow 3\,MgSiO_3 + SiO_2 + H_2O,$$

thus avoiding the transformation of protoenstatite to clinoenstatite.

3.2.2.3 $CaO–SiO_2$

Phase Diagram Calcium silicates are the most important constituents of hydraulic Portland cements (see Chapter 5), as well as of basic and acidic blast furnace slags and stabilized refractories based on dolomite; they also occur as devitrification products of some technical CNS (calcium sodium silicate) glasses. The phase diagram is complicated, as shown in Figure 3.24.

The two calcium silicates pseudowollastonite, α-$Ca_3(SiO_3)_3$, and dicalcium silicate, α-Ca_2SiO_4, melt congruently at 1544 ± 2 °C and 2130 ± 15 °C, respectively. Rankinite, $Ca_3Si_2O_7$ (S.G. *P2₁/a*) dissociates at 1460 ± 5 °C. The temperature of

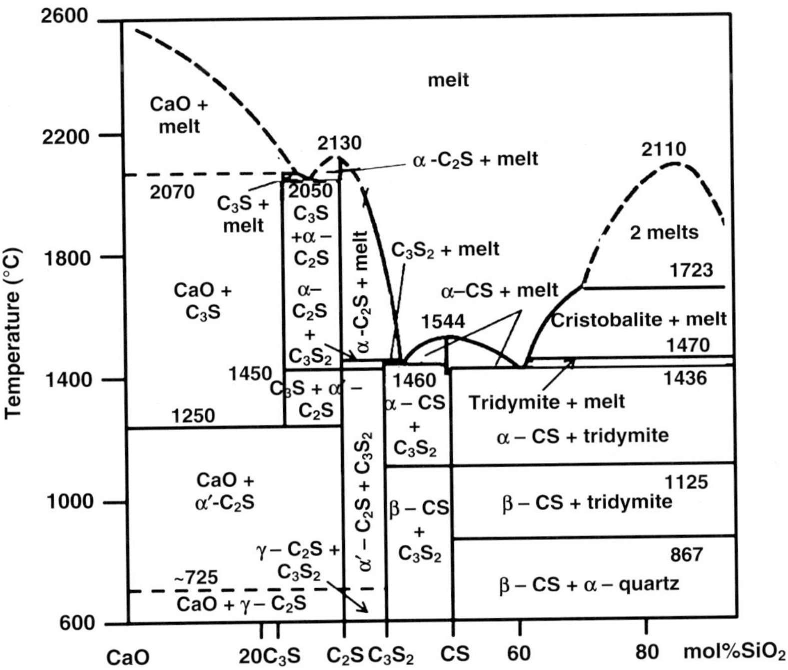

Figure 3.24 Binary system CaO–SiO_2. Phases are as follows: C_3S: tricalcium silicate, alite; C_2S: dicalcium silicate, hydraulic phase (α), larnite (α′), calcio-olivine (γ); C_3S_2: rankinite; CS: pseudowollastonite (α), wollastonite (β) (Phillips and Muan, 1959). Reprinted with permission from Wiley-Blackwell, Oxford, UK. Modifications introduced by Niesel and Thormann (1967) and Taylor (1997) are discussed in the text.

transition from wollastonite, β-$Ca_3(SiO_3)_3$ (S.G. $P\bar{1}$) to pseudowollastonite, α-$Ca_3(SiO_3)_3$ (S.G. $C2/c$) is 1125 ± 2 °C, while that of the transition from α′-Ca_2SiO_4 to α-Ca_2SiO_4 is 1450 ± 15 °C.

Tricalcium silicate, $Ca_3(O/SiO_4)$ is stable in the temperature interval 1250–2070 °C, and melts incongruently under the formation of CaO + melt (Welch and Gutt, 1959). In cement chemistry, it is known as "alite," with some Ca being substituted by Mg, and Si by Al. Ca may also be replaced by Cr, Cd, Sr, Cu, Mn, Ti, Ba, Y, and other heavy-metal ions, and consequently it can serve as a host for the immobilization of industrial toxic heavy-metal waste (see Section 5.4.5; Heimann *et al.*, 1992; Pöllmann, 2007). Tricalcium silicate is also found naturally as the mineral *hatrurite* (S.G. $R3m$). Within the interval of its stability several polymorphic transitions occur, from low-temperature triclinic structures (600–980 °C) to monoclinic (980–1070 °C) to the trigonal, high-temperature form. In the crystal structure, the isolated SiO_4 tetrahedra and $Ca^{[6]}$ and $Ca^{[7]}$ polyhedra share corners to form a framework.

Dicalcium silicate shows equally complex phase relations that are still under discussion. According to Niesel and Thormann (1967) and Taylor (1997), α-Ca_2SiO_4 melts at 2130 °C; however, cooling the melt to 1420 ± 5 °C leads reversibly to high

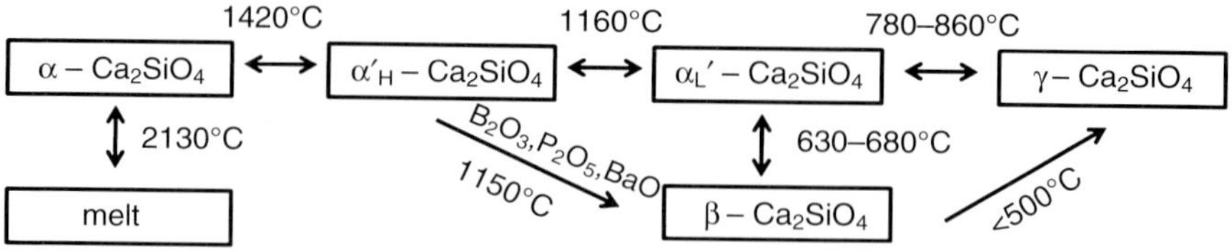

Figure 3.25 Transformation sequence of dicalcium silicate (Niesel and Thormann, 1967; Taylor, 1997).

α'_H-Ca_2SiO_4 that, at 1160 ± 10 °C, transforms to low α'_L-Ca_2SiO_4. The addition of mineralizers such as boron, phosphorus or barium oxides stabilizes β-Ca_2SiO_4 ("belite") at 680–630 °C, whereas in the pure system between 860 and 780 °C a transformation of low α'_L-Ca_2SiO_4 to nonhydraulic γ-Ca_2SiO_4 (calcio-olivine, S.G. *Pnma*) with olivine structure occurs. In calcio-olivine, the M1 and M2 sites are occupied by $Ca^{[6]}$ ions that are large compared to Mg, Fe, and Mn found in olivines of the forsterite series; hence, lattice strain makes the structure rather unstable. The β-Ca_2SiO_4 changes to γ-Ca_2SiO_4 at <500 °C. Formation of the γ-phase must be controlled during formation of cement clinker by rapid quenching to avoid transformation to this nonhydraulic polymorph of dicalcium silicate. As the density of the β-phase (3.28 Mg m^{-3}) is much higher than that of the γ-phase (2.97 Mg m^{-3}), disintegration of the β-type grains during stress-induced transformation (Nettleship *et al.*, 1992) occurs; this imparts deleterious properties (e.g., a low hydraulic activity) to the cement and hence also to the concrete produced from it. Figure 3.25 shows, schematically, the transformation sequence of dicalcium silicate.

Recently, an additional dicalcium silicate phase, κ-Ca_2SiO_4 with S.G. $P2_1/c$ was identified during the dehydration of hydrothermally synthesized α-$Ca_2(SiO_4H)OH$ at 800 °C (Yamazaki and Toraya, 2001).

3.2.3
Three-Component Systems

In this section, only two three-component systems will be briefly discussed, namely, $CaO–Al_2O_3–SiO_2$ and $K_2O–Al_2O_3–SiO_2$.

The $CaO–Al_2O_3–SiO_2$ system serves to describe in its SiO_2-poor part the phase relations and transformations in Portland and aluminate cements; in its SiO_2-rich/intermediate CaO part earthenware pottery including, bricks and stoneware (see Figure 3.26); and in its Al_2O_3-rich part ceramics with refractory properties (mullite, chamotte).

The $K_2O–Al_2O_3–SiO_2$ system is crucial to describe the microstructure of China ware (porcelain).

3.2.3.1 $CaO–Al_2O_3–SiO_2$

In Figure 3.26, for clarity, only the cotectic triangles are shown formed by conodes that connect the positions of primary ternary and binary compounds. In the

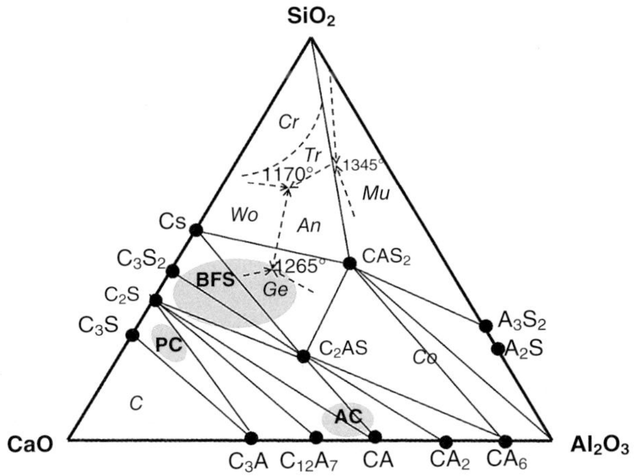

Figure 3.26 Cotectic (compatibility) triangles (solid lines) in the ternary diagram $CaO–Al_2O_3–SiO_2$ (mass%). In the SiO_2-rich portion, three important triple points are indicated at the intersection of the phase boundaries (dashed lines). The arrows show the direction of falling temperatures according to the Alkemade theorem (see Section 3.1.4.3). *An* = anorthite; *Co* = corundum; *Cr* = cristobalite; *Ge* = gehlenite; *Mu* = mullite; *Tr* = tridymite; *Wo* = wollastonite. AC, PC, and BFS indicate the compositional areas of aluminate cement, Portland cement, and blast furnace slags, respectively.

ternary system $CaO–Al_2O_3–SiO_2$ there exist only two ternary compounds: anorthite (An, $CaAl_2Si_2O_8$) and gehlenite (Ge, $Ca_2Al_2SiO_7$). Gehlenite is the Al-rich end member of the melilite ($Ca_2(Mg,Al)[Si_2O_7]$) solid solution series. In the presence of MgO, the Mg-rich end member åkermanite ($Ca_2MgSi_2O_7$) is stable, and hence the composition of the melilites is located in the quaternary system $CaO–MgO–Al_2O_3–SiO_2$ (see Figure 3.9).

The binary compounds are located along the sides of the triangle shown in Figure 3.26. The calcium aluminate phases C_3A (and $C_{12}C_7$), and the calcium silicate phases C_3S and C_2S are important clinker phases of cement (see Chapter 5), forming "aluminate," "alite," and "belite". The iron-rich calcium aluminate, "celite" or brownmillerite (C_4AF) is a compound in the ternary system $CaO–Al_2O_3–Fe_2O_3$. The triple point between An, Wo, and Tr is the ternary eutectic, with a melting point at 1170 °C; other eutectics with low melting points are located at the triple points An, Ge and Wo (1265 °C), and An, Tr and Mu (1345 °C). Another important triple point at 1335 °C exists among C_2S, CA and $C_{12}A_7$ (not shown in Figure 3.26).

Figure 3.27 shows the location of some common, coarse-grained, nonrefractory ceramics in a subsection of the ternary phase diagram.

3.2.3.2 $K_2O–Al_2O_3–SiO_2$

This system is important to describe the technology of porcelain and fine stoneware, and also the origin and evolution of some rock-forming minerals. Figure

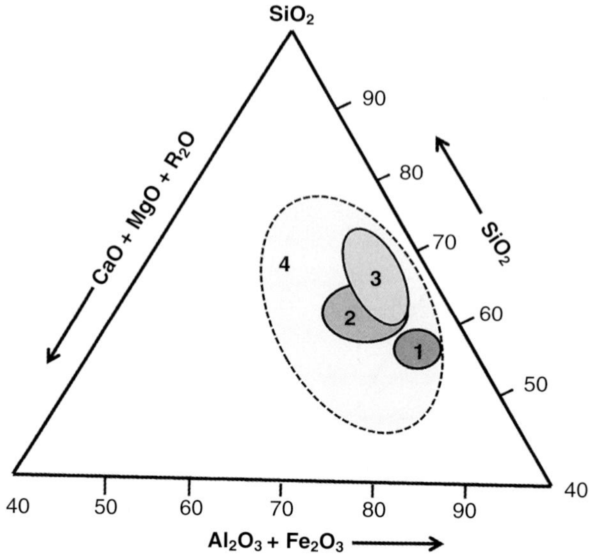

Figure 3.27 Position of nonrefractory coarse ceramics in the (multinary) system (RO + R_2O)–(Al_2O_3 + Fe_2O_3)–SiO_2. 1: Tiles and terracotta; 2: bricks; 3: stoneware pipe; 4: range of coarse ceramics.

3.28 shows a simplified version of the SiO_2-rich part of the ternary phase diagram. Potassium feldspar melts incongruently at 1170 °C, slightly above the ternary eutectic Kfs-Leu-Mu at 1140 °C.

The typical classic raw materials mixture for manufacturing porcelain consists of 50 mass% kaolin, 25 mass% K-feldspar, and 25 mass% quartz ("triaxial" porcelain), corresponding to point "A" in Figure 3.28 (69.3 mass% SiO_2, 26.2 mass% Al_2O_3, 4.5 mass% K_2O). This composition is located in the stability field of mullite, with a liquidus temperature >1700 °C. On cooling from this temperature, mullite crystallizes along the path A–B that meets at 1280 °C the phase boundary mullite–tridymite. Further crystallization occurs while following the phase boundary along the path B–C until, at 985 °C, the ternary eutectic K-feldspar–tridymite–mullite will be reached. The results of calculation of the theoretical (equilibrium) composition (Salmang *et al.*, 2007) are shown in Figure 3.29. Crystallization on cooling from high temperatures leads to the formation of mullite, reaching a maximum amount of 27.4 mass% at 1280 °C. Below this temperature tridymite crystallization starts, reaching 22.9 mass% at 985 °C, besides 47.4 mass% melt phase, and an additional amount of 2.3 mass% mullite. Further cooling to ambient temperature causes crystallization of the melt to form 26.6 mass% K-feldspar and an additional amount of 20.8 mass% tridymite. Hence, the equilibrium composition of the final porcelain body is 29.7 mass% mullite + 43.7 mass% tridymite + 26.6 mass% K-feldspar (Figure 3.29). However, in reality the microstructure of a typical porcelain body is

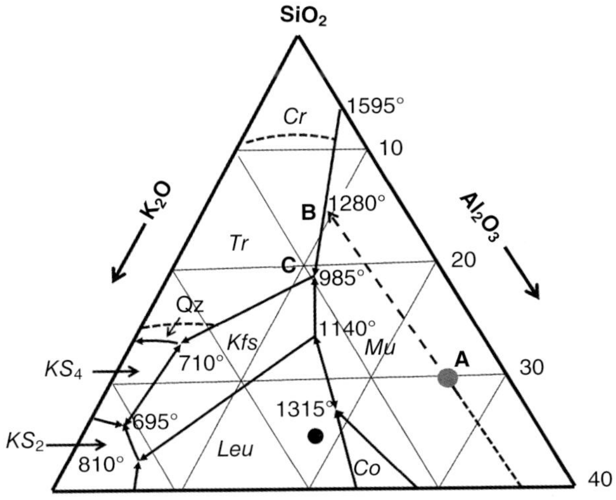

Figure 3.28 Phase boundary and coexisting phases in the SiO_2-rich section of the ternary system K_2O–Al_2O_3–SiO_2 (after Schairer and Bowen, 1947). The path A–B–C characterizes the crystallization history of a typical porcelain body (see text). *Co* = corundum; *Cr* = cristobalite; *Kfs* = potassium feldspar; KS_2 = potassium disilicate; KS_4 = potassium tetrasilicate; *Leu* = leucite; *Mu* = mullite; *Qz* = quartz; *Tr* = tridymite. The ternary eutectic is at the triple point *Kfs-Mu-Tr* with the melting temperature of 985 °C. This is also the equilibrium phase composition of porcelain.

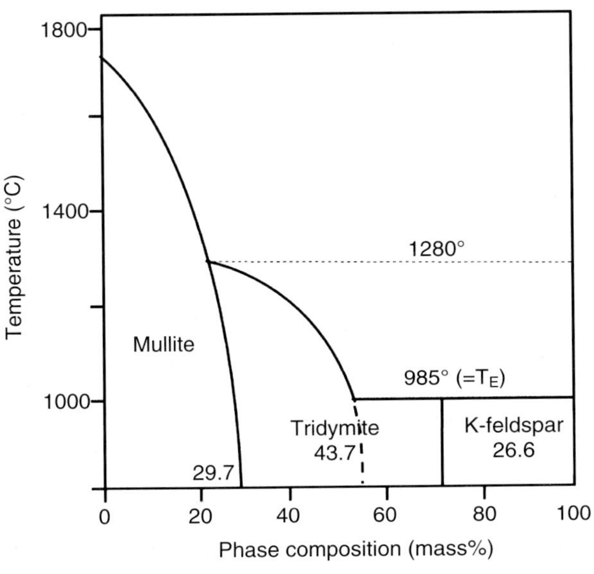

Figure 3.29 Temperature-dependence of the equilibrium phases of a triaxial porcelain raw materials mixture (50 mass% kaolin, 25 mass% quartz, 25 mass% K-feldspar). Adapted from Salmang *et al.* (2007).

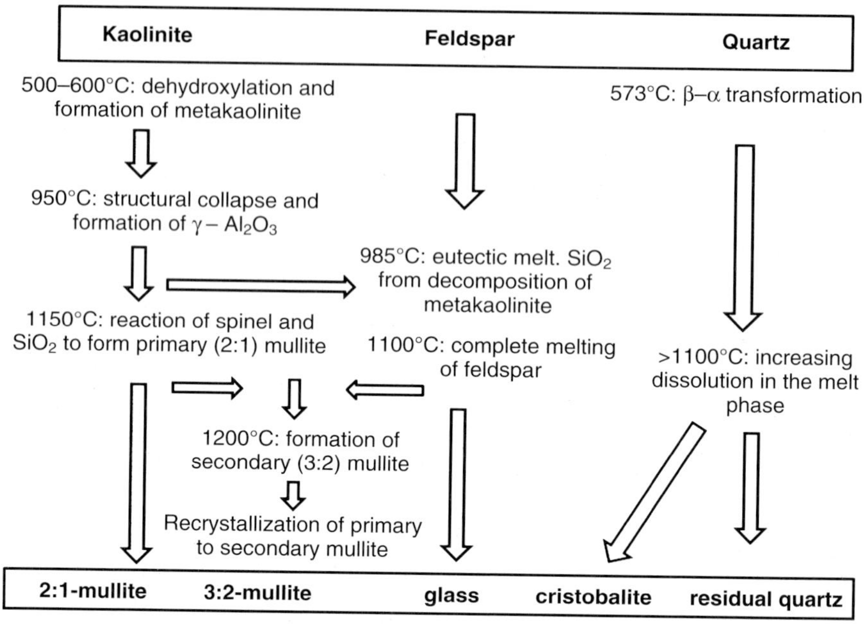

Figure 3.30 Simplified reaction scheme of formation of the microstructure of triaxial porcelain. Adapted from Schüller (1982).

very different, consisting of about 30 mass% mullite and 70 mass% $K_2O–Al_2O_3–SiO_2$ glass, since the highly viscous melt formed freezes without crystallization. Consequently, neither tridymite nor K-feldspar will be observed. Quartz grains present in the raw mass stabilize mechanically the green body, but tend to dissolve only slowly in the melt and hence are retained, since the firing time of porcelain is insufficient to lead to complete resorption (see Figure 3.30). It must also be considered that the firing temperature of porcelain (1200–1400 °C) is well below the liquidus (>1700 °C), and hence the amount of melt available will be limited. Quartz grains show rounded features and also frequently cracks that are formed as a result of residual stresses. In porcelain fabricated from mixtures containing large amounts of quartz, some cristobalite can be observed. Recently, Iqbal and Lee (1999) reinvestigated the microstructure of porcelain, and reported the typical final phase composition of a porcelain body to be 10–25 mass% mullite, 5–25 mass% quartz, and 0–8 vol% pores, dispersed in 65–80 mass% glass.

The complete reaction scheme leading to the microstructure of porcelain is shown in Figure 3.30.

The microstructure of typical porcelain-like stoneware is shown in Figure 3.31. The mineralogical equilibrium composition is expected to be mullite + quartz (tridymite) + K-feldspar (see Figure 3.29). However, all alkali ions are locked up

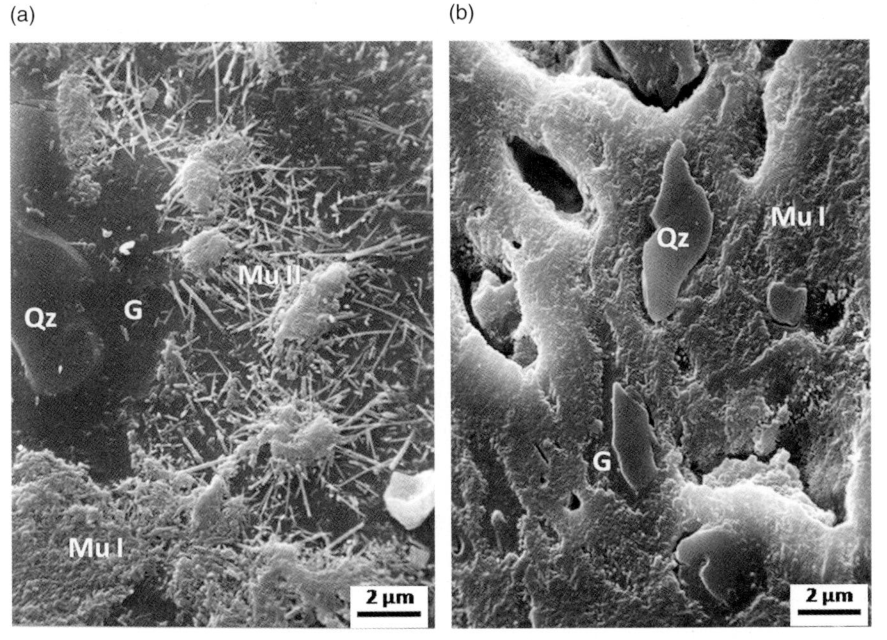

Figure 3.31 Scanning electron microscopy images of porcelain-like stoneware. (a) Vitreous China ware from Sawankhalok, Thailand (15th century C.E.), showing primary platy mullite (Mu I), secondary needle-shaped mullite (Mu II), and a partially dissolved residual quartz grain (Qz) surrounded by glass (G); (b) Early Siegburg stoneware (1300–1400 C.E.), showing predominantly primary platy mullite (Mu I), partially dissolved quartz grains with glassy reaction rims, and elongated larger pores (Heimann, 1989).

in the glass phase and consequently no feldspar forms as such. Mullite is shown to be present in its primary (2:1 mullite, Mu I) and secondary (3:2 mullite, Mu II) modifications. Moreover, quartz grains can be seen dissolving, surrounded by a pool of glass that constitutes about 70 mass% of the ceramic body (Figure 3.29). The firing temperature of the Sawankhalok ceramics shown in Figure 3.31a has been estimated to between 1200 and 1250 °C, well within the stability field of (secondary) mullite. In contrast, the early Siegburg stoneware shown in Figure 3.31b shows elongated large pores, attesting to a somewhat lower firing temperature. Moreover, the absence of any larger amounts of fluxing feldspar in the raw materials mix precluded the formation of the secondary needle-like mullite so abundant in the Thai ware, despite a similar chemical composition. This underscores the fact that the overriding criterion for the microstructure development is not the global chemical composition of the precursor materials, but rather its mineralogical composition (Kingery, 1987).

References

Abbott, R.N. (1984) $KAlSiO_4$ stuffed derivatives of tridymite: phase relationships. *Am. Mineral.*, **69**, 449–457.

Aksay, I.A. and Pask, J.A. (1975) Stable and metastable equilibrium in the system SiO_2-Al_2O_3. *J. Am. Ceram. Soc.*, **58**, 507–512.

Andersen, D.J., Lindsley, D.H., and Davidson, P.M. (1993) QUILF: a Pascal program to assess equilibria among Fe-Mg-Mn-Ti oxides, pyroxenes, olivine, and quartz. *Comp. Geosci.*, **19**, 1333–1350.

Angel, R.J. and Prewitt, C.T. (1986) Crystal structure of mullite: a reexamination of the average structure. *Am. Mineral.*, **71**, 1476–1482.

Aramaki, S. and Roy, R. (1962) Revised phase diagram for the system Al_2O_3-SiO_2. *J. Am. Ceram. Soc.*, **45**, 229–242.

Belonoshko, A.B., Dubrovinsky, L.S., and Dubrovinsky, N.A. (1996) A new high-pressure silica phase obtained by molecular dynamics. *Am. Mineral.*, **81**, 785–788.

Bowen, N.L. and Greig, J.W. (1924) The system Al_2O_3-SiO_2. *J. Am. Ceram. Soc.*, **7**, 238–254.

Buerger, M.J. (1951) Crystallographic aspects of phase transformations, in *Phase Transformations in Solids* (eds R. Smoluchowski, J.E. Mayer, and W.A. Weyl), John Wiley & Sons, Inc., New York.

Burnham, C.W. (1963) The crystal structure of mullite. *Carnegie Inst. Wash. Year Book*, **62**, 158–165.

Carpenter, M.A. and Wennemer, M. (1985) Characterization of synthetic tridymites by transmission electron microscopy. *Am. Mineral.*, **70**, 517–528.

Carpenter, M.A., Hemley, R.J., and Mao, H.K. (2000) High-pressure elasticity of stishovite and the $P4_2/mnm \leftrightarrow Pnnm$ phase transition. *J. Geophys. Res.*, **105**, 10807–10816.

Chao, E.C.T., Shoemaker, E.M., and Madsen, B.M. (1960) First natural occurrence of coesite. *Science*, **132** (3421), 220–222.

Chatterjee, N.D., Krüger, R., Haller, G., and Olbricht, W. (1998) The Bayesian approach to an internally consistent thermodynamic database: theory, database, and generation of phase diagrams. *Contrib. Mineral. Petrol.*, **133**, 149–168.

Chopin, C. (1984) Coesite and pure pyrope in high-grade blueschists of the western Alps: a first record and some consequences. *Contrib. Mineral. Petrol.*, **86**, 107–118.

Choudhury, N. and Chaplot, N.S. (2000) Free energy and relative stability of the enstatite $Mg_2Si_2O_6$ polymorphs. *Solid State Commun.*, **114** (3), 127–132.

Clark, S.P., Schairer, J.F., and de Neufville, J. (1962) Phase relations in the system $CaMgSi_2O_6$-$CaAl_2Si_2O_8$-SiO_2 at low and high pressure. *Carnegie Inst. Wash. Year Book*, **61**, 59–68.

Coes, L. (1953) A new dense crystalline silica. *Science*, **118**, 131–132.

Cruickshank, D.W.J. (1961) The role of 3d-orbitals in π-bonds between (a) silicon, phosphorus, sulphur, or chlorine and (b) oxygen or nitrogen. *J. Chem. Soc. (London)*, **1961**, 5486–5504.

Dolino, G. (1990) The α-β transition of quartz: a century of research on displacive phase transitions. *Phase Trans.*, **21**, 59–72.

Donnay, J.D.H. and Le Page, Y. (1978) The vicissitudes of the low-quartz crystal setting or the pitfalls of enantiomorphism. *Acta Crystallogr.*, **A34**, 584–594.

Dorner, B., Boysen, H., Frey, F., and Grimm, H. (1981) On the Si-O-Si bond angle in α- and β-quartz. *J. Phys.*, **42** (12), C6752–C6754.

Dove, M.T. (1999) Order/disorder phenomena in minerals: ordering, phase transitions and solid solutions, in *Microscopic Properties and Processes in Minerals*, vol. **543** (eds C.R.A. Catlow and K. Wright), NATO ASI Series C, Kluwer Academic Publishers, Dordrecht, pp. 451–475.

Dove, M.T., Giddy, A.P., and Heine, V. (1993) Rigid unit mode model of displacive phase transition in framework silicates. *Trans. Am. Crystallogr. Assoc.*, **27**, 65–74.

Dove, M.T., Keen, D.A., Hannon, A.C., and Swainson, I.P. (1997) Direct measurement of the Si-O bond length and orientational disorder in β-cristobalite. *Phys. Chem. Miner.*, **24**, 311–317.

Dove, M.T., Bosenick, A., Myers, E.R., Warren, M.C., and Roberts, S.A.T. (2000a) Modelling in relation to cation ordering. *Phase Trans.*, **71**, 205–226.

Dove, M.T., Pryde, A.K.A., and Keen, D.A. (2000b) Phase transitions in tridymite studied using "Rigid Unit Mode" theory, Reverse Monte Carlo methods and molecular dynamics simulations. *Mineral. Mag.*, **64**, 267–283.

Dove, M.T., Trachenko, K.O., Tucker, M.G., and Keen, D.A. (2000c) Rigid Unite Modes in framework structures: theory, experiment and application, in *Transformation Processes in Minerals*, vol. **39** (eds S.A.T. Redfern and M.A. Carpenter), Reviews in Mineralogy & Geochemistry, The Mineralogical Society of America, Washington, D.C., pp. 1–33.

Dove, M.T. (2001) Computer simulation of solid solutions, in *Solid Solutions in Silicate and Oxide Systems*, vol. **3** (ed. C.A. Geiger), EMU Notes in Mineralogy, Eötvös University Press, Budapest, Hungary, pp. 225–250.

Eitel, W. (1957) Structural anomalies in tridymite and cristobalite. *Am. Ceram. Bull.*, **36**, 142–148.

Fenner, C.N. (1913) The stability relations of the silica minerals. *Am. J. Sci.*, **36**, 331–384.

Flörke, O.W. (1956) Über das Einstoffsystem SiO_2. *Die Naturwiss.*, **43**, 419–420.

Flörke, O.W. (1967) Die Modifikationen von SiO_2. Eine Zusammenfassung. *Fortschr. Mineral.*, **44**, 181–230.

Flörke, O.W., Jones, J.B., and Segnit, E.R. (1975) Opal-CT crystals. *Neues Jahrb. Mineral., Mh.*, 369–377.

Flörke, O.W., Jones, J.B., and Schmincke, H.-U. (1976) A new microcrystalline silica from Gran Canaria. *Z. Kristallogr.*, **143**, 156–165.

Frost, D.J. (2008) The upper mantle and transition zone. *Elements*, **4**, 171–176.

Gibbs, J.W. (1876) On the equilibrium of heterogeneous substances. Part 1. *Trans. Conn. Acad. Arts Sci.*, **3**, 108–248.

Gibbs, J.W. (1878) On the equilibrium of heterogeneous substances. Part 2. *Trans. Conn. Acad. Arts Sci.*, **3**, 343–524.

Götze, J., Nasdala, L., Kleeberg, R., and Wenzel, M. (1998) Occurrence and distribution of "moganite" in agate/chalcedony: a combined micro-Raman, Rietveld, and cathodoluminescence study. *Contrib. Mineral. Petrol.*, **133**, 96–105.

Hammonds, K.D., Dove, M.T., Giddy, A.P., Heine, V., and Winkler, B. (1996) Rigid unit phonon modes and structural phase transitions in framework silicates. *Am. Mineral.*, **81**, 1057–1079.

Heaney, P.J. (2000) Phase transitions induced by solid solution, in *Transformation Processes in Minerals*, vol. **39** (eds S.A.T. Redfern and M.A. Carpenter), Reviews in Mineralogy & Geochemistry, The Mineralogical Society of America, Washington, D.C., pp. 135–174.

Heimann, R.B. (1975) *Auflösung von Kristallen. Theorie und Technische Anwendung*, Applied Mineralogy, vol. **8** Springer, Wien, New York, 270 pp.

Heimann, R.B. (1977) High-temperature and high-pressure polymorphs of silica: a review. *Miner. Sci. Eng.*, **9** (2), 57–67.

Heimann, R.B. (1979) Ätzerscheinungen an Kristallen. Spuren der aggression. *Lapis*, **4** (3), 27–31.

Heimann, R.B. (1989) Assessing the technology of ancient pottery: the use of ceramic phase diagrams. *Archeomaterials*, **3**, 123–148.

Heimann, R.B. (2008) *Plasma Spray Coating. Principles and Applications*, 2nd edn, Wiley-VCH Verlag GmbH, Weinheim, 427 pp.

Heimann, R.B., Conrad, D., Florence, L.Z., Neuwirth, M., Ivey, D.G., Mikula, R.J., and Lam, W.W. (1992) Leaching of simulated heavy metal waste stabilized/solidified in different cement matrices. *J. Hazard. Mater.*, **31**, 39–57.

Heuer, A.H. (1975) A classification scheme for polymorphic phase transitions in ceramics. *J. Am. Ceram. Soc.*, **58**, 529–530.

Hill, V.B. and Roy, R. (1958) Silica structure studies VI. On tridymites. *Trans. Br. Ceram. Soc.*, **57**, 496–510.

Hirose, K. and Lay, T. (2008) Discovery of post-perovskite and news views on the core-mantle boundary region. *Elements*, **4**, 183–189.

Holmquist, S.B. (1968) The thermal stability of purified tridymite: discussion. *Am. Mineral.*, **53**, 501–503.

Honess, A.P. (1927) *The Nature, Origin and Interpretation of Etch Figures on Crystals*, John Wiley & Sons Inc., New York.

Iqbal, Y. and Lee, W.E. (1999) Fired porcelain microstructure revisited. *J. Am. Ceram. Soc.*, **82** (12), 3584–3590.

Jones, J.B. and Segnit, E.R. (1975) Nomenclature and the structure of natural disordered (opaline) silica. *Contrib. Mineral. Petrol.*, **51** (3), 231–234.

Kattner, U.R. (1997) The thermodynamic modeling of multicomponent phase equilibria. *JOM*, **49** (12), 14–19.

Keat, P.P. (1954) A new crystalline silica. *Science*, **129**, 328–330.

Keith, M.L. and Tuttle, O.F. (1952) Significance of variation in the high-low inversion of quartz. *Am. J. Sci.*, Bowen Volume, 203–280.

Kihara, K. (1978) Thermal change in unit-cell dimensions and a hexagonal structure of tridymite. *Z. Kristallogr.*, **148**, 237–253.

Kingery, W.D. (1987) Microstructure analysis as part of a holistic interpretation of ceramic art and archaeological artifacts. *Archeomaterials*, **1** (2), 91–99.

Kingery, W.D., Bowen, H.K., and Uhlmann, D.R. (1976) *Introduction to Ceramics*, 2nd edn, John Wiley & Sons, Inc., New York, Chichester, Brisbane, Toronto, Singapore, p. 274.

Kitchin, S.J., Kohn, S.C., Dupree, R., Henderson, C.M., and Kihara, K. (1996) In situ ^{29}Si MAS NMR studies of structural phase transitions of tridymite. *Am. Mineral.*, **81**, 550–560.

Klug, F.J., Prochazka, S., and Doremus, R.H. (1987) Alumina-silica phase diagram in the mullite region. *J. Am. Ceram. Soc.*, **70**, 750–759.

Kubo, Y., Yamaguchi, G., and Kasahara, K. (1966) Inverted phase relation in the formation of nepheline and carnegieite from the system kaolinite-sodium carbonate. *Am. Mineral.*, **51**, 516–521.

Levin, E.M., Robbins, C.R., and McMurdie, H.F. (1964–1990) *Phase Diagrams for Ceramists*, vol. **I–VIII**, The American Ceramic Society, Westerville, OH.

McGreevy, R.L. and Pusztai, L. (1988) Reverse Monte Carlo simulation: a new technique for the determination of disordered structures. *Mol. Simul.*, **1**, 359–367.

McPherson, R. (1980) On the formation of thermally sprayed alumina coatings. *J. Mater. Sci.*, **115**, 3141–3149.

Madden, G.J. and van Vlack, L.H. (1967) Transformation of quartz to tridymite in the presence of binary silicate liquids. *J. Am. Ceram. Soc.*, **50**, 414–418.

Marcovic, S., Dondur, V., and Dimitrijevic, R. (2003) FTIT spectroscopy of framework aluminosilicate structures: carnegieite and pure sodium nepheline. *J. Mol. Struct.*, **654** (1–3), 223–234.

Miehe, G. and Graetsch, H. (1992) Crystal structure of moganite: a new structure type for silica. *Eur. J. Mineral.*, **4**, 693–706.

Moore, W.J. (1972) *Physical Chemistry*, 5th edn, Prentice-Hall, Englewood Cliffs, New Jersey.

Murakami, M., Hirose, K., Kawamura, K., Sata, N., and Ohishi, Y. (2004) Post-perovskite phase transition in $MgSiO_3$. *Science*, **304**, 855–858.

Nettleship, I., Slavick, K.G., Kim, Y.J., and Kriven, W.M. (1992) Phase transformation in dicalcium silicate: I, Fabrication and phase stability of fine-grained ß phase. *J. Am. Ceram. Soc.*, **75** (9), 2400–2406.

Niesel, K. and Thormann, P. (1967) The stability field of dicalcium silicate modifications. *Tonind.-Zeitung*, **91**, 362–369.

Niggli, P. (1954) *Rocks and Mineral Deposits*, Freeman & Company, San Francisco, Ch 9, pp. 347–427.

Osborn, E.F. and Tait, D.B. (1952) The system diopside-forsterite-anorthite. *Am. J. Sci.*, Bowen Volume, 413–433.

Phillips, B. and Muan, A. (1959) Phase equilibria in the system CaO-iron oxide-SiO_2 in air. *J. Am. Ceram. Soc.*, **42**, 313–423.

Pöllmann, H. (2007) *Immobilisierung von Schadstoffen durch Speichermineralbildung*, Shaker-Verlag, Aachen, ISBN: 978-3-8322-6377-5, 428 pp.

Putnis, A. (1992) *Introduction to Mineral Science*, Cambridge University Press, Cambridge, pp. 411–412.

Rankin, G.A. and Wright, F.E. (1915) The ternary system CaO-Al_2O_3-SiO_2. *Am. J. Sci.*, **39**, 1–79.

Richet, P. and Mysen, B.O. (1999) High temperature dynamics in cristobalite (SiO_2) and carnegieite ($NaAlSiO_4$): a Raman spectroscopy study. *Geophys. Res. Lett.*, **26** (15), 2283–2286.

Risbud, S.H. and Pask, J.A. (1977) Calculated thermodynamic data and metastable immiscibility in the system SiO_2-Al_2O_3. *J. Am. Ceram. Soc.*, **60**, 419–424.

Rockett, T.J. and Foster, W.R. (1967) The thermal stability of purified tridymite. *Am. Mineral.*, **52**, 1233–1240.

Rosenberger, F. (1979) *Fundamentals of Crystal Growth I*, Springer, Berlin.

Routschka, G. and Wuthnow, H. (2007) *Feuerfeste Werkstoffe*, 4th edn, Vulkan-Verlag GmbH, Essen.

Salmang, H., Scholze, H., and Telle, R. (2007) *Keramik*, 7th edn, Springer, Berlin.

Schairer, J.F. and Bowen, N.L. (1947) Melting relations in the system Na_2O-Al_2O_3-SiO_2 and K_2O-Al_2O_3-SiO_2. *Am. J. Sci.*, **245**, 193–204.

Schneider, H. and Komarneni, S. (2005) *Mullite*, Wiley-VCH Verlag GmbH, Weinheim, ISBN: 978-3527-3097-4-0, 487 pp.

Schneider, H., Okada, K., and Pask, J.A. (1994) *Mullite and Mullite Ceramics*, John Wiley & Sons, Ltd, Chichester.

Schulle, W. (1972) Neue Silikasteintypen zur Auskleidung von Koksöfen. *Silikattechn.*, **22**, 312–314.

Schüller, K.H. (1982) The development of microstructures in silicate ceramics. 1. Whitewares. *J. Educ. Modules Mater. Sci. Eng.*, **4** (3), 529–565.

Schultz, A. and Chang, Y.A. (1985) Computer graphics for ternary phase diagrams. *J. Metals*, **37** (11), 10–13.

Sosman, R.B. (1955) New and old phases of silica. *Trans. Br. Ceram. Soc.*, **54**, 655–670.

Sosman, R.B. (1965) *The Phases of Silica*, Rutgers University Press, New Jersey.

Stishov, S.M. and Popova, S.V. (1961) A new dense modification of silica. *Geokhimiya*, **10**, 837–839 (in Russian).

Strunz, H. and Nickel, E.H. (2001) *Strunz Mineralogical Tables. Chemical-Structural Mineral Classification System*, E. Schweizerbart, Stuttgart.

Tanaka, Y., Takeuchi, T., Lovesey, S.W., Knight, K.S., Chainani, A., Takata, Y., Oura, M., Senba, Y., Ohashi, H., and Shin, S. (2008) Right-handed or left-handed? Forbidden x-ray diffraction reveals chirality. *Phys. Rev. Lett.*, **100**, 145502.

Taylor, H.F.W. (1997) *Cement Chemistry*, 2nd edn, Thomas Telford, London.

The American Ceramic Society (1992–1996) *Phase Equilibria Diagrams*, Vol. **IX** (ed. G.B. Stringfellow), vol. **X** (ed. A.E. McHale), vol. **XI** (ed. R.S. Roth) and vol. **XII** (eds A.E. McHale and R.S. Roth), The American Ceramic Society, Westerville, Ohio.

Toropov, N.A. and Galakhov, F.Y. (1953) The mullite problem. *Vopr. Petrogr. Mineral.*, **2**, 245–255.

Weaver, J.S., Chipman, D.W., and Takahashi, T. (1979) Comparison between thermochemical and phase stability data for the quartz-coesite-stishovite transformation. *Am. Mineral.*, **64**, 604–614.

Weiss, A. and Weiss, A. (1954) A new silicon dioxide modification with fibrous structure and chain molecules. *Die Naturwiss.*, **41**, 12.

Welch, J.H. and Gutt, W. (1959) Tricalcium silicate and its stability within the system CaO-SiO_2. *J. Am. Ceram. Soc.*, **42** (1), 11–15.

West, D.R.F. (1982) *Ternary Phase Diagrams*, 2nd edn, Chapman & Hall, London, New York.

Wilson, M.J., Russell, J.D., and Tate, J.M. (1974) An electron diffraction study of the polymorphs of tridymite. *Phys. Chem. Miner.*, **21**, 421–433.

Yamazaki, S. and Toraya, H. (2001) X-ray powder data for a new phase of dicalcium silicate, κ-Ca_2SiO_4. *Powder Diffr.*, **16** (2), 110–114.

4
Mineralogy and Chemistry of the Ceramic Firing Process

4.1
Introduction

Ceramic firing of the green body (Chapter 2) establishes, and fixes permanently, a set of properties that are usually associated with ceramics, namely chemical durability, mechanical resilience, and hard and dense microstructure. *Densification* is achieved by sintering with or without the presence of a liquid phase. Owing to the variability of clay composition, the mineralogical processes occurring during the thermal transformation of clay minerals can be very complex, not in the least due to the fact that most reactions occur far removed from thermodynamic equilibrium, and hence are both time- and temperature dependent. A simplified model of these phase transformations is shown in Figure 4.1, which traces the composition of a ceramic body composed of a mixture of kaolin, quartz, feldspar and pores as a function of temperature (see also Section 3.2.3.2).

As shown in Figure 4.1, the decrease in pore volume causes almost always a (linear) shrinkage that may be in excess of 20%. Densification of the green body prior to firing, for example, by uniaxial or isostatic pressing, can reduce this large shrinkage. This is important when near-net-shaped ceramic parts are desired, thereby reducing the requirements for expensive post-firing machining with diamond tools. The properties of the ceramic body are closely related to the firing temperature and the type of atmosphere prevalent in the kiln. *Oxidizing firing* is achieved when the partial pressure of oxygen exceeds the thermodynamically defined decomposition pressure of oxides with different valencies, such as magnetite ($Fe^{2+}Fe_2^{3+}O_4$) or hausmannite ($Mn^{2+}Mn_2^{3+}O_4$). On the other hand, *reducing firing* requires a partial pressure of oxygen below the decomposition pressure of these spinels. In the case of iron oxides, the color of the fired ceramics is dependent on the valency state of the iron ions. A mastery of the redox behavior of iron oxides is at the heart of the impressive artistic achievements of ancient potters throughout the ages in different regions of the world (see for example, Maggetti and Radzewski, 1975; Heimann and Franklin, 1979; Kingery and Vandiver, 1986; Walton *et al.*, 2009).

Since the chemical and mineralogical composition of the clay mixture play such a decisive role, development of the phase composition and microstructure of the

Classic and Advanced Ceramics: From Fundamentals to Applications. Robert B. Heimann

ISBN: 978-3-527-32517-7

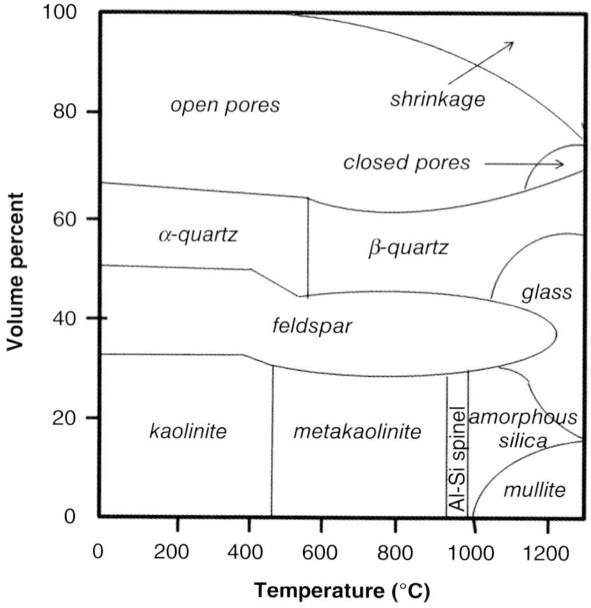

Figure 4.1 Thermal history of transformation of kaolinite, quartz and feldspar, and development of the composition of ceramic bodies as a function of firing temperature. Adapted from Norton (1952).

fired ceramics are notoriously complex. In this chapter, only the thermal transformation of kaolinite during ceramic firing will be discussed in greater detail. Here, a brief survey of phase development during the oxidizing and reducing firing of illitic clays with low and high calcite contents is presented, as well as phase relations within the technically important quinary system:

$$MgO{-}CaO{-}Al_2O_3{-}Fe_2O_3{-}SiO_2.$$

4.2 Crystallography of the Thermal Transformation of Kaolinite

The heating of pure kaolinite leads to a slight densification so as to yield a microstructure with a rather low strength. Owing to the lack of mineralizers such as alkali and alkaline earth ions, up to a temperature of 1600 °C only solid-state reactions occur, and these proceed sluggishly. However, the situation changes for natural kaolinitic clays, which invariably contain feldspars, the mineralizing alkali and alkaline earth ions contents of which lead to rapid sintering at moderate temperatures. A pure K-feldspar begins to melt incongruently at 1150 °C, but the presence of albite will reduce this temperature to 1090 °C. A mixture of kaolinite,

K-feldspar and quartz, present in porcelain paste (see Chapter 3), begins to melt at the ternary eutectic at 985 °C, and even small amounts of albite will lower this melting temperature by another 60 °C, to 925 °C. At these low temperatures, however, the amount of liquid phase formed is small, and its viscosity is very high, thus impeding diffusion and hence the progress of the ceramic reactions. In order to achieve reasonable reaction rates, the firing temperature must be raised to a value that allows the feldspar to enter the glass phase formed. As shown in Figure 4.1, the dissolution of feldspar occurs beyond a temperature of 1200 °C.

When firing a mixture of kaolinite and feldspar, the first traces of mullite are obtained at about 1000 °C. The onset of transformation can be detected in kaolinite single crystal relics, where the formation of crystallographically oriented mullite laths is aided by the presence of potassium ions.

The exact mechanism of transformation from kaolinite to mullite has been the subject of a longstanding controversy. It has long been known that the first step in the reaction sequence is the formation of a more or less amorphous dehydration product, termed *metakaolinite*, that manifests itself in differential thermal analysis (DTA) studies as an endothermic reaction peak at about 550 °C or below, depending on the presence of impurities in the clay. These studies also generally show a second, strong exothermic peak at 980 °C, the origin of which has been the key issue in a flurry of research reports between 1930 and 1990. These studies have centered around two main explanations. The first explanation assumes that the exothermic reaction has its origin in the formation of a γ-Al_2O_3-type defect spinel phase (Insley and Ewell, 1935; Brindley and Nakahira, 1959; Percival *et al.*, 1974; Leonard, 1977; Brown *et al.*, 1985; Sonuparlak *et al.*, 1987). Whilst agreeing on the central issue, the individual approaches differ in their interpretation of the composition of the spinel phase and the question as to whether also mullite, which is formed in a parallel reaction, will contribute to the exothermic event. The second explanation concerns the assumption that mullite is formed without involvement of a spinel-type phase (Comefore *et al.*, 1948; Bradley and Grim, 1951; Roy *et al.*, 1955). However, this hypothesis has been disproved in the light of more recent experimentation (see Percival *et al.*, 1974; Leonard, 1977; Chakraborty and Ghosh, 1978; Sonuparlak *et al.*, 1987).

Investigations of the transformation mechanism of kaolinite showed that in the metakaolinite phase, which is stable between about 480 and 950 °C, the tetrahedral Si–O layers expand but maintain their structural integrity, whereas in the octahedral Al–O layers a loss of OH ions occurs (Figure 4.2). The dehydroxylated kaolinite loses its *c* periodicity, but retains its *a* and *b* periodicities to a large extent. The distorted Al–O layers transform at about 950 °C (Figure 4.1) to a spinel-type phase (Brindley and Nakahira, 1959). The formation of the spinel phase occurs topotaxially with respect to the lattice metric of the parent kaolinite, whereby $[001]_{Kao} \parallel [111]_{Sp}$ and $[010]_{Kao} \parallel [110]_{Sp}$.

Measurement of the Si–O and $Al^{[6]}$-O bond lengths of kaolinite by radial electron density (RED) mapping (see Appendix C) revealed values of 161.1 pm and 192.2 pm, respectively; these values were in good agreement with those of Iwai *et al.* (1971) and Newnham (1961). Figure 4.3 shows the radial electron density of a well-

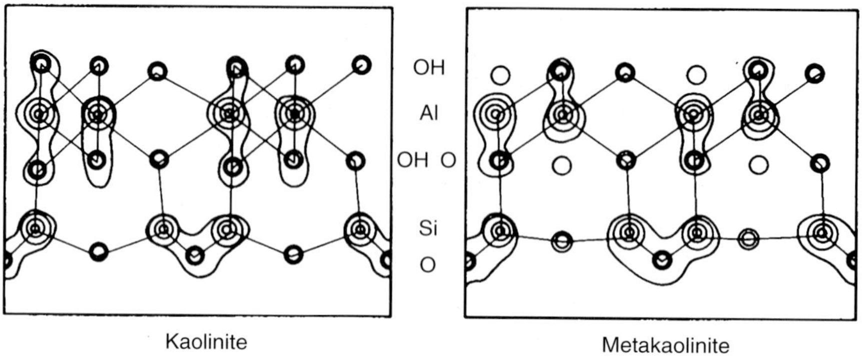

Figure 4.2 Change of the electron density distribution of kaolinite during dehydroxylation to the highly defective metakaolinite structure (Iwai *et al.*, 1971). Reprinted with permission from the International Union of Crystallography.

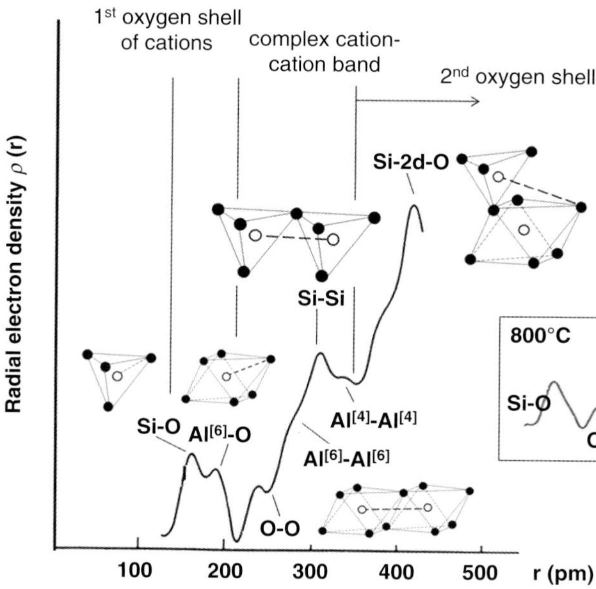

Figure 4.3 Experimental radial electron density distribution in kaolinite at 100 °C. The inset shows the metakaolinite structure at 800 °C with a negligible contribution of $Al^{[6]}$–O but a strong $Al^{[4]}$–$Al^{[4]}$ band.

ordered kaolinite, where the peaks correspond to characteristic structural distances in the kaolinite lattice, and several important connectivities are indicated. Average distances are as follows: Si–O, 162 pm; $Al^{[6]}$–O, 194.4 pm; $Al^{[6]}$–$Al^{[6]}$ 279.3 pm; Si–Si, 312.0 pm; $Al^{[4]}$–$Al^{[4]}$, 342.1 pm, and Si-2d–O, 415.0 pm (Leonard, 1977).

In metakaolinite, the Si–O distance in the tetrahedral layer increases to 165.0 pm, whereas the Al–O bond length remains essentially constant. At first glance this is rather surprising, since removal of the OH groups should be accompanied by a reorganization of the octahedral gibbsite-type layer into a tetrahedral configuration. However, even though the octahedral layer attains an extremely high degree of lattice defects (Freund, 1967), it does not collapse immediately but rather retains a truncated octahedral structure with Al ions in a fourfold coordination (Figure 4.2, right). Hence, whilst the Al configuration is loosely tetrahedral, perturbation of the electronic shells of the Al atom freed from the two hydroxyl ions is reduced in such a way that unusually large Al–O distances are maintained. Indeed, a Gaussian–Lorentzian deconvolution of the electron density distribution curve shows up to a temperature of 800 °C Al–O distances of 194.4 pm corresponding to $Al^{[6]}$–O, whereas at 900 °C in addition Al–O distances of 176.7 pm were observed akin to $Al^{[4]}$–O and indicative of the onset of the formation of γ-Al_2O_3 (Leonard, 1977).

The metakaolinite phase collapses beyond 900 °C into a spinel-type lattice, the nature of which has been vigorously disputed for 120 years. Many years ago, Le Chatelier (1887) had noted that metakaolinite appeared to behave like a mixture of silica and alumina ("mixed-oxide theory"), and that from metakaolinite alumina could be dissolved completely with dilute hydrochloric acid. Subsequently, the alumina phase was identified as having a γ-alumina spinel structure whereby two options were put forward: (i) Si–Al spinel $Si_8^{[4]}(Al_{10.67}\square_{5.33})^{[6]}O_{32}$ (or $3SiO_2 \cdot 2Al_2O_3$); or (ii) pure γ-Al_2O_3, $Al_8^{[4]}(Al_{13.33}\square_{2.67})^{[6]}O_{32}$ defect spinel. It was assumed that the large proportion of the vacancies □ in the Si–Al spinel structure rendered it unstable, and prone to collapse at higher temperature to 3 : 2-mullite (Brindley and Nakahira, 1959). Indeed, the idea of Si–Al spinel occurring as an intermediate structure was quite appealing as this would have explained the formation of mullite by a simple solid-state exsolution of silica beyond 1100 °C according to:

$$3(3SiO_2 \cdot 2Al_2O_3) \rightarrow 2(3Al_2O_3 \cdot 2SiO_2) + 5SiO_2 \qquad (4.1)$$

Si–Al spinel   3:2-mullite   cristobalite

Leonard (1977) traced the transformation sequence kaolinite–metakaolinite–spinel–mullite by studying the radial electron density (RED) distribution of kaolinite and its transition products as a function of temperature, and correlated the results with chemical information obtained with X-ray fluorescence spectroscopy. While the presence of some Si–Al spinel could not completely be ruled out, most of the Al atoms were found in the configuration of γ-Al_2O_3 beyond a firing temperature of 900 °C. This assumption was corroborated by Sonuparlak *et al.* (1987), who showed by transmission electron microscopy (TEM) in conjunction with energy-dispersive X-ray (EDX) spectroscopy that the spinel phase formed at 980 °C, indicated by a strong exothermic DTA peak, contains <10 mass% SiO_2, if any.

The formation of mullite at temperatures beyond 1100 °C has been described as the result of reaction of γ-Al_2O_3 with amorphous silica that is present in pores of the spinel structure. This solid-state reaction occurs at the interface between crystalline spinel and the amorphous silica phase. The Si^{4+} ions are more mobile than the Al^{3+} ions owing to their higher charge and smaller ionic radius, so that the predominant direction of ion movement will be towards the spinel crystallites. On the other hand, the high viscosity of the amorphous silica phase counteracts the destruction of the spinel lattice and inhibits the diffusion of Al^{3+} ions. Thus, a large number of needle-like mullite crystals will be formed with a preferred orientation whereby $[002]_{mullite} \parallel [440]_{spinel}$ (Fieger, 1989).

4.3
Thermal Transformations in Illitic Clays

Most structural clay products such as bricks, tile, pipe, earthenware pottery, and architectural terracotta, are produced from red-firing clays and shales, the main clay mineral content of which is illite. Minor amounts of kaolinite, montmorillonite and chlorite might also be present, but are considered nonessential for the intended properties of the fired product (Brownell, 1976). The presence of feldspars and, in particular, calcite provides fluxing agents that lower the sintering temperature considerably. It is therefore not surprising that ancient pottery was produced predominantly from calcareous illitic clays that, on firing to temperatures well below 1100 °C, yielded a dense and hence impervious body. Corinthian and Attic ware, most Roman Terra sigillata – notably those produced in southern Gaul (La Graufesenque, Banassac) – and most medieval earthenware products were produced from calcareous illitic clays. However, as such clays possess a rather narrow softening interval, great care had to be taken so as not to "overfire" the ware and thus induce any undesirable large-scale melting of the clay. Clear evidence of such failure has been demonstrated at many ancient pottery production sites, where copious amounts of bloated, deformed and otherwise misfigured wastes have been excavated. Yet, with an increasing control over the physicochemical parameters existing in a pottery kiln, a judicious choice of appropriate clays, and a shift to less calcareous clays, gains in quality control were subsequently achieved. This was very evident in the change from the highly calcareous Neolithic pottery of Crete to the much less calcareous clays in Middle and Late Minoan Kamares wares. Presumably, such improvements were achieved by blending clays from different sources with high (northern coast of Crete, Knossos) and low (southern coast, Phaistos) lime contents, so as to yield a consistent product (Noll, 1982; Heimann, 1989).

Figure 4.4 shows the X-ray traces of several size fractions of a typical calcareous illitic clay from Otterbach, near Jockgrim, Palatinate, Germany, that had been used to manufacture provincial Roman Terra sigillata ware between the first and third centuries C.E. in the Eastern Gaulish settlement *Tabernae rhenanae* (today's Rheinzabern). The coarse, granular nonclay constituents of quartz, K-feldspar, plagioclase, and calcite appear in the as-mined clay (Figure 4.4, trace a), whereas in the coarse silt (63 μm to 2 μm) (trace b) and clay fractions (<2 μm) (trace c), the clay minerals illite, as well as one or more members of the mixed layer (ML)

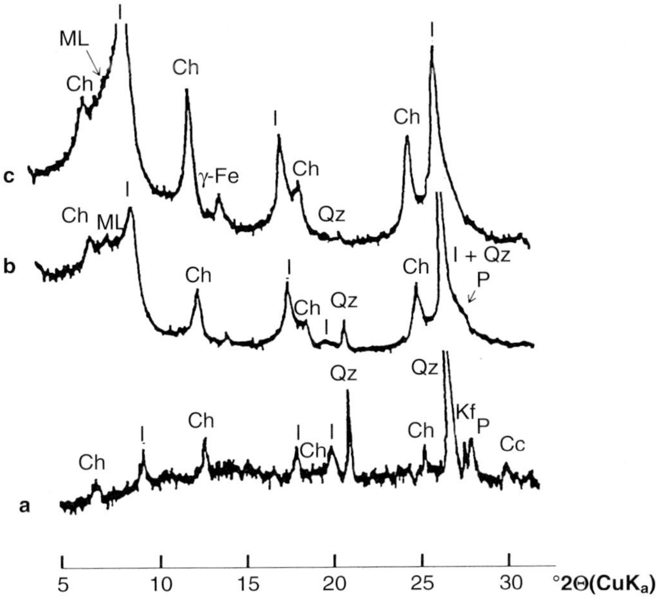

Figure 4.4 X-ray pattern of an archeological calcareous illitic clay from Otterbach, near Jockrim, Palatinate, Germany. Trace (a): The as-mined clay; Trace (b): Coarse silt fraction (63–2 μm); Trace (c): Clay fraction (<2 μm). Cc = calcite; Ch = chlorite; I = illite; Kf = K-feldspar; ML = mixed layer mineral; P = plagioclase; Qz = quartz; γ-Fe = lepidocrocite (Heimann *et al.*, 1980).

minerals and chlorite can be found. The finest fraction also contains lepidocrocite (γ-FeOOH). This fraction was used as a slip, applied to the leather-hard green body prior to firing. During oxidizing firing it produced, together with the iron content of illite and chlorite, the telltale red color of the Terra sigillata pottery.

The chemical composition of the clay shown in Figure 4.4 (in mass%) is 61.7 SiO_2, 19.3 Al_2O_3, 0.8 TiO_2, 5.5 Fe_2O_3, 7.0 CaO, 2.7 MgO, 0.8 Na_2O, 3.5 K_2O, and 0.2 P_2O_5, in addition to 514 ppm Ba, 134 ppm Cr, and 197 ppm Rb. This composition, normalized for silica, alumina and the sum of calcia and magnesia, is depicted in Figure 4.5 as a circle, located approximately at the triple point anorthite–tridymite–mullite with a melting point of 1345 °C (see also Figure 3.25). However, the considerable iron content (5.5 mass% Fe_2O_3) shifts the phase composition of the clay away from the triple point towards the phase field of cordierite and iron cordierite (sekaninaite), respectively. Consequently, the crystalline compounds of the ceramic formed by oxidizing firing to 1010 °C shows diopside, calcium-rich plagioclase, quartz and hematite, whereas the same clay fired under strongly reducing conditions ($f_{O2} < 10^{-4}$ atm; Ni/NiO buffer) at 1035 °C exhibits a phase assembly consisting of anorthite (Figure 4.6a), tridymite, iron cordierite (Figure 4.6b), magnetite, and glass (Heimann *et al.*, 1980).

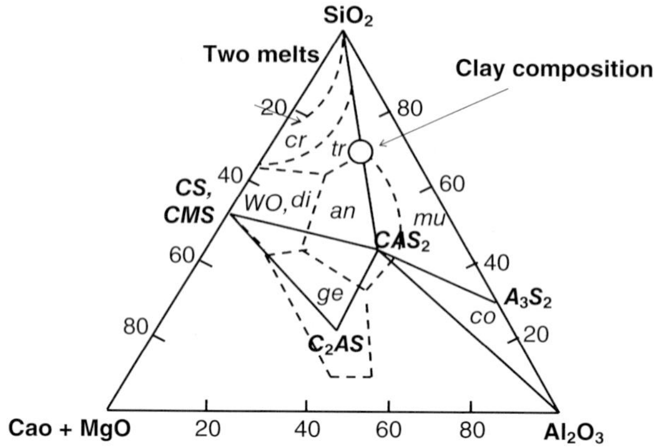

Figure 4.5 Simplified phase diagram CaO+MgO–Al_2O_3–SiO_2 with the composition of the Otterbach clays indicated (circle). The solid lines connecting primary phases are conodes delineating compatibility triangles; the dashed lines delineate phase stability regions. *an* = anorthite; *co* = corundum; *cr* = cristobalite; *di* = diopside; *ge* = gehlenite; *mu* = mullite; *tr* = tridymite; *wo* = pseudowollastonite.

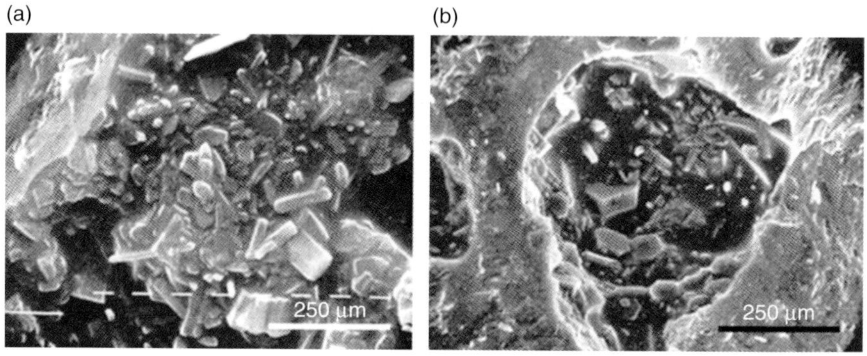

Figure 4.6 Calcareous illitic clay fired under strongly reducing conditions ($f_{O_2} < 10^{-4}$ atm; Ni/NiO buffer) at 1035 °C. (a) Anorthite crystallized from glassy phase; (b) Orthorhombic pseudohexagonal crystals of iron cordierite (Heimann *et al.*, 1980).

If greater amounts of CaO are present in the clay, its composition shifts into the compatibility triangle wollastonite (diopside)–anorthite–gehlenite. Indeed, such clays, when fired to 850–1050 °C, contain considerable amounts of gehlenite, $Ca_2Al^{IV}[AlSiO_7]$ (Peters and Jenni, 1973; Maggetti and Küpfer, 1978). Gehlenite is metastable and reacts at higher temperature (>1100 °C), with silica being released during the decomposition of metakaolinite or illite to anorthite + wollastonite (or diopside, in the presence of MgO) according to:

$$Ca_2Al[AlSiO_7] + 2SiO_2 \rightarrow CaAl_2Si_2O_8 + CaSiO_3. \qquad (4.2)$$

Since ancient earthenware ceramics were generally fired below 1100 °C, gehlenite should indeed have been formed during oxidizing firing. Although this phase is rarely detected in such ceramics, some notable exceptions have occurred when the ceramic shards were buried under strongly arid conditions. It appears that the gehlenite had dissolved under humid conditions when in contact with soil solutions, and was transformed to zeolites such as wairakite, garronite or, in the presence of phosphate ions, to brushite (Maggetti and Heimann, 1979; Heimann and Maggetti, 1980). The mechanism of formation of gehlenite outside its field of thermodynamic stability, and its occurrence in and disappearance from calcareous illitic clays, have been considered an analytical challenge as a *thermodynamic problem*. However, the solution to this problem may lie in *kinetics*, since well-processed clays with a narrow grain size distribution and, in particular, an absence

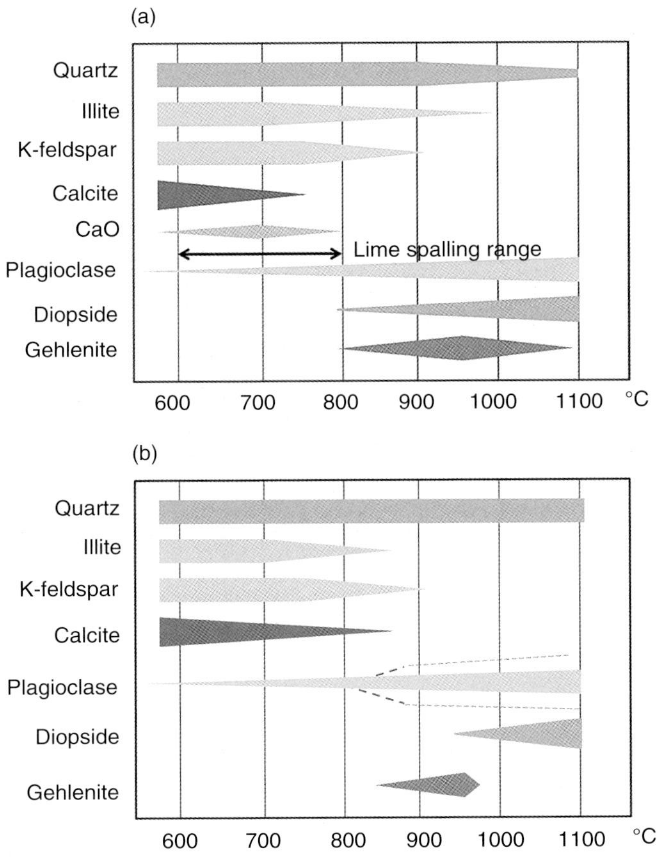

Figure 4.7 Experimentally determined phase stability in a calcareous illitic clay fired in (a) oxidizing and (b) reducing atmospheres. Under oxidizing conditions, illite is found to be stable to higher temperatures, while decarbonization of calcite is impeded under reducing conditions. The onset of diopside formation occurs at a lower temperature in the presence of a high oxygen partial pressure. After Maggetti (1982).

of larger calcite grains (such as those utilized in fine Roman Terra sigillata) do not show gehlenite, whereas coarse utilitarian ware made from identical clays contain sizeable amounts (Maggetti and Küpfer, 1978). Evidently, the reaction rate is a function of grain size which, in turn, is influenced by the degree of processing of the clay, and a clear measure of the technological skill of the potter.

High CaO contents in clay lead, during oxidizing firing, to a pronounced color change towards yellow and buff, or even white hues. For example, the highly calcareous (22 mass% CaO) and iron-rich (8 mass% Fe_2O_3) brown clay of the classic Corinthian pottery changes in color from pink (fired at 700 °C) to red (at 900 °C) to pale yellow (at 1080 °C). An originally gray clay (15 mass% CaO, 7 mass% Fe_2O_3) from Corfu changes from buff-cream (700 °C) to buff (900 °C) to white (1080 °C) (Maniatis *et al.*, 1982). The reason(s) for these color changes are still under discussion. As with the gehlenite problem, they may be related to the grain size of calcite that controls the ratio of diopside to anorthite in the fired product (Nöller and Knoll, 1985). Diopside, wollastonite and mullite all are able to incorporate Fe^{3+} ions into their crystal lattices; consequently, no free iron oxide phase such as hematite can be formed, and the yellow coloration results from a solid-state charge-transfer reaction caused by Fe^{3+} substituting Ca^{2+} for Mg^{2+} (Marfunin, 1979). On the other hand, as anorthite is unable to incorporate sizeable amounts of Fe^{3+} ceramic bodies with a low diopside/anorthite ratio it will remain red. At still higher firing temperatures beyond 1200 °C, calcareous and iron-bearing clays change to brown colors, thought to be formed by a superposition of the greenish-black hue of fayalite formed from Fe^{2+} exsolved during decomposition of trioctahedral illite, and the red color of hematite (Maggetti, 1975).

Figure 4.7 summarizes the phase changes of several typical minerals comprising a calcareous illitic clay that occur under oxidizing (Figure 4.7a) and reducing (Figure 4.7b) conditions.

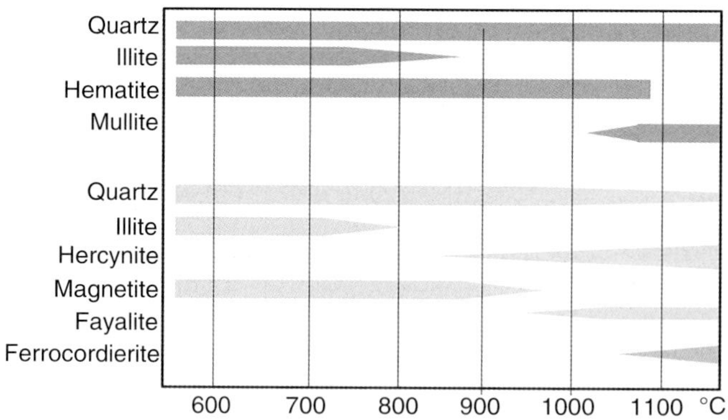

Figure 4.8 Experimentally determined phase stability in a noncalcareous, iron-rich illitic clay under oxidizing (top) and reducing (bottom) firing atmospheres. Under oxidizing conditions, illite is found to be stable to somewhat higher temperatures. The onset of mullite formation occurs at a lower temperature in the presence of a high oxygen partial pressure. After Heimann *et al.* (1980) and Letsch and Noll (1983).

In contrast to this, the mineralogical changes of a noncalcareous iron-rich clay are shown in Figure 4.8 under oxidizing (top, dark bars) and reducing (bottom, light bars) conditions. Here, the main differences lie in the stability range of illite to higher temperatures, and the earlier onset of mullite formation under oxidizing conditions.

4.4
Thermal Transformations and Phase Formation in the System $MgO–CaO–Al_2O_3–SiO_2$

This system is of great importance for producing the refractory ceramics utilized in steel-making to line the Linz–Donawitz (LD) converters of the basic oxygen-furnace process, to construct the open-hearth roofs of Siemens regenerative furnaces, and to build transport and mixing ladles for molten steel. In steel-casting, the tar-soaked magnesia refractories provide protection against corrosive attack by slags and liquid metals. These ceramic bricks must withstand not only high temperatures but also steep temperature gradients, leading to thermomechanical stresses, as well as deleterious reactions with gaseous, solid and liquid compounds, including attacks by molten iron. The ceramics can be subdivided into magnesia, magnesia-chromia, sintered dolomite, forsterite, and spinel refractories. This large range of products requires raw materials of a likewise large compositional range, such as magnesite, dolomite, chromite, olivine, serpentine, and talc. Increasingly, magnesite is synthesized by precipitating $MgCl_2$ dissolved in seawater with lime or dolomite, thus obtaining $Mg(OH)_2$ that will be dewatered and sintered at temperatures up to 2000 °C. In Israel, a special process (the Aman process) uses concentrated brine from the Dead Sea and converts it at 600–800 °C in a reactor to yield MgO (DSP; Dead Sea periclase) of exceptional purity (>99%). When starting from natural magnesite, the raw material is prefired at 350–580 °C to remove CO_2, and then sintered at >1700 °C. Dolomite requires a somewhat higher temperature, since decarbonization is complete only beyond 900 °C and final sintering is carried out at between 1800 and 2000 °C.

In producing MgO–CaO-based refractories from natural raw materials, fluxing oxides such as SiO_2 and Fe_2O_3 are always present that react at the high sintering temperatures with MgO and CaO so as to produce a wide variety of binary and ternary silicate or ferrite phases. Although these phases on the one hand act as binder phases for MgO grains, on the other hand they limit the temperature stability of the refractory. Hence, one important quality criterion of magnesia raw materials is their CaO/SiO_2 ratio (Table 4.1), which determines the stability ranges of the reaction products in the sintered refractory (Figure 4.9).

Although, the low-melting eutectics monticellite ($CaMgSiO_4$)–merwinite ($Ca_3MgSi_2O_8$)–periclase (MgO) ($T_e = 1490$ °C) and monticellite–forsterite (Mg_2SiO_4)–periclase ($T_e = 1502$ °C) assist in densification of the magnesite refractories, they reduce their hot bending strength. Hence, in industrial practice it is desirable to

Table 4.1 Dependence of silicate phases on the CaO/SiO_2 ratio in basic magnesia refractories (ternary system CaO–MgO–SiO_2).

CaO/SiO_2	Phases (in cement chemistry notation)	Mineral names
<0.94	M_2S + CMS	Forsterite + monticellite
0.94	CMS	Monticellite
0.94–1.40	CMS + C_3MS_2	Monticellite + merwinite
1.40	C_3MS_2	Merwinite
1.40–1.87	C_3MS_2 + C_2S	Merwinite + belite
1.87	C_2S	Belite
>1.87	C_2S (C_3S)	Belite (alite)

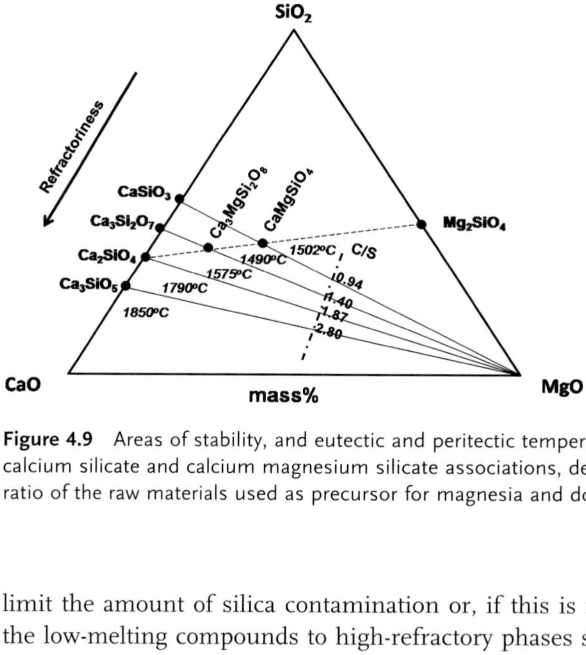

Figure 4.9 Areas of stability, and eutectic and peritectic temperatures, respectively, of calcium silicate and calcium magnesium silicate associations, depending on the CaO/SiO_2 ratio of the raw materials used as precursor for magnesia and dolomite refractories.

limit the amount of silica contamination or, if this is not feasible, to transform the low-melting compounds to high-refractory phases such as forsterite or dicalcium silicate by using appropriate additives (see Trojer *et al.*, 1981).

The addition of Al_2O_3 to the ternary system CaO–MgO–SiO_2 leads to other high-refractory phases, such as spinel (MA, $MgAl_2O_4$). As in the ternary system, in the quaternary system CaO–MgO–Al_2O_3–SiO_2 the equilibrium compositions are dependent on the CaO/SiO_2 ratio (Table 4.2).

For CaO-rich compositions the spinel phase (MA, $MgAl_2O_4$) no longer forms part of the equilibrium composition, and in this case the phase association in the quaternary system for a C/S ratio = 1.87 (see Table 4.2) is shown in Figure 4.10. For this particular composition, a peculiarity exists in that the phase association "$C_{12}A_7$" + C_2S + CA is not in equilibrium with MgO. Therefore, depending on the

Table 4.2 Equilibrium compositions in the quaternary system $CaO–MgO–Al_2O_3–SiO_2$ as a function of the CaO/SiO_2 ratio.

CaO/SiO_2	Phase assembly (in cement chemistry notation)
<0.94	$M + MA + M_2S + CMS$
0.94–1.40	$M + MA + CMS + C_3MS_2$
1.40–1.87	$M + MA + C_3MS_2 + C_5MS_3$ (<1372 °C)
	$M + MA + C_3MS_2 + C_2S$ (>1372 °C)
>1.87	$M + MA + C_2S + CA$ (lower T)
	$M + C_3A + C_3S + C$ (high T)

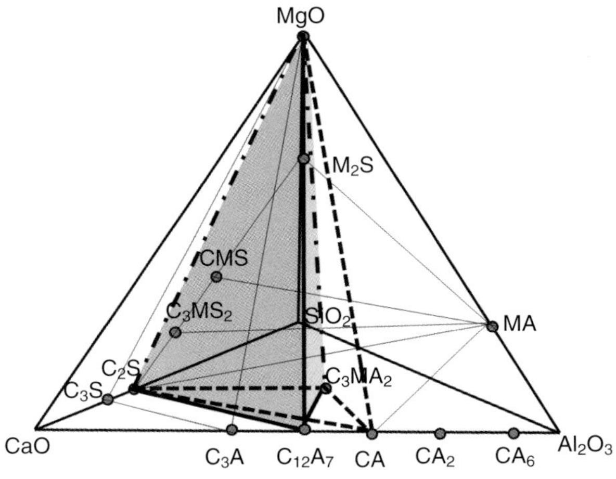

Figure 4.10 Phase association in the (Ca-rich) quaternary system $CaO–MgO–Al_2O_3–SiO_2$ for $C/S = 1.87$. The coexisting phases are either $C_{12}A_7 + C_2S + C_3MA_2 + M$ (solid conodes, shaded tetrahedron) or $CA + C_2S + C_3MA_2 + M$ (dashed conodes), depending on the firing temperature.

firing temperature, two phase associations compete–either "$C_{12}A_7$" + C_2S + C_3MA_2 + M (solid lines and shaded tetrahedron in Figure 4.10), or CA + C_2S + C_3MA_2 + M (dashed lines in Figure 4.10). The chain lines denote C_2S–M and C_3MA_2–M conodes shared by both phase tetrahedra

The mechanical wear and chemical corrosion of refractory ceramics is shown schematically in Figure 4.11. The molten slag, liquid metals, and corrosive gases will each be transported by infiltration of the pore structure and diffusion along grain boundaries, and interact with the silicate and aluminate phases. At the refractory surface facing the hot zone, a crud layer is formed which consists of a low-melting phase, followed by an infiltrated zone several millimeters thick that, with time, tends to chip off–a process called *slabbing*. Mechanically weak areas

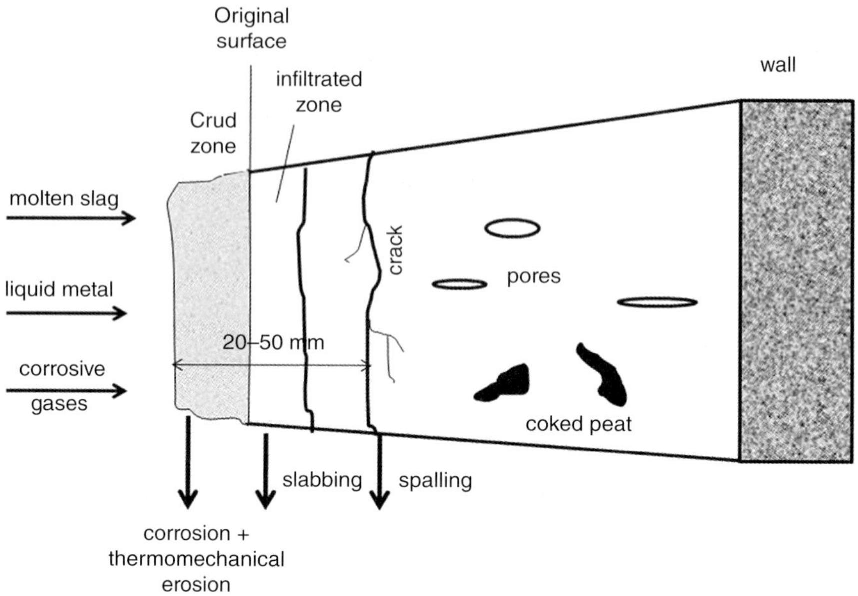

Figure 4.11 Wear and corrosion model of a basic refractory.

such as cracks are prone to spalling, especially if they are subjected to steep temperature gradients that develop during the heating and cooling of the furnace. These processes result in material losses in the range of 20 to 50 mm, that occur rather quickly. The resistance of the refractories to these deleterious mechanisms can be considerably improved by adding coked peat to seal the pores, by impeding the absorption of water that might otherwise react with free CaO present in the refractory, and by reducing wetting by liquid metals and slags. There are limitations, however, in an environment with a very high oxygen partial pressure. In an LD converter, liquid raw iron from a blast furnace will be subjected to oxygen blasting at about 1600 °C, so as to reduce its carbon content from about 4% (cast iron) to 0.1% (steel). However, trace amounts of Si, Mn, S and P present will also be oxidized and become bound into a Ca-rich slag formed by the addition of lime or dolomite (see below). Historically, phosphates locked up in the slag (Thomas slag) have been much sought-after as agricultural fertilizers.

During oxygen blasting of the iron, a series of complex reactions occur with the constituents of the refractories, and this leads to an evolutionary sequence of slag compositions that can be expressed in sections of the quinary system: CaO–MgO–Al_2O_3–SiO_2–Fe_2O_3. The first element to be oxidized is silicon. Depending on the amount of CaO present, either ferromonticellite (Ca[Mg,Fe]SiO_4; CMFS) or olivine [$(Mg,Fe)_2SiO_4$; $(M,F)_2S$] + wustite (Fe,Mg,Mn)O will be formed. These early slags have a low viscosity and so can easily infiltrate the pore structure of the refractory; but, they are also chemically very aggressive, and tend predominantly to dissolve the MgO. Consequently, slabs of dolomite or dolomitic limestone must be added to provide sufficient CaO to reduce the solubility of MgO in the slag, and this in turn leads to the formation of increasingly CaO-rich compounds such as merwinite ($Ca_3MgSi_2O_8$) and dicalcium silicate (Ca_2SiO_4, belite). The process culminates in a final slag composition that is characterized by the phase association tricalcium silicate (C_3S) + magnesiowustite (M,F) + calciowustite (C,F) + dicalcium ferrite (C_2F). These final slags contain 15–20% FeO and attack predominantly dolomite-based refractories (Trojer *et al.*, 1981).

4.5 Thermal Transformations and Phase Formation in the System MgO–(Fe$_2$O$_3$)–Al$_2$O$_3$–SiO$_2$

The technically interesting quaternary phases in this system are orthorhombic cordierite and ferrocordierite (sekaninaite), $(Mg,Fe^{2+})_2(Al_2Si)^{[4]}[Al_2Si_4O_{18}]$. Their structure consists of pseudo-hexagonal $[Al_2Si_4O_{18}]$ rings with twofold rotational symmetry and a mirror plane normal to the rings that are superimposed in [001] direction, and alternate rings are rotated by approximately 32°. These rings are connected by two $Mg^{[6]}$ and/or $Fe^{[6]}$ octahedra and irregular $(Al_2Si)^{[4]}$ tetrahedra that generate a 3-D-heteropolyhedral network (Strunz and Nickel, 2001; Figure 4.12).

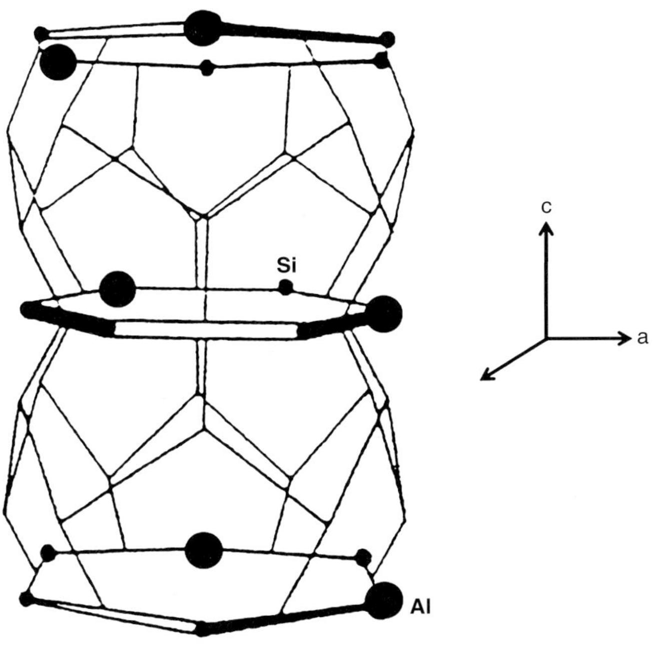

Figure 4.12 Schematic structure of cordierite. Thermal expansion occurs predominantly in the a-plane. A horizontal deformation of the charge fields of the oxygen anions causes contraction in the c-direction.

Hence, there exists a pronounced anisotropy in the coefficient of thermal expansion: the expansion in the plane of the strongly bonded tetrahedral $[Al_2Si_4O_{18}]$ rings (a-axis) is $2.9 \times 10^{-6}\,K^{-1}$ (25–1000 °C), whereas perpendicular to the rings (c-axis) – that is, within the more weakly bonded octahedral planes – the expansion is negative with $-1.1 \times 10^{-6}\,K^{-1}$ (25–1000 °C).

With increasing temperature, the weakly charged Mg^{2+} cations require a lower activation energy to oscillate around their equilibrium lattice positions, and consequently expansion occurs predominantly in the a-plane. A horizontal deformation of the charge fields of the oxygen anions causes contraction in the c-direction. The superposition of a positive expansion in the a-direction and a negative expansion in the c-direction results in a very slight volume expansion of about 0.3% between room temperature and 800 °C. The fact that cordierite ceramics are quite insensitive to thermal shock and thermally imposed deformation means that they are often used in applications such as catalyst supports (see below), as high-precision mirror supports for astronomical optical instruments, and as glass ceramics for household appliances (such as cooking ranges), as well as cooking vessels and dinner plates for microwave heating.

The production of cordierite ceramics can follow several reaction pathways, two of which are indicated by the dotted mixing lines in Figure 4.13. During the heating of a mixture of talc, kaolinite and mullite, first dehydroxylation and thus the formation of metatalc (MT)[1)] and metakaolinite (MK) occurs. The mixing line MT–MK crosses the lowest ternary eutectic protoenstatite–tridymite–cordierite at 1355 °C (point A in Figure 4.13a). At this composition and temperature, a contact melt forms between the talc and the clay mineral platelets. However, at a much lower temperature (ca. 1100 °C), a solid-state reaction occurs such that layers of cordierite composition are formed between the reacting particles; this isolates the reaction partners such that the reaction ceases at the initial firing. Whilst, during an initial slow heating, small amounts of cordierite are formed by an interfacial reaction between talc and clay, the second step (which is conducted at a higher temperature) leads to the formation of greater amounts of cordierite with a large activation energy, as ion diffusion must proceed across the inhibiting cordierite layer.

As indicated by the dotted mixing lines in Figure 4.13a, 26% metakaolinite (line MT–A), 42% metatalc (line A–MK) and 32% mullite (line A–$M_2A_2S_5$) react to form 100% cordierite (note that the line $M_2A_2S_5$–A_3S_2 corresponds to 68%). An alternate reaction pathway is shown in Figure 4.13b, where 43% metakaolinite (line MT–B), 42% metatalc (line B–MK) and 15% corundum (line B–$M_2A_2S_5$) react. A third option (not shown in Figure 4.13) is the so-called "sierralite" (chlorite) pathway, during which 62% MK, 13% MT, and 25% chlorite will react to form 100% cordierite. Based on their low coefficient of thermal expansion and high thermal shock resistance, cordierite ceramics find technical applications as spark protection for high-power switches, as heating coil supports for electric furnaces and cooking

1) Actually "metatalc" and "metakaolinite" consist of mixtures of protoenstatite and cristobalite, and of mullite and cristobalite, respectively.

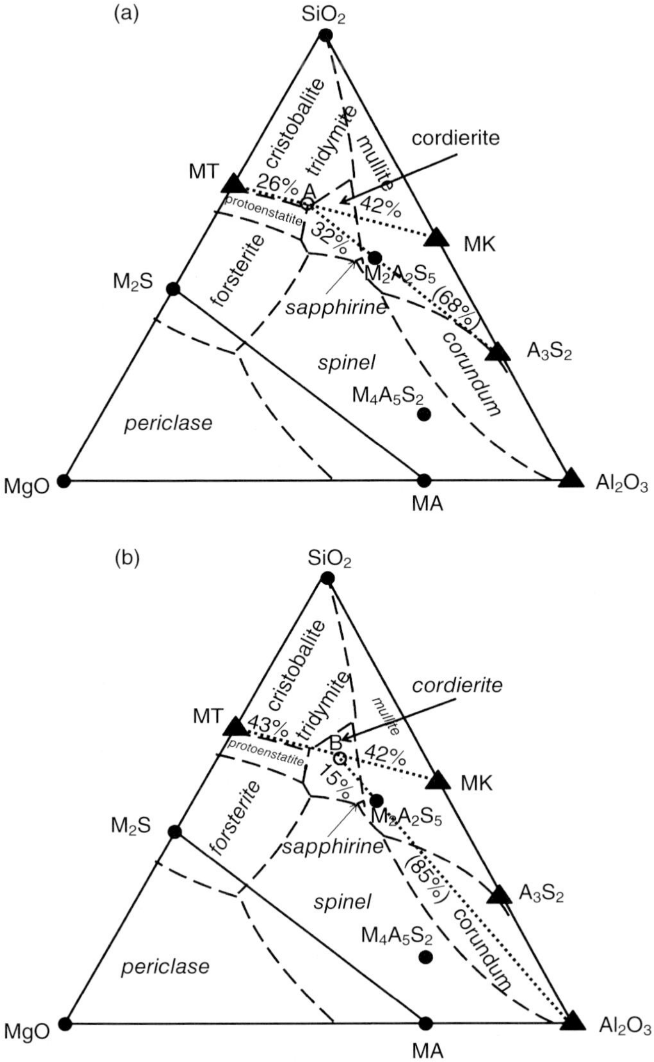

Figure 4.13 Ternary phase diagram $MgO–Al_2O_3–SiO_2$. Stability fields of primary phases are labeled in italics, binary and ternary phases are shown in cement chemical notation. The dotted lines indicate mixing lines leading to cordierite ($M_2A_2S_5$) either from mixtures of mullite + talc + kaolinite (a) or corundum + talc + kaolinite (b). For details, see the text.

ranges, as cordierite glass ceramics, cordierite-based porcelain, as chemically resistant materials utilized in chemical reactors, and as supports for three-way ceramic monolithic catalysts in automotive engines. A typical mass used for cordierite porcelain comprises 20–25% kaolin, 10–15% clay, 25–35% soapstone, 20–25% bauxite and AlOOH, respectively, and 10–15% potassium feldspar.

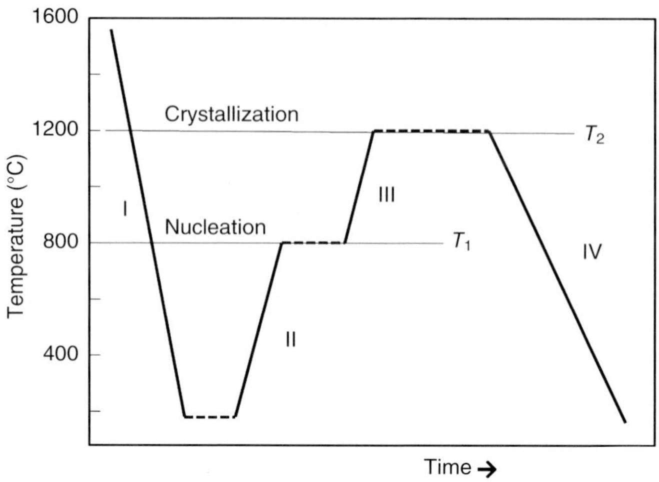

Figure 4.14 Schematics for the production of cordierite glass ceramics. The temperatures shown are approximate.

Cordierite glass ceramic is formed by a controlled heat treatment in two stages (Figure 4.14). Stage 1 follows essentially the production path of a normal technical glass involving raw materials mixing, smelting, fining, settling, forming and cooling (I). Stage 2 (II to IV in Figure 4.14) involves the "ceramization" of the glass – that is, a controlled reheating to the temperature T_1 of maximum nucleation (II), holding, heating to the temperature T_2 of sufficient crystallization conditions (III), holding, and controlled cooling to ambient temperature (IV). The result is a polycrystalline glass ceramic.

Ceramic monolithic catalyst supports are produced in form of cordierite honeycomb bodies coated with a highly disperse aluminum oxide layer embedded with fine particles of catalytically active noble metals such as platinum, rhodium and palladium, and mixtures thereof. A three-way catalyst functions to: (i) oxidize CO to CO_2; (ii) oxidize unburned hydrocarbons to CO_2 and water; and (iii) convert NO_x to nitrogen by the reaction $2NO + 2CO \rightarrow N_2 + 2CO_2$. The maintenance and control of an appropriate oxygen partial pressure is guaranteed by a stabilized zirconia-based, so-called lambda (λ)-probe (see Section 7.2.6.2). The monolithic cordierite support body has a very low average coefficient of thermal expansion in the range of $0.5 \times 10^{-6}\,K^{-1}$ between 25 °C and 1000 °C, parallel to the long axis of the body. This is caused by the forced directional orientation of the plate-like raw material particles along the decreasing pressure gradient in the nozzle of the extruder. Consequently, during firing of the green body the cordierite crystals formed grow with their c-axes in the direction of the extrusion of the honeycomb structure. Since the low thermal expansion correlates directly with the texture characteristics, X-ray diffraction can be used as a quality control tool; here, the ratio $I_{110}/(I_{110} + I_{002})$ is the so-called *texture coefficient* that represents a measure of the orientation function of the cordierite crystals.

References

Bradley, W.F. and Grim, R.E. (1951) High temperature thermal effects of clay and related minerals. *Am. Mineral.*, **36**, 182–201.

Brindley, G.W. and Nakahira, M. (1959) The kaolinite-mullite reaction series: I-III. *J. Am. Ceram. Soc.*, **42** (7), 311–324.

Brown, I.W.M., MacKenzie, K.J.D., Bowden, M.E., and Meinhold, R.H. (1985) Outstanding problems in the kaolinite-mullite-reaction sequence investigated by ^{29}Si and ^{27}Al solid-state nuclear magnetic resonance: II, High temperature transformation of metakaolinite. *J. Am. Ceram. Soc.*, **68** (6), 298–301.

Brownell, W.E. (1976) *Structural Clay Products*, Applied Mineralogy, vol. **9**, Springer, Wien, New York. 231 pp.

Chakraborty, A.K. and Ghosh, D.K. (1978) Reexamination of the kaolinite to mullite reaction series. *J. Am. Ceram. Soc.*, **61** (3–4), 170–173.

Comefore, J.E., Fischer, R.B., and Bradley, W.F. (1948) Mullitization of kaolinite. *J. Am. Ceram. Soc.*, **31** (9), 254–259.

Fieger, U. (1989) Elektronenmikroskopische Untersuchungen zur Phasen- und Gefügeentwicklung in einer aluminiumsilikatischen Keramik zur Fixierung transuran-elementhaltiger Abfälle. Dissertation, TH Darmstadt, Report KfK 4628, PWA 52/89.

Freund, F. (1967) Kaolinite-metakaolinite, a model of a solid with extremely high lattice defect concentrations. *Ber. Dtsch. Keram. Ges.*, **44** (1), 5–13.

Heimann, R.B. (1989) Assessing the technology of ancient pottery: the use of ceramic phase diagrams. *Archeomaterials*, **3**, 123–148.

Heimann, R. and Franklin, U.M. (1979) Archaeo-thermometry: the assessment of firing temperatures of ancient ceramics. *J. Int. Inst. Conserv., Can. Group*, **4** (2), 23–45.

Heimann, R. and Maggetti, M. (1980) Experiments on simulated burial of calcareous Terra Sigillata (mineralogical changes). *Br. Mus. Occas. Papers*, **19**, 163–169.

Heimann, R.B., Maggetti, M., and Einfalt, H.C. (1980) Zum Verhalten des Eisens beim Brennen eines kalkhaltigen illitischen Tones unter reduzierenden Bedingungen. *Ber. Dtsch. Keram. Ges.*, **57**, 145–152.

Insley, H. and Ewell, R.H. (1935) Thermal behavior of kaolin minerals. *J. Res. Nat. Bur. Stand.*, **14** (5), 615–627.

Iwai, S., Tagai, H., and Shimamune, T. (1971) Procedure for dickite structure modification by dehydration. *Acta Crystallogr.*, **B27** (1), 248–250.

Kingery, W.D. and Vandiver, P.B. (1986) *Ceramic Masterpieces. Art, Structure, and Technology*, The Free Press, New York.

Le Chatelier, H. (1887) De l'action de la chaleur sur les argiles. *Soc. Franc. Mineral. Bull.*, **10**, 204–211.

Leonard, A.J. (1977) Structural analysis of the transition phases in the kaolinite-mullite thermal sequence. *J. Am. Ceram. Soc.*, **60** (1–2), 37–43.

Letsch, J. and Noll, W. (1983) Phasenbildung in einigen keramischen Teilsystemen bei 600 °C–1000 °C in Abhängigkeit von der Sauerstofffugazität. *cfi/Ber. Dtsch. Keram. Gesell.*, **60** (7), 259–267.

Maggetti, M. (1975) Mineralogische Untersuchungen zur Farbe gebrannter Tone. *Bull. Soc. Fribourg Sci. Nat.*, **64** (1), 46–62.

Maggetti, M. (1982) Phase analysis and its significance for technology and origin, in *Archaeological Ceramics* (eds J.S. Olin and A.D. Franklin), Smithsonian Institution, Washington, pp. 121–133.

Maggetti, M. and Heimann, R. (1979) Bildung und Stabilität von Gehlenit in römischer Feinkeramik. *Schweiz. Mineral. Petrogr. Mitt.*, **59** (3), 413–417.

Maggetti, M., and Küpfer, T. (1978) Composition of the Terra Sigillata from La Péniche (Vidy, Lausanne, Switzerland). *Archaeometry*, **20** (2), 183–188.

Maggetti, M., and Radzewski, O.E. (1975) Zur Brennfarbe eisenhaltiger Tone. *Keram. Z.*, **27**, 1–5.

Maniatis, Y., Simopoulos, A., and Kostikas, A. (1982) The investigation of ancient ceramic technologies by Mössbauer spectroscopy, in *Archaeological Ceramics* (eds J.S. Olin and A.D. Franklin), Smithsonian Institution, Washington, DC, pp. 97–108.

Marfunin, A.S. (1979) *Spectroscopy, Luminescence and Radiation Centres in Minerals*, Springer, Berlin.

Newnham, R.E. (1961) Refinement of the dickite structure and some remarks on polymorphism in kaolin minerals. *Mineral. Mag.*, **32** (252), 683–704.

Noll, W. (1982) Mineralogie und Technik der Keramiken Alt-Kretas. *Neues. Jahrb. Mineral. Abh.*, **143** (2), 150–199.

Nöller, R., and Knoll, H. (1985) Diopside ($CaMgSi_2O_6$) as a host lattice for Fe-O. *Cryst. Latt. Defects Amorph. Mater.*, **11**, 159–163.

Norton, F.H. (1952) *Elements of Ceramics*, Addison-Wesley Press Inc., Cambridge, Mass.

Percival, H.J., Duncan, J.F., and Foster, P.K. (1974) Interpretation of the kaolinite-mullite reaction sequence from infrared absorption spectra. *J. Am. Ceram. Soc.*, **57** (2), 57–61.

Peters, T. and Jenni, J.P. (1973) *Mineralogische Untersuchungen Über Das Brennverhalten Von Ziegeltonen*. Beiträge zur Geologie Schweiz, Geotechnische Serie, vol. **50**, Kümmerly und Frey, Bern.

Roy, R., Roy, D.M., and Francis, E.E. (1955) New data on thermal decomposition of kaolinite and halloysite. *J. Am. Ceram. Soc.*, **38** (6), 198–205.

Sonuparlak, B., Sarikaya, M., and Aksay, O.S. (1987) Spinel phase formation during to 980°C exothermic reaction in the kaolinite-to-mullite reaction series. *J. Am. Ceram. Soc.*, **70** (11), 837–842.

Strunz, H. and Nickel, E.H. (2001) *Strunz Mineralogical Tables, Chemical-Structural Mineral Classification System*, 9th edn, E. Schweizerbart, Stuttgart.

Trojer, F., Obst, K.H., and Münchberg, W. (1981) *Mineralogie Basischer Feuerfest-Produkte*, Applied Mineralogy, vol. **12**, Springer, Wien, New York.

Walton, M.S., Diehne, E., Trentelman, K., Chiari, G., Maish, J., and Buxbaum, A. (2009) Characterization of coral-red slips on Greek Attic pottery. *Archaeometry*, **51** (3), 383–396.

5
Mineralogy and Chemistry of Cements

5.1
Historical Development

In the hierarchical materials triangles (see Figure 1.6), cements are shown to belong to the group of *hydraulic ceramics*, also termed chemically bonded ceramics (CBCs), to indicate that they obtain their properties not by thermal processes (as do ceramics, *sensu strictu*), but rather by their reaction with water. Moreover, the product of this reaction is stable against further reaction with water. This type of hydraulic reaction was first utilized many centuries ago by the Romans, who mixed hydraulic volcanic ash (pozzolan) with lime, sand and crushed rock to yield "*opus caementicium*," from which expression the term "cement" was derived. Artificial hydraulic "Roman cement" was first developed in Britain by James Parker in 1796. For this, the *septarian* concretions of organically derived calcite nodules, embedded in marine mudstones, were burned; subsequently, when the burned product was ground to a fine powder and mixed with sand, the mixture set hard in about 15 minutes. Earlier observations by John Smeaton, in relation to the construction of the famous Eddystone Lighthouse in the English Channel between 1755 and 1759, revealed that the "hydraulicity" of the then-available hydraulic lime mortars was directly proportional to the clay content of the precursor limestones. Armed with this knowledge, Smeaton selected those mortars for the lighthouse construction that would set and develop a reasonable strength during the 12-h period between successive high tides.

In 1824, Joseph Aspdin patented a hydraulic material, obtained by burning of an intimate mix of chalk and clay, that he called "Portland cement," on the basis that the set product was similar in appearance to the prestigious Portland limestone. The strength of the mortar made from early Portland cement was rather low, however, owing to the fact that Aspdin had fired the mixture at a temperature well below 1250 °C. Such a temperature was too low to obtain alite (Ca_3SiO_5) which, in modern Portland cements, is responsible for early strength (see Figure 5.2); instead, the reaction yielded essentially belite (β-Ca_2SiO_4), which develops its strength more slowly owing to a delayed hydrolysis. The situation was remedied in 1844, simultaneously by Isaac Johnson, Louis Vicat, and Joseph Aspdin's son, William, when the high-temperature Portland cement that they produced

Classic and Advanced Ceramics: From Fundamentals to Applications. Robert B. Heimann

ISBN: 978-3-527-32517-7

contained large amounts of alite and consequently developed the much desired early strength. Unfortunately, the innovation of high-temperature Portland cement required the addition of extra lime to the mix, as well as a much higher kiln temperature that led to a need for more fuel for the production process. In addition, the resultant clinker was very hard and rapidly wore down the millstones which, at the time, was the only available grinding technology. Although the manufacturing costs were considerably higher, the product set reasonably slowly and developed strength quickly, and this in turn opened up a wide market for the use of cement in concrete production. Moreover, from 1850 onwards, as the use of concrete in the construction industry expanded rapidly, this soon became the predominant market for cements.

While these attempts to optimize the strength and durability of cement were more or less unsystematic and empirical, the exact details of the chemistry of cement were first elucidated by Le Chatelier (1904). Later developments included the invention of reinforced concrete by Wilkinson and Lambot in 1855, and of blast furnace cement by Emil Langen in 1862. Thereafter, the twentieth century witnessed the invention and optimization of sulfate-resistant alumina cements (1908), the addition of plasticizers such as lignosulfonic acid or hydroxylated polysaccharides and superplasticizers such as sulfonated naphthalene–formaldehyde condensate, and the advent of macro-defect-free (MDF) and polymer fiber-reinforced cements, to name only a few.

Although cement represents a construction material with a long and colorful tradition, research into the product still provides a rich field of endeavor, with modern cements about to enter the realm of high-tech materials, including applications for lightweight and durable motor blocks, barges, counter tops, and many more. Further details on the history and future prospects of cement and concrete can be found in the reports of Gani (1997), Aitkin (2000), Bensted and Barnes (2002), Hewlett (2004), and Locher (2006).

5.2 Portland Cement

5.2.1 Introduction

Portland cement is fabricated by heating limestone with small quantities of materials carrying silica and alumina (such as clay) to about 1500 °C in a rotary kiln; this process is known as *calcination*. The resultant hard, sintered "clinker" is then ground together with a small amount of gypsum (ca. 3%) into a fine powder to produce so-called "Ordinary Portland Cement" (OPC), the most commonly used type of cement.

Portland cement is a basic ingredient of concrete, mortar, and most nonspecialty grouts. The most common use for Portland cement is in the production of concrete, a composite material consisting of aggregate (gravel and sand), cement, and water.

Table 5.1 Typical material properties compared on a production energy basis.

Material	Density ($t\,m^{-3}$)	Elastic modulus (GPa)	Tensile strength (MPa)	Energy per meter (GJ)	Energy per meter/unit tensile strength ($GJ\,MPa^{-1}$)
Al	2.8	70	100	360	3.6
Cu	8.9	130	200	530	2.6
Mild steel	7.8	210	300	300	1.0
Cast iron	7.9	150	150	360	2.4
Glass	2.5	65	60	50	0.8
Concrete	2.4	30	6	3.4	0.6
Reinforced concrete	2.5	35	30	3.9	0.13

As a construction material, concrete can be cast in almost any shape desired and, once hardened, it can become a structural (i.e., load-bearing) element, especially when reinforced with steel rods or meshes that act to take up the tensile forces (Table 5.1). The phase composition of OPC – that is, the relative amounts of alite, belite, tricalcium aluminate and ferrite – depends not only on the composition of the starting limestone/clay mix, but also on the cooling history of the clinker (see below). In addition, the "set time" can be manipulated by the addition of *accelerators* such as chlorides of calcium, magnesium, nickel or cobalt, or *retarders* such as nitrates of lead and zinc, tartaric acid, citric acid, and sucrose or raffinose. The addition of mineral (pumice, slags, fly ash, condensed silica gel, etc.) and chemical admixtures [sodium dodecylsulfate, lignosulfonates, hydrocarboxyl acids, poly(vinyl alcohol), etc.] leads to modified products with improved strengths and durabilities. Hence, Portland cement and the concrete produced from it emerge as the complex results of a set of parameters that can be widely varied to arrive at a broad palette of specified products for an ever-increasing spectrum of applications.

In 2002, global cement production amounted to 1.8×10^9 metric tons, and with a strongly increasing tendency. In particular, today's growing economies such as China and India utilize over-proportional shares; for example, in 2006 cement production in China amounted to 1.2×10^9 metric tons, equal to 44% of the total global production (NEAA, 2007).

5.2.2
Typical Composition and Materials Properties

The average chemical composition of Portland cement is 60–67 mass% CaO, 17–25 mass% SiO_2, 3–8 mass% Al_2O_3, 0.5–6 mass% Fe_2O_3, 0.1–4 mass% MgO, 0.5–1.3 mass% $Na_2O + K_2O$, and 1–3 mass% SO_3 (Roy, 1981). The sulfate content results from the addition of gypsum to the quenched clinker prior to grinding. The function of gypsum is twofold: (i) it acts as retarder during setting of the cement paste by inhibiting the early hydration of C_3A; and (ii) during grinding it releases water vapor, thus improving the result of grinding in terms of fineness and energy expenditure.

The general equation of clinker formation can be described by:

$$\text{Limestone} + \text{clay} \rightarrow CaO + \text{“metaclay”} + CO_2 + H_2O \quad (5.1a)$$

$$CaO + (Al, Fe)_2O_3 \cdot xSiO_2 \rightarrow C_3S + C_2S + C_3A + C_4AF \quad (5.1b)$$

The average phase composition of Portland cement (ASTM Type I, CSA 10) is ~50% alite, C_3S;[1] ~25% belite, β-C_2S; ~12% tricalcium aluminate, C_3A; ~8% ferrite, C_4AF[2]; and ~3.5% gypsum (Mindness, 1983). The structural order and crystallinity of the cement phases–and hence their hydraulic reactivities–are heavily dependent on the rate of cooling of the clinker:

- During *slow cooling* (~4 °C min^{-1}), MgO is exsolved as periclase, C_3A and C_4AF occur as crystalline phases, and alite and belite are relatively ordered and hence possess low hydraulic reactivity. Consequently, the clinker is coarse-grained and easy to grind. However, the cement is hydrolytically less active, develops a high early strength but a low 28-day strength, and is less sulfate-resistant. Slow cooling also leads to the destabilization of C_3S, forming C_2S and CaO.
- *Fast cooling* (~20 °C min^{-1}) produces disordered alite and belite that are highly reactive, whereas C_3A, C_4AF and also MgO remain in a glassy quench phase. The grain size in the clinker is small, which causes difficulties during grinding. The cement product then shows an increased hydraulic reactivity, a higher sulfate resistance, and a low early strength that, however, is strongly increased after 28 days.

Traditionally, the clinker composition is calculated from a chemical analysis of the clinker, using the time-honored Bogue equations (Bogue, 1929) and the very empirical relationships:

$$C_3S = (4.07 \cdot \%CaO) - (7.60 \cdot \%SiO_2) - (6.72 \cdot \%Al_2O_3) - (1.43 \cdot \%Fe_2O_3) - (2.85 \cdot \%SO_3) \quad (5.2a)$$

$$C_2S = (2.876 \cdot \%SiO_2) - (0.754 \cdot \%C_3S) \quad (5.2b)$$

$$C_3A = (2.650 \cdot \%Al_2O_3) - (1.692 \cdot \%Fe_2O_3) \quad (5.2c)$$

$$C_4AF = 3.043 \cdot Fe_2O_3 \quad (5.2d)$$

If some of the lime remains uncombined, it must be subtracted from the total lime content before performing the calculation; it is for this reason that a clinker analysis normally contains a value for uncombined free lime. It should also be stressed that the Bogue calculation does not give the “true” amounts of the four main clinker phases present, but differs from them principally because the actual mineral compositions differ from those assumed in the calculation, particularly

1) “Alite” is defined as C_3S with about 4% C_3A in solid solution.
2) Ferrite is not a true ternary phase but a (equimolecular?) solid solution of C_2A and C_2F.

for ferrite but also for alite. This problem is caused primarily by the fact that, under industrial conditions with limited firing and reaction progress times, the formation of clinker minerals does not reach equilibrium. In modern developments, a Rietveld refinement of the X-ray diffraction (XRD) pattern of the reaction products allows a direct measurement of the amount of cement phases present.

Other parameters are used to characterize the composition of Portland cement, most prominently the silica module (SM), the lime saturation factor (LSF), and the alumina-to-iron ratio (alumina module), (AR):

$$\mathrm{SM} = (\mathrm{SiO_2}/\mathrm{Al_2O_3} + \mathrm{Fe_2O_3}) \tag{5.3a}$$

$$\mathrm{LSF} = [100\,\mathrm{CaO}/(2.8\,\mathrm{SiO_2} + 1.18\mathrm{Al_2O_3} + 0.65\,\mathrm{Fe_2O_3})] \tag{5.3b}$$

$$\mathrm{AR} = \mathrm{Al_2O_3}/\mathrm{Fe_2O_3} \tag{5.3c}$$

The SM is typically between 2.3 and 3.5, and determines the amount of molten phase in the high-temperature zone of the rotary kiln: a high module indicates a low amount of melt, and *vice versa*. If the module is too high, then the reaction progress is slow. A large amount of fine grains remains unreacted and hence unconsolidated, and may block the movement of the charge through the kiln.

The LSF determines the proportion of free, that is, unreacted CaO. A value >100% indicates a high proportion of free lime that, however, can be reduced by appropriately long firing.

The AR value is typically around 2 and is determined by the melting temperature. The lowest melting temperature is reached if AR ~1.6, corresponding to the quaternary eutectic C_3S–C_2S–C_3A–C_4AF at 1338 °C. The amount of melt as a function of temperature and chemical composition has been estimated by Lea and Parker (1934) to yield:

$$\%\ \text{melt at } 1340\,^{\circ}\mathrm{C} = 6.1\mathrm{F} + \mathrm{M} + \mathrm{N} \tag{5.4a}$$

$$\%\ \text{melt at } 1400\,^{\circ}\mathrm{C} = 2.95\mathrm{A} + 2.2\mathrm{F} + \mathrm{M} + \mathrm{N} \tag{5.4b}$$

$$\%\ \text{melt at } 1450\,^{\circ}\mathrm{C} = 3.0\mathrm{A} + 2.25\mathrm{F} + \mathrm{M} + \mathrm{N}, \tag{5.4c}$$

where A = Al_2O_3, F = Fe_2O_3, M = MgO, and N = Na_2O (cement chemical notation).

Optimum values of the chemical parameters with regard to plant operation and clinker quality are LSF = 92–96%, SM = 2.3–2.8, and AR = 1.6–2.0 (Roy, 1981).

Table 5.1 lists several mechanical properties of common technically used construction materials. While concrete and reinforced concrete lack stiffness and tensile strength compared to metals, they are superior in their light weight and on the basis of production energy. This becomes even more apparent when comparing the ratio of the production energy expended per meter and the unit tensile strength in GJ MPa^{-1}.

Portland cement is one of the most energy-efficient construction materials, owing to the low cost of raw materials and high production figures. The energy content of technical products per unit volume relative to that of Portland cement is shown in Table 5.2. Even though the energy expenditure of Portland cement is considerable (4600–5800 kJ kg^{-1}), it is still much lower than that of other technical

Table 5.2 Energy content of technical products per unit volume relative to Portland cement.

Product	Relative energy content	Product	Relative energy content
Portland cement	**1.0**	Poly(styrole), PS	6
Sheet glass	3.0	Steel	19.2
Poly(vinyl chloride)	3.8	Stainless steel	28.2
Poly(ethylene), LDPE	4.2	Aluminum	31.8
Poly(ethylene), HDPE	4.4	Zinc	34.8

HDPE = high-density poly(ethylene); LDPE = low-density poly(ethylene).

products, and hence there is a strong incentive to replace more traditional (notably metallic) materials with high-value-added cement products. Yet, on the other hand, environmental considerations – and in particular the high CO_2 emissions from the cement kilns – call for changes in the available technology (see Section 5.4).

The typical ranges of the modulus of elasticity (Young's modulus), flexural strength and fracture toughness of selected structural materials are displayed in Figure 5.1.

5.2.3
Phase Composition

Figure 5.2 shows, in schematic form, the simplified reaction sequence and the relative amounts of reacting phases and their spatial distribution during heating to about 1500 °C a mixture of limestone and clay. The low-temperature intermediate product spurrite begins to form at about 630 °C (Kurdowski and Sobón, 1999) and decomposes, to belite, on further heating between 850 and 920 °C (Goswami *et al.*, 1989). The latter continues to react at a still higher temperature with silica released earlier by the decomposition of clay, to form the highly hydraulic clinker mineral, alite. The liberated alumina then reacts with free CaO to form tricalcium aluminate, $Ca_3Al_2O_6$.

Thus, Portland cement is characterized by the association of alite, belite, tricalcium aluminate and, in the presence of iron, ferrite (also called *celite* or *brownmillerite*). The composition of Portland cement is confined to a narrow compatibility triangle, shown as a shaded wedge in the ternary compositional diagram CaO–Al_2O_3–SiO_2 (Figure 5.3). However, in reality ubiquitously present iron and magnesium oxides add a fourth major phase ferrite ($Ca_4Al_2Fe_2O_{10}$) and, in slowly cooled clinker, accessory periclase MgO. The clay component in the mix is shown in Figure 5.3 as "metakaoline," AS_2. The intersections of the tie line AS_2–CaO (a–c) with the conodes C_3S–C_3A (y) and C_2S–C_3A (x) indicate the maximum molar fraction cx/ac = 0.31 and minimum molar fraction cy/ac = 0.26. Compositions outside this range will lead to the formation of additional, generally undesirable, components in the Portland cement, such as $C_{12}A_7$.

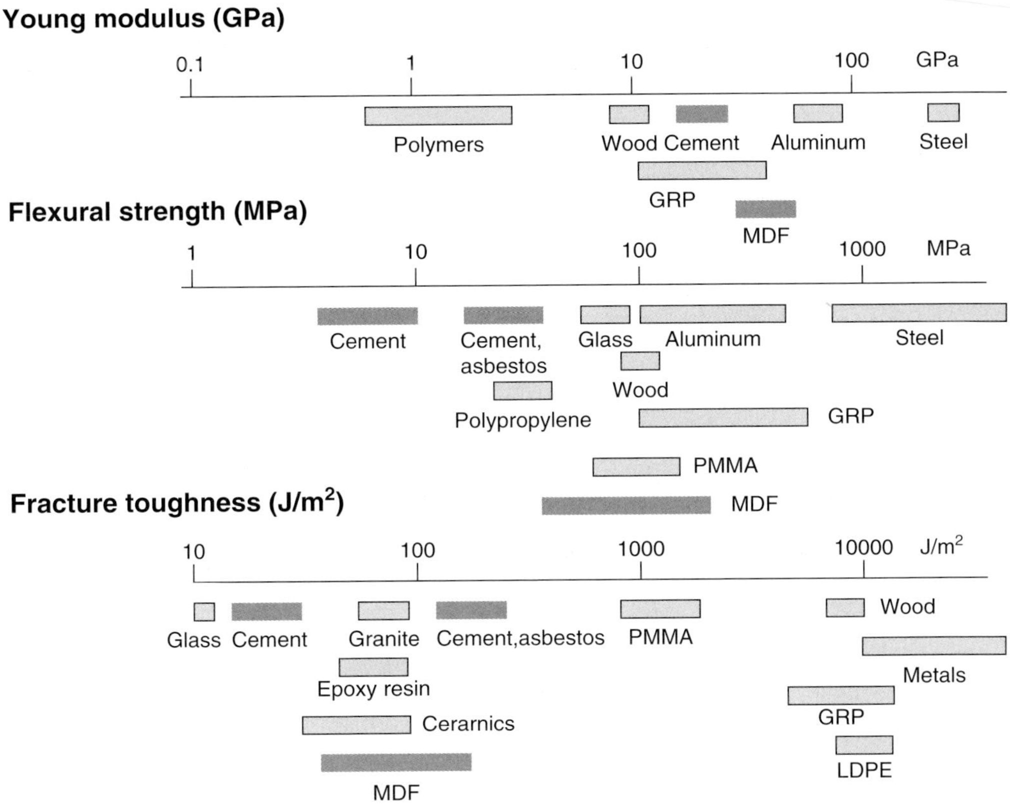

Figure 5.1 Selected mechanical properties of metals, engineering polymers, and ceramics. Cement-related materials are indicated by dark shading. GRP = glass-fiber reinforced polymers; LDPE = low-density polyethylene; MDF = macrodefect-free cement; PMMA = poly(methylmethacrylate) (plexiglass).

For reasons of clarity, the phase relations shown in Figure 5.3 neglect the existence of iron and magnesium oxides.

The crystallization history of Portland cement is shown schematically in Figure 5.4. Here, point "a" marks the approximate limiting composition of a lime-rich Portland cement, consistent with point y in Figure 5.3. This composition is actually in the stability field of CaO, so that the first product of crystallization should be CaO. The melt is continuously depleted in CaO, and hence the composition moves along the backward extension (chain line) a-CaO until it reaches the phase boundary CaO–C_3S at a′. The C_3S then starts to crystallize, while CaO redissolves in the melt. At point a″, the dissolution of CaO ceases and the composition of the melt moves along the line a″–b. At point "b" at the phase boundary C_3S–C_3A, the phase C_3A begins to crystallize and the crystallization path changes direction and moves towards the peritectic point "c"; that is, the triple point C_3S–C_3A–C_2S. Now, C_2S appears in the phase assembly and continues to crystallize until all melt has been solidified. If the starting composition contains more SiO_2 (point "p"), then the first

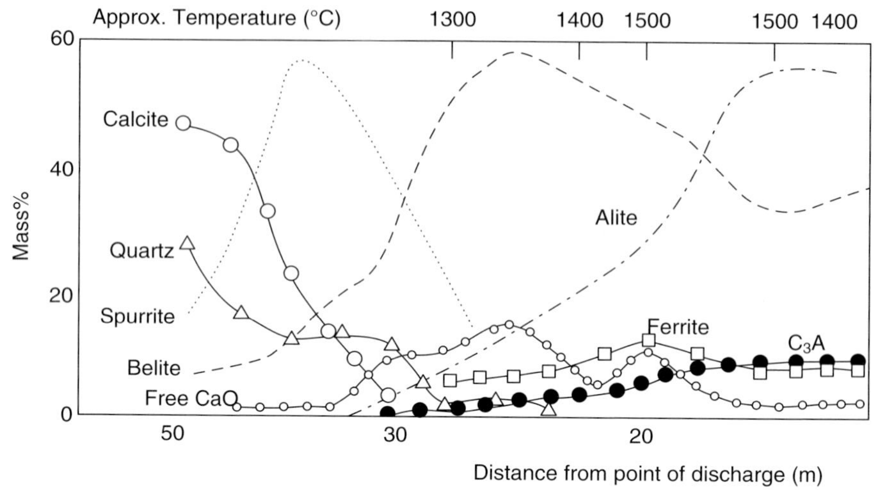

Figure 5.2 Phase composition of Portland cement clinker as a function of temperature and distance from discharge point, respectively. Alite Ca_3SiO_5, belite β-Ca_2SiO_4, ferrite (celite) $Ca_4Al_2Fe_2O_{10}$, spurrite $Ca_5[CO_3/(SiO_4)_2]$.

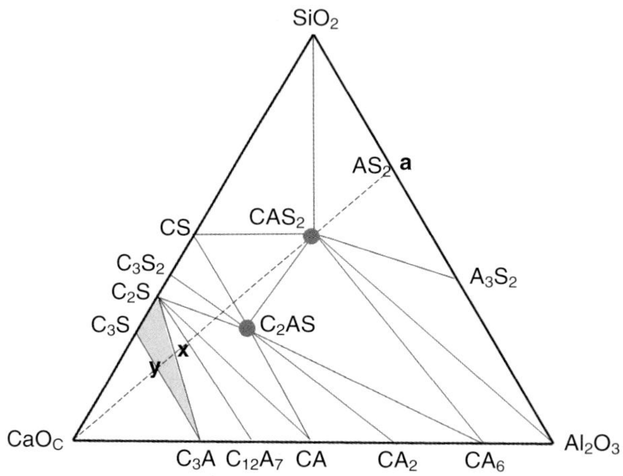

Figure 5.3 Compatibility triangles in the ternary system $CaO(C)$–$Al_2O_3(A)$–$SiO_2(S)$. The phase compositions of Portland cement are located in the shaded area. The limiting molar fractions of metakaoline AS_2 and CaO are indicated as the intersections of the tie line CaO–AS_2 with the conodes C_3S–C_3A (y) and C_2S–C_3A (x) (see also Figure 3.26).

crystallization product will be C_2S. The crystallization path moves away from C_2S towards the phase boundary C_2S–C_3S and intersects it at point p′, at which the melt composition C_3S and C_2S crystallize simultaneously. This continues until the peritectic point "c" has been reached, at which now C_3A begins to appear. At a still higher silica content (point "q"), corresponding to point "x" in Figure 5.3, the

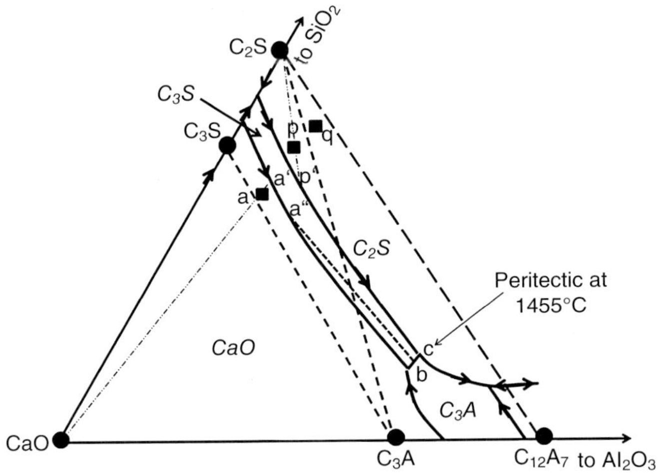

Figure 5.4 Phase relations and crystallization paths of Portland cement. Solid lines indicate phase boundaries; dashed lines indicate cotectic lines between primary phases; dotted lines indicate crystallization paths; chain lines indicate backward extensions of crystallization paths a–a′ and p–p′, respectively.

Table 5.3 Sequence of formation of clinker minerals (see also Figure 5.2).

Reaction #	Temperature (°C)	Reaction	Type of reaction
1	500	Clay →"metaclay" + H_2O	Decomposition
2	900	$CaCO_3$→ CaO(s) + CO_2(g)	Calcination
3	900–1200	CaO + "metaclay"→ C_2S(s) + C_3A	Solid state
4	1250–1280	C_2S + CaO + melt I → C_3S(s) + melt II	Liquid phase
5	>1280	CaO + SiO_2 + melt II → C_2S(s)	Liquid phase

starting composition is within the cotectic triangle C_2S–C_3A–$C_{12}A_7$. This is unfavorable, since $C_{12}A_7$ hydrolyzes much too rapidly to yield a strong cement product.

In the presence of iron oxide there is a tendency for the replacement of alumina by iron oxide. Consequently, at $Al_2O_3/Fe_2O_3 \geq 0.64$, C_3A will be replaced by C_4AF in the phase assembly.

If the starting mixture of limestone and clay is heated to 1338 °C, the first traces of melt begin to appear at the quaternary eutectic C_3S–C_2S–C_3A–C_4AF, the amount of which can be estimated with Lea and Parker's equation [Eq. (5.4a)] to be between 15% and 25%. Since the lowest melting point in the iron-free ternary system CaO–Al_2O_3–SiO_2 is still at the peritectic at 1455 °C, the melting point-suppressing action of iron oxide is obvious. Cooling the melt leads first to the crystallization of C_3A + C_4AF, before C_2S and C_3S solidify.

In conclusion, the sequence of reactions occurring during formation of the clinker minerals can be summarized as shown in Table 5.3 (Roy, 1983). It should be noted that C_2S can be formed by different reaction paths and types, as also indicated in Figure 5.2.

For details and up-to-date information the reader is referred to the excellent summary by Pöllmann (2002).

5.2.4
Hydration of Clinker Minerals

The setting and hardening of cement and concrete are the results of chemical and physical processes – that is, hydration taking place between the clinker minerals of Portland cement and water. As these reactions are complex, they can be best unraveled by studying the hydration chemistry and kinetics of the individual compounds separately. Hence, it will be tacitly assumed that the hydration of each clinker compound proceeds independently of the others present in Portland cement. Even though this assumption is not completely valid it is a reasonable first approximation (Young, 1981). Many studies have been conducted on this subject, using modern analytical techniques to follow the microscopic reaction that occurs at the surface of the individual clinker grains. In the present context, a more pragmatic approach will be followed by simplifying the processes ongoing in the C–S–H gel and relating them to the various factors that affect the strength of the set product (Young, 1981; Taylor, 1981, 1997; Gartner *et al.*, 2002).

5.2.4.1 Calcium Silicates

The hydration reactions of C_3S and C_2S are stoichiometrically very similar, but differ in the amount of calcium hydroxide (portlandite, CH) formed and the negative enthalpy of hydration. They can be described by:

$$2C_3S + 7H_2O \rightarrow C_3S_2H_4 + 3CH; \Delta H = -114\ \text{kJ mol}^{-1} \quad (5.5a)$$

$$2C_2S + 5H_2O \rightarrow C_3S_2H_4 + CH; \Delta H = -43\ \text{kJ mol}^{-1} \quad (5.5b)$$

Figure 5.5 shows a typical setting curve for C_3S. When cement powder is brought in contact with water, initially a large amount of heat is released (stage I). This so-called "early hydrolysis" lasts for about 15 min, and leads into stage II, the induction period. Here, negligible amounts of heat will be released as the reaction kinetics is dominated by the nucleation of portlandite. During this time (1–3 h) the cement paste is still pliable. Stage III is initiated by a tearing of the gel layer formed around the C_3S grains. This period of acceleration is characterized by the formation of C–S–H hydrolysis products and the production of considerable amounts of heat. During the following stage IV, heat production ceases and the reaction kinetics is slowed down as it now requires the diffusion of ions through the thickening gel layers around the hydrolyzing C_3S grains. The last stage V, reached after 12–24 h, is very slow as it is dominated by diffusion. The reactions continue for long periods of time, such that a set cement or concrete, respectively, will continue to gain strength over time.

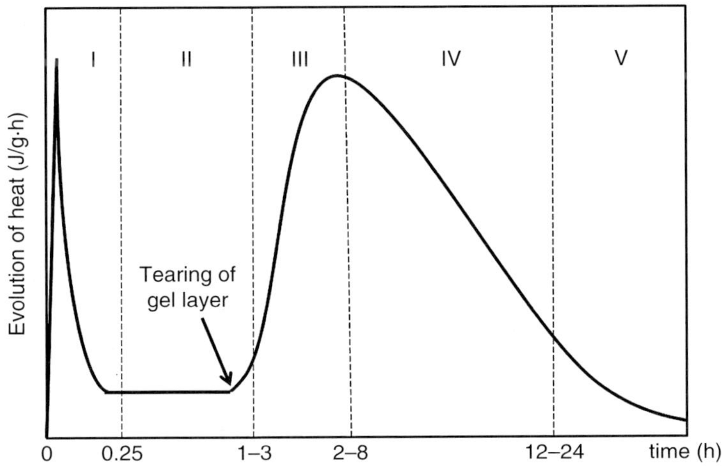

Figure 5.5 Calorimetric curve of the hydrolysis of C_3S.

Table 5.4 Interpretation of the calorimetric curve shown in Figure 5.5.

Reaction stage	Reaction kinetics	Chemical process	Property
I: Early hydrolysis	Chemical (fast)	Hydrolysis, ions pass into solution	
II: Induction period	Nucleation (slow)	Ions pass into solution	Determines early setting
III: Acceleration	Chemical (fast)	Formation of C–S–H	Determines rate of late setting
IV: Retardation	Chemical + diffusion (slow)	Formation of C–S–H	Determines rate of early strength development
V: Steady state	Diffusion (very slow)	Slow formation of hydration products	Determines rate of late strength development

The type of reaction, the ongoing chemical processes, and the effect of the individual reaction stages on the development of the properties of the set product are summarized in Table 5.4.

The hydrolysis of C_2S occurs in similar fashion, but proceeds much more slowly owing to its lower hydration enthalpy [Eq. (5.5b)].

5.2.4.2 Calcium Aluminate

In the presence of gypsum ($CaSO_4 \cdot 2H_2O$; $C\bar{S}H_2$), the hydrolysis reaction of C_3A proceeds under formation of ettringite according to:

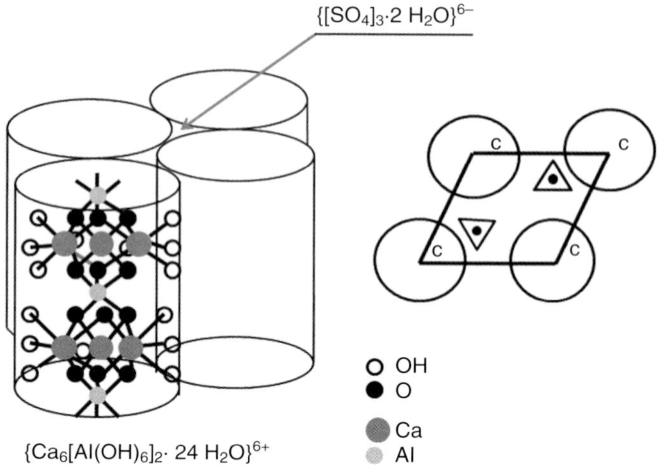

Figure 5.6 Structural model of ettringite.

$$C_3A + 3C\bar{S}H_2 + 26H_2O \rightarrow C_6A\bar{S}_3H_{32} \quad (5.6a)$$

Ettringite ($Ca_6Al_2[(OH)_{12}/(SO_4)_3] \cdot (24+2)H_2O$) is a natural mineral, in the structure of which $Al(OH)_6$ octahedra and clusters of three $Ca(OH)_4(H_2O)_4$ polyhedra alternate along the trigonal axis [00.1] to form chains by sharing edges (Figure 5.6). These chains are linked by three SO_4 tetrahedra, and an additional two H_2O may be present between the chains. Ettringite belongs to the general group of AFt phases (Al_2O_3–Fe_2O_3 trisulfate); here, the sulfate group can be partially replaced by carbonate to yield *thaumasite* and *jouravskite*, or by borate to yield *charlesite* and *sturmanite* (Strunz and Nickel, 2001). This property is at the heart of the use of aluminate-rich cement to immobilize toxic heavy-metal ions such as CrO_4^{2-} or VO_4^{3-} (e.g., Pöllmann, 2007; see also Buhlert and Kuzel, 1971; Heimann *et al.*, 1992).

If the sulfate concentration in the solution decreases, ettringite will become unstable and transform to monosulfate, a member of the AFm phases (Al_2O_3–Fe_2O_3 monosulfate) according to:

$$2C_3A + C_6A\bar{S}_3H_{32} + 4H_2O \rightarrow 3C_4A\bar{S}H_{12}. \quad (5.6b)$$

The complete hydrolysis reaction is

$$C_3A + C\bar{S}H_2 + 10H_2O \rightarrow C_4A\bar{S}H_{12};\ \Delta H = -362\,\text{kJ}\,\text{mol}^{-1}. \quad (5.6c)$$

Since ettringite forms a diffusion barrier around the C_3A grains, the hydration reaction is initially delayed. Only after about 12 h will this barrier be broken by the onset of transformation of ettringite into monosulfate (Figure 5.7). This reaction is reversible, as the monosulfate can retransform to ettringite in the presence of sulfate ions in the groundwater, leading to the structural destruction of concrete, as can be observed on many urban concrete structures. In modern sulfate-resistant

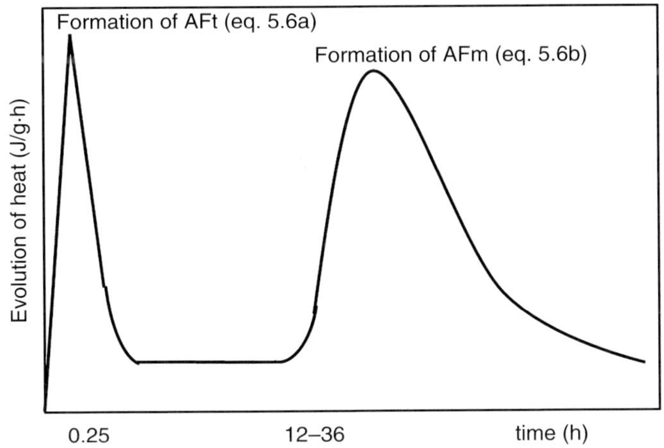

Figure 5.7 Calorimetric curve of the hydrolysis of C_3A.

cements, much of the alumina is replaced by iron oxide, thus inhibiting any sulfate-based corrosion of concrete.

The formation of ettringite from monosulfate can be described by:

$$C_4A\bar{S}H_{12} + 2C\bar{S}H_2 + 16H_2O \rightarrow C_6A\bar{S}_3H_{32} \tag{5.7}$$

Without gypsum, the hydrolysis of C_3A occurs immediately via the formation of two AFm phases:

$$2C_3A + 21H_2O \rightarrow C_4AH_{13} + C_2AH_8;\ \Delta H = -340\,kJ\,mol^{-1} \tag{5.8a}$$

that are structurally very similar to monosulfate. These phases are unstable and are converted, via a type of condensation reaction, to hydrogarnet:

$$C_4AH_{13} + C_2AH_8 \rightarrow 2C_3AH_6 + 9H_2O \tag{5.8b}$$

Hydrogarnet crystallizes in the cubic system and is structurally analogous to the silicate garnet grossularite:

$$Ca_3Al_2(OH)_{12} \rightarrow Ca_3Al_2\mathit{H}_{12}O_{12}$$

$$Ca_3Al_2Si_3O_{12} \rightarrow Ca_3Al_2\mathit{Si}_3O_{12}$$

At an elevated temperature (>80 °C), hydrogarnet forms directly according to:

$$C_3A + 6H_2O \rightarrow C_3AH_6;\ \Delta H = -240\,kJ\,mol^{-1}. \tag{5.8c}$$

However, which stable reaction products actually form will depend on the molar ratio gypsum/C_3A, as shown in Table 5.5.

5.2.4.3 **Ferrite (Brownmillerite)**

Ferrite forms the same sequence of hydration products as C_3A in the presence of gypsum, but the reactions are slower and produce less heat. With an increasing

Table 5.5 Stable hydration products of C_3A.

Molar ratio gypsum/C_3A	Stable hydration products
3	Ettrigite (AFt)
3–1	Ettringite + Monosulfate (AFm)
<1	Monosulfate–C_4AH_{13} solid solutions (AFm)
0	Hydrogarnet (C_3AH_6)

iron content, the reaction rate decreases. Pure ferrite has an insufficient calcium content to form AFt and AFm phases; this occurs only if iron and aluminum hydroxides can be formed simultaneously, according to

$$3C_4AF + 12C\bar{S}H_2 + 110H_2O \rightarrow 4\{C_6(A,F)\bar{S}_3H_{32}\} + 2(A,F)H_3 \text{ and} \tag{5.9a}$$

$$3C_4AF + 2\{C_6(A,F)\bar{S}_3H_{32}\} + 14H_2O \rightarrow 6\{C_4(A,F)\bar{S}H_{12}\} + 2(A,F)H_3 \tag{5.9b}$$

The thermodynamically stable phase is again a hydrogarnet solid solution, $C_3(A,F)H_6$.

5.2.4.4 Kinetics of Hydration

The kinetics of setting of cement depends on the rate of hydration of the individual clinker minerals; that is:

$$C_3A > C_3S > C_4AF > C_2S$$

as shown in Figure 5.8. These hydration curves pertain to pure clinker minerals. Technical clinker phases such as alite, ferrite, and belite hydrate at a somewhat

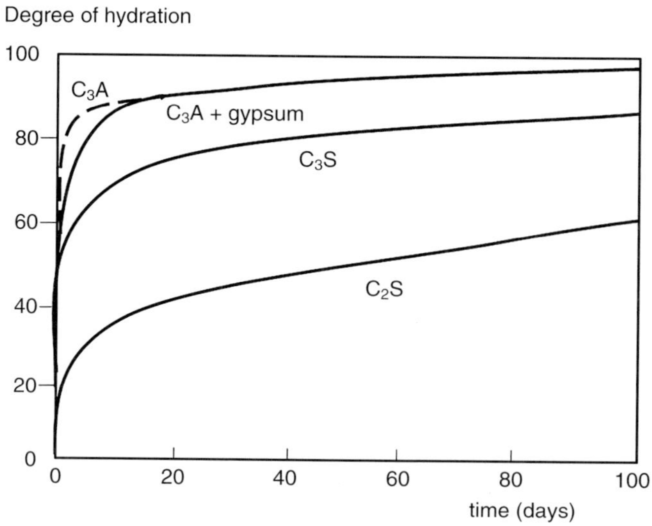

Figure 5.8 Time dependence of hydration of pure clinker phases.

faster rate. For example, pure C_2S shows a degree of hydration of about 60% after 100 days, whereas belite hydrates to about 80% within the same time. This is most likely related to the presence of impurities acting as accelerators, and affecting the reaction rate through: (i) a statistical replacement of ions in the crystal lattice that generate point defects and dislocations; (ii) polymorphic transformation induced by ion substitution; and (iii) exsolution of secondary phases at grain boundaries.

As mentioned above, the assumption that clinker minerals hydrate independently of each other cannot be upheld. For example, C_2S hydrates more quickly in the presence of C_3S since the Ca^{2+} and OH^- concentrations in the pore solution have been substantially increased by the faster-hydrolyzing C_3S. These changes also influence the rate of hydration of C_3A and C_4AF. On the other hand, C_3A and C_4AF compete for the sulfate ions introduced by gypsum. Hence, the C_3A with a higher hydrolytic reactivity tends to consume more sulfate ions than C_4AF. This, however, increases the reactivity of C_4AF as it produces less ettringite than would be expected stoichiometrically. Gypsum also increases the rate of hydration of the calcium silicates that likewise fight for the available sulfate ions, since the C–S–H phases incorporate large amounts of not only of SO_4 ions but also of Al and Fe. This situation has been nicknamed "... the fight for survival of the primary clinker phases."

Hence, gypsum plays a crucial role during the hydration history of cement. The addition of too much gypsum will lead to the formation of large amounts of ettringite which, after setting, results in an unfettered expansion and destruction of the microstructure. Yet, insufficient gypsum is also detrimental, as it will lead to the formation of AFm phases before the second stage of C_3S hydration has been completed. The available Ca ions will be used up and the nucleation of the hydration products of C_3S delayed, thus prolonging the induction period. A concrete of this type would be susceptible to secondary ettringite formation during any reaction with groundwater.

Figure 5.9 summarizes the processes that occur during the hydration of anhydrous clinker minerals (left-hand column) to form the products shown in the right-hand column. The rectangular areas occupied by the individual phases correspond approximately to their volumetric proportions in Portland cement.

5.2.5 Models of Hydration

Popular models of hydration of clinker minerals can be divided into the delayed nucleation model and the protective layer model.

5.2.5.1 Delayed Nucleation Model

This model refers to the delayed nucleation of portlandite (calcium hydroxide; CH). The end of the induction period (stage II in Figure 5.5) is marked by a maximum of supersaturation of Ca^{2+} ions in solution. Normally, this supersaturation would be removed by nucleation and a later precipitation of portlandite. However, nucleation is delayed by "poisoning" of the embryos by the adsorption

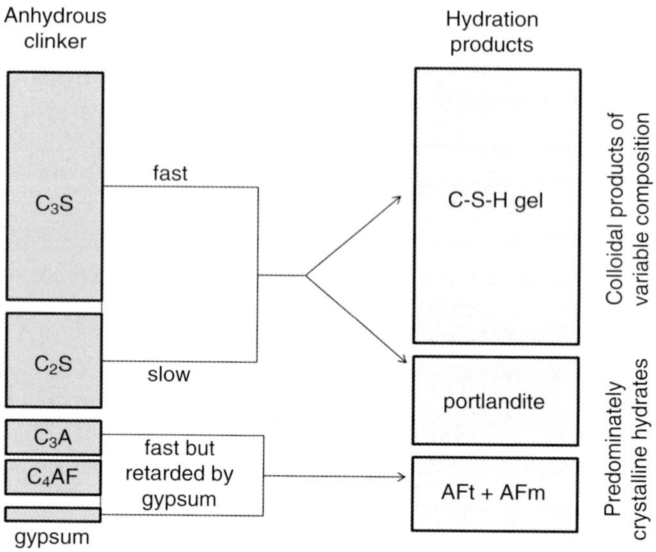

Figure 5.9 Sequence of hydration of clinker minerals.

of minute amounts of silicate from the solution before they can reach the critical size that would enable their growth. Hence, further increasing the Ca concentration in solution would inhibit the hydration of calcium silicates and thus prolong the induction period. This situation can only be resolved when, eventually, the portlandite crystallizes. Arguments against this model stem from the observation that the hydration of C_3S can be accelerated neither by the presence of an already saturated CaO solution, nor by the addition of crystalline portlandite nuclei. In addition, the formation of crystalline portlandite has been observed long before the Ca^{2+} supersaturation peak and the end of the induction period.

5.2.5.2 **Protective Layer Model**

In contrast to the previous model, the protective layer model considers the growth of portlandite as the effect–not the cause–of the accelerated hydration. Figure 5.10 shows, schematically, the sequence of events. Immediately after the addition of water to the cement powder, a colloidal layer of C–S–H gel forms around the grains (Figure 5.10b) and acts as a barrier towards any further reaction (this explains the existence of an induction period). At the end of the induction period, the protective layer would rupture and an accelerated hydration and growth of the secondary C–S–H gel would result (Figure 5.10c); this corresponds to the acceleration period (stage III). At a still later stage, an infilling of the microstructure occurs by the fine-grained C–S–H gel and by a continuing growth of crystalline portlandite (Figure 5.10d) (i.e., period of retardation; stage IV).

The rupture of the protective layer (Figure 5.10c) has been explained by two competing mechanisms:

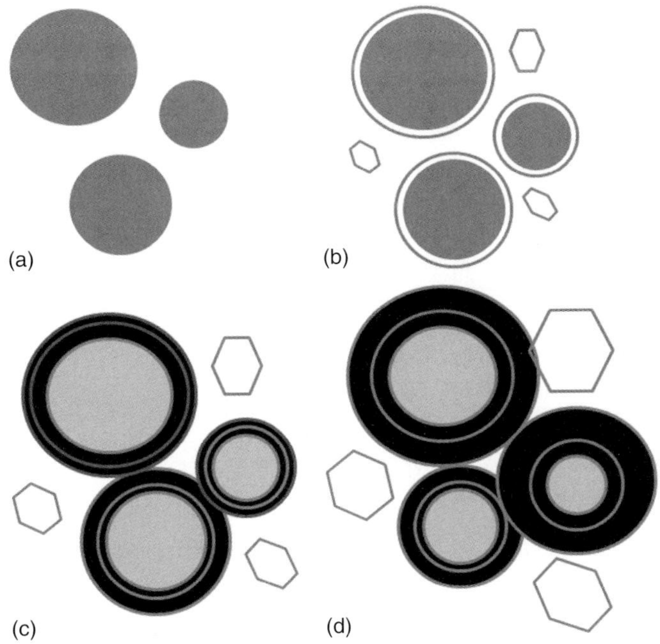

Figure 5.10 Schematics of the sequence of hydration of clinker minerals. (a) Alite grains in water; (b) Formation of a protective colloidal C–S–H layer; (c) Rupture of layer and secondary growth of C–S–H; (d) Infilling of the microstructure by fine-grained C–S–H gel and growth of crystalline portlandite.

- The first mechanism assumes a spontaneous change of the chemical and physical structure of the C–S–H gel by transformation to a secondary C–S–H phase that is more permeable to water. These semicrystalline C–S–H (I) and C–S–H (II) phases vary in composition: C–S–H (I) has a C : S ratio between 0.8 and 1.5, whereas C–S–H (II) has a C : S ratio >1.5. Within increasing time and temperature, the C–S–H crystallizes and develops a structure similar to tobermorite with C : S ratios close to 0.8 and a variable water content; that is, 0.9 nm-tobermorite (riversideite, $C_5S_6H_{0...2}$), 1.1 nm-tobermorite ($C_5S_6H_3$), and 1.4 nm-tobermorite (plombièrite, $C_5S_6H_7$). Tobermorite forms three-periodic single chains of SiO_4 tetrahedra that are held together by edge-sharing CaO_7 polyhedra and water molecules (Figure 5.11). A reevaluation of the structure has been provided by Merlino *et al.* (2001) recently. In addition to tobermorite-like phases at elevated temperatures, a variety of other crystalline C–S–H phases occur. Up to about 150 °C, tobermorite, afwillite ($C_3S_2H_3$) and α-C_2S hydrate are stable; however, beyond this temperature (and up to about 500 °C) the hydrate phases cease to be stable and will be replaced by hydrosilicates such as xonotlite (C_6S_6H) (Heimann, 1988b) and foshagite ($C_4S_3H_{1.5}$). At still higher temperatures, water-free structures dominate such as β-wollastonite (CS), rankinite (C_3S_2), and bredigite (C_2S).

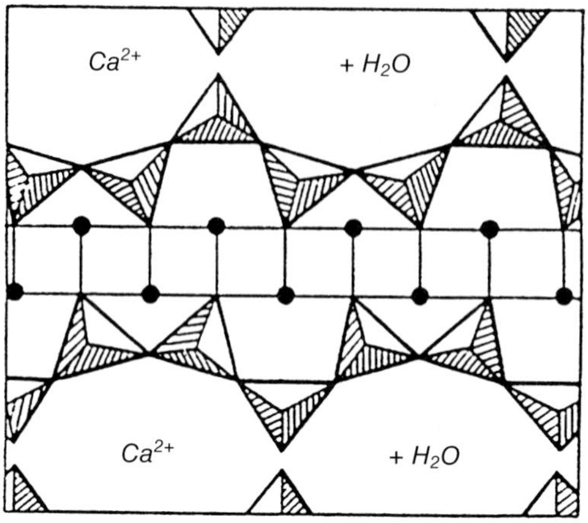

Figure 5.11 Structure of 1.1 nm tobermorite (after Megaw and Kelsey, 1956).

- The second mechanism assumes that the water-swelled C–S–H gel layer will eventually rupture by a build-up of internal osmotic pressure that is generated in response to the selectively permeable nature of the colloidal gel. Even though it allows the inward diffusion of water and the outward diffusion of Ca^{2+} and OH^- ions, it blocks the outward diffusion of the larger, bulky $H_2SiO_4^{2-}$ ions. These hydrate silicate ions will be squeezed out after rupture of the layer, extruded into the surrounding Ca^{2+}-rich solution, and immediately precipitated as a secondary C–S–H gel that grows at the surface of the original C–S–H gel (Figure 5.10c) in the form of a calcium silicate membrane.

5.2.6 Setting and Hardening of Cement

The early setting of cement corresponds to the end of the induction period (stage II in Figure 5.5). Although this suggests that it is predominantly determined by the rate of hydration of C_3S, there are indications that the recrystallization of ettringite likewise plays a role in the rate of early setting. The early compressive strength is determined by the hydration history of C_3S, whereas the contributions of C_3A and C_4AF are minor. Only during the later phases of the hardening of a cement paste does C_2S contribute to the development of strength.

In order to accelerate or retard the setting of cement, water-soluble admixtures may be used (see Section 5.3.1). Whilst many inorganic salts, such as calcium chloride, can act as accelerators, lead and zinc salts, borates and phosphates, as well as hydrocarbonic acids (e.g., citric acid or tartaric acid) and, most importantly, sugar derivatives such as raffinose or sucrose, will tend to retard the setting process. As the rate of setting can be determined by recording the heat evolved during hydration, a plot of Q_{max} (in $W\,kg^{-1}$) versus $1/t$ will yield a close to linear relationship.

For cations, the following series are obtained:

$Ca^{2+} > Ba^{2+} > Fe^{3+} >> NH_4^+ > Li^+ > Na^+ > Co^{2+} > Zn^{2+} > Pb^{2+}$

← **accelerators** | **retarders** →

This ranking pertains to chlorides and nitrates. For anions associated with calcium the ranking is:

$$Cl^-, Br^- > SCN^- > I^- > NO_3^- > ClO_4^-.$$

Whilst the rationale for these rankings is not clear, attempts to explain it have considered the diffusivities of the ions or the relative solubilities of the hydroxides. Indeed, lead and zinc salts may form dense, nonpermeable layers of hydroxide gels around the clinker grains (Figure 5.12c). On the other hand, the retarding action of organic compounds may be related to adsorption at the clinker mineral surfaces, to an inhibition of the nucleation and crystallization of the products of hydration, or to the precipitation of complex species together with colloidal C–S–H gel. Among aliphatic compounds, those with a high molecular mass and a larger number of carboxyl groups, as well as the presence of α- and β-hydroxycarboxyl groups capable of forming chelate structures with Ca ions, show the most strongly retarding activities. Among aromatic compounds, the number(s) and position(s) of the hydroxyl group(s) are important. For example, the retarding action increases from phenol to hydroquinone to catechol to pyrogallol.

As the rate of hydration of clinker grains appears to be determined by the rate of diffusion of water molecules through the colloidal C–S–H gel layer, the rate of reaction is a function of the permeability and cohesion of the layer.

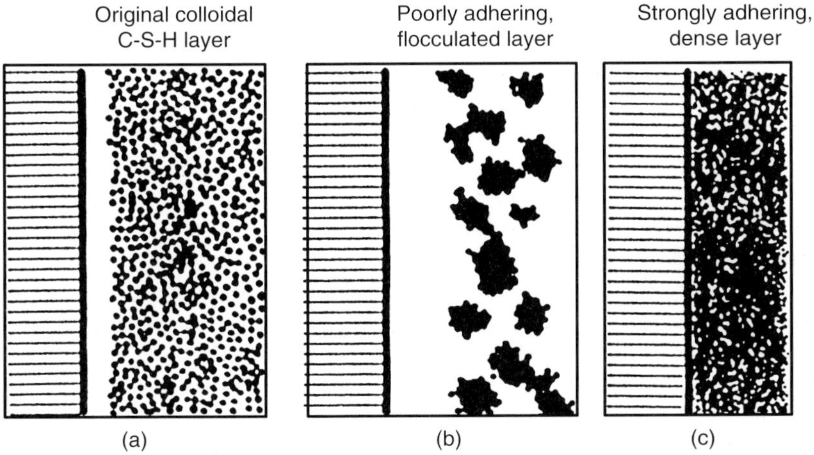

Figure 5.12 Models of changes of the structure of the colloidal C–S–H layer (a) in response to the action of accelerating (b) and retarding (c) species.

Transformation of the original colloidal layer (Figure 5.12a) towards an open, flocculated structure favors the acceleration of hydration (Figure 5.12b). Likewise, coagulation and condensation of the layer to form a dense, less-permeable and more strongly adhering coating will promote retardation (Figure 5.12c).

During the setting of a cement, a large amount of added water will be used to form hydrate phases such as C–S–H, portlandite, and AFt and AFm phases. Yet, even after complete hydration some free water will remain confined within the porous C–S–H gel and the capillary pore spaces. Notably, this occurs when surplus water is added so that the water/cement (W/C) ratio exceeds 0.6. A lack of water (W/C ratio <0.2) results in an incomplete hydration of clinker minerals. The optimum W/C ratio for Portland cement, leading to maximum strength, is $0.3 < W/C < 0.5$.

5.3
High-Performance Concretes (HPC)

When, in the production of concrete, OPC is mixed with an aggregate (e.g., gravel and/or sand), the resultant construction material demonstrates a high compressive strength but a rather low tensile strength. Thus, in order to obtain a material with a sufficient tensile strength, reinforcing iron bars and meshes are used to produce a ceramic–metal composite "concrete". Today, the increasing use of mineral and chemical admixtures has resulted in the creation of specialty cements and high-performance concretes (HPCs) with vastly improved performance and fields of application (Chatterjee, 2002).

5.3.1
Mineral and Chemical Admixtures

At this point, it is important to distinguish admixtures from additives, the latter being materials that are added during the manufacture of a cement. Additives include gypsum as a grinding aid and set controllers that function by inhibiting the hydration of calcium aluminate (see Section 5.2.4.2).

Admixtures, classified according to their function, are shown in Figure 5.13 (Hewlett and Young, 1983).

5.3.1.1 Mineral Admixtures

In addition to acting as water-proofing agents, solid mineral admixtures also: (i) increase the corrosion resistance of concrete against aggressive reagents such as chlorides and sulfates; (ii) reduce any undesired expansion through alkali–aggregate reactions; and (iii) reduce the corrosion of any steel reinforcements. These advantages are achieved: (i) by the creation of a finer pore structure that in turn reduces the permeability and thus the ion diffusion; and (ii) by reducing the amounts of calcium hydroxide present.

During the hydration of cement, large capillary pores will be filled with hydration products such that the number of large open pores will be reduced as a

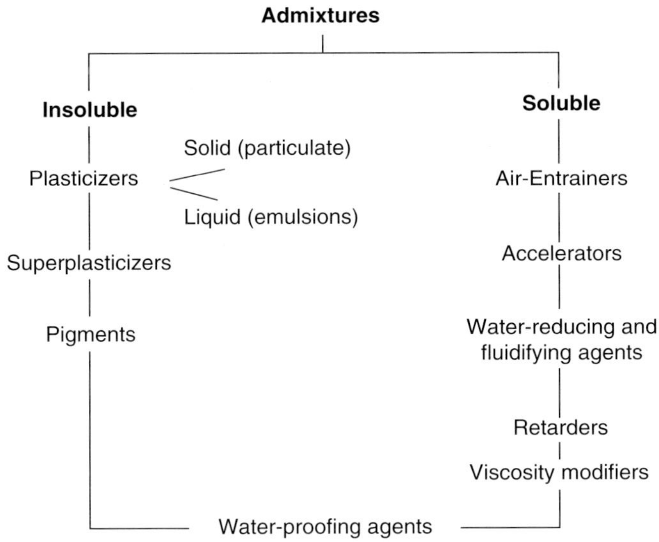

Figure 5.13 Functional classification of solid and liquid admixtures (after Hewlett and Young, 1983).

consequence. Simultaneously, the number of very fine gel pores will be increased. Mineral admixtures – and, in particular, silica fume – cause an additional shift of the pore distribution function towards smaller pore radii, as they react readily with free calcium hydroxide forming C–S–H (Huang and Feldman, 1985). For example, the addition of 20% silica fume to sulfate-resistant Portland cement (SRPC) (Bogue composition 54% C_3S, 25% C_2S, 2% C_3A, 11% C_4AF; Heimann and Hooton, 1986) yields, after 90 days, a cement with a compressive strength of 98 MPa, a permeability of $<3 \times 10^{-14}\,m\,s^{-1}$, a volume portion of pores >25 nm of 1%, and a portlandite content (measured using XRD) of zero (Burnett *et al.*, 1985). Such a virtually impermeable cement (DSP cement; see Section 5.3.2) would have the potential for use in a nuclear fuel waste disposal vault, and in particular for the grouting of rock fractures, the construction of concrete bulkheads, and the safe containment of nuclear fuel waste.

5.3.1.2 **Chemical Admixtures**

Air-Entraining Agents During the process of cement hydration, different pore populations will be formed, ranging from very small gel pores (~2 nm) to capillary pores (~500 nm) to rather coarse pores that originate from air entrapped during mixing. The coarse pores are usually filled with water that imparts a sensitivity against freeze–thaw cycling, as freezing water expands in volume and hence will destroy the microstructure. A cement with an increased frost resistance will contain many small air bubbles that act akin to a "safety valve," since water can

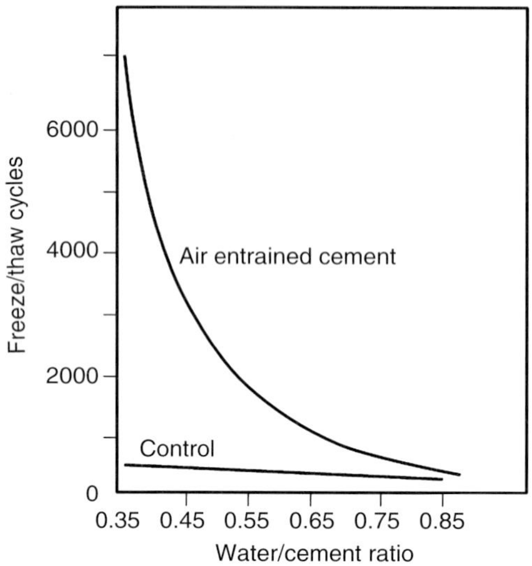

Figure 5.14 Number of freeze–thaw cycles required to cause a 25% mass loss (after Hewlett and Young, 1983).

move into these pores under the pressure gradient generated by expansion during freezing. In order to be effective, the air bubbles must be closely spaced apart (<0.4 mm, preferably 0.05 mm; Hewlett and Young, 1983) and be very small (0.02–1 mm).

Figure 5.14 shows the role of air entrainment in reducing frost damage to concrete. It is desirable to stabilize the air-entrainment pores by adding surfactants that will reduce the surface tension at the air–water interface, and hence stabilize and permanently fix the fine foam generated during high-velocity mixing. Such surface-active molecules possess polar groups such as carboxylate or sulfonate (Figure 5.15); these groups dissociate in water, with the negatively charged hydrophilic anions dissolving into the water surface and the positively charged hydrophobic moieties pointing away from the interface and forming a film that reduces the surface tension and stabilizes the air pores (Figure 5.15b). The longer the hydrophobic molecular chain, the more efficiently are the air pores stabilized. Examples of efficient surfactants include sodium dodecylsulfonate, $[CH_3–(CH_2)_{10}–CH_2–SO_3H]^- Na^+$, and sodium oleate, $[CH_3–(CH_2)_{16}–CH_2–COO]^- Na^+$.

As the surfactant molecules can also be adsorbed at the surfaces of cement (Figure 5.15c) and of small sand grains, the stabilized air bubbles can become attached to the grains so as to form "bubble bridges" (Figure 5.15d); these will increase the cohesion of the cement matrix, and simultaneously reduce their internal friction by acting like tiny "ball-bearings" (Hewlett and Young, 1983).

Water-Reducing Agents These compounds reduce the amount of water required to produce a workable cement–water paste and concrete, respectively. Hence,

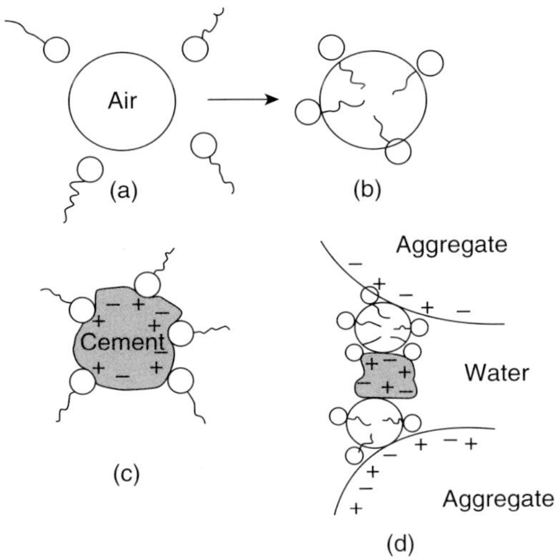

Figure 5.15 The action of air-entraining surfactants. The round hydrophilic "head" of the molecule is negatively charged, while the wiggly hydrophobic "tail" is positively charged. (a) Unstable air bubble; (b) Stable air bubble; (c) Neutralization of surface charges of cement by adsorption of surfactant molecules; (d) Formation of "bubble bridges" aggregate–air–cement–air–aggregate. Modified from Hewlett and Young (1983).

either the W/C ratio can be reduced (increase of strength) or the proportion of cement can be reduced at a maintained strength (reduction of cost). This effect is sometimes referred to as *plastification*. Typical plasticizers include lignosulfonate, hydroxycarboxylic acids (gluconic acid, citric acid, tartaric acid), or hydroxylated polymers such as chains of polysaccharides. There also exist a group of superplasticizers based on sulfonated melamine units polymerized by formaldehyde linkages, and a group based on sulfonated naphthalene that are similarly condensed with formaldehyde (Malhotra, 1981; Yoshioka *et al.*, 2002; Termkhajornkit and Nawa, 2004).

Although, like air-entraining agents, plasticizers and superplasticizers are surface-active compounds, they function in a completely different manner, owing to the large number of polar groups attached to their carbon backbone chains. Consequently, adsorption occurs at the polar backbone and not at the negatively charged end of the chain; as a result, the cement grains will acquire a higher negative charge that will, in turn, cause the individual particles to repel each other more strongly and be distributed uniformly in water. Following the addition of a superplasticizer, the zeta potential (see Section 2.4.2.4) changes from about +10 mV to −38 mV (Hattori, 1978). The adsorption of plasticizers and superplasticizers occurs predominantly at the C_3S and C_2S grains, whereas C_3A and C_4AF will be less affected.

5.3.2
DSP Cement

Many mechanical and corrosion challenges of concrete are related to the high porosity of ordinary cement and concrete, respectively. As detailed above, the ceramic microstructure of a hydraulic cement is the result of a chemical consolidation – that is, the hydration of clinker minerals accompanied by the formation of a colloidal C–S–H gel with a high surface area (10^2–$10^3\,m^2\,g^{-1}$) and a high gel porosity. A certain surplus of tempering water is required to guarantee the easy flow of concrete during casting; however, this surplus water will in time evaporate and leaves in its wake a macroscopic open porosity in the range of 25–35 vol%. Although tiny gel pores comprise the majority of the total porosity, the strength of the final cast product is controlled by the few larger pores present. An ideal concrete fabric should consist of densely packed grains of aggregate (sand, gravel) with a minimum amount of hydrated gel between the grains, and no macroscopic defects such as large pores. One way of achieving such a structure would involve a reduction of the tempering water content and the addition of superplasticizers (see above). Although this will increase the tensile (bending) strength from less than 10 MPa to about 20 MPa, it is still insufficient for the creation of a high-performance product, the tensile strength of which should be in the realm of 100 MPa.

In order to achieve better mechanical properties and a higher corrosion resistance, the maximum particle density must be combined with a minimum total porosity and, in particular, a virtual absence of large pores. There are several options towards achieving this goal, including:

1) Mechanical treatment by oscillatory or vibrational compaction, high-pressure densification, or shear mixing.
2) The addition of dispersing agents with a simultaneous reduction in the amount of tempering water.
3) The filling of large pores with sulfur ("sulfcrete") or resin.
4) The mixing of cements with variable grain size distributions.

In this case, method 4, in conjunction with method 2, results in the creation of so-called *DSP cements* (densified systems containing homogeneously arranged ultrafine particles). Basically, DSP cement consists of Portland cement with added microsilica (silica fume) and a superplasticizer. The microsilica particles, which are spherical and have diameters in the region of 0.1 μm, fill the pore spaces between the much larger clinker grains. This leads to the elimination of any large pores that would act as stress-concentration centers and crack-initiation points, and consequently the compressive strength of DSP cement is high (>150 MPa). The addition of reinforcing fibers can increase the compressive strength to values beyond 250 MPa. The DSP material is brittle and has a linear stress–strain curve up to the point of failure (Figure 5.16), which in turn leads to construction materials with a very favorable strength/density ratio. Owing to the dense structure, freezable water cannot penetrate into these materials and, for the same reason,

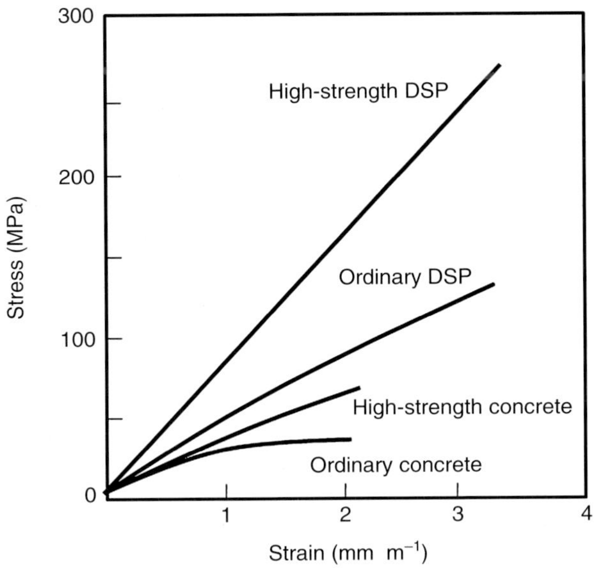

Figure 5.16 Stress–strain relations for ordinary concrete, high-strength concrete, and DSP materials.

the diffusion coefficient of chloride ions is more than one order of magnitude below that of OPC. With calcined bauxite as aggregate, an extremely high-abrasion-resistant composite can be created for use as flooring materials in chemical factories and parking garages.

5.3.3
Macro-Defect-Free (MDF) Cement

The addition of water-soluble organic polymers, such as hydroxyl(propylmethyl) cellulose or hydrolyzed poly(vinyl alcohol), to the tempering water/cement mix (ca. 100 parts cement/7 parts polymer/10 parts water), followed by intense shear mixing, permits the creation of an easily deformable dough. In this case, the volume fraction of the cement particles (ca. 0.65) is very close to that of the closest body-centered spherical packing (0.68). The polymer adheres strongly to the cement particles, with the hydrophilic regions pointing into the aqueous phase and causing crosslinking of the long chains (Figure 5.17a). The intense shear mixing causes the particles to separate (Figure 5.17b), such that the adsorbed polymer can act as an effective lubricant, allowing the particles to glide over each other and achieve a dense packing. The polymer strands remaining at the grain boundaries are able to crosslink, via Al ions (Bonapasta *et al.*, 2000), with the hydration products of the clinker minerals (Figure 5.17c). The hardened cement has a flexural strength of up to 150 MPa, a modulus of elasticity of 40–60 GPa (see

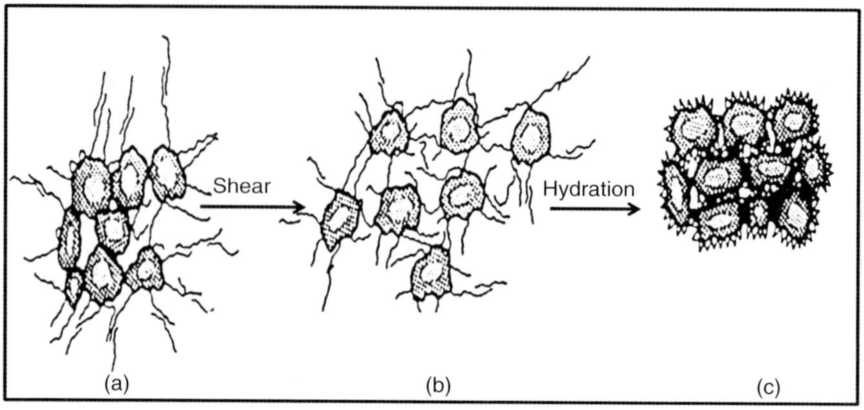

Figure 5.17 Action of polymer during formation of MDF cement dough. For an explanation, see the text.

Figure 5.1), and also demonstrates *R*-curve behavior (Mai *et al.*, 1990). The addition of Kevlar® or polyamide (Nylon 6®) fibers yields still higher values of flexural strength and fracture toughness, as well as imparting major improvements in impact strength (Di Maggio *et al.*, 1998). These fiber-reinforced composite materials rate almost as high as aluminum in terms of their flexural strength, fracture toughness, and tensile strength. Whilst the compositions of the cements may vary considerably, those with a high proportion of calcium aluminate (aluminate cement) are particularly suitable for high-strength MDF cements (Drábik *et al.*, 2002; Donatello *et al.*, 2009).

The intercalation of poly(acrylamide)–montmorillonite superabsorbing polymer composites (SAPCs) into an aluminate cement matrix leads to a noticeable improvement in the mechanical properties of the hardened cement paste. The addition of as little as 0.6% SAPCs increases the compressive strength from 36.1 to 44.4 MPa, the modulus of elasticity from 7.8 to 11.1 GPa, and the split tensile strength from 3.3 to 7.0 MPa, for a W/C ratio of 0.4 (Gao *et al.*, 1997).

5.3.4
Gas Concrete (Autoclaved Aerated Concrete; AAC)

To prepare an AAC, salt-free quartz sand, fly ash, and blast furnace slag are mixed with calcined limestone and/or Portland cement, and a paste of finely dispersed metallic aluminum particles. After milling, water is added and the slurry is cast into molds into which the steel reinforcement had previously been mounted. The reinforcement is protected against corrosion by a coating of bitumen or cement slurry. Within the molds, the aluminum particles react with the water at pH > 12 so as to generate hydrogen, and this is responsible for formation of the uniformly distributed macropores (0.5–1.5 mm diameter). During AAC production the release of hydrogen causes the raw mixture to double in volume.

Although, when the product is removed from the mold it has a solid texture, it is still soft enough to be cut into either blocks or panels that are placed into an autoclave chamber. During the steam pressure-hardening process (190 °C, 8–12 bar, 6–12 h), the quartz sand reacts with calcium hydroxide to form calcium silica hydrates such as 1.1 nm-tobermorite and gyrolith, and also xonotlite (see Section 5.2.5.2); these materials account for the AACs' great strength and other unique properties.

Since up to 80% of the AACs' volume consists of air, these materials have very low densities, which in turn accounts for their low compression strengths of, at most, 10 MPa (compared to about 100 MPa for ordinary concrete). Despite such a low compressive strength, AACs are used at construction sites worldwide for non-load-bearing applications. In particular, AACs are seen as an environment-friendly materials, for reasons that include:

1) Their production is virtually waste-free, as byproducts can easily be recycled.
2) They are produced in large industrial buildings, without toxic emissions or noise pollution; neither is there any emission of fine dusts or polluted waters. The hydrogen is replaced by air via the process of diffusion and rapidly dispersed so as not to constitute an occupational hazard.
3) The raw materials are cheap, and easy available; the quartz sand is usually mined in the immediate vicinity of the production site.
4) The primary energy consumption amounts to about 300 kWh for each cubic meter of AAC produced, which is much more economical than for similar materials. Owing to the high thermal insulation capacity of AACs, the ratio of primary energy consumption and heating energy consumption is particularly attractive.
5) The demolition of AACs at construction sites yields materials that can be either recycled to produce new AACs or deposited at normal disposal sites, without any environmental restrictions.
6) Demolished AACs can be used as a granular material for cat litter, as an oil and humidity barrier, and as insulating materials and water-retention agents in hydroculture.

5.4 Environmental Impact and Concrete Recycling

The manufacture of Portland cement can cause environmental problems at all stages of the process, including emissions of airborne pollution in the form of dust, gases, noise and vibration when operating machinery and during blasting in quarries, the consumption of large quantities of fuel during manufacture, the release of CO_2 from the raw materials during manufacture, and damage to countryside from quarrying. Equipment to reduce dust emissions during quarrying and manufacture of cement is widely used, whilst methods of trapping and separating exhaust gases are increasingly employed. Environmental protection also includes the recultivation of former quarried areas so as to recreate the original nature of the countryside.

5.4.1
CO_2 Emissions

Sources of CO_2 associated with Portland cement manufacture include: (i) the decarbonization of limestone; (ii) the exhausts of kiln fuel combustion; and (iii) the exhausts of the vehicles used in cement plants and distribution. Of these sources, the first produces a minimum of about 0.47 kg CO_2 kg^{-1} cement, whilst production via the second source varies with the plant efficiency. For example, an efficient precalciner plant will produce 0.24 kg CO_2 kg^{-1} cement, while a low-efficiency wet process may produce up to 0.65 kg CO_2 kg^{-1}. The production of CO_2 via the third source is almost insignificant (0.002–0.005 kg CO_2 kg^{-1} cement). Hence, the typical total CO_2 footprint is around 0.80 kg CO_2 kg^{-1} finished cement. This leaves aside the CO_2 associated with electric power consumption, which varies according to the local generation type and efficiency. Typical electrical energy consumption is of the order of 90–150 kWh per metric ton of cement; this is equivalent to 0.09–0.15 kg CO_2 kg^{-1} finished cement if the electricity is coal-generated. All of this amounts to about 7% of the total CO_2 generated worldwide (Malhotra, 1999).

Whilst this situation is unlikely to be affected by any enhancement in efficiency, the replacement of some cements with supplementary cementing materials not associated with CO_2 emissions may substantially reduce the problem.

5.4.2
NO_x Emissions

The emissions of oxides of nitrogen (NO_x) are associated with the burning of gasoline, coal, or other fossil fuels. *Ozone* is formed when NO_x and volatile organic compounds (VOCs) derived from sources ranging from industrial solvents to volatile resins in trees are mixed in sunlight. At ground level, ozone may cause a number of health problems, including asthma attacks, sore throat, coughing, and other health difficulties. In addition, nitrous oxide, CO_2 and methane represent the most important *"greenhouse gases."*

The NO_x emissions from cement kilns range from 1.5 to 9.5 kg ton^{-1} of clinker produced. A reduction in NO_x production is normally achieved by lowering the burning temperature, or by injecting ammonium compounds into the high-temperature exhaust stream (Klein and Rose, 1998). However, despite such action seeming to be effective, when used to reduce NO_x emissions from coal-fired electricity-generating stations it adversely affects the quality of the fly ash, which must first be treated to remove any unburned coal and ammonia gas before being used in concrete mixtures (Bremner, 2001).

5.4.3
Particulate Emissions and Visual Pollution

Levels of *particulate emissions* from exhaust gases range from 0.3 to 1.0 kg ton^{-1}, with much of this kiln dust being first collected into fabric filter baghouses and then reintroduced into the kiln feed. The dust is normally very rich in sodium and potassium chlorides that have vaporization temperatures of only 883 °C and 774 °C, respectively. In the past, before any concerted efforts were made to capture the particulate emissions, the sodium and potassium plumes from cement plant chimneys settled over the countryside, helping to combat acid rain and also acting as fertilizers in the soil. Today, however, the dust is mainly carried out in the clinker stream, where it creates problems with alkali aggregate reactions (Bremner, 2001).

Visual pollution resulting from unsightly quarries used to acquire the raw materials for cement production, or for obtaining sand and gravel, can be sculptured to meet the natural topography and, when abandoned, can be recultivated. Unfortunately, however, most quarries have a very long working life, and attempts to sculpture the topography for visual effect may be counter-productive to the efficiency of the quarrying process. The most effective end use might be for educational or recreational purposes, with special attention being paid to public safety (Bremner, 2001).

5.4.4
Water Pollution

On average, each day, each transport cement truck (in the U.S., Ready Mix) will return to the plant about 0.5 m^3 of unused cement; moreover, even when this concrete has been discharged, there is still about 300 kg of solids (cement, sand, and stones) that must be washed out of the truck with about 1000 liters of water. In the past, the returned concrete and solids were simply dumped into a pit at the job site or at the plant. However, considering that this may represent 2–4% of the total concrete produced, it is now considered too valuable to waste and can be recycled or reclaimed as sand and gravel. In order to reclaim the sand and gravel, a "reclaimer" is used; this involves adding water to the returned concrete, agitating the mixture, followed by a wet screening phase.

In the past, the cement–water slurry from the reclaimer, the wash-out water, the water used to clean the outside of the truck, plus any stormwater, was usually directed first into somewhat inefficient settling basins, and then into a local water course. Today, in Canada and the U.S. the "process water" from settling ponds or from a reclaimer can be used as the tempering water for subsequent concrete mixtures. Recent developments have enabled the concrete retained and any concrete clinging to the inside of the truck drum, plus any wash-out water, to be stabilized overnight or over the weekend by the addition of a hydration-stabilizing admixture. This stabilized concrete, with an accelerating admixture, may then be used as part of the proportions for the next load such that there is essentially no water pollution (Ruhlin, 2000).

5.4.5 Environmental Benefits

Today, cement production sites are often also used for the disposal or processing of toxic wastes, and hence are of considerable benefit to the environment. Due to their typically high internal temperatures, combined with the oxidizing atmospheres and long residence times, cement kilns have been used as a viable processing option for various types of waste streams. Indeed, the waste streams may themselves contain combustible materials that can be substituted as part of the fossil fuel normally used in the process. Waste materials used in cement kilns as a fuel supplement include car and truck tires, waste solvents and lubricants, hazardous waste (PCBs, dioxins), animal cadavers, waste plastics, sewage sludge, rice hulls, and other agricultural waste.

Portland cement manufacture also has the potential to remove industrial byproducts from the waste stream, effectively sequestering some environmentally damaging wastes. These include slag, fly ash from coal-fired power plants, silica fume from steel mills, and synthetic gypsum from the desulfurization of fuel gases (FGD gypsum). On the other hand, these materials are increasingly utilized as mineral admixtures to reduce the amount of clinker needed, and hence reduce the CO_2 footprint of cement production by a substantial margin.

In spite of these environmental advantages, significant pollution problems may result from using toxic waste as fuels in cement kilns. Since the quantity of toxic fuel use is often maintained below local regulatory requirements by blending, most of these kilns will employ significantly less stringent air pollution control devices than other industrial or waste-burning incinerators. This may lead to significant increases in local air pollution and has, on occasion, caused agricultural produce from surrounding farms to become contaminated with high levels of heavy metals, PCBs, dioxins, and other toxic compounds.

A second area of the application of cements for environmental purposes is that of the stabilization/immobilization of liquid toxic industrial waste, and in particular that of hazardous heavy-metal waste streams containing chromium, vanadium, cadmium, and other metals. As noted in Section 5.2.4.2, both Cr and V can enter the inter-chain spaces of ettringite to replace SO_4^{2-} groups (Buhlert and Kuzel, 1971), but they can also be adsorbed by C–S–H phases. In addition, Cr^{3+} has been found to replace Si^{4+} ions in C–S–H (Ivey *et al.*, 1990). Experimental evidence exists that the retention and release, respectively, of heavy metals is controlled by the solution chemistry of aluminum, and hence the composition of the cement matrix (Heimann *et al.*, 1992). Both, Cr^{6+} and V^{5+} exhibit strong mobility responses to varying levels of Al_2O_3 that in turn are related to interactions associated with varying pH levels in leachates. The upper portion of Figure 5.18 shows the response surface of the normalized mass loss (NML) of Cr^{6+} as a function of the cement composition (expressed in % Al_2O_3) and the concentration of the Cr^{6+} waste load (in mg l^{-1}). The isopleths of NML are shown in the lower part of Figure 5.18. The strongly parabolic relationship points to a strong matrix effect that is manifest in the control of the leachability of the Cr by the level of Al in the cement. The

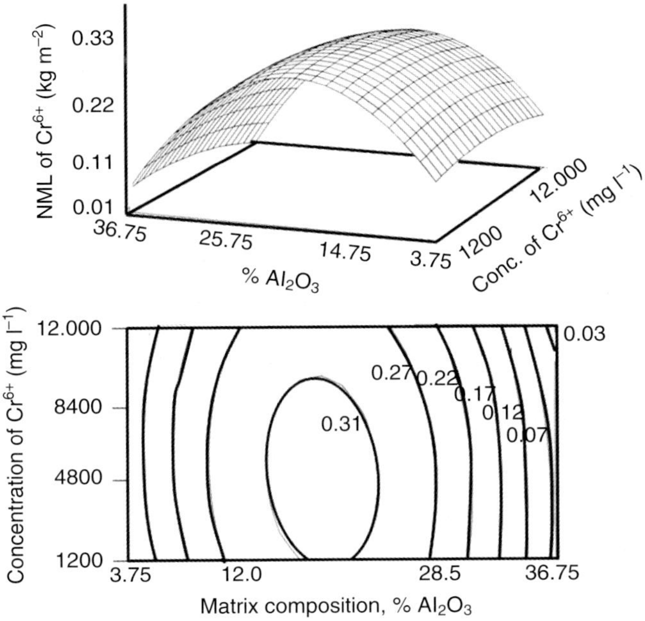

Figure 5.18 Relationship between normalized mass loss (NML) of Cr^{6+} (in kg m^{-2}) and concentration of Cr^{6+} in the cement mix (in mg l^{-1}) and composition of matrix cement expressed by % Al_2O_3. Upper diagram: Parabolic response surface. Lower diagram: isopleths of NML(Cr^{6+}) (Heimann *et al.*, 1992).

cements investigated were OPC (Canada Cement Lafarge Ltd) with 3.75% Al_2O_3, and aluminous cement (Ciment Fondu T1; Canada Cement Lafarge Ltd) with 36.75% Al_2O_3; intermediate compositions were obtained by mixing sulfate-resistant Portland cement (Kalicrete, type 50; Canada Cement Lafarge Ltd) with 7.11% Al_2O_3, as well as aluminous cement with varying amounts of fly ash (Western Canada Fly Ash). The normalized mass loss was calculated as $NML_i = (c_i \cdot V/c_i^0)/SA$ where c_i is the concentration of element i in the solution (in mg l^{-1}), c_i^0 is the concentration of element in the solid cement matrix (in mg kg^{-1}), V is the volume of leach solution (in liters), and SA is the surface area of the solid (in m^2) (Mendel, 1983).

Studies of the immobilization behavior of Cd, Cr, and V in the cement matrices shown above revealed that: (i) Cd is effectively retained in all cements and shows no sensitivity to the leachate pH; and (ii) both NMLs of Cr and V are maximized at intermediate levels of alumina in the cement matrix and minimized at high and low alumina contents, presumably related to the amphoteric nature of Al (Heimann *et al.*, 1992; Ivey *et al.*, 1990). Multicomponent systems tests of the interaction of cement and radioactive waste forms (used UO_2 fuel, fuel recycle waste glass) have provided evidence that actinides (plutonium 239 + 240, americium 241, curium 244) would be efficiently adsorbed onto cement (Heimann, 1988a), presumably related to the presence of apatite structures (Beall and Allard, 1982).

5.5 Future Developments and Outlook

Future developments in the field of classic cement production and utilization may be characterized by:

- The development of computer-aided mathematical models of cement and concrete properties and performance (e.g. Kadaschewitsch *et al.*, 2005).
- The development of computer-aided expert systems to assess specific properties and areas of novel application.
- A linking of knowledge bases in other disciplines of materials science (metallurgy, ceramics) with those of cement chemistry, so as to apply the hydraulic properties of cement to new fields of application.
- An improvement in the understanding of factors that influence the performance of cement, the relationship between process variables and cement performance, and an optimization of cement production in terms of energy base, product properties, and energy economy.
- An improvement of the mechanical properties of concrete, cement–polymer composite materials, and MDF cements.
- Investigations into the role of hot and warm pressing, and the addition of fly ash, blast furnace slags, and microsilica.
- An improvement of the long-term survival probability of cement and concrete materials, with the goal of reaching Weibull moduli >10.

Today, many of these aspects are being investigated under the umbrella of Nanoscale Research on Cements and Concrete (Nanocem, 2009), a consortium of European academic and industrial partners, founded in 2004.

5.5.1 Reduction of the Emission of Greenhouse Gases

The environmental concerns voiced in Section 5.4 call urgently for a radical paradigm change in selecting cement raw materials. This will guarantee a sustainable novel technology with drastically reduced emissions of greenhouse gases, reductions in energy expenditure, and prospects of closing the materials' cycle by an increased recycling of demolished concrete structures.

Hence, the thrust of innovation for the future is to reduce greenhouse gases such as CO_2 and NO_x by modification of the chemistry of cement by replacing clinker minerals by fly ash or blast furnace slag, by the use of wastes, and by adopting more energy-efficient processes (see Section 5.4.5). The use of composite cements and concretes, and the addition of pozzolans, will also help to achieve the goal of reducing CO_2 emissions.

Indeed, the easiest and most effective way to reduce greenhouse gas production would be to increase the use of such silica-rich byproducts such as fly ash, slag and silica fume, thus reducing the amount of cement used per cubic meter of concrete. Concrete generates about 7% of the total CO_2 emitted worldwide. In 2006, a minimum of 2.2 billion metric tons of cement was produced. If a ton of cement produces 0.8 tons of CO_2, and if only 20% of the cement can be replaced with slag or fly ash, then the CO_2 reduction would amount to 350 million tons per year worldwide. Over the past decade, the average annual increase in CO_2 emissions was 1.3% or nearly 300 million tons a year worldwide. The cement industry alone could easily offset global warming, and at the same time enhance the properties of the concrete produced.

5.5.2
Recycling of Concrete

The demolition and recycling of concrete are high on the agenda of developing sustainable technologies. The per capita total annual consumption of concrete in Germany is presently about 2300 kg (ca. 1 m^3), and concrete waste generated by selective demolition amounts to about 20% of this value. At present, waste concrete is comminuted by attrition milling and used as low-quality aggregates for concrete or as filler materials at construction sites. However, the costs of milling are high, owing to high wear of the machinery and high energy expenditure; moreover, the tolerance of the milling equipment for steel reinforcement is low and the particulate pollution due to dust is considerable. Hence, there is a need for improved technologies that would result in the efficient separation of concrete into its constituents of gravel, sand and cement flour, as well as the reinforcements. At the Forschungszentrum Karlsruhe (FZK), a pilot plant is currently in operation that utilizes the cavitation introduced by a high-voltage pulsed power to selectively fragment concrete that has been submerged in water.

FRANKA is a semi-industrial prototype plant for the selective decomposition of up to 1 t h^{-1} of pre-crushed materials such as concrete, rocks, or other polycrystalline materials. Its main components are a high-voltage pulse generator that delivers a 300 kV, 15 kA pulse with a 12 Hz repetition rate, and a water-filled reaction chamber. In the reaction chamber the materials to be treated are exposed for a well-defined time interval to the high-voltage discharges. In addition, a transport system continuously feeds the materials to be treated into and out of the reaction chamber. There are additional components for the separation of crushed materials from water, material classification, and for water cleaning and reprocessing (Hoppe *et al.*, 2002). A schematic of an earlier version of the prototype plant is shown in Figure 5.19.

The physical basis of electrodynamic fragmentation is the initiation of a high-voltage discharge through a dielectric solid material to form a thin plasma channel with high energy density. The deposition within a few microseconds of an energy of 10–100 J cm^{-1} leads to temperatures of about 10 000°C and pressures up to

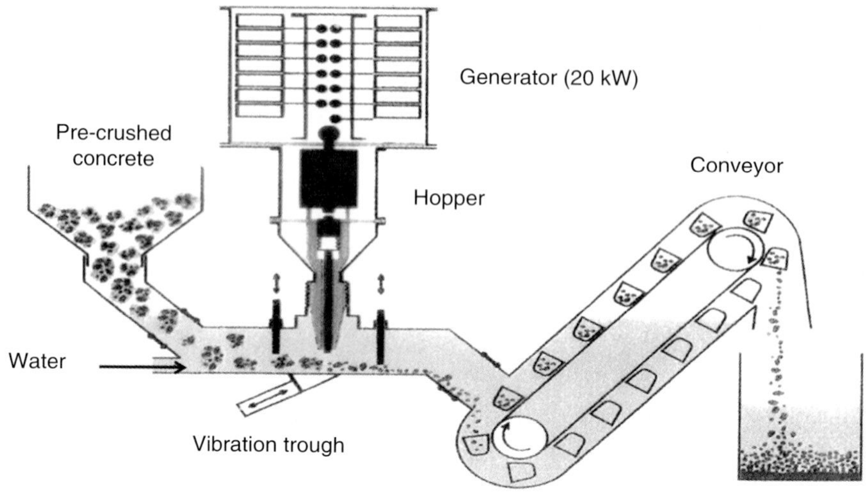

Figure 5.19 A schematic representation of the FRANKA prototype plant for the selective fragmentation of concrete. Illustration courtesy of Dr P. Stemmermann, FZK.

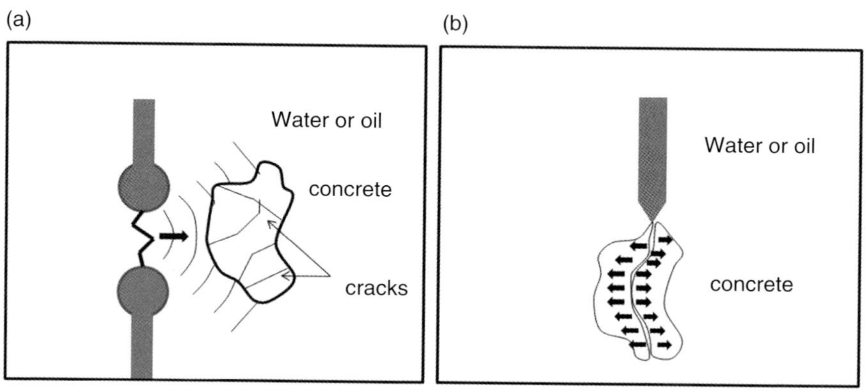

Figure 5.20 Comparison between destruction by compression (a) and explosion (b). The former delivers an energy density of 6–10 kJ per pulse, but the latter only 0.4 kJ per pulse. Illustration courtesy of Dr P. Stemmermann, FZK.

10^{10} Pa. The channel, which initially is only 10–50 μm wide, expands and emits a pressure pulse into the surrounding material. Through this pressure pulse, and through radially propagating cracks, inclusions can be separated from the matrix and the material can be ripped apart at the material interfaces (Figure 5.20). In order to force the high-voltage discharge through the solid, the latter must be embedded in a dielectric liquid of a higher electric breakdown strength. For a steep voltage ramp, water is a suitable liquid since under these conditions it has a higher breakdown field strength than most solids.

From an economics standpoint, the cement flour generated by the selective fragmentation of concrete is chemically and physically almost identical to raw cement flour, and hence can be used to replace the latter by approximately 50%. Accordingly, substantial savings of raw materials (limestone, clay) and energy, the preservation of the natural environment (disposal sites, resources), and the reduction of CO_2 emissions will add to the benefits of intelligent concrete recycling.

References

Aitkin, P.C. (2000) Cements of yesterday and today: concrete of tomorrow. *Cem. Concr. Res.*, **30** (9), 1349–1359.

Beall, G.W. and Allard, B. (1982) Chemical aspects governing the choice of backfill material for nuclear waste repositories. *Nucl. Technol.*, **59** (3), 405–408.

Bensted, J. and Barnes, P. (2002) *Structure and Performance of Cements*, 2nd edn, Spon Press, London, ISBN: 0-419-23330-X, 565 pp.

Bogue, R.H. (1929) Calculation of the compounds in Portland cement. *Ind. Eng. Chem. Anal. Ed.*, **1**, 192–197.

Bonapasta, A.A., Buda, F., and Colombet, P. (2000) Cross-linking of poly(vinyl alcohol) chains by Al ions in macro-defect-free cement: a theoretical study. *Chem. Mater.*, **12** (3), 738–743.

Bremner, T.W. (2001) Environmental aspects of concrete: problems and solutions. Proceedings 1st All-Russian Conference on Concrete and Reinforced Concrete, 9–14 September, 2001.

Buhlert, R. and Kuzel, H.J. (1971) Über den Einbau von Cr^{3+} und Fe^{3+} in Ettringit. *Zem. Kalk Gips*, **24**, 83–85.

Burnett, N.C., Hooton, R.D., Heimann, R.B., and Onofrei, M. (1985) The development of durable cementitious materials for use in a nuclear fuel waste disposal facility. *Mater. Res. Soc. Symp. Proc.*, **50**, 461–468.

Chatterjee, A.K. (2002) Special cements, in *Structure and Performance of Cements*, 2nd edn (eds J. Bensted and P. Barnes), Spon Press, London, pp. 186–236.

Di Maggio, R., Franchini, M., Guerrini, G., Migliaresi, C., and Poli, S. (1998) MDF cement compositions with improved impact strength. US Patent 5,814,146 (29 September 1998).

Donatello, S., Tyres, M., and Cheeseman, C.R. (2009) Recent developments in macro-defect-free (MDF) cements. *Constr. Build. Mater.*, **23** (5), 1761–1767.

Drábik, M., Mojumdar, S.C., and Slade, R.C.T. (2002) Prospects of novel macro-defect-free cements for the new millennium. *Ceramic-Silikáty*, **46** (2), 68–73.

Gani, M.S.J. (1997) *Cement and Concrete*, Chapman & Hall.

Gao, D., Heimann, R.B., and Alexander, S.D.B. (1997) Box-Behnken design applied to study the strengthening of aluminate concrete modified by a superabsorbent polymer/clay composite. *Adv. Cement Res.*, **9** (35), 93–97.

Gartner, E.M., Young, J.F., Damidot, D.A., and Jaws, I. (2002) Hydration of Portland cement, in *Structure and Performance of Cements*, 2nd edn (eds J. Bensted and P. Barnes), Spon Press, London, pp. 57–113.

Goswami, G., Padhy, B.P., and Panda, J.D. (1989) Thermal analysis of spurrite from a rotary cement kiln. *J. Therm. Anal.*, **35** (4), 1129–1136.

Hattori, K. (1978) Experience with mighty superplasticizer in Japan. Proceedings, International Symposium on Superplasticizers in Concrete, Ottawa, Canada, 29–31 May, vol. 1, 48–86.

Heimann, R.B. (1988a) Interaction of cement and radioactive waste forms in multicomponent system tests at 200 °C. Part 1: Leaching and sorption of cesium, strontium and actinides. *Cem. Concr. Res.*, **18**, 389–400.

Heimann, R.B. (1988b) Interaction of cement and radioactive waste forms in multicomponent system tests at 200 °C.

Part 2: mineralogical changes of cement. *Cem. Concr. Res.*, **18**, 554–560.

Heimann, R.B. and Hooton, R.D. (1986) Mineralogical changes of various cement formulations during reaction with groundwater in the presence of Ca- and Na-bentonite at 150 °C. *Can. Mineral.*, **24**, 289–302.

Heimann, R.B., Conrad, D., Florence, L.Z., Neuwirth, M., Ivey, D.G., Mikula, R.J., and Lam, W.W. (1992) Leaching of simulated heavy metal waste stabilized/solidified in different cement matrices. *J. Hazard. Mater.*, **31**, 38–57.

Hewlett, P.C. and Young, J.F. (1983) Physico-chemical interactions between chemical admixtures and Portland cement. *J. Mater. Educ.*, **9** (4), 389–435.

Hewlett, P.C. (2004) *Lea's Chemistry of Cement and Concrete*, 4th edn, Elsevier. ISBN: 10-07-506-6256-5.

Hoppe, P., Singer, J., Bluhm, H., Frey, W., Giehse, H., Massier, H., Edinger, W., and Schweike, U. (2002) FRANKA-stein: design, operation and industrial application. Proceedings 25th International Power Modulator Symposium and High Voltage Workshop, Hollywood, CA, 30 June–3 July 2002, pp. 559–562, ISSN 1076-8467.

Huang, C.Y. and Feldman, R.F. (1985) Influence of silica fume on the microstructural development in cement pastes and mortars. *Cem. Concr. Res.*, **15**, 285–289.

Ivey, D.G., Heimann, R.B., Neuwirth, M., Shumborski, S., Conrad, D., Mikula, R.J., and Lam, W.W. (1990) Electron microscopy of heavy metal waste in cement matrices. *J. Mater. Sci.*, **25**, 5055–5062.

Kadaschewitsch, I. and Sloyan, D. (2005) Aerated autoclaved concrete: stochastic structure model and elastic properties. *Proc. Appl. Math. Mech. (PAMM)*, **5** (1), 419–420.

Klein, M., and Rose, D. (1998) Development of CME National Emission Guidelines for cement kilns, in *CANMET/ACZ International Symposium on Sustainable Development of the Cement and Concrete Industry* (ed. V.M. Malhotra), ACI Special Publications, Ottawa, pp. 16–30.

Kurdowski, W. and Sobón, M. (1999) Mineral compositions of build-up in cement kiln preheater. *J. Therm. Anal. Calorim.*, **55** (3), 1021–1029.

Le Chatelier, H.L. (1904) *Recherches Expérimentales sur les Constituants des Mortiers Hydrauliques*, 2nd edn, Vve C. Dunod, Paris.

Lea, M. and Parker, F.W. (1934) Investigations on a portion of the quaternary system $CaO\text{-}Al_2O_3\text{-}SiO_2\text{-}Fe_2O_3$. *Philos. Trans. R. Soc. London*, **234** (731), 1–41.

Locher, F.W. (2006) *Cement: Principles of Production and Use*, Verlag Bau + Technik GmbH, Düsseldorf, ISBN: 3-7640-0420-7.

Mai, Y.W., Barakat, B., Cotterell, B., and Swain, M.V. (1990) R-curve behaviour in a macro-defect-free cement paste. *Philos. Mag. A*, **62** (3), 347–361.

Malhotra, V.M. (1981) Superplasticisers: their effect on fresh and hardened concrete. *Concr. Int.*, **3**, 66–81.

Malhotra, V.M. (1999) Making concrete "greener" with fly ash. *Concr. Int.*, **21** (5), 61–66.

Megaw, H.D. and Kelsey, Ch. (1956) Crystal structure of tobermorite. *Nature*, **177**, 390–391.

Mendel, J.E. (1983) *Nuclear Waste Materials Handbook of Test Methods*, DOE/TIC-11400, Pacific Northwest Laboratories, Richland, WA, pp. 1–11.

Merlino, S., Bonaccorsi, E., and Armbruster, T. (2001) The real structure of tobermorile 11 Å: Normal and anomalous forms, OD character and polytypic modifications. *Eur. J. Mineral.*, **13**, 577–590.

Mindness, S. (1983) Concrete materials. *J. Educ. Modules Mater. Sci. Eng.*, **5** (6), 983–1046.

Nanocem (2009) Nanoscale Research on Cement and Concrete, www.nanocem.org (accessed 29 September 2009).

Netherlands Environmental Assessment Agency (2007) China now no. 1 in CO_2 emissions; USA in second position.

Pöllmann, H. (2002) Composition of cement phases, in *Structure and Performance of Cements*, 2nd edn (eds J. Bensted and P. Barnes), Spon Press, London, pp. 25–56.

Pöllmann, H. (2007) *Immobilisierung von Schadstoffen durch Speichermineralbildung*, Shaker-Verlag, Aachen, ISBN: 978-3-8322-6377-5, 428 pp.

Roy, D.M. (1981) Portland cement: constitution and processing. Part I: Cement manufacture. *J. Educ. Modules Mater. Sci. Eng.*, **3** (4), 625–647.

Roy, D.M. (1983) Portland cement: constitution and processing. Part II: Cement constitution and kiln reactions. *J. Educ. Modules Mater. Sci. Eng.*, **5** (3), 471–490.

Ruhlin, D. (2000) Compliance matters – stormwater permitting for concrete producers. *Concrete Products Magazine*, May, 28–34.

Strunz, H. and Nickel, E.H. (2001) *Strunz Mineralogical Tables. Chemical- Structural Mineral Classification System*, E. Schweizerbart, Stuttgart, ISBN: 3-510-65188-X.

Taylor, H.F.W. (1981) Portland cement: hydration products. *J. Educ. Modules Mater. Sci. Eng.*, **3** (3), 429–449.

Taylor, H.F.W. (1997) *Cement Chemistry*, 2nd edn, Academic Press.

Termkhajornkit, P. and Nawa, T. (2004) The fluidity of fly ash-cement paste containing naphthalene sulfonate superplasticizer. *Cem. Concr. Res.*, **34** (6), 1017–1024.

Yoshioka, K., Tazawa, E., Kawai, K., and Enokata, T. (2002) Adsorption characteristics of superplasticizers on cement component minerals. *Cem. Concr. Res.*, **32** (10), 1507–1513.

Young, J.F. (1981) Hydration of Portland cement. *J. Educ. Modules Mater. Sci. Eng.*, **3** (3), 403–428.

6
Introduction to Advanced Ceramics

6.1
General Properties of Advanced Ceramics

According to a joint study conducted by VAMAS (Versailles Project on Advanced Materials and Standards; Early and Rook, 2004), the ASTM Committee C-28 on Advanced Ceramics, and the USACA (United States Advanced Ceramics Association) the term "advanced ceramics" (in Japan: fine ceramics) refers to ceramic materials that are:

- highly specialized by exploiting unique electric, magnetic, optical, mechanical, biological, and environmental properties;
- performing well under extreme conditions such as high temperature, high pressure, high stress, high radiation, and high corrosive exposure;
- predominantly inorganic–nonmetallic;
- relatively expensive, with properties and failure mechanisms not yet fully understood;
- capable of solving current manufacturing and use problems;
- high value-added products owing to their sophisticated processing technology;
- not presently profitable in terms of return of investment (ROI), but which offer great promise for the future; and
- positioned at the beginning of the development cycle and not yet widely used with respect to their potential.

In particular, advanced ceramics excel in terms of their high-temperature stability (up to 2500 °C and beyond), high hardness, high corrosion resistance, low thermal expansion, and a variety of electrical properties ranging from insulators to semiconductors to highly conductive materials (Table 6.1). Although the fracture toughness and impact resistance of these materials are low compared to metals and some engineering plastics, modern processing technologies – and, in particular, the development of particulate-reinforced advanced ceramics – have led to the creation of transformation-toughened ceramics such as partially stabilized zirconia and modern nanostructured ceramics which provide materials that perform well, even under adverse mechanical conditions.

Classic and Advanced Ceramics: From Fundamentals to Applications. Robert B. Heimann

ISBN: 978-3-527-32517-7

Table 6.1 Comparison of important properties of metals, engineering plastics, and advanced ceramics.

Property	Metals	Engineering plastics	Advanced ceramics
Maximum temperature for continuous use (°C)			
Typical	1000	250	1200
Maximum	1500	350	2500
Hardness	Medium–high	Low–medium	High
Toughness			
Flexibility	Medium–high	High	Low
Impact resistance	High	Medium–high	Low
Corrosion resistance	Low–medium	Medium	High
Coefficient of thermal expansion	High	Medium	Low
Electrical properties	Conductive	Insulative to conductive (with fillers)	Insulative to conductive
Density	High	Low	Medium

One useful classification scheme divides advanced ceramics into *structural* and *functional* categories:

- *Structural advanced ceramics:* These are used for applications where a component of an engineering system is subjected to high mechanical, tribological, thermal, or chemical loads. Typical structural ceramics are alumina, partially and fully stabilized zirconia, cordierite, mullite, spinel, silicon nitride and SiAlONs, silicon carbide, boron nitride, titanium nitride, and titanium boride. The performance-application diagram (Figure 6.1) for zirconia, silicon carbide, silicon nitride and alumina shows that zirconia ceramics excel in high-temperature applications such as thermal barrier coatings where high density, high bending strength, high thermal expansion and low thermal conductivity are required. Silicon carbide ceramics possess high thermal conductivity and high microhardness, but are deficient in strength and thermal shock resistance. Hence, the area of application of such ceramics is in heating elements for furnaces operating up to 1600 °C in ambient atmospheres. Silicon nitride on the other hand shows a high thermal shock resistance, sufficient strength and hardness, and intermediate values of density and thermal conductivity which suggest applications in automotive engines

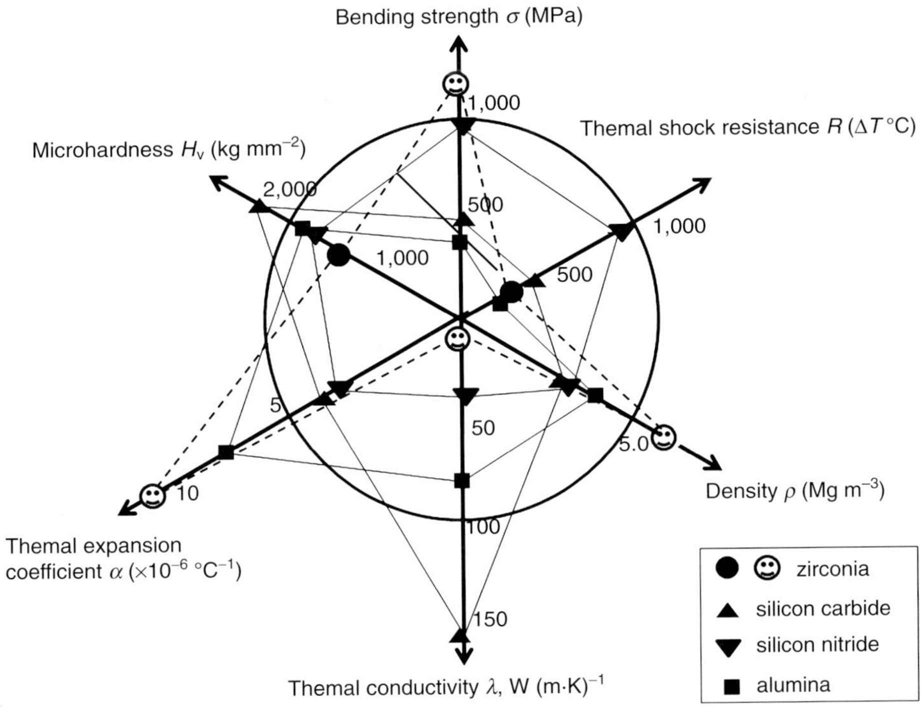

Figure 6.1 Performance–application diagram of four typical advanced structural ceramics: zirconia; silicon carbide; silicon nitride; and alumina. Owing to the dependency of the properties on processing conditions, the values given are only tentative. Zirconia ceramics are characterized by a combination of useful properties unmatched by other oxide and nonoxide ceramics.

and heat exchangers, as well as for ceramic cutting tools for the high-speed machining of hard metals. Lastly, alumina is hard and shows a rather high thermal expansion, but performs badly in terms of bending strength and thermal shock resistance. Frequently, design engineers supplement performance–application diagrams with diagrams that match the required mechanical stress profile with the property profile of an envisaged material, as shown in Figure 6.2. This shows the required mechanical stress profile for drilling mud pumps (chain line) juxtaposed to the property profile of Cr–Ni steel (solid line). As long as the former stays within the contour line of the latter, the steel material is sufficient to cope with the mechanical stresses that occur during operation of the mud pump. However, the requirements for superior abrasion and erosion resistance are not met by the austenitic steel, and this requires either a different base material or a resilient ceramic coating, such as tungsten carbide or chromium carbide.

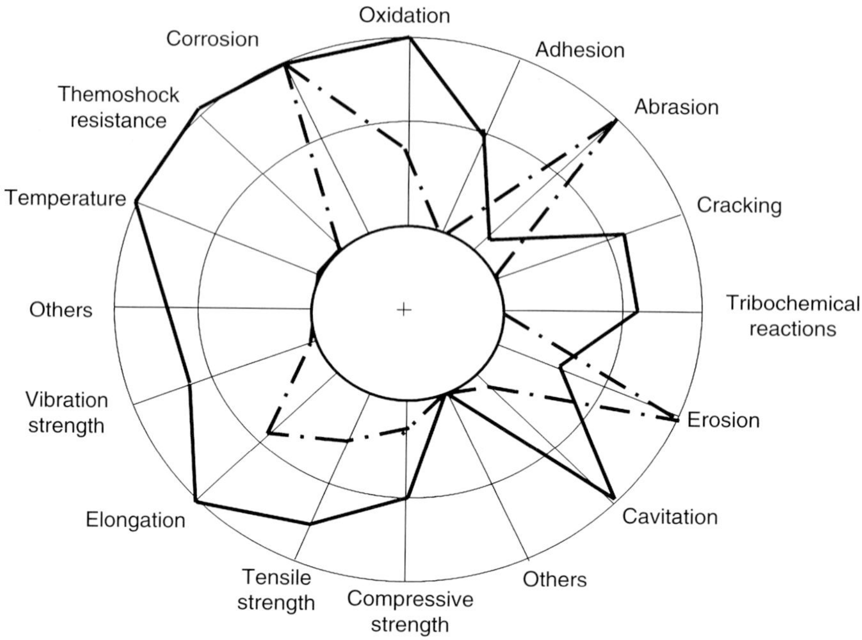

Figure 6.2 Matching the stress profile of a drilling mud pump (chain line) and the wear property profile of corrosion-resistant austenitic Cr–Ni steel (solid line).

- *Functional advanced ceramics:* These, in contrast to structural ceramics, utilize microstructural effects localized within the volume, at grain boundaries or at the surfaces of conducting or nonconducting ceramics. Such effects encompass semiconducting, varistor, piezoelectric, pyroelectric, ferroelectric and superconducting properties. Novel ceramic technologies for the automotive industry include knock and oxygen sensors, exhaust gas catalysts and fuel cells, as well as future ceramic gas turbines and adiabatic turbo-compound diesel engines (Okada, 2008).
- *Bioceramics:* Bioceramic materials such as alumina and zirconia exploit exceptional structural stability in a highly corrosive body environment and hence biological inertness, whereas hydroxyapatite and tricalcium phosphate provide functional osseoinductive and osseoconductive properties (Heimann, 2007).

Table 6.2 shows some application fields, typical components and some advanced ceramics solutions to fulfill the requirements for optimum component performance. Higher-performance ceramics and their future role have been reviewed by Nicholson (1996).

In providing a brief summary of the complete field of modern advanced ceramics, Figure 6.3 illustrates, collectively, the properties and areas of application of

Table 6.2 Application fields, selected components and some advanced ceramics alternatives.

Application field	Component	Properties utilized	Advanced ceramics
Processing technology	Chemical apparatus	Corrosion resistance	Aluminum oxide
	Sliding rings	Wear resistance	Silicon carbide
	Thread guides		Titanium oxide
	Wire-drawing dies		Zirconium oxide
Shaping of materials	Cutting tools	Wear resistance	Aluminum oxide
	Grinding disks	Hardness	Silicon nitride
	Sand-blasting nozzles		Silicon carbide
			Boron carbide
High temperature	Burner nozzles	Heat resistance	Silicon nitride
	Welding nozzles	Corrosion resistance	Silicon carbide
	Heat exchanger	Heat conductivity	Aluminum oxide
	Crucibles		Carbon
	Heat pipes		Boron nitride
Engines	Valve seats	Heat resistance	Aluminum titanate
	Turbocharger	Corrosion resistance	Silicon carbide
	Gas turbine	Heat conductivity	Silicon nitride
	Catalyst support		Cordierite
	A-probe		Zirconium oxide
Electronics, electrical engineering, optics	Substrates	Special electrical and magnetic properties	Aluminum oxide
	Capacitors		Titanate perovskites
	Sensors		Ferrites
	Laser materials		
	Magnets		
Energy technology	Nuclear fuel	Radiation resistance	Uranium oxide
	Solid-state electrolyte	Ionic conductivity	Zirconium oxide
			β-Aluminum oxide
Medical technology	Hip joints	Mechanical resistance	Aluminum oxide
	Bone replacement	Surface finish	Calcium phosphate
		Biocompatibility	

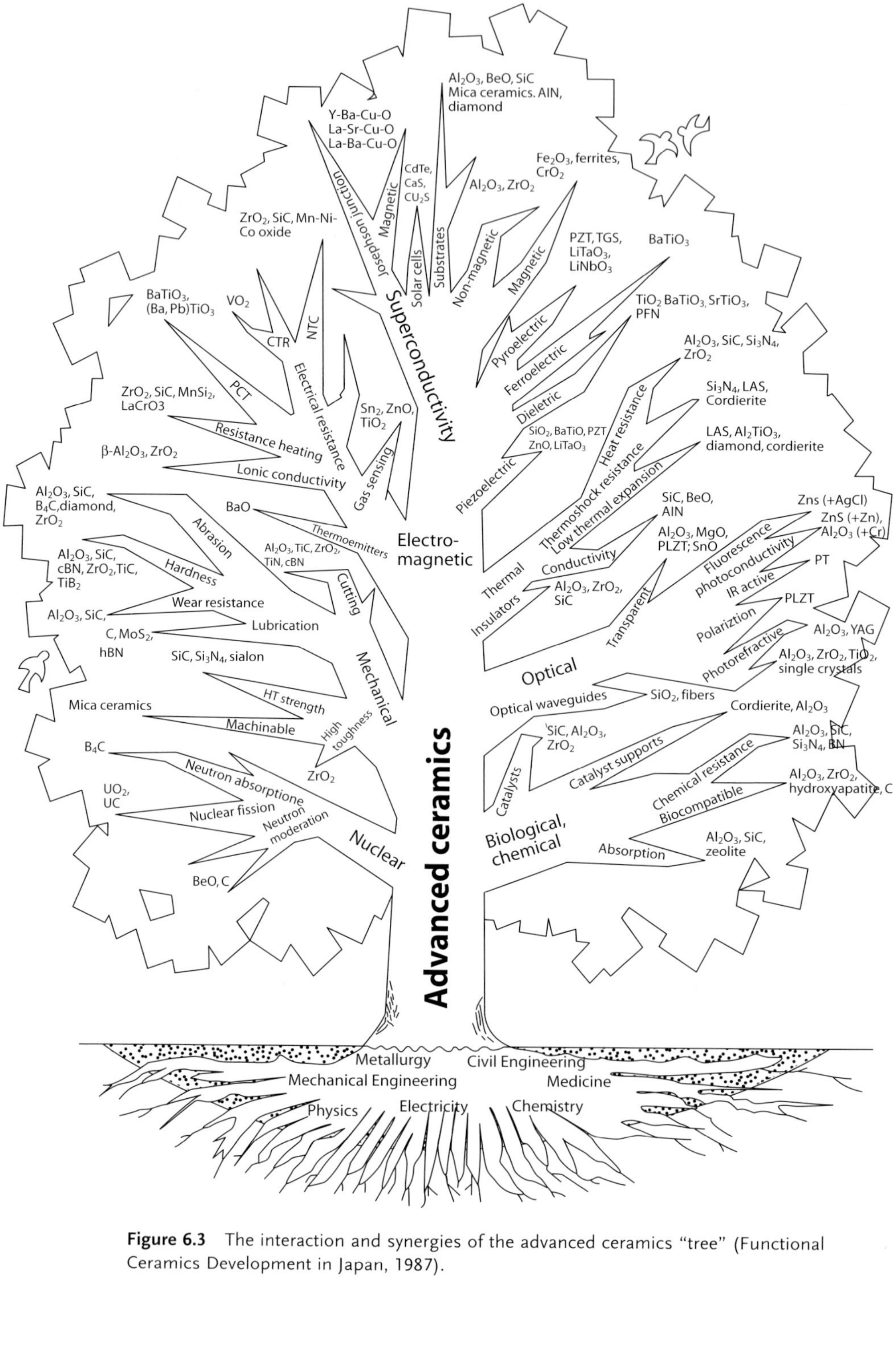

Figure 6.3 The interaction and synergies of the advanced ceramics "tree" (Functional Ceramics Development in Japan, 1987).

advanced ceramics. In this intuitive and compelling figure, the advanced (fine) ceramic "tree" is nourished in terms of knowledge and expertise through its roots in physics, mechanical engineering, metallurgy, chemistry, civil engineering, medicine, and so on. From the trunk of the tree main branches emerge that depict the general properties such as mechanical, nuclear, electric, thermal, optical, and biological. Smaller side branches refer to a mix of secondary properties and applications, while the leaves of the tree symbolize the individual ceramics compounds. Whilst this is, of course, a typical Japanese view, it shows very clearly the interactions and synergetic relations of the subject under discussion, namely *advanced ceramics*.

6.2 The Current World Market Situation

Compared to mass-produced traditional structural ceramics, the production volume of advanced functional ceramics is very small. However, their high value-added nature yields impressive global sales figures, and even more impressive average annual growth rates (AAGRs). A somewhat outdated forecast by Reh (1998) predicted the global advanced ceramics volume to reach €18 billion in 2000, corresponding to an AAGR of 6% (see Table 6.3). This places advanced ceramics among the classical ceramic heavyweights such as floor tiles, bricks and roof tiles, even though the estimated AAGR is rather conservative compared to other predictions shown below.

According to an extensive market analysis conducted by Abraham (2005), the worldwide "(advanced) ceramic fever" which began during the late 1980s had largely abated during the 1990s. Consequently, in the United States the growth of structural markets projected to reach billions of dollars within a short period of

Table 6.3 Global sales predictions of ceramic products (in € billion; original values in DM) (Reh, 1998).

Sector	Year			AAGR% (1990–1995)	AAGR% (1995–2000)
	1990	1995	2000		
Tiles	8	12	17	10	5
Tableware	7	9	9	7.7	4
Sanitary ceramics	4	5	5	7	4
Refractories	13	11	10	−2	−2
Bricks/roof tiles	12	17	22	9	9
Advanced ceramics	10	13	18	5	6
Total	54	67	81		

time simply failed to occur. The reason for such failure originated in part from the shelving of some defense programs, including ceramic armor, as a result of massive cuts in military spending by the government. The anticipated upswing of the automotive engine markets also failed to occur, and other ceramic market segments lost ground when newer technologies emerged that would no longer require ceramics. Furthermore, overly optimistic forecasts of the impact of the then newly discovered high-temperature superconductors led to a rather sluggish development of the global advanced ceramics market. In the US, the advanced ceramics market declined by as much as 10% from 2000 to 2001 as the result of the 9/11 terrorist attack and the continuing economic downturn. The damage done to the market by the 2008 financial crash must still be assessed.

These problems notwithstanding, in 2004 the large-scale commercialization of advanced structural ceramics finally began to shift in a new direction, highlighted by the emergence of new technologies to produce ultrapure fine ceramic powders with tailored size distributions that considerably helped to improve the performance of ceramic components. Advanced ceramics suppliers scaled up the production processes, and new applications emerged for high-end ceramic products, especially within the military realm. In addition, nanosized ceramic powders have recently been commercialized and are finding increasing applications in the emerging field of ceramic nanotechnology (see Chapter 12).

These developments point to new growth opportunities for structural ceramics, electronic ceramics, high-performance coatings and chemical/environmental ceramics during the years ahead.

6.2.1 Advanced Structural Ceramics

Structural ceramics are seeing a strong growth rate due to the increased demand for ceramic armor for personal protection and military vehicles. In 2004 alone, US companies such as Ceradyne, Cercom, Armor Holdings/Simula and ArmorWorks delivered armor worth over US$200 million to the US military. With still more orders expected in the future, the Business Communications Co. (BCC) forecasts that the structural advanced ceramics market will double in size, from US$650 million in 2004 to US$1.3 billion by 2009, due largely to the continued boom in armor demand (see Table 6.4).

6.2.2 Advanced Electronic Ceramics

As shown in Table 6.4, whilst electronic ceramics constitute a mature market, some of its segments are also growing strongly, such as piezoelectric ceramics, hard and soft magnets, and semiconductor packages. Overall, the electronic ceramics market segments maintained about a 6% growth rate from 2003 to 2004, and BCC projects that these markets will grow 7.5% annually from 2004 to 2009.

Table 6.4 US market forecast for advanced ceramic components (in US$ million) (Abraham, 2005).

Segment	Year 2004	Year 2009	AAGR (%)
Structural ceramics	650	1300	15.0
Electronic ceramics	6060	8700	7.5[a)]
Ceramic coatings	1108	1600	7.6
Chemical processing and environment-related	1660	2330	7.0
Total	**9478**	**13 930**	**8.0**

AAGR = average annual growth rate.
a) 2003–2004: 6%.

The use of electronic ceramics is highly dependent on the health of the electronic component and electrical equipment industries, which accounted for a combined 52% of total demand in 2005. The largest outlets for advanced ceramics in the electrical equipment market are insulators and permanent magnets, which accounted for a combined 38% of total demand in 2005. Other advanced ceramic electrical equipment includes piezoelectric igniters, heating elements, heat-shielding components, connectors, and seals. All advanced ceramic use is included in this market, such as those instances where electrical equipment is utilized in the assembly of machinery and transportation equipment (Prokop, 2007).

Demand for microwave dielectric ceramics used in telecommunications, as well as in cable communications using optical fibers, is increasing rapidly. Whereas, in the past microwave communications were used primarily for military purposes such as radar, weapon guidance systems and satellite communications, more recently microwaves have been utilized extensively in communications devices such as mobile radios and phones, and in satellite broadcasting (see Chapter 8). The market volumes for ceramic dielectric materials for chip capacitors and multilayer capacitors (MLCCs) were estimated in a study by Paumanok Publications Inc. (2006).

While electronic ceramics continue to hold the largest share of the market at about 60% (see Table 6.4), structural ceramics are estimated to yield the fastest growth rate, with 15% annually between 2004 and 2009. The total value of the US advanced ceramic components market for 2004 is estimated to be US$ 9.5 billion, and this was expected to increase to US$ 13.9 billion by 2009, with an exceptional growth rate of 8%. These figures, as presented by Abraham (2005), tied in reasonably well with an earlier prediction by the Gorham Advanced Materials Institute (see also Abraham, 2000) which indicated that the world demand for advanced ceramics would increase beyond 7% annually to US$ 34 billion in 2004, of which electronic components would account for about 75%. These components included portable computing devices, cellular phones, pagers, electronic games systems, consumer electronics, electronic components in motor vehicles, aircraft, industrial machinery and instrumentation, household appliances, and other durable goods.

6.2.3 Advanced Ceramic Coatings

The US market for high-performance ceramic coatings and services was estimated at US$ 1.1 billion in 2004, and is expected to increase to US$ 1.6 billion by 2009, at an AAGR of 7.6% (Abraham, 2005; see Table 6.4); however, a somewhat lower estimate was provided recently by Prokop (2007). Thermal spray coatings dominate the total ceramic coatings market, and will continue to do so in 2009, as this area is estimated to grow at an AAGR of 8.6% through the forecast period. Such a high growth rate is due not only to an increased demand for wear- and corrosion-resistant coatings, but also to a need for thermal barrier coatings in gas turbines for commercial and military aircraft over the next five years. Currently, many airlines in the US are acquiring new aircraft, and demand is also increasing in Europe and Asia. The share of techniques such as physical vapor deposition (PVD) and chemical vapor deposition (CVD) in the total market will vary little, as their markets are expected to grow by 5.8% and 5.0%, respectively through the forecast period. The market for other coating techniques, including spraying/dip-coating, sol–gel deposition, micro-arc oxidation (MAO) and laser-assisted thermal spraying, is expected to grow at an AAGR of 8.6% (Abraham, 2005). In particular, although the sol–gel processing of ceramics and glasses enjoyed a global sales figure of US$ 1 billion in 2006, this was proposed to rise to US$ 1.4.billion in 2011, corresponding to an AAGR of 6.3% (Abraham, 2006).

According to a study conducted by the Gorham Advanced Materials Institute (1990), the global sales of advanced ceramics were predicted to exceed US$ 4 billion in 1995, 80% of which was proposed to relate to ceramic coatings. Yet, whilst these predictions overestimated the actual growth rate by close to 100%, the real growth was still impressive. For example, the global thermal spray market for the year 2003 consisted of the following market segments: original equipment manufacturers and end users US$ 1.4 billion; large coating service companies US$ 0.8 billion; small coating service companies US$ 0.6 billion; and powder and equipment sales US$ 0.7 billion. Although this total was US$ 3.5 billion, the figure increased to US$ 5 billion in 2004, of which 45% was accounted for by atmospheric d.c. plasma spray technology (Fauchais *et al.*, 2006). Recent advances in the areas of wear- and corrosion-resistant ceramic coatings, as well as thermal barrier coatings, were reviewed by Heimann and Lehmann (2008) and Lehmann and Heimann (2008).

Besides these high-technology applications, a major market for ceramic coatings also exists within the resource industries, including oil and gas, mining, forestry, pulp and paper and agricultural industries, construction, manufacturing and electronics, and the automotive and aerospace industries. In particular, those industries threatened by international competition, the erosion of raw materials prices, and shifts to new materials and technologies must radically improve their operational efficiency, industrial diversification and environmental compatibility in order to survive. One important component of this struggle is constituted by the coatings that combat the wear, erosion and corrosion that occur at all levels of operations in industry, that impart new functional properties, extend the service life of machinery, and contribute in general to a sustainable development required for the increasingly environment-conscious world of the next few decades. To give an example of economic scale, in Germany alone the wear and corrosion of technical systems account annually for the equivalent of 4.5% of the country's gross national product (GNP), corresponding to € 35 billion (Henke *et al.*, 2004; Heimann, 2008).

In the United States, the consumption of *advanced ceramic coatings* is forecast to increase by 6.4% annually to US$1.1 billion in 2010. It is suggested that engine parts will show the greatest growth potential, based largely on recovery in the transportation equipment sector, and particularly in aerospace equipment. Potential growth applications include building insulation, smart self-repairing anticorrosion (Andreeva and Shchukin, 2008) and fire protection coatings, as well as nanocrystalline coatings (Prokop, 2007). However, the market volume prediction appears to be surprisingly low, and estimates provided by other analysts, such as the Global Industry Analyst, Inc. (2009), may be more realistic.

6.2.4
Chemical Processing and Environment-Related Ceramics

Due to new stringent regulations for automobile emissions and hot industrial gases, ceramic automotive catalyst supports and filters are being used increasingly to reduce pollutants from these emissions. Along with ceramic filters and membranes for cleaning industrial and domestic waste water, these markets are expected to remain steady. Overall, the market for chemical processing and environment-related ceramics reached about US$1.7 billion in 2004, and was projected to reach US$ 2.3 billion by 2009, with an annual growth rate of 7% (Abraham, 2005) (Table 6.4).

So-called "green materials" are replacement materials for products that are deemed environmentally unfriendly. Given the worldwide attention being paid to a healthy and sustainable economy, it is not surprising that the global market for green materials is expected to show an AAGR of 7.4%, increasing the market volume of US$ 6.1 billion in 2005 to an estimated US$ 8.7 billion in 2010 (Gagliardi, 2006b). The lion's share of this market, at 75%, would account for replacement materials for polybrominated biphenyls (PBBs) and polybrominated diphenyl ethers (PBDEs), such as metal hydrate-based resins. Lead-free soldering materials and coatings showed a growth rate of about 10% between 2003 and 2005, with a market volume of US$ 460 million expected in 2007. In this context, the quest for lead-free ferroelectric and piezoelectric ceramics for sensor and actuator applications takes on a particular urgent role (see Section 8.3.3).

6.2.5
Bioceramics, Bioactive Glasses, and Coatings

One field of application with a large potential is that of bioceramic coatings based on hydroxyapatite and second-generation bioceramics. Such biocompatible coatings for bone implants have promising applications to solve some of the health problems of an aging population (Hench, 1991; Heimann, 2007; Heimann 2008; see also Chapter 10).

According to a market study conducted by Gagliardi (2006a), the world market for bioactive glasses, ceramics, and composites is estimated to exceed US$ 1 billion in 2011, up from US$ 378 million in 2004, with a huge AAGR of 17.2%. The largest proportion, at 60%, is that of bioglasses and glass ceramics, with an AAGR of 16.5% to reach a market volume of US$ 600 million in 2011. The highest annual growth rate, at 18.5%, is expected for bioactive composite materials and tissue engineering scaffolds (market volume in 2011: US$ 423 million).

6.3
Recent US and Global Forecasts

The US advanced ceramics industry forecast shown in Table 6.4 (Abraham, 2005) is well supported by the findings of a very recent study conducted by Prokop (2007), which estimates that the demand for advanced ceramics in the US will increase by 7.0% annually, from a weak 2005 base to over US$12 billion in 2010 (Table 6.5). Following a few years of retracted demand, the industry began to gather momentum in 2003, gained strength in 2004 and 2005, and is poised to continue growth through 2010. New or improved uses for advanced ceramics offer growth potential to ballistic armor for personal protection and shielding of military vehicles, for ceramic composite automotive brakes, for diesel particulate filters, and

Table 6.5 US advanced ceramic demand by market (in US$ million) (Prokop, 2007).

Sector	Year			AAGR (2000–2005) (%)	AAGR (2005–2010) (%)
	2000	2005	2010		
Electronic components	3750	2820	4130	−5.5	7.9
Electrical equipment	1670	1680	2370	0.1	7.1
Industrial machinery	1124	1160	1500	0.6	5.3
Transportation equipment	1060	1135	1600	1.4	7.1
Other markets	1446	1830	2500	4.8	6.4
Total	**9050**	**8625**	**12100**	**−1.0**	**7.0**

AAGR = average annual growth rate.

for a wide variety of biomedical joint replacement products (see Chapter 10) and piezoceramic sensors and actuators (see Chapter 8).

It is likely that cordierite, titanate and zirconate ceramics will record the most rapid rates of growth, due to their uses in environmental systems, medical products, electronic components, and household appliances. A continued expansion in electronic component shipments will provide opportunities for titanate perovskites and other ceramics. Nonetheless, cordierite, titanates and other ceramics will undoubtedly benefit from a continued, environmentally driven trend to reduce the amounts of particulates, nitrogen oxides (NO_x), and sulfur oxides that are released into the atmosphere. Technological advances in the medical product market will also provide many opportunities, notably for monolithic ceramics such as alumina and zirconia used for femoral balls in hip endoprostheses, as well as biocompatible hydroxyapatite and tricalcium phosphate coatings for the metal stems of hip implants (see also Chapter 10). Likewise, dental ceramics will continue to experience high growth rates through 2010.

In contrast to the above proposals, it is likely that other advanced ceramics–notably alumina, beryllia, silicon carbide and silicon nitride (see Chapter 11)–will demonstrate below-average growth rates owing to environmental concerns, competition from other ceramics (see Table 6.4), and an increasing reliance on slower-growing market segments such as cutting tools that are based on ceramic alloys (e.g., modified alumina and SiAlONs). In this situation, whilst alumina will surely remain the prominent material, its market share will be eroded by ferrites, and by beryllia- and zirconia-based ceramics.

The demand for advanced ceramics in the *industrial machinery* market is projected to increase by 5.3% each year through 2010, from a weak 2005 base. In this case, the primary driving force will be the cyclical recovery in industrial markets following the downturn of the early 2000s.

As the users of machine tools and other industrial machinery begin to benefit from the generally stronger macroeconomic environment worldwide, the need to expand capacity, to bring idling capacity back online, and to perform upgrades to existing equipment will provide market opportunities for the producers of ceramic wear parts. In the case of cutting tools, which are more of a consumable item, economic recovery will mean a greater demand for the tools necessary to cut and form metal.

Transportation equipment will also post healthy gains, with growth resulting from an acceleration in the production of both aerospace and motor vehicle equipment, as well as through the increased use of ceramics in catalytic converter substrates, engine bearings, ceramic armor for military vehicles, diesel engine particulate filters, and ceramic matrix composite brakes.

Monolithic ceramics represent the dominant and best-established segment of the industry. These products will experience more rapid gains, due primarily to their use in the key electronic components market that is expected to rebound from the low growth rates experienced during the period between 2000 and 2005.

The fastest growth area for *ceramic matrix composites* will also be in engine parts, while cutting tool and wear part applications will lag in their overall growth. In

Table 6.6 US advanced ceramics demand by product group (in US$ million) (Prokop, 2007).

Sector	Year			AAGR (2000–2005) (%)	AAGR (2005–2010) (%)
	2000	2005	2010		
Monolithic ceramics	7980	7530	10 600	−1.2	7.1
Ceramic coatings	740	770	1 050	0.8	6.4
Ceramic matrix composites (CMCs)	330	325	450	−0.3	6.7
Total demand	**9050**	**8625**	**12 100**	**−1.0**	**7.0**

AAGR = average annual growth rate.

general, growth will derive from the increased use of ceramic matrix composites in a variety of extreme environment applications, where the enhanced strength and durability of these ceramics provide performance advantages.

Tables 6.5 and 6.6 list the US advanced ceramics demand by market sectors and product groups, respectively, projected to the year 2010 (Prokop, 2007).

Market research conducted by the Freedonia group (Freedonia Custom Research, 2006) indicates a US advanced ceramics demand of more than US$ 12 billion in 2010, of which 87% will be for monolithic ceramics, 9% for ceramic coatings, and 4% for ceramic matrix composites (CMCs). The Freedonia group study estimated the worldwide demand for automotive electronics to be US$ 117 billion in 2009, showing a growth rate of 7% annually–much more rapid than that for vehicle production itself. The greatest success, however, was proposed to relate to ceramic armor, with the structural ceramics market being expected to double in size, from US$ 650 million in 2004 to US$ 1.3 billion in 2009, as the demand for ceramic armor continued.

Currently, high-performance diamond coatings as well as electrode and electrolyte coatings for solid oxide fuels cells (SOFCs) are on the verge of creating technological breakthroughs in the microelectronics industry (Heimann, 1991; Müller *et al.*, 2002).

6.3.1
Advanced Ceramic Powders

It is proposed that advanced ceramic and nanosized ceramic powders will record a healthy AAGR of 8.9% between 2006 and 2011, boosting the market volume in the US from US$ 2.2 billion in 2006 to US$ 3.4 billion in 2011 (McWilliams, 2006).

Table 6.7 World demand for carbon nanotubes (in US$ million) (Freedonia Custom Research, 2006).

Sector	Year			AAGR (2009–2014) (%)	AAGR (2014–2020) (%)
	2009	2014	2020		
Electronic components	90	395	4530	34	50
Motor vehicles	31	165	1130	39	38
Aerospace and defense	10	65	640	46	46
Other markets	84	445	3100	40	38
World total	**215**	**1070**	**9400**	38	44

AAGR = average annual growth rate.

6.3.2 Carbon Nanotubes

One particularly interesting, and likewise promising, market with unprecedented growth rates appears to exist for carbon nanotubes (CNTs), the global demand for which in electronic, automotive and aerospace and defense applications is summarized in Table 6.7.

CNTs have demonstrated major potential in both micro- and nanoelectronic components, in lifestyle products and sports equipment, as electrode materials for batteries, as supercapacitors, as fuel cells and novel actuator devices as well as electronic displays and, in particular, as novel drug delivery systems (McWilliams, 2006; Brand *et al.*, 2007). Unfortunately, however, the large-scale biomedical application of CNTs is still pending as serious doubts must first be resolved in relation to their potential biotoxicity.

According to a study conducted by Oliver (2007), the market volume of CNTs in the US jumped from US$ 51 million in 2006 to US$ 79 million in 2007. An AAGR of 74%, which was even more impressive than that predicted by Freedonia Custom Research (2006) (see Table 6.7), should result in a market volume of US$ 807 million in 2011. The application fields of CNTs include energy storage (batteries), electronics, and composite materials, with the latter role being expected to account for 80% of global CNT production.

6.3.3 Ceramic Superconductors

During the years following the discovery of high-temperature superconducting ceramics in 1986 (Bednorz and Müller, 1986), highly exaggerated sales figures were promoted, in some cases with the clear expectation for scientists and

Table 6.8 Estimated global market shares for superconductor technology (in € million) (Conectus, 2009).

Field of business	Year			
	2004	2007	2010	2013
Research & Technology Development (RTD)	600	660	835	955
Magnetic resonance imaging (MRI)	2950	3300	3410	3525
Total of RTD and MRI	**3550**	**3960**	**4245**	**4480**
New large-scale applications	35	65	150	410
New electronic applications	65	60	125	210
Total of emerging new businesses	**100**	**125**	**275**	**620**
Total market	**3650**	**4085**	**4520**	**5100**
Market share for LTS	3610	4025	4350	4600
Market share for HTS	40	60	170	500

HTS = high-temperature superconductor; LTS = low-temperature superconductor.

research institutions to acquire huge funding from industry and government-owned utilities for continuing such research. Today, however, the predictions are more sober and realistic. Table 6.8 lists the data published recently by the Consortium of European Companies Determined to Use Superconductors (Conectus, 2009).

In concluding this brief review of published market forecasts for advanced ceramic materials, it should be emphasized that the close-to-catastrophic downturn of the US and world economies, highlighted by a series of stunning bank failures during the summer and fall of 2008, may well make many – if not all – of the forecast sales figures obsolete. However, within this context guidelines have been provided by using technology foresight methods, in particular roadmapping (Da Costa *et al.*, 2003), to recommend promising applications and areas of scientific research on advanced ceramics between 2010 and 2025 (Rödel *et al.*, 2008). These roadmaps are described in greater detail in Chapter 12.

References

Abraham, Th. (2000) U.S. advanced ceramics growth continues. Ceramic Industry. BMP Media, 1 August 2000. Available at: www.ceramicindustry.com/Archives (accessed 8 December 2009).

Abraham, Th. (2005) New opportunities for advanced ceramics. Ceramic Industry. BMP Media, 1 June 2000. Available at: www.ceramicindustry.com/Archives (accessed 8 December 2009).

Abraham, Th. (2006) Sol-gel processing of ceramics and glasses. BCC Research Report, AVM016E, June 2006. Available at: www.bccresearch.com/report/index (accessed December 9, 2009).

Andreeva, D.V. and Shchukin, D.G. (2008) Smart self-repairing protective coatings. *Mater. Today*, **11** (10), 24–30.

Bednorz, J.G. and Müller, K.A. (1986) Possible high T_c superconductivity in the

Ba-La-Cu-O system. *Z. Phys.*, **B64** (1), 189–193.

Brand, L., Eickenbusch, H., Hoffknecht, A., Krauß, O., Zweck, A., and Pohle, D. (2007) *Innovations- und Marktpotenzial neuer Werkstoffe. Monitoringbericht 2007.* Zukünftige Technologien Consulting, VDI-Technologiezentrum GmbH, Düsseldorf, Germany.

Conectus (2009) Consortium of European Companies Determined to Use Superconductors. Available at: www.conectus.org (accessed 12 August 2009).

Da Costa, O., Boden, M., Punic, Y., and Zappacosta, M. (2003) Science and Technology Roadmapping: From Industry to Public Policy. Institute for Prospective Technological Studies (IPTS) Report, Joint Research Centre, European Commission, issue 73, April 2003.

Early, J.G. and Rook, H.L. (2004) Versailles project on advanced materials (VAMAS). *Adv. Mater.*, **8** (1), 9–12.

Fauchais, P., Montavon, G., Vardelle, M., and Cedelle, J. (2006) Development in direct current plasma spraying. *Surf. Coat. Technol.*, **201** (5), 1908–1921.

Freedonia Custom Research (2006) Advanced Ceramics to 2010. Study # 2134, December 2006. The Freedonia Group, 767 Beta Drive, Cleveland, OH. Available at: www.freedoniagroup.com (accessed 9 December 2009).

Functional Ceramics Development in Japan (1987) Report to the Canadian Embassy, Tokyo.

Gagliardi, M. (2006a) Bioactive glasses, ceramics, composites, other advanced materials. BCC Research Report, AVM054A, October 2006. Available at: www.bccresearch.com/report/index (accessed 9 December 2009).

Gagliardi, M. (2006b) Green materials for electrical, electronic and other applications. BCC Research Report, AVM051A, February 2006. Available at: www.bccresearch.com/report/index (accessed 9 December 2009).

Global Industry Analyst, Inc. (2009) High performance ceramic coatings. Report. Available at: www.reportlinker.com/p0131236/High-Performance-Ceramic-Coatings.html (accessed 9 December 2009).

Gorham Advanced Materials Institute (1990) Thermal spray market to grow . Gorham, Maine, USA. Available at: www.highbeam.com/doc/1G1-41702725.html (accessed 8 December 2009).

Heimann, R.B. (1991) Advanced ceramic coatings by plasma spraying: a technology for the 1990s and beyond. *Process. Adv. Mater.*, **1**, 181–192.

Heimann, R.B. (2007) Novel approaches towards design and biofunctionality of plasma-sprayed osteoconductive calcium phosphate coatings for biomedical implants: the concept of bond coats, in *Trends in Biomaterials Research* (ed. P.J. Pannone), Nova Science Publishers Inc., Hauppauge, New York, Ch 1, pp. 1–80.

Heimann, R.B. (2008) *Plasma Spray Coating. Principles and Applications*, 2nd edn, Wiley-VCH Verlag GmbH, Weinheim. 427 pp.

Heimann, R.B. and Lehmann, H.D. (2008) Recently patented work on thermally sprayed coatings for protection against wear and corrosion of engineered structures. *Recent Pat. Mater. Sci.*, **1** (1), 41–55.

Hench, L.L. (1991) Bioceramics: from concept to clinic. *J. Am. Ceram. Soc.*, **74** (7), 1487–1510.

Henke, H., Adam, D., Köhler, A., and Heimann, R.B. (2004) Development and testing of HVOF-sprayed tungsten carbide coatings applied to moulds for concrete roof tiles. *Wear*, **256**, 81–87.

Lehmann, H.D. and Heimann, R.B. (2008) Thermally sprayed thermal barrier coating (TBC) systems: a survey of recent patents. *Recent Pat. Mater. Sci.*, **1** (2), 140–158.

McWilliams, A. (2006) Advanced ceramics and nanoceramic powders. BCC Research Report, NAN015E, December 2006. Available at: www.bccresearch.com/report/index (accessed 9 December 2009).

Müller, M., Bouyer, E., von Bradke, M., Branston, D.W., Heimann, R.B., Henne, R., Lins, G., and Schiller, G. (2002) Thermal induction plasma processes for the synthesis of SOFC materials. *Materialwiss. Werkstofftech.*, **33**, 322–330.

Nicholson, P.S. (1996) Higher-performance ceramics – towards 2000. *Key Eng. Mater.*, **122–124**, 3–14.

Okada, A. (2008) Ceramic technologies for automotive industry: current status and perspectives. *Mater. Sci. Eng. B*, **161**, 182–187.

Oliver, J. (2007) Carbon nanotubes: technology and commercial products. BCC Research Report, NAN024C, March 2007. Available at: www.bccresearch.com/report/index (accessed 9 December 2009).

Paumanok Publications Inc. (2006) Ceramic dielectric materials: world markets, technologies and opportunities: 2006–2010. Available at: www.marketresearch.com/vendors/viewvendors.asp (accessed 9 December 2009).

Prokop, P. (2007) Advanced ceramics in demand. Ceramic Industry, BMP Media, June 1, 2007. Available at: www.ceramicindustry.com/Archives (accessed 8 December 2009).

Reh, H. (1998) Die Welt der Keramik: Zulieferer, Produzenten und Verbraucher. *cfi/Ber. DKG*, **75** (1/2), 51.

Rödel, J., Kounga, A.B.N., Weissenberger-Eibl, M., Koch, D., Bierwisch, A., Rossner, W., Hoffmann, M.J., Danzer, R., and Schneider, G. (2008) Development of a roadmap for advanced ceramics: 2010–2025. *J. Eur. Ceram. Soc.*, **29** (9), 1549–1560.

7
Oxide Ceramics: Structure, Technology, and Applications

7.1
Alumina

7.1.1
General Properties and Application

Alumina is the most cost-effective and widely used structural engineering material in the family of advanced ceramics. The raw materials from which this high-performance, technical grade ceramic is produced are readily available and reasonably priced, resulting in good value for the cost in fabricated alumina shapes.

Hence, alumina (corundum, α-Al_2O_3) is considered the "workhorse material" of the structural ceramics industry. Its high hardness, abrasion resistance, and chemical inertness makes it an ideal material to perform well in a variety of aggressive environments, ranging from mining industry to chemical industry to metal manufacturing and processing, to ceramic armor and biomedical applications. Its electrically insulating nature, coupled with its moderate thermal conductivity, reasonably low dielectric permittivity, and low dielectric loss, offer a plethora of applications that include electronic substrates for integrated circuits (ICs) and automotive spark plugs.

However, these advantageous properties are partially offset by the material's low tensile and flexural strengths and fracture toughness, and low thermal shock resistance (see Figure 6.1). Hence, the decision to include alumina-based wearing parts in engineering designs must be assessed judiciously so as to avoid catastrophic failure in service under harsh conditions. An overview of the important properties of alumina ceramics is provided in Table 7.1. The wide range of values attests to the fact that alumina ceramics depend on processing parameters, including the sintering temperature, sintering atmosphere, impurity content, grain size, and other extrinsic and intrinsic factors. For example, the extremely pure and fine-grained alumina specifications used in femoral heads for hip endoprostheses demonstrate noticeably higher strengths, toughness, and hardness characteristics (see Tables 10.2 and 10.3).

High-purity alumina parts and devices can be utilized in both oxidizing and reducing atmospheres to 1925 °C. The mass loss in vacuum between 1700° and

Classic and Advanced Ceramics: From Fundamentals to Applications. Robert B. Heimann

ISBN: 978-3-527-32517-7

Table 7.1 Mechanical, thermal, elastic and electrical properties of high-alumina ceramics.

Property	Range
Relative density ($Mg\,m^{-3}$)	3.4–3.7
Thermal expansion coefficient ($\times 10^{-6}\,°C^{-1}$)	7.5–8.5
Compressive strength (MPa)	1000–2800
Tensile strength (MPa)	140–170
Flexural strength (MPa)	280–420
Wear strength (MPa·m)	550–600
Fracture toughness ($MPa \cdot m^{1/2}$)	3–4
Thermal conductivity ($W \cdot m^{-1}\,K^{-1}$)	30–40
Specific heat ($J\,kg^{-1}\,K^{-1}$)	880
Elastic modulus (GPa)	350–400
Shear modulus (GPa)	140–160
Bulk modulus (GPa)	210–250
Microhardness ($kg\,mm^{-2}$)	1400–1800
Dielectric strength ($kV\,mm^{-1}$)	10–17
Dielectric permittivity	9.8 (1 MHz)
Volume resistivity (ohm·cm)	$>10^{14}$ (RT)

RT = room temperature.

2000 °C ranges from 10^{-5} to $10^{-7}\,g\,cm^{-2} \cdot s^{-1}$. Alumina resists attack by all gases except for wet fluorine, and is resistant to all common reagents except for hydrofluoric and hot orthophosphoric acids. An elevated temperature corrosion occurs in the presence of alkali metal vapors, particularly at lower purity levels (<90% Al_2O_3 content).

The additions of either chromium oxide or manganese oxide is known to improve the material's hardness and toughness (see Section 10.5.1.1). Other additions can be made to improve the ease and consistency of metal films fired to alumina ceramics for subsequently brazed and soldered assemblies.

7.1.2
Processing of Alumina

The process of producing pure alumina from bauxite ore (Bayer process) has changed very little since its inception in 1893, and can be divided into three stages of extraction, precipitation, and calcination.

7.1.2.1 Extraction

The aluminum hydroxide (gibbsite, γ-$Al(OH)_3$) and oxyhydroxide (boehmite, γ-AlOOH; diaspore α-AlOOH) minerals[1] in bauxite ore are selectively extracted from the insoluble components (mostly quartz, clay minerals, and iron and tita-

1) Minor amounts of transitional alumina phases such as η-alumina (Tilley and Eggleton, 1996) or χ-alumina (Singh and Gilkes, 1995) may also be present in natural bauxite ore.

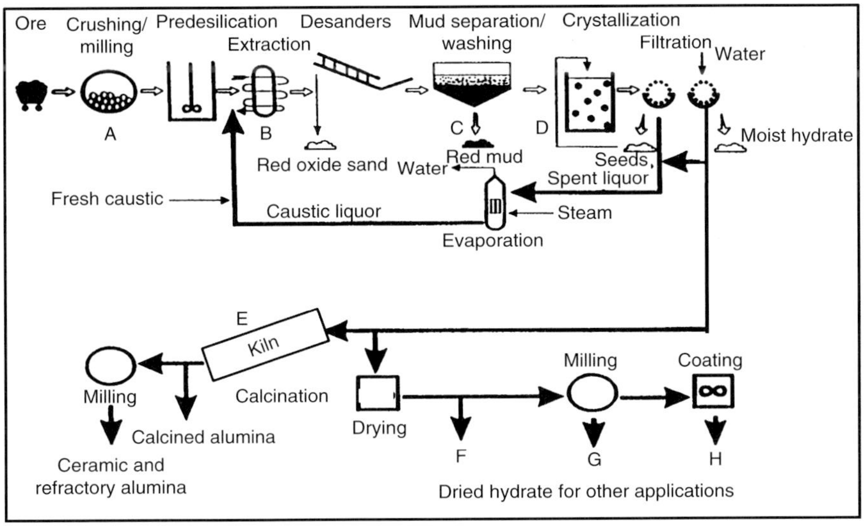

Figure 7.1 Process flow sheet of the Bayer process. Modified after Evans (1996).

nium oxides) by dissolving the ore in a solution of sodium hydroxide (caustic soda) according to:

$$Al(OH)_3 + OH^- \rightarrow [Al(OH)_4]^- \qquad (7.1)$$

The ore is washed, crushed and milled (Figure 7.1, step A) to reduce the particle size and make the minerals more available for extraction. It is then combined with the process caustic liquor and sent as slurry to a heated pressure digester for extraction (step B).

The concentration, temperature and pressure within the digester are set according to the properties of the bauxite ore. Ores with a high gibbsite content can be processed at 140 °C, whereas the processing of boehmite requires temperatures between 200 and 240 °C. The steam pressure at 240 °C is approximately 3.5 MPa. Although the application of higher temperatures would, in theory, be advantageous, there are several disadvantages that include corrosion problems and the possibility of oxides other than the alumina dissolving into the caustic liquor.

Following the extraction stage, the insoluble oxide residue must be separated from the aluminum-containing liquor by a process known as *settling* (step C). Here, the liquor is purified by filtering before being transferred to the precipitators. The insoluble, so-called "red mud" from the first settling stage is thickened and washed to recover the caustic soda, which is then recycled back into the main process.

7.1.2.2 **Precipitation**

Crystalline pure aluminum hydroxide (gibbsite) is then precipitated from the so-called "pregnant" digestion liquor after cooling and diluting, and adding seed crystals (step D) according to:

$$[Al(OH)_4]^- \rightarrow Al(OH)_3 + OH^- \tag{7.2}$$

This is essentially the reverse of the extraction process, except that the product's nature is carefully controlled by the plant conditions, including seeding or selective nucleation, precipitation temperature, and the cooling rate. The gibbsite crystals are then classified into size fractions and fed into a rotary or fluidized-bed calcination kiln (step E). Undersized particles are fed back into the precipitation stage.

7.1.2.3 Calcination

The gibbsite crystals are calcined (step E) to form high-purity alumina (α-Al_2O_3) as the precursor material for a variety of alumina ceramics according to:

$$2Al(OH)_3 \rightarrow Al_2O_3 + 3H_2O \tag{7.3}$$

The calcination process must be carefully controlled since it dictates the properties of the final product. A secondary process stream containing dried gibbsite can be separated to produce aluminates, zeolites, filler materials for toothpaste, fire retardants, and others (steps F–H). As will be shown below, the stepwise removal of water and OH groups, respectively, leads to a plethora of transitional alumina polytypic structures that eventually will determine the performance of the end products.

The stability of aluminum hydroxides shows also a pronounced pH dependence. Hydroxides precipitated from an aqueous solution of $AlCl_3$ show the structures of gibbsite at pH 5–7, bayerite at pH 7–11, and nordstrandite at pH 11–13. This means that the controlled dehydration of aluminum hydroxides can lead to alumina products with tailored properties for a large variety of industrial applications. In particular, the size and morphology of the crystalline alumina particles can be adjusted in such a way that, on subsequent sintering, ceramics with predetermined and reproducible properties can be obtained (see also Figure 7.1, steps F to H).

The worldwide production of alumina was estimated to be 62.4 million tons in 2006, an increase from the 1995 value of 38 million tons (Evans, 1996), with Australia (17.7 million tons), China (8.6 million tons), Brazil (5.3 million tons) and USA (5.2 million tons) as the leading producers (Source: USGS Mineral Resources Program). While 90% of the alumina is used to extract metallic aluminum by electrolysis, approximately 10% of the total is non-metallurgical alumina used for advanced ceramic allocations (see Section 7.1.4.3 below).

7.1.3 Structure of Alumina Polymorphs

7.1.3.1 Stable Alumina Polymorphs

Under ambient conditions, only two stable modifications of Al_2O_3 exist. The thermodynamically stable polymorph between room temperature and the melting point at 2050 °C is α-Al_2O_3 (corundum), which forms a lattice of hexagonally close-packed oxygen atoms with stacking order AB-AB..., in which two-thirds of the

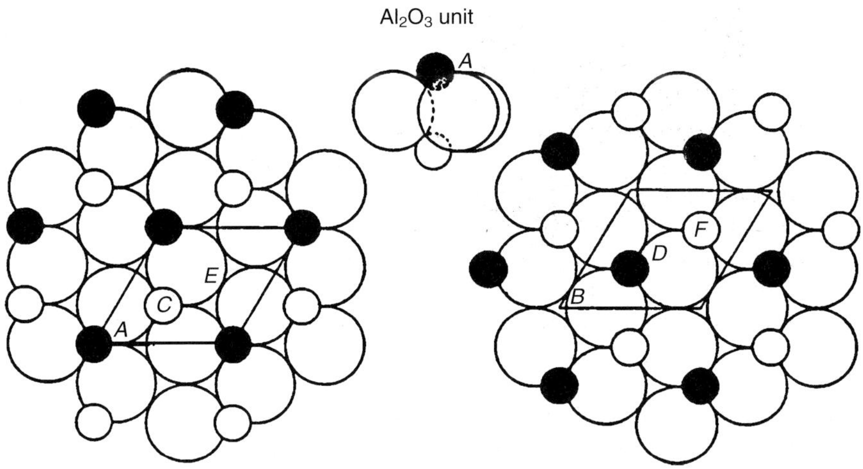

Figure 7.2 Structure of α-Al_2O_3. The framework is composed of AB-AB...-stacked gibbsite-like sheets of oxygen octahedra.

octahedral sites are occupied by Al cations (Figure 7.2). Corundum crystallizes in the symmetry group $\bar{3}m$ [space group (S.G.) $R\bar{3}c$] with ditrigonal scalenohedra as the general forms and a lattice constant a = 513 pm, α = 55°16′. The oxygen octahedra share edges to form six-membered rings that are linked into gibbsite-like sheets parallel (0001). The sheets are stacked into a framework structure by sharing the faces and corners of octahedra. In the hexagonal cell (a = 475 pm, c = 1298 pm), six gibbsite-like sheets are stacked in such a way that positions ABCDEF indicated in Figure 7.2 lie on one common line parallel [0001].

The γ-Al_2O_3 polymorph is a product of dehydroxylation of gibbsite (γ-$Al(OH)_3$) or boehmite (γ-AlOOH) at 450 °C. Gibbsite has a slightly deformed quasi-hexagonal close-packed oxygen arrangement with a stacking order ABB-AAB... (Figure 7.3), whereby the layer stacks are separated by mirror planes and held together by hydrogen bonds. This arrangement provides the structural motif of the defect spinel lattice of γ-Al_2O_3 composed of cubic close-packed oxygen atoms with stacking order ABC-ABC...

Recently, the structure of gibbsite was re-evaluated by Balan *et al.* (2006) using *ab initio* density functional theory (DFT) calculation with local density approximation (LDA), and the following unit cell constants were reported: a = 874.2 pm, b = 511.2 pm, c = 980.1 pm, β = 94.54°; S.G. $P2_1/n$ (14).

The cubic γ-Al_2O_3 (S.G. $Fd\bar{3}m$ has a lattice constant a = 790 pm, which is close to $c_{gibbsite}/(\sqrt{3}/\sqrt{2})$. The transformation of gibbsite into the defect spinel involves a lateral shifting of one oxygen layer to form the f.c.c. stacking order ABC-ABC... Above 1000 °C, γ-Al_2O_3 converts to stable α-Al_2O_3. In the presence of impurities and during hydrothermal treatment, this transformation is more complex as it imparts the occurrence of transient phases such as $\bar{\delta}$- and θ-Al_2O_3 (see below).

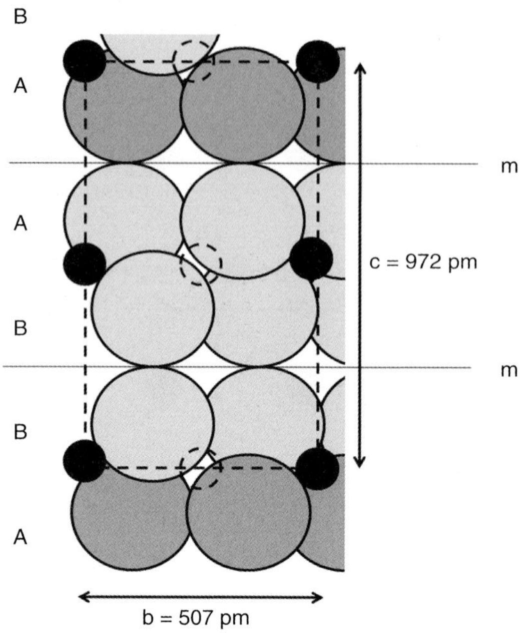

Figure 7.3 Layered structure of monoclinic gibbsite, γ-$Al(OH)_3$ with stacking order AB-BA-AB ...

Figure 7.4 shows a simplified scheme of the distribution of cations between the two types of interstices of the spinel lattice. Al ions occupy both tetrahedral (squares) and octahedral (circles) interstices. However, for reason of charge balance of the 16 available octahedral sites, only 13⅓ are filled, leaving 2⅔ vacancies per 32 oxygen ions in the spinel unit cell.

However, the true relative partial occupancy of Al in each lattice position (Wyckoff parameter) remains a matter of dispute. According to the results of recent X-ray diffraction (XRD) studies (Paglia *et al.*, 2003; Smrcok *et al.*, 2006), the refined occupancy parameters for Al were found to be 0.83(3) in 8a, 0.818(13) in 16d, 0.066(14) in 16c, and 0.044(18) in 48f Wyckoff positions. Hence, the Al ions are distributed over octahedral and tetrahedral sites in a ratio of 63:37. Some 6% of all ions occupy non-spinel positions (Wyckoff parameters 16c and 32e) to yield a formula $Al_{2.67}O_4$ or $[Al\ Al_{5/3}\square_{1/3}]O_4$ when recast as a spinel-type structure. Molecular modeling of the structure of γ-Al_2O_3 and a Rietveld refinement of the neutron diffraction data suggested the existence of a supercell of cubic and tetragonal subcells, with a significant number of cations in c-symmetry positions (Paglia, 2004). Hence, a new structure was proposed for γ-alumina with a tetragonal space group $I4_1/amd$ (141). Based on DFT and LDA-relaxed calculations, Menéndez-Proupin and Gutiérrez (2005) suggested a tetragonally distorted cubic lattice with a = c = 784.8 pm, b = 787.7 pm, α = 89.8°, β = 88.8°, γ = 89.8°. Based on these considerations, Ching *et al.* (2008) argued that the traditional view of stoichiomet-

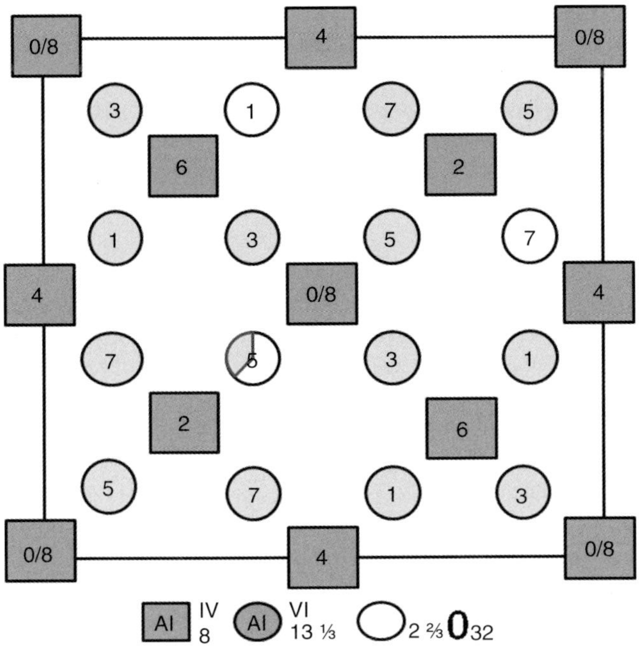

Figure 7.4 Distribution of Al ions between two types of interstices in γ-Al_2O_3 with defect spinel structure. The numbers refer to height along the third dimension, expressed in eighths of cubic unit cell parameters (790 pm). The squares correspond to Al in tetrahedral interstices, and the lightly shaded circles to Al in octahedral interstices. The blank circles correspond to lattice vacancies.

ric γ-Al_2O_3 being a defective spinel with cation vacancies should be modified. Rather, it may be better described as an amorphous network-like structure such that the ratio of tetrahedrally to octahedrally coordinated Al is close to 0.6, and the O ions are bonded to Al in either a threefold or fourfold configurations in about equal proportions.

It should also be mentioned that a "tetragonal γ-alumina" (a = 796 pm, c = 780 pm) described earlier by Saalfeld (1958) was reinvestigated by Rooksby and Rooymans (1961) and renamed δ-alumina, with a larger tetragonal unit cell (a = 796 pm, c = 1170 pm) and cubic closed-packed ABC-ABC... stacking of oxygen atoms. Alternatively, δ-alumina could be described as a superstructure of γ-alumina with ordered cation vacancies. According to Levin and Brandon (1998), δ-alumina has a tetragonal metric when formed during the dehydration of boehmite, but an orthorhombic metric (symmetry group: *222*, S.G. either $P2_12_12$ (18) or $P2_12_12_1$ (19)) when crystallized from a melt. The rapid quenching of molten alumina, as realized during atmospheric plasma spraying, yields a mixture of γ- and δ-alumina that, on annealing to 1100 °C, transforms to θ-alumina (Figure 7.7) (Dubsky *et al.*, 1997).

The so-called β-Al_2O_3 is, strictly speaking, not an alumina polymorph but rather an alkali aluminate (see Section 7.1.4.2).

7.1.3.2 Transitional Alumina Polymorphs

The kinetics and structural relationships of the dehydration of gibbsite, boehmite, and diapore occurring as the main minerals in bauxite ore are quite complex, and in several aspects remain controversial. Depending on the grain size, heating rate and water partial pressure, three main pathways of dehydration have been described (Wefers and Misra, 1987; Ingram-Jones *et al.*, 1996). The general structural principle of these dehydration sequences is that the stacking order of oxygen atoms is maintained during transformation within the series of the transitional aluminas (i.e., gibbsite $\rightarrow$ boehmite $\rightarrow \chi \rightarrow \gamma \rightarrow \delta \rightarrow \theta$), all of which have a cubic close-packed ABC-ABC... stacking sequence. On the other hand, rapid quenching during the chemical vapor deposition (CVD) of alumina for coatings of cemented carbide cutting tools leads to the sequence $\kappa \rightarrow \alpha$, with hexagonal close-packed layers of oxygen atoms (Prengel *et al.*, 1994). All metastable alumina structures are ordered or partially ordered cation arrays at the interstitial sites of either cubic f.c.c. (γ-, η-, θ-, and δ-alumina) or hexagonally close-packed (χ-, κ-, and α-alumina) oxygen atoms (Levin and Brandon, 1998).

Dehydration/Calcination: Pathway 1 When slowly heated in air, small grains of gibbsite, γ-$Al(OH)_3$, convert initially at about 270 °C to χ-Al_2O_3, that is stable to about 850 °C (Fondeur *et al.*, 2004). The structure consists of disordered AB-AB... layers and still contains as many as 0.5 mol water per mol Al_2O_3. The lattice metric is hexagonal, with a = 556 pm, and c = 864 pm (Brindley, 1961) or 1344 pm (Saalfeld, 1960), depending on the number of stacked gibbsite-like layers – that is, either four or six layers of an average thickness 212 pm. During heating beyond 850–900 °C, χ-alumina transforms to κ-alumina with an h.c.p. oxygen atom arrangement (Brindley, 1961). The stacking order is ABAB-BABA-...; that is, the second layer stack is rotated by 180° relative to the first. At about 1200 °C, κ-alumina transforms to the stable α-alumina phase.

However, recent investigations have identified a different structure for κ-alumina (Halvarsson, 1994; Ollivier *et al.*, 1997). According to these studies, κ-alumina crystallizes in an orthorhombic structure with symmetry group *mm2*, S.G. $Pna2_1$, and the lattice constants a = 483.4 pm, b = 831.0 pm, c = 893.7 pm. The crystal lattice can be visualized as composed of two layers alternatively stacked along the direction of the [001] (Figure 7.5). The first layer consists of single chains of edge-sharing AlO_6 octahedra connected through single chains of corner-sharing AlO_4 tetrahedra, while the second layer shows double zigzag chains of edge-sharing AlO_6 octahedra parallel to [100], with vacant octahedra in between (Gross and Mader, 1997).

A recent investigation by Whittington and Ilievski (2004) charted somewhat different pathways of the thermal dehydration of gibbsite. At the high heating rates and temperatures relevant to the Bayer process, the sequence gibbsite $\rightarrow \chi \rightarrow \gamma \rightarrow \theta \rightarrow \alpha$ was found.

Dehydration/Calcination: Pathway 2 Starting from larger gibbsite grains and during rapid heating, locally hydrothermal conditions may occur. However, for these conditions, a different transition pathway has been observed experimentally (Figures 7.6 and 7.7).

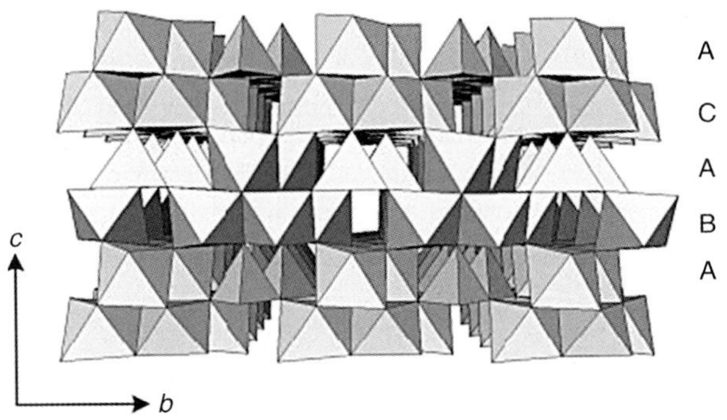

Figure 7.5 Structure of κ-alumina, showing the pseudo-close-packed AB-AC... stacking of oxygen atoms in octahedral and tetrahedral sites (Ollivier *et al.*, 1997). Reproduced with permission from The Royal Society of Chemistry.

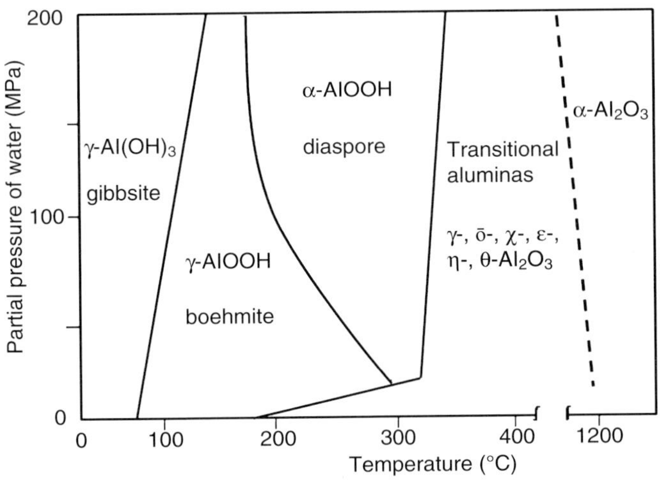

Figure 7.6 Tentative phase relations during hydrothermal dehydration of gibbsite.

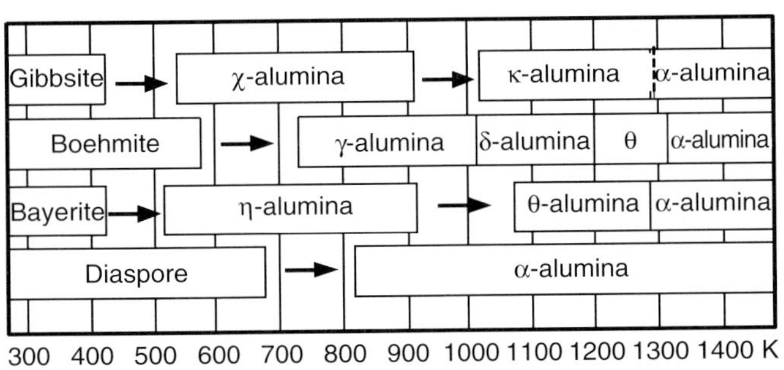

Figure 7.7 The dehydroxylation/calcinations sequences of aluminum hydroxides. Modified after Wefers and Misra (1987).

Gibbsite first transforms to boehmite, whereby the layered h.c.p. oxygen arrangement is maintained. Boehmite transforms above 300 °C to γ-alumina and ō-alumina which, at 900 °C, forms θ-alumina and eventually α-alumina (corundum) beyond 1150 °C (Messing *et al.*, 1986). Minor phases during the transformation of gibbsite include monoclinic bayerite, α-$Al(OH)_3$ with a hexagonal AB-AB-... stacking order, as well as triclinic nordstrandite with a somewhat disordered gibbsite lattice (Wall *et al.*, 1962). As opposed to gibbsite in nordstrandite, this disordering causes lengthening of the c- and a-axes by about 0.5% and 1.0%, respectively (Strunz and Nickel, 2001).

Increasing water pressure causes the stability field of gibbsite to increase somewhat (Figure 7.6). However, boehmite, γ-AlOOH becomes metastable whereas the stable phase at higher temperature is diaspore, α-AlOOH.

Dehydration/Calcination: Pathway 3 Bayerite, α-$Al(OH)_3$ occurring as a minor mineral constituent in bauxite, dehydrates to form initially γ-Al_2O_3, which in turn transforms on continued heating via η-alumina to θ-Al_2O_3 to α-Al_2O_3. This reaction sequence may be identical to that described by Whittington and Ilievski (2004) for gibbsite dehydrated under high heating rates, and also appears to be in accord with the results of an earlier investigation conducted by Saalfeld (1958). The η-alumina has, presumably, a cubic close-packed spinel-type structure with a = 794 pm (Stumpf *et al.*, 1950; Brown *et al.*, 1953) or a = 791.4 pm (Zhou and Snyder, 1991), and may even be a derivative structure of γ-alumina. The crystal structure of θ-alumina is monoclinic [S.G. *C2/m* (12)], with a distorted spinel-type structure that can be thought of as a transient stage between the low-temperature f.c.c structure of the transitional aluminas and the high-temperature h.c.p. structure of corundum (Wefers and Misra, 1987; Zhou and Snyder 1991; Mo and Chang, 1998).

Recent studies have indicated that the mechanical activation of bayerite by soft grinding accelerates the phase transition to stable α-Al_2O_3. The onset of phase transition occurs in the presence of α-Al_2O_3 seeds already at 300 °C, some 800 °C lower than in the non-ground state. The transition sequence is:

$$\alpha\text{-Al(OH)}_3 \rightarrow \gamma\text{-AlOOH} \rightarrow \gamma\text{-Al}_2\text{O}_3 \rightarrow \alpha\text{-Al}_2\text{O}_3 \quad (7.4)$$

which suggests that the formation of the η- and θ-Al_2O_3 transitional phases observed by other authors is suppressed under these conditions (Du *et al.*, 2008).

Summary The transformation sequences described above, as first provided by Wefers and Misra (1987) and subsequently reconfirmed by Souza Santos *et al.* (2000) and Ingram-Jones *et al.* (1996), are shown in Figure 7.7. The latter authors investigated the phase formation of fine and coarse gibbsite particles at slow (soak) and fast (flash) calcinations conditions. *Fine gibbsite particles* under slow calcinations conditions follow the reaction pathway I, with $\chi \rightarrow \kappa \rightarrow \alpha$. During flash calcinations, however, the dehydroxylation follows the sequence $\chi \rightarrow \gamma \rightarrow \bar{o} \rightarrow \theta \rightarrow \alpha$. *Coarse gibbsite particles* during slow heating transform by two pathways: (i) gibbsite → boehmite $\rightarrow \gamma \rightarrow \bar{o} \rightarrow \theta \rightarrow \alpha$ (reaction pathway II); or (ii) gibbsite → χ → κ → α (minor pathway). During flash calcination at about 800 °C a cross-over χ → γ takes place, without formation of the κ-phase.

7.1.4
Specific Properties and Applications

7.1.4.1 Duplex Al_2O_3–ZrO_2 Ceramics

The addition of polycrystalline metastable tetragonal Y-partially stabilized zirconia (Y-PSZ) aggregates to a fine-grained alumina matrix results in so-called "duplex" structures (see Figure 7.32), with a high fracture strength >700 MPa and fracture toughness up to 12 MPa · $m^{1/2}$. Since Y-PSZ has a lower elastic modulus (ca. 210 GPa) than alumina (350–400 GPa), cracks introduced by external loads tend to move towards the zirconia particle aggregates. The crack energy will then be dissipated by forcing the t-ZrO_2 to transform to m-ZrO_2. This mechanism will be augmented by the compressive stress generated, that tends to counteract crack movement.

The considerably improved mechanical properties of such duplex ceramics are currently being exploited in novel developments in the field of femoral heads for hip endoprostheses that rely on high-purity alumina (BIOLOX® delta), with the addition of 17 vol% tetragonally stabilized zirconia and 1.4 vol% chromia particles. Whilst the former addition provides mechanical strengthening by transformation toughening (see Section 7.2.4), the latter addition acts as a reinforcement by dissipating crack energy via the deflection of crack paths. By applying these mechanisms, the four-point bending strength is increased three- to fourfold compared to that of unalloyed alumina (Table 7.1), to reach 1400 MPa, while the fracture toughness is increased to 6.5 MPa · $m^{1/2}$ (CeramTech, 2006) (see also Chapter 10).

7.1.4.2 Stuffed Alumina Ceramics

β-Al_2O_3 (β-alumina) is not an aluminum oxide, but rather a "stuffed" alumina compound, $MAl_{11}O_{17}$, where M is a monovalent ion such as Na, K, Rb, Ag, Te, or Li. Increasing amounts of M ions lead to related structures such as β′-alumina (MAl_7O_{11}) and β″-alumina (MAl_5O_8), with noticeably increased conductivities. β-alumina is thought to crystallize in the hexagonal system with a = 558 pm and c = 2245 pm.

Owing to the mobility of the monovalent ions in response to an electric field, β-aluminas show remarkable two-dimensional (2-D) ionic conductivity (see Section 7.2.6.2), and hence can serve as a semipermeable membrane for NaS batteries (see below). Structurally, β-alumina is composed of spinel-like blocks of cubic close-packed oxygen ions with the stacking order ABC-ABC..., in which the Al ions occupy both tetrahedral and octahedral interstices (as shown in Figure 7.4). These blocks are joined together by rather open oxygen layers within which the mobile ions can freely move. As shown in Figure 7.8, this Na–O layer forms a mirror plane within the structure of β-alumina.

The electrical conductivity of β-alumina depends on the size of the M cation. If the ion radius of the M cation becomes larger, its mobility is impeded and hence its transference number is decreased. On the other hand, if the M cation is too

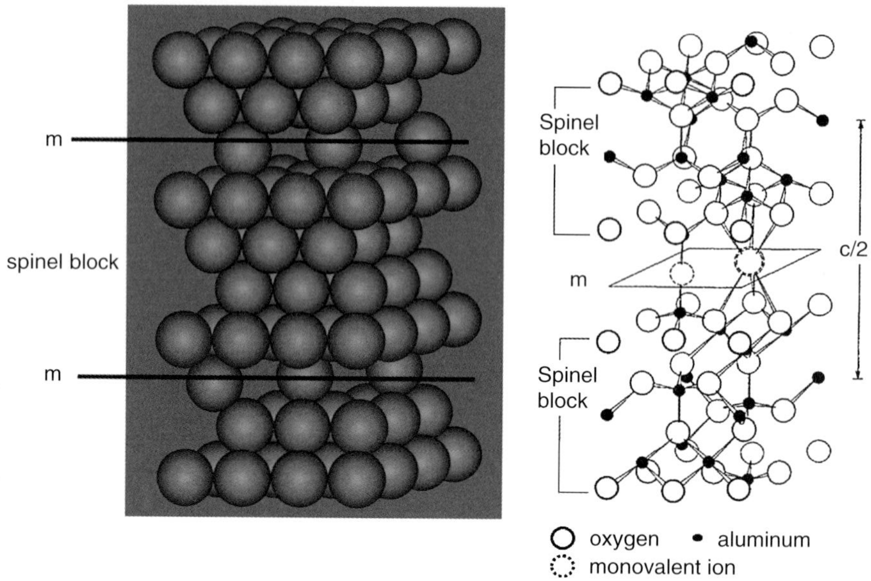

Figure 7.8 Structure of β-alumina. The spinel blocks with cubic close-packed oxygen ordering are separated by open layers of oxygen ions loosely bound to monovalent ions, forming a mirror plane *m* (Felsche, 1968). Reprinted with permission from Oldenbourg Wissenschaftsverlag, München, Germany.

small (as is the case for Li^+), it will "rattle around" within the conduction channels of the β-alumina structure and its mobility will be reduced. Details of the theory of ionic conductivity and numerical conductivity values can be obtained from Kingery *et al.* (1976) (see also May, 1978).

7.1.4.3 **Selected Applications of Alumina Ceramics**

Of the approximately 6 million tons of non-metallurgical alumina produced worldwide in 2006, about 50% was used to produce refractory and wear ceramics, 20% abrasives, 15% whiteware and spark plugs, and 15% other technical ceramics.

Owing to its favorable mechanical, electrical and chemical properties, the typical uses of polycrystalline α-alumina ceramics include:

- Wear pads
- Sealing rings and washers (e.g., for hot water mixing batteries and automotive water pumps)
- High-temperature electrical insulators
- High-voltage insulators, as in spark plugs
- Surge arrester sleeves
- Rectifier housings for thyristors
- Gas laser tubes
- Sheaths for high-voltage sodium vapor lamps (Lucalox™)

- Furnace liner tubes
- Thread and wire guides
- Abrasion- and erosion-resistant pipe and elbow liners
- Liners for Venturi nozzles
- Filters for molten metals
- Cutting tool bits
- Grinding media
- Ballistic armor
- Electronic substrates
- Nose cones and radomes for missiles, infrared detector windows for short-range attack missiles

Applications of single-crystal corundum include:

- Laser crystals
- Dental roots
- Bone screws (only in Japan)
- Optical windows for high-temperature–high pressure reactors
- Scratch-proof watch glasses
- Ruby bearings for watches
- Artificial gemstones

Single crystals of pure and doped alumina can be grown using well-established techniques such as flame fusion (Verneuil process), Czochralski crystal pulling, and top-seeded solution growth (TSSG), as well as under hydrothermal conditions and from the vapor phase (Franke *et al.*, 1968). Some of these applications will be discussed in greater detail below.

Alumina-Lined Pipes Extraordinary amounts of coal are burned worldwide, not only in conventional coal-fired power plants but also in modern developments, including cogeneration power plants. The burners of these units must be fed with pulverized coal or coal slurry that is transported pneumatically to the burner nozzles at high velocity and throughput. However, as most coals contain a percentage of highly abrasive quartz grains, severe wear problems of the feeder pipes may ensue. This wear is particularly destructive in areas where the feeding stream changes direction or is constricted, such as in elbows, pipe bends and Venturi tubes used to measure flow rates and Venturi scrubbers to clean flue gas emissions. Currently, alumina is being applied successfully not only to combat mechanical wear and erosion of the coal powder transport piping, but also for the handling of abrasive fly and bottom ash, for example, as liners for ash slurry lines.

Other applications include lined pipes to transport acidic wood chip slurries in pulp and paper mills, as well as mixtures of sand, tailings, cements and water for backfilling shafts and drifts in ore mines. Furthermore, alumina protection is applied to transport pipes of sand tailings in Canadian tarsand extraction facilities in northern Alberta, and for the lining of cyclones to clean and classify ores in all fields of wet mineral processing (Foster, 1996).

Filters for Molten Metal Inclusions in molten metals, such as oxide skins, mold sand particles, inoculation reaction byproducts or furnace slags, can be efficiently removed by filtering through pressed cellular, extruded cellular, or foam ceramic filters manufactured from high-temperature-resistant, chemically inert alumina, and also mullite, silicon carbide, and stabilized zirconia. Alumina is particularly useful when filtering liquid aluminum alloys in the range of 750–850 °C, and copper-based alloys at 1000–1200 °C (Matthews, 1996). These ceramic filters are designed for use in either "batch" or "in-mold" filtration:

- In *batch filtration*, large quantities of molten metal may be filtered while being transferred from the melting kiln to the holding furnace, or from the holding furnace to the casting ladle.
- For *in-mold filtration*, the ceramic filter is placed across one of the channels of the mold. Since, in this design, the melt is filtered immediately before solidifying inside the mold, the cleaning efficiency is very high.

Spark Plugs Spark plug insulators, which typically contain 88–95% alumina, were first developed during World War II, as the then-existing mullite insulators (with a high silica content) were particularly susceptible to corrosive attack from the lead tetraethyl contained in the vehicles' fuel, and also from additives in the lubricants. Alumina provides to the spark plug insulators not only chemical inertness but also a wide range of temperature stability (from subzero to over 1000 °C), a reasonably high thermal conductivity required to avoid preignition, a high mechanical strength, and an ability to withstand voltages of the order of 20 kV at high temperatures.

Cutting Tool Bits The way in which the cutting speed of metal milling has been increased over time as a response to the development of hard tool and specialty steels requiring tool materials with ever-increasing hardness, toughness, and wear resistance, is shown graphically in Figure 7.9. Cutting tool inserts manufactured from alumina were first developed in Germany during World War II, when tungsten carbide as the then predominant material for such purposes was in scarce supply. Today, alumina is alloyed with other ceramics such as zirconia, silicon nitride, and titanium carbide ("black" alumina) and nitride, so as to increase the cutting speed beyond 2000 m min^{-1}. The basic properties of three different cermet cutting tools, produced by Sandvik Coromant, are compared in Table 7.2.

The key to producing functional and longlasting alumina cutting tools is to maximize the impact strength and to minimize not only the abrasive, adhesive, and solution wear but also the plastic deformation. Apart from normal wear and tear, ceramic cutting tools undergo a special type of wear caused by solid-state welding reactions between the tool material and the metal turnings. This leads to a zone of intense shear within the turning, to high local temperatures exceeding 1000 °C, and to chemical reactions between the tool materials and the milled metal. This leads in turn to: (i) abrasive wear of the metal surface, caused by hard particles being broken off the cutting tool; (ii) adhesive wear among tools and turned or

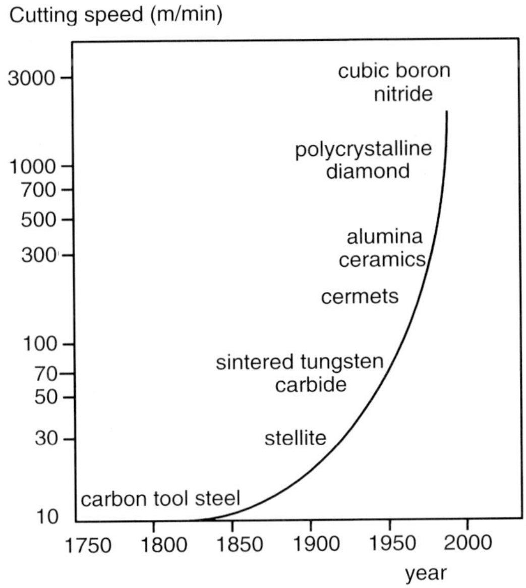

Figure 7.9 Trends in the speed of cutting tool materials and inserts. Adapted from Tesaki and Tanaguchi (1984).

Table 7.2 Characteristics of selected ceramic cutting tool materials.

	Oxide ceramics	**Mixed ceramics**	**SiAlON**
Sandvik grade	CC620	CC650	CC680
Composition	$Al_2O_3 + ZrO_2$	$Al_2O_3 + ZrO_2 + TiN + TiC$	$Si_3N_4 + Al_2O_3 + Y_2O_3$
Hardness at RT (HV1.0)	1700	1900	1800
Hardness at 1000 °C (HV1.0)	650	800	950
Flank wear resistance	High	Very high	Medium
Solution wear resistance	Very high	High	Low
Bulk toughness	High	Medium	Very high
Edge chipping resistance	Medium	Low	High
Thermal shock resistance	Low	Medium	High

RT = room temperature.

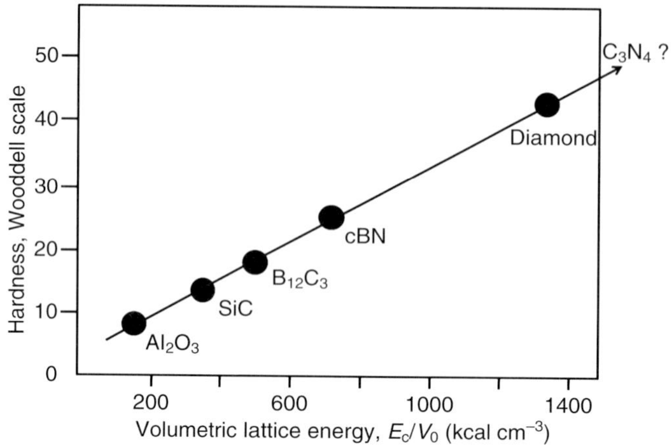

Figure 7.10 Energy required to break down ceramic cutting materials.

milled metals; (iii) diffusional or solution wear by chemical reactions occurring between the ceramic and the metal; and (iv) a general plastic deformation by the combined action of stress and temperature and, in turn, induced fracture processes. The typical countermeasures taken to overcome this situation include: (i) increasing the abrasive hardness of the cutting tool material (Figure 7.10); (ii) improving the cohesive bonding within the polycrystalline cutting tool so as to prevent strong adhesive bonding to the turnings; and (iii) selecting ceramic materials with a low solubility in the metal, or coating them with TiC or TiN.

Specific routes to realize improvements include: (i) the use of a smaller grain size by adding magnesia so as to suppress grain growth during sintering; (ii) alloying with zirconia; and (iii) adding TiC to reduce attrition wear at the tool flanks and also to increase the thermal conductivity and hence improve the thermal shock resistance. This involves the incorporation of SiC whiskers and the addition of TiN, BN, or SiAlON (see Section 11.3.4.4).

Several ceramic cutting materials are compared, on the basis of the resistance they show against abrasive wear, in Figure 7.10. The Wooddell abrasion hardness measures the amount of work required to break down a unit volume of material. The data in Figure 7.10 indicate that, although alumina is certainly not the ceramic material of choice in terms of its abrasive stability, for economic reasons it remains a strong contender for the cutting tool market.

Grinding Materials Owing to its high abrasion (Rosival) hardness, alumina has been used since antiquity in the form of smirgel (emery) to cut and polish less-resistive materials, including rocks, steel, nonferrous metals and jewelry. Archeological evidence exists of its use in Neolithic China (ca. 4000–3500 B.C.; Lu *et al.*, 2005) and in Bronze Age Minoan Crete (ca. 1700–1500 B.C.; Heimpel *et al.*, 1988). In order to produce alumina for loose or cemented grinding materials, the raw material is first melted in an electric furnace using graphite elec-

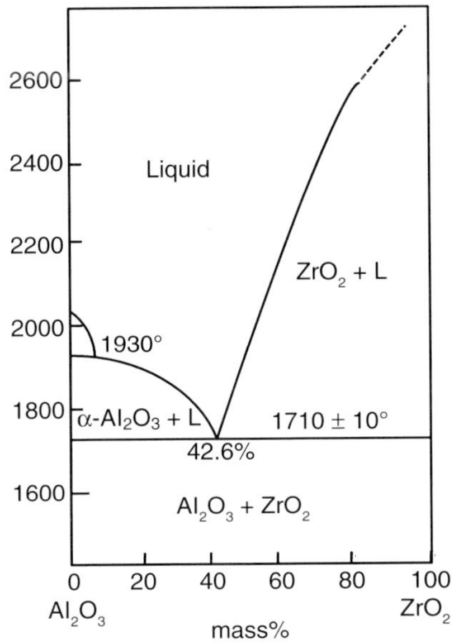

Figure 7.11 Binary phase diagram Al_2O_3–ZrO_2 (after Alper, 1967).

trodes. The solidified melt is then broken up into grains of several sizes and, after classification by sieving, is bonded with resin or a glass-bonding agent to produce grinding wheels.

Since the 1960s, the expansion of the steel industry has triggered efforts to improve the quality of the surface finish, to lower the cost of cutting and grinding, and to create new grinding applications. In particular, when steel was cast into ingots that were treated in holding furnaces before roll milling, it was first necessary to remove any slag and scale imperfections by spot grinding and increasingly, also to condition all of the ingot surfaces. This required not only the development of more durable abrasives in coarse grit sizes, but also the use of stronger grinding wheels and automated machines (Patchett, 1996). One means of producing grinding abrasives with higher durabilities was to alloy alumina with zirconia, by melting together the precursor ceramics and rapidly quenching the resultant melt. In this way, unique grinding properties were achieved with a eutectic composition of 57.4 mass% Al_2O_3 + 42.6 mass% ZrO_2 (Figure 7.11). The microstructure developed during rapid solidification consisted of dendrites of alumina in an alumina–zirconia matrix. The amount of tetragonal zirconia present was also shown to have a critical influence on the properties of the end product.

Ballistic Armor Alumina is used to produce lightweight armor systems for personal and vehicular protection against ballistic projectiles. Currently, the tiles are produced by hot-pressing or slip-casting, with subsequent sintering. While alumina has the same inherent advantages, such as easy availability, a well-known

Table 7.3 Critical properties of ceramic armor materials (Matchen, 1996).

Property	Al_2O_3	SiC	$B_{12}C_3$
Density ($Mg\,m^{-3}$)	3.5–3.9	3.13	2.45
Hardness KN ($kg\,mm^{-2}$)	2260	2800	2850
Elastic modulus (GPa)	370	400	450
Flexural strength (MPa)	275	460	345
Sonic velocity ($km\,s^{-1}$)	5	11.5	13

processing technology, and comparatively low cost, it faces competition from other ceramics such as silicon carbide (and, in particular boron carbide) that exhibit a lower areal density, a greater hardness, and a substantially higher sonic velocity (Table 7.3).

The projectile defeat system consists of sandwiching the ceramic plates between a ballistic Nylon™ spall cover and a Kevlar™ or glass-fiber-reinforced plastic backup plate material (see Figure 11.5). The mechanism of energy dissipation of the impacting projectile follows a three-step event (Matchen, 1996): (i) destruction of the projectile tip within 9 μs and yielding of the backup polymeric material at the conoid interface; (ii) erosion of the projectile by the ceramic fragments within 9–15 μs, with 40% of the impact energy carried off by eroded projectile material; and (iii) absorption by dynamic deflection of the remaining 60% of impact energy after 15 μs by the backup material.

Electronic Substrates High-quality alumina substrates are used ubiquitously in IC devices as a substrate onto which the electronic components (i.e., the source, gate and drain) of a transistor for modern very large-scale integration (VLSI) applications can be built. It can also be used as a thermal management ceramic in radiofrequency (RF) power transistors, when the substantial waste heat generated by the electronic switching processes must be dissipated both rapidly and efficiently. The manufacture of alumina substrates begins with the mixing of ultrapure alumina powder with additives and organic binders, and the formation of a "green sheet" into which holes are drilled. After having applied a conducting line network by screen printing, the green sheets are laminated into thicker sheets. The organic binder is then removed by heating at low temperature, after which the sheets are sintered at temperatures >1500 °C in a hydrogen atmosphere.

Both, insulated gate bipolar transistor (IGBT) and metal-oxide-semiconductor (MOS) modules for high-power applications require the use of direct copper-bonded (DCB) alumina substrates. In this case, thin copper sheets are bonded to one or both sides of a thin alumina or AlN plate, by heating to a temperature slightly above the melting point of copper. Conducting paths are then etched into the copper sheets, using a technology adapted from the manufacture of printed circuit boards. The substrate is then attached to a heat sink by its being soldered to the bottom copper layer. The suitability of DBC substrates in power electronics

Table 7.4 Potential ceramic materials for electronic substrates (Matweb, 2008; Accuratus Corp., 2008). Data for Cu, Al and Si are given for comparison. All data are subject to fluctuations depending on purity and processing conditions.

Material	λ ($W m^{-1} K^{-1}$)	ε (1 MHz)	α ($\times 10^{-6} K^{-1}$)	R ($\Omega \cdot cm$)	$\tan\delta$ (1 MHz)	U_c ($kV mm^{-1}$)	σ (MPa)
Al_2O_3 [a]	35	9.8	8.4	$>10^{14}$	0.0002	16.9	380
Si_3N_4 [b]	30	10.3	3.3	$>10^{7}$	0.143	10–15	830
BN	70–120	4 [c]	−0.7	$>10^{14}$	0.0012	370	15–25
AlN	140–180	9	4.5	$>10^{11}$	0.004–0.01	14–30	320
BeO	285	6.7	6.8	$>10^{13}$	0.0012	10–14	200
SiC–BeO	270		3.7	$>10^{14}$			450
SiC	120		4.0	10^{2}–10^{6}			550
Cu	390		20	$1.7 \cdot 10^{6}$			
Al	230		25	$2.7 \cdot 10^{6}$			
Si	125	11.8	3.5–4.4	0.01			
Diamond	2000	5.7	1.1	10^{13}–10^{16}		1000	

a) 99.5%.
b) 8.8 GHz.
c) Cubic high-pressure (HP) modification.

is based especially on the high thermal conductivity of Al_2O_3 (up to $35\,W m^{-1} K^{-1}$) and AlN ($140–180\,W m^{-1} K^{-1}$), and the high thermal capacity and heat dissipation of the thick copper layer (200–600 μm).

Electronic substrate materials should possess a high thermal conductivity, λ, a high electrical resistivity, R, a high flexural strength, σ, a low dielectric permittivity, ε, a low loss tangent, δ, a high dielectric strength, U_C, and a coefficient of thermal expansion, α, which is as close as possible to that of the semiconductor material used in the device. Two important figures of merit are $\lambda > 200\,W m^{-1} K^{-1}$ and $\varepsilon < 4$. While several ceramic materials are available which meet the λ requirement (AlN, BeO, SiC–BeO; Table 7.4), there is none that complies with the ε condition, this being met only by polymers that, unfortunately, demonstrate thermal conductivities several orders of magnitude below the target value. Although some glasses, as well as cordierite and steatite, have permittivities in the range of 5–6, these again fail to meet the λ condition.

As is evident from the data in Table 7.4, alumina is not the most suitable material to fulfill the requirements for an ideal electronic substrate. In particular, its thermal conductivity is among the lowest listed, the rather high dielectric permittivity may lead to inductive crosstalk and noise generation as well as signal delay in the electronic circuit, and the high coefficient of thermal expansion does not match that of silicon, so that differential stresses are to be expected if the heat generated is not efficiently dissipated. On the other hand, the dielectric strength, U_c, the electrical resistivity, R, and the flexural strength, σ, are suitably high for application in DCB power transistors. However, the most important reason to use alumina is its unmatched cost effectiveness when compared to the other ceramic options.

Owing to their high thermal conductivities and comparatively low dielectric permittivities, AlN (and in particular BeO; Schaetzle, 2007) appear to be good candidates. Indeed, they are already used in a variety of applications, even though the toxicity of BeO is a matter of concern. AlN is an excellent substrate material for thermal management, with a high λ and a coefficient of thermal expansion close to that of silicon to ensure good thermal cycling performance. However, it shows limited signal delay management owing to the high value of ε. Such a signal delay management is provided by polyimide with $\varepsilon < 4$ but a very low value of λ. AlN/polyimide composites might represent a suitable compromise, however, and research in this area is currently being conducted (e.g., Akiyama *et al.*, 2007). The ultimate substrate material is diamond, with $\lambda = 2000\,\mathrm{W\,m^{-1}\,K^{-1}}$, $\alpha = 1.1 \cdot 10^{-6}\,\mathrm{K^{-1}}$, $\varepsilon = 5.7$, and an unmatched dielectric strength of about $1\,\mathrm{MW\,mm^{-1}}$ (Table 7.4). The Sumitomo Electric Corporation, in Japan, already produces synthetic diamond substrates of small size and suitable thickness as heat sink material for ICs, and CVD diamond film are currently being investigated for thermal management purposes (Schneider *et al.*, 2007).

High-Temperature Sodium–Sulfur Battery This battery employs the ionic conductivity of β-alumina, that acts as a diaphragm containing liquid sodium which serves as the negative pole of the battery (Figure 7.12). The steel casing, which usually is

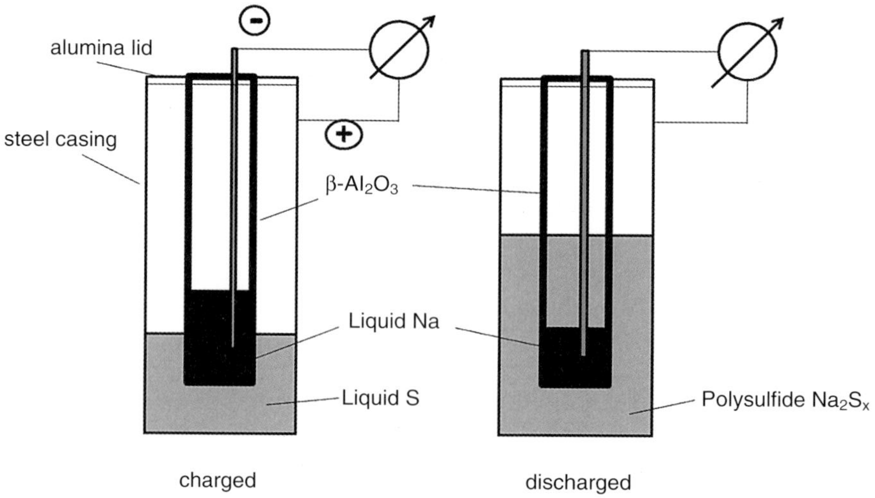

Figure 7.12 Schematic representation of a high-temperature, sodium–sulfur battery.

protected at the inside by a thin coating of molybdenum or chromium to counteract corrosion by liquid sulfur, is sealed off hermetically from the atmosphere by an airtight lid made from dense alumina. This outer steel container serves as the positive pole of the cell.

During the discharge phase, molten elemental sodium donates electrons to the external circuit, and hence serves as the anode. The liquid sodium is separated from the outer steel cylinder (which contains the sulfur absorbed into a carbon sponge) by a cylinder made from ion-conducting β-alumina. As β-alumina is a good conductor of sodium ions, but a poor conductor of electrons, self-discharge is effectively avoided. The electrons given off by the sodium atoms drive an electric current through the molten sodium to the contact, through the electrical load, and back to the sulfur container. Another electron reacts with sulfur to form S_n^{2-}, sodium polysulfide, such that the discharge process can be represented by the equation:

$$2Na^+ + 4S^{2-} \rightarrow Na_2S_4 + 6e \quad (7.5)$$

The cell voltage is approximately 2 V and, as the cell discharges, the sodium level drops. However, during the charging phase the reverse process takes place; that is, on external electric charging field the polysulfide will be decomposed, releasing sodium such that the sodium level rises. Once the cell is running, the heat produced by the charging and discharging cycles is sufficient to maintain the operating temperatures, and usually no external heat source is required.

Despite one failed attempt to use NaS batteries to power electrically operated automotive vehicles (Ford's "Ecostar" in 1971), sodium sulfur batteries are currently under development (Oshima *et al.*, 2004) as a possible energy storage application to support renewable energy plants, specifically wind farms and solar generation plants. In the case of a wind farm, the battery would store energy during times of high wind but low power demand; the stored energy could then be discharged from the batteries during peak load periods. Recently, NGK Insulators Ltd of Japan supplied the Japanese Futamata wind farm with seventeen NaS batteries having a total storage capacity of 34 MW, and which will be used to store the electricity generated during night-time operation. During the day time, the battery output is combined with wind turbine-generated electricity and supplied to the power grid operated by Tohoku Electric Power Co. Inc. of Japan (Yomogita, 2008).

In addition to this "power shifting," it is conceivable that NaS batteries could be used throughout the day to assist in stabilizing the power output during wind fluctuations. These types of battery present an option for energy storage in locations where other storage options are not feasible due to remoteness or terrain constraints.

Because of its high energy density (Table 7.5), the NaS battery has also been proposed for outer-space applications, as these cells can be made space-hardened. In fact, in November 1997 a test sodium sulfur cell was flown on the STS-87 Space Shuttle mission to demonstrate its operation in space. The experimental battery had a mass energy density of 150 W·h kg^{-1} (threefold the energy density of a nickel hydrogen battery and about eightfold that of a conventional lead-acid battery) and, when operated at 350 °C, performed satisfactorily over a 10-day period whilst in orbit.

Table 7.5 Comparison of a typical NaS battery and a conventional lead-acid battery (Kamibayashi, 2001).

Property	NaS battery	Lead-acid battery
Volume energy density ($kWh\,m^{-3}$)	170	40
Mass energy density ($kWh\,ton^{-1}$)	117	20
Charge/discharge efficiency (%)	>86	>84
Maintenance demand	None	Regular
Number of life cycles	>2500	>1200

Although concerns have been expressed in relation to the high temperature of operation (300–350 °C) and the risk of explosion on contact with moisture if the sealed steel container were to be breached, recent developments have proven the operational safety of the device (Oshima *et al.*, 2004). Concerns of failure due to the brittleness of the β-alumina electrolyte cylinder can be alleviated by reinforcing its structure with zirconia particles.

Activated Alumina Historically, all transitional polymorphs of alumina obtained by dehydroxylation of aluminum hydroxide below 1000 °C were termed "active" or "activated" alumina, since they served as an important carrier for catalysts. For example, catalysts used in the oxychlorination of ethylene (Degussa, 1999), and noble metal (Pt, Pd) catalysts used in the purification of exhaust gases (ICT, 2002), rely on porous, reactive support materials that can be provided by γ/η-alumina surfaces. Other applications include adsorbents to remove water from organic liquids, drying media of gases, the removal of fluoride from drinking waters with excessively high natural fluoride levels, color and odor removal from industrial effluents, chromatographic uses in the laboratory, the use of catalysts in their own right (e.g., a catalyst to remove hydrogen sulfide from natural gas; the Claus process), reforming catalysts to boost the octane numbers of gasoline, and automotive exhaust catalysts to oxidize unburned hydrocarbons and toxic carbon monoxide to water and carbon dioxide. Further details on the production routes and uses of activated alumina can be found in Evans (1996).

7.2
Zirconia

7.2.1
Introduction

Over the past forty years, zirconia has been developed into a highly sophisticated advanced ceramic material that is utilized in many important applications via the exploitation of its superior mechanical and unique functional properties.

The exceptional *mechanical properties* of zirconia are mainly based on the phenomenon of transformation toughening (this is described in detail below). The delayed martensitic transformation of the tetragonal high-temperature modification into the monoclinic low-temperature modification by adding stabilizing oxides leads to increased stress intensity factors K_{Ic} and also *R*-curve behavior. Hence, appropriately stabilized zirconia has been dubbed "ceramic steel" (Garvie *et al.*, 1975).

The *functional properties* of fully stabilized zirconia (FSZ) and partially stabilized zirconia (PSZ) are based largely on the material's ionic conductivity at elevated temperatures – a property which is utilized in gas sensors and electrolytes for high-temperature solid oxide fuel cells (SOFCs). High-temperature ionic conductivity is based on the existence of vacancies in the oxygen ion lattice caused by the addition of stabilizing cations with valencies lower than that of the zirconium matrix cation. Furthermore, the low thermal conductivity of zirconia has led to applications in thermal barrier coatings for gas turbine vanes and blades. The bioinert properties of yttria-stabilized tetragonal zirconia polycrystal (Y-TZP), coupled to its high strength and toughness, are currently utilized in femoral heads for hip endoprostheses (see Section 10.6).

7.2.2 Processing of Zirconia

The starting raw materials for the production of zirconia are the minerals zircon ($ZrSiO_4$) and, to a lesser extent, baddeleyite (β-ZrO_2), all of which are mined predominantly in Australia, South Africa, and the USA. The chemical syntheses of zirconia include thermal "demixing" via the carbothermal reduction of zirconium silicate above 1750 °C, the direct chlorination by a mixture of coke and chlorine below 1200 °C, and the "alkoxide" route by reacting zirconium silicate with sodium hydroxide or sodium carbonate and treating the resultant complex hydroxide with sulfuric acid to form substoichiometric zirconyl sulfate ($Zr_5O_8(SO_4)_2 \cdot xH_2O$) that will be calcined above 800 °C.

Carbothermal reduction. Thermal demixing by *carbothermal reduction* at temperatures >1750 °C generated by an electric arc is governed by the chemical reaction:

$$ZrSiO_4 + C \rightarrow ZrO_2 + SiO\uparrow + CO\uparrow. \quad (7.6a)$$

In a second processing step, the zirconia is purified by treating it with chlorine and hydrogen according to:

$$ZrO_2 + 2Cl_2 + 2H_2 \rightarrow ZrCl_4 + 2H_2O. \quad (7.6b)$$

The zirconium chloride is subsequently converted to zirconia (see below).

Direct chlorination. A variant of this process is the *direct chlorination*, which is conducted at temperatures between 800 and 1200 °C according to:

$$ZrSiO_4 + 4C + 4Cl_2 \rightarrow ZrCl_4\uparrow + SiCl_4\uparrow + 4CO\uparrow \quad (7.6c)$$

$$ZrCl_4 + H_2O \rightarrow ZrOCl_2 + 2HCl. \quad (7.6d)$$

Zirconium chloride evaporates and is subsequently condensed at between 150 and 180 °C, and thus is separated from $SiCl_4$ (which condenses only at the much lower temperature of−10 °C). $ZrCl_4$ is hydrolyzed with water and forms a saturated solution of $ZrOCl_2$ [Eq. (7.6d)]. On cooling, this zirconyl chloride crystallizes at between 65 and 20 °C, and the crystals are separated and dried at 85 °C; eventually, calcination at >800 °C yields zirconia. Alternatively, ammonia can be added to the zirconyl chloride solution to yield:

$$ZrOCl_2 + 2NH_3 + 3H_2O \rightarrow Zr(OH)_4 + 2NH_4Cl. \quad (7.6e)$$

Subsequently, the zirconium hydroxide is calcined to form zirconia.

Alkoxide demixing. The *alkoxide demixing process* is based on the following sequence of reactions:

$$ZrSiO_4 + 4NaOH \rightarrow Na_2ZrO_3 + Na_2SiO_3 + 2H_2O (T > 600\,°C) \quad (7.7a)$$

$$\text{or } ZrSiO_4 + 2Na_2CO_3 \rightarrow Na_2ZrO_3 + Na_2SiO_3 + 2CO_2 (T > 1000\,°C) \quad (7.7a')$$

$$Na_2ZrO_3 + H_2O \rightarrow \text{“complex hydroxide”} \quad (7.7b)$$

$$\text{“complex hydroxide”} + 2H_2SO_4 + xH_2O \rightarrow Zr_2O_2(SO_4)_2 \cdot 3ZrO_2 \cdot yH_2O \quad (7.7c)$$

$$Zr_2O_2(SO_4)_2 \cdot 3ZrO_2 \cdot yH_2O \rightarrow 5ZrO_2 + 2SO_2 + yH_2O (T > 800\,°C) \quad (7.7d)$$

Other possible processing routes include the calcia processing route (reaction of zirconium silicate with CaO at 1600 °C, and leaching the calcium silicate formed by treatment with HCl), and the plasma route (thermal dissociation of zirconium silicate in a plasma reactor above 2100 °C, and leaching the glassy silica formed with NaOH).

Calcia processing. The *calcia processing* reaction can be formulated as follows:

$$ZrSiO_4 + CaO \rightarrow CaZrSiO_5 (T > 1100\,°C) \quad (7.8a)$$

$$\text{or } ZrSiO_4 + CaO \rightarrow ZrO_2 + CaSiO_3 (T > 1600\,°C). \quad (7.8b)$$

The calcium silicate will subsequently be dissolved in HCl.

Plasma route. The *plasma route* can be expressed by the dissociation reaction:

$$ZrSiO_4 \rightarrow (ZrO_2 + SiO_2)(T > 2100\,°C) \quad (7.9)$$

On cooling, the silica solidifies in a vitreous state, and the crystalline zirconia suspended in the glassy matrix is liberated by dissolving the silica glass in NaOH. This process yields very fine-grained pure crystalline zirconia, since any natural impurities of zircon (such as titanium and iron oxides) will effectively be removed by incorporation into the glassy silica phase.

In order to produce Y_2O_3-stabilized zirconia with homogeneous distribution, sol–gel processes or coprecipitation have been adopted, based on the hydrolysis of stoichiometric mixtures of zirconyl chloride and yttrium trichloride (see, e.g., Carter *et al.*, 2009). The hydroxides formed are subsequently subjected to aceotropic distillation, drying, calcining between 850 and 950 °C, wet milling, and spray-drying.

7.2.3
Structure of Zirconia

The crystallographic data of the most important polymorphic structures of zirconia are listed in Table 7.6.

On heating, the monoclinic β-phase (natural mineral baddeleyite) of pure zirconia is stable up to 1170 °C (see Figure 7.15), after which it transforms to the tetragonal form that is stable up to 2370 °C. Beyond that temperature, the cubic phase exists up to the melting point at around 2700 °C. In the *monoclinic structure*, Zr^{4+} has the unusual coordination number of 7, where four oxygen atoms (O_{II}) form with the central zirconium atom an almost symmetrical tetrahedral arrangement (Figure 7.13a); the three remaining oxygen atoms (O_I) are strongly disordered. It is for this reason that there is a tendency for twinning. The twin plane (001) is composed of O_{II} atoms that are rather mobile and can easily be moved out of their equilibrium positions. In the projection onto the (010) plane (Figure 7.13b), the boundary of the twin plane (001) is shown as the dotted line. This is formed by rotation around a twofold axis through $x = ½$ and $y = ½$, and displaces one half of the twin against the other by ½ a.

Even though the *tetragonal form* has a coordination number of [8] its structure is likewise distorted. Four oxygen atoms surround the central zirconium atom in the form of a flattened tetrahedron at a distance of 206.5 pm, while the other four oxygen atoms are found at a distance of 245.5 pm at the vertices of an elongated tetrahedron rotated against the former by 90° (Figure 7.14a). The *cubic zirconia* (Figure 7.14b) has the face-centered cubic (f.c.c.) fluorite-type structure in which each Zr^{4+} atom has an eightfold symmetry against the oxygen atoms that are arranged in two ideal tetrahedra.

Table 7.6 Crystallographic data of zirconia.

Modification	Monoclinic	Tetragonal	Cubic
Space group	$P2_1/c$	$P4_2/nmc$	$Fm3m$
Lattice parameters (pm)	a = 516	a = 509	a = 512
	b = 519	c = 518	
	c = 530		
	ß = 98.9°		
Density ($Mg\,m^{-3}$)	5.83	6.10	6.09

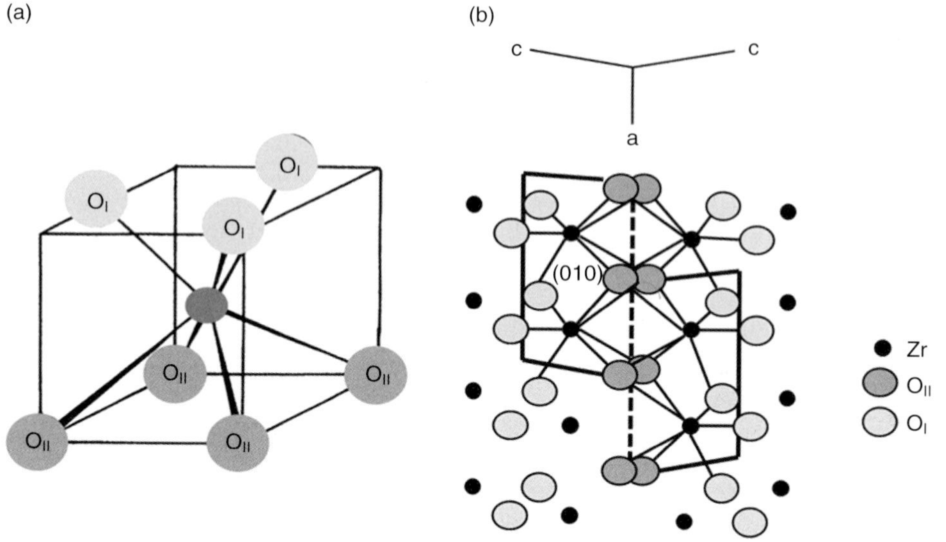

Figure 7.13 Structure of monoclinic zirconia. (a) Idealized ZrO_7 polyhedron, showing the two types of oxygen (O_I and O_{II}); (b) Projection onto the (010) plane, showing the plane of twinning (dotted line).

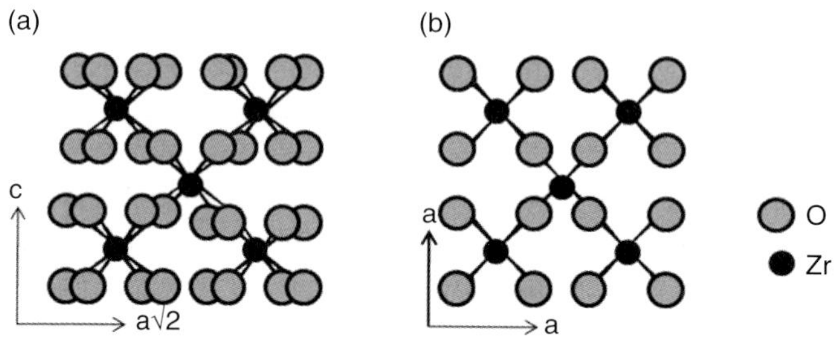

Figure 7.14 Structure of (a) tetragonal and (b) cubic zirconia.

The phase diagram of the binary system ZrO_2–Y_2O_3 is shown in Figure 7.15, where the various compositions and coating properties for different applications are indicated. The phase relationships of the binary system ZrO_2–Y_2O_3 exhibit several compositions, for which the grain sizes of the tetragonal (t)-phase, bending strengths, and fracture toughness are displayed. In general, the fracture toughness increases with a decreasing amount of yttria, whereas the thermal stability increases with an increasing yttria content. This requires trade-offs that depend on the application of the material. For 8 mass% (2.65 mol%) yttria, the phase assembly at ambient temperature should consist of a mixture of cubic and monoclinic zirconia. However, extensive XRD studies on plasma-sprayed coatings have demonstrated a tetragonal phase and only a small amount of monoclinic phase

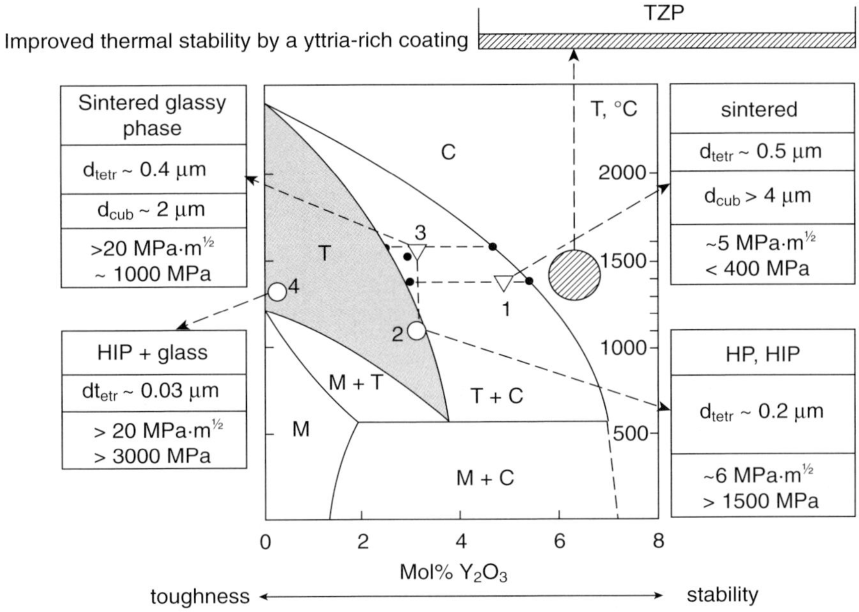

Figure 7.15 Phase diagram of the binary system ZrO_2–Y_2O_3 (for details, see the text). TZP = tetragonal zirconia polycrystal.

(e.g., Jasim *et al.*, 1992). This tetragonal (t′) phase does not transform to the monoclinic (m) phase because of the rapid splat cooling during plasma spraying. Rather, it is a tetragonal solid solution with a high yttria content that, on annealing at high temperatures (1300 °C), is slowly transformed by a diffusion-controlled mechanism to equilibrium tetragonal (t) zirconia with a low yttria content, and cubic (c) zirconia (Wu *et al.*, 1989).

7.2.4
Transformation Toughening of Zirconia Ceramics

The tetragonal–monoclinic phase transformation in zirconia shows a *martensitic* nature; that is, a diffusionless shear process at close to the velocity of sound that is characterized by a transformation hysteresis during the heating and cooling cycles. The width of this hysteresis (see Figure 7.16) is dependent on the processing history of the material, the amount and type of stabilizing ions, and also (heavily) on the grain size. Smaller zirconia particles transform at lower temperatures compared to larger particles. Excellent accounts of the martensitic transformation in zirconia have been provided by Evans (1984) and Behrens (1993).

The rate of the martensitic transformation and its mechanism are analogous to those encountered during the transformation of α-iron (b.c.c.) to γ-iron (f.c.c.) shown in Figure 7.17. However, there is one very important difference between the martensitic transformation in inherently ductile materials such as iron, and

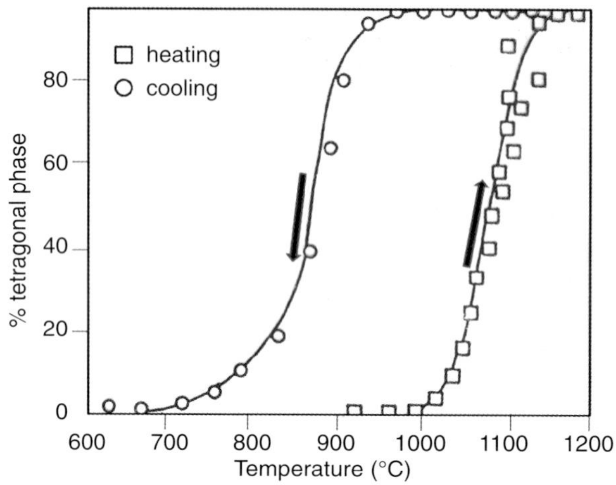

Figure 7.16 Hysteresis during the monocline–tetragonal phase transformation of undoped zirconia.

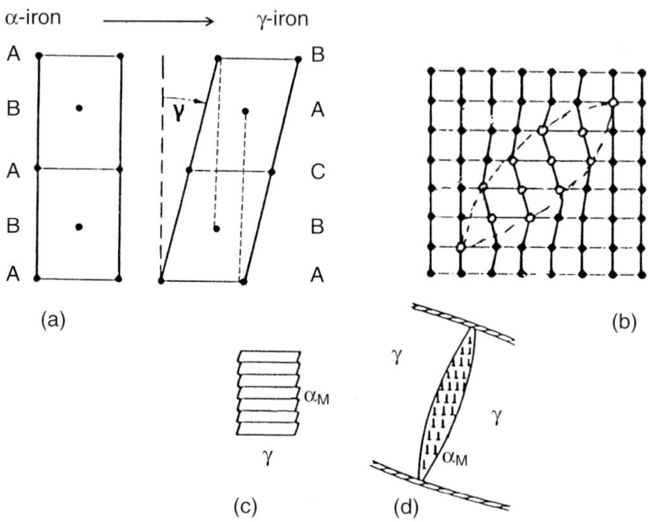

Figure 7.17 Mechanism of martensitic transformation of α- to γ-iron. (a) Shearing, during which the b.c.c stacking order ABABA... is transformed to the f.c.c. stacking order BACBA...; (b) Shearing within the matrix lattice; (c) Inner plastic deformation by sideways sliding of small crystalline slabs to maintain the volume of the martensite α_M phase within the matrix γ-iron phase; (d) Lamella of α_M within a γ-crystal at a grain boundary. The adjustment of strain is facilitated by edge dislocations (after Hornbogen, 2008). Reprinted with permission from Springer, Science and Business Media.

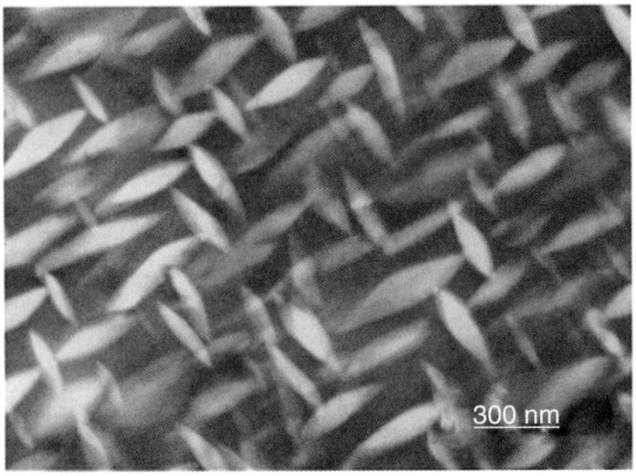

Figure 7.18 Transmission electron microscopy image showing spindle-shaped tetragonal zirconia precipitates within a monoclinic zirconia matrix phase in Mg-partially stabilized zirconia (Birkby and Stevens, 1996). Reprinted with permission of Trans Tech Publication, Millersville, DA, USA.

in brittle materials such as zirconia. In the latter case, the *von Mises criteria* for plastic deformation are not satisfied and hence the deformation process is accommodated by a change in shape that manifests as a serrated surface (akin to that shown in Figure 7.17c on a microscopic scale) and the formation of microcracks at the grain boundaries of monoclinic twins. Whilst these microcracks are very small, and well within the order of the martensitic twin thickness, they can link up to generate cracks of a critical size (Birkby and Stevens, 1996). For example, Ce-stabilized tetragonal zirconia polycrystals with $\sigma_f = 125$ MPa and $K_{Ic} = 35\ \text{MPa} \cdot \text{m}^{1/2}$ yield a critical flaw size of 2.5 mm (Tsukuma *et al.*, 1985).

Oblate spheroidal precipitates of the tetragonal zirconia phase within the matrix of monoclinic zirconia are shown in Figure 7.18.

Since this transformation results in a large change in volume (ca. 5%; see Table 7.6), in pure zirconia the elastic limit and yield strength will be exceeded, thus causing the formation of cracks and an eventual structural disintegration of the monolithic zirconia bodies. The addition of yttria, calcia, magnesia, ceria, or scandia will delay the transformation. As the charge of these aliovalent cations is generally less than that of four-valent zirconium, O^{2-} vacancies will form in the zirconia lattice, as shown in Figure 7.19 (these vacancies are responsible for the ionic conductivity of stabilized zirconia; see below). The aliovalent replacement reactions, in Kröger–Vink notation, can be described as:

$$ZrO_2 + Y_2O_3 \rightarrow 2Y'_{Zr} + V_O^{**} + 3O_O^X \tag{7.10a}$$

$$ZrO_2 + CaO \rightarrow Ca''_{Zr} + V_O^{**} + O_O^X \tag{7.10b}$$

Nucleation and the growth of spindle-shaped tetragonal particles occur that are stable against transformation, even at room temperature. The metastable

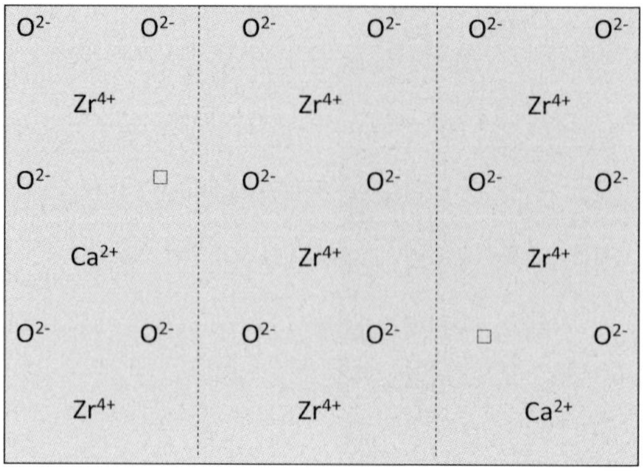

Figure 7.19 Replacement of Zr^{4+} ions by Ca^{2+} ions in the zirconia lattice creating vacancies □ in the oxygen lattice.

tetragonal particles induce tangential stresses and hence produce microcracks in the surrounding matrix that are subcritical in terms of the Griffith–Orowan fracture mechanics (Figure 7.20, upper left).

The classic Griffith–Orowan theory describes the relationship between the strength and toughness of brittle materials such as ceramics (Griffith, 1920; Orowan, 1949). In the simple basic equation of the theory, the stress to fracture σ_f is related to Young's elastic modulus E, the fracture energy γ, and the critical crack length c, by:

$$\sigma_f = \sqrt{(2E\gamma/\pi c)} \tag{7.11}$$

As the critical stress intensity factor (= fracture toughness) $K_{Ic} = \sqrt{2E \cdot \gamma}$, the fracture stress becomes

$$\sigma_f = \frac{1}{\sqrt{\pi}} \cdot K_{1C} \cdot c^{-1/2} \tag{7.11a}$$

Thus, a crack of critical length is able to move through the stress field, but will be deflected by the particles that transform spontaneously to the monoclinic form in the vicinity of the progressing crack (Figure 7.20, upper right). The crack energy will be absorbed, and the volume change will prevent the advancing crack from spreading further. This is expressed in an increase of the fracture toughness of the partially stabilized zirconia by this so-called *transformation toughening* mechanism (Figure 7.20, lower). This implies that the mechanical properties of such an engineered ceramic zirconia material are no longer limited by any process-induced flaws, since damage tolerance is available owing to the pseudo-plastic deformation behavior. Furthermore, the extension of the transformation zone (as shown in Figure 7.20) will keep a check on the population of the flaws, since each flaw will now be limited by the length of the transformation zone, and not by its individual length and the wide variability of processing flaws inherent in all ceramic materials. Hence, the reproducibility of the critical fracture stress will be considerably

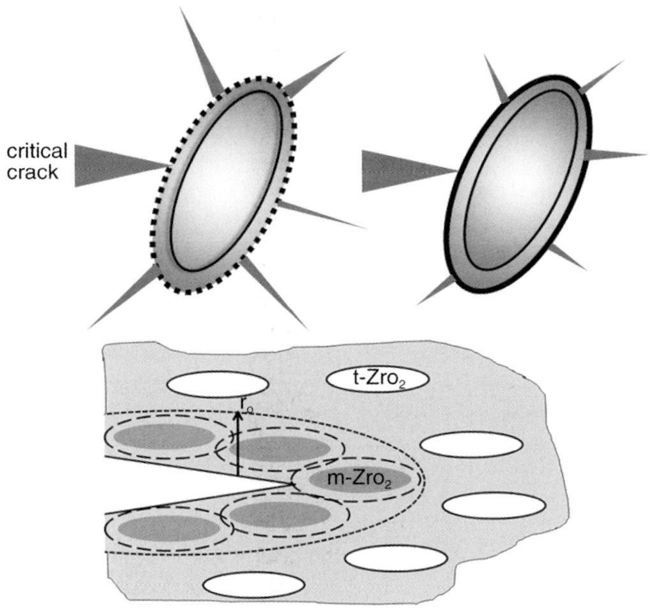

Figure 7.20 Schematics of the mechanism of transformation toughening of stabilized zirconia. Upper: Formation of subcritical microcracks around a transformed zirconia grain (left) and deflection of an arriving (critical) crack by the strain field around the transformed grain (right). Lower: A crack penetrating monoclinic zirconia with embedded untransformed tetragonal zirconia ($t\text{-}ZrO_2$) particles generates around its tip a stress field with radius r_0 that triggers a martensitic transformation t → m by releasing the matrix pressure on the t-phase. The volume expansion (3–5%) and existing shear stresses (1–7%) cause compressive stresses around the advancing crack tip, and halt its further movement.

increased, as is reflected in high values of the Weibull modulus *m* for stabilized zirconia ceramics.

Indeed, the K_{Ic} values of PSZ are very much higher than those of other engineering ceramics, varying between 7 (CaO–PSZ with 3.3 mass% CaO), 9 (MgO–PSZ with 3.4 mass% MgO), 10 (Y_2O_3–PSZ with 8 mass% Y_2O_3) and 11 $MNm^{-3/2}$ (CeO_2–PSZ). It was this unique feature that prompted the authors of one of the first reports describing the mechanism of transformation toughening of zirconia to refer to "ceramic steel" (Garvie *et al.*, 1975).

The bending strengths are in the range of 600 to 800 MPa, and reach a maximum of 1000 MPa for the Y-TZP used for femoral ball heads (see Section 10.6; also Figure 6.1). Mixed oxide ("duplex") ceramics can achieve high values of fracture toughness, typically 12 $MNm^{-3/2}$ (Si_3N_4 + 22 vol% PSZ) or 15 $MNm^{-3/2}$ (Al_2O_3 + 16 vol% PSZ) (see Figure 7.30).

In addition to improved mechanical strength and fracture toughness, transformation-toughened PSZ exhibits *R*-curve behavior; that is, the resistance against

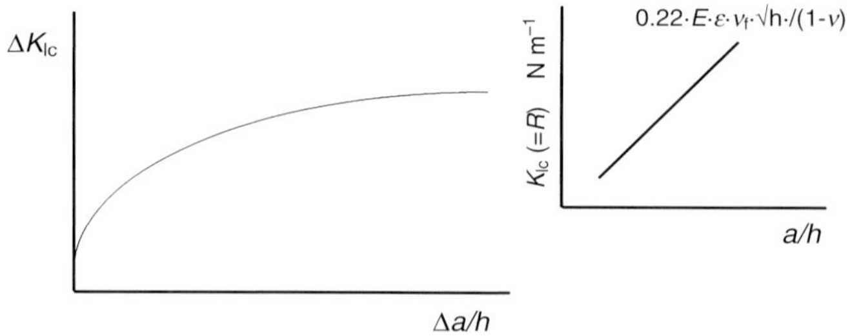

Figure 7.21 Schematic rendering of the *R*-curve behavior of stabilized zirconia. The resistance of zirconia against crack propagation increases with increasing relative crack length, a/h (a = crack length; h = length of transformation zone).

crack propagation increases with increasing crack length according to $K_{Ic} \propto E \cdot \varepsilon \cdot v \cdot \sqrt{h}/(1-\nu)$, where E is the modulus of elasticity, ε is strain induced by the phase transformation, v is the volume fraction of tetragonal zirconia, h is the length of the transformation (interaction) zone, and ν is the Poisson number (Figure 7.21).

While the use of the stress intensity factor K_{Ic} as a single-value measure of the toughness is appropriate for many materials, in some cases – notably in zirconia but also in cortical bone (Nalla *et al.*, 2005) – the fracture resistance actually increases with crack extension, promoting stable crack growth and requiring a resistance-curve (*R*-curve) fracture mechanics approach. In particular, *R*-curves are necessary to describe the fracture resistance of materials toughened by crack-tip shielding (Ritchie, 1988, 1999; Evans, 1990), that is, by extrinsic toughening mechanisms such as crack bridging, constrained microcracking, or *in-situ* phase transformations that develop in the wake of a crack as it extends. In such instances, while crack extension commences at a crack-initiation toughness, K_0, sustaining further crack extension requires higher driving forces until typically a "plateau" or steady-state toughness is reached (Figures 7.21 and 7.22). The corresponding slope of the *R*-curve can be considered a measure of the crack growth toughness.

Figure 7.22 shows that, in magnesia-PSZ, very high stress intensity levels can be obtained compared to alumina. Indeed, this was one of the reasons why Y-TZP was thought of as a superior ceramic material to supersede alumina for application in femoral heads for hip endoprosthetic implants (see Chapter 10). The difference in *R*-curve behavior appears to be based on the fact that transformation toughening observed in stabilized zirconia is much more effective at arresting crack movement than the existence of grain boundary bridges and grain interlocking mechanisms that are prevalent in alumina.

Compressive surface layers can form during processing of stabilized zirconia (see Figure 7.23), owing to the absence of a hydrostatic pressure near the free surface of a zirconia body.

Gentle abrasion and polishing of the surface causes more t-ZrO_2 grains to transform so that the thickness of the surface layer under compressive stress can be

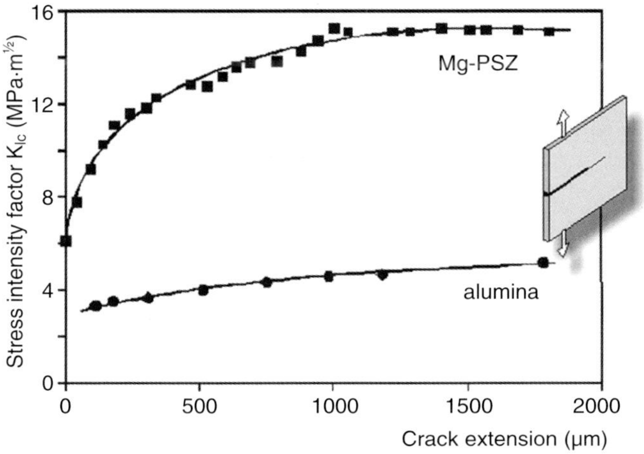

Figure 7.22 *R*-curve behavior of Mg-partially stabilized zirconia (Mg-PSZ) compared to that of alumina (Bleise and Steinbrech, 1994; Inghels *et al.*, 1990).

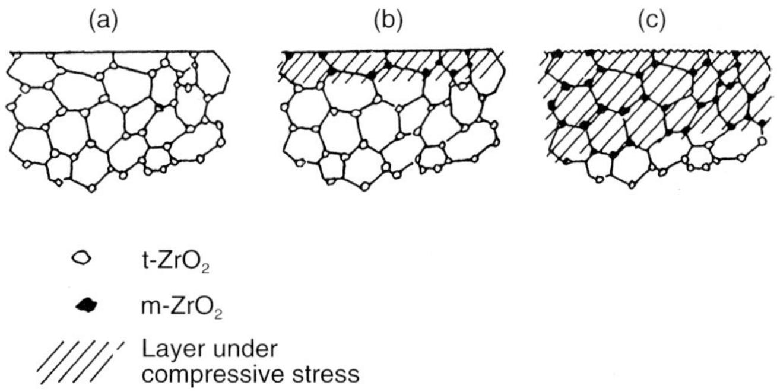

Figure 7.23 (a) Free surface of stabilized zirconia at sintering temperature; (b) During cooling, the t-ZrO_2 particles near the surface transform to m-ZrO_2 and thus induce a compressive stress in the matrix; (c) The thickness of the compressive layer can be increased by abrading/polishing the surface.

increased to 100 μm, and beyond, with the proportion of monoclinic phase decreasing exponentially as increasing amounts of material are removed. The maximum strengthening effect is reached if the thickness of the transformation zone subjected to compressive stress is higher than the critical crack length, but is still small compared to the cross-section of the ceramic body. This fact is very important: the stabilized zirconia material reacts much less sensitively to small surface defects than do other ceramics and glasses, since each service-imposed wear loss with a thickness less than the critical crack length is transformed to compressive stress. This effect is exploited to strengthen the surface of zirconia femoral heads for hip endoprostheses, by thorough polishing.

7.2.5 Binary Phase Systems

The technologically most important binary systems to achieve stabilization are based on the addition of various amounts of MgO, CaO, Y_2O_3, CeO_2, and Sc_2O_3 to zirconia. The fact that the addition of small amounts of dopant oxides allows quenching of the cubic phase from high temperature directly to a metastable tetragonal phase, thus avoiding the deleterious monoclinic phase, was first reported by Lefèvre (1963).

7.2.5.1 System ZrO_2–MgO

As shown in Figure 7.24, there is almost no solid solution of MgO in zirconia until the transformation temperature m → t has been reached at 1240 °C. The solubility increases slowly with temperature, but beyond 1400 °C a cubic solid solution with a eutectic composition of 87 mol% ZrO_2 + 13 mol% MgO becomes stable. Quenching of the material causes nucleation and the growth of oblate spheroidal precipitates of the tetragonal zirconia phase within the matrix of cubic zirconia (Figure 7.18). On further cooling below 1240 °C the tetragonal particles transform to the monoclinic phase, causing transformation toughening (as described in detail above).

7.2.5.2 System ZrO_2–CaO

This binary system is more complex owing to the existence of the stoichiometric compounds calcium tetrazirconate, $CaZr_4O_9$ at 20 mol% CaO and calcium zirconate, $CaZrO_3$ at 50 mol% CaO (Figure 7.25).

Below 2400 °C the tetragonal solid solution is stable up to a CaO content of about 6 mole%, but on cooling the system passes through region B, in which the tetragonal and monoclinic solid solutions are simultaneously stable; continued cooling results in a monoclinic solid solution + $CaZr_4O_9$. At between 6 and about 17 mol% CaO there exists, above the eutectic transformation temperature of 1140 °C, a two-phase field tetragonal ss + cubic ss, indicating the phase stability of CaO–PSZ. Quenching below 1140 °C leads to the transformation of t_{ss} to m_{ss} within a matrix that is still cubic. In contrast, slow cooling to 1000 °C results in the phase assembly t_{ss} + $CaZr_4O_9$. Within the cubic phase field (17–20mol% CaO) a similar reaction takes place, showing transformation from c_{ss} to t_{ss} between 1140 and 1000 °C, and from t_{ss} to m_{ss} below 1000 °C.

7.2.5.3 System ZrO_2–Y_2O_3

The phase relations of the very ZrO_2-rich side of the binary system have already been shown in Figure 7.15. The most significant characteristic of this technically enormously important system is the decease of the t → m transformation temperature with increasing yttria content. With increasing temperature, there exists

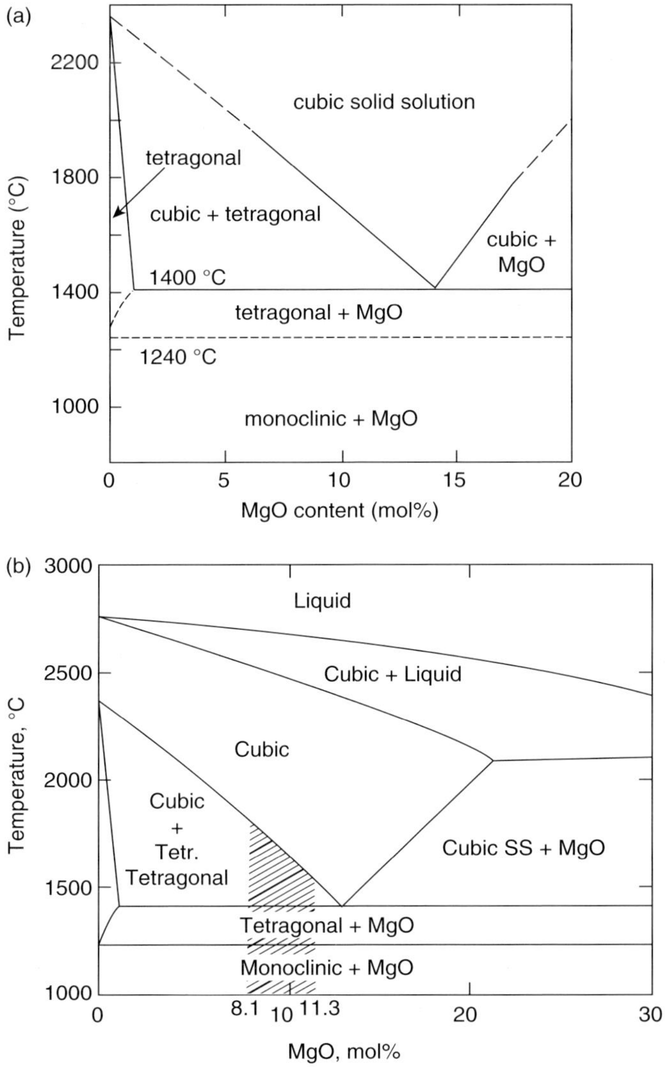

Figure 7.24 (a) Subsolidus and (b) high-temperature fields of the binary phase diagram ZrO_2–MgO (Grain, 1967). Reprinted with permission from Wiley-Blackwell, Oxford, UK.

a small area in which, above the transformation temperature, the two-phase assembly t_{ss} + m_{ss} is stable, but beyond this phase field the transformable tetragonal phase is stabilized between 0 and about 4 mol% Y_2O_3. An increasing yttria content leads to the wide stability field of t_{ss} + c_{ss} such that, eventually, a homogeneous cubic solid solution will be formed that exists between the melting point and ambient temperature. This is the field of existence of *fully stabilized zirconia*.

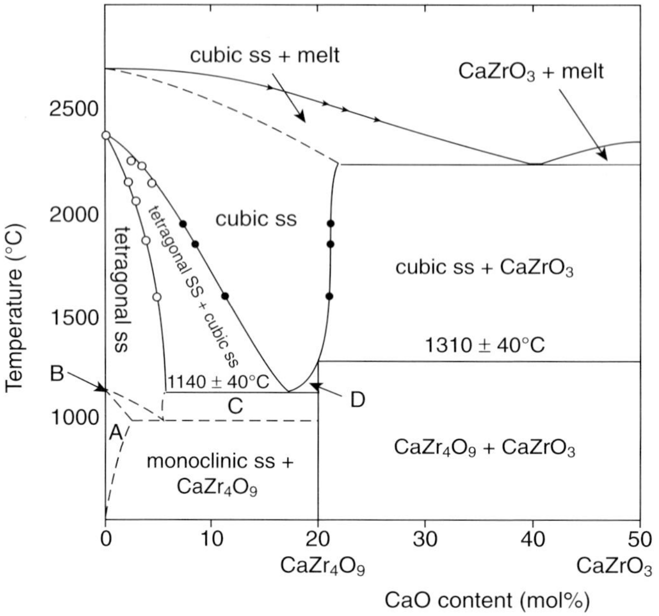

Figure 7.25 Equilibrium binary phase diagram ZrO_2–CaO. A = monoclinic solid solution; B = tetragonal and monoclinic solid solutions; C = tetragonal solid solution + $CaZr_4O_9$; D = cubic solid solution + $CaZr_4O_9$ (Stubican *et al.*, 1984). Reprinted with permission from Wiley-Blackwell, Oxford, UK.

Table 7.7 Transformation of nontransformable t′-zirconia on annealing (Jasim, Rawlings, and West, 1992).

Temperature (°C)/time (h)	Zirconia phases	Y_2O_3 (mol%)			
		t′	t	c	m
1050/1000	t′ ≫ c ≫ m				
1200/ 170	t > t′ ≫ c ≫ m	7.5	4.0		
1300/ 500	t > c > m		5.0	15.0	
1400/ 90	c > m ≫ t			15.0	4–6
1500/210	c > m > t		4.9	14.5	5.5

Very rapid quenching of molten zirconia, as experienced in plasma-sprayed zirconia coatings, leads to a stabilization of t′-zirconia (so-called "non-transformable tetragonal zirconia") between about 3 and 6 mol% yttria that contains high amounts of yttria (>7 mol%) dissolved in its lattice (Figure 7.27, b). On annealing, this yttria will be exsolved by transforming t′-zirconia to yttria-poor t-zirconia (4–5 mol%) and yttria-rich c-zirconia (>14 mol%) (see Table 7.7).

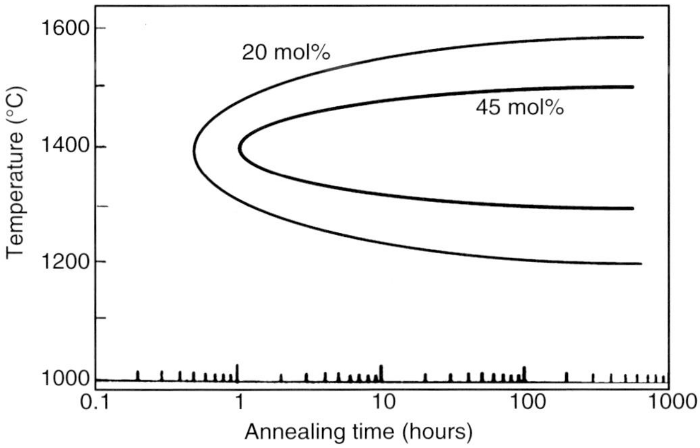

Figure 7.26 Temperature–time–transformation curves for Y-PSZ.

The fracture toughness of t′-zirconia was found to be less (~4 MN · $m^{-3/2}$) than that of t-zirconia (>7 MN · $m^{-3/2}$). The maximum K_{Ic} value was obtained in a product annealed at 1200 °C for 170 h, with a composition of 43% t + 34% t′ + 15% c + 8% m (Table 7.7) (Jasim *et al.*, 1992).

This exsolution of yttria from t′-zirconia follows a time–temperature-transformation (TTT) curve, as shown in Figure 7.26. The "nose" of the TTT curve at 1400 °C characterizes the least time for a given volume of t′-zirconia (20 mol%: 30 min; 45 mol%: 1 h) to transform to (t+c)-zirconia. This results from the competition between the driving force for the exsolution of yttria, which increases with increasing temperature, and the atomic mobility, which decreases with decreasing temperature. For details on the construction of TTT curves, see Grange and Kiefer (1941).

Figure 7.27a shows the full ZrO_2–Y_2O_3 system at high temperature, as reported by Rouanet (1971) and Skaggs *et al.* (1972). Here, the liquid maximum was found at 77 mol% ZrO_2/23 mol% Y_2O_3 and 2805 °C. Two major discrepancies pertained to: (i) the nature of the invariant point at 18 mol% ZrO_2/82 mol% Y_2O_3 (a eutectic at 2380 °C according to Rouanet, or a peritectic at 2483 °C according to Skaggs); and (ii) the existence of a two-phase region C_1 (cubic fluorite-type) + C_2 (cubic Tl_2O_3-type). The high temperature stability of a binary phase $Zr_3Y_4O_{12}$ with hexagonal structure was also debated (Ray and Stubican, 1977; Goff *et al.*, 1999).

7.2.5.4 **System ZrO_2–CeO_2**

As shown in Figure 7.28 for the binary system ZrO_2–CeO_2, the amount of dopant to be added to achieve a t → c transition is much higher than in the other systems discussed in the context of stabilization. At room temperature, a wide miscibility gap separates the monoclinic from the cubic phase field. Instead, the metastable tetragonal t′-phase transform to the cubic phase via an intermediate t″-phase, the existence of which has been confirmed by Raman investigations (Yashima *et al.*, 1994) to occur for a Ce molar fraction $0.65 < x < 0.85$. The intermediate phase is

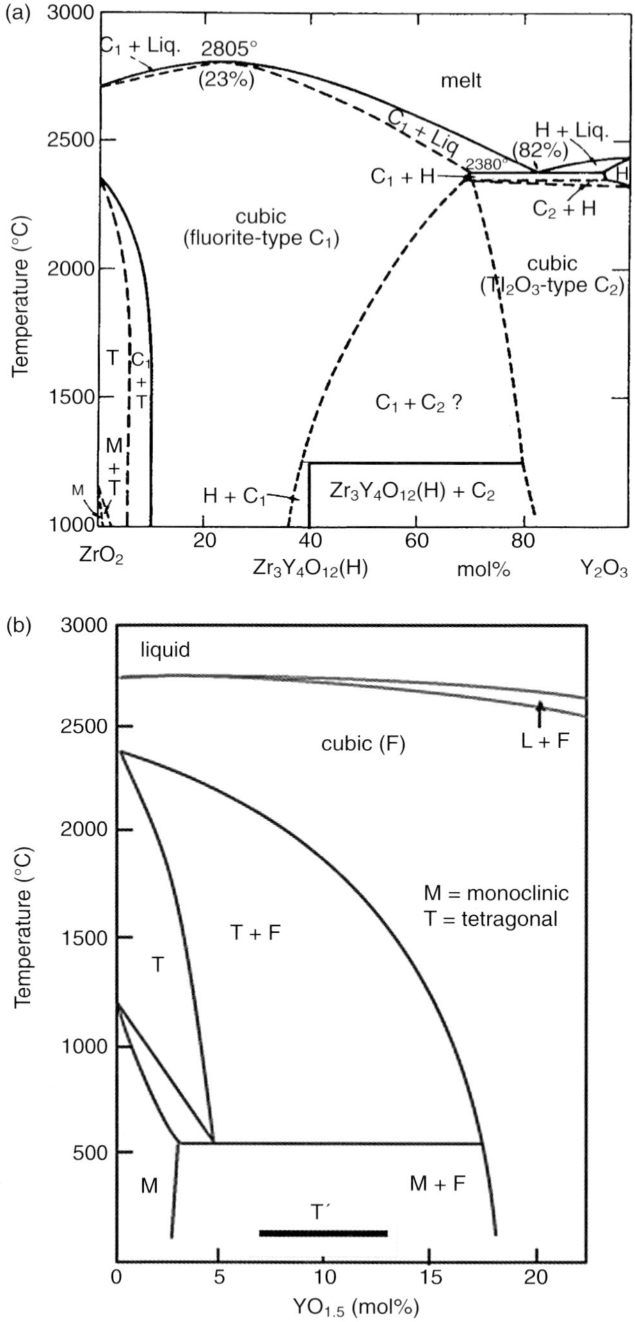
(a)
3000
2500
2000
1500
1000
Temperature (°C)
C1 + Liq.
2805°
(23%)
melt
C1 + Liq
H + Liq.
(82%)
2380°
C1 + H
H
C2 + H
cubic
(fluorite-type C1)
cubic
(Tl2O3-type C2)
T
C1
+
T
M
+
T
M
C1 + C2 ?
H + C1
Zr3Y4O12(H) + C2
ZrO2
20
40
60
80
Zr3Y4O12(H)
mol%
Y2O3
(b)
3000
2500
2000
1500
1000
500
Temperature (°C)
liquid
cubic (F)
L + F
M = monoclinic
T = tetragonal
T + F
T
M
M + F
T′
0
5
10
15
20
YO1.5 (mol%)

Figure 7.27 (a) High-temperature portion of the binary phase diagram ZrO_2–Y_2O_3 (Rouanet, 1971; Skaggs *et al.*, 1972). The region showing coexistence of the two cubic phases C_1 and C_2 is doubtful, as are the eutectic (or peritectic) relations at the high yttria site of the system; (b) Zirconia-rich side of the complete phase diagram. T′ depicts the range of existence of nontransformable tetragonal zirconia obtained during rapid quenching from the melt, as realized in plasma-sprayed coatings (Heimann, 2008). Reprinted with permission from Wiley-Blackwell, Oxford, UK.

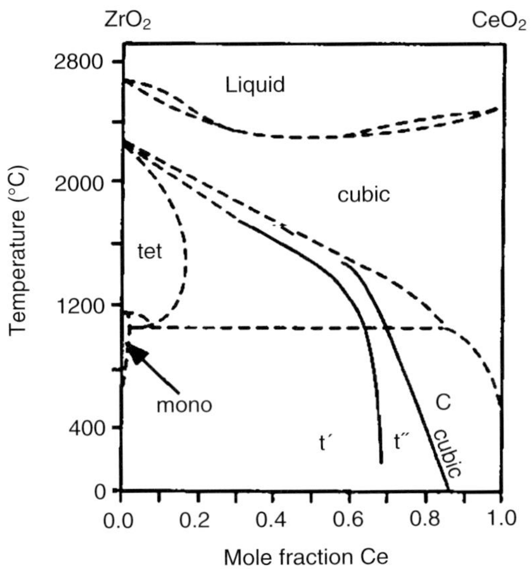

Figure 7.28 Binary phase diagram ZrO_2–CeO_2 (Heaney, 2000, after Yashima *et al.*, 1994). Reprinted with permission from The American Mineralogical Society and Wiley-Blackwell, Oxford, UK.

metrically cubic but symmetrically tetragonal (Heaney, 2000). In contrast to the systems ZrO_2–Y_2O_3 (Figure 7.27) and ZrO_2–Sc_2O_3 (Figure 7.29), there appears to exist a liquid minimum at about $x = 0.5$ and below 2400 °C.

7.2.5.5 System ZrO_2–Sc_2O_3

This system is complicated by the occurrence of several rhombohedral binary phases (β, γ, δ). The phase diagrams (Figure 7.29) shows that, with increasing scandia content, the monoclinic m-phase (which is stable between 0 and <2.5 mol% Sc_2O_3) transforms to a tetragonal t*-phase (that is stable between 5 and 8 mol% Sc_2O_3). However, beyond 8 mol% the binary phases $Sc_2Zr_7O_{17}$ (β, $9 < x < 13$ mol% Sc_2O_3), $Sc_2Zr_5O_{13}$ (γ, $15 < x < 23$ mol% Sc_2O_3), $Sc_4Zr_3O_{12}$ (δ, $23 < x < 40$ mol% Sc_2O_3) and Sc_2O_3 can be observed. The cubic zirconia phase (c) is obtained between approximately 7.5 and 10 mol% Sc_2O_3. The monoclinic phase m and the tetragonal phase t* convert with increasing temperature to a

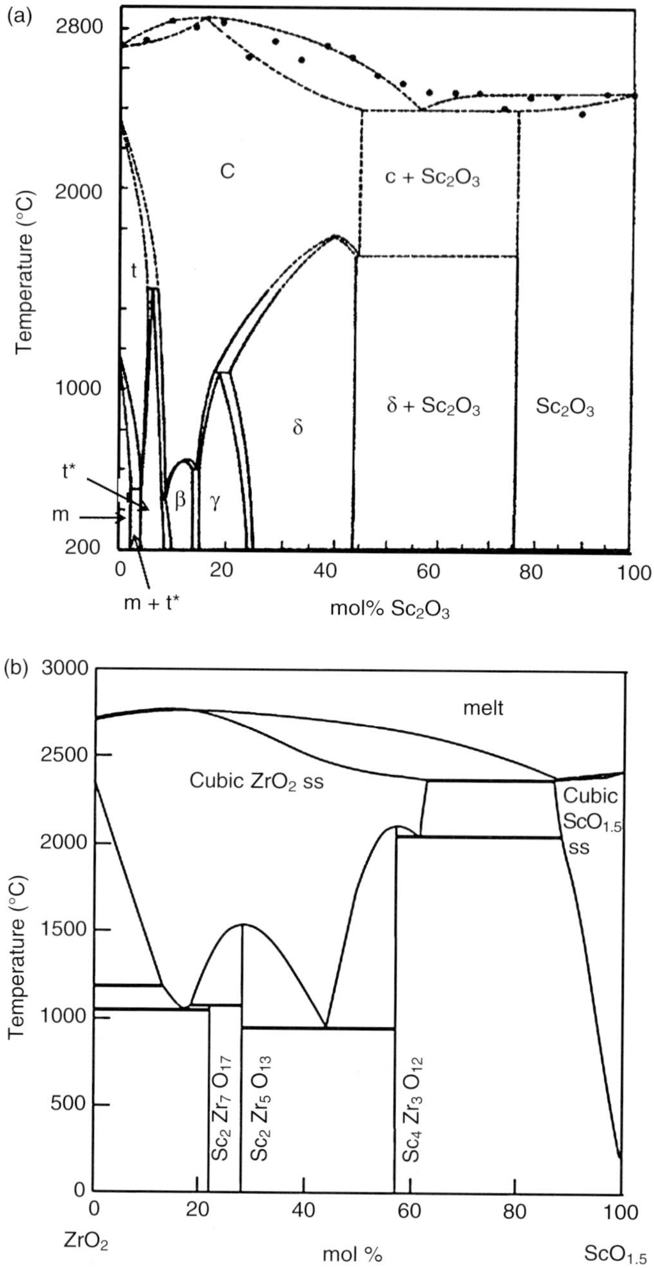

Figure 7.29 (a) Experimental (after Ruh *et al.*, 1977) and (b) calculated (Jacobson *et al.*, 2001) binary phase diagram ZrO_2–Sc_2O_3. Reprinted with permission from Wiley-Blackwell, Oxford, UK.

second tetragonal phase, t, and pass through a phase field in which m + t* simultaneously occur. The binary compounds β, γ, and δ decompose at higher temperatures to form cubic zirconia. The maximum melting point has been found to be at 2870 °C (10 mol% scandia), and there appears to exist a eutectic with 55 mol% scandia at 2400 °C; however, these phase relations and transformation temperatures require experimental re-evaluation. For example, the calculated phase diagram shown in Figure 7.29b suggest the existence of a true eutectic between the γ- and δ-phases at about 45 mol% $ScO_{1.5}$ and slightly below 1000 °C, instead of a "buried" peritectic point, as obtained experimentally by Ruh *et al.* (1977) and by Spiridonov *et al.* (1970). The calculated stability field of cubic Sc_2O_3 is much smaller than was obtained experimentally.

Since the electric conductivity of Sc-doped zirconia exceeds that of Y- or Ca-stabilized materials (see Figure 7.33), investigations have been undertaken to develop this material as an electrolyte for SOFCs operating at high temperature (Biedermann, 2002; see also Section 7.2.6.2).

7.2.6
Selected Applications of Zirconia

7.2.6.1 Structural Applications

The range of dense *structural ceramics* includes the processing of custom-blended ceramic alloy powders by isostatic pressing or extrusion to meet customer specifications. Both, Mg- and Y-stabilized zirconia alloys provide excellent erosion, corrosion and abrasion resistance, along with temperature resistance, fracture toughness, and strength. These ceramic alloy components are used in extreme service applications that take advantage of their superior wear- and corrosion-resistant properties. The numerous applications of these materials in chemical- and oil-producing operations continue to provide an economic payback by virtue of their reduced maintenance requirements and less down time.

As a result of the partial stabilization of zirconia (PSZ) ceramics, including mixed-oxide zirconia–alumina composite ceramics (see Section 7.1), these materials exhibit a high structural resilience that is utilized in devices as diverse as hard and tough knives for the high-speed cutting of paper and polymers, wear-resistant roller bearings, ultra-tough femoral ball heads for hip endoprostheses, wire-drawing dies, valve seats, piston heads and exhaust parts of automotive internal combustion engines, and scrapers and screw-type winding engines in mining. Other structural applications include precision ceramic balls, press tooling, dental blanks, extrusion tooling, and wear plates. Specialty products cover the range of porous ceramic filters, diffusers and adsorbent filter/drier cores for refrigeration applications, dispersion parts, vacuum chucks, and inkpads.

Wear- and Erosion-Resistant Applications Zirconia's exceptional resistance to erosive wear and corrosion has led to its application in pump component parts subjected to high stress, such as shafts, couplings, or thrust parts. These types of pump are currently utilized as high-performance sludge and process pumps for the chemical industry, while applications abound in the petrochemical industry as valves, seats, sleeves, liners and nozzles for pumps operating under rough environmental conditions, including sour gases, highly abrasive sand, and elevated temperatures (Ogando, 2005). As an example, Mg–PSZ liners provide a superior abrasion resistance to tungsten carbide, and a much better impact resistance than silicon carbide. The dense structural ceramic material is usually precision ground using conventional diamond cutters and the wheels of surface grinders, computer numerical controlled (CNC) milling machines, jig grinders, center-less grinders, CNC lathes, ID grinders, and so on.

The high-temperature stability of zirconia (melting point >2700 °C) and its reasonable thermoshock resistance (ΔT = 300–400 °C), coupled with an exceptionally low thermal conductivity (<2 W m^{-1} K^{-1}) and a high thermal expansion (>10 × 10^{-6} K^{-1}) (see Figure 6.1), leads to applications in plasma-sprayed thermal barrier coatings for aerospace gas turbine blades, tundishes and ladles to handle liquid metals, as refractory fibers for high-temperature insulation (see below), as lining bricks for glass smelting tanks, and as cutting tools for the high-speed machining of hard Ni-based superalloys such as Inconel® and Hastelloy®.

One specific property of zirconia is its excellent edge retention, which has led to applications as cutting blades for the paper industry, as scalpels for precision cutting in the operation theater, as scissors for hairdressing, as cutting tool for Kevlar™ fibers, and even as sushi knives.

Zirconia parts with superior diamond tool-generated surface finishes displaying tolerances as low as 1 μm and high strength have also been used as split coupling devices for optical fibers in telecommunication systems.

As zirconia displays high stability against most aggressive gases, liquids and melts, an important application is in refractory bricks and liners within environments where the combined effect of erosion and thermal shock would cause the failure of other high-temperature ceramic materials such as alumina or mullite. Such applications can be found as reinforcement for alumina parts in sliding gate valves for continuous steel casting, as nozzles and stoppers for steel holding and transfer ladles, and as the liners of (optical) glass smelting tanks. While zirconia resists acid and basic slags and glasses at high temperatures, destabilization occurs on prolonged contact with silicate and aluminum silicate refractories above 1400 °C.

Refractory Fibers Recently, zirconia-based insulating material with a low density and a low thermal conductivity has been developed in the form of fibers, paper, felt, board and shaped articles. The material is a cubic zirconia solid solution stabilized with yttria, and has a maximum usable temperature of >2100 °C. The innovative fabrication technique involves the use of an organic precursor fiber as a structural template, impregnated with an aqueous solution of zirconium chloride and yttrium chloride. The metallic salts are deposited within the organic fiber, which can subsequently be burned off by a controlled oxidation. The hollow remainder is then fired at a sufficiently high temperature (800–1300 °C) so as to induce crystallization, after which the oxide particles are sintered to develop a ceramic bond. Other techniques to produce refractory fibers involve phase inversion and sintering using a mixture of *N*-methyl-2-pyrrolidone (NMP) and water as an internal coagulant (Yin *et al.*, 2008).

These materials offer outstanding properties, including thermal stability to 2600 °C and corrosion resistance to hot alkalis and many aggressive chemicals and melts; neither are they wetted by a range of molten metals. The applications of zirconia refractory fibers range from highly efficient thermal insulation to separators in aerospace batteries, hot gas filters and electrolysis diaphragms (see below). One particularly topical application may be its use as a chemical and thermal safety barrier since, when placed beneath the core of a nuclear reactor it would serve as both a chemical and thermal shroud in the event of a meltdown (Birkby and Stevens, 1996).

Strengthening of Zirconia Ceramics During recent years, research and development efforts have resulted in engineered ceramics which can be strengthened by: (i) reinforcement by mixing with zirconia; (ii) heat treatment of partially stabilized zirconia; (iii) chemical reactions to produce a matrix with dispersed zirconia particles; (iv) duplex structures; and (v) whisker-reinforced zirconia.

- **Zirconia-reinforced ceramics:** This involves the addition of 15–25 vol% ZrO_2 to ceramic matrix materials such as alumina (see Section 7.1), silicon nitride, or silicon carbide (Figure 7.30). The fracture toughness of these mixed ceramics will be considerably increased, so that applications become possible which require high toughness and durability. Examples include cutting tool for the high-speed machining of hard metals, or the sealing parts for internal

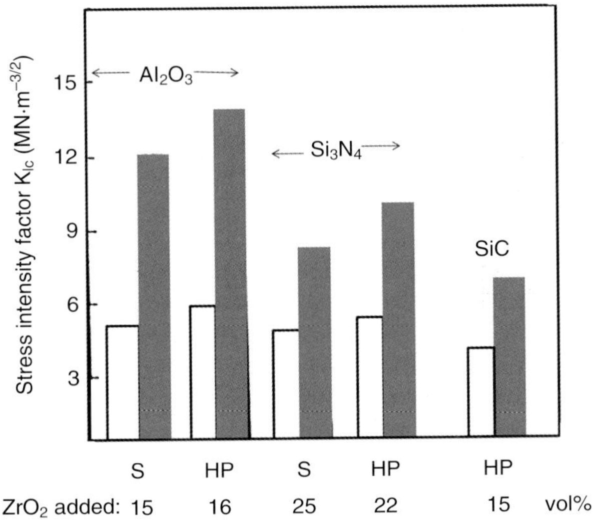

Figure 7.30 Increase of the fracture toughness of alumina, silicon nitride and silicon carbide by the addition of 15–25 vol% zirconia. S= sintered; HP = hot-pressed. Empty bars indicate the matrix ceramic; filled bars indicate ceramics with zirconia added.

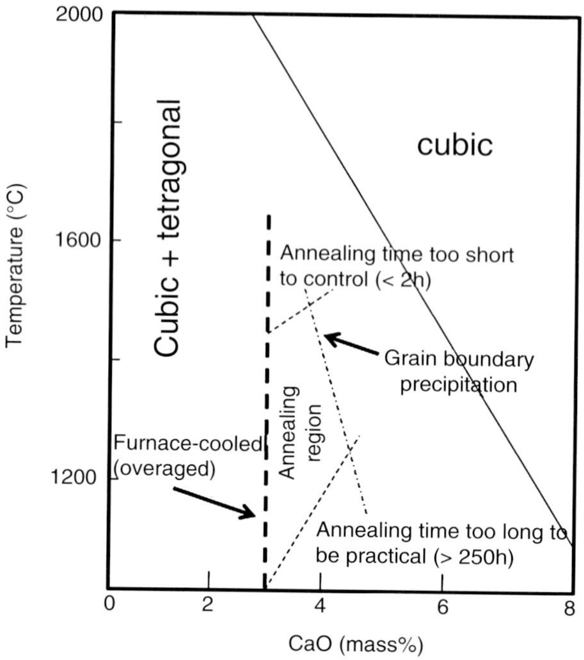

Figure 7.31 Processing phase diagram to optimize the annealing conditions of Ca-stabilized zirconia.

combustion engines operated at high temperature, in order to optimize fuel performance and economy.

- **Heat treatment of stabilized zirconia:** This can dramatically increase the strength of the ceramics; for example, annealing Ca–PSZ with 3.3% CaO at 1300 °C for 10 h increases its modulus of rupture, from 300 to 650 MPa. This effect is thought to be related to a precipitation of CaO-rich zirconia clusters at grain boundaries that act as crack-branching centers. The processing phase diagram (Figure 7.31) shows how the appropriate selection of annealing temperatures (1200–1450 °C) and times (10–200 h) lead to ceramic materials with different properties, and hence applications.

- **In situ dispersion of zirconia particles in a mullite matrix:** This can be achieved by reacting zirconium silicate with alumina according to:

$$2ZrSiO_4 + 3Al_2O_3 \rightarrow 2ZrO_2 + 3Al_2O_3 \cdot 2SiO_2. \quad (7.12)$$

- Hot-pressed mixtures develop strength values up to 400 MPa. A second example is the improvement of the strength of ceramic electrolytes such as β-Al_2O_3 for high-temperature Na–S batteries by ZrO_2 particle reinforcement according to:

$$Na_2ZrO_3 + xAl_2O_3 \rightarrow ZrO_2 + Na_2O \cdot xAl_2O_3 \quad (7.13)$$

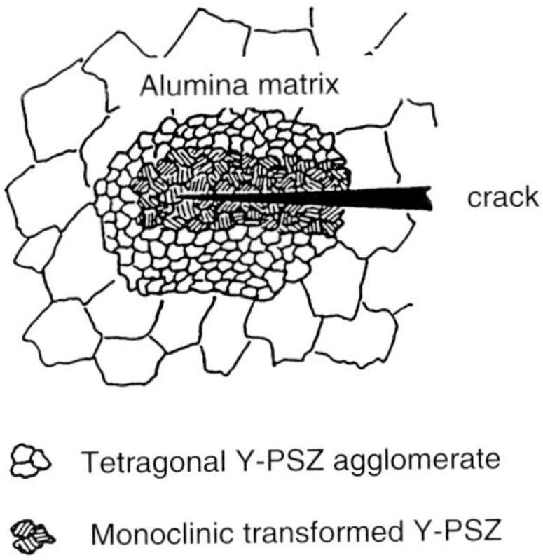

Figure 7.32 Reinforcement mechanism of an alumina–zirconia duplex ceramics with agglomerated Y-PSZ particles.

- **Duplex structures:** These are able to exploit the properties of zirconia in an optimum manner by adding polycrystalline metastable tetragonal YSZ aggregates to a fine-grained alumina matrix (Figure 7.32). These duplex structures exhibit a fracture strength >700 MPa and fracture toughness up to $12\,\mathrm{MPa \cdot m^{1/2}}$. Since zirconia has a lower modulus than alumina, cracks introduced by external loads tend to move towards the zirconia particle aggregates. Their crack energy will then be dissipated by forcing the t-ZrO_2 to transform to m-ZrO_2. This mechanism will be augmented by the compressive stress generated that tends to counteract crack movement.
- **Whisker-reinforced zirconia:** This has been obtained by adding SiC whiskers to TZP material. The ceramic, when hot-pressed at 1450 °C, shows a substantial increase in strength at 1000 °C, and significantly improved cyclic fatigue strength (Zhan *et al.*, 1997). However, during long-term service in an oxidizing atmosphere, the SiC whiskers tend to be changed by transforming into SiO_2, with a concurrent decrease in strength. The addition of alumina whiskers instead might alleviate this negative effect (Roberts *et al.*, 2008).

7.2.6.2 **Functional Applications**

Applications of the ion-conducting properties of zirconia include electrolytes for so-called λ-probes (oxygen sensors to measure the oxygen partial pressure in carburetors equipped with a three-way catalytic converter), probes to measure oxygen levels in molten steel, water electrolyzers, and SOFCs. Since, at very high temperatures, the ionic conductivity through vacancies in the oxygen sublattice of YSZ will be superseded by electronic conductivity (presumably via an electron-hopping mechanism), applications abound such as heating elements for high-temperature kilns in conjunction with a gas or metallic resistance pre-heater, as electrodes for magneto-hydrodynamic generators with gas temperatures exceeding 1800 °C, and as cathodes for air plasma burners. Reviews of additional applications have been provided by Stevens (1993) and Birkby and Stevens (1996).

Ionic Conductivity The O^{2-} anions are highly mobile in the stabilized zirconia lattice, owing to the existence of 10–20% vacancies. This high oxygen diffusivity results in a high anion conductivity, whereby the oxygen vacancies act like positive charge carriers. Efficient ionic conductors should have the following properties (West, 1992):

- There must exist many mobile ions in the structure.
- Many lattice vacancies must be present that can be occupied by mobile ions.
- The empty and filled positions must have a comparable energy, and the activation energy for site exchange must be low.
- The crystal structure should have a three-dimensional (3-D) lattice that shows open channels for ion movement.

Each of these requirements is fulfilled in the case of fully or partially stabilized zirconia. The conductivity will vary by several orders of magnitude, and depend on various intrinsic (type and concentration of stabilizing oxides, impurity concentration) and extrinsic (temperature, surrounding atmosphere, service time) parameters. Some general effects of these parameters are discussed briefly below.

Type of dopant cation. The ionic conductivity depends on the ionic radius of the dopant cation. With decreasing radius, the conductivity increases (Table 7.8), since the then smaller difference to the ionic radius of Zr (79 pm) decreases the lattice strain and reduces the association of the dopant cations with the oxygen vacancies. This results in a higher mobility of the oxygen anions and, thus, a higher conductivity.

Concentration of dopant cation. Independent of the type of dopant cation, the conductivity is maximized for a concentration corresponding to that required to stabilize the cubic modification of ZrO_2 given by the corresponding phase diagram. Lesser concentrations do not provide sufficient vacancies for optimum oxygen ion mobility, while higher concentrations give rise to the formation of ordered defects,

Table 7.8 Conductivities at 1000 °C of ZrO_2 doped with different oxides (Minh and Takahashi, 1995).

Oxide	Ionic radius of cation (pm)	Concentration (mol%)	Conductivity ($10^{-2}\,\Omega^{-1}\,cm^{-1}$)	Activation energy ($kJ\,mol^{-1}$)
Nd_2O_3	104	15	1.4	104
Sm_2O_3	100	10	5.8	92
Y_2O_3	92	8	10.0	96
Yb_2O_3	86	10	11.0	82
Sc_2O_3	81	10	25.0	62

vacancy clusters and electrostatic interactions, all of which tend to decrease the conductivity.

Concentration of impurities. The electric conductivity of polycrystalline zirconia is the sum of the volume and the grain boundary conductivities. In particular, the grain boundary conductivity depends on the grain size and the level of impurities that tend to concentrate at grains boundaries (see Figures 8.18 and 8.19). Such impurities can be introduced when sintering aids such as alumina or silica are being added during the processing of zirconia.

Temperature dependence. Since doped zirconia can be considered a solid-state electrolyte, its ionic conductivity increases with increasing temperature according to:

$$\sigma_i = A\exp(-U/kT) + B\exp([U^* + E_d]/kT) \quad (7.14)$$

where U is the activation energy of oxygen ion mobility, U^* is the activation energy of transport by electron hopping, and E_d is the energy of disordered ions. Hence, the first term of Eq. (7.14) describes the contribution of ionic conductivity, and the second term that of electronic conductivity.

Figure 7.33 illustrates the temperature dependence of the ionic conductivity of ceramic electrolytes relevant for applications in SOFCs.

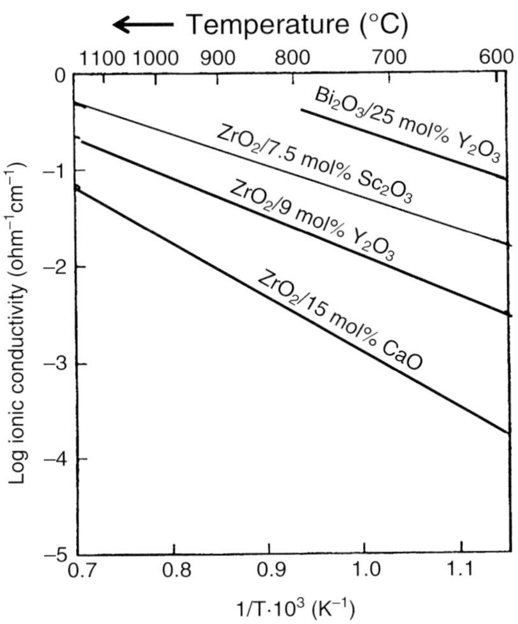

Figure 7.33 Temperature dependence of the ionic conductivity of ceramic electrolytes relevant for application in solid oxide fuel cells. Modified after Haering (2001).

Surrounding atmosphere. Under oxygen partial pressures (0.21 atm to 10^{-17} atm) typical for the operation of SOFCs with a stabilized zirconia electrolyte, the ionic conductivity is virtually independent of the oxygen partial pressure, and the electronic conductivity is negligible.

Service time. While the volume conductivity of a solid-state electrolyte such as fully stabilized (cubic) zirconia (FSZ) reaches a steady state within a comparatively short time, its grain boundary conductivity changes with service time, mostly due to the accumulation of impurities. In PSZ, similar effects occur in relation to the precipitation of tetragonal ZrO_2 and associated deterioration of conductivity. Because of this aging behavior of PSZ, FSZ is preferred in SOFC applications (Subbarao, 1981; Minh and Takahashi, 1995).

Galvanic Gas Sensors A typical example is the galvanic oxygen concentration cell with a zirconia solid-state electrolyte, applied in the so-called λ-probe to measure and control the oxygen fugacity in a carburetor. Figure 7.34 shows the functional principle of a zirconia oxygen sensor. Yttria-stabilized zirconia is an almost pure oxygen ion conductor between 400 and 1000 °C, the charge-carrying ability of which is supported by a high concentration of vacancies in the oxygen sublattice. The incorporation of Y^{3+} ions at regular Zr^{4+} sites creates these anion vacancies, that are electron-donor states. At elevated temperatures, these can be lifted into the conduction band and hence they can contribute to the electrical conductivity. According to the Nernst equation, the electromotoric force generated by the galvanic cell of the type $O_2(pI),Pt/(ZrO_2)_{0.85}(CaO)_{0.15}/Pt,O_2(pII)$ is

$$E = (kT/4e)\ln[p_{O2}(I)/p_{O2}(II)], \tag{7.15}$$

where $p_{O2}(I)$ and $p_{O2}(II)$ are, respectively, the oxygen partial pressures in the reference cell (air, 0.21 atm) and in the measuring cell (i.e., the carburetor). Whilst this measuring principle is well established, problems occur owing to the frequent

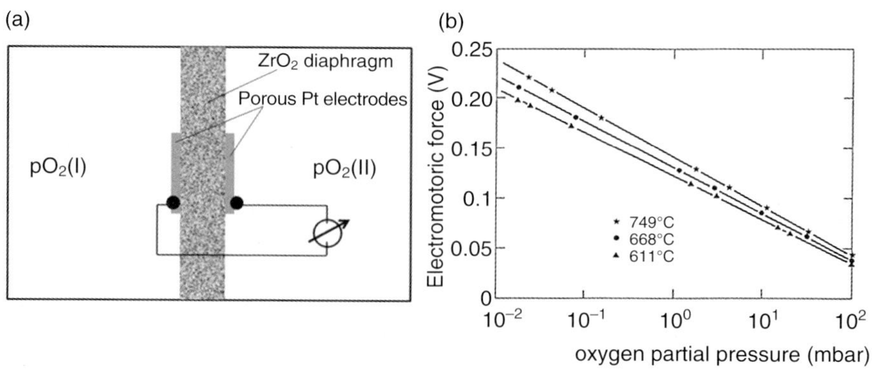

Figure 7.34 ZrO_2 oxygen sensor. (a) Functional principle; (b) Linear dependence of electromotoric force on oxygen partial pressure and temperature (Heywang, 1984).

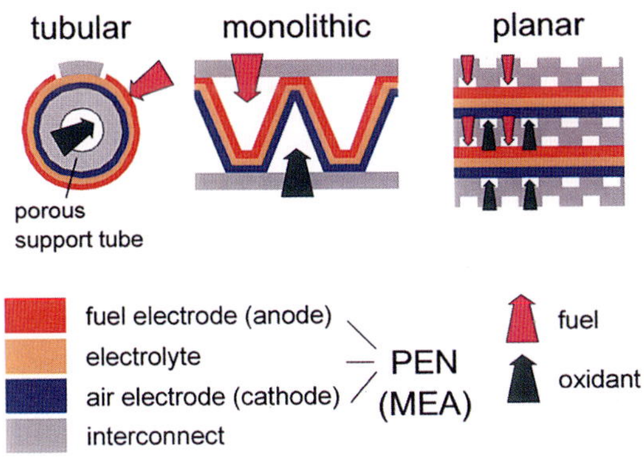

Figure 7.35 Tubular, monolithic and planar geometries of solid oxide fuel cells. Illustration courtesy of M. Müller (Müller, 2001).

presence of other gases in the car exhaust, such as carbon monoxide, carbon dioxide, nitrogen, and NO_x.

Solid Oxide Fuel Cells (SOFCs) Fuel cells convert chemical energy stored in the fuel directly into electrical energy through an electrochemical reaction, and are thus considered to be one of the most advantageous energy conversion technologies of the immediate future. They combine high conversion efficiency, flexibility in fuel use, and cogeneration capability with very low chemical or acoustic pollution of the environment. Figure 7.35 shows three different designs with tubular, monolithic, and flat-plate geometries. In all cases, the anode (fuel electrode) consists of Ni or a Ni–Al_2O_3 or Ni–ZrO_2 composite material (e.g., see Benyoucef *et al.*, 2009). The cathode (air electrode) is frequently made from perovskite-type ceramics, such as Sr-doped lanthanum composite oxides ($LaMnO_3$, $LaCoO_3$, $LaSrMnO_3$). The two working electrodes are separated by a gastight YSZ electrolyte diaphragm.

The working principle of a SOFC is shown in Figure 7.36. At the air electrode (cathode), oxygen combines with electrons, and the negatively charged oxygen ions are transported through the stabilized zirconia electrolyte via oxygen vacancies. In order to obtain reasonable transport rates, the temperature of the electrolyte must be at least 950 °C. At the fuel electrode (anode), a mixture of hydrocarbons and water reacts to produce hydrogen and CO, by internal reforming. The oxygen ions carried through the electrolyte oxidize the hydrogen and CO to CO_2 and water, releasing electrons that are transported back via an external circuit to the cathode.

From this working principle, it follows that the p- and n-electrodes must have a high proportion of open porosity that will ensure fast and effective gas transport. On the other hand, the electrolyte transports only charged species (i.e., electrons) and hence must be gastight. Recent studies have focused increasingly on producing the intricate shapes and compositions required for the PEN ("positive-electrolyte-negative") by applying plasma-spraying technology (Figure 7.37). This includes the PSZ electrolyte, the cathode and anode, as well as the total PEN structure. The

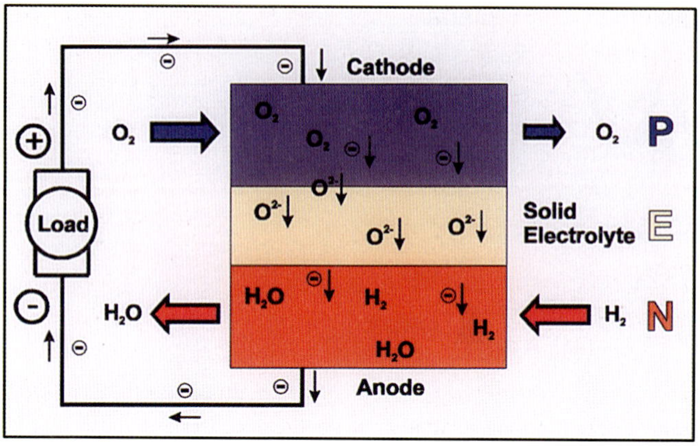

Figure 7.36 Operating principle of a solid oxide fuel cell. Illustration courtesy of M. Müller (Müller, 2001).

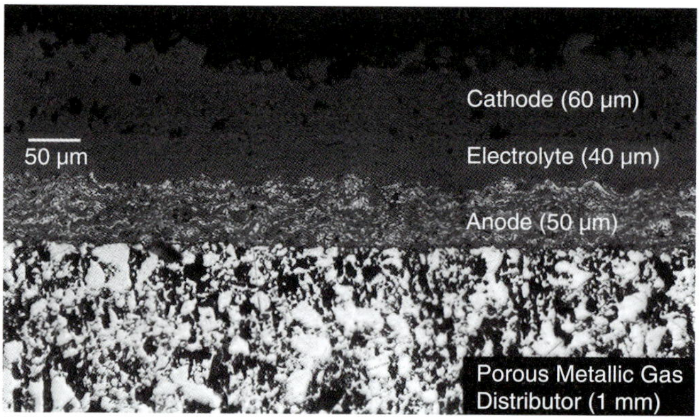

Figure 7.37 PEN assembly (anode: Ni; electrolyte: Y-stabilized zirconia; cathode: (La,Sr) MnO_3) deposited by r.f. induction plasma spraying on a porous Ni support. Illustration courtesy of Dr M. Müller, Bosch AG, Stuttgart, Germany.

interconnects of the planar SOFC (i.e., the bipolar plates) which, until now have been hot-pressed and rolled to a thickness of 3 mm and finished by electrochemical machining (ECM), can also be plasma-sprayed. Preliminary experiments have been conducted with RF inductively-coupled plasma-sprayed $94Cr5Fe1Y_2O_3$ powders. The major challenge for the economic production of bipolar plates (interconnects) by plasma spraying relates to the production of a system of orthogonal grooves to carry the natural fuel/water mixture and air, respectively. Orthogonality is required to ensure effective cooling and a uniform temperature distribution that would considerably reduce thermal stresses across the PEN. A second problem is *poisoning* of the active cathode with nickel or chromium that has diffused from the bipolar plate at the working temperature of 950 °C. In order to prevent this situation, a protective layer of (La,Sr)CrO_3 could be sprayed onto the cathode using a low-pressure plasma spraying (LPPS) technique. Further details and reference material relating to this subject are available in Heimann (2008) (see also Müller, 2001).

At present, intense research is continuing worldwide to develop novel, more efficient electrode materials and interconnects so as to increase the electrochemical conversion rate (fuel efficiency), the active surface area and the electrical efficiency, and to reduce the operating temperature and system degradation. Advanced production technologies, including the use of plasma spray techniques to fabricate the SOFCs, are also currently under development (Table 7.9).

Recently, the Swiss company Sulzer Hexis AG has developed the concept of an SOFC for small-scale domestic application to provide heating on demand and to cover peak power requirements. This so-called "Hexis" (Heat Exchanger Integrated Stack) cell consists of circular zirconia electrolyte disks of about 12 cm diameter that carry the anode and cathode (Figure 7.38). The individual cells are

Table 7.9 Systems parameters and development goals of low-pressure plasma (LPP)-sprayed solid oxide fuel cells.

Parameter	Present	Future
System power (kW)	1	7
Cell diameter (mm)	120	200
Active surface (cm^2)	100	300
Operating temperature (°C)	920	850
Power density ($mW\,cm^{-2}$)	175	175
Fuel efficiency (%)	43	63
Electrical efficiency (%)	23	35
Systems degradation per 1000 h (%)	5	1

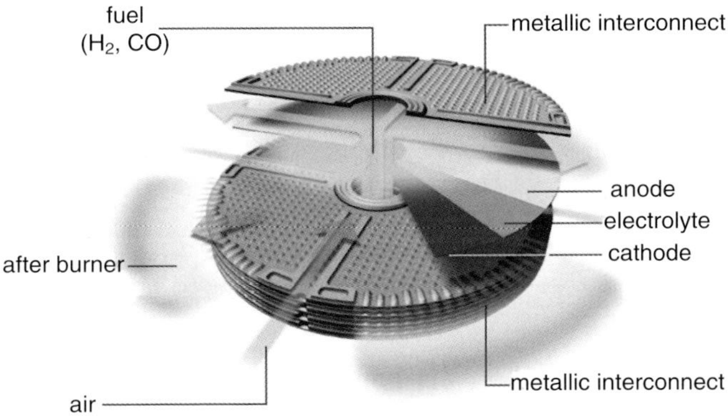

Figure 7.38 Solid oxide fuel cell Hexis (Heat Exchanger Integrated Stack) (Schuler, 2007) Reprinted with permission from Schweizerischer Verein des Gas- und Wasserfachs.

stacked via metallic interconnects to provide a power output of several kW. At present, field tests are ongoing with the prototype Galileo 1000 N system (Schuler, 2007), the aim being to solve any operational challenges associated with the system's stability during long-term use under peak service conditions.

Thermal Barrier Coatings (TBCs) The combustion temperatures of car engines and aerospace gas turbines are severely limited by the inability of common metallic construction materials to withstand temperatures in excess of approximately 1000 °C. However, the application of thermal barrier coatings based on stabilized zirconia represents one way of increasing the combustion temperature, and hence the thermodynamic efficiency and environmental compatibility by reducing the amount of hydrocarbons expelled in the exhaust gases.

In general, two methods are used to deposit TBCs onto a superalloy substrate: (i) atmospheric plasma spraying (APS); and (ii) electron beam-physical vapor deposition (EB-PVD).

The TBCs deposited by plasma spraying consist (adjacent to the metallic substrate; Inconel, Hastelloy, Ti6Al4V) of a dense MCrAlY (M = Ni, Co) alloy bond coat which is laid down by low-pressure plasma spraying (LPPS), followed by a porous stabilized zirconia top coat in the thickness range of 300 to 1000 μm (Cernuschi *et al.*, 2004), as shown schematically in Figure 7.39. During operation,

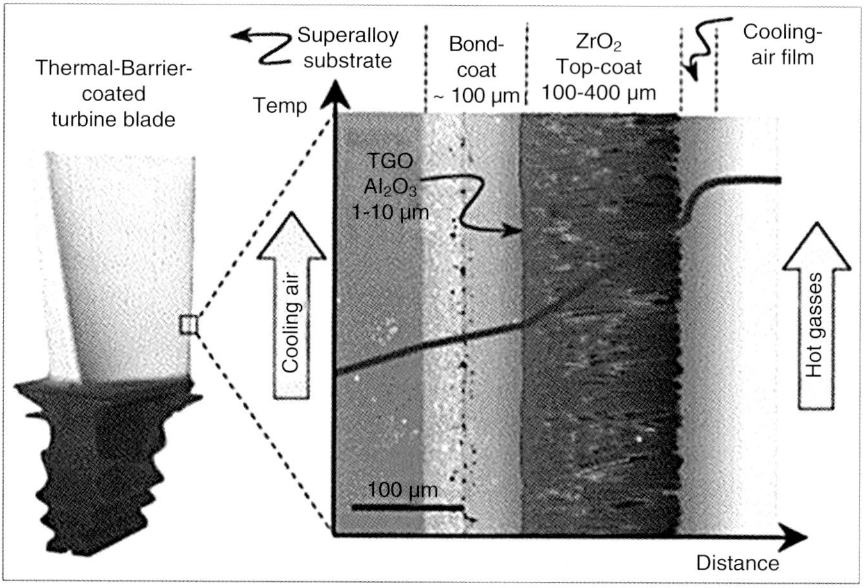

Figure 7.39 Design of thermal barrier coatings (TBCs) applied to the Ni superalloy substrate of a gas turbine blade, showing the typical microstructure of an electron beam-vapor deposited coating (EB-PVD) as well as the thermally grown oxide (TGO) layer at the substrate/bond coat interface (Padture *et al.*, 2002). Reprinted with permission from the American Association for the Advancement of Science.

oxygen diffuses through the porous zirconia layer and reacts with the bond coat to form a thermally grown oxide (TGO) interface consisting of alumina that is proposed to inhibit any further oxygen diffusion down to the superalloy substrate level. The growth of the TGO layer generally follows a parabolic rate law, and is influenced by the surface roughness of the LPPS MCrAlY coating (Seo *et al.*, 2007).

The thermal and mechanical property differences of YSZ-TBC coatings prepared via the two processing routes result from differences in the morphology of the porosity present within the TBC layer (Figure 7.40). In APS layers, inter-splat pores result from the impingement of molten droplets that splash across the substrate surface (Heimann, 2008). These pores are roughly aligned parallel to the substrate surface, and are accompanied by microcracks and fine-grain boundaries (Figure 7.40a; see also Figure 7.41). The pores provide a high impedance to heat flow through the thickness of the coating, and this results in a layer with a low thermal conductivity. The spallation life of these coatings is believed to be governed by a combination of the disc-like coating "defects" and the significant coefficient of thermal expansion (CTE) mismatch between the TBC layer and the underlying substrate.

In contrast, the TBC layers produced by EB-PVD have a columnar microstructure with elongated intercolumnar pores that become predominantly aligned

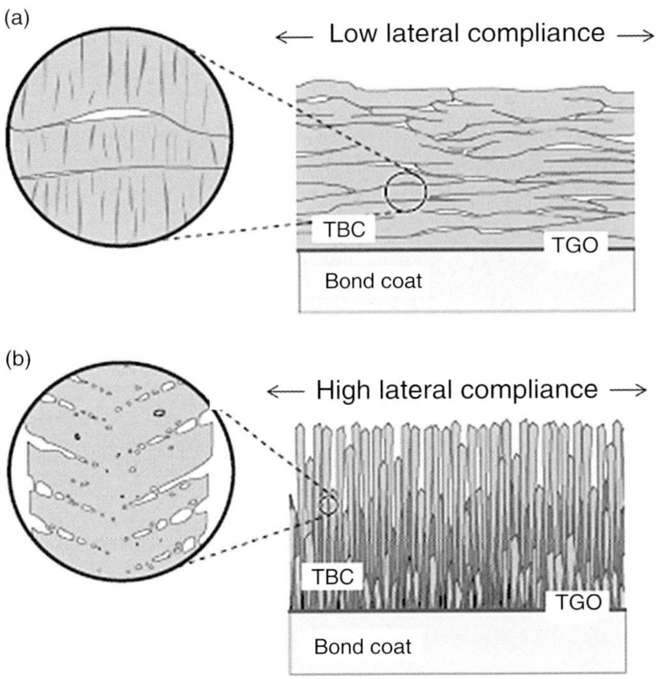

Figure 7.40 Pore morphologies in (a) APS and (b) EB-PVD-deposited thermal barrier coatings.

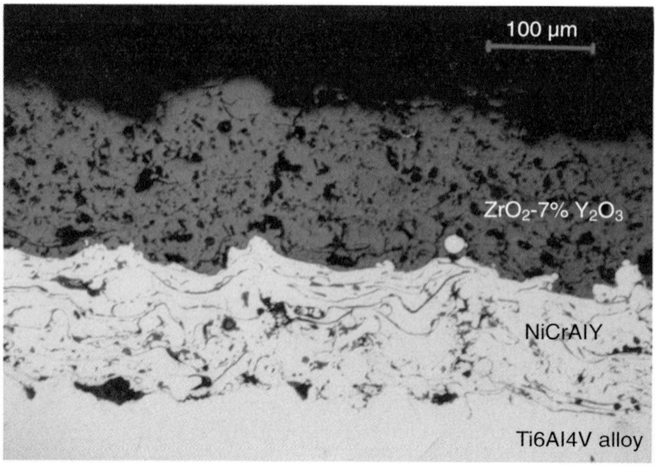

Figure 7.41 Porous Y-stabilized zirconia thermal barrier coating (TBC) on top of a dense NiCrAlY bond coat covering the Ti6Al4V alloy substrate of a gas turbine blade (Heimann *et al.*, 1991).

perpendicular to the plane of the coating as its thickness increases (Figure 7.40b). A finer distribution of intracolumnar pores is also seen to exist. The elongated intercolumnar pores increase the compliance of the coating in the plane of the substrate, and this leads to the improved spallation lifetimes of the TBC systems.

Since, in contrast to the rather dense electron beam-assisted coatings, the plasma-sprayed PSZ-based thermal barrier coatings are inherently porous (Figure 7.41), corrosive gases may find their way through any connected porosity to interact negatively with the austenitic superalloy substrate. However, this high degree of porosity will reduce the thermal conductivity of the coatings to values $<1.0\,W\,m^{-1}\,K^{-1}$, whereas that of EB-PVD coatings are in the range of $2.0\,W\,m^{-1}\,K^{-1}$, depending on the roughness of the substrate and the deposition conditions.

Today, the development and improvement of zirconia-based thermal barrier coatings are carried out in two main areas (Heimann, 2008; Lehmann and Heimann, 2008). The first area relates to the aerospace industry, and is concerned with the coating of austenitic superalloy blades and vanes of gas turbine engines so as to increase the lifetime of components subjected to a variety of degrading processes. As shown in Figure 7.42, the lifetime of an uncoated Ni-superalloy turbine component part is limited by mechanical fatigue below 800 °C, by hot corrosion between 800 and 900 °C, and by thermal fatigue between 900 and 1050 °C. Above this temperature, oxidation and creep are lifetime-controlling factors. Zirconia-based thermal barrier coatings increase the thermal stability of the turbine blades to beyond 1200 °C by improving the blades' cooling efficiency and, in conjunction with (Ni,Co)CrAlY bond coats, preventing hot corrosion by molten salts (sodium sulfate, sodium vanadate) and corrosive gases. In particular,

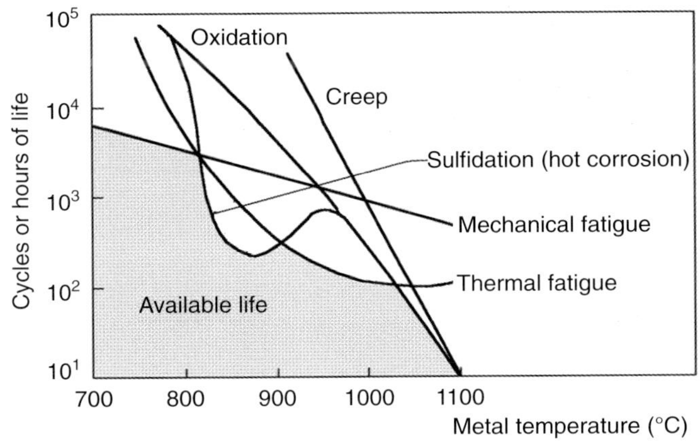

Figure 7.42 Factors influencing gas turbine component life (Spera and Grisaffe, 1973).

V_2O_5 as a possible constituent of low-quality kerosine may react with Y-stabilized TBCs to form ZrV_2O_7 and YVO_4, thus delaying the formation of m-ZrO_2 (Chen *et al.*, 2008).

The second area of application is as a potential material for inclusion in reciprocating internal combustion engines, principally diesel engines. The goals here are to insulate components such as pistons, valves, intake manifolds and exhaust ports, and to protect the moving parts from wear and corrosion.

Currently, the bulk of zirconia precursor powders produced are used for gas turbine applications. However, in the future this market segment is estimated to decrease, whereas diesel engine applications will increase (for past estimates, see Heimann, 1991). Figure 7.43 shows the NASA scenario of development trends in the in-service temperatures of gas turbines for aerospace applications. The first step in the development cycle has been already reached, by increasing the turbine blade surface temperatures to 1100 °C. However, the second – somewhat visionary – step will require the development of completely new ceramic structural materials and systems that can withstand surface temperatures of up to 1400 °C so that fuels can be burned at considerably higher efficiencies, with concurrent reductions in the production of polluting gases. Bridging between these two steps is the current coating technology, which has a temperature limit of approximately 1200 °C (BMFT, 1994).

Yet, this limit must be overcome in highly loaded blades and vanes of high-pressure gas turbines, and should potentially reach 1500 °C. The resultant increase in efficiency may derive from reduced cooling and/or increased inlet temperatures.

Although, thick thermal barrier coatings (TTBCs) satisfy thermal efficiency requirements, they create problems by accumulating residual stresses. Hence, an optimization of the ceramic materials has been achieved in two areas: (i) a *chemical*

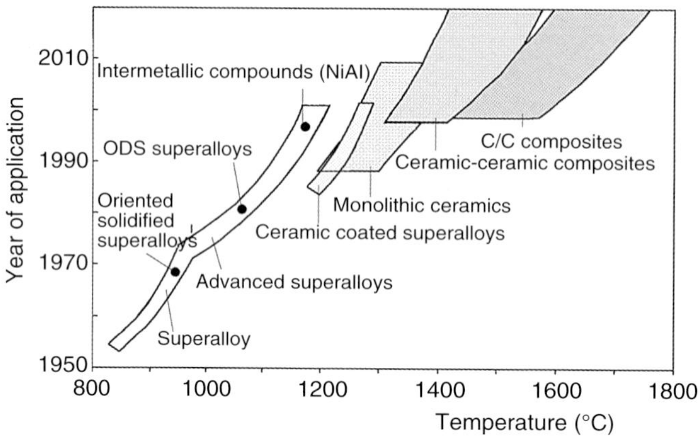

Figure 7.43 NASA scenario of development trends in in-service temperature of aerospace gas turbines (BMFT, 1994).

modification, by optimizing the amount of stabilizing yttria; and (ii) a *microstructural modification*, by optimizing the coating density.

As a result of such *chemical modifications*, the lifetime of a TBC has been maximized by varying the yttria content in zirconia. An yttria concentration of 6–8 mass% coincides with the maximum amount of the nonequilibrium, "nontransformable" tetragonal zirconia polymorph (t′-zirconia) (see Section 7.2.5.3; see also Figure 7.27b).

The present state-of-the-art technology consists of applying zirconia coatings, partially stabilized with 7–8 mass% yttria, over a metallic bond coat of Ni- or CoCrAlY alloy (Figure 7.41). The bond coat must be applied using vacuum plasma spray (VPS) technology so as to prevent the formation of alumina scale that is suspected of reducing adhesion strength to the superalloy substrate. The bond coat adhesion also appears to be controlled by the yttrium concentration, as this reduces the interface stress by minimizing the difference in valence electron densities between the substrate and the bond coat. The addition of yttrium can also increase the valence electron density ρ_{hkl} of the interface, thereby increasing the interface cohesion force.

Currently, the PSZ top layer is generally sprayed in APS mode, since VPS experiments have shown that in this case the PSZ layer develops Young's moduli threefold larger than are observed for APS coatings. Because of this, greater thermal stresses are expected to be generated in VPS-PSZ coatings. In Figure 7.15, the bending strengths and fracture toughness are displayed above the phase relationships of the binary system ZrO_2–Y_2O_3 with several compositions, grain sizes of the t-phase. In general, fracture toughness increases with decreasing amounts of yttria, whereas the thermal stability increases with increasing yttria content. This requires trade-offs to be made, depending on the application of the material as explained in detail in Section 7.2.3.

A second aspect of chemical modification concerns the replacement of the stabilizing yttria by other elements, but studies in this area have been relatively few in number. By using ceria instead of yttria, the microhardness and crack resistance could be markedly improved for a PSZ stabilized with 15 mol% CeO_2 and beyond. In order to reduce the thermal conductivity of ZrO_2-based thermal barrier coatings, doping was carried out with Al_2O_3, NiO, Nd_2O_3, Er_2O_3 and Gd_2O_3 (Nicholls *et al.*, 2002), as well as La_2O_3 (Matsumoto *et al.*, 2006).

The *microstructural modifications* relate to either porosity control or the development of graded coating systems. Porosity and microcrack distribution impart to the material a tolerance to both thermal and residual stress; thus, the lifetime of a TBC is sensitive to density variations that can be controlled by the correct adjustment of the plasma spray parameters by statistical experimental design methodology (Heimann, 2008, 2009). The effect of porosity and splat interfaces on the effective properties of zirconia coatings were monitored with small-angle neutron scattering (SANS) and scanning electron microscopy (SEM), using an object-oriented finite element (OOFE) method (Wang *et al.*, 2003) as well as various nondestructive testing (NDT) methods (Rogé *et al.*, 2003).

An additional important quest is for stress control, since coated turbine section parts are subject to extremely steep temperature gradients. Residual stresses in the as-sprayed PSZ coating reduce the required thermal shock resistance, and hence the coating lifetime by cracking and delamination. These stresses arise from nonequilibrium states with respect to the composition, structure and microstructure of the PSZ, as well as from the difference in the CTE between the superalloy substrate and the 7% yttria-PSZ coatings. The adjustment of these parameters has been attempted through changes of the ceramic composition, by the application of ceramic/bond coat gradient layers, as well as substrate preheating or cooling. In the past, although much effort has been expended to predict residual coating stresses by modeling the life expectancy of the TBCs, problems were encountered by the assumption of a continuum theory and the non-consideration of elastic finite elements, nonlinear processes, and the general fractal nature of plasma-sprayed coatings (Heimann, 2008).

The future performance profile of TBCs will include improved thermal barrier function and oxidation resistance, as well as resistance to both corrosion and erosion. With improved TBCs, economic and environmental requirements will be addressed such as higher combustion temperatures resulting in better fuel efficiency, lower cooling air requirements and hence higher compressor effectivity, and lower thermal stresses on materials that will result in a longer service life and maintenance cycles. There are, however, problems with TBCs that manifest themselves in increased spalling and chipping at higher operational temperatures and times. This is caused by higher residual stresses in the coatings, by bond coat oxidation, and also by volume changes induced by phase transformations. Here, improvements are urgently needed to attain performance levels anticipated by NASA (BMFT, 1994) for airborne gas turbines (10 000 h at turbine blade surface temperatures up to 1400 °C), for stationary gas turbines (25 000 h at up to 1000 °C), and also for large stationary diesel engines (5000 h at cycling loads). Quality control protocols are needed to more accurately predict the lifetime and reliability, as well as to obtain a deeper insight into the failure mechanisms of TBCs by applying techniques such as acoustic emission (AE) and the implementation of real-time and nondestructive tests for the on-line characterization of coatings.

Recently, investigations have also been performed to apply detonation spraying, the aim being to deposit suitable TBCs based on YSZ over a NiCrAlY bond coat on a Ni-base superalloy (M38G) (Ke *et al.*, 2006). These coatings had low thermal conductivities of between 1.0 and 1.4 $W\,m^{-1}\,K^{-1}$ (at 200–1200 °C), and showed excellent resistance against thermal shock. Moreover, thermal cycling up to 400 times (heating to 1050 °C, cooling to room temperature by forced air cooling), followed by 200 cycles to 1100 °C and cooling to room temperature by forced water quenching, introduced extensive cracking and eventual failure by spalling. At an early stage during the cycling process, a continuous alumina layer was formed at the ceramic/bond coat interface, and eventually a Ni/Co-rich TGO layer also developed; in addition, the transformation of t-ZrO_2 to m-ZrO_2 resulted in a final, catastrophic spallation.

Several degradation factors have been identified that limit the lifetime of a TBC, as shown schematically in Figure 7.44.

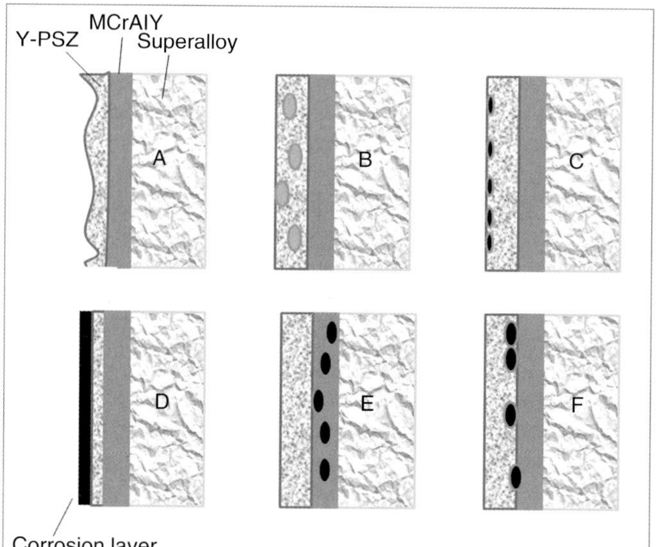

Figure 7.44 Several potential ceramic failure mechanisms of Y-PSZ thermal barrier coatings. (A) Spalling due to stresses introduced by thermal cycling; (B) Destabilization of zirconia due to cubic–monoclinic phase transformation; (C) Erosion and microspalling due to particulate-hot gas impingement; (D) Chemical corrosion by reaction with fuel impurities; (E) Oxidation of MCrAlY bond coat by hot corrosion; (F) Infiltration by gaseous impurities and condensation of high-stress phases.

To prevent the in-service failure of TBCs by the mechanisms shown in Figure 7.44 during long-term, high-temperature exposure and thermal cycling, respectively, several approaches were tested including:

- Infiltration of the porous coating with hot erosion-resistant materials, such as Ni_3Al or CVD-SiC.
- Reduction of surface porosity by applying a top layer of very fine zirconia powder.
- Treatment of plasma-sprayed PSZ with a slurry of 0.1 mm zirconia, and subsequent thermal densification.
- Addition of Si to the MCrAlY bond coat to improve its corrosion and erosion resistances.
- Laser surface remelting ("laser glazing").
- Reactive laser remelting with alumina.
- Plasma-spraying of PZT with a low yttria content, the expansion of which during the $t \rightarrow m$ transformation counteracts the thermal shrinkage.

Bioceramic Applications The performance requirements of yttria-stabilized tetragonal zirconia polycrystal (TZP) to form biocompatible, strong components for use as hip, knee, and dental prostheses, and which demonstrate long-term resistance against aggressive body fluids and mechanical wear and tear, during a predicted lifetime of 15–20 years in the human body, include:

- a high chemical resistance;
- an absence of effects that negatively affect the body's metabolism;
- a long lifetime under physiological conditions; and
- high strength, and in particular a high fatigue strength.

For details of these prostheses, the reader is referred to Chapter 10.

Synthetic Gemstones The optical properties of cubic zirconia phase resemble those of diamond in terms of the refractive index (2.15–2.18), while its dispersion (0.058–0.066) exceeds that of diamond (0.044). Hence, cubic zirconia [tradenames: CZ, CZ diamond, Diamonique™ (QVC), Absolut™ (HSN) and many others] has its market niche as a diamond simulant. The melting of zirconia powder requires temperatures in excess of 2800 °C; these can be achieved by "skull melting," using a device consisting of a cylinder of water-cooled copper "finger" containing the zirconia powder, and which is surrounded by a copper coil that can be energized at 4 MHz, with a power consumption of up to 100 kW (Birkby and Stevens, 1996) (Figure 7.45). As zirconia at low temperature is virtually electrically insulating, it does not carry any current, and the current flow must be initiated by introducing small pieces of metallic zirconium to the powder charge. When a temperature above 1100 °C has been reached, the zirconia becomes electronically conducting; heating by Ohmic current flow is then continued until all the powder has melted, with the exception of a thin, thermally insulating shell ("skull") adjacent to the water-cooled copper cylinder. The melt container is then lowered slowly through

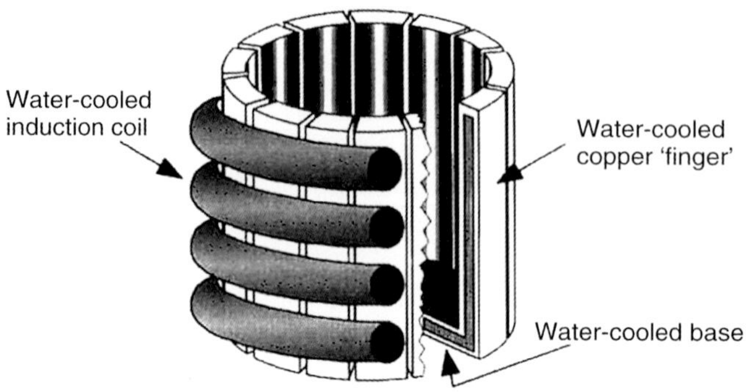

Figure 7.45 Schematic rendering of an induction skull melting crucible. Illustration courtesy of Castings Technology International; www.castingstechnology.com/ismtialloys.asp.

a temperature gradient to initiate crystallization. Doping of the powder with transition metal oxides leads to colored gemstones that are sold as fashion jewelry.

Other technical applications based on the optical properties of cubic zirconia include high-temperature optic pressure sensors and components for laser optics.

Immobilization of Radioactive Waste Owing to its high hydrolytic and radiation stability, and an ability to incorporate large amounts of actinide ions, cubic zirconia has been considered a promising matrix for the immobilization and/or transmutation of radioactive waste (Beall and Rittler, 1982; Heimann and Vandergraaf, 1988). Hence, cubic zirconia was studied as a model system for radiative effects in ceramic oxide matrices (Thomé *et al.*, 2009), to be utilized as advanced waste forms for radioactive waste streams from fuel reprocessing operations containing high levels of actinides such as Np, Pu, Cm, and Am isotopes. For this, Y-stabilized cubic zirconia samples doped with Hf and Nb and irradiated with a thermal neutron flux of 10^{14} neutrons $cm^{-2} s^{-1}$ for 36 days were leached under mild hydrothermal conditions (85–150 °C) in a saline groundwater solution (SCSSS; Standard Canadian Shield Saline Solution) in the presence of granite and a Ca-exchanged montmorillonite to simulate the conditions in a future radioactive waste repository deep within a granite pluton of the Canadian Shield (Heimann and Vandergraaf, 1988). From the decay-correlated activities of ^{95}Nb and ^{181}Hf in the leachate, the leach rate of the cubic zirconia matrix was calculated, and found to be very low, ranging from 1.4×10^{-12} to $7.0 \times 10^{-12} kg m^{-2} s^{-1}$ at 25 °C and 85 °C, respectively. At 150 °C, in the presence of Ca-montmorillonite, an approximately 100 nm-thick hydrated aluminosilicate reaction layer was formed on the surface of zirconia that was enriched in radioactive species.

The radiation stability of cubic zirconia was recently probed by irradiation with ions in the MeV to GeV energy range (Thomé *et al.*, 2009). At low energies the ballistic effects dominated, creating lattice damage which peaked around the ion projected range. At high energies, however, electronic excitation occurred, and the damage was manifest predominantly through the creation and overlap of ion tracks. As these findings may be generalized, however, many nuclear oxide ceramics are thought to possess exceptional stability against radiation-induced damage.

7.3 Titania

7.3.1 General Properties and Applications

The important physical and mechanical properties of sintered titania are summarized in Table 7.10.

Today, the predominant application of titania is still in pigments, owing to its optical properties and stability against ultraviolet (UV) radiation. However, other applications have emerged including oxygen sensors, antimicrobial coatings, photocatalysts to clean waste gases and liquids, and potential photovoltaic cells with a high degree of energy conversion. These applications will be discussed in more detail below.

7.3.2 Processing of Titania

Pure primary titanium dioxide does not occur naturally, but is derived by weathering from ilmenite ($FeTiO_3$), perovskite ($CaTiO_3$), and titanite (sphene) ($CaTiSiO_5$). The weathering products form leuxocene ores with up to 68% rutile content. The

Table 7.10 Typical physical and mechanical properties of titania.

Property	Value
Density ($Mg\,m^{-3}$)	4.0
Modulus of rupture (MPa)	140
Compressive strength (MPa)	680
Fracture toughness ($MPa \cdot m^{-3/2}$)	3.2
Shear modulus (GPa)	90
Elastic modulus (GPa)	230
Microhardness (HV0.5) ($kg\,mm^{-2}$)	880
Resistivity at RT (ohm·cm)	10^{12}
Resistivity at 700 °C (ohm·cm)	$2.5 \cdot 10^4$
Dielectric permittivity @ 1 MHz	85
Dissipation factor @ 1 MHz	$5 \cdot 10^{-4}$
Dielectric strength ($kV\,mm^{-1}$)	4
Thermal expansion (RT-1000 °C) (K^{-1})	$9 \cdot 10^{-6}$
Thermal conductivity ($W\,m^{-1}\,K^{-1}$)	11.7

RT = room temperature.

most prevalent occurrence of titania is in rutile or ilmenite mineral sands mined in Canada, Australia, the USA, and South Africa.

These ores are the principal raw materials used in the manufacture of titanium dioxide pigment. The first step to purify the ore is basically a refinement, using either sulfuric acid (sulfate process; Gesenhues *et al.*, 2003) or chlorine (chloride process) as an extraction agent.

Sulfate process. The sulfate process yields both anatase and rutile grades, depending on the particular processing routes. Ilmenite or titanium slag is first digested with sulfuric acid, and the product is diluted with water or dilute acid. This causes most of the titanium dioxide from the ore to be dissolved as a titanium oxosulfate, $TiO(SO_4)$, and the iron is present in its two-valent oxidation state. The resultant liquor is clarified by sedimentation to remove any insoluble residues, such as silica. The iron is removed by crystallization as its sulfate salt ($FeSO_4 \cdot 7H_2O$), followed by filtration. To produce the anatase form, a small portion of the clarified liquor is neutralized with alkali to produce anatase microcrystals; these microcrystals are then introduced into the mother liquor, which is hydrolyzed under carefully controlled conditions to produce crystals of anatase. The crystals are subsequently filtered, washed, calcined, and micronized. During calcination, the final temperature reaches about 800–850 °C.

To produce the rutile form of titanium dioxide, the clarified liquor is hydrolyzed in the presence of a rutile seeding agent, obtained by neutralizing a small portion of the mother liquor in the presence of hydrochloric acid. The resultant crystals are filtered, washed, and calcined at temperatures between 900 and 930 °C.

Chloride process. The chloride process yields the rutile form of titanium dioxide. At temperatures between 800 and 1200 °C, chlorine is reacted in a fluidized-bed reactor with a titanium-containing mineral under reducing conditions (in the presence of coke) to form anhydrous $TiCl_4$. Further purification requires separation by fractional condensation. The conversion of $TiCl_4$ to TiO_2 may be accomplished by either direct thermal oxidation, or by reaction with steam in the vapor phase at temperatures in the range of 900–1400 °C. A small amount of $AlCl_3$ is generally added to promote formation of the rutile form. The titanium dioxide product is then washed, calcined, and packaged (Kuznetsof, 2006).

Alternatively, titanium-containing minerals can be reacted with concentrated hydrochloric acid to form solutions of $TiCl_4$ which are then further purified. Hydrolysis of the tetrachloride will yield the dioxide which is filtered off, washed, calcined, and packaged.

Both, the sulfate and chloride processes encounter problems if larger amounts of impurities such as alkaline earth oxides are present in the titanium oxide materials. Such impurities can be selectively sulfated and subsequently leached out by the action of sulfur trioxide gas. In a particularly important aspect, titania slag from the smelting of ilmenite is reacted with sulfur trioxide gas; this causes the calcium and magnesium oxide impurities to be converted into double sulfates that can then be leached from the slag. Furthermore, the addition of oxygen to the sulfur trioxide gas results in a concurrent transformation of the titanium-bearing phases to rutile (Elger *et al.*, 1978).

Ilmenite can be upgraded to synthetic rutile (>90% TiO_2) by removing its iron content, using the *Becher process*. In this process, an ilmenite concentrate containing 55–65% TiO_2 is fed to a rotary kiln to oxidize the two-valent iron in ilmenite to three-valent iron according to:

$$4FeTiO_{3(s)} + O_{2(g)} \rightarrow 2(Fe_2O_3 \cdot TiO_{2(s)}) + 2TiO_{2(s)} \quad (7.16)$$

In a second step, the pseudobrookite, $Fe_2^{3+}TiO_5$ is reduced with coal (and sulfur) according to:

$$Fe_2O_3 \cdot TiO_{2(s)} + 3CO_{(g)} \rightarrow 2Fe_{(s)} + TiO_{2(s)} + 3CO_{2(g)} \quad (7.17)$$

and heated to beyond 1200 °C. The metallic iron formed is removed by aeration in large tanks that contain 1% NH_4Cl solution, with air being pumped through the tank. With continuous agitation, the iron will "rust" and precipitate out of solution in the form of a slime:

$$4Fe_{(s)} + 4O_{2(g)} + H_2O \rightarrow 4FeOOH \cdot nH_2O \quad (7.18)$$

An important modification of the Becher process is the development of the Synthetic Rutile Enhancement Process (SREP) route, using various leaching methods to reduce the level of radioactivity in the synthetic rutile product to internationally acceptable levels.

In a process variant (Enhanced Acid Regeneration System; EARS process), the ilmenite is roasted to convert the titanium component into the insoluble rutile form and to condition the iron component for leaching. The product is then rapidly leached at atmospheric pressure in hydrochloric acid to remove the iron, leaving rutile crystals in the former ilmenite grain. This "synthetic" rutile (typically 96–98% TiO_2) is then washed, filtered, and calcined. The iron chloride leach liquors are processed to regenerate the acid, and the iron oxide pellets can be sold for use by the steel or cement industries.

7.3.3 Structure of Titania

Titanium dioxide occurs as a mineral in three polymorphic modifications: tetragonal rutile (*P4₂/mnm* (136)); tetragonal anatase (*I4₁/amd* (141); and orthorhombic brookite (*Pbca* (61)). The crystal structure of the thermodynamically stable rutile consists of edge-sharing TiO_6 octahedra that form straight chains parallel [001]. The chains are linked by sharing corners of the octahedra. As evident from Figure 7.46a, this configuration resembles an f.c.c. lattice. However, the close-packed oxygen sheets are slightly kinked (by about 20°) to permit only a tetragonal symmetry with a 4_2 screw axis. This causes also a distortion of the individual TiO_6 octahedra.

In anatase, the TiO_6 octahedra share four edges to form a pseudo-tetragonal framework structure, and in brookite the octahedra share three edges. Each

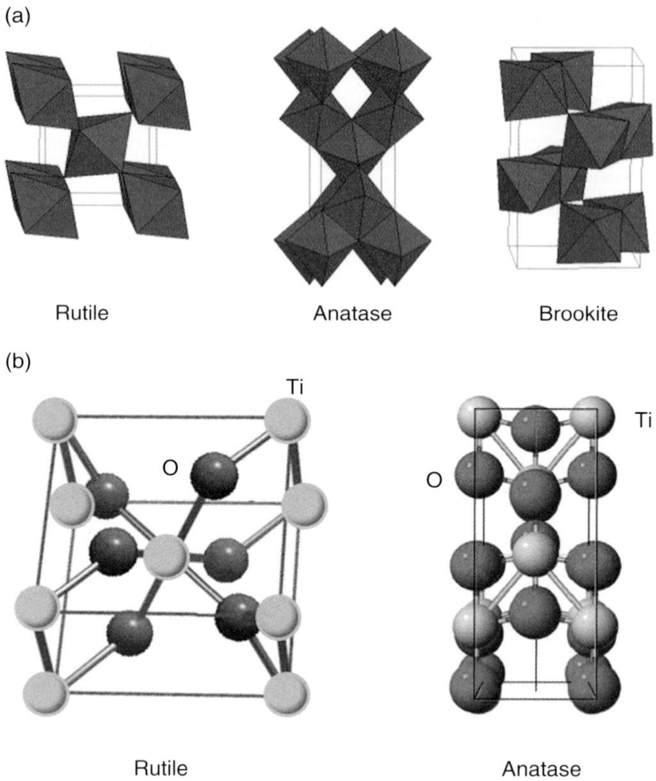

Figure 7.46 (a) Arrangement of TiO_6 octahedra in the three titania modification (Smyth, 2008); (b) Representation of the structures of rutile and anatase.

distorted octahedron is linked with three others to form zigzag chains, with each octahedron sharing one edge with an octahedron of the neighboring chain thus forming nets parallel (100) (Figure 7.46a,b).

7.3.4
Selected Applications

The structural applications of sintered titania are limited by its relatively poor mechanical properties, in particular its low modulus of rupture, compressive strength, and fracture toughness (Table 7.10). However, there are a number of uses including pigments, fillers for paper and plastics, gas sensors, and photocatalytic and antimicrobial applications.

7.3.4.1 **Pigments**
The most important application of titanium dioxide is in powder form as a widely used pigment for providing whiteness and opacity to products such as paints and

coatings (including glazes and enamels), plastics, papers, inks, fibers, and food and cosmetics. Titania is very white, with a high refractive index of >2.4 that is surpassed only by diamond. The refractive index determines the opacity that the material confers to the matrix in which the pigment is suspended. Hence, owing to its high refractive index, only relatively low levels of titania pigment are required to achieve a white opaque coating.

The high refractive index and bright white color of titanium dioxide make it also an effective opacifier in glass and porcelain enamels, replacing the tin oxide used previously. One of the major advantages of the material for exposed applications is its resistance to discoloration under UV light.

7.3.4.2 **Semiconductor Gas Sensors**

Semiconducting n-type metal oxides, including TiO_2, demonstrate a reproducible decrease in electrical resistivity in the presence of certain chemical species, usually those associated with oxygen deficiency or excess electrons at the surface (Newnham and Ruschau, 1991). The major requirements of any semiconductor sensor are high sensitivity, high selectivity, long-term stability, high reproducibility, and short response and recovery times. In general, such sensors consist of a suitable semiconducting metal oxide, such as SnO_2, TiO_2, ZnO or γ-Fe_2O_3, together with an embedded catalyst (usually a noble metal such as Pt, Pd, Ru, Rh) and a heater (Heywang, 1984).

The operational principle of a semiconductor gas sensor is concerned with the detection of changes of space and surface charges in response to the exposure of a gas. For example, a gas molecule X arriving at a sensor surface will react with a reversibly adsorbed oxygen atom $O_{r,ad}$, thereby being oxidized to XO. When XO desorbs from the surface, an oxygen vacancy V^+ remains that contributes one electron to the conduction band of the semiconducting oxide, thus increasing the electrical conductivity according to:

$$X + O_{r,ad}\, \overleftrightarrow{k_1 k_{-1}} XO\uparrow + V^+ + e^-, \tag{7.19}$$

where k_1 and k_{-1} are the rate constants of the adsorption and desorption reactions, respectively.

The temporal change of the surface electron density $n_s^{(x)}$ in response to the presence of a reducing gas is:

$$dn_s^{(x)}/dt = k_1[X]\cdot[O_{r,ad}] - k_{-1}[XO]\cdot[V^+]\cdot n_s^{(x)}. \tag{7.20}$$

Under equilibrium conditions the excess electron density is:

$$n_s^{(x)} = \sqrt{k_1/k_{-1}}\cdot\sqrt{[O_{\eta ad}]/[XO]}\cdot\sqrt{[X]}. \tag{7.21}$$

Calculating the total electrical surface conductivity σ_s of the titanium oxide as a function of the gas concentration [X] yields:

$$\sigma_s = \sigma_s^{(o)} + \sigma_s^{(s)} + \sigma_s^{(X)} = e\cdot\mu_e\left[n_s^{(o)} + n_s^{(s)} + n_s^{(X)}\right], \tag{7.22}$$

where the superscripts o, s, and X refer to the partial surface conductivities of the oxygen-independent, oxygen-dependent, and gas concentration-dependent reactions, respectively. Finally, the relative change of the conductivity in the presence of a gas yields:

$$\frac{\Delta\sigma_s}{\sigma_s} = \frac{\sigma_s^{(X)}}{\sigma_s^{(O)} + \sigma_s^{(s)}} = \frac{\left(\sqrt{\frac{k_1}{k_{-1}}}\right)\left(\sqrt{\frac{[O_{\eta ad}]}{[XO]}}\right)}{n_s^{(O)} + \left(\sqrt[4]{\frac{k_{-1}}{k_1}}\right)\cdot\left(\sqrt{[O_{\eta ad}]}\right)\cdot\left([O_2]^{-1/4}\right)}\sqrt{[X]} \tag{7.23}$$

As is evident from Eqs (7.21) and (7.23), the excess surface conductivity is directly proportional to the square root of the partial pressure [X] of the gas if, per mole of the gas, only one oxygen atom at the surface will be consumed; that is, only one lattice vacancy V^+ will be created. However, if two oxygen atoms are being consumed per mole of the monitored gas (V^{++}), then the excess surface conductivity will be increased with the cube root of [X] (Heywang, 1984). Also, the signal of the sensor will be maximized by minimizing the surface electron density $n_s^{(o)}$ that is related to defects at the surface of the sensing metal oxide.

Many possible types of sensor have been developed that respond to a variety of reducing, oxidizing, and combustible gases. A unique application of a TiO_2-based gas sensor has been designed in Japan (Egashira *et al.*, 1988), to monitor the freshness of fish (the "sushi sensor"). In this case, the adenosine triphosphate (ATP) in fish flesh is converted on spoiling, into inosine and hypoxanthine, in addition the odorous gaseous trimethylamine (TMA) and dimethylamine (DMA) being released. The presence of these gases can be monitored by a solid-state ruthenium-doped TiO_2 sensor. Here, ruthenium catalyzes the decomposition reaction of TMA and DMA, and acts as an electron acceptor in the space charge region of the TiO_2 semiconductor. The ensuing decrease in resistance can be directly related to the ATP content, and hence the freshness of the fish (Newnham and Ruschau, 1991).

Titania-based gas sensors may also be developed for "breathalyzer" purposes to detect alcohol in the breath of inebriated drivers. For example, Nb–Pt co-doped TiO_2 thin films can detect ethanol at concentrations below 300 ppm, with a response and recovery time of less than 2 minutes (Le *et al.*, 2005).

As the sensitivity of a semiconductor gas sensor depends on the efficiency with which electrons are exchanged, maximization of the surface area by using nano-sized particles of the sensing metal oxide, such as TiO_2, SnO_2 or ZnO, is an important prerequisite. This has led to the design of integrated ultrafine particle gas sensors with large surface areas (Chan *et al.*, 2002; Eranna *et al.*, 2004). Since their operating temperatures are substantially lower (130–150 °C) than those of standard gas sensors (≥350 °C), integration with a single-chip, monolithic type microprocessor is feasible (Figure 7.47).

7.3.4.3 **Photocatalysis**

Titania acts as an electrode or electrode coating in photoelectrolysis cells to enhance the efficiency of electrolytic splitting of water into hydrogen and oxygen (the Honda–Fujishima effect; see below). The most important use of the photocatalytic functional performance of titania (in particular anatase) has been found in the

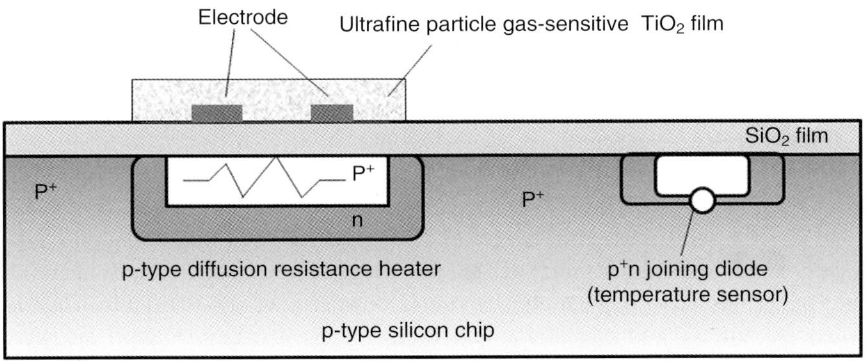

Figure 7.47 Schematic representation of an integrated ultrafine particle gas sensor.

design and development of many new effective reactors for purifying industrial and domestic waste water and gases. Other applications include self-cleaning and self-sterilizing surfaces, anti-fogging surfaces, anticorrosion systems, photocatalytic lithography, and photochromics (Fujishima *et al.*, 2008).

The outstanding photocatalytic activity of the anatase polymorph of titanium dioxide (TiO_2) was discovered and studied by Fujishima and Honda (1972). The use of n-type TiO_2 as an anode and Pt as a cathode, and irradiation by sunlight or UV light, led to the evolution of hydrogen at the Pt electrode and oxygen at the TiO_2 site. A comprehensive account on the surface science of titanium dioxide, based on almost 800 references, has been produced by Diebold (2003).

The basic mechanism of photocatalysis is as follows. Light of sufficient energy will be absorbed by the semiconducting TiO_2, and electrons gain energy sufficient to be elevated from the valence to the conduction band; this is akin to the formation of an electron-hole pair. The threshold energy of this process is given by the band gap of the semiconducting material: the energy of the photons must exceed the band gap of anatase at 3.2 eV. This energy corresponds to a wavelength of 390 nm, thus requiring UV radiation. However, in modern developments anatase is doped with oxides of transition metals of Group V of the Periodic System of Elements (PSE), such as Ta (Sobczynski and White, 1985), Nb (Cui *et al.*, 1995; Burlacov *et al.*, 2005) or V (Gao *et al.*, 2000), that shift down the conduction band to lower energies by injecting new electronic states into the wide-band gap TiO_2 host. Organic dye coatings on TiO_2 grains can also be used for this purpose (see the *Grätzel cell* below; Yao *et al.*, 2008). The electrons and holes created by irradiation with photons of energy >3.2 eV will be spatially separated and diffuse to the surface of the TiO_2, where the electrons will trigger reduction reactions, and the holes oxidation reactions. The latter includes the oxidation of OH^- ions to form highly reactive OH^* radicals. These radicals are able to oxidize and split very stable and difficult-to-handle organic contaminants, such gestagenic hormones, steroids, antibiotics, PCB, dioxins, and related chemical species.

In the past, many techniques have been explored to bring effluents to be decontaminated into close contact with the photoactive titania. These have included the use of a pumped slurry of anatase particles, the deposition of thin titania layers by a wet precipitation/sol–gel technique onto glass or metal substrates, the screen printing of porous anatase layers consisting of TiO_2 nanoparticles (20–30 nm), the CVD of titania by pyrolysis, the PVD of titania including RF and d.c. magnetron sputtering, and the oxidation of metallic titanium by the PLASMOCER process (for references, see Burlacov *et al.*, 2006). However, many of these techniques have experienced serious limitations, related predominantly to problems with filtration and recycling or the regeneration of titania powder after use, long-term coating integrity, surface contamination, and many other obstacles. Hence, alternate approaches have been sought based on the immobilization of photoactive titania powder by thermal spraying on different types of technological support, for example, metal sheet, ceramics, glass, and polymers. *Thermal spraying* appears to be an appropriate technology to produce coatings with desirable photocatalytic characteristics and with additional promising economic prospects for industrial scale-up (see for example, Burlacov *et al.*, 2006; Burlacov *et al.*, 2007). As the photocatalytic reactions occur only in the very thin top layer of the titania coating, much attention must be paid to the morphological and microstructural evolution and quality of the coating surfaces.

As the immobilization of a TiO_2 photocatalyst in the form of a coating has many practical advantages over suspended powder, the coating techniques used in the past have included thermal spraying (Berger-Keller *et al.*, 2003; Ye *et al.*, 2003; Toma *et al.*, 2005; Burlacov *et al.*, 2006; 2007), CVD (O'Neill *et al.*, 2003) and, in particular, sol–gel deposition (Atashbar *et al.*, 1998), all of which have been tested to produce well-adhering coatings with favorable surface properties. However, in order to achieve mechanical stability of the sol–gel-derived TiO_2 coatings, and high crystallinity during its subsequent calcination treatment, it is essential to apply high temperatures in the range of 450 to 750 °C. Unfortunately, this may have dramatic consequences, since at elevated temperatures an irreversible polymorphic transformation occurs of the anatase phase to the photocatalytically less-active rutile phase. Hence, there is a need to process TiO_2 (anatase) coatings at temperatures substantially below the onset of phase transition to rutile. Cold gas dynamic spray (CGDS) deposition on pliable polymer substrates appears to be a promising technology to achieve this goal (Burlacov *et al.*, 2007).

Anatase powder was deposited, using CGDS, onto poly(sulfone), poly(ethersulfone) and poly(phenylsulfone) sheets at 450 and 500 °C gas temperature, and 15–30 bar nitrogen gas pressure. A soft substrate provided by a polymeric surface allows anchoring of the highly accelerated hard and brittle titania particles solidly within the surface asperities produced by impact with high kinetic energy. Raman spectroscopy confirmed that the phase composition of the precursor TiO_2 powder had been conserved during the deposition process. Moreover, the coatings demonstrated superior adhesion to the polymeric substrates, with surface roughness R_a values between 1.8 (450 °C) and 3.4 (500 °C) μm, and photocatalytic efficiencies equal to or higher than that of the reference coating Degussa P-25. In particular, coatings deposited onto poly(ethersulfone) (450 °C, 20–30 bar) showed initial reaction rates (rate of pH decrease) of between 0.6 and 1.0 pH h^{-1}, whereas those deposited under comparable conditions onto poly(phenylsulfone) and poly(sulfone) showed values of 0.3–0.6 pH h^{-1} and 0.25–0.40 pH h^{-1}, respectively. The initial reaction rate of the control (standard Degussa P-25 coating) was about 0.29 pH h^{-1} (Burlacov *et al.*, 2007). As some thermoplastic polymers exhibit properties that are conducive to their use in photocatalytic reactors (e.g., light weight, reasonable transmissivity in UV-VIS-NIR light, mechanical flexibility, reasonable chemical stability against UV-A radiation, cost efficiency, and easy processability), the combination of a homogeneous, thin, well-adhering titania (anatase) coating deposited by CGDS with a polymer substrate with satisfactory UV transmittance and mechanical and chemical stability, might provide a paradigm shift for liquid and gaseous waste treatment by photocatalysis. The results of these attempts have been protected by patent law (Ballhorn *et al.*, 2007).

7.3.4.4 **Antimicrobial Coatings**

The photocatalytic activity of titania is applied in thin coatings of materials to exhibit self-cleaning and bactericidal properties under exposure to UV radiation. These properties make the material a candidate for applications such as medical devices, food preparation surfaces, air-conditioning filters, sanitary ware surfaces, and also for materials used to perform ultrafiltration in water treatment plants (e.g., Alem *et al.*, 2008). The growth of phytopathogenic bacteria such as *Enterobacter cloacae* and *Erwinia carotovora* (Yao *et al.*, 2008), and fish pathogens such as *Streptococcus iniae* and *Edwardsiella tarda* (Cheng *et al.*, 2008), has been effectively inhibited by dye-modified TiO_2 films under UV-A irradiation.

TiO_2-coated medical-grade AISI 304 stainless steel provides an effective antimicrobial surface coating method for medical implements, thereby reducing the risk of hospital-acquired infection, and in particular infections caused by *Staphylococcus aureus* and *Escherichia coli* (Chung *et al.*, 2008).

Surfaces coated with titania exhibit superhydrophilicity. Water molecules adsorbed onto the TiO_2 surfaces desorb during irradiation by sunlight or UV light, owing to the heating effect of the irradiation source. Whilst the amount of H_2O adsorbed onto the TiO_2 surfaces decreases, the distribution of the hydrogen bonds within the H_2O molecules also decreases, which in turn results in a reduction of the surface tension of the H_2O clusters. This was found to be an important driving force by which H_2O clusters on the TiO_2 surface spread out thermodynamically, to form thin H_2O layers with an extremely small contact angle (Takeuchi *et al.*, 2005). As a consequence, extremely efficient wetting of the coated surface takes place, that gives rise to the *self-cleaning* or *lotus effect*. Today, these coatings are applied to exterior walls and tiles, as well as to tiles in swimming pools, windows, fabrics, and any other conceivable surface.

7.3.4.5 **Photovoltaic Applications**

The particular electronic properties of the semiconducting TiO_2 surface have been utilized to design experimental dye-sensitized solar (photovoltaic) cells (DYSC).

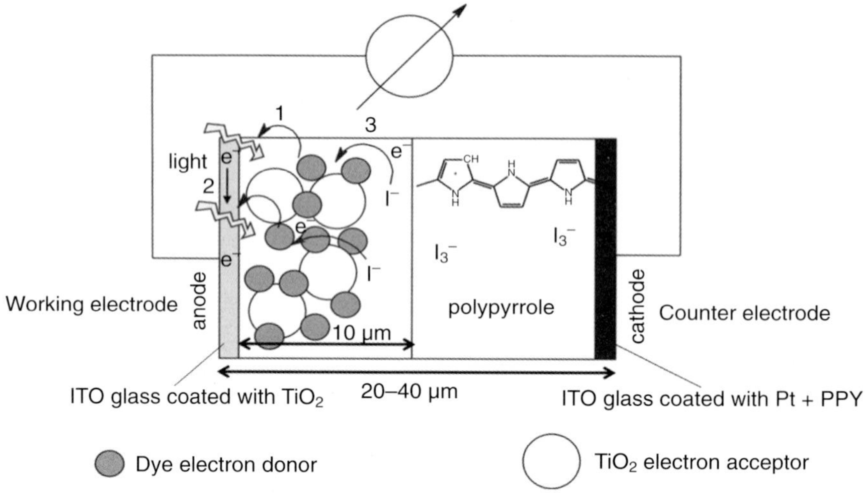

Figure 7.48 The operating principle of the Grätzel cell (http://en.wikipedia.org/wiki/Dye-sensitized_solar_cell, 2008). For details, see the text.

One of these realizations is the so-called *Grätzel cell* (Figure 7.48), that exploits the basic principles of photosynthesis (O'Regan and Grätzel, 1991). This consists of two planar electrodes fashioned from glass, typically separated by a small distance of 20–40 μm. Both electrodes are coated with a thin (0.5 μm) transparent, electrically conductive layer such as indium tin oxide (ITO) or fluorine-doped tin oxide (FTO). The working electrode (anode) carries an approximately 10 μm nanoporous layer of TiO_2 (anatase), onto which surface an organic dye such as an anthocyanoside or Ru-polypyridine is adsorbed. The counterelectrode (cathode) is coated with a thin catalytic Pt layer. An additional layer of a conducting polymer such as polypyrrole (PPY) provides better current-carrying properties to the cell by establishing a scaffold to increase the surface area. The free space between the two electrodes is filled with a redox electrolyte consisting of a solution of iodine (I_2) and potassium iodide (KI), such that the redox system I^-/I_3^- acts as an electron-carrying liquid.

As TiO_2 is not photocatalytically active when irradiated with visible light, but rather requires light with a wavelength <390 nm to elevate electrons from the valence to the conduction band, a dopant is required to inject electrons into the wide bandgap of 3.2 eV of anatase. The dye, having been adsorbed via OH groups at the TiO_2 surface, is excited by the irradiation and transfers energy, thus sensitizing the semiconductor in the visible region according to:

$$2D + h \cdot \nu \rightarrow 2D^* (\text{step 1}) \tag{7.24a}$$

The excited dye molecule D* transfer electrons into the conduction band of TiO_2 (step 2) by:

$$2D^* + 2TiO_2 \rightarrow 2D^+ + 2(TiO_2 + e^-). \quad (7.24b)$$

The atomic iodine, I, generated at the anode (step 3) combines to molecular iodine, I_2, that reacts with iodide ions, I^- to I_3^-. At the cathode, I_3^- will be decomposed to re-form three iodide ions, I^-. Not all details of these reaction chains have been elucidated in detail, with much remaining obscure, due partly to the extreme speed of some of the transfer reactions. For example, the electron injection process was found to last less than 25 fs, whereas the return of an electron from the TiO_2 to the ionized dye molecule D^+ requires some milliseconds. However, in the presence of the I_3^-/I^- redox system the dye molecule will be regenerated already within 100 ns. It is important that the ionized dye molecule recovers its neutral stage very quickly by stripping the required electron from the iodide ion, I^-, and oxidizing it to triiodide, I_3^-, instead of using the injected electron for recombination since, in the latter case, the reaction would short-circuit the cell. The triiodide then recovers its missing electron by diffusing to the counterelectrode.

In recent developments (Bai *et al.*, 2008), the somewhat problematic liquid electrolyte has been replaced by an ionic liquid consisting of a solvent-free ternary eutectic melt as the redox electrolyte. Further information on the history of the Grätzel cell, and details of the latest breakthroughs, can be found at: http://en.wikipedia.org/wiki/Dye-sensitized_solar_cell (2008).

Such DYSCs have demonstrated clear promise for covering demands for electricity generated from renewable resources, with their price/performance ratio (in $kWh\,m^{-2}\,year^{-1}$) appearing to be sufficiently high to compete successfully with fossil fuel-generated electricity. Consequently, commercial application has been forecast in the European Union Photovoltaic Roadmap (De Moor *et al.*, 2002), and prospects are being evaluated to develop this technology into a potentially significant contributor to renewable electrical energy generation by the year 2020.

References

Accuratus Corp. (2008) www.accuratus.com/materials.html (accessed 10 December 2009).

Akiyama, M., Morofuji, Y., Kamohara, T., Nishikubo, K., Ooishi, Y., Tsubai, M., Fukuda, O., and Ueno, N. (2007) Preparation of oriented aluminum nitride thin films on polyimide films and piezoelectric response with high thermal stability and flexibility. *Adv. Funct. Mater.*, **17** (3), 458–462.

Alem, A., Sarpoolaky, H., and Keshmiri, M. (2008) Titania ultrafiltration membrane: preparation, characterization and photocatalytic activity. *J. Eur. Ceram. Soc.*, **29**, 629–635.

Alper, A.M. (1967) Interrelationship of phase equilibria, microstructure and properties of fusion-cast ceramics, in *Science of Ceramics*, vol. **3** (ed. G.H. Stewart), Academic Press, London, pp. 335–369.

Atashbar, M.Z., Sun, H.T., Gong, B., Wlodarski, B., and Lamb, R. (1998) XPS study of Nb-doped oxygen sensing TiO_2 thin films prepared by sol-gel method. *Thin Solid Films*, **326** (1–2), 238–244.

Bai, Y., Cao, Y., Zhang, J., Wang, M., Li, R., Wang, P., Zakeeruddin, S., and Grätzel, M. (2008) High-performance dye-sensitized solar cells based on solvent-free electrolytes produced from eutectic melts. *Nat. Mater.*, **7**, 626–630.

Balan, E., Lazzeri, M., Morin, G., and Mauri, F. (2006) First-principle study of the OH-stretching modes of gibbsite, Locality: hypothetical structure calculated with DFT. *Am. Mineral.*, **91**, 115–119.

Ballhorn, R., Peterka, F., Kreye, H., Burlacov, I., Stoltenhoff, T., and Jirkovský, J. (2007) Verfahren zur Herstellung photokatalytisch aktiver Polymere. DE 10 2004 038 795 B4 (19.07.2007).

Beall, G.H. and Rittler, H.L. (1982) Highly refractory glass-ceramics suitable for incorporating radioactive wastes. US Patent 4,314,909 (February 9, 1982).

Behrens, G. (1993) The martensitic transformation in zirconia. Ph.D. dissertation. Case Western Reserve University.

Benyoucef, A., Klein, D., Rapaud, O., Coddet, C., and Beyoucef, B. (2009) Thermal stability of atmospheric plasma sprayed (Ni,Cu,Co)-YSZ and Ni-Cu-Co-YSZ anodes cermets for SOFC application. *J. Phys. Chem. Solids*, **70** (12), 1487–1495.

Berger-Keller, N., Bertrand, G., Coddet, C., and Meunier, C. (2003) *Thermal Spray 2003: Advancing the Science & Applying the Technology* (eds C. Moreau and B. Marple), ASM International, Materials Park, OH, USA, p. 1403.

Biedermann, K. (2002) Synthese von Schichten aus kubisch-stabilisiertem Zirkonoxid mit Hilfe plasmaunterstützter Abscheidung aus der Dampfphase. Unpublished Master Thesis, Technische Universität Bergakademie, Freiberg.

Birkby, I. and Stevens, R. (1996) Applications of zirconia ceramics. *Key Eng. Mater.*, **122–124**, 527–552.

Bleise, D. and Steinbrech, R.W. (1994) Flat R-curve from stable propagation of indentation cracks in coarse-grained alumina. *J. Am. Ceram. Soc.*, **77** (2), 315–322.

BMFT (1994) *Neue Materialien Für Schlüsseltechnologien Des 21. Jahrhunderts*, MaTech, Bundesmin. für Forschung und Technologie (BMFT), Bonn, August 1994.

Brindley, G.W. (1961) The reaction series, gibbsite → chi alumina → kappa alumina → corundum. II. *Am. Mineral.*, **46**, 1187–1190.

Brown, J.F., Clark, D., and Elliott, W.W. (1953) The thermal decomposition of the alumina trihydrate, gibbsite. *J. Chem. Soc. (London)*, 84–88.

Burlacov, I., Müller, M., and Heimann, R.B. (2005) Group V metal oxides (V, Nb and Ta) doping of titania coatings by means of suspension plasma spraying. Proceedings 17th International Symposium on Plasma Chemistry, 7–12 August 2005, Toronto, Ontario, Canada.

Burlacov, I., Jirkovský, J., Müller, M., and Heimann, R.B. (2006) Induction plasma-sprayed photocatalytically active titania coatings and their characterization by micro-Raman spectroscopy. *Surf. Coat. Technol.*, **201**, 255–264.

Burlacov, I., Jirkovský, J., Kavan, L., Ballhorn, R., and Heimann, R.B. (2007) Cold gas dynamic spraying (CGDS) of TiO_2 (anatase) coatings on polymer substrates: microstructural characterisation and photocatalytic efficiency. *J. Photochem. Photobiol. A*, **187** (2–3), 285–292.

Carter, G.A., Hart, R.D., Rowles, M., Ogden, M.I., and Buckley, C.E. (2009) Industrial precipitation of yttrium chloride and zirconyl chloride: effect of pH on ceramic properties for yttria partially stabilised zirconia. *J. Alloys Compd.*, **480** (2), 639–644.

CeramTech (2006) *BIOLOX® Delta – Nanoverbundwerkstoff Für Die Endoprothetik*, CeramTech AG, Plochingen, Germany.

Cernuschi, F., Ahmaniemi, S., Vuoristo, P., and Mäntylä, T. (2004) Modelling of thermal conductivity of porous materials: application to thick thermal barrier coatings. *J. Eur. Ceram. Soc.*, **24**, 2657–2667.

Chan, P.C.H., Yan, G.Z., Sheng, L.Y., Sharma, R.K., Tang, Z., Sin, J.K.O., Hsing, I.M., and Wang, Y. (2002) An integrated gas sensor technology using surface micro-machining. *Sens. Actuators B*, **82** (2), 277–283.

Chen, Z., Speakman, S., Howe, J., Wang, H., Porter, W., and Trice, R. (2008) Investigation of reactions between vanadium oxide and plasma-sprayed yttria-stabilized zirconia coatings. *J. Eur. Ceram. Soc.*, **29** (8), 1403–1411.

Cheng, T.C., Chang, C.Y., Chang, C.I., Hwang, C.J., Hsu, H.C., Wang, D.Y., and Yao, K.S. (2008) Photocatalytic bactericidal effect of TiO_2 film on fish pathogens. *Surf. Coat. Technol.*, **203** (5–7), 925–927.

Ching, W.Y., Ouyang, L., Rulis, P., and Yao, H. (2008) *Ab initio* study of the physical properties of γ-Al_2O_3: Lattice dynamics, bulk properties, electronic structure, bonding, optical properties, and ELNES/XANES spectra. *Phys. Rev. B*, **78** (1), 014106.

Chung, C.J., Lin, H.I., Tsou, H.K., Shi, Z.Y., and He, J.L. (2008) An antimicrobial TiO_2 coating for reducing hospital-acquired infection. *J. Biomed. Mater. Res.*, **85** (1), 220–224.

Cui, H., Dwight, K., Soled, S., and Wold, A. (1995) Surface acidity and photocatalytic activity of Nb_2O_5/TiO_2 photocatalysts. *J. Solid State Chem.*, **115** (1), 187–191.

De Moor, H., Jäger-Waldner, A., Herrero, J., *et al.* (2002) PVNET. European Roadmap for PV R & D. Draft version 12/2002, 34 pp.

Degussa (1999) Supported catalyst, process for its production as well as use for oxychlorination of ethylene. EP 19980121259 A1.

Diebold, U. (2003) The surface science of titanium dioxide. *Surf. Sci. Rep.*, **48**, 53–229.

Du, X., Su, X., Wang, Y., and Li, J. (2008) Thermal decomposition of grinding activated bayerite. *Mater. Res. Bull.*, **44** (3), 660–665.

Dubsky, J., Neufuss, K., and Kolman, B. (1997) Phase composition changes in annealed plasma-sprayed zircon-alumina coatings, in *Thermal Spray: A United Forum for Scientific and Technological Advances* (ed. C.C. Berndt), ASM International, Materials Park, OH, USA, pp. 473–476.

Egashira, M., Shimizu, Y., and Taka, Y. (1988) Enhancement of trimethylamine sensitivity of semiconductor gas sensors by ruthenium. *Chem. Lett.*, **17**, 389–392.

Elger, G.W., Stafler, R.A., and Sanker, P.E. (1978) Process for purifying a titanium-bearing material and upgrading ilmenite to synthetic rutile with sulfur trioxide. US Patent 4,120,694 (October 17, 1978).

Eranna, G., Joshi, B.C., Runthala, D.P., and Gupta, R.P. (2004) Oxide materials for development of integrated gas sensors: a comprehensive review. *Crit. Rev. Solid State Mater. Sci.*, **29** (3–4), 111–188.

Evans, A.G. (1984) Toughening mechanisms in zirconia alloys, in *Advances in Ceramics*, vol. **12** (eds N. Claussen, M. Rühle, and A.H. Heuer), The American Ceramic Society, Columbus, OH, pp. 193–212.

Evans, A.G. (1990) Perspective on the development of high toughness ceramics. *J. Am. Ceram. Soc.*, **73**, 187–206.

Evans, K.A. (1996) The manufacture of alumina and its use in ceramics and related applications. *Key Eng. Mater.*, **122–124**, 489–526.

Felsche, J. (1968) The alkali problem in the crystal structure of ß-alumina. *Z. Kristallogr.*, **127**, 94–100.

Fondeur, F.F., Hobbs, D.T., and Fink, S.D. (2004) Aluminum Leaching of "Archived" Sludge from Tanks 8F, 11H and 12 H. WSRC-TR-2004-00180, Westinghouse Savanna River Company, Aiken, SC 29808, USA.

Foster, J. (1996) Ceramic applications for wear protection of pipe liners and cyclones. *Key Eng. Mater.*, **122–124**, 247–278.

Franke, W., Heimann, R., and Willgallis, A. (1968) Bildung von Mn-dotierten Einkristallen von Korund aus der Gasphase. *J. Cryst. Growth*, **2**, 160–164.

Fujishima, A. and Honda, K. (1972) Electrochemical photolysis of water at a semiconductor electrode. *Nature*, **238**, 37–38.

Fujishima, A., Zhang, X., and Tryk, D.A. (2008) TiO_2 photocatalysis and related surface phenomena. *Surf. Sci. Rep.*, **63** (12), 515–582.

Gao, Y., Thevuthasan, S., McCready, D.E., and Engelhard, M. (2000) MOCVD growth and structure of Nb- and V-doped TiO_2 films on sapphire. *J. Cryst. Growth*, **212**, 178–190.

Garvie, R.C., Hannik, R.H.J., and Pascoe, R.T. (1975) Ceramic steel? *Nature (London)*, **258** (5537), 703–704.

Gesenhues, U., Optehostert, R., and Philipps, K. (2003) Production of titanium dioxide. EP 20000916956 (June 25, 2003).

Goff, J.P., Hayes, W., Hull, S., Hutchings, M.T., and Clausen, K.N. (1999) Defect structure of yttria-stabilized zirconia and its influence on the ionic conductivity at elevated temperatures. *Phys. Rev. B*, **59** (22), 14202–14219.

Grain, C.F. (1967) Phase relations in the ZrO_2-MgO system. *J. Am. Ceram. Soc.*, **50** (6), 288–290.

Grange, R.A. and Kiefer, J.M. (1941) Transformation of austenite on continuous cooling and its relation to transformation at constant temperature. *Trans. ASM*, **29**, 85–116.

Griffith, A.A. (1920) The phenomenon of rupture and flow in solids. *Philos. Trans. R. Soc. London*, **A221**, 163–198.

Gross, H.L. and Mader, W. (1997) On the crystal structure of κ-alumina. *Chem. Commun.*, 55–56.

Haering, C. (2001) Degradation der Leifähigkeit von stabilisiertem Zirkoniumdioxid in Abhängigkeit von der Dotierung und den damit verbundenen Defektstrukturen. Unpublished Ph.D. Thesis, University Erlangen, Germany.

Halvarsson, M. (1994) The microstructure and stability of CVD alumina coatings. Ph.D. Thesis, Chalmers University of Technology, Göteborg, Sweden.

Heaney, P.J. (2000) Phase transformation induced by solid solution, in *Transformation Processes in Minerals*, vol. **39** (eds S.A.T. Redfern and M.A. Carpenter), Reviews in Mineralogy and Geochemistry, Mineralogy Society of America, Geochemical Society, Washington, D.C., pp. 135–174.

Heimann, R.B. (1991) Advanced ceramic coatings by plasma spraying: a technology for the 1990s and beyond. *Proc. Adv. Mater.*, **1**, 181–192.

Heimann, R.B. (2008) *Plasma Spray Coating. Principles and Applications*, 2nd edn, Wiley-VCH Verlag GmbH, Weinheim, Ch 8.1.

Heimann, R.B. (2009) Better quality control: stochastic approaches to optimize properties and performance of plasma sprayed coatings. *J. Therm. Spray Technol.*, doi: 10.1007/s11666-009-9385-3.

Heimann, R.B. and Vandergraaf, T.T. (1988) Cubic zirconia as a candidate waste form for actinides: dissolution studies. *J. Mater. Sci. Lett.*, **7**, 583–586.

Heimann, R.B., Lamy, D., and Merchant, V.E. (1991) Post-plasma spraying laser densification of partially stabilized zirconia coatings, in *Ceramic Coatings–A Solution Towards Reducing Wear and Corrosion*, Transactions, 17th CUICAC Workshop (ed. R.B. Heimann), Laval University, Québec City, Québec, Canada.

Heimpel, W., Gorelick, L., and Gwinnett, A.J. (1988) Philological and archaeological evidence for the use of emery in the Bronze Age Near East. *J. Cuneiform Stud.*, **40**, 195–210.

Heywang, W. (ed.) (1984) *Sensorik, Halbleiter-Elektronik*, Bd. **17**. Springer, Berlin, p. 171.

Hornbogen, E. (2008) *Werkstoffe. Aufbau und Eigenschaften von Keramik-, Metall-, Polymer- und Verbundwerkstoffen*, 9th edn, Springer, Berlin, 594 pp.

ICT Co. Ltd. (JP), International Catalyst Technology (2002) Exhaust gas purifying catalyst. EP 20010114935.

Inghels, E., Heuer, A.H., and Steinbrech, R.W. (1990) Fracture mechanics of high-toughness magnesia-partially stabilized zirconia. *J. Am. Ceram. Soc.*, **73** (7), 2023–2031.

Ingram-Jones, V., Slade, R.C.T., Davies, T.W., Southern, J.C., and Salvador, S. (1996) Dehydroxylation sequences of gibbsite and boehmite: study of differences between soak and flash calcinations and of particle size. *J. Mater. Chem.*, **6**, 73–79.

Jacobson, N.S., Copland, E.H., and Kaufman, L. (2001) Thermodynamic database for the NdO1.5-YO1.5-YbO1.5-ScO1.5-ZrO2 system. NASA/TM-2001-210753. NASA Glenn Research Center.

Jasim, K.M., Rawlings, R.D., and West, D.R.F. (1992) Stability of ZrO_2-Y_2O_3 t′-phase formed during laser sealing. *Mater. Sci. Technol.*, **8** (1), 83–91.

Kamibayashi, M. (2001) Advanced Sodium-Sulfur Battery Systems. Tokyo Electric Power Company (TEPCO), Japan.

Ke, P.L., Wang, Q.M., Gong, J., Sun, C., and Zhou, Y.C. (2006) Progressive. damage during thermal shock cycling of D-gun sprayed thermal barrier coatings with hollow spherical ZrO_2-$8Y_2O_3$. *Mater. Sci. Eng. A.*, **435/436**, 228–236.

Kingery, W.D., Bowen, H.K., and Uhlmann, D.R. (1976) *Introduction to Ceramics*, 2nd edn, John Wiley & Sons, Inc., New York, Ch 17.

Kuznetsof, P.M. (2006) Titanium Dioxide. Chemical and Technical Assessment. Report UN Food and Agricultural Organization (FAO), pp. 1–8.

Le, D.T., Vuong, D.D., Duy, N.V., Hieu, N.V., and Chien, N.D. (2005) Preparation and characterization of nanostructured TiO_2 and SnO_2 materials for gas sensor applications. Proceedings, 8th German-Vietnamese Seminar in Physics and Engineering, 3–8 April, Erlangen, Germany.

Lefèvre, J. (1963) Contribution à l'étude de différentes modifications structurales des phases de type fluorine dans les systems à base de zircone ou d'oxyde de hafnium. I. Étude d'une deformation discontinue de la maille fluorine. La transformation allotropique de la zircone. *Ann. Chim.*, **8**, 118–124.

Lehmann, H.D. and Heimann, R.B. (2008) Thermally sprayed thermal barrier coating (TBC) systems: a survey of recent patents. *Recent Pat. Mater. Sci.*, **1** (2), 140–158.

Levin, I. and Brandon, D. (1998) Metastable alumina polymorphs: crystal structures and transition sequences. *J. Am. Ceram. Soc.*, **81** (8), 1995–2012.

Lu, P.J., Yao, N., So, J.E., Harlow, G.E., Lu, J.F., Wang, G.F., and Chaikin, P.M. (2005) The earliest use of corundum and diamond in prehistoric China. *Archaeometry*, **47** (1), 1–12.

Matchen, B. (1996) Application of ceramics in armor products. *Key Eng. Mater.*, **122–124**, 333–342.

Matsumoto, M., Takayama, H., Yokoe, D., Mukai, K., Matsubara, H., Kagiya, Y., and Sugita, Y. (2006) Thermal cycle behavior of plasma sprayed La_2O_3, Y_2O_3 stabilized ZrO_2 coatings. *Scr. Mater.*, **54**, 2035–2039.

Matthews, A.L. (1996) Ceramic filters for the cast metal industry. *Key Eng. Mater.*, **122–124**, 293–304.

Matweb (2008) www.matweb.com/ (accessed 10 December 2009).

May, G.J. (1978) Ionic conductivity and crystal structure of Mg- and Ca-doped Na-β-alumina. *J. Mater. Sci.*, **13**, 261–267.

Menéndez-Proupin, E. and Gutiérrez, G. (2005) Electronic properties of bulk γ-Al_2O_3. *Phys. Rev. B*, **72**, 035116.

Messing, G.L., Kumagai, M., Shelleman, R.A., and McArdle, J.L. (1986) Seeded transformations for microstructural control in ceramics, in *Science of Ceramic Chemical Processing* (eds L. Hench and D. Ulrich), John Wiley & Sons, Inc., New York, pp. 259–272.

Minh, N.Q. and Takahashi, T.T. (1995) Electrolyte, in *Science and Technology of Ceramic Fuel Cells* (eds N.Q. Minh and T.T. Takahashi), Elsevier, Amsterdam, pp. 69–116.

Mo, S.D. and Chang, W.Y. (1998) Electronic and optical properties of θ-Al_2O_3 and comparison to α-Al_2O_3. *Phys. Rev. B*, **57** (24), 15219–15228.

Müller, M. (2001) Entwicklung elektrokatalytischer Oxidschichten mit kontrollierter Struktur und Dotierung aus flüssigen Precursoren mittels thermischer Hochfrequenzplasmen. Unpublished Ph.D. Thesis, Technische Universität Bergakademie, Freiberg.

Nalla, R.K., Kruzic, J.J., Kinney, J.H., and Ritchie, R.O. (2005) Mechanistic aspects of fracture and R-curve behavior in human cortical bone. *Biomaterials*, **26**, 217–231.

Newnham, R.E. and Ruschau, G.R. (1991) Smart electroceramics. *J. Am. Ceram. Soc.*, **74**, 463–480.

Nicholls, J.R., Lawson, K.J., Johnstone, A., and Rickerby, R.S. (2002) Methods to reduce the thermal conductivity of EB-PVD TBCs. *Surf. Coat. Technol.*, **151/152**, 383–391.

O'Neill, S.A., Clark, R.J.H., and Parkin, I.P. (2003) Anatase thin films on glass from the chemical vapor deposition of titanium (IV) chloride and ethyl acetate. *Chem. Mater.*, **15**, 46–50.

O'Regan, B. and Grätzel, M. (1991) A low-cost, high-efficiency solar cell based on dye-sensitized colloidal TiO_2 films. *Nature*, **353** (6346), 737–740.

Ogando, J. (2005) Zirconia pump liners last longer, lower total coast. Design News, 7 March 2005.

Ollivier, B., Retoux, R., Lacorre, P., Massiot, D., and Férey, G. (1997) Crystal structure of κ-alumina: an x-ray powder diffraction, TRM and NMR study. *J. Mater. Chem.*, **7** (6), 1049–1056.

Orowan, E. (1949) Fracture and strength of solid. *Rep. Progr. Phys.*, **12**, 185–232.

Oshima, T., Kajita, M., and Okuno, A. (2004) Development of sodium-sulfur batteries. *Int. J. Appl. Ceram. Technol.*, **1**, 269–276.

Padture, N.P., Gell, M., and Jordan, E.H. (2002) Thermal barrier coatings for gas-turbine engine applications. *Science*, **296** (5566), 280–284.

Paglia, G. (2004) Determination of the structure of γ-alumina using empirical and first principle calculations combined with supporting experiments. Ph.D. dissertation, Curtin University of Technology.

Paglia, G., Buckley, C.E., Rohl, A.L., Hunter, B.A., Hart, R.D., Hanna, J.V., and Byrne, L.T. (2003) A new structural model for boehmite-derived γ-alumina. *Phys. Rev. B*, **68**, 144110.

Patchett, J.E. (1996) The use of advanced ceramics in abrasive applications. *Key Eng. Mater.*, **122–124**, 325–332.

Prengel, H.G., Heinrich, W., Roder, G., and Wendt, K.H. (1994) CVD coatings based on medium temperature CVD κ- and α-Al_2O_3. *Surf. Coat. Technol.*, **68/69**, 217–220.

Ray, S.P. and Stubican, V.S. (1977) Fluorite related ordered compounds in the ZrO_2-CaO and ZrO_2-Y_2O_3 systems. *Mater. Res. Bull.*, **12** (5), 549–556.

Ritchie, R.O. (1988) Mechanisms of fatigue crack propagation in metals, ceramics and composites: role of crack-tip shielding. *Mater. Sci. Eng.*, **103**, 15–28.

Ritchie, R.O. (1999) Mechanisms of fatigue-crack propagation in ductile and brittle solids. *Int. J. Fract.*, **100**, 55–83.

Roberts, J.M., Singh, J.P., and Scattergood, R.O. (2008) Microstructure and properties of alumina-whisker-reinforced tetragonal zirconia polycrystal matrix, in *Ceramics Engineering and Science Proceedings*, vol. **12** (7-8) (ed. J.B. Wachtman, Jr), The American Ceramic Society, Inc., Columbus, OH, pp. 1193–1200.

Rogé, B., Fahr, A., Giguère, J.S.R., and McRae, K.I. (2003) Nondestructive measurement of porosity in thermal barrier coatings. *J. Therm. Spray Technol.*, **12** (4), 530–535.

Rooksby, H.P. and Rooymans, C.J.M. (1961) The formation and structure of delta alumina. *Clay Miner. Bull.*, **4**, 234–238.

Rouanet, A. (1971) Zirconium dioxide-lanthanide oxide systems close to the melting point. *Rev. Int. Hautes Temp. Refract.*, **8** (2), 161–180.

Ruh, R., Garrett, H.J., Domagala, R.F., and Patel, V.A. (1977) The system zirconia-scandia. *J. Am. Ceram. Soc.*, **60** (9–10), 399–403.

Saalfeld, H. (1958) Dehydration of gibbsite and the structure of a tetragonal gamma-Al_2O_3. *Clay Miner. Bull.*, **3**, 249–256.

Saalfeld, H. (1960) Struktur des Hydrargillits und der Zwischenstufen beim Entwässern. *Neues Jahrb. Mineral. Abh.*, **95**, 1–87.

Schaetzle, J.E. (2007) BeO takes the heat in RF transistor packages. Microwaves & RF. Available at: www.mwrf.com/issues/Index.cfm?IssueID=602 (accessed 10 December 2009).

Schneider, H., Locatelli, M.L., Achard, J., Scheid, E., Tounsi, P., and Ding, H. (2007) Study of CVD diamond films for thermal management in power electronics. Proceedings, 2007 European Conference on Power Electronics and Applications, 2–5 September, Aalborg, Denmark, pp. 1–8.

Schuler, A. (2007) Das Brennstoffzellen-Projekt Hexis. *GasWasserAbwasser (gwa)*, **2**, 93–98. See also www.hexis.com.

Seo, D., Ogawa, K., Tanno, M., Shoji, T., and Murata, S. (2007) Influence of heat exposure time on isothermal degradation of plasma sprayed CoNiCrAlY coatings. *Surf. Coat. Technol.*, **201**, 7952–7960.

Singh, B. and Gilkes, R.J. (1995) The natural occurrence of chi-alumina in lateritic pisolites. *Clay Mineral.*, **30**, 39–44.

Skaggs, S.R., Richardson, N.L., and Nelson, L.S. (1972) Phase relationships in the zirconia-yttria system. *Bull. Am. Ceram. Soc.*, **51**, 324–326.

Smrcok, L., Langer, V., and Krestan, J. (2006) γ-alumina: a single crystal x-ray diffraction study. *Acta Crystallogr. C*, **62**, i83–i84.

Smyth, J. (2008) Mineral Structure and Property Data: TiO2 group. Available at: http://ruby.colorado.edu/~smyth/min/tio2.html (accessed 10 December 2009).

Sobczynski, A. and White, J.M. (1985) Photodecomposition of water vapor over

TiO_2 catalysts doped with Ta_2O_5. *J. Mol. Catal.*, **29**, 379–391.

Souza Santos, P., Souza Santos, H., and Toledo, S.P. (2000) Standard transition aluminas. Electron microscopy studies. *Mater. Res.*, **3** (4), 104–114.

Spera, D.A. and Grisaffe, S.J. (1973) *Life Prediction of Turbine Components: On-Going Studies at the NASA-Lewis Research Center*, NASA Technical Memorandum TMX-2664. National Technical Information Service, Springfield, VA.

Spiridonov, F.M., Popova, L.N., and Popilskii, R.Y. (1970) On the phase relations and the electrical conductivity in the system ZrO_2-Sc_2O_3. *J. Solid State Chem.*, **2** (3), 430–438.

Stevens, R. (1993) Zirkonoxidkeramik, in *Technische Keramische Werkstoffe* (ed. J. Kriegesmann), Deutscher Wirtschaftsdienst, ch. 4.2.2.0.

Strunz, H. and Nickel, E.H. (2001) *Strunz Mineralogical Tables*, 9th edn, E. Schweizerbart, Stuttgart.

Stubican, V.S., Corman, G.S., Hellmann, J.R., and Senft, G. (1984) Phase relationships in some ZrO_2 systems, in *Advances in Ceramics*, vol. **12** (eds B. Claussen, M. Rühle, and A.H. Heuer), The American Ceramic Society, Columbus, OH, pp. 96–106.

Stumpf, H.C., Russell, A.S., Newsome, J.W., and Tucker, C.M. (1950) Thermal transformation of aluminas and alumina hydrates – Reaction with 44% technical acid. *Ind. Eng. Chem.*, **42** (7), 1398–1403.

Subbarao, E.C. (1981) Zirconia – an overview, in *Advances in Ceramics* (eds A.H. Heuer and L.W. Hobbs), The American Ceramic Society, Columbus, OH, pp. 1–24.

Takeuchi, M., Sakamoto, K., Martra, G., Coluccia, S., and Anpo, M. (2005) Mechanism of photoinduced superhydrophilicity on the TiO_2 photocatalyst surface. *J. Phys. Chem. B*, **109** (32), 15422–15428.

Tesaki, M. and Tanaguchi, H. (1984) High-speed cutting tools: sintered and coated. *Kogyo Zairyo (Ind. Mater.)*, **32**, 64–71.

Thomé, L., Moll, S., Sattonnay, G., Vincent, L., Garrido, F., and Jagielski, J. (2009) Radiation effects in cubic zirconia: a model system for ceramic oxides. *J. Nucl. Mater.*, **389** (2), 297–302.

Tilley, D.B. and Eggleton, R.A. (1996) The natural occurrence of eta-alumina (η-Al_2O_3) in bauxite. *Clays Clay Mineral.*, **44** (5), 658–664.

Toma, F.L., Bertrand, G., Chwa, S.O., Meunier, C., Klein, D., and Coddet, C. (2005) Comparative study on the photocatalytic decomposition of nitrogen oxides using TiO_2 coatings prepared by conventional plasma spraying and suspension plasma spraying. *Surf. Coat. Technol.*, **200** (20–21), 5855–5862.

Tsukuma, K., Ueda, K., and Shiomi, M. (1985) Mechanical properties of isostatically hot-pressed zirconia (yttria)/alumina composites. Proceedings, 38th Annual Pacific Coast Regional Meeting of the American Ceramic Society, Irvine, CA, October 1985.

Wall, J.R.D., Wolfenden, E.B., Beard, E.H., and Deans, T. (1962) Nordstrandite in soil from West Sarawak, Borneo. *Nature*, **196**, 264–265.

Wang, Z., Kulkarni, A., Deshpande, S., Nakamura, T., and Herman, H. (2003) Effect of pores and interfaces on effective properties of plasma-sprayed zirconia coatings. *Acta Mater.*, **51**, 5319–5334.

Wefers, K. and Misra, C. (1987) Oxides and Hydroxides of Aluminium. Alcoa Technical Papers 19, Alcoa Laboratories.

West, A.R. (1992) *Grundlagen der Festkörperchemie*. VCH, Weinheim, pp. 329–362.

Whittington, B. and Ilievski, D. (2004) Determination of the gibbsite dehydration pathway at conditions relevant to Bayer refineries. *Chem. Eng. J.*, **98** (1/2), 89–97.

Wu, B.C., Chang, E., Chang, S.E., and Tu, D. (1989) Degradation mechanisms of ZrO_2-8 wt% Y_2O_3/Ni22Cr10Al1Y thermal barrier coatings. *J. Am. Ceram. Soc.*, **72** (2), 212–218.

Yao, K.S., Cheng, T.C., Li, S.J., Yang, L.Y., Tzeng, K.C., Chang, C.Y., and Ko, Y. (2008) Comparison of photocatalytic activities of various dye-modified TiO_2 thin films under visible light. *Surf. Coat. Technol.*, **203** (5–7), 922–924.

Yashima, M., Arashi, H., Kakihana, M., and Yoshimura, M. (1994) Raman scattering study of cubic-tetragonal phase transition in $Zr_{1-x}Ce_xO_2$ solid solution. *J. Am. Ceram. Soc.*, **77**, 1067–1071.

Ye, F.X., Ohmori, A., and Lee, C.J. (2003) *Thermal Spray 2003: Advancing the Science & Applying the Technology* (eds C. Moreau and B. Marple), ASM International, Materials Park, OH, USA, p. 169.

Yin, W., Meng, B., Meng, X., and Tan, X. (2008) Highly asymmetric yttria stabilized zirconia hollow fibre membranes. *J. Alloys Compd.*, **476**, 566–570.

Yomogita, H. (2008) Can batteries save embattled wind power? Tech-On, 28 October.

Zhan, G.D., Zhang, Y.Z., Shi, J.L., and Yen, T.S. (1997) Cyclic fatigue-crack propagation in SiC-whisker-reinforced Y-TZP composites. *J. Eur. Ceram. Soc.*, **17** (8), 1011–1017.

Zhou, R.S. and Snyder, R.L. (1991) Structures and transformation mechanisms of the η, γ and θ transition aluminas. *Acta Crystallogr. B*, **47**, 617–630.

8
Electroceramic Materials

8.1
Introduction

The complexity of electroceramic materials cannot be covered exhaustively in the present short chapter; instead, only basic information on the various dielectric effects will be provided. Examples of ceramic and single crystal materials, and their properties and applications will be given that are intended to stand *pars pro toto* for the completeness of information.

Many oxide ceramic materials display properties that are conducive to important modern technical applications as sensors, actuators, capacitors, thermistors, varistors, solid electrolytes, ionic conductors, superconductors, permanent magnets with soft and hard characteristics, optoelectronic shutters, and many others. Frequently, such ceramics have distorted perovskite or spinel structures that impart ferro-, piezo- or pyroelectric "smart" properties (Newnham and Ruschau, 1991). "Ferroics" are oxide ceramics with moveable domain walls that can be shifted in response to electrical, magnetic, temperature, and stress field gradients. They include ferrimagnetics ($Y_3Fe_5O_{12}$, YIG), ferroelectrics ($BaTiO_3$), superconductors ($YBa_2Cu_3O_{7-\delta}$), piezoelectrics ($Pb(Zr,Ti)O_3$, PZT), PTC thermistors ($(Ba,La)TiO_3$), hard (permanent) ferrimagnetic materials ($BaFe_{12}O_{19}$, magnetoplumbite), soft ferrimagnetic (transformer) magnets ($(Mn,Zn)Fe_2O_4$), nonlinear electro-optics [$(Pb,La)(Zr,Ti)O_3$; PLZT], electrostrictive ceramics [$(Pb(Mg,Nb)O_3$; PMN], and many others. The current estimated worldwide market share of these materials is approximately US$ 25 billion (see Chapter 6).

In contrast to polycrystalline *ceramic* materials with ferroic properties, there exist nonferroic *single crystal* piezoelectrics such as α-quartz or materials with a calcium gallium germanate (CGG) structure, as well as single crystal pyroelectrics with perovskite structure such as lithium tantalate ($LiTaO_3$).

Classic and Advanced Ceramics: From Fundamentals to Applications. Robert B. Heimann

ISBN: 978-3-527-32517-7

8.1.1
Definition and Properties of Ferroic, Smart, and Intelligent Materials

Most advanced polycrystalline ceramics described in this chapter belong to the so-called “ferroic materials” with moveable domain walls[1] (see Section 8.2). The driving forces (input parameters or stimuli) for realignment of the domain walls can be electrical (ferroelectric materials), magnetic (ferro/ferrimagnetic materials), stress field (ferroelastic materials) or temperature gradients (pyroelectric materials that are simultaneously ferroelectric). Changes of these gradients cause linear or nonlinear responses (outputs), such as the generation of electrical charges, magnetization, strain, or temperature whereby input and output parameters are coupled by physical effects, as shown in Table 8.1. “Trivial” materials responses are arranged along the diagonal of the matrix, whereas off-diagonal effects are characteristic features of “smart” or even “intelligent” materials (Newnham and Ruschau, 1991, 1993). Frequently, an effect and its inverse effect are utilized in coupled sensor-actuator applications, for example piezoelectric effect (sensor) and inverse piezoelectric effect or (quadratic) electrostriction (actuator), pyroelectric and electrocaloric effects, or photovoltaic and electro-optic effects. As shown in Table 8.1, these effects are arranged symmetrically to the diagonal of the matrix.

It is important to state that the combination of a sensor and an actuator mimics two very basic and unique functions of a living organism – being aware of its sur-

Table 8.1 Stimulus (input) – response (output) matrix of ferroic materials. Adapted from Uchino (2000).

Input	Output				
	Charge, current	Magnetization	Strain	Temperature	Light
Electrical field	Electrical permittivity	Electromagnetic effect	Inverse piezoelectric effect	Electrocaloric effect	Electro-optic effect
Magnetic field	Magneto-electric effect	Permeability	Magneto-striction	Magneto-caloric effect	Magneto-optic effect
Stress	Piezoelectric effect	Piezomagnetic effect	Elastic constant	–	Photoelastic effect
Entropy	Pyroelectric effect	–	Thermal expansion	Specific heat	–
Light	Photovoltaic effect	–	Photostriction	–	Refractive index

1) Ferroic crystals possess two or more orientation states or domains, and under a suitably chosen driving force the domain walls move, switching the crystal from one domain state to another (Newnham, 2005). Domain walls are defined as topological solitons, that is, 2-D membranes formed by spontaneously breaking a discrete symmetry at a phase transition.

rounding (sensing) and being able to respond by motion to the sensation received (acting). However, a *smart material* is not simply a sensor that receives a stimulus and responds with a signal. It is not simply an actuator that produces a useful motion either. Smart materials function both ways – as a sensor *and* as an actuator. Some very smart materials have a *control system*; that is, they analyze the sensing signal received and make a choice as what type of response to generate. In particular, they tune their properties in time and space to optimize future behavior, and hence move along a learning curve. Such actively smart materials may rightly be called *intelligent* (Dry, 1993; Rogers, 1995). They may mimic biological systems that evolve into complex life forms by relying on persistent disequilibrium to optimize dynamic behavior (Kelly, 1994; Newnham, 2000). This disequilibrium triggers rapid responses that enable the ferroic material to operate near an instability and to easily cross the morphotropic phase boundary (see Section 8.3.2). The distinction between smart and intelligent materials is essentially one between linear and nonlinear properties, with the latter being adjusted by bias fields or forces to control the response (Newnham and Ruschau, 1991).

There are four major families of smart materials, exemplified by: *piezoelectric* $Pb(Zr,Ti)O_3$ (PZT); *electrostrictive* $Pb(Mg_{1/3}Nb_{2/3})O_3$ (PMN); and *magnetorestrictive materials* $(Tb,Dy)Fe_2$ (Terfenol™),[2] as well as *shape memory alloys* such as NiTi (Nitinol™) (Newnham, 2000). Shape memory alloys have a broad phase transition from a partially ordered (austenitic) CsCl-like structure to a martensitic phase of lower symmetry. In PMN, a similar diffuse phase transition from a partially ordered paraelectric high-temperature state to a ferroelectric low-temperature relaxor state exists. In this regard, PMN is a ferroelectric analog to ferroelastic Nitinol™. In PZT, which arguably is the most important actuator material, there are similar changes in symmetry (see Section 8.3.2). At high temperature, PZT is paraelectric with a cubic symmetry that changes to ferroelectric rhombohedral and tetragonal phases at lower temperature (see Figure 8.16). Useful actuators based on PZT operate near the morphotropic phase boundary (MPB), where maximum ferro- and piezoelectric properties can be expected. Terfenol™ is the magnetic analog to ferroelectric PZT. Magnetocaloric materials such as metallic Gd and $Gd_5(Si_{4-x}Ge_x)$ have interesting potential applications in magnetic refrigeration (Gscheidner, 2008).

All of these materials have at least two phase transitions that can be described in terms of thermodynamic functions with two ordering parameters (see Appendix B). Ferroic materials are operated near an instability to make domain walls with their associated dipoles and strains moveable, as encountered in PZT or Terfenol™. On the other hand, a second type of material involves a partially ordered phase, as in PMN or the shape memory alloys. These materials are operated near a diffuse phase transition with two coexisting phases, a high-temperature austenite-like phase and a low-temperature martensite-like phase. A third type of smart

2) The names Terfenol™ and Nitinol™ relate respectively, to **Ter**bium **Fe**rro **N**aval **O**rdnance **L**aboratory and **Ni**ckel **Ti**tanium **N**aval **O**rdnance **L**aboratory and hence refer to their R&D origin.

material involves composite materials with coupled phase transformations that may combine polymeric materials with phase transitions, causing large changes in elastic properties and ferroelectric ceramics with an associated instability of the dielectric properties.

Other typical ferroic ceramic materials are ferromagnetics such as Fe_3O_4, ferrimagnetics such as doped Fe_2O_3 ("ferrites"), ferroelectrics such as $BaTiO_3$, piezoelectrics such as lead zirconate/lead titanate (PZT), and ferroelastics such as $Pb_3[O/GeO_4]$ (Otto, 1979; see also Müller-Lierheim *et al.*, 1977). Ferroelectric materials are dielectrics, the spontaneous polarization of which, caused by charge imbalances within the unit cell of the crystal structure, can be altered by the application of an outside electric field (see Section 8.3). Pyroelectrics are confined to the 10 crystallographical point groups with polar axes (see Table 8.3) that impose spontaneous polarization (see Section 8.4.1). Piezoelectric materials are confined to the 20 point groups without a center of symmetry (see Section 8.4.2); in these cases, a mechanical stress causes polarization through shifting of charges, whereas the application of an electric field induces a mechanical deformation (strain) (see Table 8.1).

Both, ferroic and nonferroic electroceramic materials have important industrial applications, some of which are outlined in Table 8.2 (further details are provided below). The very broad field of ferrimagnetic ceramics (ferrites) will not be described in this treatise; for this information, the reader is referred to more specialized literature (Krupicka, 1973; Goldman, 2006).

8.1.2
Historical Development of Dielectric Ceramics

The historical development of dielectric ceramics, and in particular for applications as efficient capacitors, is shown in Figure 8.1 (see also Table 8.4). The first dielectric capacitors based on titania were invented during the 1920s by Siemens in Germany, and further developed during the 1930s. Concurrently, capacitors based on magnesium titanate and silicate (steatite) were investigated and utilized until the 1940s, when the development of ferroelectrics such as barium titanate was begun. Pure stoichiometric $BaTiO_3$ is an excellent material for the construction of capacitors, owing to its very high dielectric constant ($\varepsilon > 7000$). At this time, $BaTiO_3$ was the epitome of a dielectric, and therefore in Germany was given the tradename "Epsilon". Unfortunately, $BaTiO_3$ is not applicable to fashion electronic devices that require temperature stability, as the temperature coefficient of the resonance frequency (τ_f) has a large negative value. Moreover, since its polarization mode is based on spontaneous dipole polarization produced by distortion of the oxygen coordination octahedra surrounding the titanium ion, the dispersion of ε occurs in the microwave region with a concomitant substantial reduction of the electrical quality ("Q") factor (see Sections 8.2.1 and 8.2.2).

Much later, it was found that barium titanates with a large surplus of TiO_2 in the lattice, such as $BaTi_4O_9$ and $Ba_2Ti_9O_{20}$, have good properties as microwave

Table 8.2 Application and examples of some typical electroceramics.

Function/effect	Application	Examples
Ferrimagnetic domains	Microwave magnetics	$Y_3Fe_5O_{12}$ (YIG)
Ferroelectric domains	Multilayer capacitors	$BaTiO_3$ (BT)
Ionic conduction	Gas sensors, electrolytes	$(Y,Ca)ZrO_2$, β-Al_2O_3
Superconduction	SQUIDs	$YBa_3Cu_3O_{7-\delta}$
Piezoelectricity	Piezoelectric oscillators	α-SiO_2
Ferrimagnetic domains	Magnetic tape	γ-Fe_2O_3 (maghemite)
Ferroelectric domains	PTC thermistors	$(Ba,La)TiO_3$
Electrostriction	Actuators	$Pb(Mg,Nb)O_3$ (PMN)
Piezoelectricity	Piezotransducers	$Pb(Zr,Ti)O_3$ (PZT)
Ferrimagnetic domains	Hard ferrimagnetics	$BaFe_{12}O_{19}$ (magnetoplumbite)
Electronic conduction	NTC thermistors	VO_2, Mn-, Ni-, Co-, Fe-oxides
Ferrimagnetic domains	Soft ferrimagnetics	$(Mn,Zn)Fe_2O_4$ (ferrites)
Electro-optic effect	Non-linear optics	$(Pb,La)(Zr,Ti)O_3$ (PLZT)
Magnetostriction	Magnetostrictive material	$Tb_{0.3}Dy_{0.7}Fe_{1.9}$ (Terfenol)
Electrical insulation	Electronic substrates	Al_2O_3, SiC–BeO–AlN
Surface conduction	Chemical sensors	SnO_2, TiO_2
Grain boundary phenomena	Varistors, BLCs	$(Zn,Co)O$–Bi_2O_3
Metal-semiconductor transition	Critical T-NTC thermistors	VO_2

dielectrics (see Section 8.4), even though they are no longer ferroelectrics. To date, barium titanate doped with oxides of strontium, bismuth, neodymium, samarium, and tungsten, as well as complex barium–zinc-tantalum oxide perovskites, fulfill the dielectric requirements in terms of permittivity, *Q*-factor, and temperature coefficients of the resonance frequency and permittivity to a large degree. Giant dielectric permittivity was rather unexpectedly discovered in $CaCu_3Ti_4O_{12}$ (CCTO; Ramirez *et al.*, 2000; Whangbo and Subramanian, 2006; Barbier *et al.*, 2008), with values of ε exceeding 10^4 at low frequency. There is some evidence, however, that this effect may not be related to true ferroelectricity, but may instead involve the existence of highly polarizable relaxational modes with a characteristic gap energy of 28 meV. The Internal Barrier Layer Capacitance (IBLC) model was also invoked to explain the experimentally observed fact that ε increases with the sintering time of CCTO, due to the incorporation of an intergranular CuO phase into the structure of CCTO (Romero *et al.*, 2009).

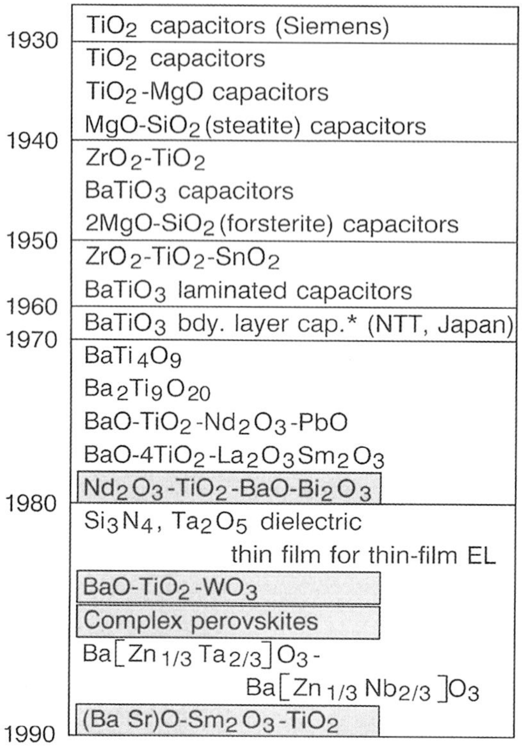

Figure 8.1 Historical development of dielectric ceramics.

8.2 Physics of Dielectric Materials

8.2.1 Dielectric Effects

Dielectric materials can be subdivided according to their physical effects (Kleber, 1998). *Normal dielectrics* undergo polarization in response to an outer electric field; that is, a shift of charge carriers relative to their equilibrium positions. *Pyroelectrics* are materials in which a temperature gradient creates a macroscopic spontaneous polarization as the result of singular; that is, polar axes in the crystal structure (see Section 8.5.1). All pyroelectrics are also piezoelectric (see below). A special case of (nonlinear) pyroelectricity is *ferroelectricity*, the orientation of spontaneous polarization of which can be reversed by the application of an outer electric field. Within ferroelectric crystals, domains are being established in which the orientation of the polarization of individual crystallites is identical but is different from that of the neighboring domains. The domains are separated by walls that, by definition, are topological solitons – that is, two-dimensional (2-D) membranes formed by the spontaneous breaking of a discrete symmetry at a phase transition (see Section 8.3). While in ferroelectrics the electric dipoles within a domain all point in the same direction, *antiferroelectrics* are characterized by an interpenetration of sublattices of antiparallel ordered arrays of electric dipoles. Since the adjacent dipole moments cancel each other out, the macroscopic spontaneous polarization is zero, but dielectric anomalies exist (Dalal and Bussmann-Holder, 2007).

Another special case of pyroelectricity is *ferroelasticity*. In these materials, the direction of the spontaneous polarization can be changed by applying a mechanical stress. In some crystals such as gadolinium molybdate [$Gd_2(MoO_4)_3$; GMO] the ferroelastic effect is coupled with a ferroelectric effect (Bohm and Kürten, 1973).

Piezoelectrics are materials in which a mechanical stress results in a shift of charge carriers, and hence a polarization (see Section 8.5.2). The basic requirement of piezoelectricity is the existence of a crystal structure lacking a center of symmetry. Conversely, this asymmetry causes a strain which is linearly proportional to the electric field strength, an effect termed the *reverse piezoelectric effect*. In a symmetric crystal, the strain is proportional to the square of the electric field strength, known as the *electrostrictive strain effect*.

Ferroelectric and piezoelectric ceramics, in particular, play an ever-increasing role as materials for electrical and electronic applications that include multilayer capacitors (MLCs), bypass capacitors, dielectric resonators for frequency stabilization of microwave circuits, low-noise oscillators and low-insertion loss bandpass filters for microwave communication components, dielectric waveguide resonators, piezoelectric transducers and sensors, piezomechanical actuators and motors, PTC thermistors, and a large variety of novel emerging utilizations. Some of these applications are described later in the chapter (see Section 8.7).

8.2.2
Electric Polarization in Dielectric Materials

The dielectric and electromechanical coupling properties of ceramics are generally dependent on three figures of merit: (i) the dielectric permittivity ε; (ii) the (electrical) quality factor $Q = (\tan\hat{o})^{-1}$, where $\hat{o}$ = dielectric loss angle; and (iii) the temperature coefficients of both resonance frequency, τ_r and dielectric permittivity, τ_ε (for a more detailed description, see Section 8.2.3). These parameters in turn depend crucially on phase purity, content of impurities and dopants, respectively, and grain size (see Figure 8.7), and the grain size distribution of the ceramic starting powder used to fabricate monolithic devices. Powder synthesis is customarily achieved by classical ceramic processing routes, such as solid-state reactions of constituent oxides, carbonates, or nitrates. However, increasingly sophisticated methods are being employed to avoid the repeated mixing, grinding and calcination steps that are known to introduce impurities. This involves synthesizing the desired and frequently nanosized (see Section 12.5) compositions from true or quasi-homogenous solutions by the coprecipitation of hydroxides or catecholates (Maison *et al.*, 2003), sol–gel phase transition (Kirby, 1988), or hydrothermal methods (Phanichphant and Heimann, 2004).

In contrast to metals, with their typically high electrical and thermal conductivities, ceramics and most polymeric materials are dielectrics. Whilst in metals, the electrons are delocalized and distributed randomly within the crystal lattice (the "electron cloud" model) and hence can move freely in response to an outer electric field, in predominantly covalent and ionic bonded ceramics the electrons are tightly bound to the atomic core. Hence, in an ideal dielectric material there is no long-range transport of charge carriers but only a displacement of electrons, ions, or dipoles relative to their equilibrium positions in the lattice. Consequently, in ceramics the electronic conductivity takes a distant second place to ionic conductivity. The degree of displacement corresponds to the restoring force exerted by the outer electric field that may be compared in a mechanical analog to the restoring force of an extended spring. Since positive and negative charges will be displaced in opposite directions, the crystal acquires a *dipole momentum* $\mathbf{p}$; the macroscopic dipole momentum per unit volume is the *polarization*

$$\mathbf{P} = \varepsilon_0 \chi^{e} \mathbf{E}, \tag{8.1}$$

with the dielectric permittivity of the vacuum ε_0 [=8.854 × 10^{-12} As/(Vm)], the dielectric susceptibility χ^e, and the macroscopic electric field strength in the dielectric, $\mathbf{E}$. The susceptibility is an anisotropic material constant, and links the two vectors $\mathbf{P}$ and $\mathbf{E}$. Hence, it can be described by a (polar) symmetric tensor of second rank ($\chi^{e}_{ij} = \chi^{e}_{ji}$). In the triclinic crystal system the tensor of the dielectric susceptibility has six independent components. With increasing crystal symmetry, this number reduces and shows for tetragonal, rhombohedral or hexagonal crystals, only two independent tensor components (χ^{e}_{11} and χ^{e}_{33}).

Polarization and susceptibility can be determined by measuring and comparing the capacities C of a capacitor with a dielectric (C) and an empty capacitor (C_0, in vacuum). The ratio C/C_0 determines the *relative dielectric permittivity* $\varepsilon_r = \varepsilon/\varepsilon_0$, where ε is the permittivity of the dielectric and ε_0 is the absolute dielectric constant of the vacuum. For the isotropic case, the susceptibility χ^e and the relative dielectric permittivity ε_r are related by

$$\chi^{e} = \varepsilon_r - 1 = \varepsilon/\varepsilon_0 - 1. \tag{8.2}$$

Hence, both ε_r and ε are second rank tensors and link the vector of the electric field strength $\mathbf{E}$ and the dielectric displacement vector $\mathbf{D}$ by

$$\mathbf{D} = \varepsilon_0 \varepsilon_r \mathbf{E} = \varepsilon \mathbf{E} \tag{8.3a}$$

and with Eqs (8.1) and (8.2)

$$\mathbf{D} = \varepsilon_0 \mathbf{E} + \mathbf{P}. \tag{8.3b}$$

From Eq. (8.3a) it follows that the dielectric displacement $\mathbf{D}$ and hence the polarization $\mathbf{P}$ both reduce to zero when the electric field disappears. However, there are crystals that display a so-called *spontaneous polarization* $\mathbf{P}_s$ already in the absence of an electric field, caused by the presence of singular (polar) directions in their crystal structures. This happens in the ten symmetry groups *1, m, 2, mm2, 3, 3m, 4, 4mm, 6,* and *6mm* (Table 8.3). It is important for any further

Table 8.3 Classification of the 32 crystallographic point groups with respect to crystal centrosymmetry and polarity (Uchino, 1994).

Polarity	Centro symmetry	No. of point groups	Cubic		Hexagonal		Tetragonal		Rhombohedral		Orthorhombic	Monoclinic		Triclinic
Nonpolar (22)	Yes (11)	11	$m\bar{3}m$	$m\bar{3}$	$6/mmm$	$6/m$	$4/mmm$	$4/m$	$\bar{3}m$	$\bar{3}$	mmm	$2/m$		$\bar{1}$
			432		622		422							
	No (21)	11		23		$\bar{6}$		$\bar{4}$	32		222			
			$\overline{43}m$		$\bar{6}m2$		$\overline{42}m$							
Polar (pyro) (10)		10			$6mm$	6	$4mm$	4	$3m$	3	$2mm$	2	m	1

considerations to note that spontaneous polarization can also occur in polycrystalline ceramics with a polar symmetry group if an oriented arrangement exists of crystallites aligned by a preferred texture to yield a polarization distribution function (Jones, 2010).

The magnitude of the spontaneous polarization is temperature-dependent

$$d\mathbf{P}_s = \mathbf{p} * dT, \tag{8.4}$$

where **p*** is the "true" pyroelectric coefficient. The change of spontaneous polarization causes a charge displacement in response to changing temperature; that is, the opposite ends of a polar axis will be differently charged, a phenomenon termed the "*pyroelectric effect*." True pyroelectricity (see Section 8.5.1) occurs in crystals that lack a center of symmetry and have unique polar axes, as realized in the ten point groups at the bottom of Table 8.3.

On the other hand, all point groups without a center of symmetry show the piezoelectric effect, with the exception of *432*, the high overall symmetry of which appears to prevent charge delocalization. The 20 piezoelectric point groups are enclosed by a solid line (see Table 8.3).

Some pyroelectrics have an additional property, in that the direction of the spontaneous polarization can be changed by applying either an outside electric field or a mechanical stress. The former materials are termed *ferroelectrics*, and the latter *ferroelastics*. In this regard, both ferroelectricity and ferroelasticity are special cases of pyroelectricity.

The changes in polarization directions can be explained by invoking the concept of the polarizability, α, of ions. This quantity relates the dipole momentum **p** to the local field strength $\mathbf{E}_L$ that is associated with the deformation of the electron orbital shells of individual ions:

$$\mathbf{p} = \alpha \cdot \mathbf{E}_L \tag{8.4a}$$

According to Heywang (1951), the effective polarizability in crystals of (cubic) perovskite structure can be expressed by

$$\alpha = \alpha_A + \alpha_B + 3\alpha_O + \gamma\alpha_O, \tag{8.5}$$

where the subscripts A, B, and O refer, respectively to the combined electronic and displacement polarizabilities of the A-, B-, and oxygen ions. The complex factor γ (Lorentz factor) considers the long-range Coulombic interactions in structures with a noncubic environment, and contains the scaling factors $p = 0.69$ and $q = 2.39$ to yield

$$\gamma = [q\,B* + p(O*\text{-}A*)]^2 / [(2/3)[1 + pO*]\text{-}O*(q^2B* + p^2A* + p^2O*/3), \tag{8.6}$$

with $A* = \alpha_A/\varepsilon_0 v$, $B* = \alpha_B/\varepsilon_0 v$, and $O* = \alpha_O/\varepsilon_0 v$ (where v is the volume of the unit cell). For structures with cubic environments, the interaction is given by the well-known Clausius–Mosotti equation.

Owing to the high value of q, the polarizability of the B-ions, combined with the likewise high polarizability of the oxygen, ions contribute to a large extent to the effective (total) polarizability α. For the case of barium titanate, $A* = 0.20$ leads to

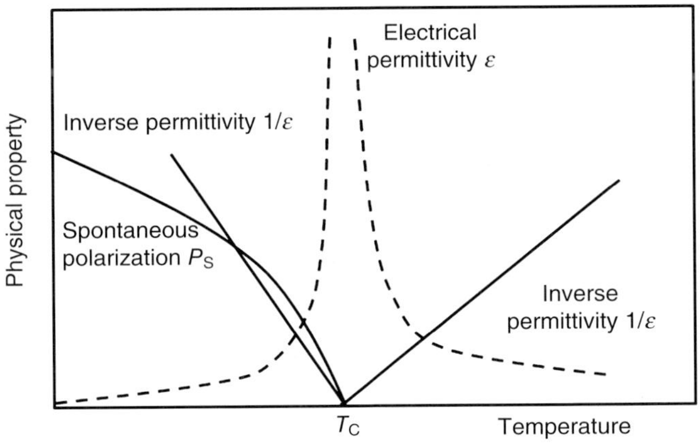

Figure 8.2 Temperature dependence of spontaneous polarization $\mathbf{P_S}$ and dielectric permittivity ε_r of a ferroelectric ceramics in the vicinity of the Curie temperature, T_C.

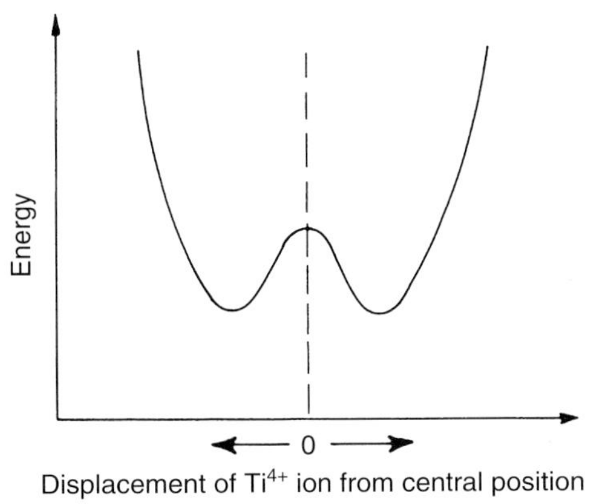

Figure 8.3 Double-energy well of Ti^{4+} along the z-axis.

$\gamma = 1.38$ and $\varepsilon_r(0) = 26.6$. However, a very slightly larger $A^* = 0.21$ yields $\gamma = 1.73$ and hence $\varepsilon_r(0) \rightarrow \infty$; that is, a polarization catastrophe associated with a second-order ("λ"-type) ferroelectric–paraelectric phase transition at T_C (see Figure 8.2).

This behavior will be augmented by the displacement of oxygen ions, causing a dynamic covalency. The latter occurs when B-sites are replaced by small transition metal ions with unfilled d-orbitals that "rattle" around within the oxygen coordination polyhedron "cage." This will lead to asymmetric approaches of the B-ions to oxygen ions, and hence an increase in the proportion of the covalent bonding force. The system reacts to that situation by a partial charge transfer, and consequently an additional dipole momentum appears, causing the B-ions to move in a double-potential energy well structure (Figure 8.3), leading to an order–disorder (OD)

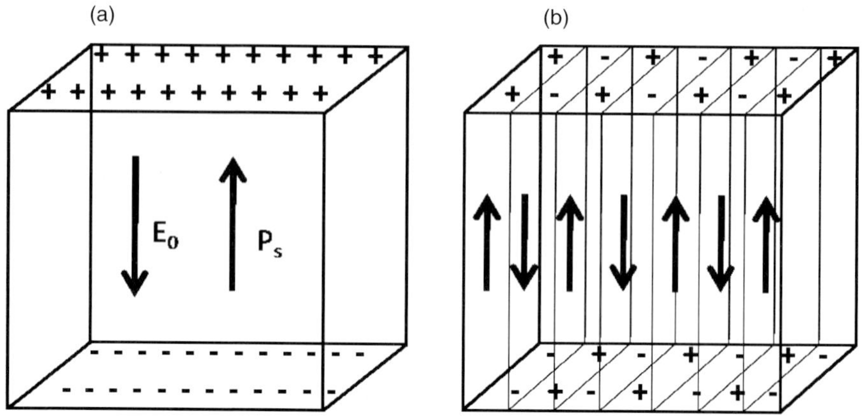

Figure 8.4 Surface charges of a single-domain ferroelectric ceramic material associated with spontaneous polarization $\mathbf{P}_S$ (a) and formation of 180°-multidomains by twinning to minimize electrostatic energy (b).

phase transition. In Ti-based perovskites, an applied electric field enables Ti^{4+} ions to overcome the shallow energy barrier "hump" between the two equivalent energy states, and consequently to move from one well into the other well, thus reversing the direction of polarity in this lattice region. By cooperative coupling the energy increment required for a neighboring unit cell to make the same transition will be reduced, and eventually the entire domain will switch to the new direction in lock-step. Within a domain the direction of spontaneous polarization is identical, but differs from domain to domain.

As the orientation of domains can be deduced from the (higher) symmetry of the paraelectric "aristotype" phase, the domains are structurally related to each other by twin planes (domains) obeying appropriate twin laws, usually 180°-domains (Figure 8.4). In some ferroelectric materials (notably barium titanate), there exist also 90°-domains that originate from mechanical stress associated with the phase transition by domain movement, and hence are induced by the ferroelastic effect (Nord, 1994). Not only the high polarizability of the B-ions but also a higher polarizability of A-ions favor a ferroelectric behavior. For example, the Curie temperature of $PbTiO_3$ is considerably higher (T_C = 490 °C; see Figure 8.16) than that of $BaTiO_3$ (T_C = 130 °C), since $\alpha_{A(Pb)} = 1.4\ \alpha_{A(Ba)}$. The anharmonicity of the perovskite lattice leads to a decrease of polarizability with increasing temperature ($\delta\varepsilon/\delta T < 0$), and in turn to a disappearance of ferroelectricity at the Curie temperature (Figure 8.2). The material becomes paraelectric, such that the relative dielectric permittivity ε_r obeys the linear Curie–Weiss law, $\varepsilon_r = C/(T - T_C)$.

The fact that an electric field can reverse the direction of spontaneous polarization by the cooperative movement of domain walls – that is, the collective ordering of local dipole moments – requires an energy contribution that the dielectric material takes from the surrounding electric field. In an oscillating a.c. field, the phase shift between current and voltage in the absence of a dielectric is exactly 90° (see Figure 8.8). The presence of a ferroelectric material results in a slightly smaller phase shift of 90°-δ, where δ is the angle of dielectric loss, and $1/\tan\delta \equiv Q$, the "electrical" quality factor. The Q-factor plays an important role in electrical resonance, as the sharpness of tuning increases with increasing Q-factor by decreasing the damping function. The application of a strong static electric field to a polycrystalline ferroelectric ceramic at an elevated temperature can permanently fix the direction of the 180°-domain orientation – that is, the direction of polarization. This process is termed "poling."

The existence of 180° domains is a consequence of minimizing the electrostatic energy by twinning (see Figure 8.4). On the other hand, a multidomain state can be transformed to a single-domain state by applying a field parallel to one of the polar directions; that is, [100] in the tetragonal point group *4mm*, [110] in the orthorhombic point group *2mm*, and [111] in the trigonal point group *3m* (see Figure 8.15). If a mechanical stress is superposed on the electric field, 90°-domains can be created in tetragonal materials, and 71°- and 109°- domains in rhombohedral materials that reduce the internal strain. Conversely, the applied external field will affect the domain orientation, and hence the internal strain, to yield a *ferroelastic effect*. Depoling can be achieved by raising the temperature to well above the Curie point, followed by rapid cooling without a field present.

Plotting the polarization against the electric field strength yields the typical hysteresis loop of a ferroelectric material (Figure 8.5; see also Figure 8.10). After switching off the field, a part of the preferential domain orientation will be retained (remanent polarization). To remove this remanent polarization, a reverse electric

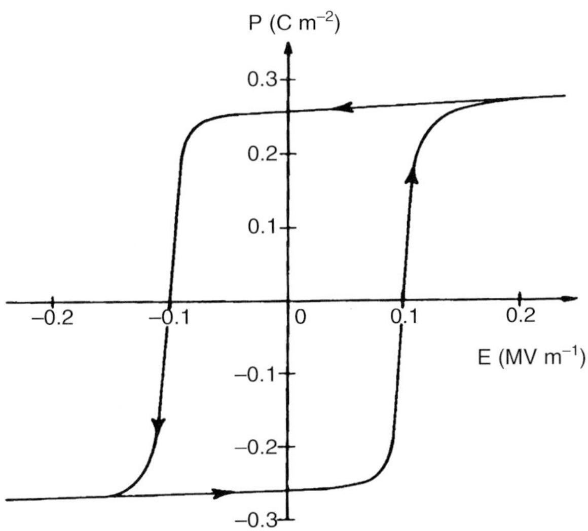

Figure 8.5 Hysteresis loop for a single-domain single crystal of barium titanate (Moulson and Herbert, 2003). Reprinted with permission from Wiley-Blackwell Ltd, Oxford, UK.

field with a coercitive strength E_C must be applied in order to reproduce the original (statistical) domain orientation (P = 0). This behavior is analogous to the magnetization (*B*)–magnetic field strength (*H*) hysteresis loop of ferromagnetic materials, and indeed was the original reason for Valasek (1924) to coin the term "ferroelectricity."

The near-vertical portions of the loop in Figure 8.5 correspond to the reversal of the spontaneous polarization when opposite 180°-domains nucleate and grow. The near-horizontal portions represent saturated single-domain states. The remanent polarization (E = 0) shown is about $0.27\,\mathrm{C\,m^{-2}}$, the coercitive field strength (P = 0) is (–) $0.1\,\mathrm{MV\,m^{-1}}$.

The ease with which ferroelectric ceramics can be poled suggests only weakly stable structural states that enable distortive phase transitions from low-temperature ferroelectric to high-temperature paraelectric, or antiferroelectric to paraelectric states (Carpenter, 1992). In single crystals such as $LiTaO_3$, the displacive phase transition can be described mechanistically as a swinging motion of the small B ion through the plane of the four oxygen atoms in the center of the oxygen coordination polyhedron in the perovskite lattice (see Figure 8.13).

Although, $SrTiO_3$ is *per se* not ferroelectric even at low temperatures, it is very close to the ferroelectric threshold so that an isotopic replacement of oxygen ($^{16}O \rightarrow {}^{18}O$) or partial cation substitution reduces quantum fluctuations and makes the material ferroelectric (Itoh *et al.*, 1999; Choudhury *et al.*, 2008). Hence the phonon dynamics as a function of temperature will be discussed using $SrTiO_3$ as a model substance.

The frequency of the "soft" (= low-frequency) mode phonons decreases by approaching the transition temperature and "freeze" at the Curie temperature in a polar, that is, ferroelectric configuration (Figure 8.6).

As shown in Figure 8.6, the antiferrodistortive modes consist of an A_{1g} nondegenerate branch and an E_g doubly degenerate branch. These branches collapse to zero energy at the phase transition from the low-temperature antiferroelectric to the high-temperature paraelectric phase near 106 K. In contrast to this, the ferroelectric modes consist of a nondegenerate A_{2u} branch and a doubly degenerate E_u branch that are very close together in frequency. These ferroelectric modes approach zero frequency as the temperature is decreased to zero Kelvin, but remain finite as pure strontium titanate remains paraelectric at all temperatures, so that its extrapolated Curie temperature is below zero (Scott and Ledbetter, 1997).

The high polarizabilities of the TiO_6-octahedra results in anomalously high values of the dielectric permittivities (see Figure 8.2), as well as the electro-optical, nonlinear optical, piezoelectric and electromechanical coupling coefficients (Cross, 1993). This is at the very heart of the technical application of ferroelectric materials as ultrasonic oscillators, acoustic and optical frequency multipliers, dielectric amplifiers, acoustic and optical frequency modulators, switches, sensors, actuators, and many more. For detailed accounts on these applications, see Uchino (1994, 1996), Cross (1993), and Heywang *et al.* (2009).

Experimental evidence exists that the grain size of a polycrystalline ferroelectric ceramic strongly influences the crucial dielectric properties such as permittivity,

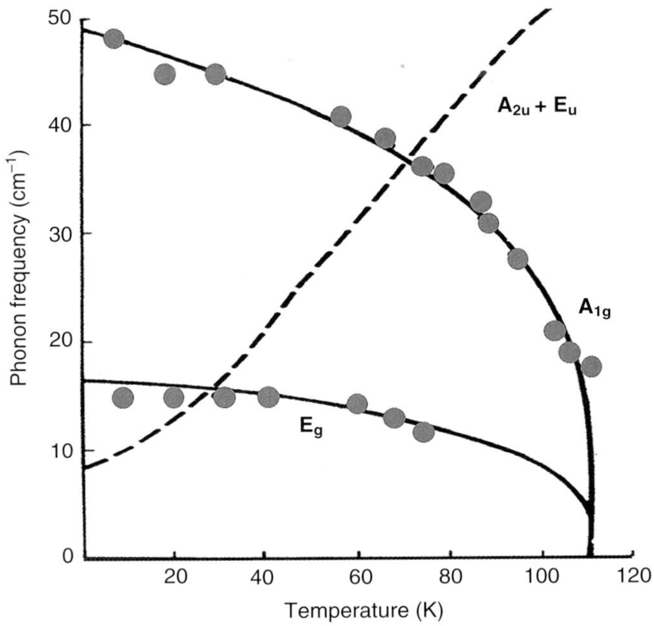

Figure 8.6 Soft optical phonon frequencies in antiferroelectric $SrTiO_3$ versus temperature. Adapted from Worlock *et al.* (1969).

electromechanical coupling coefficient (i.e., the efficiency of conversion of mechanical into electrical energy and *vice versa*) and, in particular, the electro-optical behavior. Figure 8.7a shows an example of the dependency of the relative dielectric permittivity ε_r of $BaTiO_3$ on the grain size between 1 and 50 μm. Clearly, ε_r is seen to strongly increase with decreasing grain size at T_C, and also increases in the vicinity of the tetragonal–orthorhombic phase transition (MPB) around 0 °C, thus raising the permittivity level below T_C. For very small crystals, however, the T_C decreases (Kniepkamp and Heywang, 1954; Anliker *et al.*, 1952). To attain small grain sizes during conventional solid-state powder sintering, doping plays an important role, as substitution with higher valency ions (e.g., La^{3+} for Ba^{2+}, and Nb^{5+} for Ti^{4+}) inhibits grain growth. Figure 8.7b shows that the tetragonality (c/a) of $BaTiO_3$ decreases with decreasing particle size until in the vicinity of 100 nm, when the cubic phase is stable and ferroelectricity ceases to exist.

8.2.3 Characteristic Dielectric Parameters

The useful properties of a dielectric ceramic material – that is, their figures of merit, and in particular for microwave applications (see Section 8.4) – can be described by:

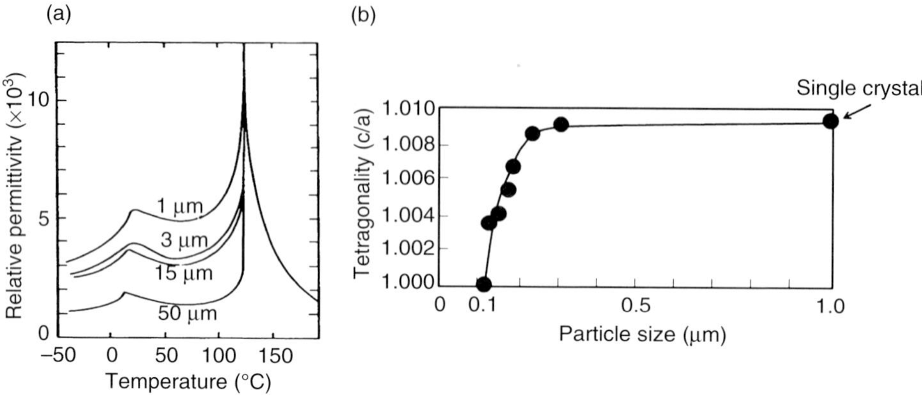

Figure 8.7 (a) Dependency of the relative dielectric permittivity ε_r of $BaTiO_3$ on grain size and temperature (after Kinoshita and Yamaji, 1976); (b) Change of tetragonality of $BaTiO_3$ with particle size (after Uchino, 1995).

- the dielectric permittivity, expressed by the dielectric constant ε;
- the angle of dielectric loss ô, expressed by the quality ("*Q*") factor ($Q = 1/\tan$ ô); and
- the temperature coefficient of the resonance frequency τ_f.

Dielectric loss is caused by the interaction of an outer electric field with the dielectric ceramic material. The reversal of the direction of spontaneous polarization by a cooperative movement of the domain walls requires an energy contribution that the dielectric material takes from the surrounding electric field. In an oscillating a.c. field, the phase shift between current I and voltage U in the absence of a dielectric is exactly 90° (Figure 8.8). The presence of a dielectric material results in a slightly smaller phase shift of (90° – δ), where δ is the angle of dielectric loss, and $1/\tan\delta \equiv Q$. The *Q*-factor plays an important role in electrical resonance, as the sharpness of tuning increases with increasing *Q*-factor by decreasing the damping function.

A desirable dielectric ceramic should have:

- a high dielectric constant exceeding $\varepsilon = 100$, since the required size of a resonator is proportional to $1/\sqrt{\varepsilon}$ (Figure 8.36); that is, a high value of ε leads to a miniaturization of the device;
- a high *Q*-factor; that is, a low dielectric loss; and
- a zero temperature coefficient of the resonance frequency to attain temperature stability of the device.

When an electric field is applied to a dielectric ceramic material, the electrical charges within the material will shift, and a polarization is induced that is

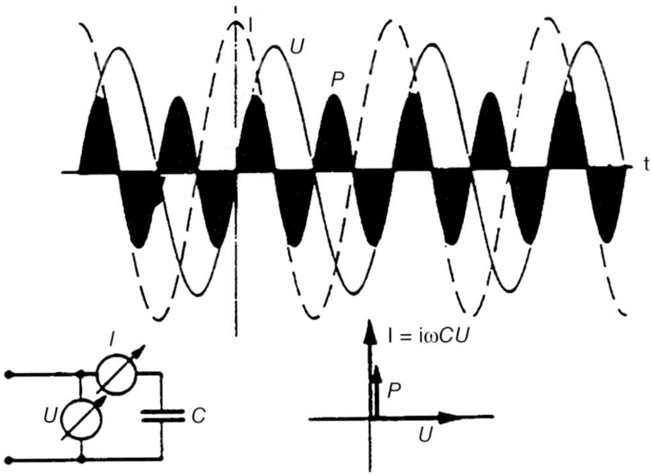

Figure 8.8 Current I, voltage U and power P of a capacitor. Without a dielectric, the phase shift between I and U is exactly 90°, but in the presence of a dielectric it is (90° − ô), where ô is the dielectric loss angle.

proportional to the electric field strength. Depending on the size of the region affected, five different modes of polarization can be distinguished: electron polarization, ε_e; ion polarization, ε_i; dipole polarization, ε_d; space and surface charge polarization, ε_s; and orientation polarization, ε_o . While the first four polarization modes are temperature-invariant, the magnitude of the orientation polarization – that is, the alignment of dipoles in the direction of an electric field – is inversely proportional to the temperature.

In the presence of an alternating current, the polarization of the material can follow the changes of the electric field only in the lower-frequency region. As the frequency increases, polarization is no longer able to follow the changes of the field, and consequently a dispersion of the dielectric permittivity occurs with increasing frequency (Figure 8.9).

8.3 Ferroelectric Ceramics

The physical phenomenon of ferroelectricity – initially termed Seignette electricity – was first discovered in sodium potassium tartrate tetrahydrate (Rochelle or Seignette salt), and later in analogy to ferromagnetic behavior coined ferroelectricity by Valasek (1924). Its history is listed in Table 8.4, which shows the impressive change from a curious isolated property to a widespread and economically enormously important and promising ceramic engineering material (Cross and Newnham, 1987).

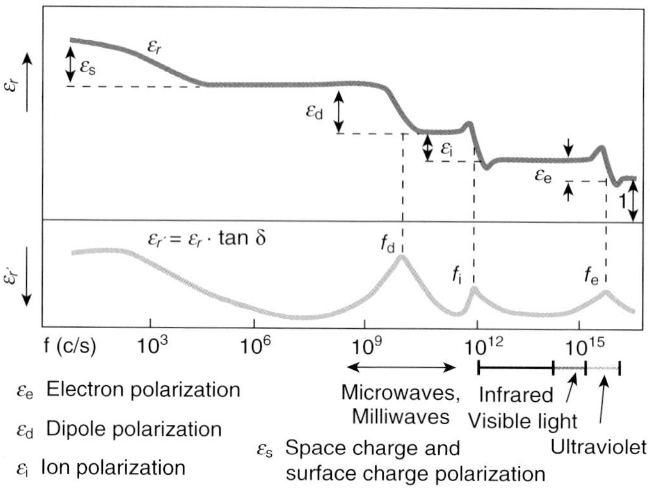

Figure 8.9 Dispersion of polarization of dielectric materials.

Table 8.4 History of research and development in the field of ferroelectrics (Cross and Newnham, 1987).

	Important R&D events in ferroelectricity
1920–1930	Rochelle salt period: discovery of ferroelectricity
1930–1940	KDP age: thermodynamic and atomistic models of ferroelectricty
1940–1950	Early barium titanate era: high-K capacitors developed
1950–1960	Period of proliferation: many new ferroelectrics discovered
1960–1970	Age of high science: soft modes and order parameters
1970–1980	Age of diversification: ferroics, electro-optics, thermistors
1980–1990	Age of integration: packages, composites, and integrated optics
1990–2000	Age of miniaturization: size effects, manipulated modes and dipoles

Intensive research and development efforts conducted during the past fifty years have revealed four main types of structure with ferroelectric properties:

- *Corner-sharing octahedral structures* (true perovskites such as $BaTiO_3$, BT; $Pb(Zr_xTi_{1-x})O_3$, PZT; $Pb(Mg_{1/3}Nb_{2/3})O_3$, PMN; tungsten bronze-type structures, such as $PbNb_2O_6$; bismuth oxide layer structures (Aurivillius phases) such as $Bi_4Ti_3O_{12}$ and $PbBi_2Nb_2O_9$; and perovskite-related structures such as $LiNbO_3$ and $LiTaO_3$.

- *Compounds containing hydrogen-bonded radicals,* such as KH_2PO_4 (KDP), triglycine sulfate (TGS) and Rochelle salt.
- *Organic polymers,* such as polyvinylidene fluoride [PVDF; $(CH_2CF_2)_n$] and its copolymers with trifluoroethylene.
- *Ceramic polymer composites* (see Newnham *et al.*, 1980; Janas and Safari, 1995).

The crystal structures and functional principles of these ferroelectric materials have been described by Safari *et al.* (1996).

By definition, ferroelectrics – that is, materials showing one or more ferroelectric phases in a realizable range of pressure and temperature – are materials in which, in the ferroelectric phase, the crystal is spontaneously electrically polarized and the polarization has more than one possible equilibrium orientation that results in a domain structure. To establish true ferroelectricity in a material, it must be confirmed that the polarization can be reoriented between the different orientation (domain) states by a realizable electric field.

The realignment of domains by an outside electric field is accompanied by a deformation. The main axis of deformation is parallel to the field; that is, a crystal rod shows elongation by a small percentage parallel, and shortening perpendicular to the direction of the field. The application of a strong electric field to a polycrystalline ferroelectric ceramics results in domain reorientation. However, after removing the field, a part of the preferential orientation will be retained by the material (remanent polarization), an effect which is entirely analogous to the remanent magnetization of a ferromagnetic material; in other words, a hysteresis loop exists (Figure 8.10). In order to remove the domain orientation, a reverse field with a coercive field strength $\mathbf{E}_c$ must be applied that reproduces the original (statistical) domain orientation and reduces the polarization to $\mathbf{P} = 0$.

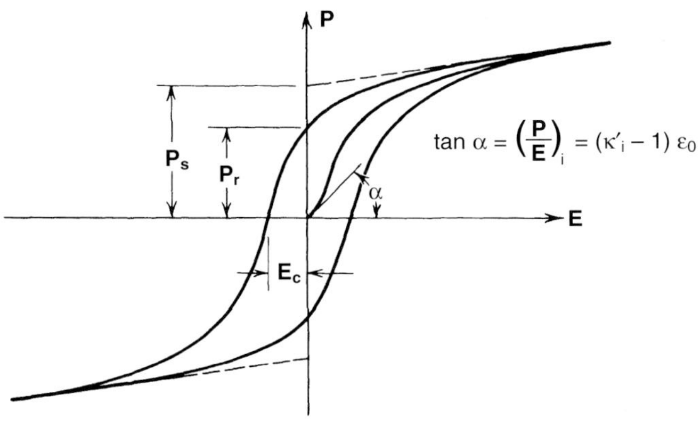

Figure 8.10 Polarization of a ferroelectric ceramic in response to the change of an outer electric field strength, **E**.

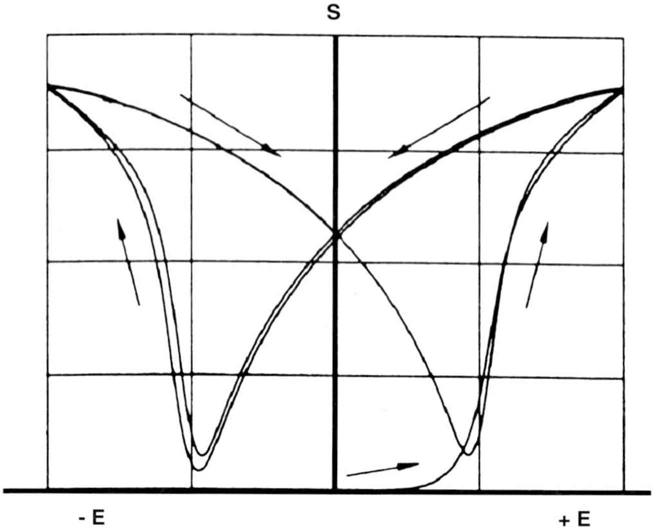

Figure 8.11 Butterfly loop of a piezostrictive material such as PZT, showing the strain S associated with a reversal of the polarization direction.

The total deformation $S = \Delta l/l$ of a ferroelectric material undergoing a hysteresis loop can be described by the so-called "butterfly loop" (Figure 8.11). At high field strengths **E** there is saturation; that is, the ferroelectric domains show maximum orientation and hence saturation polarization $\mathbf{P}_s$. The application of a very strong d.c. field at elevated temperature, as realized during poling, induced piezoelectric properties in the polycrystalline ferroelectric ceramics by permanently fixing the direction of polarization.

The strain change associated with the domain reversal in a ferroelectric ceramic is shown schematically in Figure 8.12, where the arrows delineate the direction of polarization. It should be noted that after a full loop 1 to 5, the direction of polarization of the individual domains has exactly reversed.

Recently, extensive investigations have been conducted with the aim of developing lead-free ferroelectric and piezoelectric ceramic materials; this was in response to legislation introduced by several countries to phase out PZT and other lead-based ceramics. For example, <001>-textured $(K_xNa_{1-x})NbO_3$ (KNN) with a perovskite structure shows a field-induced strain comparable to typical actuator-grade PZT, with a high piezoelectric constant of up to $d_{33} = 416$ pC N^{-1} (Saito *et al.*, 2004). The addition of sintering aids such as $K_{5.4}Cu_{1.3}Ta_{10}O_{29}$ to KNN results in a material with $Q_m > 680$, an electromechanical coupling coefficient $k_p = 0.45$, $d_{33} = 160$ pC N^{-1}, and a T_C around 420 °C (Matsubara *et al.*, 2005). The doping of KNN with Li, Sb, Ta, Ag, Sr, Ba and Ca allows tailoring of the dielectric properties and electromechanical performance.

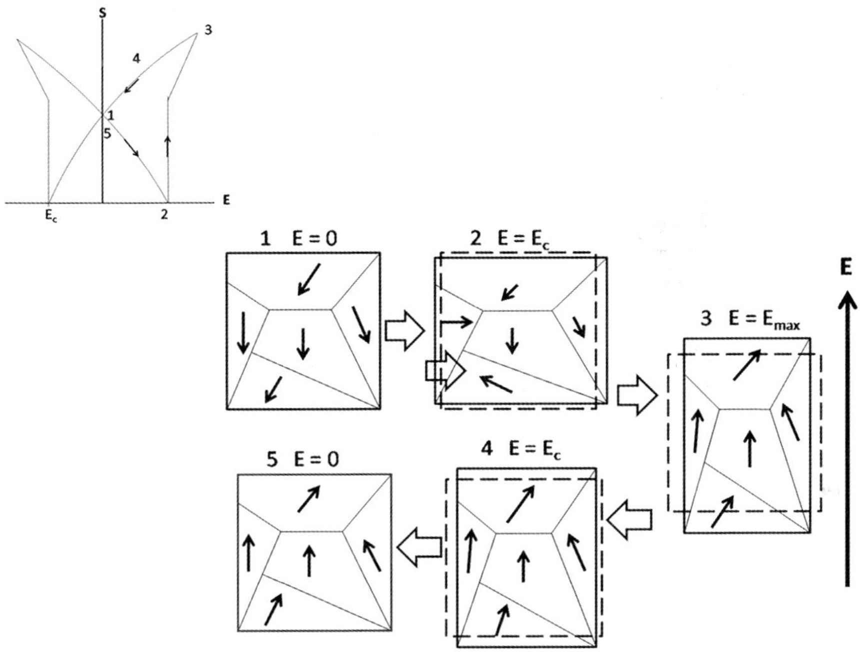

Figure 8.12 Strain change associated with the domain reversal in a ferroelectric ceramic (modified after Uchino, 1995).

8.3.1 Barium Titanate (BT)

Many important ferroelectric ceramics are based on the (pseudo)cubic perovskite structure with the general formula ABX_3 (Figure 8.13), with X = O (oxygen). This structure consists of large A-type cations with low valency (K^+, Na^+, Ca^{2+}, Sr^{2+}, $\mathbf{Ba^{2+}}$, Pb^{2+}, La^{3+}, Bi^{3+} and others) situated at the corners of the cubic unit cell, and small highly charged B-type cations (Nb^{5+}, Ta^{5+}, $\mathbf{Ti^{4+}}$, Zr^{4+}, Sn^{4+}, Ga^{3+} and others) in the center of the cubic unit cell, surrounded octahedrally by six oxygen ions that occupy the centers of the faces of the cubic unit cell. In titanate perovskites, the $[TiO_6]$ octahedra form chains parallel to the crystallographic z-direction (Figure 8.14), while the A- and B-sites can be occupied by cations of different valencies, thus leading to complex perovskites. For example, replacement of the B-site cation Ti^{4+} by $1/3Mg^{2+} + 2/3Nb^{5+}$ and the A-site cation by Pb^{2+} leads to the important ferroelectric relaxor ceramic PMN (lead magnesium niobate; see Section 8.3.3).

On the other hand, the A positions can be shared by dissimilar ions such as Na^+ and Bi^{3+}, yielding the complex ferroelectric relaxor ceramic BNT (bismuth sodium titanate) (Hosono *et al.*, 2001; Pookmanee *et al.*, 2001; Pookmanee *et al.*, 2003; see Section 8.3.3).

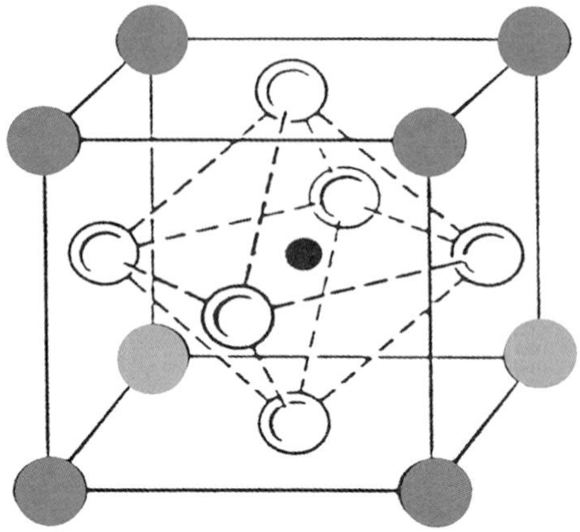

Figure 8.13 Ionic positions in ideal perovskite (S.G. *Pm3m*).

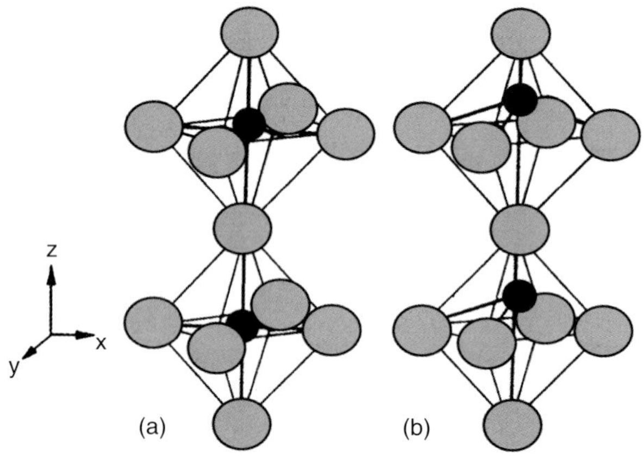

Figure 8.14 (a) [TiO_6] octahedra chain with a symmetry center above the Curie temperature, T_C; (b) Ti–O-chain with spontaneous polarization in z-direction below T_C. The large circles represent oxygen ions, the small circles represent the B-type Ti^{4+} ions.

The stability of the perovskite structure depends on Goldschmidt's tolerance factor $(r_A + r_O)/\sqrt{2}(r_B + r_O)$, with $0.7 < t < 1.2$. For tolerance factors outside this range nonperovskite structure are stable, including pyrochlore-type structures $A_2B_2X_6Z$, Aurivillius phases (Aurivillius, 1949), or tetragonal tungsten bronze (TTB)-type structures $Ba_2MTi_2X_3O_{15}$ (M = Ln,Bi; X = Nb,Ta) (Stennett *et al.*, 2005).

Aurivillius phases of the type $Bi_2A_{n-1}B_nO_{3n+3}$ (n = 1–5) are layered bismuth oxides that may be described structurally as intergrowths of fluorite-type $(Bi_2O_2)^{2+}$ layers with perovskite $(A_{n-1}\ B_nO_{3n+3})^{2-}$ layers (Blake *et al.*, 1997). Typical examples are Bi_2WO_6 (n = 1), $Bi_2SrTa_2O_9$ (n = 2), and $Bi_4Ti_3O_{12}$ (n = 3).

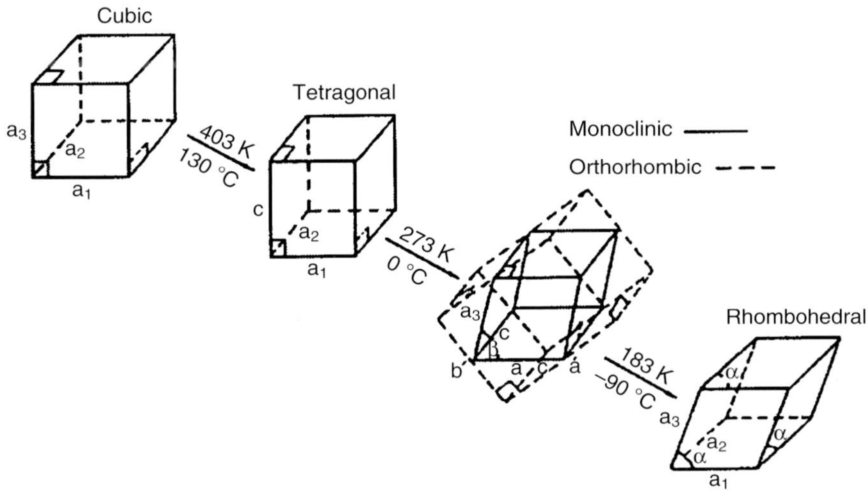

Figure 8.15 Unit cell distortions of single crystals of barium titanate (Kay and Vousden, 1949). Reprinted with permission from Taylor & Francis Ltd. (http://www.informaworld.com).

As is evident from Figure 8.14b, displacement of the B-cation from the center of the oxygen octahedron in the z-direction at a temperature below T_C introduces an asymmetric charge distribution within the coordination polyhedron, and thus induces spontaneous polarization, the direction of which can be reversed by an outside electric field.

Figure 8.15 shows the temperature-dependent symmetry relations of barium titanate, one of the most intensely investigated ferroelectric ceramics. Displacement of the Ti^{4+} ion from the center of the coordination polyhedron in z-direction <001> causes a distortion of the originally cubic symmetry $m\bar{3}m$ (space group $Pm\bar{3}m$, No. 221) to a tetragonal symmetry *4mm* (space group *P4mm*, No. 99) below 130 °C (see Figure 8.14b). However, still other phase transitions have been observed in $BaTiO_3$. At about 0 °C, distortion along the face diagonal <110> changes the symmetry group to orthorhombic (pseudo-monoclinic) *2mm* (space group *A2mm*, No. 38), whilst at −90 °C distortion along the space diagonal <111> changes the overall symmetry group to rhombohedral *3m* (space group *R3m*, No. 160). Since the space groups mentioned are subgroups of $Pm\bar{3}m$ (the "aristotype" according to Megaw, 1973), a so-called perovskite "family tree" emerges (Bärnighaus, 1979). The electric nature of the cubic high-temperature phase with $m\bar{3}m$ symmetry above T_C = 130 °C is paraelectric.

8.3.2
Lead Zirconate/Lead Titanate (PZT)

The structural relationships in ferroelectric ceramics can be explained more clearly using the ferroelectric (and piezoelectric) solid solution system of $Pb(Zr_xTi_{1-x})O_3$

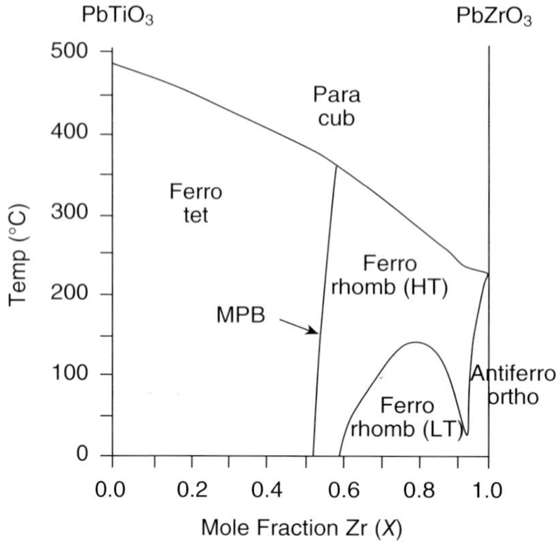

Figure 8.16 Phase diagram of the solid solution series $Pb(Ti_{1-x}Zr_x)O_3$.The ferroelectric tetragonal and rhombohedral phase fields are separated by the morphotropic phase boundary (MPB) (Heaney, 2000). Reprinted with permission from The Mineralogical Society of America.

(lead zirconate titanate, PZT). The two end members are: $PbTiO_3$, which undergoes, at 490 °C, a simple phase transition from a paraelectric cubic to a ferroelectric tetragonal phase; and $PbZrO_3$, which behaves in a more complex fashion, having a transition from cubic to an antiferroelectric orthorhombic phase. The replacement of Zr by Ti produces low- and high-temperature rhombohedral ferroelectric phases. Figure 8.16 shows the phase diagram of the solid solution system $Pb(Zr_xTi_{1-x})O_3$. Below T_C, the phase field of the ferroelectric states is separated by a near-vertical line, the MPB.

The term "morphotropism" was introduced by the German crystallographer Ritter von Groth as early as 1870, and has been defined as a structural change introduced by chemical substitution (von Groth, 1870). The morphotropic phase diagram shown in Figure 8.16 belongs to type IV (Heaney, 2000), in which the space groups of the low-temperature ferroelectric phases are subgroups of the paraelectric high-temperature phase. Extensive studies have confirmed that, in the vicinity of the MPB, properties such as electric susceptibility χ^e – and thus the dielectric permittivity ε – reach anomalously high values (cf. Figure 8.2). This is usually explained by the large number of possible directions of polarization. Whereas, in the region of the ferroelectric tetragonal structure only six possible domain orientations exist (conforming to the six <100> directions) and in the region of the rhombohedral structure eight orientations commensurate with the eight <111> directions exist, close to the MPB the ferroelectric domains adopt fourteen orientations (6+8) for alignment during poling (Heywang, 1965; Heywang *et al.*, 2009). Since, at the phase boundaries, a natural tendency exists for atomic rearrangement, ferroelectric materials will interact strongly with an external electric field, such that maximum values of ferroelectric and piezoelectric properties are obtained along the MPB (see below).

8.3.3 Ferroelectric Relaxor Ceramics

In contrast to classic ferroelectrics, with a sharp phase transition at T_C and along the MPB, respectively relaxor ceramics are characterized by a diffuse OD phase transition without any change in macroscopic symmetry, a slim hysteresis loop (as opposed to the wide hysteresis in classical ferroelectrics; see Figures 8.5 and 8.10), and associated small coercive fields E_C as well as small remanent (P_R) and spontaneous polarizations (P_S). Most importantly, the polarization of relaxor ceramics does not vanish at T_C, but rather retains finite values up to a higher temperature, termed the *Burns temperature* (T_B), as shown schematically in Figure 8.17. The dielectric permittivity of relaxor ceramics attains a maximum value at a temperature T_{max} for a particular frequency and, as this frequency increases, T_{max} also increases. The temperature dependence of ε_R does not obey the Curie–Weiss law just above T_{max}, but only beyond T_B where $T_B > T_{max}$. In contrast to the displacive phase transformation in classical ferroelectrics, the diffuse transition in relaxor ceramics does not involve any change in macroscopic symmetry, and is of neither first nor second order.

The reason for these differences is attributed to the existence of nanosized domains (so-called "polar regions") in which indeed a hysteresis loop as well as symmetry changes can be observed. In perovskites of type ABO_3, the substitution of ions of different valencies, sizes and polarizabilities at the A and B sites produces a variety of dipolar effects that are able to introduce a sufficiently high degree of disorder. This disorder breaks the transitional symmetry and thus prevents the formation of long-range ordered domain states.

For example, in the typical relaxor ceramic lead magnesium niobate (PMN; $Pb[Mg_{1/3}Nb_{2/3}]O_3$), disorder is generated by the differences in valency, ionic radii,

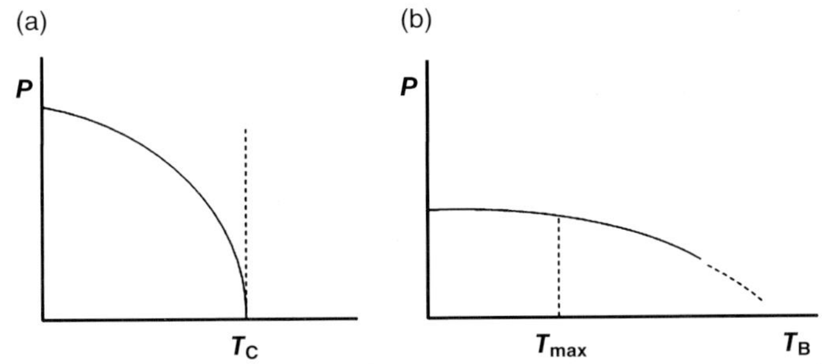

Figure 8.17 Temperature dependence of the polarization in classic ferroelectric (a) and ferroelectric relaxor ceramics (b). T_C: Curie temperature; T_B: Burns temperature.

and electronegativities of the B site cations Mg and Nb. This results in nanometer-scale domains of chemically different nature. Owing to thermal motions close to the diffuse ferroelectric OD phase transition, polarization fluctuations also exist among these domains. Since tightly coupled dipoles within niobium-rich nano-sized domains act like superparaelectrics, their reorientation under the influence of temperature variations, caused by the surrounding electric field, produces large dielectric permittivities and large electrostrictive effects. This is a hallmark of ferroelectric relaxor ceramics.

Likewise, in lead lanthanum zirconate titanate (PLZT) relaxors, the substitution of La^{3+} for Pb^{2+} at A sites produces randomly distributed Pb^{2+} vacancies. In bismuth sodium titanate (BNT; $[BiNa]Ti_2O_6$), the replacement of Bi^{3+} by Na^+ creates charge imbalances and vacancies. Ceramics with compositions $Ba_{1-x}Na_xTi_{1-x}Nb_xO_3$ are either classical ferroelectrics (for $0 \leq x \leq 0.075$), ferroelectrics or antiferroelectrics (for $0.055 \leq x \leq 1$) or relaxor ferroelectrics (for $0.075 \leq x \leq 0.055$) with a diffuse transition temperature without any frequency dispersion. The relaxor behavior increases with increasing compositional deviation from both $BaTiO_3$ and $NaNbO_3$ (Khemahem *et al.*, 2000).

In conclusion, complex perovskite relaxor ceramics are characterized by a very diffuse range of the ferroelectric–paraelectric OD phase transition, owing to nanoscopic compositional fluctuations. The minimum domain size that still sustains cooperative phenomena leading to ferroelectric behavior is the so-called *Känzig region* (Känzig, 1951), and is on the order of 10 to 100 nm in PMN. In contrast to normal ferroelectric ceramics, relaxor ceramics show a frequency dependence of the dielectric permittivity as well as the dielectric loss tangent, which presumably is caused by the locally disordered structure that creates shallow, multipotential wells.

The properties of the ferroelectric relaxor material BNT, $(BiNa)Ti_2O_6$ are described in more detail below. This material is of substantial interest, as the replacement of lead-containing electroceramics would constitute an environmental advantage. In addition, it would alleviate the need to control the lead oxide partial pressure during the mixed-oxide synthesis of lead-based perovskites such as PMN or PZN (Moulson and Herbert, 2003). However, these advantages are limited by the high coercitive field strength E_C of 73 kV cm^{-1} (Nagata and Takenaga, 1997) that is known to cause problems during poling (Herabut and Safari, 1997). There is also a strong tendency for the compound to decompose during sintering by evaporative losses of volatile Na and Bi components (see below).

X-ray powder diffraction analyses of hydrothermally synthesized BNT powders (Pookmanee *et al.*, 2004) have revealed a rhombohedral perovskite phase with symmetry group *R3m* (Pookmanee *et al.*, 2003). The diffraction patterns of undoped BNT powder sintered at 1000 °C for 1, 2, and 3 h are shown in Figure 8.18 (traces a, b, and c, respectively), and those of BNT doped with 1 at% La 2 at% La sintered at 1000 °C for 3 h in Figure 8.19 (traces d and e, respectively). Sintering caused a substantial decomposition of the undoped BNT by the thermal evaporation of Na_2O and Bi_2O_3 according to:

$$4(BiNa)Ti_2O_6 \rightarrow Bi_2Ti_2O_7 + BiNaTi_6O_{14} + 3/2\,Na_2O\uparrow + \tfrac{1}{2}Bi_2O_3\uparrow \tag{8.7}$$

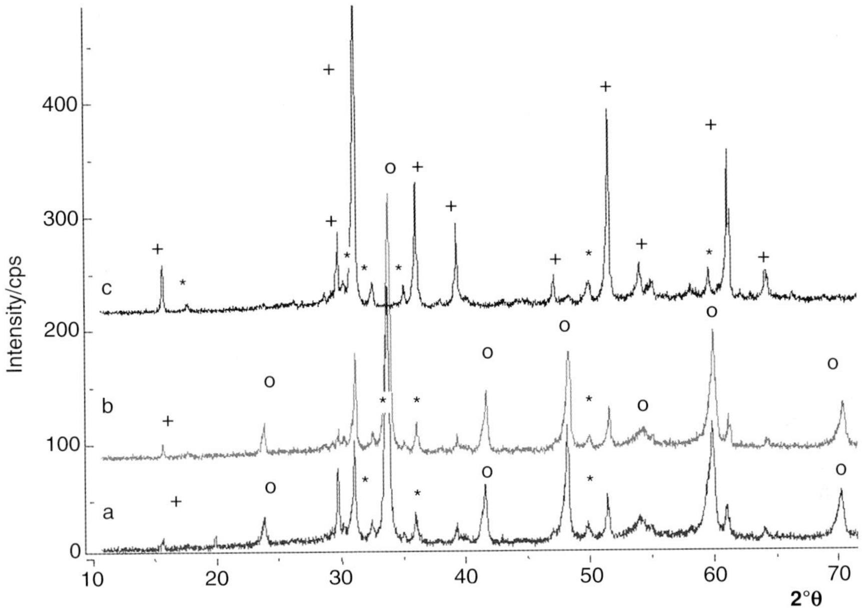

Figure 8.18 X-ray diffraction patterns of hydrothermally synthesized BNT powder sintered at 1000 °C for 1 h (trace a), 2 h (trace b), and 3 h (trace c). The interplanar spacings are marked as: o = ferroelectric rhombohedral phase of BNT, + = paraelectric pyrochlore-type phase $Bi_2Ti_2O_7$, and * = paraelectric $BiNaTi_6O_{14}$ (Pookmanee *et al.*, 2003, 2004).

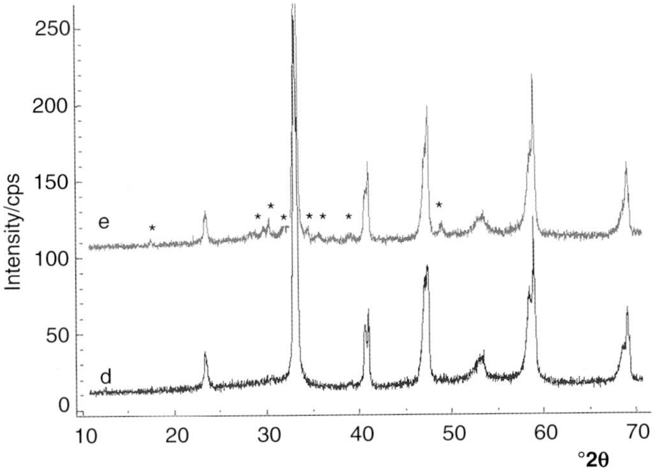

Figure 8.19 X-ray diffraction patterns of hydrothermally synthesized BNT powder doped with 1 at% La (trace d) and 2 at% La (trace e) sintered at 1000 °C for 3 h. * = $BiNaTi_6O_{14}$ (Pookmanee *et al.*, 2003).

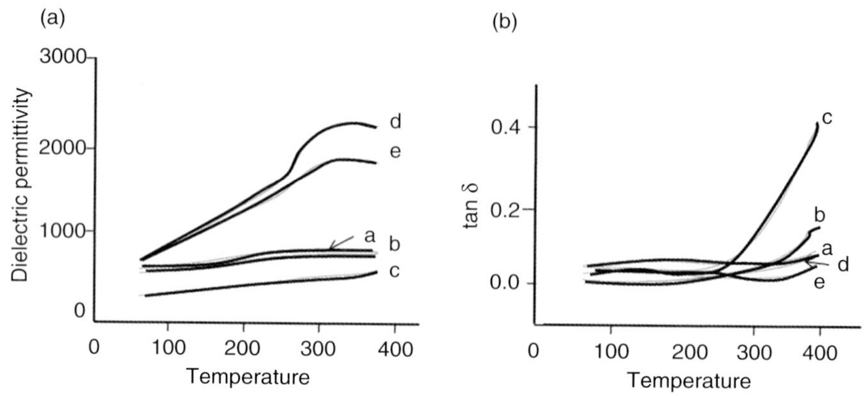

Figure 8.20 Dielectric permittivities (a) and dielectric loss tangent δ (b) of sintered BNT (curves a–c) and La-doped BNT (curves d and e) measured at 10 kHz (Pookmanee *et al.*, 2003).

In fact, sintering for 3 h at 1000 °C completely destroys the ferroelectric perovskite phase $(BiNa)Ti_2O_6$, and the product shows only the interplanar spacings of the paraelectric pyrochlore phase $Bi_2Ti_2O_7$ and the "stuffed" titanium oxide $BiNaTi_6O_{14}$ (Figure 8.18, trace c). In contrast, the doping of BNT with 1 at% La appears to stabilize the perovskite phase. Trace d in Figure 8.19 shows a mixture of rhombohedral ferroelectric and tetragonal antiferroelectric $(BiNa)Ti_2O_6$ phases, with the latter appearing at the low angle flank of the interplanar spacings of the rhombohedral phase. However, a sample doped with 2 at% La revealed additionally small (but noticeable) amounts of $BiNaTi_6O_{14}$ (trace e in Figure 8.19), indicating the onset of thermal decomposition.

Figure 8.20a shows the temperature variation of the dielectric permittivity of undoped BNT samples (curves a–c), as well as of BNT doped with 1 at% La (curve d) and 2 at% La (curve e), all sintered at 1000 °C. For this, the measurement frequency was 10 kHz. The dielectric permittivities of the undoped samples varied approximately linearly with temperature, and hence followed the Curie–Weiss law. The low values of dielectric permittivity, and their near-linear variation with temperature, could be assigned to the deviation from the ferroelectric perovskite composition, and the increasing presence of paraelectric contributions from the decomposition products that cause an increase in electrical conductivity. On the other hand, in concurrence with the diffuse OD phase transition from the antiferroelectric to the paraelectric phase at T_C, the dielectric permittivity of the La-doped samples reached a maximum at 350 °C. The dielectric permittivity of BNT doped with lanthanum was more than twice that of undoped BNT, and was larger for 1 at% La (~2300) than for 2 at% La (~2000). The lower value at a higher La concentration was presumed to be related to the superposition of an increasing deformation of the rhombohedral lattice of BNT towards a (pseudo) cubic symmetry with increasing dopant concentration, and the onset of thermal decomposition, producing small amounts of paraelectric $BiNaTi_6O_{14}$ (as confirmed by trace e in Figure 8.19). The slightly larger La^{3+} ions (ionic radius 101 pm) were substituting for Bi^{3+} ions (ionic radius 96 pm) at the A sites of the perovskite lattice to yield a compound of the type $(BiNa)_{1-(3/2)x}La_{2x}Ti_2O_6$ $(0.01 < x < 0.02)$. Beyond a dopant concentration of 2 at% La, the rhombohedral ferroelectric structure was seen to transform to a (pseudo)cubic paraelectric structure, because of the increasing lattice strain (Nagata and Takenaga, 1997). Figure 8.20b shows the dielectric loss tangent of the samples measured at a frequency of 10 kHz. For the undoped BNT sintered for 1 and 2 h (curves a and b), the loss tangent increased somewhat with temperature beyond 350 °C. However, the loss tangent increased strongly for undoped BNT sintered for 3 h (curve c), in concurrence with an increasing thermal decomposition that led to the disappearance of the ferroelectric phase according to Eq. (8.7). On the other hand, the loss tangents of BNT doped with La (curves d and e in Figure 8.20b) and sintered at 1000 °C for up to 3 h remained low, indicating a structural stabilization of the ferroelectric perovskite phase.

More detailed recent information on ferroelectric relaxor ceramics can be obtained from Ye (2008). Lead-free $BaFe_{0.5}Nb_{0.5}O_3$ ceramics with multiferroic properties were developed by Bochenek *et al.* (2009).

8.4 Microwave Ceramics

8.4.1 Grain Boundary Engineering

Without exception, dielectric ceramics are semiconductors, which means that they possess a high resistance within the grain boundary region but a low resistance within the crystal grains. When an external field is applied, most of it will concentrate in the high-resistance region; hence, the grain boundaries control the movement of electrons, with the consequence that any alteration of the composition and the microstructure of the grain boundary region is very important as it may lead to dielectric ceramics with improved permittivities, *Q*-factors, time constants, and temperature coefficients of both the permittivity and the resonance frequency.

Figure 8.21 shows a typical ceramic microstructure (a) with impurities and pores within the microcrystal grain and along grain boundaries as well as grain boundary divided deposits (b).

Grain boundary-divided deposits of thickness between 20 Å and about 1 μm consist of an ionic stratification of impurities separated along the grain boundary. As impurities are easily dissolved in the grain boundaries, the composition and crystal phases, respectively, in the grain boundary are very different from those inside the crystal grain. When the amount of impurities increases beyond the limit of solid solution, a separate crystalline phase precipitates at the grain boundary.

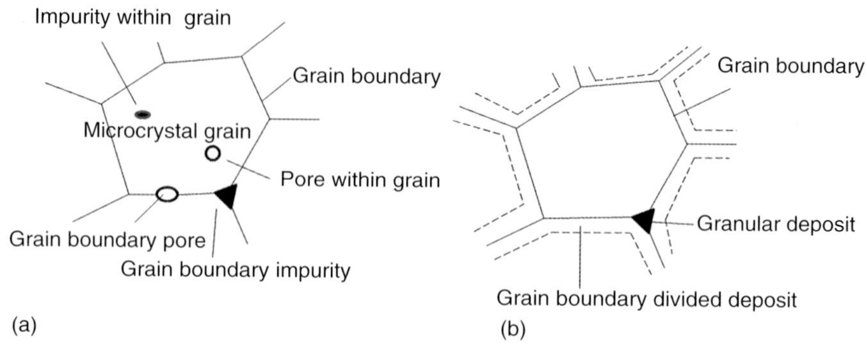

Figure 8.21 (a) Typical ceramic microstructure and (b) formation of grain boundary-divided deposits by dissolution of impurities in the grain boundary.

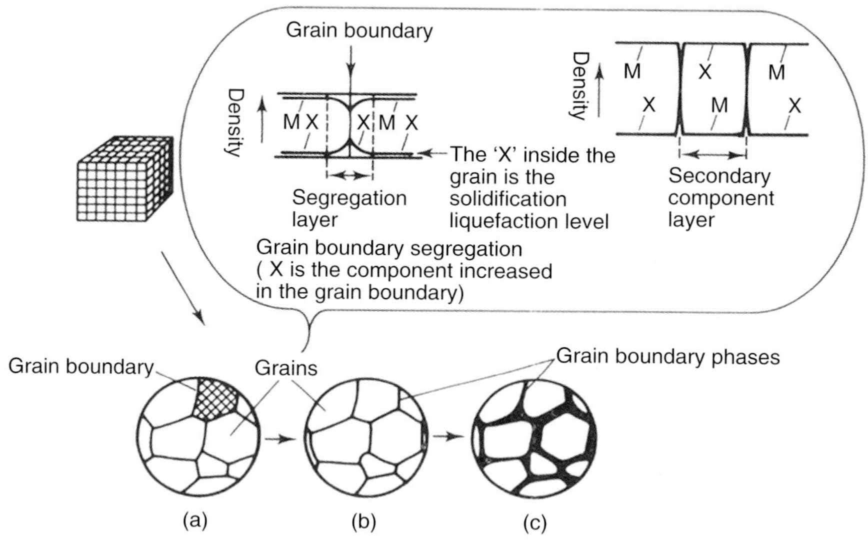

Figure 8.22 Segregation of an impurity at the grain boundaries and development of the microstructure pattern with increasing impurity level (a–c) (Ichinose, 1987). Reprinted with permission from Wiley-Blackwell Ltd, Oxford, UK.

Such diffuse deposits have thicknesses of between 100 Å and 10 μm, and are formed by liquid-phase sintering (LPS), which occurs when the melting point of the precipitate is lower than the sintering temperature of the bulk grain. Finally, if the amount of impurities exceeds the saturation concentration of the solid solution by a large margin, then particulate impurities will precipitate as granular deposits between the grains (Figure 8.21b).

Figure 8.22 shows the segregation of impurities at grain boundaries, the region with increased energy that provides the driving force for the segregation reaction.

At low concentration, the impurities "X" dissolve in solid solution in the matrix phase "M" (Figure 8.22a); however, with increasing concentration the solubility limit will be exceeded so that a second (or third) phase exists at the grain boundaries (Figure 8.22b and c).

A typical early example of grain boundary engineering has been provided by n-type doping of the inner part of $BaTiO_3$ crystallites with Sb, Nb, La, or Bi; and p-type doping of their surface layers with Cu, Co, Ni, Fe, or Mn. In addition, a TiO_2-rich intermediate layer separating the doped crystallites leads to a capacitor ceramic (SIBATIT H™, Siemens) with improved time constant, resistance to high electrical field strength, and reproducibility (Kniepkamp and Heywang, 1954; Schmelz and Schwaen, 1978).

8.4.2
Design of Microwave Ceramics

In order to design low-noise oscillators and narrow band microwave filters with low insertion losses, the resonator material should have a high Q-value (details of this are provided in Section 8.7.1). While the Q-factor of coaxial resonators (see Figure 8.36) is usually limited to values of less than 1500, owing to the finite conductivity of the silver or palladium–rhodium electrode coating, substantially higher Q-factors can be achieved with solid or strip-line resonators using ceramic materials with high intrinsic Q-values. There is a conundrum, however, in that whereas the dielectric permittivity remains constant with respect to frequency in the microwave region (see Figure 8.9), tan ô increases with increasing frequency; that is, Q decreases. A consequence of this is that the product of Q and f is approximately a constant in the microwave band.

Unfortunately, two fundamental relationships work against the existence of an "ideal" dielectric microwave ceramics:

- The inverse ε–Q relationship; that is, if ε increases, Q decreases rapidly (Figure 8.23). This is related to the fact that Q is not just a generic material constant but depends heavily on the microstructure that, in turn, is a function of the grain size of the precursor ceramic powder and the processing of the ceramics. Impurities often form grain boundary phases and lattice defects that increase the dielectric losses through either a relaxation of space charge polarization at the grain boundaries of multiphase dielectric ceramics (Figure 8.24a), or phonon scattering by anharmonic 3-phonon coupling involving the decay of one transversal optical (TO) into two thermal acoustic (TA) phonons in single-phase materials (Figure 8.24b),
- The well-known Cockbain relation (Cockbain and Harrop, 1968) between ε and τ_ε; that is, as ε increases the temperature coefficient of the dielectric permittivity decreases to strongly negative values.

Hence, strategies are required to balance ε, τ_ε, and Q by grain boundary engineering. Several routes towards materials with improved characteristic dielectric parameters are outlined in Section 8.4.1 (see also Wersing, 1996).

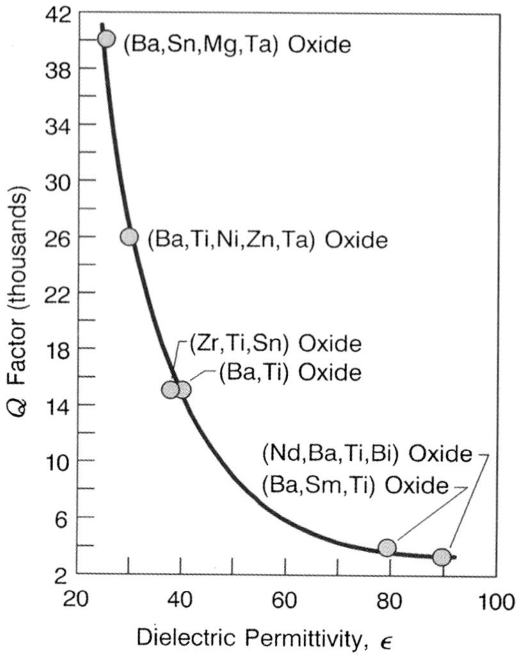

Figure 8.23 Inverse relationship between the mechanical Q-factor and the dielectric permittivity.

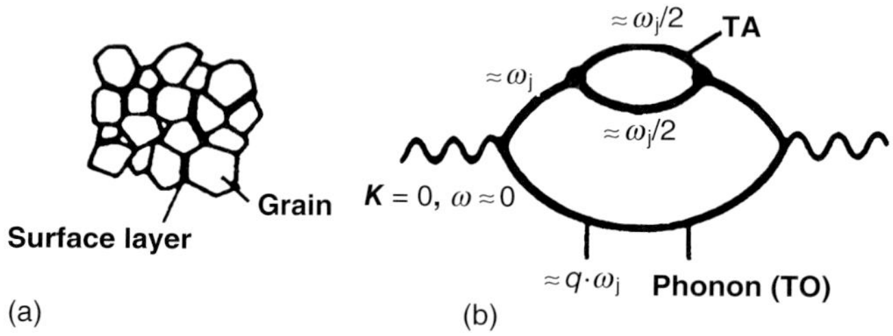

Figure 8.24 (a) Relaxation of space charge polarization at grain boundaries in multiphase dielectrics; (b) Phonon scattering by anharmonic three-phonon coupling in single-phase dielectrics (b).

As indicated in Section 8.2.2 and shown in Figure 8.24, dielectric losses occur in single as well as multiphase ceramics by phonon scattering and a relaxation of the space charge polarization at grain boundaries, respectively. A significant decrease of scattering losses can be achieved by a reduction of the grain size, as well as the porosity of the polycrystalline ceramics by appropriate technologies such as sol–gel precipitation, hot isostatic pressing (HIP), thin film deposition, or by the addition of sintering aids in conjunction with an optimization of the sinter temperature, pressure, and time. For example, complex perovskites such as $Ba(Zn_{1/3}Ta_{2/3})O_3$ when fired at 1350 °C for 120 h develop a *Q*-factor of 15 000 but a very low values of *Q* when fired for only 2 h. This is caused by the fact that the Zn and Ta atoms at B sites of the perovskite lattice form a superlattice that requires time to develop through comparatively slow solid-state diffusion. On the other hand, sintering at 1650 °C for only 5 min leads to a perfectly ordered structure with an extremely high *Q*-factor of 35 000 at 10 GHz (Kawashima *et al.*, 1983).

Since oxygen vacancy defects also tend to decrease the *Q*-factor, the firing atmosphere and post-firing annealing must be carefully considered. This is reminiscent of the severe processing problems of high-temperature superconductors (see Section 9.7). In fact, the dielectric loss in ceramics is analogous to the resistance experienced by flux penetration in type II superconductors.

Attempts were made to improve the dielectric parameters ε, Q, τ_f, and τ_ε by adding TiO_2 to $BaTiO_3$ to yield $BaTi_4O_9$ and $Ba_2Ti_9O_{20}$ (Figure 8.25).

The dielectric permittivities ε_r are in the range of 35 to 40, with Q around 9000 at a resonance frequency of 4 GHz. There is a severe drawback, however, in that the temperature coefficients of the dielectric permittivity τ_ε are about 50 ppm °C^{-1} for $BaTi_4O_9$ and 40 ppm °C^{-1} for $Ba_2Ti_9O_{20}$, and further increase with increasing TiO_2 content to very high positive values. In contrast, the Q-factors tend to decrease to about 5000, much lower than those of the complex perovskites developed during the 1980s. The temperature coefficients of the resonance frequency τ_f were 20 ppm °C^{-1} for $BaTi_4O_9$ (Masse *et al.*, 1971) and 2 ppm °C^{-1} for $Ba_2Ti_9O_{20}$ (O'Bryan and Thomson, 1974).

Grain boundary engineering by doping $BaTi_4O_9$ with as little as 2 mol% tungsten oxide (WO_3) yields ceramics with temperature coefficients of the resonance

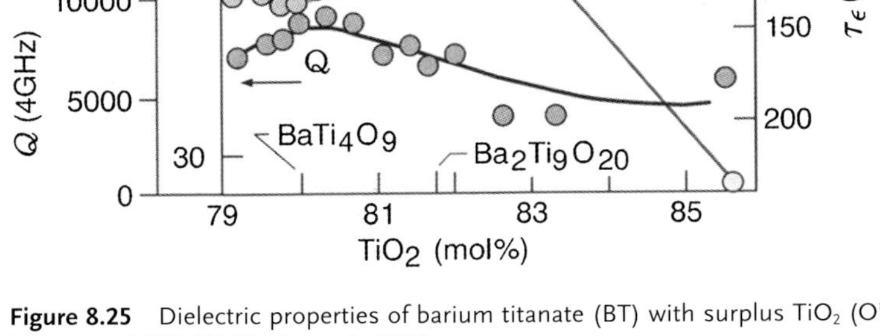

Figure 8.25 Dielectric properties of barium titanate (BT) with surplus TiO_2 (O'Bryan and Thomson, 1974; Nishigaki *et al.*, 1986).

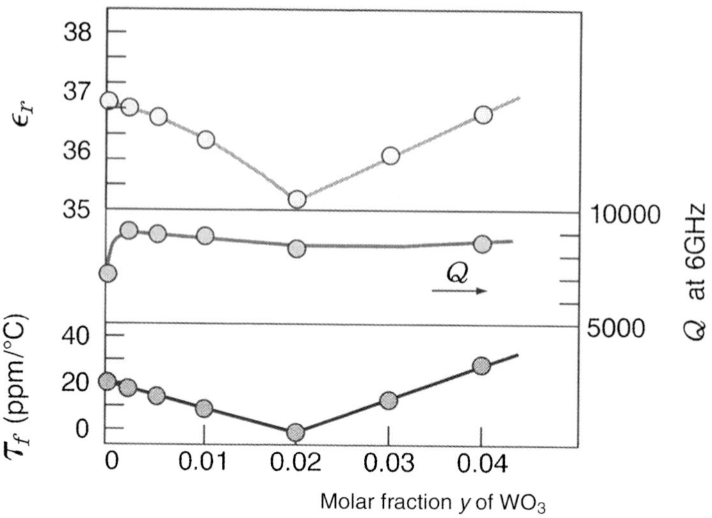

Figure 8.26 Dielectric properties of $BaTi_4O_9$ doped with WO_3 (Nishigaki *et al.*, 1988).

frequency τ_f close to zero (Figure 8.26), accompanied by a slight decrease in the dielectric permittivity and the Q-factor (Nishigaki *et al.*, 1988). Microstructural studies revealed that for $0 < y < 0.02$ (where y is the molar fraction of WO_3), the phases $BaTi_4O_9$, $Ba_2Ti_9O_{20}$ and $BaWO_4$ coexist, while for $0.02 < y < 0.04$ $Ba_2Ti_9O_{20}$, $BaWO_4$ and TiO_2 exist; this indicates that the dopant WO_3 does not dissolve in the barium titanate structure, but rather forms grain boundary deposits of barium tungstate, $BaWO_4$. Apparently, the large positive temperature coefficient of the resonance frequency of $BaTi_4O_9$ (+20 ppm °C^{-1}) will be compensated by the large negative τ_f values of $BaWO_4$ (−33 ppm °C^{-1}). This is expressed by the excellent correspondence of the measured coefficients with those calculated following a volume ratio mixing rule $\tau_f = \Sigma v_i \cdot \tau_{fi}$.

Another instructive example of successful grain boundary engineering concerns the doping of $BaTiO_3$ with strontium and samarium oxides. Figure 8.27a shows the ternary phase diagram $BaTiO_3$–Sm_2TiO_5–TiO_2 with the area of interest highlighted; two stoichiometric ternary compounds are shown, $BaO \cdot Sm_2O_3 \cdot 3TiO_2$ and $BaO \cdot Sm_2O_3 \cdot 5TiO_2$. Along the conode between these two phases a composition has been found (Kawashima *et al.*, 1985) corresponding to $(Ba_{1-x}Sr_x)O \cdot Sm_2O_3 \cdot 4.7TiO_2$ ($x = 0.05$) with $\tau_f = 0$, $\varepsilon = 80$, and $Q = 3700$ at 3 GHz resonance frequency (Figure 8.27b). The microstructure of this phase does not show any segregation of strontium, but rather a solid solution of Sr in a $BaSm_2Ti_5O_{14}$ phase with small amounts of secondary TiO_2 and $Ba_2Ti_9O_{20}$.

The introduction of neodymium into barium titanate leads to a product with a high dielectric permittivity of 80, which is attributed to a single-phase $BaNd_2Ti_5O_{14}$ (NBT), analogous to the phase described above in samarium-doped barium

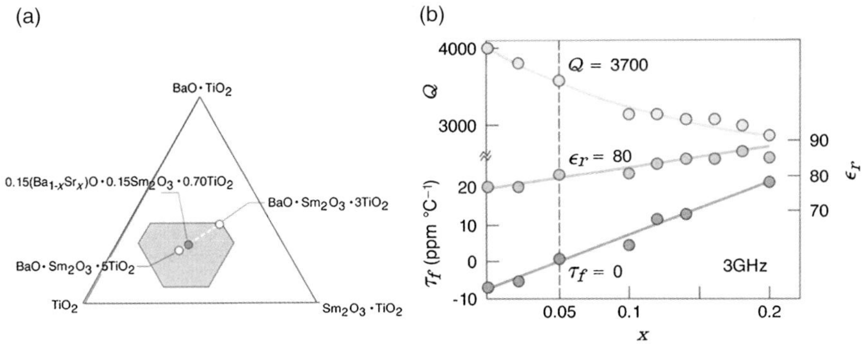

Figure 8.27 (a) Ternary phase diagram for $BaTiO_3$–Sm_2TiO_5–TiO_2; (b) Dielectric properties of phase $(Ba_{1-x}Sr_x)O{\cdot}Sm_2O_3{\cdot}4.7TiO_2$ ($x = 0.05$) (Kawashima *et al.*, 1985).

Table 8.5 Strategies of designing dielectric ceramics.

Microstructure	Examples	ε	$Q{\cdot}f$ (GHz)	τ_f (ppm K^{-1})
Solid solution	Ba $(Mg_{1/3}Ta_{2/3})O_3$	25	200 000	4.4
With cation ordering	Ba $(Zn_{1/3}Ta_{2/3})O_3$	30	100 000	0.0
Solid solution	$(CaSr)ZrO_3$	30	30 000	0.4
Without ordering	$(ZrTiSn)O_2$	40	50 000	0.0
Mixed ceramic phases	$CaZrO_3$–$BaZrO_3$	30	8 000	0.0
	Nd_2O_3–BaO–TiO_2–Bi_2O_3	90	6 000	0.0
Ceramic—glass mixtures	$CaTiO_3$–$CaTiSiO_3$	80	500	>0.0
Combined	$Pb(ZrCe)O_3$–$Sr(TiNiNb)O_3$	170	4 000	87
Ceramic resonators	($\tau f \approx -1100$, ≈ 890 ppm K^{-1})			
?	?	>300	>6 000	0.0

titanate. To improve τ_ε and to lower the sintering temperature, bismuth titanate is usually added, which results in a two-phase structure with a bismuth-rich and a titanium-rich phase at the interface of the NBT grains (Kolar *et al.*, 1981; Wersing, 1991). By the substitution of Bi for Ti, two Ti ions in $BaTiO_3$ will be replaced by one three-valent and one five-valent Bi ion to yield $Ba_2[Bi^{3+}Bi^{5+}]O_6$ that separates as the Bi-rich phase at grain boundaries together with TiO_2. This doping results in a substantial lowering of the temperature coefficient of the resonance frequency that was found to reach $\varepsilon_f \approx 0$ for a degree of substitution $x = 0.34$ (Wersing, 1991).

Several strategies for the design of dielectric microwave ceramics with high permittivities, high Q-factors, and a close to zero temperature coefficient of the resonance frequency, are listed in Table 8.5.

For permittivities >25, the highest Q-factor can be obtained with partially ordered complex perovskites such as $Ba(Mg_{1/3}Ta_{2/3})O_3$. Single-phase $(ZrTiSn)O_2$, as well as solid solutions such as $(Ca,Sr)ZrO_3$ or $(Ca,Sn)(Li,Nb,Ti)O_3$, have much lower $Q{\cdot}f$-values but τ_f values of zero. Mixtures of phases with large negative and positive τ_f values such as $CaTiO_3$ and $BaZrO_3$, or $LaAlO_3$ and $BaZrO_3$ have high permittivities and τ_f values close to zero, but drastically reduced Q-factors. Antiferroelectric $PbZrO_3$ doped with CeO_2 ($\tau_f = -1100\,ppm\,K^{-1}$) laminated with paraelectric $SrTiO_3$ doped with NiO and Nb_2O_5 ($\tau_f = +890\,ppm\,K^{-1}$) yield a compound dielectric material with $\varepsilon = 170$, $\tau_f = +87\,ppm\,K^{-1}$, and $Q = 1300$ at 3 GHz.

Future developments will focus on improved ceramic processing technology to increase the quality factor, as well as on the development of mixed crystal systems rather than homogeneous systems. The insertion losses of filters might also be drastically reduced by the application of thin-film technology, including HT superconducting films that operate at liquid nitrogen temperatures (see Section 9.8.6).

8.5 Pyroelectric and Piezoelectric Ceramics

The pyroelectric and piezoelectric effects can be visualized by the interaction of thermal and electric energies, and kinetic and electric energies, respectively (Figure 8.28). Here, the side connecting the electrical and thermal energy corners represents the pyroelectric effect and produces no kinetic energy, whereas the side connecting the kinetic and electrical energy corners represents the piezoelectric effect and produces no heat (Figure 8.28, left).

Both, piezoelectric and pyroelectric behavior is possible only in ferroelectric ceramics, or in otherwise polar materials that are deposited as textured thin films.

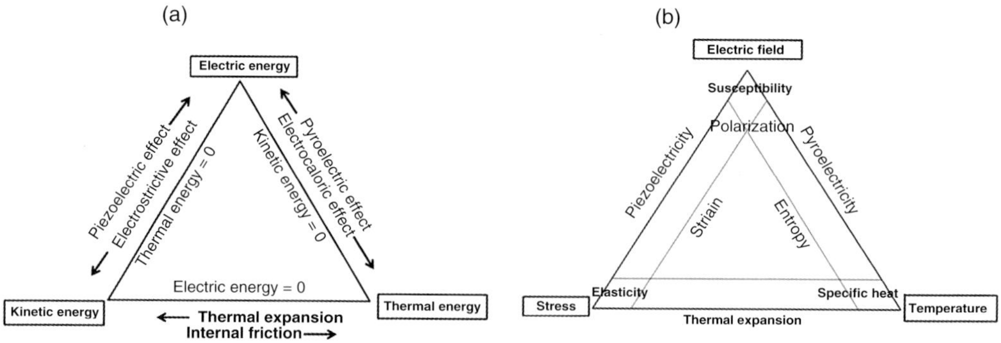

Figure 8.28 (a) Visualization of the pyroelectric and piezoelectric effects as two sides of the same coin. The directions of the arrows symbolize the stimulus–response chain (see Table 8.1); (b) Linear relationships among mechanical, electrical, and thermal variables (Heckmann diagram; Newnham, 2005).

Other piezoelectric materials require either utilization as single crystal slices cut in fixed relation to the electrical axes, as in α-quartz (Bradaczek *et al.*, 1990) and langasite (Bohm *et al.*, 1999), or as epitaxially deposited layers on suitable substrates (Heywang *et al.*, 2009).

8.5.1 Pyroelectric Ceramics

Pyroelectricity is a property inherent in crystals with unique polar axes that are, consequently, without a center of symmetry. However, as opposed to piezoelectricity not all symmetry groups (point groups) lacking a center of symmetry are pyroelectric, whereas all pyroelectrics are also piezoelectrics. Most pyroelectrics belong to the linear dielectrics as the polarization is linearly dependent on the electric field. However, some – for example, KDP and Seignette salt – are nonlinear pyroelectrics – that is, they are ferroelectrics.

As shown in Table 8.3 above the 21 point groups without a center of symmetry only the 10 groups *1, 2, m, mm2, 3, 3m, 4, 4mm, 6,* and *6mm* are pyroelectric. The prototype of a pyroelectric crystal is tourmaline with point group *3m*, the c-direction of which is the direction of the polar axis. Figure 8.29 shows a ditrigonal pyramid as the general form {hkil} of the point group *3m* (left) and a crystal of tourmaline (right). Pyroelectric behavior as displayed by tourmaline is characterized by the development of opposite charges at both ends of the polar axis during heating or cooling of the crystal (see Table 8.1), whereby the pole becoming electrically positive during heating is called the *analogous* pole, and the other the *antilogous* pole. The reason for this behavior can be explained as follows.

An individual volume element of the tourmaline crystal shows, even in the absence of an electric field, a spontaneous electric polarization owing to the polar axis. This polarization creates apparent surface charges at opposite ends of the polar axis that, however, are compensated by an adsorbed water film (and the

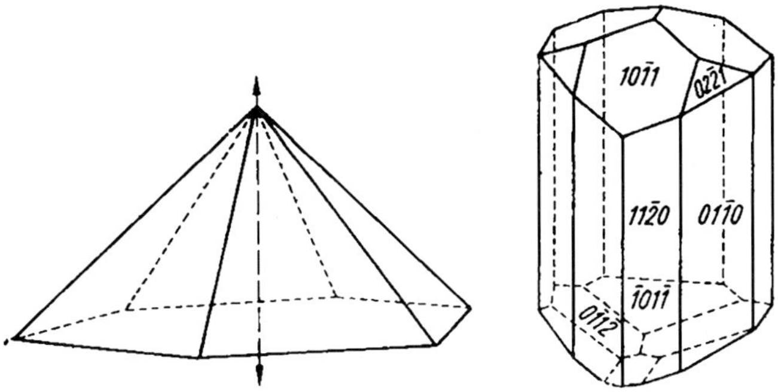

Figure 8.29 (a) General form {hkil} (ditrigonal pyramid) of the pyroelectric point group *3m*; (b) A typical crystal of tourmaline.

conductivity of the surrounding air), so that at ambient temperature these polarization-induced surface charges cannot be observed. Changing the temperature changes also the polarization state [see Eq. (8.4)] and, in turn, the displacement density of the apparent surface charges. Only if the temperature change is so rapid that the charge balance by free charges is delayed (the water film and the air are poorly conducting) can the charge separation be observed.

While crystals of the 10 point groups without a center of symmetry and unique polar axes are called "true" pyroelectrics, there exists the phenomenon of "false" pyroelectricity in which crystals not belonging to one of the 10 point groups can develop charge separation during heating or cooling. For example, on rapid heating or cooling, crystals with point group *32* (such as piezoelectric α-quartz) will, owing to the absence of a center of symmetry, generate elastic tensions within the crystal that in turn produce pyroelectric charges.

As outlined above in Section 8.2.1, special cases of pyroelectricity include ferroelectricity ("nonlinear pyroelectricity") and ferroelasticity; some pyroelectrics demonstrate the additional property, that the direction of the spontaneous polarization can be changed by applying either an outside electric field or a mechanical stress. The former materials are termed *ferroelectrics*, and the latter *ferroelastics*. A correlation between ferroelectric and pyroelectric behavior can, for example, be observed in lead zirconate titanate (PZT, $Pb(Zr,Ti)O_3$) that is ferroelectric with a spontaneous polarization below the Curie temperature at 490 °C (Figure 8.16), as well as piezoelectric. As the polarization $\mathbf{P}_S$ is temperature-dependent (see Figure 8.2), even subtle changes in temperature can be detected by likewise subtle changes in voltage. Based on this principle, passive infrared (PIR) detectors can be designed for remote temperature sensing by satellite or aircraft for the environmental surveillance of land and ocean areas, for measurements of CO and SO_2 concentrations in air, for medical applications such as no-touch skin measurements and cancer detection, and the remote sensing of temperature under adverse conditions such as high-radiation fields encountered in nuclear reactors, as well as infra-red (IR) burglar alarms. Since the recorded voltages are very low, the resistance of the devices must be $10^{12}\,\Omega$ or higher, in order to detect small changes of the intensity of IR radiation that stand out against an only very slightly different background. Such devices are highly sensitive and allow the detection of IR radiation power down to about 10^{-11} W.

Studies are currently in progress to develop functional transducers to transform thermal energy in electric energy with a power output of several $W\,cm^{-3}$ working volume, and an efficiency of >20%.

Modern pyroelectric sensing materials include gallium nitride, cesium nitrate or lithium tantalate, as well as organic compounds such as cobalt phthalocyanine or polyvinylidene fluoride (PVDF) (Lang, 2005). Displacement sensors based on PZT (Lozinski *et al.*, 1998; De Cicco *et al.*, 1999) or PLZT (lead lanthanum zirconate titanate) (Kongtaweelert *et al.*, 2004) show high sensitivity and low dielectric loss, combined with moderate permittivity but a large pyroelectric coefficient, $\mathbf{p}^*$. While many pyroelectric materials are single crystals, polycrystalline ferroelectric thin films may also possess pyroelectric sensor properties.

The inverse effect – that is, the dependence of temperature on an external electric field ($\Delta T = \mathbf{q} \cdot \mathbf{E}$; where $\mathbf{q}$ is the electrocaloric coefficient) – is used today in devices for electrostatic cooling to achieve temperatures close to the absolute temperature, as required for experiments conducted in the Large Hadron Collider.

8.5.2 Semiquantitative Model of Piezoelectricity

The brothers Jacques and Pierre Curie are credited with the discovery of piezoelectricity in a number of hemiedric crystals (Curie and Curie, 1880). Today, piezoelectrics are utilized in acousto-electronic devices and sensors based on bulk and surface acoustic waves, piezomechanical sensors to monitor pressure, power, and acceleration, as actuators for micropositioning devices, band pass filters with low insertion losses, as electro-optic devices for optical memories, displays for high-definition televisions, and possibly as transparent piezoelectric speaker membranes as well as miniaturized piezoelectric transformers and motors. As the classic piezoelectric material is α-quartz, the basic relationships are detailed below using it as a model structure. Further details on the piezoelectric properties of quartz, and of its history, discovery and utilization, are available elsewhere (Ballato, 2009).

As shown in Table 8.1, the piezoelectric effect causes the creation of charges in a dielectric and ferroic material, respectively, in response to an applied stress field. The opposite effect – that is, the induction of strain (deformation) by applying an outside electric field – is called the *inverse piezoelectric effect*. Piezoelectricity requires that no symmetry center exists in the crystal structure. The piezoelectric properties of ceramic materials are described by four parameters: (i) the dielectric displacement $\mathbf{D}$; (ii) the electric field strength $\mathbf{E}$; (iii) the applied stress $\mathbf{X}$; and (iv) the strain (deformation) $\mathbf{x}$. These are related by two equations that apply to the (direct) piezoelectric effect: $\mathbf{D} = e \cdot \mathbf{x}$ and $\mathbf{E} = h \cdot \mathbf{x}$, and two equations that apply to the inverse piezoelectric effect: $\mathbf{x} = g^* \cdot \mathbf{D}$ and $\mathbf{x} = d^* \cdot \mathbf{E}$. The four coefficients e, h, g^*, and d^* are termed the *piezoelectric coefficients*.

The *direct effect coefficients* are defined by the derivatives $(\delta\mathbf{D}/\delta\mathbf{X}) = d$ (piezoelectric strain coefficient), $(\delta\mathbf{D}/\delta\mathbf{x}) = e$, $-(\delta\mathbf{E}/\delta\mathbf{X}) = g$ (piezoelectric voltage constant) and $-(\delta\mathbf{E}/\delta\mathbf{x}) = h$. The *converse-effect coefficients* are defined by the derivatives $(\delta\mathbf{x}/\delta\mathbf{E}) = d^*$, $(\delta\mathbf{x}/\delta\mathbf{D}) = g^*$, $-(\delta\mathbf{X}/\delta\mathbf{E}) = e^*$, and $-(\delta\mathbf{X}/\delta\mathbf{D}) = h^*$. As the piezoelectric coefficients are higher-rank tensors, their mathematical treatment is rather tedious. Fortunately, in higher symmetric crystals the number of tensorial components will be drastically reduced due to symmetry constraints. An example is shown below.

Materials-dependent figure of merits in piezoelectric ceramics are:

- the piezoelectric strain constant d (for actuators): $\mathbf{x} = d \cdot \mathbf{E}$
- the piezoelectric voltage constant g (for sensors): $\mathbf{E} = g \cdot \mathbf{X}$ ($g = d/\varepsilon$, where ε = dielectric permittivity)
- the electromechanical coupling factor k (= stored mechanical or electrical energy/input electrical or mechanical energy), with $k^2 = d^2/\varepsilon \cdot \varepsilon_0 \cdot s$, where s = elastic compliance)

- the energy transmission coefficient λ (= output mechanical or electrical energy/input electrical or mechanical energy), with $\lambda \leq (1/3\text{–}1/4) \cdot k^2$
- the efficiency η (= output mechanical or electrical energy/consumed electrical or mechanical energy), and
- the acoustic impedance Z, with $Z = \rho \cdot v_s$, where ρ is density and v_s is the velocity of sound, or acoustic energy transfer $Z = (\rho \cdot c)^{1/2}$, where c is the elastic stiffness, with $c = 1/s$.
- the mechanical impedance $Z_m = F/v$, where F is the force and v is the velocity.

Barium titanate (point group *4mm* below the Curie temperature of 130 °C) has the (reduced) piezoelectric strain constant (d) matrix

$$\begin{bmatrix} 0 & 0 & 0 & 0 & d_{15} & 0 \\ 0 & 0 & 0 & d_{15} & 0 & 0 \\ d_{31} & d_{31} & d_{33} & 0 & 0 & 0 \end{bmatrix} \tag{8.8}$$

The induced strains $\mathbf{x}_i$ are calculated by $\mathbf{x} = d \cdot \mathbf{E}$ as

$$\begin{bmatrix} x_1 \\ x_2 \\ x_3 \\ x_4 \\ x_5 \\ x_6 \end{bmatrix} = \begin{bmatrix} 0 & 0 & d_{31} \\ 0 & 0 & d_{31} \\ 0 & 0 & d_{33} \\ 0 & d_{15} & 0 \\ d_{15} & 0 & 0 \\ 0 & 0 & 0 \end{bmatrix} \cdot \begin{bmatrix} E_1 \\ E_2 \\ E_3 \end{bmatrix} \tag{8.9}$$

Matrix multiplication yields $x_1 = x_2 = d_{31} \cdot \mathbf{E}_3$, $x_3 = d_{33}.\mathbf{E}_3$, $x_4 = d_{15}.\mathbf{E}_2$, $x_5 = d_{15}.\mathbf{E}_1$, and $x_6 = 0$. Consequently, an induced strain under an electric field $\mathbf{E}_3$ applied along the c-axis (x_3) leads to an elongation in the c-direction ($x_3 = d_{33}.\mathbf{E}_3$, $d_{33} > 0$) but to a contraction in a-direction ($x_1 = x_2 = d_{31}.\mathbf{E}_3$, $d_{31} < 0$). Figure 8.30 shows the induced shear strain under an electric field $\mathbf{E}_1$. In this particular case, $x_5 = d_{15}.\mathbf{E}_1$ with $d_{15} > 0$ and $x_5 > 0$.

Based on the unit cell of α-quartz, the mechanism of the piezoelectric effect can be visualized by a simple semiquantitative model. As a consequence of a mechanical deformation, the silicon and oxygen atoms change their position slightly with respect to each other. However, the distance of directly neighboring atoms remains (almost) constant owing to the strong binding forces as also expressed by the structural approach based on the RUM (rigid unit mode) in silicon framework structures (Dove *et al.*, 2000). As shown in Figure 8.31 (left), in the undeformed state the Si atoms and the O_2 atom pairs are at the same distance to each other on a circle circumscribing the unit cell. Applying a mechanical load in y-direction a displacement dx occurs that deforms the circle to an ellipse (Figure 8.31, right). Note that the original distances r between Si and O atoms are still retained.

As there is an electrical charge compensation in the undeformed state, the polarization $\mathbf{P}$ is zero:

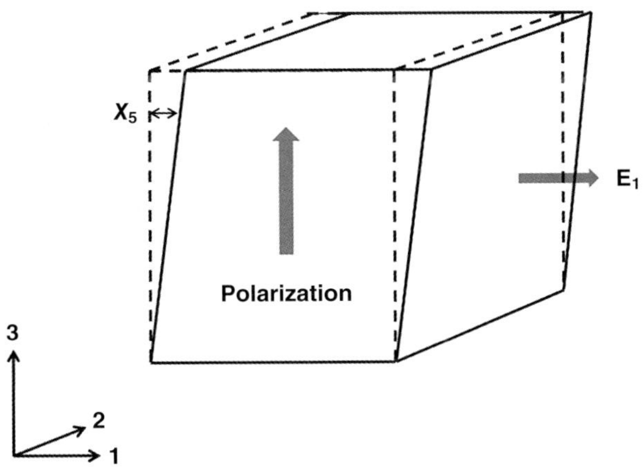

Figure 8.30 Induced shear deformation in the presence of an electric field vector component, E_1.

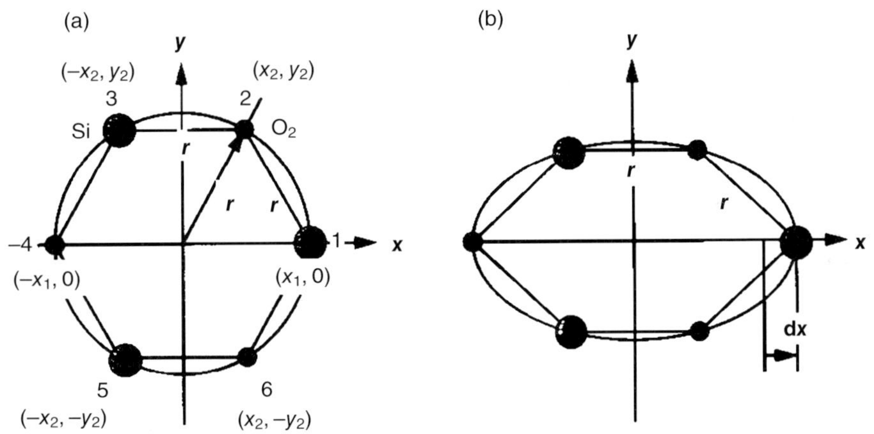

Figure 8.31 Schematic model of the deformation of a unit cell of α-quartz in y-direction. (a) Without mechanical loading; (b) After mechanical loading. For reasons of clarity, the degree of deformation has been grossly exaggerated.

$$\mathbf{p} = \frac{1}{V}\sum_{i=1}^{6} qi \cdot \mathbf{r}_i = \frac{M}{V} = 0, \tag{8.10}$$

where V is the volume of the unit cell, M the dipole momentum, q the electric charge (+4e for Si, −4e for the pair of oxygen atoms), and $\mathbf{r}$ the atomic distance (displacement vector).

The electrical dipole momentum in x-direction M_x can be calculated using the following geometric relations (Nowotny *et al.*, 2008):

$$2x_2 = r,\ (x_1 - x_2)^2 + y_2^2 = r^2 \tag{8.11a}$$

$$dx_2 = 0,\ 2(x_1 - x_2)dx_1 - 2(x_1 - x_2)dx_2 + 2y_2dy_2 = 0,\ dx = dx_1 \tag{8.11b}$$

to yield

$$M_x = 8e \cdot dx. \tag{8.12}$$

As the volume of the unit cell is proportional to the product of r^2 and the height L in z-direction, $V \propto r^2 \cdot L$, and the dilatation in x-direction is $S_{xx} = dx/x = dx_1/x_1 = dx/r$, the polarization can be expressed by:

$$\mathbf{P}_x \propto \frac{8edx}{r^2L} = \frac{8e}{rL}S_{xx}. \tag{8.13a}$$

As polarization and dilatation have the same direction, the effect is called the *longitudinal* piezoelectric effect. Alternatively, since dilation in y-direction and direction of polarization are perpendicular to each other, the effect is called the *transversal* piezoelectric effect, and the polarization can be expressed by:

$$\mathbf{P}_y \propto -\frac{8edx}{r^2L} = -\frac{8e}{rL}S_{xx}. \tag{8.13b}$$

8.5.3
Novel Piezoelectric Single Crystals with CGG Structure

In order to operate devices that utilize piezoelectric materials at elevated temperatures, the absence of any structural or electric phase transitions is mandatory. It is for this reason that why, at high temperature, the "standard" piezoelectric material α-quartz no longer functions because, at 573 °C the piezoelectric α(low)-quartz with symmetry group *32* undergoes a displacive phase transformation to β(high)-quartz with symmetry group *622*.[3] On the other hand, PZT fails to retain its piezoelectric properties at temperatures above 500 °C, because of a ferroelectric–paraelectric phase transformation at the Curie temperature (see Section 8.3.2).

The materials group with calcium gallium germanate (CGG; $Ca_3Ga_2Ge_4O_{14}$) structure is based on ternary oxides such as langasite (LGS; $La_3Ga_5SiO_{14}$) and its homologous niobium ("langanite", LGN) and tantalum ("langataite", LGT) compounds, as well as quaternary oxides such as strontium niobium gallium silicon oxide (SNGS) and strontium tantalum gallium silicon oxide (STGS) (Heimann and Hengst, 2004). Most importantly, these compounds have piezoelectric strain coefficients d that are considerably larger than those of α-quartz, as well as high electromechanical coupling coefficients k^2 (Table 8.8), but do not show any

3) While the piezoelectric tensor of the point symmetry group *622* contains indeed four non-zero components, the longitudinal piezoeffect is missing.

structural phase transition up to their melting points around 1470 °C. These properties provide excellent advanced functional materials for high-temperature applications. For example, since phase-matched optical second harmonic generation (SHG) is also possible in several compounds with CGG structure (such as LGN and LGT), intense studies are currently being conducted to develop materials with optimized properties in terms of a reduced electrical conductivity at high temperature, a smaller anisotropy of thermal expansion, and improved temperature-compensated orientations (Heimann *et al.*, 2003a). However, there is a pronounced increase in the dielectric loss above 350 °C (Stade *et al.*, 2002) that is most likely induced by a significant rise in electric conductivity, and this limits piezoelectric applications at higher temperatures.

8.5.3.1 Structure of CGG Compounds

The crystal structure of langasite and its homologous compounds belongs to the structure type of CGG with point symmetry group (crystal class) *32*, to which α-quartz also belongs. However, in contrast to α-quartz with the enantiomorphic space groups of, respectively $P3_121$ (left-handed) and $P3_221$ (right-handed), the space group of langasite is of lower symmetry, that is, *P312* (Kaminski *et al.*, 1983). This "empty" space group is not enantiomorphic in its own right, and consequently left- and right-handed forms have been identified in the same space group by appropriate asymmetric arrangements of the atoms (Heimann *et al.*, 2003a). Consequently, CGG-type compounds do not form a framework structure akin to quartz with a helical arrangement of SiO_4-tetrahedra along the three-fold screw axes 3_1 or 3_2, but rather a layered structure of GeO_4 tetrahedra in CGG or $(Ga,Si)O_4$ tetrahedra in LGS perpendicular to the c-axis, connected by their vertices (Figure 8.32a and b). These layers are further connected by octahedrally coordinated Ge and Ga ions, respectively as well as Ca and La ions, respectively in eightfold coordination. Consequently, in contrast to α-quartz, the basis planes

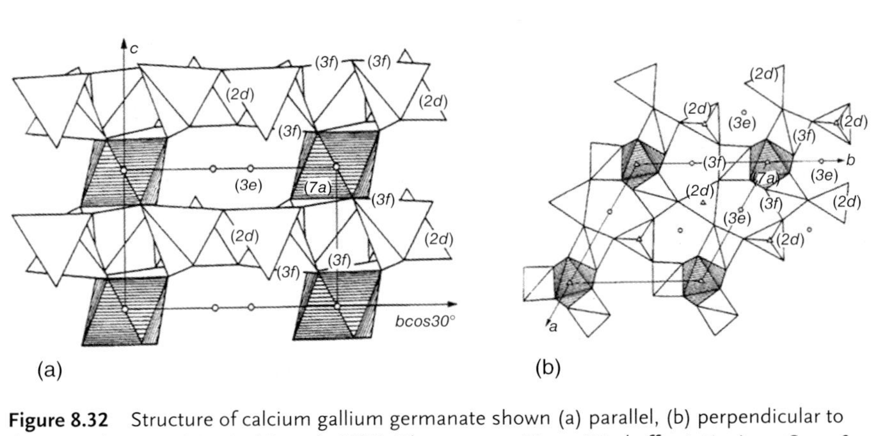

Figure 8.32 Structure of calcium gallium germanate shown (a) parallel, (b) perpendicular to the [00.1] direction (Kaminskii *et al.*, 1983). The atom positions (Wyckoff notation) are Ca → 3e, Ge → 1a, 2d, 1/3 3f, Ga → 2/3 3f. Reprinted with permission from Wiley-VCH Weinheim.

{00.1} ought to be part of the set of equilibrium form faces. Etching experiments carried out with hot 85% orthophosphoric acid on *Z*-cuts of langasite display well-developed triangular and six-sided etch pits, confirming the notion of {00.1} being an F-face in the Hartman–Perdockian sense (Hartman and Perdock, 1955; Heimann, 1975).

The general chemical formula of these compounds is $A_3BC_3D_2O_{14}$. The coordination polyhedron A is a decahedron (coordination number [8], point symmetry group 2, Wyckoff position 3e) and occupied by Ca atoms in CGG, Sr atoms in SNGS, and La atoms in LGS. The coordination polyhedron B is an octahedron (coordination number [6], point symmetry group 32, Wyckoff position 1a) and occupied by Ge in CGG, Nb in SNGS, and Ga in LGS. The coordination polyhedron C is a tetrahedron (coordination number [4], point symmetry group 3, Wyckoff position 3f) and occupied by 1/3 Ge + 2/3 Ga in CGG, and Ga in SNGS and LGS. The final coordination polyhedron D is a (smaller) tetrahedron (coordination number [4], point symmetry group 3, Wyckoff position 2d) and occupied by Ge in CGG, Si in SNGS, and 1/2 Ga + 1/2 Si in LGS. Hence, CGG can be written as $Ca_3Ge[(Ga_2Ge)Ge_2O_{14}]$ and LGS as $La_3Ga[Ga_3(GaSi)O_{14}]$, whereby the square brackets denote the tetrahedrally coordinated part of the structure. The occupation of the cation positions C in CGG are statistically shared among 2/3 Ga and 1/3 Ge ions, and the cation positions D in LGS are statistically shared among 1/2 Ga and 1/2 Si ions (Silvestrova *et al.* 1987). In SNGS, $Sr_3Nb[Ga_3Si_2O_{14}]$ an ordered symmetry exists with all structurally distinct sites occupied by dissimilar atoms (Takeda *et al.*, 2000). This structural difference presumably imparts important differences in the dielectric and elastic properties of the two groups, as shown below.

The ternary and quaternary oxides with CGG structure provide ample opportunities for isoelectronic (aliovalent) substitution of cations. Hence, it is not surprising that to date some sixty different compounds have been identified, all of which belong to the CGG structural family.

8.5.3.2 Czochralski Growth of Single Crystals

Single crystals of LGS, LGN, LGT, SNGS and STGS can today be grown with high crystallographic perfection using the conventional Czochralski technique (Bohm *et al.*, 1999; Figure 8.33). Most of the crystals grown to date have been right-handed. Whilst most crystals boules of LGS, LGN and LGT were grown in *c*(*Z*)-direction, and hence show a symmetry consistent with the trigonal metric, those crystals grown along the electric *X* [10.0]- and *Y* [12.0]-axes are asymmetric, as the (morphological) crystal symmetry is lower in the latter directions compared to the former. The crystals are orange to yellow (LGS, LGN, LGT) or yellow (SNGS, STGS) in color, presumably owing to the incorporation of iridium or platinum impurities from the crucible material, or other elements from the precursor chemicals. Along the circumference of the cylindrical part of the *Z*-grown boules, as well as in their shoulder region, six symmetrical $\{10\bar{1}0\}$ facets appear. In the center of the cylindrical part a triangular core is visible that penetrates the entire crystal parallel to the *c*-direction, and is thought to originate from the concurrent

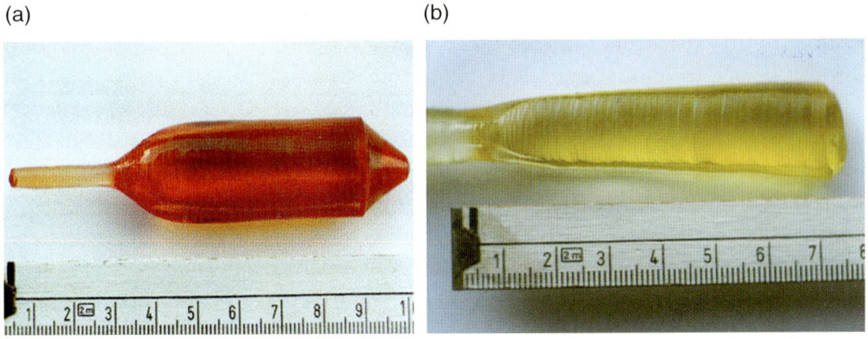

Figure 8.33 Czochralski-grown single crystal boules of (a) langasite (LGS) and (b) strontium niobium gallium silicate (SNGS) (Heimann and Hengst, 2004). Reprinted with permission from Wiley-VCH Weinheim.

growth of a morphologically important F-type (0001) facet at the phase boundary between the melt and the growing crystal.

It is noteworthy to mention that LGS, LGN and LGT can be easily grown along the *Z*-axes, but with much more difficulty along the Cartesian *X*- and *Y*-axes, when SNGS and STGS grow exclusively in the *X*- or *Y*-direction. The reason for this is not yet clear, but may be related to the structural differences in terms of the perfect cation ordering mentioned above.

The as-grown STGS crystals have small {0001} and large $\{2\bar{1}\bar{1}0\}$ facets. Two faces of the latter orientation intersect at the phase boundary (growth front) in a sharp edge along the *c*-(*Z*-) direction. In this way, the crystals grow with an atomically smooth phase boundary, in contrast to the growth mode in the other directions mentioned above of crystals of the CGG or LGS structure family. This exceptional growth behavior in *Y*-direction results in straight growth striations along the entire length of the crystal (Figure 8.33b; Heimann *et al.*, 2003b).

8.5.3.3 Selected Properties of Single Crystals with CGG Structure

Specific Electric Conductivity The specific electric conductivities σ of LGS, LGN and LGT were measured using a four-point technique at temperatures between 795 and 1000 K on crystal slices cut perpendicular to the *Z*-axes. The temperature dependence of the conductivities in dielectric materials with a limited defect density are caused by localized electronic states, and thus the data yield a straight line in a plot of ln σ versus $1/T$, the slope of which is a measure of the activation energy ΔE of the charge carriers in a Mott-type transition model. These activation energies were found to be 1.1 (LGS), 1.0 (LGN) and 0.9 (LGT) eV, respectively. The rather strong increase in electric conductivity (LGS: $10^{-4}\,\Omega^{-1}\text{cm}^{-1}$ at 530 °C, $5 \times 10^{-3}\,\Omega^{-1}\text{cm}^{-1}$ at 730 °C; LGT $10^{-4}\,\Omega^{-1}\text{cm}^{-1}$ at 600 °C, $8 \times 10^{-4}\,\Omega^{-1}\text{cm}^{-1}$ at 730 °C) appears to limit the high-temperature applications of these materials.

Optical Activity Since compounds with CGG structure belong to the acentric symmetry group *32* they display optical activity; that is, rotation of the plane of linearly polarized light. Figure 8.34 shows the specific rotation (° mm^{-1}) of LGS, LGN, LGT, and of SNGS and STGS as a function of the wavelength. The rotation angles increase nonlinearly towards shorter wavelengths, consistent with the effect of normal rotation dispersion, and follows a hyperbolical relationship as expressed by Boltzmann's equation. These rotation angles are particularly large for SNGS and STGS. All crystals investigated rotate clockwise, and hence are right-handed in accordance with the IEEE notation (IEEE Standard on Piezoelectricity, 1988). The curves for LGN and LGT are almost identical. The specific rotation dispersion $\rho(\lambda)$ can be fitted to Boltzmann's equation $\rho(\lambda) = A_1/\lambda^2 + A_2/\lambda^4$ (Table 8.6).

Indices of Refraction The principal refractive indices $n_{(1)}$ $(=n_{(o)})$ and $n_{(3)}$ $(=n_{(e)})$ and their dispersions determined between 365 and 2325 nm wavelengths, can be fitted to a four-parametric Sellmeier equation of the form $n^2(\lambda) = A + B/(\lambda^2\text{-}C) - D\lambda^2$ (Table 8.7).

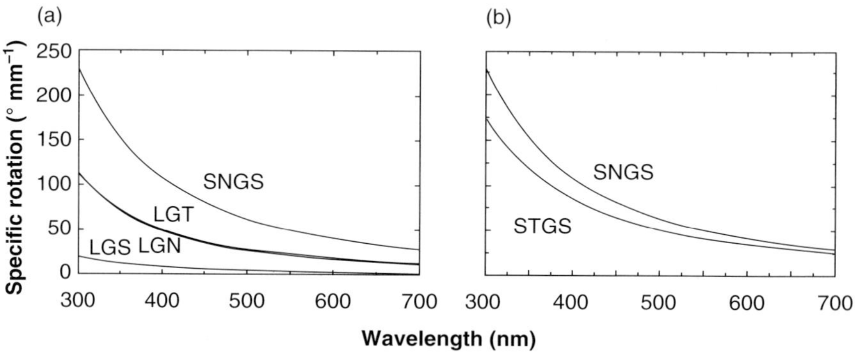

Figure 8.34 Specific rotation of LGS, LGN, LGT and SNGS (a) and SNGS and STGS (b) parallel to the *Z*(*c*)-axis as a function of the wavelength (Heimann *et al.*, 2003a, 2003b). Reprinted with permission from Wiley-VCH Weinheim.

Table 8.6 Boltzmann parameters A_1 and A_2 for LGS, LGN, LGT, SNGS, and STGS (Heimann *et al.*, 2003a, 2003b).

	LGS	LGN	LGT	STGS	SNGS
A_1 [° nm]	0.9491	4.526	5.543	10.863	12.570
A_2 [° nm^3] × 10^{-6}	0.06996	0.5207	0.4136	0.4407	0.7428

Table 8.7 Sellmeier coefficients for LGS, LGN, and LGT (χ^2 = sum of squares of the residuals; Stade et al., 2002).

	LGS		LGN		LGT	
	$n_{(1)}$	$n_{(3)}$	$n_{(1)}$	$n_{(3)}$	$n_{(1)}$	$n_{(3)}$
A	3.5129	3.5558	3.6842	3.7947	3.6487	3.7467
B	0.0383	0.0387	0.0466	0.0508	0.0443	0.0475
C	0.0236	0.0225	0.0297	0.0325	0.0267	0.0272
D	0.0096	0.0098	0.0096	0.0104	0.0094	0.0101
χ^2 ($\times 10^8$)	1.86	1.77	2.00	7.32	3.57	3.48

Piezoelectric Coefficients, Dielectric Permittivities, and Electromechanic Coupling Coefficients As already described in detail in Section 8.4.2 the piezoelectric properties of a non-centrosymmetric crystal may be described by their piezoelectric coefficients (piezoelectric moduli, piezoelectric strain constants), d_{ijk}. These coefficients are components of a third-rank tensor with eight non-vanishing components. The tensor can be written in a 6×3 matrix form with only five non-vanishing components for the trigonal symmetry group *32*. For reasons of symmetry, only two components are independent, denoted usually as d_{11} ($=d_{111}$) and d_{14} ($=2d_{123}$) (Voigt's notation). Thus, the piezoelectric strain tensor of the symmetry group *32* is:

$$\begin{matrix} d_{11} & -d_{11} & 0 & d_{14} & 0 & 0 \\ 0 & 0 & 0 & 0 & -d_{14} & -2d_{11} \\ 0 & 0 & 0 & 0 & 0 & 0 \end{matrix}$$

While d_{11} of LGS, LGN and LGT are more than twice, and d_{14} more than six times that of α-quartz, SNGS behaves differently as the value of d_{14} is very close to that of α-quartz (Table 8.8). On the other hand, the high effective electromechanical coupling coefficient k^2 of SNGS (Chou *et al.*, 2001), which is almost twice as high as those of LGS, LGN or LGT, and five times as high as that of α-quartz, makes the material an interesting substitute of quartz or even lead zirconate titanate (PZT; k^2 typically around 0.5) in applications such as band pass wave filters and transducers. As the coupling coefficient determines the efficiency of conversion of mechanical energy to electrical energy normalized to the input mechanical energy and *vice versa*, the high value of 0.64% can readily pass bands up to approximately 25% of the resonant frequency. The piezoelectric coefficients d_{11} and d_{14} decrease between −200 °C and +250 °C, possibly in relation to the increase in electric conductivity with increasing temperature (Stade *et al.*, 2002). The temperature dependence of the capacitance (and thus the dielectric permittivities) displays anomalous behavior, in particular at low frequencies. The dielectric losses increase strongly at low frequencies beyond about 350 °C; this again suggests an increase of the material's conductivity with temperature.

Table 8.8 Piezoelectric coefficients in p(As)/N, relative dielectric permittivities, and effective electromechanical coupling coefficients in percent of LGS, LGN, LGT (Bohm *et al.*, 2000), STGS (Takeda *et al.*, 2000; Heimann and Hengst, 2004) and SNGS (Heimann and Hengst, 2004; Chilla *et al.*, 2003). Quartz is given for comparison. The signs of d_{ij} conform to the IEEE standard.

	LGS	LGN	LGT	STGS	SNGS	α-Quartz
d_{11}	6.15 ± 0.2	7.41 ± 0.1	7.06 ± 0.1	4.89[a]	6.7 ± 0.2	2.91
d_{14}	−6.01 ± 0.7	−6.16 ± 0.5	−4.32 ± 0.5	–	−0.8 ± 0.1	−0.73
$\varepsilon_{11}/\varepsilon_0$	19.2	20.7	19.6	13.1	12.4	4.53
$\varepsilon_{33}/\varepsilon_0$	50.7	79.0	76.5	17.9	21.9	4.68
k^2_{eff} (%)	0.38	0.38	0.39	–	0.64[b]	0.134

a) Takeda *et al.* (2000).
b) Chou *et al.* (2001).

Elastic Coefficients The elastic stiffness coefficients c_{ij} can be calculated from the measured velocity of propagation of bulk acoustic ultrasonic waves, according to the Papadakis method (quartz transducer with center frequency of 20 MHz) (Papadakis, 1967), on differently oriented bar-shaped samples using the equations given by Truell *et al.* (1969) and corrected for the piezoelectric contributions (Ljamov, 1983; Ikeda, 1990). The samples were oriented in axial directions *XYZ*, and 45° rotated against the *X*- and *Y*-axes, respectively. In order to obtain optimized values for the elastic materials parameters, the elastic stiffness coefficients c_{ij} were used to calculate and critically compare the results of surface acoustic wave (SAW) measurements.

The surface-guided acoustic waves (SAW) can be thermoelastically generated by a laser beam focused by a cylindrical lens onto the surface of the material, and their ultrasonic velocities measured using a piezoelectric foil of poly(vinylidene fluoride). To increase the ultra-violet (UV) absorption efficiency, the samples are coated with a 50 nm Al layer, as they are almost transparent at the laser wavelength chosen. Angular phase velocity dispersion data for SAWs on an *X*-cut and for pseudo ("leaky")-surface acoustic waves (PSAWs) on a *Z*-cut were measured for LGS, LGN, and LGT by Bohm *et al.* (2000) and Chilla *et al.* (2002), and for SNGS by Heimann *et al.* (2002) and Chilla *et al.* (2003).

The elastic stiffness components are considerably smaller for SNGS compared to LGS, LGN and LGT. In particular, the very small tensor component c_{14} for SNGS has been confirmed by several independent measurements. Although the reason for this is not yet clear, it may concern the ordered arrangement of the atoms in the quaternary SNGS structure, as opposed to the statistical structure of the ternary compounds LGS, LGN, and LGT.

Synopsis Single crystals of ternary and quaternary compounds with CGG structures share with the "standard" piezoelectric material α-quartz the acentric symmetry group *32*, but with different space group (*P312* as opposed to $P3_121$ and $P3_221$, respectively). Today, high-quality single crystals of LGS, LGN and LGT, as well as of SNGS and STGS, can be grown using the Czochalski method. The piezoelectric strain constants d_{11} and d_{14} of these compounds are considerably larger than those of α-quartz. Together with moderately high electromechanical coupling coefficients and mechanical *Q*, as well as the absence of a destructive phase transformation between −200 °C and the melting points at about 1470 °C, these compounds represent potential novel materials for sensors and actuators operating at elevated temperatures. The large effective electromechanical coupling coefficient of SNGS points to a very efficient conversion of mechanical energy to electrical energy, and hence suggests an application for narrow band pass filters with low insertion losses (see Section 8.6.1). Likewise, the unusually high values of the specific optical rotation, coupled with a high transparency, high piezoelectric coefficients, electro-optic properties, and the existence of a phase-matched optical SHG in LGN and LGT, suggest unique applications that include the doubling and mixing of frequencies applied to devices operated at elevated temperatures. The angular dispersion of the phase velocities of SAWs (Inoue and Sato, 1998) and a moderate Faraday effect make the investigated members of the CGG family interesting contenders for novel sensors in environmentally challenging applications.

8.6 Electro-Optic Ceramics

Many dielectric crystals are currently applied as optical transducers by utilizing the Faraday effect (rotation of the plane of polarization of light by a magnetic field), the electro- and magnetogyration (change of optic activity by a constant or time-varying electric or magnetic field; Zheludev and Vlokh, 1983), the linear electro-optic (Pockels effect) and the quadratic electro-optic (Kerr) effect, in addition to effects related to frequency multiplication via higher harmonics (Scheuerer *et al.*, 2003). Only the electro-optic effects will be briefly discussed in the following sections.

8.6.1 Linear Electro-Optic (Pockels) Effect

The application of an external electric field deforms the optical indicatrix (Fresnel ellipsoid) of crystals lacking a center of symmetry in such a way that its birefringence is changed. The dependence of the birefringence on the electric field **E** is linear, and can be analytically described by a change of the impermeability tensor $a = (\varepsilon_\infty)^{-1}$ by the electric field **E** [see Eq. (8.4a)] and the polarization **P**, respectively:

$$\Delta a_{ij} = r^{\sigma}_{ijk} \cdot \mathbf{E}_k; \qquad \Delta a_{ij} = m^{\sigma}_{ijk} \cdot \mathbf{P}_k, \tag{8.14}$$

where r^{σ}_{ijk} and m^{σ}_{ijk} are the components of the electro-optic tensor of third rank at constant stress σ ("unclamped"), and $\mathbf{E}_k$ are the components of the electric field.

Since the electro-optic tensor has the same symmetry as the tensor of the inverse piezoelectric effect, the linear electro-optic (Pockels) effect is confined to the symmetry groups in which piezoelectricity occurs (see Table 8.3). The electro-optic coefficients of most dielectric materials are small (of the order of 10^{-12} m V^{-1}), with the notable exception of ferroelectrics such as potassium dihydrogen phosphate (KDP; KH_2PO_4), lithium niobate ($LiNbO_3$), lithium tantalate ($LiTaO_3$), barium sodium niobate ($Ba_2NaNb_5O_{15}$), or strontium barium niobate ($Sr_{0.75}Ba_{0.25}Nb_2O_6$) (Zheludev, 1990). For example, the tensorial matrix of KDP with symmetry group $\bar{4}2m$ has the form

	E_1	E_2	E_3
$\Delta\alpha_1$	0	0	0
$\Delta\alpha_2$	0	0	0
$\Delta\alpha_3$	0	0	0
$\Delta\alpha_4$	r_{14}	0	0
$\Delta\alpha_5$	0	r_{14}	0
$\Delta\alpha_6$	0	0	r_{36}

with $a_{11} = a_1$, $a_{22} = a_2$, $a_{33} = a_3$, $a_{23} = a_{32} = a_4$, $a_{31} = a_{13} = a_5$, $a_{12} = a_{21} = a_6$ (Voigt notation).

The electro-optic coefficients of the piezoelectric α-quartz (point group *32*, S.G. *P3₁21*, *P3₂21*), r_{11} and r_{25} were measured as −0.48 pm V^{-1} and −0.23 pm V^{-1}, respectively (Boháty, 1982b). For the ferroelectric KDP (point group $\bar{4}2m$), the coefficients are considerably larger, with $r_{14} = r_{25} = -8.3$ pm V^{-1} and $r_{36} = r_{123} = 10.2$ pm V^{-1} (Landolt-Börnstein, 1984, 1993) (see Table 8.9). The novel piezoelectric ceramics with CGG-type lattices (point group *32*, S.G. *P312*) were found to have intermediate values of the electro-optic coefficients (see below, Table 8.9).

8.6.2
Quadratic Electro-Optic (Kerr) Effect

This effect is even less pronounced than the linear Pockels effect in crystals. In contrast to the linear electro-optic effect that is confined to crystals lacking a center

Table 8.9 Electrooptic constants (in pm V^{-1}) of LGS, LGN, and LGT (Stade et al., 2002). Related values for α-quartz (Boháty, 1982b) and potassium dihydrogen phosphate (KDP; KH_2PO_4) (Landolt-Börnstein, 1984, 1993) are given for comparison.

	LGS	LGN	LGT	α-Quartz	KDP
r_{111} (r_{11})	−2.68(8)	−2.62(7)	−2.82(9)	−0.48	10.2 (r_{36})
r_{231} (r_{25})	1.22(6)	0.85(5)	0.75(6)	−0.23	−8.3 (r_{14})

of symmetry, the quadratic effect can occur in all symmetry groups, and even in isotropic materials such as liquids and glasses. It can be described by the equation:

$$\Delta aij = \sum\nolimits_{k,l} R_{ijkl} E_k E_l \tag{8.15}$$

where R_{ijkl} is a fourth-rank tensor corresponding in its symmetry to the piezo-optic tensor $\mathbf{q}$ (Kleber, 1998).

8.6.3
Electro-Optic Constants of Piezoelectric Single Crystals of CGG-Type

The linear electro-optic effect (Pockels effect) and its temperature dependence can be investigated using Michelson and Jamin interferometers. The measurements are based on a strategy that stimulates only one of the crystal's vibration directions, and that compensates the change of path difference, induced by an electric field in the measured crystal, by an electric field in a reference crystal positioned in a second laser beam (Boháty, 1982a, 1982b). In addition, relative changes of path difference caused by the applied voltage and the temperature can be measured using a Sénarmont compensator (Wood and Glazer, 1980). Table 8.9 lists the room-temperature values of the two independent electro-optical constants r_{111} and r_{231} (renumbered according to Voigt's notation r_{11} and r_{25}, respectively). The former is a normal component of the tensor; the latter is a shear component.

8.6.4
Electro-Optic Lead Lanthanum Zirconate Titanate (PLZT)

Functional electro-optic ceramics can be obtained by the partial substitution of lead by lanthanum in PZT. The resultant lead lanthanum zirconate titanate (PLZT) was developed in 1971 by Sandia National Laboratory, and has the general formula $[Pb_{1-x}La_x][(Zr_yTi_z)_{1-x/4}]O_3$ ($y + z = 1$). The ceramic is highly transparent with very low light scattering when hot-pressed, and shows a high optical anisotropy (birefringency) in an electric field, and a fast switching time in the range of microseconds. For technical applications, depending on the composition of the solid solution, three major types of electro-optic characteristics may be exploited: (i) memory characteristics (composition $x/y/z = 0.8/0.65/0.35$); (ii) the linear electro-optic (Pockels) effect (composition $x/y/z = 0.8/0.60/0.40$); and (iii) the quadratic electro-optic (Kerr) effect (composition $x/y/z = 0.95/0.65/0.35$) (Moulson and Herbert, 2003).

The fabrication of PLZT is achieved in four general ways: (i) hot pressing at 1000–1300 °C and 20–60 MPa; (ii) pressureless sintering in a PbO atmosphere to account for evaporation losses of lead; (iii) hot isostatic pressing; and (iv) multistage sintering.

The working principle of devices based on the nonlinear optical properties of PLZT lays in the change of birefringence in response to an external electric field. By modification of the optical indicatrix (i.e., the indices of refraction) the optical path lengths will be altered, such that the electronically triggered phase modulation of light is possible. As this phase modulation is different for the two waves

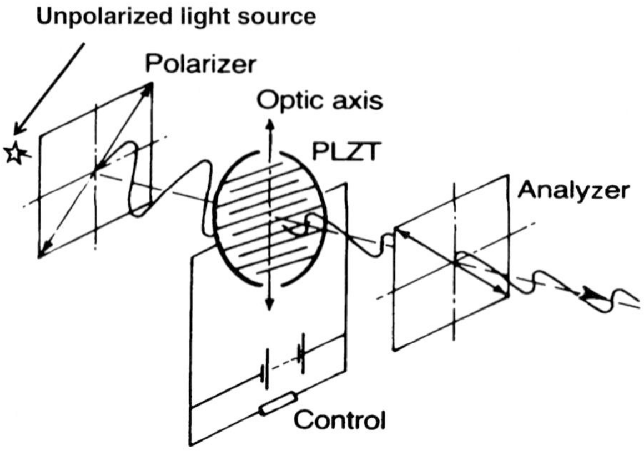

Figure 8.35 Principle of an optical shutter based on an electro-optic PLZT ceramic plate (Moulson and Herbert, 2003). Reprinted with permission from Wiley-Blackwell Ltd, Oxford, UK.

propagating through a birefringent crystal, their interference in devices with crossed polars gives rise to an associated amplitude modulation (Kaminow and Turner, 1966).

Hence, the application of PLZT in electro-optic devices such as in picture storage and display systems, including ferroelectric ceramic picture devices (FERPICs) in double-refraction mode, optical shutters for SEM stereo cameras, optical modulators, antiglare lenses, flash goggles (Figure 8.35) to protect the eyes of pilots of military aircraft from the light of a nuclear flash, light control in welding monitors, medical television stereo-viewing systems, optical memories for picture storage, stereo observation of the ocean floor, and transparent piezoelectric speakers.

Figure 8.35 shows the working principle of an optical shutter for flash goggles. In normal vision mode, the light from the polarizer incident on the PLZT plate has its plane of polarization rotated through 90° by the birefringence induced by a voltage of 800 V. As a result, the emergent light penetrates the analyzer and enters the eye of the pilot. Under flash conditions, a photodiode detects the flash and switches off the voltage. As the plane of polarization is now no longer rotated, the analyzer will stop most of the light and thus protect the eye from being burned.

8.7 Selected Applications of Electroceramic Materials and Devices

8.7.1 Microwave Resonators

Today, microwave dielectric ceramics used in telecommunications, as well as in cable communications using optical fibers, are undergoing rapid development. In the past, microwave communications were used primarily for military purposes such as radar, weapons guidance systems and satellite communications. However, more recently microwaves have been used extensively in mobile communication devices such as mobile radios and telephones, and in satellite broadcasting. In general, these communications devices contain a microwave circuitry unit that is composed of semiconductor components and resonator system components. Whilst much progress has been made in the field of semiconductors by using large bandgap materials such as gallium arsenide and tailored multilayer quaternary gallium-indium-arsenic-antimony semiconductors as resonators, dielectric ceramics are now being used with excellent microwave properties that are characterized by high output, low loss, and high selectivity in the 800–950 MHz band range, rather than conventional metal cavity devices based on standing electric waves produced in their interior. These systems include mobile, portable and cordless telephones, as well as satellite broadcast receivers in the superhigh-frequency (SHF) band at frequencies exceeding 10 GHz. Microwave dielectric ceramics are also used as resonators of bandpass filters for transmitting and receiving ultrahigh-frequency (UHF) signals in the 300 MHz to 3 GHz range, and SHF signals in the 3 GHz to 30 GHz range.

8.7.1.1 Types and Characteristics of Resonators

Figure 8.36 shows three types of operating system, namely solid dielectric resonators, coaxial resonators (hollow metal-plated ceramic cylinders), and stripline resonators on alumina or polymer substrates.

The three most important performance parameters are the insertion loss (i.e., the power dissipated by the bandpass filter), the bandwidth, and the attenuation.

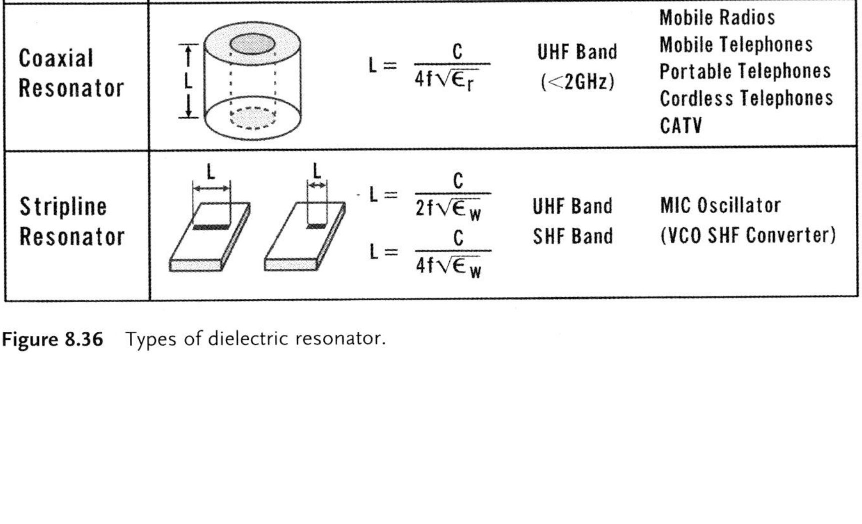

Figure 8.36 Types of dielectric resonator.

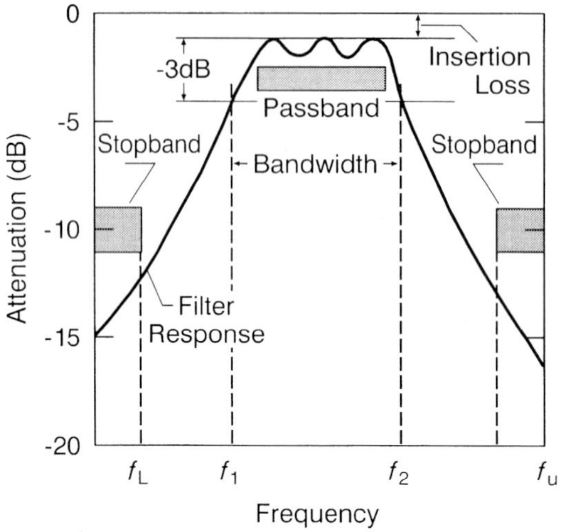

Figure 8.37 Idealized characteristic of a microwave bandpass filter.

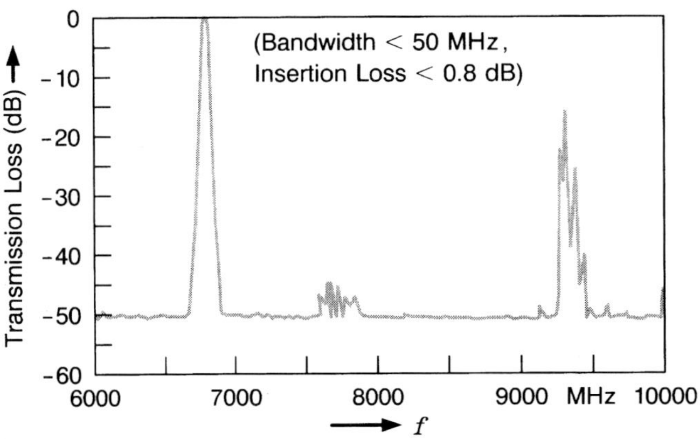

Figure 8.38 Forward characteristic of a typical dielectric resonator in the SHF band.

Both the bandwidth and attenuation are functions of the number of individual resonators in the device and their coupling, but not of the ceramic materials from which they are made, nor of the electrode coating. An idealized summary of these parameters is shown in Figure 8.37.

Figure 8.38 shows the characteristics of a typical dielectric resonator made from a $BaTiO_3$-based ceramic materials with $\varepsilon = 38$, coupled via air lines and operating in the SHF band. The small bandwidth (<50 MHz) and insertion losses (<0.8 dB) guarantee excellent frequency separation.

8.7.2
Examples of Applications of Ferroelectric Ceramic Materials

- **Multilayer capacitors (MLCs):** The volumetric efficiency (i.e., capacitance per unit volume) can be enhanced using an MLC structure with alternating layers of high dielectric constant materials such as BT and an electrically conducting electrode material. By tape-casting, a layer thickness below 20 μm can be obtained. Even better results in terms of volumetric efficiency can be achieved using ferroelectric relaxor ceramics such as PMN, which exceeds BT in terms of the value of ε and shows a much lower grain size. Both properties combine to obtain MLC devices with a high mechanical Q and low-temperature coefficient of the dielectric constant. However, there are some inherent problems in applying lead-based relaxor ceramics: the dielectric losses are highest just below the Curie point, and the permittivity is highly frequency-dependent in this temperature range. Also, lead losses by evaporation during sintering shift the structure of the PMN to that of a pyrochlore phase with low ε and, in addition, cause problematic environmental pollution.
- **Nonvolatile ferroelectric random access memory (FRAM) devices:** Dynamic random access memory (DRAM) and static random access memory (SRAM) devices based on semiconductor technology are volatile; that is, they will lose stored information when the power fails. Nonvolatile devices such as CMOS (complementary metal oxide semiconductors) and EEPROMs (electrically erasable read-only memories) are forbiddingly expensive for mass-produced electronic devices. As described above (see Section 8.3), the magnitude and direction of polarization of a ferroelectric ceramics can be reversed by applying an external electric field, and this method is used by FRAMs to store (or erase) data. As the materials have a nonlinear hysteresis curve, the polarization remains in the same state when the voltage is switched off (i.e., the information originally stored is maintained). In addition, FRAMs may be radiation-hardened for use in harsh environments such as outer space (Scott and Paz de Araujo, 1989).
- **Ferroelectric thin film application:** Ferroelectric thin films for electro-optic applications require optically highly transparent material with a high degree of crystallinity. If the propagation of light is along the plane of the film, the device can be used as an optical wave guide, whereas if the light passes through the film, then optical memories and display devices can be constructed. Hence, such a device might replace the use of bulk PLZT ceramics.
- **Other applications:** Other applications of thin films include pyroelectric detectors and surface acoustic wave (SAW) substrates. The latter devices consist of a piezoelectric substrate onto which interdigital electrodes are deposited, for example by screen printing. An elastic wave generated at the input interdigital transducer (IDT) travels along the surface and is detected by the output interdigital transducer (OIT). These devices are mainly used as delay lines and filters in microwave and television communications (see also Chilla *et al.*, 2003).

8.7.3
Examples of Applications of Piezoelectric Ceramic Materials

Piezoelectric ceramics can be found in numerous devices, ranging from such mundane applications as gas lighters to sophisticated transducers, actuators and sensors for domestic, industrial, and military uses (Uchino, 1996). A contemporary account of the history and the physics of the piezoelectric effect, including its modern applications, can be found in Heywang *et al.* (2009). Here, only a few applications will be listed.

- **Medical ultrasonic applications:** The principle of ultrasonic imaging is based on the pulse echo mode of operation. Piezoelectrics can be operated in an active mode to transmit sound, and in a passive mode as a sound receiver in which sound energy is converted into an electrical signal. In the pulse echo mode the materials performs both functions simultaneously. A sound wave is propagated into the medium (e.g., human tissue) and a faint echo is received back after a short delay which is caused by the acoustic impedance mismatch at the interface of the two dissimilar materials. This reflected echo generates a voltage signal that is used to produce the image of an internal organ. For more detail, see the reviews by Christensen (1988) or Safari *et al.* (1996).
- **Gas igniters:** The construction principles relies on two oppositely poled cylinders of a piezoelectric ceramics, generally PZT attached end-to-end so as to double the charge available to generate a spark (Figure 8.39). A compressive force is quickly applied via a spring-loaded anvil to avoid leakage of charge. The voltage V generated is given by the equation:

  $$V = g_{33} \cdot (L/A) \cdot \mathbf{X}, \tag{8.16}$$

  where $g_{33} = -(\delta \mathbf{E}/\delta \mathbf{X})$ (measured in $\mathrm{V \cdot m\,N^{-1}}$) is the piezoelectric voltage constant, L is the length of the ceramic cylinder (in meters), A is the circular surface area struck (in m^2), and $\mathbf{X}$ is the applied stress (force; in N). The voltage of the spark can reach several tens of kilovolts.

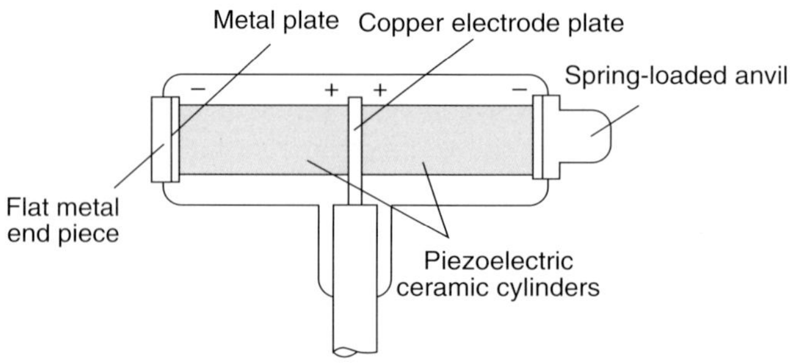

Figure 8.39 Piezoelectric ceramic gas igniter.

- **Displacement transducers and actuators:** The application of a force to a long piezoelectric cantilever beam causes it to bend, with one side of the beam in compression and the other side in tension. Hence, the piezoelectric effects cancel out and no net electrical output is obtained. However, cantilever bimorphs made from two halves of separate beams poled in opposite directions with an electrode in between (parallel configuration), as well as on the top and bottom surfaces (series configuration), generate a net voltage output when a force is applied. On the other hand, the application of a voltage to the bimorph generates a bending momentum and turns the device into an actuator for a variety of applications requiring comparatively large displacements with low forces. Examples include optical beam deflectors, printer hammers, loudspeakers, and video recording heads (Jones *et al.*, 1996).

- Novel Reduced and Internally Biased Oxide Wafer (RAINBOW) actuators (Haertling, 1994) consist of a piezoelectric layer and a chemically reduced layer created at high temperature. The bilayer system is thermally stressed, and develops on cooling a dome-shaped distortion. On the application of a voltage, a relatively large change in dome height can be attained. Such a device can be used in medical micropumps, variable focus mirrors, speakers, and microswitches. A further development has been reported concerning a miniature flextensional device consisting of a piezoelectric disc sandwiched between two crescent-shaped metallic end caps ("moonies") (Xu *et al.*, 1991). For further information on ceramic actuators, see Uchino (1993). It has been recognized that in modern multilayer piezoelectric actuators (MPAs), the combination of thermal, electrical and mechanical loads during service may affect the functional integrity of the devices. As the details of these effects and their synergistic coupling are still unknown, modeling of the nonlinear behavior of these temperature-sensitive functional properties and their implementation into finite element analysis (FEA) tools has been performed recently (Grünbichler *et al.*, 2008).

- **Accelerometers:** Accelerometers create an electrical output that is proportional to the acceleration. Several configurations are available commercially, including compression, shear, and flexural beam accelerometers. In the case of a shear mode accelerometer, the configuration consists of a cylinder of piezoelectric ceramic material joined to a fixed central post on the inside and a toroid mass on the outside (Figure 8.40). In response to an axial acceleration, the cylinder is subjected to a shear force between the outer mass and the inner post. However, any motion in the radial direction has no effect as the d_{11} of PZT is zero. Consequently, the device is highly directional and can be easily miniaturized.

- **Piezoelectric transformers:** A flat piezoelectric plate carries electrodes on half of its larger surface (drive section) and on an edge (generating section) (Figure 8.41). The regions between the larger face electrodes and the edge electrode are poled individually, so that their directions of polarization are perpendicular to each other. If an input a.c. voltage V_i with a specific resonant frequency

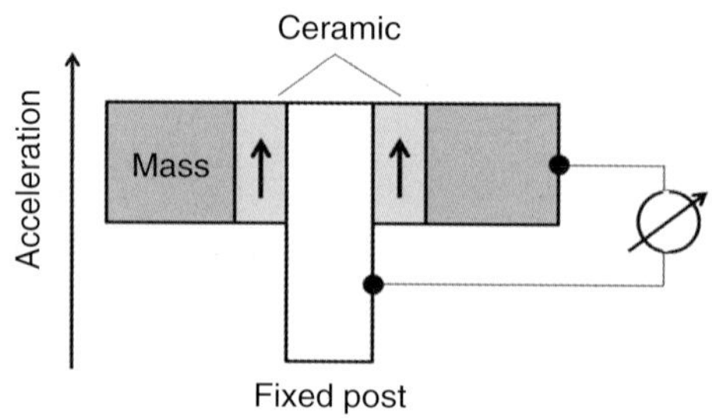

Figure 8.40 Shear mode accelerometer.

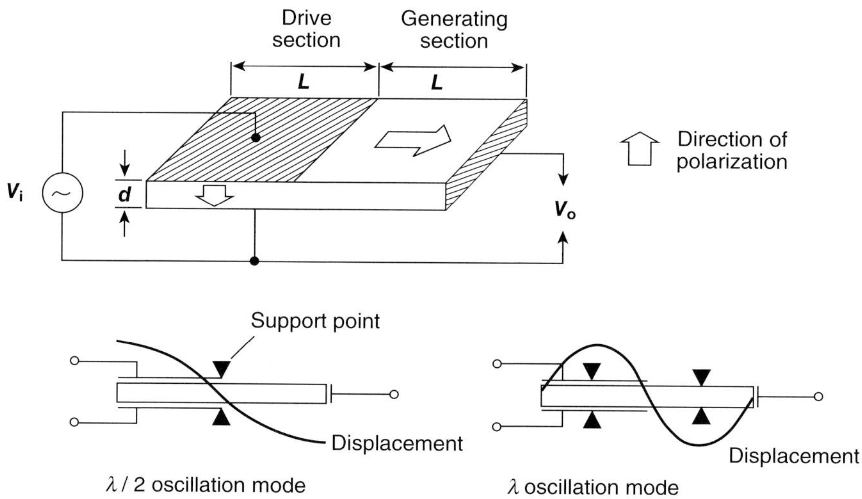

Figure 8.41 Working principle of a piezoelectric transformer (after Ichinose, 1987). Reprinted with permission from Wiley-Blackwell Ltd, Oxford, UK.

determined by the length $2L$ of the driver section is applied, a mechanical vibration is induced lengthwise by the inverse piezoelectric effect or the electrostrictive effect. This vibration triggers a high-voltage output V_o in the generating section by the piezoelectric effect. Hence, the voltage step-up process occurs through an electrical–mechanical–electrical conversion sequence. The step-up ratio, G, is given by the equation

$$G = (V_o/V_i) = (4/\pi^2)\cdot Q_m \cdot k_{31} \cdot k_{33} \cdot (L/d), \tag{8.17}$$

where Q_m is the mechanical quality factor, L is half of the length of the plate, d is the plate thickness, and k_{31} and k_{33} are the components of the electromechanical coupling coefficient. The step-up function works at either the λ or the $\lambda/2$ oscillation mode. The ratio G is proportional to the ratio of the input to output capacitance and the efficiency of conversion of mechanical to electrical energy. This principle of stepping up the voltage is currently applied in high-voltage extra-high-tension (EHT) transformers for miniature television receivers.

- **Other applications:** Today, many more devices are being developed that employ direct and converse piezoelectric effects, including buzzers for cellular telephones, high-power transducers for ultrasonic machining and cleaning, and medical electronic components such as the so-called "Lipotriptor," which is used to destroy kidney and urinary calculi by high-frequency oscillation. Typical other applications include (Uchino, 1995; Jones *et al.*, 1996):
  - Hydrophones
  - Microphones
  - Underwater transducers
  - Air transducers
  - Electrical wave filters
  - Ink jet/dot matrix printers
  - Sonar and fish finders
  - Delay line transducers
  - Surface acoustic wave (SAW) devices
  - Smart actuators for active shape control, active vibration control, active noise control
  - Load cells for force and pressure measurement
  - Piezoelectric motors (inchworm, standing or propagating wave modes, vibratory coupler ("woodpecker") or surface wave ("surfing")-type motors)
  - Voltage-driving and photo-driving bimorphs (walking machines, photophones)
  - Laser beam scanner
  - Precision positioners for VCR tape heads and fly-cutting machines
  - Activators for multilayer deformable mirrors for astrophysical observation
  - Crash sensors to release airbags in passenger cars
  - Micro-angle adjusting devices
  - Automatic camera focusing systems and camera shutters
  - Sensors for burglar alarm
  - Antiknocking sensors
  - Bulk wave crystal sensors (QCM)

Additional information on piezoelectric materials and their applications can be found in Uchino (1996), Sirisoonthorn (1997), and Moulson and Herbert (2003). For a very recent account, see Heywang *et al.* (2009).

8.7.4
Reliability of Devices

Apart from the mechanical damage of the inherently brittle dielectric ceramic materials, several breakdown mechanisms have been identified that tend to cause a deterioration in electrical performance.

Intrinsic breakdown starts when the small number of electrons in the conduction band gain kinetic energy sufficient to ionize the constituent ions. As a consequence, the number of electrons is increased such that, eventually, an "electron avalanche" forms in the materials, followed by complete failure. The intrinsic breakdown strengths of dielectric materials are on the order of $100\,\mathrm{MV\,m^{-1}}$.

Thermal breakdown is caused by the fact that d.c. conductivity results in Joule heating. Under an a.c. field, there is additional energy dissipation, with heat being generated in the dielectric materials faster than it can be dissipated to the surroundings. The subsequent rise in temperature will lead to an increase in conductivity and dielectric loss, which eventually culminates in a runaway situation and thermal breakdown. The breakdown voltage, U_b, is proportional to the thermal conductivity of the materials, λ, the function φ (which is dependent on the thickness of a flat disc and the heat transfer coefficient), and is inversely proportional to the angular frequency of the a.c. field ω, the temperature coefficient of the loss factor τ, the dielectric permittivity ε, and the loss factor tan δ:

$$U_b = \sqrt{\frac{\lambda}{\varpi \cdot \tau \cdot \varepsilon \cdot \tan\delta}\varphi} \tag{8.18}$$

Finally, *discharge breakdown* may occur when gas discharges occur in the pores of a dielectric material. The electric strength of the ceramics depends upon the specimen and pore sizes; in other words, larger pores will increase the probability of failure as the electric field in a pore, $\mathbf{E}_{\mathrm{Pore}}$ is greater than that in the surrounding dielectric $\mathbf{E}$ by

$$\mathbf{E}_{\mathrm{Pore}} = \frac{\varepsilon_r}{\varepsilon_{r(pore)}}\mathbf{E}, \tag{8.19}$$

where ε_r is the relative dielectric permittivity. For the limiting case of a spherical pore, $\mathbf{E}_{\mathrm{Pore}} = 3/2 \cdot \mathbf{E}$. For example, in alumina ceramics the electric strength decreases with increasing thickness, but increases with density.

References

Anliker, M., Känzig, W., and Peters, M. (1952) Das Verhalten sehr kleiner ferroelektrischer Teilchen. *Helv. Phys. Acta*, **25**, 474–475.

Aurivillius, B. (1949) Mixed bismuth oxides with layered lattices. *Ark. Kemi*, **1**, 463–480.

Ballato, A. (2009) Basic material quartz and related innovations, in *Piezoelectricity: Evolution and Future of a Technology* (eds W. Heywang, K. Lubitz, and W. Wersing), Springer, Berlin, pp. 9–36.

Barbier, B., Combettes, C., Guillemet-Fritsch, S., Chartier, T., Rossignol, F., Rumeau, A., Lebey, T., and Dutarde, E. (2008) $CaCu_3Ti_4O_{12}$ ceramics from co-precipitation method: dielectric properties of pellets and thick films. *J. Eur. Ceram. Soc.*, **29**, 731–735.

Bärnighaus, H. (1979) Group-subgroup relations between space groups: a useful tool in crystal chemistry. *Proceedings, Conference "Kristallographische Gruppen", University of Bielefeld, Germany, 2–16 September, 1979*, pp. 139–175.

Blake, S.M., Falconer, M.J., McCreedy, M., and Lightfoot, P. (1997) Cation disorder in ferroelectric Aurivillius phases of the type

$Bi_2ANb_2O_9$ (A = Ba,Sr,Ca). *J. Mater. Chem.*, **7** (8), 1609–1613.

Bochenek, D., Surowiak, Z., and Poltierova-Vejpravova, J. (2009) Producing the lead-free $BaFe_{0.5}Nb_{0.5}O_3$ ceramics with multiferroic properties. *J. Alloys Comp.* **487** (1–2), 572–576.

Bohatý, L. (1982a) Dynamisches Verfahren zur Messung von elektrostriktiven und elektrooptischen Effekten: Tinkalkonit $Na_2B_4O_5(OH)_4 \cdot 3H_2O$. *Z. Kristallogr.*, **158**, 233–239.

Bohatý, L. (1982b) Der lineare elektrooptische Effekt von alpha-Quarz: eine Neubestimm-ung. *Z. Kristallogr.*, **161**, 299–302.

Bohm, J. and Kürten, H.D. (1973) Domänen II. Art in ferroelektrischem Gadoliniummolybdat. *Phys. Status Solidi A*, **19**, 179–183.

Bohm, J., Heimann, R.B., Hengst, M., Roewer, R., and Schindler, J. (1999) Czochralski growth and characterization of piezoelectric single crystals with langasite structure: $La_3Ga_5SiO_{14}$ (LGS), $La_3Ga_{5.5}Nb_{0.5}O_{14}$ (LGN) and $La_3Ga_{5.5}Ta_{0.5}O_{14}$ (LGT). Part I. *J. Cryst. Growth*, **204**, 128–136.

Bohm, J., Chilla, E., Flannery, C., Fröhlich, H.-J., Hauke, T., Heimann, R.B., Hengst, M., and Straube, U. (2000) Czochralski growth and characterization of piezoelectric single crystals with langasite structure: $La_3Ga_5SiO_{14}$ (LGS), $La_3Ga_{5.5}Nb_{0.5}O_{14}$ (LGN) and $La_3Ga_{5.5}Ta_{0.5}O_{14}$ (LGT). Part II. Piezoelectric and elastic properties. *J. Cryst. Growth*, **216**, 293–298.

Bradaczek, H., Hildebrandt, G., Kuhr, A., and Nestler, B. (1990) Automatic sorting quartz oscillators. *Acta Crystallogr. A*, **46**, PS-01.04.30.

Carpenter, M.A. (1992) Thermodynamics of phase transitions in minerals: a macroscopic approach, in *The Stability of Minerals* (eds G.D. Price and N.L. Ross), Chapman & Hall, London, ch. 5, pp. 172–215.

Chilla, E., Flannery, C.M., Fröhlich, H.-J., Bohm, J., Heimann, R.B., Hengst, M., and Straube, U. (2002) Elastic constants of langasite-type crystals determined by bulk and surface guided acoustic modes. *Proceedings, IEEE Ultrasonic Symposium, Munich, Germany*, pp. 364–367.

Chilla, E., Kunze, R., Weihnacht, M., Bohm, J., Heimann, R.B., Hengst, M., and Straube, U. (2003) Acoustic waves measurements on SNGS crystals and determination of material constants. *Proceedings, IEEE Ultrasonic Symposium, 5–8 October 2003, Honolulu, Hawaii, 5A-1.*

Chou, M.M.C., Jen, S., and Chai, B.H.T. (2001) Investigation of crystal growth and materials constants of ordered langasite structure. *Proceedings, IEEE Ultrasonic Symposium 2001, Atlanta, GA, USA*, vol. 1, p. 225.

Choudhury, N., Walter, E.J., Kolesnikov, A.I., and Loong, C.K. (2008) Large phonon band-gap in $SrTiO_3$ and the vibrational signatures of ferroelectricity in $ATiO_3$ perovskites: First principles lattice dynamics and inelastic scattering of $PbTiO_3$, $BaTiO_3$ and $SrTiO_3$. *Phys. Rev.*, **B77**, 134111 (10 pages).

Christensen, D.A. (1988) *Ultrasonic Bioinstrumentation*, John Wiley & Sons, Inc., New York.

Cockbain, A.G. and Harrop, P.J. (1968) The temperature coefficient of capacitance. *J. Phys. D: Appl. Phys.*, **1**, 1109–1115.

Cross, L.E. (1993) *Ferroelectric Ceramics: Tutorial Reviews, Theory, Processing and Applications*, Birkhauser, Basel.

Cross, L.E. and Newnham, R.E. (1987) History of ferroelectrics, in *Ceramics and Civilization*, vol. **III** (ed. W.J. Kingery), The American Ceramic Society, Ohio, pp. 289–395.

Curie, J. and Curie, P. (1880) Développement, par pression, de l'électricité polaire dans les cristaux hémièdres à faces inclinées; also Sur l'électricité polaire dans les cristaux hémièdres à faces inclinées. *Comp. Rend. Acad. Sci. Fr.*, **XCI**, 294–295, 383–386.

Dalal, N. and Bussmann-Holder, A. (eds) (2007) *Ferro- and Antiferroelectricity. Order/Disorder Versus Displacive.* Structure and Bonding, vol. **124**, Springer, ISBN: 978-3-540-49602-1.

De Cicco, G., Morten, B., Dalmonego, D., and Prudenziati, M. (1999) Pyroelectricity of PZT-based thick films. *Sens. Actuators A*, **76**, 409–415.

Dove, M.T., Trachenko, K.O., and Tucker, M.G. (2000) Rigid unit modes in framework structures: theory, experiment and application, in *Transformation Processes in Minerals.* Reviews in Mineralogy and Geochemistry, vol. **39** (eds S.A.T. Redfern and M.A. Carpenter), Mineralogical Society of America, Washington, DC, ch. 1, pp. 1–33.

Dry, C.M. (1993) Passive smart materials for sensing and actuation. *J. Intell. Mater. Syst. Struct.*, **4** (3), 420–425.

Goldman, A. (2006) *Modern Ferrite Technology*, 2nd edn, Springer, ISBN-10: 0-387-28151-7.

von Groth, P. (1870) Über Beziehungen zwischen Krystallform und chemischer Constitution bei einigen organischen Verbindungen. *Ber. Dtsch. Chem. Ges.*, **3**, 449–457.

Grünbichler, H., Kreith, J., Bermejo, R., Supancic, P., and Danzer, R. (2008) Modelling of the ferroic material behaviour of piezoelectrics: characterisation of temperature-sensitive functional properties, in *Electroceramics XI* (ed. R. Freer), The University of Manchester, pp. 1–7.

Gscheidner, K.A., Jr (2008) The magnetocaloric effect, magnetic refrigeration and ductile intermetallic compounds. *Acta Mater.*, **57** (1), 18–28.

Haertling, G.H. (1994) Rainbow ceramics–a new type of ultra-high-displacement actuator. *Am. Ceram. Soc. Bull.*, **73** (1), 93–96.

Hartman, P. and Perdock, W. (1955) On the relations between structure and morphology of crystals. *Acta Crystallogr.*, **8**, 49–52.

Heaney, P.J. (2000) Phase transformations induced by solid solution, in *Transformation Processes in Minerals. Reviews in Mineralogy and Geochemistry*, vol. **39** (eds S.A.T. Redfern and M.A. Carpenter), Mineralogical Society of America, Washington, DC, pp. 135–174.

Heimann, R.B. (1975) *Auflösung von Kristallen*, vol. **8** (eds V.D. Fréchette, H. Kirsch, L.B. Sand, and F. Trojer), Applied Mineralogy, Springer-Verlag, Wien, New York, ISBN: 3-211-81278-4, 270 pp.

Heimann, R.B. and Hengst, M. (2004) Czochralski grown single crystals with acentric symmetry group 32. *Adv. Eng. Mater.*, **6** (7), 588–594.

Heimann, R.B., Hengst, M., Rossberg, M., and Bohm, J. (2003a) Giant optical rotation in piezoelectric crystals with calcium gallium germanate structure. *Phys. Status Solidi A*, **195** (2), 468–474.

Heimann, R.B., Hengst, M., Rossberg, M., and Bohm, J. (2003b) Growth and optical activity of strontium tantalum gallium silicon oxide ($Sr_3TaGa_3Si_2O_{14}$, STGS). *Phys. Status Solidi A*, **198** (2), 415–419.

Herabut, A. and Safari, A. (1997) Processing and electromechanical properties of $(Bi_{0.5}Na_{0.5})_{(1-1.5x)}La_xTiO_3$ ceramics. *J. Am. Ceram. Soc.*, **80** (11), 2954–2958.

Heywang, W. (1951) Zur wirksamen Feldstärke im kubischen Gitter. *Z. Naturforsch.*, **6a**, 219–220.

Heywang, W. (1965) Ferroelektrizität in perowskitischen Systemen und ihre technischen Anwendungen. *Z. Angew. Phys.*, **19**, 473–481.

Heywang, W., Lubitz, K., and Wersing, W. (eds) (2009) *Piezoelectricity: Evolution and Future of a Technology*, Springer, Berlin.

Hosono, Y., Harada, K., and Yamashita, Y. (2001) Crystal growth and electrical properties of lead-free piezoelectric material. *Jpn. J. Appl. Phys.*, **40**, 5722–5726.

Ichinose, N. (ed.) (1987) *Introduction to Fine Ceramics*, John Wiley & Sons, Ltd, Chichester, New York, Brisbane, Toronto, Singapore, ISBN: 0-471-91445-2, 160 pp.

IEEE Standard on Piezoelectricity (1988) 176-1987. ANSI/IEEE Standard 176-1987, Institute of Electrical and Electronic Engineers, New York, pp. 1–53.

Ikeda, T. (1990) *Fundamentals of Piezoelectricity*, Oxford Science Publications, Oxford, UK.

Inoue, K. and Sato, K. (1998) Propagation characteristics of surface acoustic waves on langasite. *Jpn. J. Appl. Phys. Part 1*, **37**, 2909–2913.

Itoh, M., Wang, R., Inaguma, Y., Yamaguchi, T., Shan, Y.J., and Nakamura, T. (1999) Ferroelectricity induced by oxygen isotope exchange in strontium titanate perovskite. *Phys. Rev. Lett.*, **82**, 3540–3543.

Janas, V.F. and Safari, A. (1995) Overview of fine-scale piezoelectric ceramic/polymer composite processing. *J. Am. Ceram. Soc.*, **78** (11), 2945–2955.

Jones, D.J., Prasad, S.E., and Wallace, J.B. (1996) Piezoelectric materials and their applications. *Key Eng. Mater.*, **122–124**, 71–144.

Jones, J.L. (2010) The effect of crystal symmetry on the maximum polarization of polycrystalline ferroelectric materials. *Mater. Sci. Eng. B*, **167** (1), 6–11.

Kaminow, J.P. and Turner, E.H. (1966) Electrooptic light modulators. *Proc. IEEE*, **54**, 1374–1612.

Kaminskii, A.A., Mill, B.V., Khodzhabagyan, G.G., Konstantinova, A.F., Okorochkov, A.I., and Silvestrova, I.M. (1983) Investigation of trigonal $(La_{1-x}Nd_x)Ga_5O_{14}$ crystals. I. Growth and properties. *Phys. Status Solidi A*, **80**, 387–398.

Känzig, W. (1951) Röntgenuntersuchungen über die Seignetteelektrizität von Bariumtitanat. *Helv. Phys. Acta*, **24**, 175–216.

Kawashima, S., Nishida, M., Ueda, I., and Ouchi, H. (1983) $Ba(Zn_{1/3}Ta_{2/3})O_3$ ceramics with low dielectric loss at microwave frequencies. *J. Am. Ceram. Soc.*, **66** (6), 421–423.

Kawashima, S., Nishida, S., Ueda, I., Ouchi, H. (1985) Dielectric properties at microwave frequencies of the ceramics in $BaO\text{-}Sm_2O_3\text{-}TiO_2$ system. *Ext. Abstracts, 87th Annual Meeting of the American Ceramic Society, Cincinnati, Electronic Division*. Paper No. 15-E-85.

Kay, H.F. and Vousden, P. (1949) Symmetry changes in barium titanate at low temperature and their relation to its ferroelectric properties. *Philos. Mag.*, Series 7, **40** (309), 1019–1040.

Kelly, K. (1994) *Out of Control. The New Biology of Machines, Social Systems, and the Economic World*, Addison-Wesley Press.

Khemahem, H., Simon, A., von der Mühll, R., and Ravez, J. (2000) Relaxor or classical ferroelectric behavior in ceramics with composition $Ba_{1-x}Na_xTi_{1-x}Nb_xO_3$. *J. Phys. Condens. Matter*, **12**, 5951–5959.

Kinoshita, K. and Yamaji, A. (1976) Grain size effect on dielectric properties in barium titanate. *J. Appl. Phys.*, **47**, 371–373.

Kirby, K.W. (1988) Alkoxide synthesis techniques for $BaTiO_3$. *Mater. Res. Bull.*, **23** (6), 881–890.

Kleber, W. (1998) *Einführung in die Kristallographie*. 18th edn, Verlag Technik, Berlin, pp. 257–270, 301.

Kniepkamp, H. and Heywang, W. (1954) Über Depolarisationseffekte in polykristallinem $BaTiO_3$. *Z. Angew. Phys.*, **6** (9), 385–390.

Kolar, D., Gaberscek, S., Volavsek, B., Parker, H.S., and Roth, R.S. (1981) Synthesis and crystal chemistry of barium neodymium titanium oxides ($BaNd_2Ti_3O_{10}$ and $BaNd_2Ti_5O_{14}$) and neodymium titanium oxide ($Nd_4Ti_9O_{24}$). *J. Solid State Chem.*, **38**, 158–160.

Kongtaweelert, S., Anuragudom, P., Nualpralaksana, S., Heimann, R.B., and Phanichphant, S. (2004) Characterization of hydrothermally synthesized PLZT for pyroelectric applications. *J. Electroceram.*, **13**, 209–214.

Krupicka, S. (1973) *Physik der Ferrite*, Vieweg, Braunschweig, ISBN: 3-528-0-8312-3.

Landolt-Börnstein (1984) *Numerical Data and Functional Relationships in Science and Technology*, (eds K.H. Hellwege and O. Madelung), New Series, vol. **III/18**, Springer, Berlin.

Landolt-Börnstein (1993) *Numerical Data and Functional Relationships in Science and Technology*, New Series, vol. **III/29b**, Springer, Berlin.

Lang, S.B. (2005) Pyroelectricity: from ancient curiosity to modern imaging tool. *Phys. Today*, **60**, 31–36.

Ljamov, V.E. (1983) *Polarization Effects and the Anisotropy of Acoustic Wave Interactions in Crystals*, University Press, Moscow (in Russian).

Lozinski, A.H., Wang, F., Uusimaki, A., and Leppavuori, S. (1998) Thick film pyroelectric linear array. *Sens. Actuators A*, **68**, 290–293.

Maison, W., Kleeberg, R., Heimann, R.B., and Phanichphant, S. (2003) Phase content, tetragonality, and crystallite size of nanoscaled barium titanate synthesized by the catacholate process: effect of calcination temperature. *J. Eur. Ceram. Soc.*, **23**, 127–132.

Masse, D.J., Purcel, R.A., Readey, D.W., Maguire, E.A., and Hartwig, C.P. (1971) New low-loss high-K temperature compensated dielectric for microwave applications. *Proc. IEEE*, **59**, 1628–1629.

Matsubara, M., Kikuta, K., and Hirano, S. (2005) Synthesis and characterization of $(K_{0.5}Na_{0.5})(Nb_{0.7}Ta_{0.3})O_3$ piezoelectric ceramics sintered with sintering aid $K_{5.4}Cu_{1.3}Ta_{10}O_{29}$. *Jpn. J. Appl. Phys.*, **44**, 6618–6623.

Megaw, H.D. (1973) *Crystal Structures: A Working Approach*, Saunders Co., Philadelphia, pp. 282–283.

Moulson, A.J. and Herbert, J.M. (2003) *Electroceramics. Materials, Properties, Applications*, 2nd edn, John Wiley & Sons, Ltd, Chichester.

Müller-Lierheim, W., Suski, T., and Otto, H.H. (1977) Factor group analysis of the Raman spectrum of $Pb_5Ge_3O_{11}$. *Phys. Status Solidi B*, **80** (1), 31–41.

Nagata, H. and Takenaka, T. (1997) Lead-free piezoelectric ceramics of $(Bi½Na½)TiO_3$–½$(Bi_2O_3 \cdot Sc_2O_3)$ system. *Jpn. J. Appl. Phys.*, **36**, 6055–6057.

Newnham, R.E. (2000) Molecular mechanisms in smart materials, in *Applied Mineralogy in Research, Economy, Technology, Ecology and Culture*, vol. **1** (eds D. Rammlmair *et al.*), Proceedings 7th International Congress on Applied Mineralogy (ICAM), Göttingen, 17–19 July 2000, Balkema, Rotterdam, Brookfield, pp. 7–9.

Newnham, R.E. (2005) *Properties of Materials: Anisotropy, Symmetry, Structure*, Oxford University Press, Oxford UK, ISBN: 0-19-852075-1, 378 pp.

Newnham, R.E. and Ruschau, G.R. (1991) Smart electroceramics. *J. Am. Ceram. Soc.*, **74**, 463–480.

Newnham, R.E. and Ruschau, G.R. (1993) Electromechanical properties of smart materials. *J. Intell. Mater. Syst. Struct.*, **4** (3), 289–294.

Newnham, R.E., Bowen, L.J., Klicker, K.A., and Cross, L.E. (1980) Composite piezoelectric transducers. *Mater. Eng.*, **92** (2), 93–106.

Nishigaki, S., Yano, S., Kato, H., and Nonomura, T. (1986) Dielectric properties of BaO-TiO_2-WO_3 system at microwave frequency. *3rd US-Japan Seminar on Dielectric and Piezoelectric Ceramics, T-4, November 9–12.*

Nishigaki, S., Yano, S., Kato, H., Hirai, T., and Nonomura, T. (1988) BaO-TiO_2-WO_3 microwave ceramics and crystalline $BaWO_4$. *J. Am. Ceram. Soc.*, **71** (1), C11–C17.

Nord, G.L., Jr (1994) Transformation-induced twin boundaries in minerals. *Phase Trans.*, **48**, 107–134.

Nowotny, H., Benes, E., and Radel, S. (2008) Piezoelectric transducers and resonators. Vienna University of Technology, Lecture SS 2008, Version 1.0. Available at: tph.tuwien.ac.at/~hnowotny/08pz/pz0.8_00.ps (accessed 15 December 2009).

O'Bryan, H.M. and Thomson, J., Jr (1974) A new $BaO \cdot TiO_2$ compound with temperature stable high permittivity and low microwave loss. *J. Am. Ceram. Soc.*, **57** (10), 450–453.

Otto, H.H. (1979) Die Kristallstruktur der ferroischen Verbindung $Pb_3[O/GeO_4]$. *Z. Kristallogr.*, **149**, 227–240.

Papadakis, E.P. (1967) Ultrasonic phase velocity by the pulse-echo overlap method. *J. Acoust. Soc. Am.*, **42**, 1045–1051.

Phanichphant, S. and Heimann, R.B. (2004) Hydrothermal synthesis of submicron- to nano-sized ferroelectric powders: properties and characterization. *Chiang Mai Univ. J.*, **3** (2), 113–132.

Pookmanee, P., Phanichphant, S., and Heimann, R.B. (2001) Synthesis and properties of bismuth sodium titanate (BNT). Part I: chemical synthesis of fine single-phase bismuth sodium titanate powders. *cfi/Ber. DKG*, **78** (7), 27–30.

Pookmanee, P., Phanichphant, S., Straube, U., and Heimann, R.B. (2003) Synthesis and properties of bismuth sodium titanate (BNT). Part II: dielectric properties of sintered BNT and La-doped BNT. *cfi/Ber. DKG*, **80** (11), E41–E44.

Pookmanee, P., Rujijanagul, G., Ananta, S., Heimann, R.B., and Phanichphant, S. (2004) Effect of sintering temperature on microstructure of hydrothermally prepared bismuth sodium titanate ceramics. *J. Eur. Ceram. Soc.*, **24**, 517–520.

Ramirez, A.P., Subramanian, M.A., Gardel, M., Blumberg, G., Li, D., Vogz, T., and Shapiro, S.M. (2000) Giant dielectric constant response in a copper titanate. *Solid State Commun.*, **115** (5), 217–220.

Rogers, C.A. (1995) Intelligent materials. *Sci. Am.*, **273** (3), 154–157.

Romero, J.J., Leret, P., Rubio-Marcos, F., Quesada, A., and Fernández, J.F. (2009) Evolution of the intergranular phase during sintering of $CaCu_3Ti_4O_{12}$ ceramics. *J. Eur. Ceram. Soc.*, **30** (3), 737–742.

Safari, A., Panda, R.K., and Janas, V.F. (1996) Ferroelectricity: materials, characteristics and applications. *Key Eng. Mater.*, **122–124**, 35–70.

Saito, Y., Takao, H., Tani, T., Nonoyama, T., Takatori, K., Homma, T., Nagaya, T., and Nakamura, M. (2004) Lead-free piezoelectrics. *Nature*, **432** (7013), 84–87.

Scheuerer, R., Häußler, M., Renk, K.F., Schomburg, E., Koschurinov, Y., Pavelev, D., Maleev, N., Ustinov, V., and Zhukov, A. (2003) Frequency multiplication of microwave radiation by propagating space-charge domains in a semiconductor superlattice. *Appl. Phys. Lett.*, **82** (17), 2826–2828.

Schmelz, H., Schwaen, W. (1978) Capacitor dielectric with inner blocking layers and method for producing the same. US Patent 4,131,903.

Scott, J.F. and Ledbetter, H. (1997) Interpretation of elastic anomalies in $SrTiO_3$ at 37 K. *Z. Phys. B*, **104**, 635–639.

Scott, J.F. and Paz de Araujo, C.A. (1989) Ferroelectric memories. *Science*, **246**, 1400–1405.

Silvestrova, I.M., Pisarevski, Y.V., Kaminskii, A.A., and Mill, B.V. (1987) Spectroscopy of disordered doped $La_3Ga_5SiO_{14}$ single crystals. *Fiz. Tverd. Tela*, **29**, 1520–1522 (in Russian).

Sirisoonthorn, S. (1997) Piezoelectric materials and their applications, in *Proceedings, International Conference on Materials Technology, Recent Developments and Future Potential*, Chiang Mai University, Thailand, pp. 224–235.

Stade, J., Bohatý, L., Hengst, M., and Heimann, R.B. (2002) Electrooptic, piezoelectric and dielectric properties of langasite ($La_3Ga_5SiO_{14}$), langanite ($La_3Ga_{5.5}Nb_{0.5}O_{14}$) and langataite ($La_3Ga_{5.5}Ta_{0.5}O_{14}$). *Cryst. Res. Technol.*, **37**, 1113–1120.

Stennett, M.C., Miles, G.C., Sharman, J., Reaney, I.M., and West, A.R. (2005) A new family of ferroelectric tetragonal tungsten bronzes. *J. Eur. Ceram. Soc.*, **25** (12), 2471–2475.

Takeda, H., Sato, J., Kato, T., Morikoshi, H., Shimamura, K., and Fukuda, T. (2000) Synthesis and characterization of $Sr_3TaGa_3Si_2O_{14}$ single crystals. *Mater. Res. Bull.*, **35**, 245–252.

Truell, R., Elbaum, C., and Chick, B. (1969) *Ultrasonic Methods in Solid State Physics*, Academic Press, New York, London.

Uchino, K. (1993) Ceramic actuators: principles and applications. *Mater. Res. Soc. Bull.*, **18**, 42–48.

Uchino, K. (1994) Ferroelectric ceramics, in *Materials Science and Technology*, vol. **11**, Structure and Properties of Ceramics (eds R.W. Cahn, P. Haasen, and E.J. Kramer), VCH, Weinheim, pp. 635–677.

Uchino, K. (1995) Ceramic Actuators. *Proceedings, Smart Materials Seminar, The Royal River Hotel, Bangkok, Thailand, 7 March 1995.*

Uchino, K. (1996) *Piezoelectric Actuators and Ultrasonic Motors*. Electronic Materials: Science & Technology (ed. H.L. Tuller), Kluwer Academic Publ., Deventer. ISBN: 978-07923-98-11-0, 364 pp.

Uchino, K. (2000) *Ferroelectric Devices*, Materials Engineering, vol. **16**, Marcel Dekker, New York. ISBN: 08-2478-1333, 320 pp.

Valasek, J. (1924) Dielectric anomalies in Rochelle salt crystals. *Phys. Rev. Bull.*, **24** (5), 560–568.

Wersing, W. (1991) High frequency ceramic dielectrics and their application in microwave components, in *Electronic Ceramics* (ed. B.C.H. Steele), Elsevier, London, New York, pp. 76–83.

Wersing, W. (1996) Microwave ceramics for resonators and filters. *Curr. Opin. Solid State Mater. Sci.*, **1** (5), 715–731.

Whangbo, M.H. and Subramanian, M.A. (2006) Structural model of planar defects in $CaCu_3Ti_4O_{12}$ exhibiting a giant dielectric constant. *Chem. Mater.*, **18** (14), 3257–3260.

Wood, G.I. and Glazer, A.M. (1980) Birefringence measurement using the rotating-analyser method: ferroelastic phase transition in $BiVO_4$. *J. Appl. Crystallogr.*, **13**, 217–223.

Worlock, J.M., Scott, J.F., and Fleury, P.A. (1969) Soft phonon modes and the 110 K phase transition of $SrTiO_3$, in *Light Scattering Spectra of Solids* (ed. G.B.

Wright), Springer, Berlin, Heidelberg, New York, p. 689.

Xu, Q.C., Belsic, J., Yoshikawa, S., Srinivasan, T.T., and Newnham, R.E. (1991) Piezoelectric composite with high sensitivity and high capacitance for use at high pressure. *IEEE Trans. UFFC*, **38** (6), 634–639.

Ye, Z.G. (ed.) (2008) *Handbook of Advanced Dielectric, Piezoelectric and Ferroelectric Materials: Synthesis, Properties and Applications*, Woodhead Publishing Ltd., Abingdon, Cambridge, UK.

Zheludev, I.S. (1990) *Kristallphysik und Symmetrie* (ed. J. Bohm), Akademie-Verlag, Berlin, pp. 165–168.

Zheludev, I.S. and Vlokh, O.G. (1983) Electro- and magnetogyration. *J. Chem. Sci. (India)*, **92** (4–5), 421–427.

9
Superconducting Ceramics

9.1
Introduction

The research and development of ceramic superconductors is of great strategic importance for a variety of emerging energy technologies. It is not only the insatiable curiosity of scientists but also the urgent quest for a sustained development of future energy technologies that is pushing research groups in universities, government research organizations and key industries towards exciting new results. Nevertheless, despite much effort a complete understanding of the theory of the superconducting quantum state, as well as the development of room-temperature superconductors and efficient processing technologies, remain challenges for the near future. Likewise, despite high expectations, the large-scale application of ceramic superconductors for electrical high-tension energy transmission cables, electric motors and generators, as well as microcircuit switch components, are still missing. Hence, the break-even point between conventional and superconducting technologies appears, at present, to have been postponed indefinitely (Heimann, 2004).

This chapter has been conceived and written in its entirety by Professor emeritus Dr Hans Hermann Otto, formerly at Technische Universität Clausthal-Zellerfeld. The author of this book is very grateful to Prof. Otto for having provided a very concise high-level introduction to the extremely complex field of superconductivity, and its challenging state-of-the-art theories, processing routes, and potential applications. More elaborate theoretical issues related to the Bardeen–Cooper–Schrieffer (BCS) approach to superconductivity, and the contrasting theories for high-T_c superconductors that are more of electronic and magnetic nature, are detailed in Appendix E.

9.2
Definitions

Certain materials exhibit a more or less abrupt drop in their electric resistivity to zero and a strong diamagnetism (expulsion of magnetic flux lines) when cooled

Classic and Advanced Ceramics: From Fundamentals to Applications. Robert B. Heimann

ISBN: 978-3-527-32517-7

to below a critical temperature (T_c). If both physical properties exist simultaneously, the material is termed a *superconductor*, characterized by a super-current flowing through the crystal lattice without any dissipation of energy. As the most interesting materials for industrial applications with high T_cs are not ductile but rather are very brittle, their fabrication processes from precursor powders are similar to those of ceramics. Therefore, these materials are classified as *ceramic superconductors*.

Other materials with potentially wide ranges of use in the near future are known as *ultraconductors*; examples include polymers such as atactic polypropylene with properties similar to superconductors – that is, high electric conductivity and current densities over a wide temperature range, even though their resistivity does not reach zero, their electric conductivity at room temperature is orders of magnitude higher than that of copper. Moreover, owing to the predominantly one-dimensional (1-D) structure of polymers compared to two-dimensional (2-D) ceramic superconductors, the Meissner effect does not exist in ultraconductors.

9.3 Historical Developments

In 1908, the famous Dutch physicist Heike Kamerlingh Onnes succeeded in liquefying helium, a feat for which he was awarded the Nobel Prize in 1913. By using this cryogenic technique, Kamerlingh Onnes (1911) discovered the superconductivity of pure metals by observing persisting currents below the superconducting transition. For example, liquid mercury could be purified by repeated distillation to show a sudden disappearance of resistivity at 4.1 K (Kamerlingh Onnes, 1911). Later, in 1933, Walter Meissner and Robert Ochsenfeld, two German researchers, discovered the strong diamagnetism of superconducting materials, which later was referred to as the Meissner–Ochsenfeld effect (Meissner and Ochsenfeld, 1933). In this case, magnetic flux lines will be expelled from a superconducting sample when cooled in the presence of a magnetic field.

In 1935, Fritz and Heinz London (London and London, 1935) provided a first phenomenological approach to the theory of superconductivity. Using the concept of a superelectron with twice the electron's mass and charge, the London equations described very well the properties of a superconductor; that is, its infinite electric conductivity as well as the decay of the magnetic field in a thin surface layer of a superconductor (the Meissner–Ochsenfeld effect) (see Appendix E).

Next came the likewise phenomenological Ginzburg–Landau theory of superconductivity, based on the Landau theory of a second-order phase transition (see also Appendix B) that predicted the coherence length and penetration depth as two characteristic parameters of a superconductor (Ginzburg and Landau, 1950). Based on this theory, Abrikosov derived the notion that the magnetic field penetrates type II superconductors in quantized flux tubes, commonly in the form of a hexagonal network (Abrikosov, 1957). The existence of this vortex lattice was

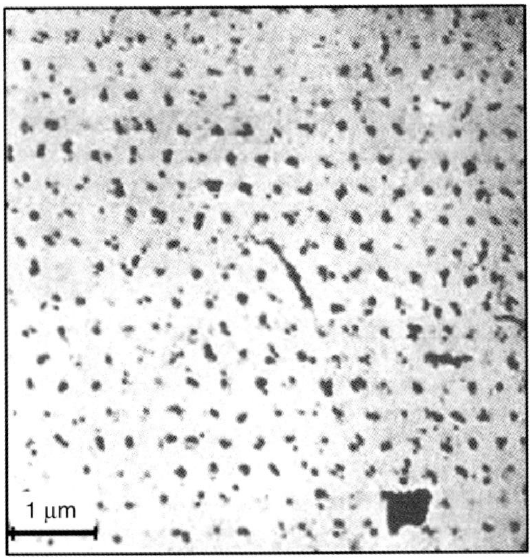

Figure 9.1 Decorated vortex lattice of superconducting $Pb_{0.96}In_{0.04}$ at 1.1 K, magnetic field 195 G (Essmann and Träuble, 1967).

verified experimentally by Essmann and Träuble (1967), who used evaporated colloidal iron particles for decoration (Figure 9.1).

Fröhlich (1954) proposed the possibility of superconductivity in 1-D metals due to spontaneously sliding charge density waves (CDWs).

In 1957, the American physicists J. Bardeen, L. N. Cooper, and J. R. Schrieffer (1957) developed a microscopic theory of superconductivity (the BCS theory), and were subsequently awarded the Nobel Prize in physics in 1972. The BCS theory is applicable to metals and alloys (conventional low-T_c superconductors). The Coulombic repulsion of electrons is thought to be overcome by a phonon-mediated mechanism, whereby two electrons with their spins aligned in opposite directions strongly attract each other (Cooper pairing). Cooper pairs condense to a remarkably stable macroscopic quantum state, described by identical wave functions (see Appendix E).

Doll and Nähbauer (1961) proved experimentally the quantization of magnetic flux in a superconducting lead ring.

A milestone towards the industrial application of superconductors was the discovery and theoretical explanation of Cooper pair tunneling through a weak link (nonconducting barrier) between two superconductors (Giaever, 1960; Josephson, 1962). The technical application of the Josephson tunneling effect forms the basis of superconducting quantum interference devices (SQUIDs), serving to precisely determine the 2 e/h ratio and voltage standards.

With the discovery of the important superconducting A15 alloys Nb_3Sn (Matthias *et al.*, 1954; $T_c = 18.1$ K) and later Nb_3Ge (Gavaler, 1973; $T_c = 23.2$ K), the road was paved to commercial applications, an example being the superconducting wires made from NbTi (T_c = 17.5 K), produced by the Westinghouse Electric Corporation in about 1962.

During the following years, although many different substance classes were found to be superconducting, the maximum critical temperature remained that of Nb_3Ge, at 23.2 K. Among the newly discovered phases the more important were the cluster compound $PbMo_6S_8$ (Chevrel *et al.*, 1971), organic superconductors such as (tetramethyltetra-selenafulvalene)$_2$ hexafluorophosphate [$(TMTSF)_2PF_6$] (Jerome *et al.*, 1980), and the heavy fermion superconductors such as $CeCu_2Si_2$ (Steglich *et al.*, 1979), UBe_{13} (Ott *et al.*, 1983), and UPt_3 (Steward *et al.*, 1984). The proposal of a possible electron-pairing mechanism along 1-D carbon chains leading to superconductivity (Little, 1964) has not yet been substantiated (see also Heimann *et al.*, 1999), although superconductivity in single-walled carbon nanotubes (SWCNTs) below 20 K has been confirmed (Tang *et al.*, 2001) (some additional ceramic superconducting compounds are listed in Table 9.2).

Although in the meantime, low-T_c superconducting compounds with perovskite structures were discovered, such as $SrTiO_3$ (T_c = 1.2 K; Schooley *et al.*, 1964), the cation-deficient phases of the molybdenum bronze type, and $BaPb_{0.75}Bi_{0.25}O_3$ (T_c = 13 K; Sleight *et al.*, 1975), it proved to be a major surprise when, in 1986, Johannes Georg Bednorz and Karl Alexander Müller found superconductivity around T_c = 35 K in Ba-doped lanthanum cuprate (Bednorz and Müller, 1986), the first of a series of high-T_c cuprate compounds discovered during the following years. One of these compounds, $YBa_2Cu_3O_{6.9}$ (Wu *et al.*, 1987), has been extensively investigated, since its transition temperature of T_c = 93 K (which is well above the boiling temperature of liquid nitrogen as a refrigerant; 77 K), its ease of formation, high current density, and magnetic field acceptance are well suited for many applications.

Subsequently, various cuprates with cations of 6*s* electron configuration (Tl, Bi, Pb, Hg) were discovered with steadily increasing transition temperatures, but still considerably distant from room temperature. A maximum transition temperature of T_c = 138 K under ambient conditions was identified for the compound $Hg_{0.8}Tl_{0.2}Ba_2Ca_2Cu_3O_{8.33}$ (Dai *et al.*, 1995), and the pure Hg cuprate of this structure type shows under high pressure even T_c = 164 K (Gao *et al.*, 1994). The important perovskite-related superconducting compounds, together with their transition temperatures and references to their discoverers, are summarized in Table 9.1.

Soon, it became evident that doping in the layered high-T_c cuprates created holes in the oxygen sublattice, forming O^{1-} instead of raising the valence of copper to the oxidation state 3+. However, a phonon-mediated (BCS-like) theory failed to fully explain the mechanism of the occurrence of superconductivity in high-T_c cuprates. To remedy this failure, many theory-based studies have been conducted and reported during the past twenty years with varied success, including the introduction of Zhang–Rice singlets, the Hubbard and t-J models, the resonating valence bond (RVB) theory of Anderson (1987), which was published within weeks of the discovery of cuprate superconductivity and recently revived, and finally the theories of Varma (1999, 2006) and Simon and Varma (2002) of quantum critical fluctuations (QCF) of an order parameter that focuses on interacting current loops in the CuO_2 plaquettes (see Figure E.6). Verification of the existence of extended

Table 9.1 General formulas of important homologous series of cuprate superconductors and derived hole concentration, p.

General formula	p	Reference[a)]
$La_{2-x}(Ba,Sr)_xCa_{n-1}Cu_nO_{2n+2-\delta}$	$x-2\delta$	Bednorz and Müller (1986)
$(Y,Ln)_{m-x}Ca_xBa_{2m}Cu_{3m+n}O_{7m+n-\delta}$	$m+x-2\delta$	Wu *et al.* (1987)
$Bi_{2-x}Pb_x(Sr,Ca)_2(Ca,Sr,Bi)_{n-1}Cu_nO_{2n+4+\delta}$	$x+2\delta$	Michel *et al.* (1987)
$Tl_{m-x}M_s$+$(Ba,Sr)_2(Ca_{1-y}Tl_y)_{n-1}Cu_nO_{m+2n+2-\delta}$	$2-m+\{x(3-s)-(n-1)y-2\delta\}$	Sheng *et al.* (1988)
$Hg_{1-x}Tl_xBa_2Ca_{n-1}Cu_nO_{2n+2+\delta}$	$2\delta-x$	Putilin *et al.* (1993)

a) Reference relates to the first discovered member of the series.

circulating orbital currents was recently achieved by polarized neutron diffraction (Fauqué *et al.*, 2006; Mook *et al.*, 2008; Li *et al.*, 2008).

To complete the history of materials discovery, further superconducting compound groups discovered have included carbon compounds such as fullerides and nanotubes, MgB_2 and, finally, the very recently described pnictides as a demonstration of things to come. Superconductivity in graphite intercalation compounds (GICs) was first observed by Hannay *et al.* (1965), but the transition temperatures of all as-yet known GICs did not much exceed 4 K. On the other hand, if alkaline ions are intercalated in C_n fullerene cluster structures (Hebart *et al.*, 1991), then a transition temperature of $T_c = 39$ K could be observed in cesium fulleride, Cs_3C_{60} (Takabayashi *et al.*, 2009), while SWCNTs reach at least $T_c = 15$ K (Tang *et al.*, 2001). Finally, MgB_2 was shown to exhibit unusually high temperature superconductivity, with $T_c = 39$ K (Nagamitsu *et al.*, 2001), and only very recently Kamihara *et al.* (2008) discovered $LaO_{1-x}F_xFeAs$ as the first member of a new family of layered superconductors with iron and pnictogen ions (see Section 9.5.3).

In completing these remarks, a quick examination of hydrogen is worthwhile. Superconductivity with $T_c = 17$ K is observed in the metallic state of silane, SiH_4, with a hexagonal close-packed structure at 96 GPa (Eremets *et al.*, 2008).

9.4 Material Classification

Superconducting materials can be classified in different ways. One common classification is based on their response to a high magnetic field, whereby pure metals with an almost perfect crystal lattice (except for V, Tc, and Nb) belong to type I superconductors. The magnetic flux lines are unable to penetrate the material, and above a critical magnetic field, B_c, the superconductivity suddenly disappears

(Figure E.7a); type I superconductors are, therefore, not suited for high-magnetic field applications. Because of the large distance allowed for electrons in a perfect lattice to be coupled, superconductivity for this class of metals can be well explained by the BCS theory. In contrast, superconducting alloys and compounds such as cuprates belong to type II superconductors. These behave differently, in that the magnetic flux lines can penetrate the material and allow, besides a higher critical temperature, far higher current densities and a greater tolerance towards stronger magnetic fields, if the magnetic flux lattice is pinned.

One different way of considering superconductors is related to their compliance with the classic BCS theory. Hence conventional (i.e., BCS-compatible) and unconventional (i.e., BCS-incompatible) materials exist. The common scientific language therefore distinguishes also between LTS (low-temperature superconductors) and HTS (high-temperature superconductors). In general, LTS are electron-doped (n-type), while HTS are hole-doped (p-type) phases.

9.5 Crystal Chemistry

9.5.1 Nb-Bearing Low-Temperature Superconductors

Commercially used type II LT-superconductors, when applied for the construction of magnets, are ductile Nb and NbTi alloys with excellent mechanical properties and deformability, as well as the brittle Nb_3Sn alloy. Apart from this, Nb and Al are the most common materials applied for superconducting tunnel junctions (see Section 9.8.6). The ductile NbTi alloy carries maximum critical currents around a composition of 47 mass% Ti; this alloy and Nb itself belong to the body-centered cubic crystal structures. The Nb_3Sn alloy crystallizes in the cubic A15 structure type (S.G. $Pm\bar{3}n$), as shown in Figure 9.2a. An important point for practical applications is the A15 stability range, from 18 to 25 atom% Sn, and the dependence of superconductivity on the critical magnetic field. The body-centered Sn lattice with lattice parameter a = 0.5293 nm contains Nb chains that alternatively bisect the cube faces (Nb–Nb distance of the chain cluster $d_{\text{Nb-Nb}} = 0.264$ nm). Tetragonal lattice distortion is observed at low temperature, showing a value of about a/c = 1.0026 at 10 K. As a rule, superconductivity is more resident in low-dimensional than in three-dimensional (3-D) structures, and extended clusters or at least layered structures favor a high transition temperature. The very recently discovered compounds MgB_2 (Section 9.5.2) and iron pnictides (Section 9.5.3) will underline these arguments.

9.5.2 Superconducting MgB_2

This metallic compound is classified as an unusual high-temperature-superconducting (UHTS) compound, because $T_c = 39$ K (Nagamitsu *et al.*, 2001).

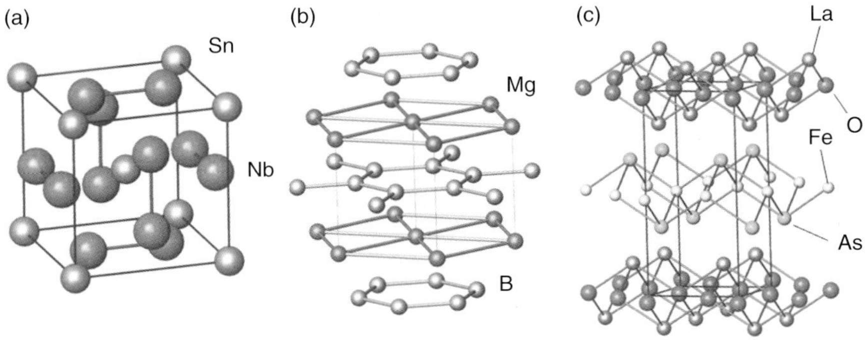

Figure 9.2 Superconductors crystallize in different crystal systems. (a) Cubic crystal structure of the A15 alloy Nb_3Sn. Space group $Pm\overline{3}n$ (No. 223), a = 0.5293 nm; (b) Crystal structure of MgB_2. Space group *P6/mmm* (No. 191), a = 0.3086 Å, c = 0.3524 nm; (c) Tetragonal layered crystal structure of the pnictide $LaO_{1-x}F_xFeAs$. Space group *P4/nmm* (No. 129/2), a = 0.402 69 nm, c = 0.870 10 nm.

MgB_2 crystallizes in a hexagonal layered structure, space group *P6/mmm* with lattice parameters of a = 0.3086 nm and c = 0.3524 nm, as shown in Figure 9.2b, with the honeycombed boron layers alternating with Mg layers. The strong interlayer σ-bonds are only partly situated in the boron layer, since fewer valence electrons are available compared to the carbon atoms of the similar graphite structure, and weak intralayer π-bonds are important for understanding the exceptional superconducting properties.

9.5.3
Iron Pnictides

In 2008, the first member of a new superconducting compound family was described with iron instead of copper and a high proportion of pnictogen ions,[1)] such as arsenic (Kamihara *et al.*, 2008; Takahashi *et al.*, 2008). The crystal structure of the first-discovered compound $LaO_{1-x}F_xFeAs$ is illustrated in Figure 9.2c. This belongs to the space group *P4/nmm*, with lattice parameters of a = 0.402 69 nm and c = 0.870 10 nm, and consists of alternating, puckered layers of LaO^{1+} and $FeAs^{1-}$ units that resemble the $Bi_2O_2^{2+}$ layers of the Aurivillius phases (Aurivillius, 1949) or the PbO layers of α-PbO. Adequate doping through the replacement of oxygen by fluorine leads to the superconducting transition of $LaO_{1-x}F_xFeAs$ at T_c = 26 K. The application of high pressure of about 4 GPa for $x = 0.11$ increases the transition temperature to T_c = 43 K (Takahashi *et al.*, 2008), but this record was soon beaten and announced as T_c = 52 K for $NdO_{1-x}F_xFeAs$ and T_c = 55 K for $SmO_{1-x}F_x$ FeAs under pressure (Ren *et al.*, 2008).

1) Pnictogens are elements of the 5th group of the PSE; pnictides are binary compounds of a pnictogen.

9.5.4
Cuprate Superconductors

The exceptional superconducting properties of the cuprates are caused by the remarkable ability of the copper atom to vary its charge and oxygen coordination, leading to 2-D and anisotropic crystal structures built up of infinite CuO_2 nets separated by large cations. The cuprate structures are related to the cubic $SrTiO_3$ perovskite prototype shown in Figure 9.3a. Large soft cations such as La^{3+}, Sr^{2+}, or Ba^{2+} at the center of the unit cell are surrounded by 12 oxygen atoms, and small cations at the corner, such as Ti^{4+} (or copper ions in the case of cuprates), show six octahedrally arranged neighbors. By removing the central oxygen layer from the perovskite structure, the coordination of the central cation is reduced to 8, and consequently smaller cations such as Ca^{2+} or Y^{3+} are more appropriate to occupy this site. The octahedral coordination of the second cation is also reduced to a half-octahedral surrounding. Further depleting the oxygen layers finally results in a very simple tetragonal all-layer or infinite layer compound (Figure 9.3b), as found in the composition $Ca_{0.86}Sr_{0.14}CuO_2$ under ambient conditions by Siegrist *et al.* (1988). The electron-doped superconductors $Sr_{0.86}Nd_{0.14}CuO_2$ (T_c = 40 K; Smith *et al.*, 1991) and $Sr_{0.9}La_{0.1}CuO_2$ (T_c = 43 K; Er *et al.*, 1991) derive from this parent structure. By combining these structural building blocks with complete or incom-

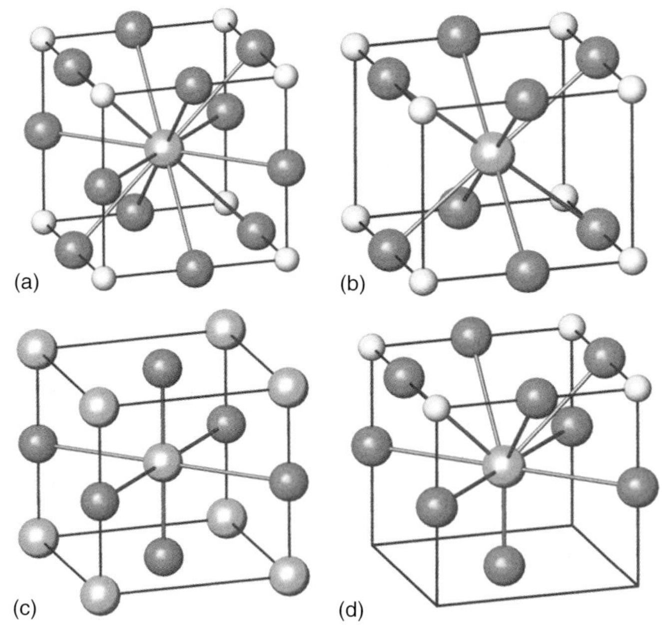

Figure 9.3 Basic structural types of cuprate superconductors. (a) Perovskite structure (cubic); (b) Infinite- layered structure (tetragonal); (c) Rocksalt prototype structure (reduced cell); (d) Composite layer between infinite layer and rocksalt building blocks.

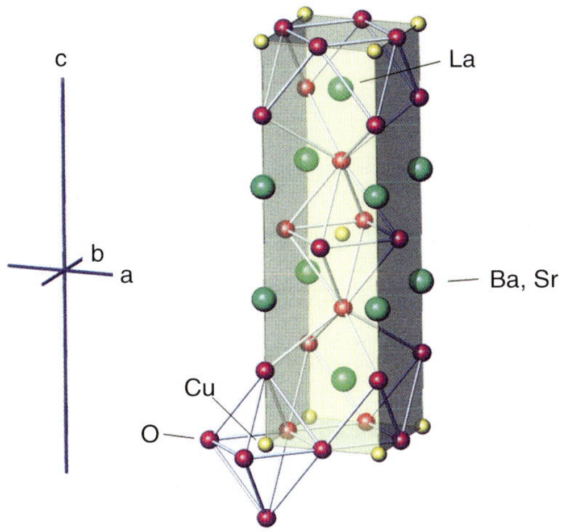

Figure 9.4 K_2NiF_4 prototypic structure as realized in $La_{2-x}(Ba,Sr)_xCuO_{4-\delta}$. Space group *I4/mmm* (No. 139), a = 0.380 nm, c = 1.323 nm.

plete rocksalt blocks (Figure 9.3c) that may serve as a charge reservoir, most superconducting cuprate structures can be represented in homologous series.

The composite layer between both structural building blocks with a 9-coordinated large cation is displayed in Figure 9.3d. For instance, the original lanthanum cuprate structure may be considered a transition between alternately stacked perovskite and rocksalt aristotype structures (Figure 9.4). The resultant crystal structures are tetragonally or slightly orthorhombically distorted. Homologous series of crystal structures with stacks of blocks along the c-axis offer the possibility of introducing a symbolic description by numbering the cations in the formula unit, but apart from that commonly termed names are also used. If the compounds contain three CuO_2 layers, optimum superconducting transition temperatures can be obtained.

From a simple equation using the block distance parameters given in Table 9.2, sufficiently accurate c-lattice parameters can be obtained. The structures themselves can be simply expressed by sequences of cationic coordination numbers along the c-direction, thereby best clarifying common motives (Otto *et al.*, 1998). Figure 9.5 illustrates examples of Tl-based superconductors and those symmetry elements that translate the structural motifs along the c-axis. Inversion operations lead to a double layer rocksalt-like charge reservoir block, thus changing the Bravais lattice from primitive (P) to body-centered (I) or A-face-centered.

Details of the nomenclature and the importance of the charge reservoir block, as well as the doping level, are explained below for the Y–Ba–Cu–O system.

In Figure 9.6, two structures of the system are compared: the nonsuperconducting tetragonal phase of $YBa_2Cu_3O_6$ (left); and the superconducting orthorhombic phase of $YBa_2Cu_3O_7$ (right), respectively. The latter is formed from the first by an

Table 9.2 Lattice parameter relations of important homologous series of superconductors deduced from structural building blocks (Otto, 1992–2002). The c-lattice parameter is given as $c = \mathbf{m_L}[(\mathbf{n_1} + 1)d_1 + \mathbf{n_2}d_2 + \mathbf{n_3}d_3]$, where $\mathbf{n_i}$ denotes the number of cations in the four-membered formula symbol and d_1 = charge reservoir distance, d_2 = composition plane (rocksalt) distance and d_3 = infinite layer distance. $\mathbf{m_L} = 1$, if $\Sigma \mathbf{n_i}$ = even, i = 1 to 4 (P-lattice), $\mathbf{m_L} = 2$, if $\Sigma \mathbf{n_i}$ = odd (I-lattice, A-lattice). $(4a^2d_2)^{1/3}$ is comparable to the lattice parameters of BaO or SrO (rocksalt-type).

$\mathbf{n_1}$	$\mathbf{n_2}$	$\mathbf{n_3}$	a (Å)	d_1 (Å)	d_2 (Å)	d_3 (Å)	$(4a^2d_2)^{1/3}$(Å)
Cu	Ba	Y	3.85	1.90	2.53	2.86	5.31
Tl	Ba	Ca	3.85	1.98	2.80	3.14	5.50
Hg	Ba	(Y,Ln)	3.86	1.96	2.80	3.17	5.51
Hg	Ba	Ca	3.88	1.96	2.80	3.00	5.52
—	Ba	Ca	3.85	1.96	2.80	3.16	5.50
—	(La,Sr)	Ca	3.80	1.81	2.41	3.16	5.18
(Tl,Pb)	Sr	Ca	3.80	2.08	2.39	3.15	5.17
Bi	Sr	Ca	3.80	2.48	2.39	3.16	5.17
Cu/Pb	Sr	Y	3.83	1.93	2.44	3.13	5.23
(Pb,Tl)	Sr	Tl	3.84	1.98	2.35	2.98	5.18

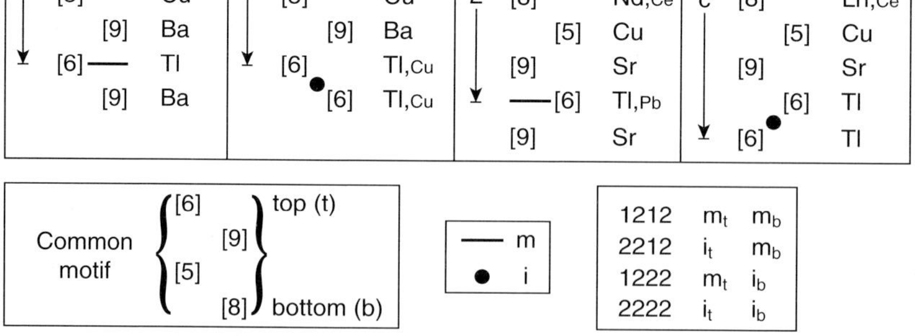

Figure 9.5 Schematic representation of Tl-cuprate structures using cationic coordination numbers. Atomic strings lined up along the c axis with origin in 0,0,0 compared to those in ½,½,0 are depicted as horizontally displaced. Common structure motifs can easily be discerned. The solid lines symbolize mirror planes m, and the solid dots centers of inversion, i.

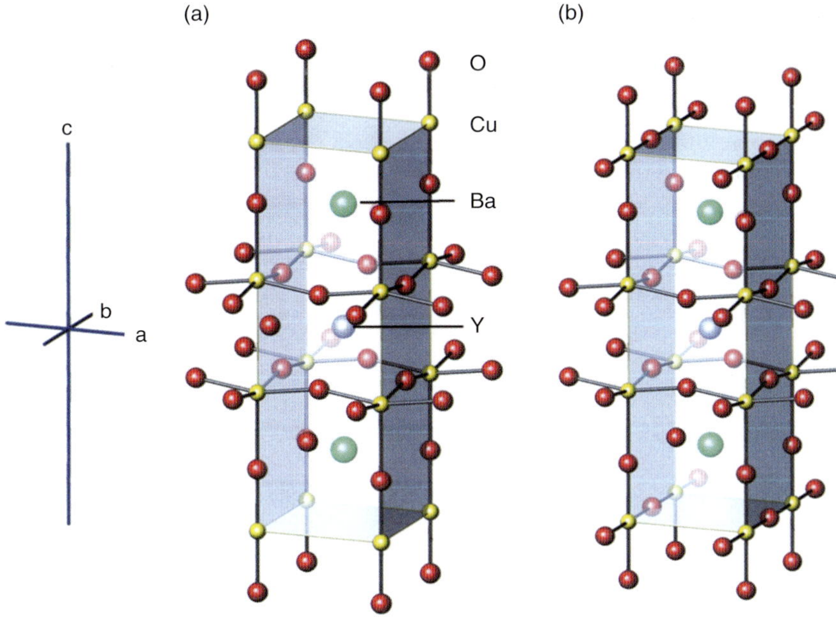

Figure 9.6 (a) Crystal structure of tetragonal $YBa_2Cu_3O_6$. Space group *P4/mmm*; (b) Crystal structure of orthorhombic $YBa_2Cu_3O_7$. Space group *P/mmm* (No. 47), a = 0.381 87 nm, b = 0.399 33 nm, c = 1.166 87 nm.

uptake of oxygen on slow cooling from 650 °C to 400 °C. The structural transformation leads to twinning of the orthorhombic phase. The tetragonal structure consists of Y^{3+} ions sandwiched between cuprate half-octahedral layers formed by Cu^{2+} ions that are connected through their apical oxygen atoms by Cu^{1+} ions with typical dumbbell coordination (Figure 9.6, left). The uptake of additional oxygen changes its coordination to four under the formation of chains, thereby raising the formal charge of the copper atoms and, concurrently, that of the layers to $Cu^{2+\delta}$ or, physically more correct, leaving Cu^{2+} ions but introducing holes in the oxygen sublattice (Figure 9.6, right). These O^{1-} ions are the most important features to explain the appearance of superconductivity in cuprates. The structural building block facilitating such hole doping is self-explanatorily dubbed the "charge reservoir block." On partial uptake of oxygen, an underdoped compound is formed that exhibits a low T_c of about 60 K compared to the maximum attainable $T_c = 93$ K for $YBa_2Cu_3O_{6.9}$. The additional substitution of Ca^{2+} for Y^{3+} ions leads to an overdoped compound. The distinct properties of both underdoped and overdoped phases contribute to the theoretical understanding of the superconducting mechanism. The superconducting compound $YBa_2Cu_3O_7$ is labeled "123" (or YBCO for short), or Y-1212 when using the four-membered symbol derived from the cation numbers of the rearranged formula $CuBa_2YCu_2O_7$ that distinguishes between both Cu sites.

Although one would expect the pure (undoped) odd-electron compounds La_2CuO_4 or $YBa_2Cu_3O_6$ to be electrically conductive, they are actually Mott insulators.[2] It is the comparable value between the ionization energy of the Cu^{2+} ion and the electron affinity of the O^{2-} ion that indeed renders the cuprates half-filled charge transfer insulators. The Cu^{2+} ions with odd d^g electron configuration order antiferromagnetically below the Néel temperature (T_N). Doping through the replacement of La^{3+} by Ba^{2+} or Sr^{2+} creates holes at O sites that are believed to couple antiferromagnetically to the holes at Cu sites, forming so-called Zhang–Rice singlets (Zhang and Rice, 1988).

From the viewpoint of the crystal chemistry, the creation of O^{1-} holes in the large cation-mediated cuprates is to be expected. Whereas, the O^{2-} ion must always be stabilized by environmental hybridization, the O^{1-} ion is a uniquely stable oxygen ion by itself. Large, soft cations such as Ba^{2+} are able to accumulate such O^{1-} ions in their coordination sphere. Due to the large size of Ba^{2+}, a high oxygen coordination is required, which means in turn that the charge of some surrounding oxygen ions can easily adopt the O^{1-} state, since the charge of the earth alkaline ion is 2+.

Still a mystery is the ability of the cuprates to self-dope. Most compounds show near-optimum doping on formation; indeed, according to Table 9.1 the Tl-bearing compounds with T_c as high as 125 K exhibit complicated self-doping expressed through the term $\{x(3 - s) - (n - 1)y - 2\delta\}$. The Tl site may be partially occupied by Cu^{1+} and the Ca site partially by Tl^{3+}. The reason for such an intrinsic cationic replacement is presumably related to the gain of lattice energy connected with the reduction of lattice strain.

The crystal chemistry of Bi^{3+} compounds is dominated by its lone electron pair that requires more space. Its dipole momentum can be compensated only by a Bi^{3+} double layer arrangement, with the weak interlayer bonding resulting in near-perfect cleavage along (001). This property is most favorable for producing highly textured superconducting samples for commercial applications, such as the orthorhombic Bi-2212 and Bi-2223 (BSCCO) phases (Table 9.3). The longer heavy-atom distance in comparison with Tl^{3+} superconductors requires the smaller Sr^{2+} ion instead of Ba^{2+} to fulfill spatial restrictions for optimum T_c. The superconducting transition temperature is maximized for Bi-2223 ($T_c = 110$ K), and falls off for higher members of the Bi series, as demonstrated up to the Bi-2267 phase.

To comply with health and environmental regulations, neither Tl nor Hg can be considered for widespread application in superconductors, owing to the risk of their accidental toxic release. Even though Tl in superconducting cuprates is in a 3+ valence state (and is therefore less toxic), its easy reduction to the 1+ oxidation state and its high cationic mobility are dangerous features. Tl^{1+} can readily replace K^{1+} due

2) Atoms with repulsive interactions in a periodic potential possess two possible ground states of the many body system. If the repulsive interactions are weak, the ground state is a superfluid Bose–Einstein condensate. However, if the interactions between the atoms are strong, then the atoms form a Mott insulating state of matter. Since the motion of the atoms through the lattice is blocked due to the repulsive interactions between the atoms, the system behaves like an insulator.

Table 9.3 Crystallographic data of important orthorhombic cuprate superconductors. The incommensurate modulation of the Bi compounds has been neglected.

Formula	Symbol	Space group	Lattice parameters (nm)			Reference
			a	b	c	
$CuBa_2YCu_2O_{6.9}$	Y-1212 YBCO, "123"	*Pmmm*	0.38187	0.38833	1.16687	Wu *et al.* (1987)
$Bi_2Sr_2CaCu_2O_{8+\delta}$	Bi-2212	*Amaa*	0.5395	0.5396	3.0665	Maeda *et al.* (1988)
$Bi_2Sr_2Ca_2Cu_3O_{10+\delta}$	Bi-2223 BSCCO	*Amaa*	0.5411	0.5420	3.7290	Tarascon *et al.* (1988)
$(Bi,Pb)_2Sr_2Ca_2Cu_3O_{10+\delta}$	(Bi,Pb)-2223	*Amaa*	0.5339	0.5483	3.772	Endo *et al.* (1988)

to its comparable charge and ion radius, and could therefore induce irreversible disturbances of the central nervous system, similar to the deleterious effect of the methylmercury ion, CH_3Hg^+. On the other hand, Bi^{3+} is quite acceptable, as demonstrated by the widespread use of bismuth subcitrate as an antimicrobial medical remedy against upset stomach ("Peptobismol"). When released, the earth alkaline ions Ba^{2+} and Sr^{2+} would be precipitated as almost insoluble carbonates or sulfates.

9.6 Theory

Common to all concepts of superconductivity theory is that electrons are able to overcome their mutual Coulombic repulsion by binding together in pairs. As these Cooper pairs (Figure 9.7) do not need to obey Pauli's exclusion principle, which allows them to condense into a single quantum state at low temperature, approximately akin to a Bose–Einstein condensate. The glue that binds electrons together is different for low-T_c (electron doped) metals and alloys, and for high-T_c (hole doped) compounds. Details of the BCS phonon-mediated theory of superconductivity and the mainstream theories of high-temperature superconductivity of electronic or magnetic origin are provided in Appendix E.

9.7 Materials Processing

The processing of ceramic superconductors is a highly challenging enterprise which has, in fact, limited their large-scale industrial applications to date. In

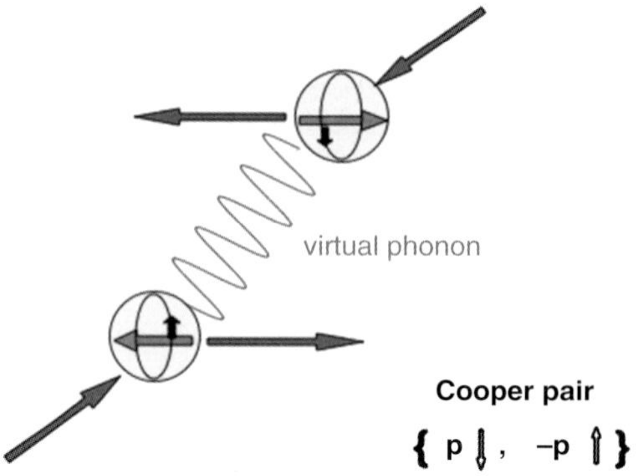

Figure 9.7 Illustration of a Cooper pair in the Feynman momentum diagram notation. The energy gain for pairing is maximized if the electrons have alternating up and down spin directions and equal but opposite momentum.

general, brittle alloys such as Nb_3Sn or ceramic high-T_c superconductors require confinement by a conducting metal matrix for thermal and shape stability reasons, and to lessen hazardous short-circuits and the release of potentially toxic substances. Hence, the production of wires requires ductile metal tubing into which fine powders of superconducting materials will be filled, heated under pressure, and hammered and/or drawn to the desired dimensions (see for example, Mamalis *et al.*, 2000). Single wires may be combined to multi-filamentary strands, tapes, and cables. The tube materials are copper for Nb_3Sn, copper or silver for YBCO, silver for BSCCO, and stainless steel for MgB_2. The filled tubes are drawn successively from millimeter-diameter tubes down to μm-size filaments. The performance is sensitive not only to composition but also to compositional uniformity.

9.7.1 Nb_3Sn Fabrication Processes

Currently, three large-scale fabrication processes are used commercially to produce conventional type II Nb_3Sn superconductors. A collection of characteristic materials parameters for Nb_3Sn is listed in Table 9.4, while an overview of some important processing details is given in Table 9.5. Because the Sn source is spatially separated from the Nb, solid-state diffusion determines the transport of Sn into Nb to form the desired A15 compound. The wires are drawn to their final diameter usually after coil winding, while the nonreacted material remains ductile. The wires need to be twisted, and their lengths exceed 1 km at a diameter of 1 mm. Superconducting coils from Nb_3Sn alloy have been considered for the ITER (International Thermonuclear Experimental Reactor) magnetic confinement fusion project at Cadarache, France. In this case, current densities of about 3000 A mm^{-2} at 12 Tesla and upon liquid helium cooling were reported (see Table 9.5).

Table 9.4 Materials parameters for Nb_3Sn.

Parameters	Value
Critical temperature T_c (K)	18.1
Cubic lattice parameter a (nm)	0.5293
Tetragonal distortion a/c at 10 K	1.0026
Martensitic transformation T_m (K)	43
Sommerfeld constant γ (mJ K^{-2} mol^{-1})	137
Debye temperature Θ (K)	234
Upper critical field $\mu_0 \cdot H_{c2}$ (T)	25
Thermodynamic critical field $\mu_0 \cdot H_c$ (T)	0.52
Lower critical field $\mu_0 \cdot H_{c1}$ (T)	0.038
Ginzburg–Landau coherence length ξ_{co} (nm)	3.6
Penetration depth λ (nm)	124
Ginzburg–Landau ratio $\kappa = \lambda/\xi_{co}$	34
Energy gap Δ (meV)	3.4
Electron-phonon interaction constant λ^{eph}	1.8

9.7.2 MgB_2 Fabrication Process

MgB_2 is an interesting new material with a hexagonal layer structure (Figure 9.2b) and a critical temperature of T_c = 39 K. In order to produce wires, a combined PIT-HIP forming technique is applied whereby MgB_2 powder is confined in a thin stainless steel tube that is coiled and exposed to a hot isostatic pressure consolidation process. This reduces the porosity and introduces additional structural defects, both of which are thought to increase the wire's current-carrying capacity. In addition, a 5% substitution of carbon for boron increases the wire's ability to withstand magnetic fields from 16 Tesla to 36 Tesla, thus exceeding the critical field of 25 Tesla of Nb_3Sn. This also lowers the critical temperature slightly, from 39 K to 35 K. The neon used as refrigerant (it is less expensive than helium) is produced industrially by the cryogenic fractional distillation of liquefied air. Low-cost magnetic resonance imaging (MRI) scanners based on MgB_2 technology may become available in the near future.

9.7.3 Processing of HTS Cuprates

The most important prerequisites for understanding the complicated phase relations of multi-element systems are exact temperature–concentration diagrams or, at least (and more practically) reduced pseudo-binary diagrams, some of which have been excellently summarized by Šesták (1992) for YBCO and Bi compounds, and by Majewski (1997) for (Bi,Pb)-2223. Since contemporaries have claimed these diagrams to be uncooperative due to their complexity, usually only minimal information is being extracted from these diagrams (e.g., the incongruent melting

Table 9.5 Large-scale fabrication for superconducting Nb_3Sn alloy.

	Bronze process	Internal Tin (IT) process	Powder-in-tube process (PIT)
Starting assembly	Nb filaments in a CuSn (bronze) solid solution matrix, Cu is protected by a barrier	Sn core surrounded by Cu embedded in Nb rods, Sn diffusion barrier[a]	Nb_3Sn embedded in Nb rods bundled in a matrix of pure Cu
Heat treatment	650–700 °C, 50–100 h	Around 700 °C	700 °C, 2 days, protective atmosphere
Main advantage	Very fine filaments. Wire fabrication is straightforward	Higher Sn to Nb ratio, drawing to final size without intermediate annealing possible	High Sn activity, inhibited grain growth and well-separated filaments
Disadvantage	Limited Sn solubility in the bronze, work hardening, repeated annealing after drawing required, Sn gradient in A15 structure	Large effective filament size and filament interconnection	Nonductility of the powder core
Filament dimensions	26 000, 1 μm	70 to 200 μm	30 to 50 μm
Current density ($A\,mm^{-2}$) at 4.2 K and 12 T	1000	3000	2300

a) Rod re-stack process (RRP).

point) that depend on oxygen pressure and, as in the case of Bi compounds, also on the lead content and the silver chosen as the tube material.

9.7.3.1 Bulk Material and Tapes from YBCO

The incongruent melting point of YBCO (123), depending on the oxygen partial pressure, is approximately 1020 °C. From a melt of 123-composition, the green Y_2BaCuO_5 phase (211) will first be formed. However, the quenching of a molten 123 mixture promotes the formation reaction at a lower temperature, due to the enhanced proximity of the constituent atoms. This melt processing technique (or melt–quench–melt growth process) employs the partial or complete melting (1040 °C) of the material, while the remaining crystallites of the 211 phase possibly act as pinning centers. Invariably, subsequent long-lasting (24 h) oxidative annealing between 600 °C and 400 °C, under an O_2 atmosphere, is required to form the superconducting phase. The melt-texturing process is used for the directional solidification of approximately 1 µm-thick layers of YBCO on substrates. For instance, the semi-continuous fabrication process of YBCO tapes of the second generation (2G) used 1 µm-thick films of YBCO on 2 µm Ag, surrounded by 20 µm of a copper stabilizer on each side. This only 50 µm-thick tape has a superior bend strain characteristic. A coil was wound from 462 m of the tape that held the world record of 26.8 Tesla (NHMFL Lab), with a critical current density of 1569 A mm^{-2} and a coil constant of 44.4 mT A^{-1}. Such a tape can carry a current of 72–82 A at 77 K and at self field produced by the transport current. High-field wires of the second generation (2G) from REBCO (RE = Y, Eu, Er) with effective pinning centers through Zr substitution can withstand high critical magnetic fields of up to 70 T.

As melt-textured samples cannot directly be produced in sufficiently large sizes for special applications, several other techniques have recently undergone successful testing to join textured blocks to larger monoliths. In contrast, bulk YBCO samples can much more easily be shaped at low oxygen pressure and temperature below 700 °C, via the formation of $YBa_2Cu^{1+}Cu_2^{2+}O_6$; this is followed by oxidation and hot-isostatic pressing (HIP) to yield very compact material in different shapes.

9.7.3.2 **BSCCO Wires and Tapes**

The relevant superconducting compounds for technical application are Bi-2212, and pure or Pb-substituted Bi-2223. At a low cooling temperature (liquid helium), Bi-2212 will perform better than Bi-2223, but the latter compound has proved to be superior after liquid nitrogen cooling. Wires produced to date belong to the first generation (1G). The Bi-2212 compound can easily be synthesized via a single-step powder process by the calcining of stoichiometrically mixed oxides or carbonates. In this case, Bi-2212 melts incongruently at 870 °C under 0.2 atm oxygen partial pressure, and at 830 °C in a pure N_2 atmosphere. The copper-rich compounds of the homologous series each also show incongruent melting, with only slightly increased temperatures; thus, the formation process of Bi-2223 is more elaborate. The substitution of lead for bismuth not only stabilizes this phase but also raises the incongruent melting point and the rate of phase formation (the preferred composition is the phase $Bi_{1.6}Pb_{0.4}Sr_2Cu_3O_{10+\delta}$). It is convenient to use a precursor mixture of Bi-2212, Ca_2PbO_4 and CuO, or the carbonates as needed, to form the desired amount of Bi-2212 in the mixture. Sintering of the pressed precursor pellets in a reducing atmosphere at approximately 820 °C for more than 20 h still resulted in a multiphase assembly, and this was also suitable for the final powder-in-tube process. In particular, the heating must be carried out carefully so as to avoid the inclusion of any remaining carbonate into the partially melted cuprate droplets, as any carbon present will be detrimental to the material's superconductivity. For the melt-textured solidification process, the presence of incongruently dissolved Ag is crucial since, on cooling, this will precipitate uniformly throughout the specimen, thus supporting the formation of (Bi,Pb)-2223 from (Bi,Pb)-2212.

The strongly anisotropic shape of the crystallites favors the development of a texture parallel to the surface of the Ag tube in powder-in-tube (PIT) processing, and this is highly desirable for the superconducting properties. The Ag content also serves as a source of pinning centers that hinder the flux line motion.

In the multi-step PIT process, silver tubes of 8 mm diameter and 2 mm wall thickness are filled with calcined powder, after which the tube diameter is first reduced through wire swaging and drawing to 1.3 mm, and then further reduced by cold rolling to 0.1 mm. This is followed by annealing to partial melting, with subsequent uniaxial cold pressing at 1 GPa and a finishing annealing process. It should be noted that the details of this process are the subjects of various patents, and consequently will differ widely.

The demands on the mechanical strength of the wires are high. Strain is produced by the coil fabrication, through thermal dilation, and by the action of the strong Lorentz force $\sigma = j \cdot B \cdot r$ during service. Assuming a current density of $j = 100\,\text{A mm}^{-2}$, magnetic field strength $B = 20\,\text{T}$, and a coil length r of only 10 cm, the Lorentz force yields a remarkably high stress of $\sigma = 100\,\text{MPa}$. Multi-filament strands and tapes are produced using more than 5000 single fibers. Both, Bi-2212 and (Bi,Pb)-2223 wires and cables are marketed by companies such as Nexans, the American Superconductor Corp. (AMSC), and VAC Hanau. The production of bulk superconducting materials from Bi cuprates is also straightforward. Here, the graphite-like material can easily be compacted hot-isostatically into large tubes with reduced porosity and pronounced texture; an example is the production of magnetic shielding panels.

9.8 Applications of Ceramic Superconductors

The range of applications of superconducting materials is potentially enormous, and includes specialized uses as thin layers and single crystals, apart from bulk ceramic materials. Some market projections are listed in Table 9.6, from which it is clear that LTS, at least for the immediate future, will command a market share far exceeding that of HTS.

Many relevant properties and already commercially available applications of ceramic superconductors are listed in Tables 9.7 and 9.8; it is important to note that LTS operate at low temperatures, close to 4 K (liquid helium), while HTS operate from the temperature of liquid helium to 77 K (liquid nitrogen). Although limited details of these materials are described in the following sections, the electronic applications which often employ thin films or type I superconducting metals have been partly omitted.

9.8.1 Cables for Power Grids

The exploitation of future technological successes, culminating in the fabrication of km-long superconducting cables and ribbons, offers the chance to install power

Table 9.6 Estimated global market shares for superconductor technology (in € million) (Conectus, 2007).

Business field	Year			
	2004	2007	2010	2013
Research & Technology Development (RTD)	600	660	835	955
Magnetic resonance imaging (MRI)	2950	3300	3410	3525
Total of RTD and MRI	3550	3960	4245	4480
New large-scale applications	35	65	150	410
New electronic applications	65	60	125	210
Total of emerging new businesses	100	125	275	620
Total market	3650	4085	4520	5100
Market share for LTS	3610	4025	4350	4600
Market share for HTS	40	60	170	500

Table 9.7 Critical superconducting properties of commercially used ceramic type II superconductors.

Superconductor	T_c (K)	J_C (A mm^{-2})	B_C (Tesla)
Bi-2223	110	>130	>30
Bi-2212	95	130	>30
REBCO[a)]	92	120	26.8 to 70
MgB_2	39	520 (self-field)	36
Nb_3Sn	18.1	100	25
NbTi	17.5	20	15

a) RE = Y, Eu, Er.

supply grids that are up to fivefold more efficient than those operated with conventional Cu wires. Moreover, the cables can be installed underground, and the current does not need to be transformed. Today, it is timely to give this technology its rightful chance. In 2008, the Holbrook Superconductor Project on Long Island, New York–the world's first large transmission voltage superconductor project–moved to its second phase. The cooperating companies [Nexans, American Superconductor Corp., Air Liquide, and local (LIGA) as well as national (DOE) US authorities] planned to utilize cost-efficient HTS tapes of the second generation and to develop solutions for cable joining. At full capacity, the HTS system would be expected to transmit up to 574 MW of electricity. With regards to the cable size, it is encouraging that the HTS ribbon-shaped wires conduct 150-fold the amount of electricity carried by similar-sized Cu wires. The Bi2223/

Table 9.8 Important contemporary and potential applications of superconductors.

Sensorics	**High-frequency techniques**	**Microelectronics**	**Ultra-fast optoelectronics**
SQUID[a] magnetometers	Passive electronic circuits	Conducting guidance for chips	Photodiodes
MIR[b] sensor coils	Oscillators	Transistors	Optical-to-FSQ[c] transducers
IR bolometers	Mixers and receivers	Digital switching	Telecommunication
X-ray detectors			
Energy techniques	**Magnetic techniques**	**Further magnetic applications**	**Medical imaging techniques**
Power cables	Accelerator magnets	Magnetic screening	MRI[d]
Generators, motors, transformers	Fusion magnets, tokamaks	Magnetic bearings	CT[e]
Current limiters, FCL[f]	Ultrahigh-field magnets NMR	High-gradient magnetic separation	
SMES[g]	Magnetic levitation trains, wind turbines, flying wheel	Ultra-low temperatures through nuclear demagnetization	
	Current connectors	MHD magneto-hydrodynamics	

a) Superconducting quantum interference device.
b) Micropower impulse radar.
c) Single flux quantum logical device.
d) Magnetic resonance imaging.
e) Computed tomography.
f) Fault current limiters.
g) Superconducting magnetic energy storage (rings).

Ag composite was chosen as the superconducting material, and liquid nitrogen cooling was planned.

9.8.2
Superconducting Magnets

Superconducting magnets are, in general, high-current systems using current densities in excess of $100\,A\,mm^{-2}$. As depicted in Figure 9.8, their applications range from low-field magnets up to ultrahigh-field NMR magnets that can deliver frequencies exceeding 1 GHz. Oxford Instruments have reported Internal-Tin route-produced strands of Nb_3Sn for superconducting magnets with an engineered critical current density of $j_e > 150\,A\,mm^{-2}$ at 23.5 Tesla. A 950 MHz instrument has already been tested.

Currently, a steadily growing market for low-field magnets lies in the huge demand for MRI instruments used widely in diagnostic medicine. Likewise, generators for automotive applications will benefit from superconducting magnets, with considerably reduced weights owing to a lack of iron cores. This allows the power of the engine to be increased remarkably, and the efficiency improved. Fault current limiters (FCLs) are devices designed to limit currents in case of failure within the power-transmitting networks. These units, which are currently under development, are produced not only from MgB_2 tapes but also from BSCCO and YBCO.

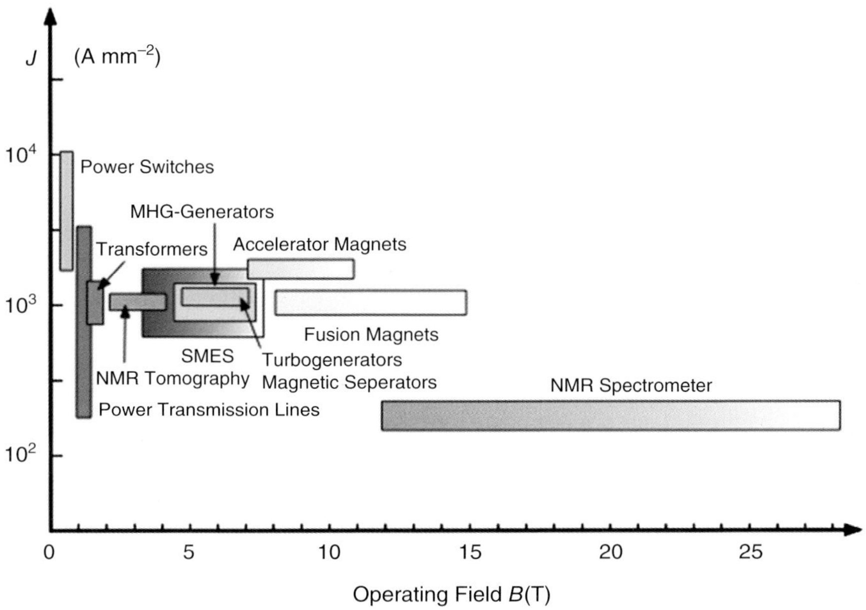

Figure 9.8 Scheme depicting superconductor current densities and the operating magnetic field for magnets used in diverse applications (this is an updated version of the illustration of Komarek (1995).

9.8.3
Magnetic Levitation and Combined Bearings

This effect has, arguably, a great technological future. In fact, the widely publicized images of levitation of a supercooled ceramic superconductor disk magically floating some distance above a strong permanent magnet has shaped the public's perception of superconductivity more than anything else. The physical background of this effect (the Meissner–Ochsenfeld effect) is an expulsion of the magnetic field from the perfectly diamagnetic material (see Figure E.1 in Appendix E). The technical application of magnetic levitation requires a high magnetic field and also a high field gradient. Already realized applications include the floating Maglev train, the "Flying Wheel" energy storage technique, a new generation of high-power (1 GW) wind turbines, and elaborate projects such as NASA's proposal for airspace catapults to transport spacecraft and satellites cost-efficiently into orbit around the Earth.

Special features are required for further applications. For instance, a large volume at high magnetic field of magnets used for nuclear demagnetization to achieve ultra-low temperatures, and for high-gradient magnetic separator (HGMS) applications a high magnetic field gradient at the surface of the separation medium. Further applications are listed in Table 9.8.

Stable magnetic bearings have also been applied favorably, using bulk YBCO material for equipment with rotary motions (e.g., generators, motors, flywheel, wind turbines).

9.8.4
Superconducting Magnetic Energy Storage (SMES)

Superconductor energy storage technology can provide very high power almost instantaneously, because of the very short delay of charge and discharge times. Although, at present, this system is only in the engineering test phase (American Superconductor Corp.), the technology may become important in the future. The energy is stored in the magnetic field created by the flow of direct currents in a superconducting coil which has been cooled to below its superconducting transition temperature. The stored energy, E, is a function of carrying current I and coil parameters such as its geometric shape parameter f, radius R, and the number of turns N, yielding $E = 0.5\, f(\xi,\delta) \cdot R \cdot N^2 \cdot I^2$.

It has been suggested that SMES rings may function in tandem with nuclear power reactors to provide a loss-less storage and retrieval of electric energy produced by nuclear fission. The cost of storage systems depends on many factors, including the level of the stored energy (1 GW·h may be achieved commercially), the mechanical support required to counteract the Lorentz force, the critical current, the coil material, and the refrigerator and power-conditioning systems. Whether, ultimately, HTS or LTS systems will be more economical will depend on developments in the near future.

Another possibility for storing energy by applying superconductor technology is the frictionless flying wheel energy storage (flywheel).

9.8.5
Magnetic Shielding

It is essential for national security that vital electronic systems are effectively shielded against destruction in the case of a nuclear weapons attack, which would be accompanied by a strong electromagnetic pulse. In fact, such a shielding technique has begun to emerge, with magnetic shielding capable of preventing the penetration of the Earth's magnetic field into sensitive measurement systems having already been developed.

9.8.6
Electronic Applications

Most electronic applications are based on the tunnel effect between two superconductors that are separated by a weak (isolating) link that is only a few atomic layers thick. Currently, Nb and Al are used preferably to build Josephson tunnel junctions (JTJs), whereas ceramic superconductors are still less common. However, the situation may soon change following the discovery of coherently emitting tera-Hertz radiation of intrinsic BSCCO JTJs (Ozyuzer *et al.*, 2007). Some aspects of electronic applications are considered in detail at this point.

The most attractive feature of LTS applications for photon detection is the very low energy level around 1 meV, this being sufficient to generate a charge carrier due to the weak bond of electrons in LTS. In contrast, in a semiconductor, approximately 1 eV is required to generate a single charge carrier. When a photon is absorbed at either electrode of a JTJ, its energy ($h \cdot \nu$) generates a cascade of broken Cooper pairs, resulting in two energetic quasiparticles per broken pair. The phase difference between the energy levels in both superconductors can be influenced by an applied direct voltage U_T to generate a Josephson a.c. current of frequency $f = 4\pi \cdot e \cdot U_T h^{-1}$. The combination of two JTJs in a superconducting ring leads to a device that can measure ultra-weak magnetic fields (10^{-14} T) via quantum interference of the branched currents resulting from an applied d.c. current (direct current SQUID).

Recently, new X-ray detectors have been fashioned from arrays of JTJs that have an astonishingly low energy resolution of 12 eV (FWHM) at 5.9 keV X-ray energy, using Pb and Al as the superconducting materials (Angloher *et al.*, 2001).

With regards to the BSCCO high-T_c material consisting of stronger insulating layers around Bi, sandwiched by metallic cuprate layers, it was shown recently (Ozyuzer *et al.*, 2007) that this structure can operate like a stack of intrinsic JTJs. In this case, a fixed voltage applied to the c-direction produces an oscillating current in the THz frequency range. Shifting the working current frequency to the resonating mode of the entire sample yields a (laser-like) in-phase emitting mode of all junctions. Clearly, the availability of coherent THz radiation will lead to the development of many new applications.

Superconducting materials allow the creation of ultra-fast and ultra-sensitive single-flux quantum logic devices. While the incoming radiation breaks the Cooper pairs, thereby heating up the electron gas, the detector responds in an ultrafast fashion before the mobile carriers can reach thermal equilibrium with the lattice. There are no bandgap limitations for superconductors, thus ensuring a uniform absorption within the entire spectrum, from infra-red to ultra-violet. When close to the critical temperature, a very fast superconducting NbN photodetector will respond with a time delay of less than 30 ps (Sobolewski, 1999).

In telecommunications technology, HTS filters (constructed from YBCO thin films) provide near-perfect filters that allow almost 100% of a desired signal to pass through, while rejecting 100% of an unwanted signal.

Finally, *spintronics* represents a new field of research, development, and applications that utilizes electron spins for high-control quantum electronics. Recently, the depairing of Cooper pairs in low-T_c superconductors such as Al metal was observed experimentally in superconductor–ferromagnet hybrid structures (Beckmann *et al.*, 2004). As the paired state can survive within a characteristic distance when soaked from the superconductor into normal metal wires, electrons with oppositely oriented spins can be sampled in adequately magnetized wires, provided that the wire separation is within, but close to, the coherence length of the Cooper pairs.

9.9 Outlook and Future Developments

The emergence of superconductor technology in the near future will help to utilize more economically the steadily decreasing conventional global energy resources so as to satisfy the ever-increasing demands for power of a rapidly growing world population. Moreover, the strongest high-field superconducting magnets ever built may soon be capable of compressing the plasma in nuclear fusion reactors for several seconds or even minutes, thereby delivering energy from this future source to the growing global population, ensuring its fair sharing and making the world "green and peaceful." The recent discovery of pnictide superconductors, derived from naturally occurring minerals, has shown very impressively that Nature has in fact beaten humankind at their own game on the road to room-temperature superconductivity.

From an historic standpoint, however, four issues at the onset of superconducting technology have hindered its proliferation:

- In the past, when a strong current was passed through a superconductor, its superconductivity was destroyed by the magnetic field created, before any appreciable values for utility applications could be achieved.
- Reliance on strategic materials such as the rare-earth elements Y, Sc, and Bi to synthesize HTS will most likely lead to problems of sustaining raw materials availability in case of political conflicts.
- Expensive refrigeration units and high power levels are required to maintain low operating temperatures.

- The existence and continued development of adequate competing technologies based on normal electric conductor technologies will tend to slow the development of HTS.

Although these issues still pose problems for superconducting applications, they are predicted to improve over time. Recent advances in terms of the performance of superconducting materials bear testimony to the vigorous developments in this field. Furthermore, the reliability and efficiency of refrigeration systems has improved significantly, to the point where some devices can now operate on hybrid electrical power systems. And further exciting developments are surely waiting in the wings? Unfortunately, the development of high-temperature superconductors may lead to product discontinuity for conventional conductors and the machinery and devices made from them, including wire and transmission cables, electric motors and generators, and microcircuit components (Heimann, 1991). Hence, the acceptance of high-temperature superconductors by industry, with resultant large-scale global applications, must first break through the technological equilibrium which was established long ago. But, for the moment, such an achievement seems to be far away (see Table 9.6).

References

Abrikosov, A.A. (1957) On the magnetic properties of superconductors of the second group. *Zh. Eksp. Teor. Fiz.*, **32**, 1442–1452 (in Russian).

Anderson, P.W. (1987) The resonating valence bond state in La_2CuO_4 and superconductivity. *Science*, **235** (4793), 1196–1198.

Angloher, G., Hettl, P., Huber, M., Jochum, J., von Feilitsch, F., and Mößbauer, R.L. (2001) Energy resolution of 12 eV at 5.9 keV from aluminum superconducting tunnel junction detectors. *J. Appl. Phys.*, **89**, 1425–1427.

Aurivillius, B. (1949) Mixed bismuth oxides with layered lattices. *Arki. Kemi*, **1**, 463–480.

Bardeen, J., Cooper, L.N., and Schrieffer, J.R. (1957) Microscopic theory of superconductivity. *Phys. Rev.*, **108**, 1175–1204.

Beckmann, D., Weber, H.B., and von Löhneysen, H. (2004) Evidence for crossed Andreev reflection in superconductor-ferromagnet hybrid structures. *Phys. Rev. Lett.*, **93**, 197003–197006.

Bednorz, J.G., and Müller, K.A. (1986) Possible high T_c superconductivity in the Ba-La-Cu-O system. *Z. Phys.*, **B64** (1), 189–193.

Chevrel, R., Sergent, M., and Prigent, J. (1971) Sur de nouvelles phases sulfurées ternaires du molybdene. *J. Solid State Chem.*, **3**, 515–519.

Conectus (2007) Consortium of European Companies Determined to Use Superconductors, www.conectus.org (accessed 14 December 2009).

Dai, P., Chakoumacos, B.C., Sun, G.F., Wong, K.W., Li, Y., and Lu, D.F. (1995) Synthesis and neutron powder diffraction study of the superconductor $HgBa_2Ca_2Cu_3O_{8+\delta}$ by Tl substitution. *Physica C*, **243**, 201–206.

Doll, R. and Nähbauer, M. (1961) Experimental proof of magnetic flux quantization in a superconducting ring. *Phys. Rev. Lett.*, **7** (2), 51–52.

Endo, U., Koyama, S., and Kwai, T. (1988) Preparation of the high-T_c phase of Bi-Sr-Ca-Cu-O superconductor. *Jpn. J. Appl. Phys.* **27**, L1476–L1479.

Er, G., Miyamoto, Y., Kanamura, F., and Kikkawa, S. (1991) Superconductivity in the infinite-layer compound $Sr_{1-x}La_xCuO_2$

prepared under high pressure. *Physica C*, **181** (1–3), 206–208.

Eremets, M.I., Trojan, I.A., Medvedev, S.A., Tse, J.S., and Yao, Y. (2008) Superconductivity in hydrogen dominated materials: silane. *Science*, **319** (5869), 1506–1509.

Essmann, U. and Träuble, H. (1967) The direct observation of individual flux lines in type II superconductors. *Phys. Lett.*, **24 A** (10), 526–527.

Fauqué, B., Sidis, Y., Hinkov, V., Pailhès, S., Lin, C.T., Chaud, X., and Bourges, P. (2006) Magnetic order in the pseudogap phase of high-T_c superconductors. *Phys. Rev. Lett.*, **96**, 197001 (4 pages).

Fröhlich, H. (1954) On the theory of superconductivity: the one-dimensional case. *Proc. R. Soc. London A*, **223**, 296–305.

Gao, L., Xue, Y.Y., Chen, F., Xiong, Q., Meng, R.L., Ramirez, D., Chu, C.W., Eggert, J.H., and Mao, H.K. (1994) Superconductivity up to 164 K in $HgBa_2Ca_{m-1}Cu_mO_{2m+2+\delta}$. *Phys. Rev. B*, **50**, 4260–4263.

Gavaler, J.R. (1973) Superconductivity in Nb-Ge films above 22K. *Appl. Phys. Lett.*, **23**, 480–482.

Giaever, I. (1960) Electron tunneling between two superconductors. *Phys. Rev. Lett.*, **5**, 464–466.

Ginzburg, V.L. and Landau, L.D. (1950) On the theory of superconductivity. *Zh. Eksp. Teor. Fiz.*, **20**, 1064–1082 (in Russian).

Hannay, N.B., Geballe, T.H., Matthias, B.T., Andres, K., Schmidt, P., and MacNair, D. (1965) Superconductivity in graphitic compounds. *Phys. Rev. Lett.*, **14**, 225–226.

Hebard, A.F., Rosseinsky, M.J., Haddon, R., Murphy, R.C., Glarum, D.W., Palstra, S.H., Ramirez, T.T., and Kortan, A.P. (1991) Superconductivity at 18 K in potassium-doped C_{60}. *Nature (London)*, **350** (6319), 600–601.

Heimann R.B. (1991) Technological progress and market penetration of advanced ceramics in Canada. *Am. Ceram. Soc. Bull.*, **70** (7), 1120–1127.

Heimann, R.B. (2004) Applied mineralogy – an important driving force towards a sustained development of future technologies, in *Applied Mineralogy. Developments in Science and Technology*, Proceedings ICAM 2004, vol. **1** (ed. M. Pecchio, *et al.*), International Council of Applied Mineralogy do Brasil, Sao Paulo, Brasil, pp. 3–11.

Heimann, R.B., Evsyukov, S.E., and Kavan, L. (eds) (1999) *Carbyne and Carbynoid Structures*, Physics and Chemistry of Materials with Low-Dimensional Structures, vol. **21**, Kluwer Academic Publishers, Dordrecht, The Netherlands, ISBN: 0-7923-5323-4, 444 pp.

Jerome, J., Mazaud, A., Ribault, N., and Bechgaard, K. (1980) Superconductivity in a synthetic organic conductor $(TMTST)_2PF_6$. *J. Phys. Lett.*, **41**, L95–L98.

Josephson, B.D. (1962) Possible new effects in superconducting tunneling. *Phys. Lett.*, **1**, 251–252.

Kamerlingh Onnes, H. (1911) The resistance of pure mercury at helium temperature. *Commun. Leiden*, **120b**, 3–5.

Kamihara, Y., Watanabe, T., Hirano, M., and Hosono, H. (2008) Iron-based layered superconductor $La[O_{1-x} F_x]FeAs$ ($x = 0.05$–0.12) with $T_c = 26$ K. *J. Am. Chem. Soc.*, **130**, 3296–3297.

Komarek, P. (1995) *Hochstromanwendungen der Supraleitung*. Teubner Studienbücher Physik/Elektrotechnik, Stuttgart.

Li, Y., Balédent, V., Barisic, N., Cho, Y., Fauqué, B., Sidis, Y., Yu, G., Zhao, X., Bourges, P., and Greven, M. (2008) Unusual magnetic order in the pseudogap region of the superconductor $HgBa_2CuO_{4+\delta}$. *Nature (London)*, **455** (7211), 372–375.

Little, W.A. (1964) Possibility of synthesizing an organic superconductor. *Phys. Rev.*, **134**, A1416–A1424.

London, F. and London, H. (1935) The electromagnetic equations of the supraconductors. *Proc. R. Soc. London A*, **149**, 71–88.

Maeda, H., Tanaka, Y., Fukutomi, M., and Asano, T. (1988) A new high-T_c oxide superconductor without a rare earth element. *Jpn. J. Appl. Phys.*, **27**, L209–L210.

Majewski, J. (1997) Phase diagram studies in the system Bi-Pb-Sr-Ca-Cu-O-Ag. *Supercond. Sci. Technol.*, **10**, 453–467.

Mamalis, A.G., Manolakos, D.E., Szalay, A., and Pantazopoulos, G. (2000) *Processing of High-Temperature Superconductors at High*

Strain Rates, CRC Press, ISBN: 9-781-566-768-788, 268 pp.

Matthias, B.T., Geballe, T.H., Geller, S., and Corenzwit, E. (1954) Superconductivity of Nb_3Sn. *Phys. Rev.*, **95**, 1435.

Meissner, W. and Ochsenfeld, R. (1933) Ein neuer Effekt bei Eintritt der Supraleitfähigkeit. *Die Naturwiss.*, **21** (44), 787–788.

Michel, C., Hervieu, M., Borel, M.M., Grandin, A., Deslandes, F., Provost, J., and Raveau, B. (1987) Superconductivity in the Bi-Sr-Cu-O system. *Z. Phys. B*, **68**, 421–423.

Mook, H.A., Sidis, Y., Fauqué, B., Balédent, V., and Bourges, P. (2008) Observation of magnetic order in a superconducting $YBa_2Cu_3O_{6.6}$ single crystal using polarized neutron scattering. *Phys. Rev. B*, **78**, 020506–020509.

Nagamitsu, J., Nakagawa, N., Muranaka, T., Zenitani, Y., and Akimitsu, J. (2001) Superconductivity at 39 K in magnesium diboride. *Nature (London)*, **410** (6824), 63–64.

Ott, H.R., Rudigier, N., Fisk, Z., and Smith, J.L. (1983) UBe_{13}: an unconventional actinide superconductor. *Phys. Rev. Lett.*, **50**, 1595–1598.

Otto, H.H. (1992–2002) "New Superconducting Materials", lectures at TU, Clausthal. Scriptum (in German) available from the author on request.

Otto, H.H., Ringshandl, S., and Baltrusch, R. (1998) (Tl, Pb, Cu)Sr_2(Tl, Pb, Cu)$_2Cu_2O_{8-\delta}$: a nonsuperconducting Tl-1222* cuprate with both heavy atom single and oxygen depleted double layer motifs. *Physica C*, **300** (3–4), 191–199.

Ozyuzer, L., Koshelev, A.E., Kurter, C., Gopalsami, N., Li, Q., Tachiki, M., Kadowaki, K., Yamamoto, T., Minami, H., Yamaguchi, H., Tachiki, T., Gray, K.E., Kwok, W.-K., and Welp, U. (2007) Emission of coherent THz radiation from superconductors. *Science*, **318**, 1291–1293.

Putilin, S.N., Antipov, E.V., Chmaissem, O., and Marezio, M. (1993) Superconductivity at 94 K in $HgBa_2CuO_{4+\delta}$. *Nature (London)*, **363**, 226–228.

Ren, Z.A., Lu, W., Yang, J., Yi, W., Shen, X.L., Li, Z.C., Che, G.C., Dong, X.L., Sun, L.L., Zhou, F., and Zhao, Z.X. (2008) Superconductivity in the iron-based F-doped layered quaternary compound Nd[$O_{1-x}F_x$]FeAs. *Europhys. Lett.*, **82**, 57002.

Schooley, J.F., Hosler, W.R., and Cohen, M.L. (1964) Superconductivity in semiconducting $SrTiO_3$. *Phys. Rev. Lett.*, **12**, 474–475.

Šesták, J. (1992) Phase diagrams in CuO_x-based superconductors. *Pure Appl. Chem.*, **64** (1), 125–136.

Sheng, Z.Z., Hermann, A.M., Vier, D.C., Schulz, S., Oseroff, S.B., George, D.J., and Hazen, R.B. (1988) Superconductivity in the Tl-Sr-Ca-Cu-O system. *Phys. Rev. B*, **38**, 7074–7076.

Siegrist, T., Zahurak, S.M., Murphy, D.W., and Roth, R.S. (1988) The parent structure of the layered high-temperature superconductors. *Nature (London)*, **334** (6179), 231–232.

Simon, M.E. and Varma, C.M. (2002) Detection and implications of a time-reversal breaking state in underdoped cuprate. *Phys. Rev. Lett*, **89**, 247003 (4 pages).

Sleight, A.W., Gillon, I.L., and Bierstedt, P.E. (1975) High temperature superconductivity in the $BaPb_{1-x}Bi_xO_3$ systems. *Solid State Commun.*, **17**, 27–28.

Smith, M.G., Manthiram, A., Zhou, J., Goodenough, J.B., and Mackert, T.J. (1991) Electron-doped superconductivity at 40 K in the infinite-layer compound $Sr_{1-y}Nd_yCuO_2$. *Nature (London)*, **351** (6327), 549–551.

Sobolewski, R. (1999) Ultrafast superconducting optoelectronics. *Proceedings, 12th Annual Meeting, Lasers and Electro-Optics Society (LEOS '99), IEEE 1999, 8–11 November, San Francisco, CA*, pp. 631–632.

Steglich, F., Aarts, J., Bredl, C.D., Lieke, W., Meschede, D., Franz, W., and Schäfer, H. (1979) Superconductivity in the presence of strong Pauli paramagnetism: $CeCu_2Si_2$. *Phys. Rev. Lett.*, **43**, 1892–1895.

Steward, G.R., Fisk, Z., Willis, J.O., and Smith, J.L. (1984) Possibility of coexistence of bulk superconductivity and spin fluctuations in UPt_3. *Phys. Rev. Lett.*, **52**, 679–682.

Takabayashi, A.Y., Ganin, P., Jeglic, D., Arcon, T., Takano, Y., Iwasa, Y., Ohishi, M., Takata, N., Takeshita, K., Prassides, K., and Posseinsky, M.J. (2009) The disorder-free non-BSC superconductor Cs_3C_{60} emerges from an antiferromagnetic insulator parent state. *Science*, **323** (5921), 1585–1590.

Takahashi, H., Igawa, K., Arii, K., Kamihara, Y., Hirano, M., and Hosono, H. (2008) Superconductivity at 43 K in an iron-based layered compound $LaO_{1-x}F_xFeAs$. *Nature (London)*, **453** (7193), 376–378.

Tang, Z.K., Zhang, L., Wang, N., Zhang, X.X., Wen, G.H., Li, G.D., Wang, J.N., Chan, C.T., and Sheng, P. (2001) Superconductivity in 4 angstrom single-walled carbon nanotubes. *Science*, **292** (5526), 2462–2465.

Tarascon, J.M., LePage, Y., Barboux, P., Bagley, P.G., Greene, L.H., McKinnon, W.R., Hull, G.W., Giroud, M., and Hwang, D.M. (1988) Crystal substructure and physical properties of the superconducting phase $Bi_4(Sr,Ca)_6Cu_4O_{16+x}$. *Phys. Rev. B*, **37**, 9382–9389.

Varma, C.M. (1999) Pseudogap phase and the quantum-critical point in copper-oxide metals. *Phys. Rev. Lett.*, **83**, 3538–3541.

Varma, C.M. (2006) Theory of the pseudogap state of the cuprates. *Phys. Rev.*, **B73**, 155113. (17 pages).

Wu, M.K., Ashburn, J.R., Torng, C.J., Hor, P.H., Meng, R.L., Gao, L., Huang, Z.Y., Wang, Y.Z., and Chu, C.W. (1987) Superconductivity at 93 K in a new mixed-phase Y-Ba-Cu-O compound system at ambient pressure. *Phys. Rev. Lett.*, **58**, 908–910.

Zhang, F.C., and Rice, T.M. (1988) Effective Hamiltonian for the superconducting Cu oxides. *Phys. Rev. B*, **37**, 3759.

10
Bioceramic Materials

10.1
Introduction

Bioceramic materials have developed into a very powerful driver of advanced ceramics research and development. For many years bioceramics, both bioinert materials such as alumina, zirconia and, to a limited extent titania (Lindgren *et al.*, 2009), and bioconductive materials such as hydroxyapatite, tricalcium phosphate and calcium phosphate cements, have been used successfully in clinical practice. In addition, applications continue to emerge that use biomaterials for medical devices. An excellent account of the wide range of bioceramics available today has recently been produced by Kokubo (2008), in which issues of the significance of the structure, mechanical properties and biological interaction of biomaterials are discussed, and their clinical applications in joint replacement, bone grafts, tissue engineering, and dentistry are reviewed. The type and consequences of cellular responses to a variety of today's biomaterials have been detailed in recent books (Di Silvio, 2008; Basu *et al.*, 2009; Planell *et al.*, 2009).

10.1.1
Scope and Socioeconomic Consequences

The number of patients receiving biomedical implants to correct skeletal defects and heal diseases continues to increase. Today, on a worldwide basis, there exists a huge demand for load-bearing hip, knee and dental endoprosthetic implants, as well as for bone replacement parts in the maxillar–mandibular area, the ossicular chain of the inner ear, and alveolar ridge and iliac crest augmentation. Currently, in the United States and in Europe, more than 1 000 000 hip and knee arthroplasties are performed annually, and the trend is increasing. By comparison, in Germany in 2007, the total number of hip and knee joint implants was 355 000.

The properties and functions of biomaterials – and in particular of bioceramics – are frequently discussed in the context of hip endoprosthetic implants. Hence, this chapter will focus on the most commonly utilized bioceramic materials such as alumina, stabilized zirconia (Y-stabilized Tetragonal Zirconia Polycrystal; Y-TZP), and calcium phosphates (notably hydroxyapatite), all of which are compo-

Classic and Advanced Ceramics: From Fundamentals to Applications. Robert B. Heimann

ISBN: 978-3-527-32517-7

nents of hip endoprosthetic implants. Some brief descriptions of bioglasses and resorbable bioceramics, including tricalcium- and tetracalcium phosphate as bone replacement materials for the treatment of large bone defects, will complement this foray into the world of inorganic biomaterials.

The increasing demand for endoprostheses is the result of the wear and tear to which the hip and knee joints are subjected during a human's lifetime. An average person walks about one million steps each year, with a frequency of about 1 Hz. Using a conservative step length of 0.64 m, this amounts – in an average lifespan of 75 years – to 4.7×10^7 load changes while walking a total distance of 47 000 km – which is about the circumference of the Earth. The load on the joints is approximately equal to the body mass whilst at rest (1 kN), but is two- to threefold the body mass during normal walking, up to fivefold while jogging, and up to eightfold when jumping. Beyond that threshold, the risk of damage to the joints increases dramatically. Today, as people tend generally to live longer and to become increasingly overweight due to overeating and lack of exercise, the protective tissue lining of the acetabular cup eventually wears away, the friction is increased, and inflammation, pain, and finally immobilization, will result. At this point, a total hip replacement (THR) is the only reasonable option to maintain mobility, freedom from pain, and hence a rewarding life in old age. Yet, with an increasing frequency younger people today require such surgery because their lifestyle might include damaging sports activities that promote premature wear on their joints. In this case, a problem can arises because younger patients will generally outlive the average lifetime of a contemporary implant; consequently, remediation surgery will likely be required, at substantial additional cost to the healthcare system of the country involved. The early onset of pain and associated mental stress on the patient adds yet another ethical dimension to this problem of the incompatible lifespans of the patient and the implant.

Today, however, an additional serious problem exists, namely that of *osteoporosis*. Increasingly, the loss of bone substances that occurs in elderly people leads to an embrittlement of the bones, with the associated risk of fracture. This effect is caused by a lack of calcium that, in turn, may be related to a nutrition that is too rich in meat, wheat and dairy products, all of which produce an acidic environment within the body. In consequence, the body will use calcium ions to maintain the pH of the blood serum within acceptable limits. The financial cost of this situation is substantial; in Germany, about 150 000 osteoporotic fractures of the neck of the femur occur annually, this being 40-fold the number occurring among the population of Thailand. It is not surprising, therefore, that in 2002 the World Health Organization (WHO) included osteoporosis among the ten most frequently encountered diseases worldwide.

10.1.2
Basic Aspects of Biomineralization

Human bone is a strong, tough, and highly durable composite material which consists of about 70% micro- to nanocrystalline biological apatite and 30% microfibrils of collagen I, all of which are organized in a hierarchical manner, as shown in Figure 10.1 (Pasteris *et al.*, 2008). Nanocrystals of bioapatite that are about $30 \times 50 \times 2\,nm^3$ in size (G and H in Figure 10.1) are arranged with their c-axes parallel to the extension of five collagen molecules which themselves consist of triple helical strands of collagen fibers to form a microfibril (F). These inorganic–organic composite microfibrils are bundled together to form larger fibrils (E) that are, in turn, grouped to form even larger mineralized fibers (D). Hence, a spatially hierarchical organization exists that forms the basic structural units of bone (Glimcher, 2006). During the formation of bone, the collagen I matrix develops first; this is followed by a second step in which the hydroxyapatite nanocrystals become embedded. This mutual arrangement is subject to dynamic processes in which bone matter is formed by cells known as *osteoblasts* and subsequently dissolved by *osteoclasts*, according to a mechanical demand that is triggered by the presence or absence of stress and according to the well-known Wolff's law. During the process, a significant role is played by the piezoelectric nature of bone, as this provides the stimuli for the growth and resorption of bone cells (Bassett *et al.*, 1974). Bones that bear weight are subjected to tensile stresses along their circumference; this in turn provide piezoelectric signals to release the bone growth proteins that will not only strengthen the bone but also lead to an increase in its diameter. In the absence of any such loading, however, the osteoclasts will dominate to dissolve the bone, and in such a situation the bones will show signs of atrophy. In particular, under low-gravity conditions (as are experienced in outer space) the deterioration in bone density is very rapid. Indeed, other than exposure to fields of strongly ionizing radiation in outer space, this effect is considered to be the most serious impediment to long-term space travel, especially as it apparently cannot be overcome even by maintaining vigorous exercise.

The intimate intergrowth of nanosized hydroxyapatite platelets and collagen microfibrils (see Figure 10.1, item F) is at the heart of the extraordinary strength, yet extreme flexibility, of bone (Currey, 2004).

10.1.3 Design of Endoprosthetic Implants

Although the stems and femoral balls of hip implants can be fashioned from austenitic surgical stainless steel or cobalt chromium molybdenum alloy, the current state of the art for hip endoprosthetic implants with extremely low wear rates (Spinelli *et al.*, 2009) is a shaft manufactured from bioinert Ti6Al4V or Ti6Al7Nb alloys and equipped with an alumina femoral ball that articulates against an acetabular cup anchored in the hip bone (Figure 10.2). This cup is made from cp-Ti or Ti alloy, and lined with either ultrahigh-molecular-weight polyethylene (UHWM-PE) or, in recent developments, with alumina so as to assure a low coefficient of friction when articulating against a femoral ball that is also made of alumina. This is desirable as the natural highly viscous synovial fluid (which is based on hyaluronic acid, specific proteins and glycoproteins, and acts as a highly efficient lubricants) is still absent from the implant system, despite

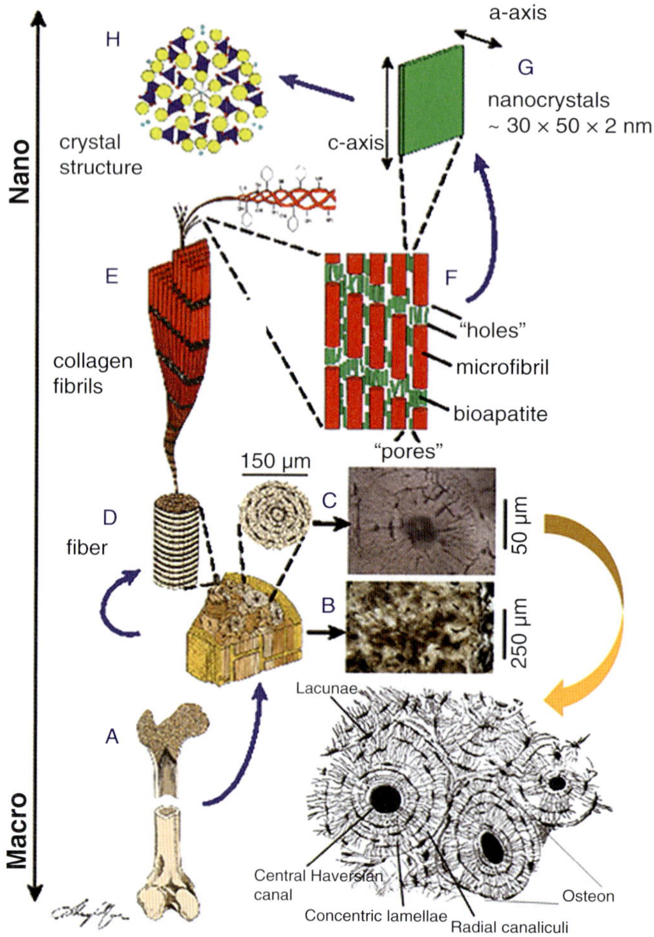

Figure 10.1 Schematics of the hierarchical architecture of cortical bone. (A) Longitudinal section of femur; (B) Enlarged cross-section of cortical bone showing cylindrical osteons; (C) Enlargement of an osteon, showing the central Haversian canal with a blood vessel, the concentric lamellas and the radial canaliculi. A more detailed view of an osteon is shown in the inset in the bottom right; (D) Collagen fiber composed of hundreds of fibrils. The evenly spaced dark spirals are periodic gaps shown in more detail in (F); (E) An array of five collagen molecules, constituting the smallest organic structural unit of the bone; (F) Enlargement of a collagen microfibril showing the oriented intergrowth with hydroxyapatite nanocrystals; (G) Individual crystalline platelet of hydroxyapatite; (H) Crystal structure of fluorapatite as a stand-in for the more complex biological apatite (modified after Pasteris *et al.*, 2008). Reprinted with kind permission from The Mineralogical Society of America.

Figure 10.2 Hip joint endoprosthesis system consisting of an alumina femoral ball attached to a Ti6Al4V stem coated with hydroxyapatite (right) and a matching acetabular cup lined with alumina (left). Illustration courtesy of Prof. Dr Gert Willmann, (deceased) CeramTec AG, Plochingen, Germany.

extensive attempts to adapt synovial fluid constituents for this purpose (Roba *et al.*, 2009).

The technology of THR has been solidly established from a clinical standpoint, with many engineering solutions having been devised to vary the form, length, and surface details of the metallic stems, as well as the diameter, neck length and type of the inside taper of the ceramic femoral balls – all of which are constructed from either alumina or Y-stabilized zirconia. In addition, a thin stable bioactive hydroxyapatite coating is frequently applied to the metallic stem; this will elicit a specific biological response at the interface of the implant material by controlling its surface chemistry through the adsorption of noncollagenous proteins. This in turn will result in a strong and lasting osseoconductive bond between the living tissue and the biomaterial. Further details of the function and performance requirements of hydroxyapatite coatings are discussed below.

In this chapter, the discussions will be limited to three typical bioceramic materials – alumina, Y-stabilized zirconia, and hydroxyapatite – all of which are applied as components of hip endoprostheses to restore the mobility and well-being of an increasing number of patients. As discussed above, there is today a tendency for the first-time recipients of such implants to be younger, and consequently there is an increasing demand for the long-term *in vivo* stability of the individual components of the implant system. This includes not only the fatigue strength and corrosion resistance of the bioinert metallic stem, but also the bending and shear strengths, as well as the surface frictional properties of the monolithic ceramics parts and the polymeric liners of the acetabular cups. Attention must also be paid to the cohesive and adhesive strengths of these materials, as well as the resorption resistance of the bioactive hydroxyapatite coatings. However, as mentioned above, it is clear that an increasing proportion of patients will outlive the expected lifetime (presently 15–20 years) of their hip endoprostheses. Consequently, investigations are being undertaken worldwide in an attempt to produce improved biomaterials with increased mechanical, chemical, biological and frictional properties that can be sustained long-term within the body's environment (Lee *et al.*, 2006).

10.2
The Concept and Definition of Biocompatibility

Any material that is incorporated into a human organism must abide by certain properties, so as to ensure that there are no negative interactions with living tissues. Biomaterials, by definition, are inorganic compounds that are designed to replace a part or a function of the human body in a safe, reliable, economic, and physiologically and esthetically acceptable manner (Hench and Ethridge, 1982). But, as *biomaterials* are inorganic structures, they do not include renewable "*biological*" materials obtained from natural sources such as wood, plant fibers, hides, sinew, bone, ivory, and others.[1)]

One of the most important properties of biomaterials is their so-called *biocompatibility*. This is not an individual property *per se*, but rather relates to the various interactions at the cell and tissue level to which the material is subjected. Hence, a "systems approach" is required (Williams, 1985). According to this modern view, biocompatibility refers to the ability of a material to perform with an appropriate host response, in a specific application. Hence, biocompatibility is neither a single event nor a single phenomenon; rather, it is meant to be a collection of processes involving different, but interdependent, interaction mechanisms between materials and living tissues (Williams, 1990). Consequently the notion of biocompatibility encompasses not only the biological safety of a material, as assessed by the ISO 10993 (2003–2006) norm, but also the physico-chemical characteristics, the design, sterilization procedures, and packaging of a medical device (Piconi, 2000).

In increasing order of biocompatibility the interaction of biomaterials with living tissue can be defined as follows (Wintermantel and Ha, 1996):

- *Incompatible materials* release to the body substances in toxic concentrations, and/or they trigger the formation of antigens that may cause immune reactions. Such reactions may range from simple allergies, to inflammation, to septic rejection, with the associated severe health consequences.
- *Biocompatible materials* also release substances, albeit in nontoxic concentrations, that may lead only to benign tissue reactions such as the formation of a fibrous connective tissue capsule, or weak immune reactions that cause the formation

1) In the case of alloplastic spongiosa replacement (see Figure 10.4) of diseased bone, however, this seemingly clear definition becomes rather blurred.

Table 10.1 Examples of metallic, ceramic and polymeric biomaterials and their applications (Willmann, 1995).

Material	Application	Biological behavior
Stainless (austenitic) steel	Osteosynthesis (bone screws)	Biotolerant
Bone cement (PMMA)	Fixation of implants	Biotolerant
cp-titanium	Acetabular cups	Bioinert
Ti6Al4V alloy	Shafts for hip implants, tibia	Bioinert
CoCrMo alloy	Femoral balls and shafts, knee implants	Bioinert (?)
Alumina	Femoral balls, inserts of acetabular cups	Bioinert
Zirconia (Y-TZP)	Femoral balls	Bioinert
HD-polyethylene	Articulation components	Bioinert
Carbon (graphite)	Heart valve components	Bioinert
CFRP	Inserts of acetabular cups	Bioinert
Hydroxyapatite	Bone cavity fillings, coatings, ear implants, vertebrae replacement	Bioactive
Tricalcium phosphate	Bone replacement	Bioactive
Tetracalcium phosphate	Dental cement	Bioactive
Bioglass	Bone replacement	Bioactive

of giant cells or phagocytes. Such materials are often termed *biotolerant*, and include austenitic stainless steels (AINSI 316L) or bone cement consisting of polymethylmethacrylate (PMMA) (see Table 10.1).

- *Bioinert materials* do not release any toxic constituents, but neither do they show any positive interactions with living tissue. The body generally responds to these materials by forming a nonadherent capsule of connective tissue around the bioinert material. In the case of bone remodeling, this manifests itself by a shape-mediated *contact osteogenesis*. Only compressive forces can be transmitted through the bone–material interface ("bony on-growth"). Typical bioinert materials include titanium and its alloys (Geetha *et al.*, 2009), ceramics such as alumina, zirconia and titania, and some polymers, as well as carbon (see Table 10.1).
- *Bioactive materials* show a positive interaction with living tissues that includes also the differentiation of immature cells towards bone cells. In contrast to bioinert materials, a chemical bonding to the bone occurs along the interface;

this is thought to be triggered by the adsorption of bone growth-mediating proteins at the biomaterials surface. Hence, there will be a biochemically mediated, strong *bonding osteogenesis*. In addition to compressive forces, to some degree tensile and shear forces can also be transmitted through the interface ("bony in-growth"). Typical bioactive materials include calcium phosphates and bioglasses (see Table 10.1). It is believed that the bioactivity of calcium phosphates is associated with the formation of hydroxycarbonate apatite (HCA), similar to bone-like apatite (LeGeros and LeGeros, 1984).

10.3
The Interaction of Implant Materials and Living Tissues: A Basic Approach

The fixation of an implant in the human body is a dynamic process that involves a remodeling of the interface zone between the implant and living tissue at all dimensional levels – from the molecular up to the cell and tissue morphology level; and at all time scales – from the first second up to several years after implantation (Kasemo and Lausmaa, 1991). This situation is represented in Figure 10.3, in which the logarithmic length and time scales indicate this complex dynamic process. Whilst immediately after implantation a space filled with biofluid exists next to the implant surface, the proteins will in time be adsorbed at the titanium oxide surface, becoming a layer that is several nanometers thick and covering the titanium alloy surface. This will, in turn, give rise to osteoinduction by the proliferation of cells and their differentiation towards bone cells, revascularization, and eventual gap closing. Ideally, this will result in a strong bond being formed between the implant and the tissue, although on occasion connective tissue is formed at the interface and results in a fibrous tissue capsule that prevents osteointegration (see the inset in Figure 10.3) and causes implant loosening. In order to prevent this undesirable situation, a calcium phosphate coating of several tens or hundreds of micrometers thickness can be applied by using a variety of surface-coating techniques (see below).

The composition of the bioliquid changes with time, in response to the chemical adsorption processes of molecules that mediate bone formation, and also in response to transport reactions by an outward diffusion of titanium atoms or ions through the thin oxide layer and a concomitant inward diffusion of oxygen to the metal–oxide interface. This diffusion process is aided by lattice defects in the titanium oxide layer such as grain boundaries, isolated vacancies or vacancy clusters, and interstitial atoms. Several other reactions may occur, including the corrosion and partial dissolution/resorption of the oxide layer. This process appears to be a limiting factor for some implants, even though the resorption rate is slow due to the high chemical stability of titanium oxide.

The chemical adsorption of biomolecules such as chondroitin sulfate at the oxide surface provides a template for the adhesion of collagen strands that precede osseointegration. The nature of interaction of a biomaterial with a living tissue is best characterized by two terms: *osteoconduction* and *osteoinduction*:

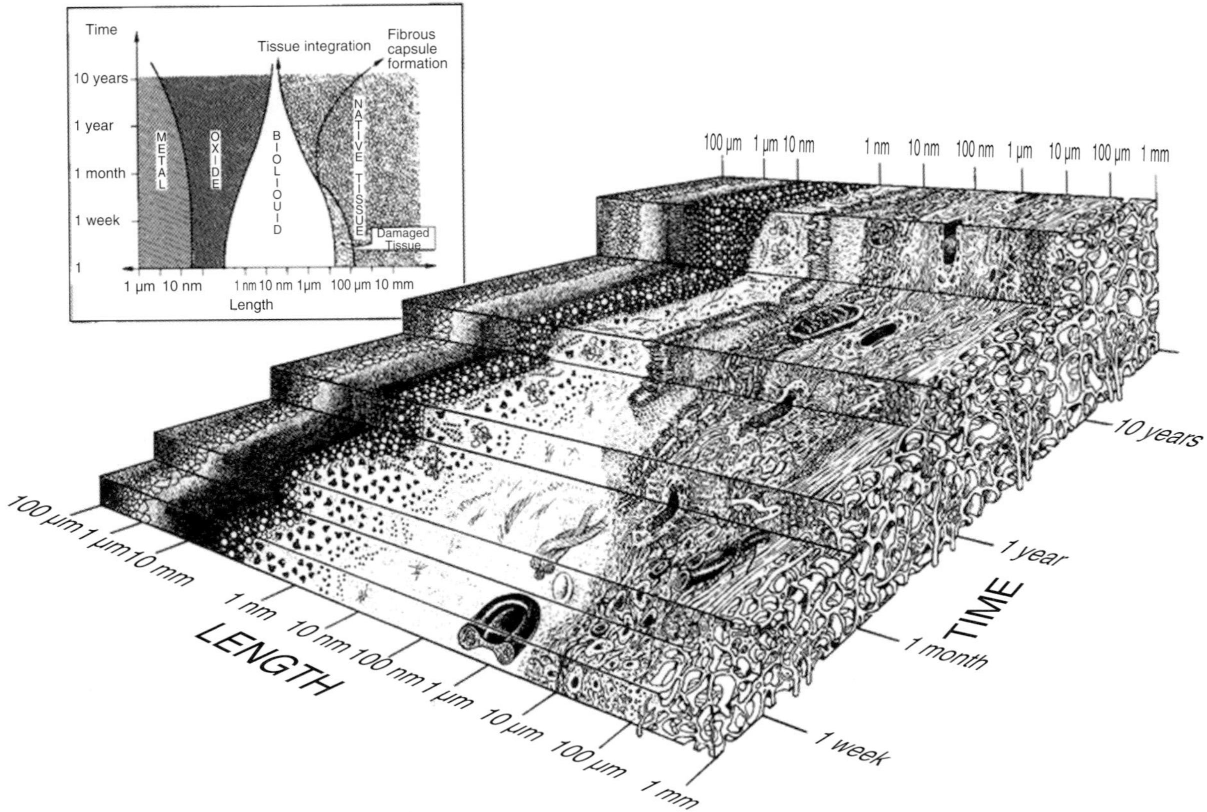

Figure 10.3 Dynamic behavior of the interface between a metallic implant (left) and bony tissue (right) (Kasemo and Lausmaa, 1991). Reprinted with permission from Prof. Bengt Kasemo, Chalmers University of Technology, Gothenburg, Sweden.

- *Osteoconductivity* is the ability of a biomaterial to support the in-growth of bone cells, blood capillaries, and perivascular tissue into the gap between implant and existing bone. Efficient in-growth is supported by interconnected pores of 150–450 μm size. Hence, the development of such a pore system in plasma-sprayed hydroxyapatite coatings is of the utmost importance, as nonporous coatings may act like bioinert materials and their eventual substitution by bone is not guaranteed.
- In contrast, *osteoinduction* refers to the transformation of undifferentiated mesenchymal precursor cells into osseoprogenitor cells preceding endochondral ossification which is either orthotopic (i.e., within the bony surrounding) or heterotopic (i.e., within muscle or fatty tissues).

In addition, the concept of *osteostimulation* by growth-supporting cytokines such as transforming growth factor-β (TGF-ß), insulin-like growth factor-1 (IGF-1), tumor necrosis factor-α (TNF-α) or recombinant human bone morphogenetic proteins (rhBMPs; Urist, 1965) continues to be the subject of extensive clinical research, and has indeed also been applied therapeutically (Marks and Popoff, 1988; Service, 2000).

Bone replacement materials for larger bone defects are designed to possess a certain degree of primary stability – that is, the ability to restore the continuity of the bone and also to provide a sufficient mechanical loading capability. Moreover, they must have an ability to be resorbed and physiologically remodeled towards new, strong bone tissue. During this process it is critical that the rate of resorption of the bone replacement material and the growth rate of new bone are essentially identical; otherwise, a gap would develop into which connective tissue would invade, thus preventing a solid osseointegration.

Regrettably, no biomaterial has yet been developed or identified that is both mechanically stable and sufficiently osteoinductive. The classic bioceramics, such as alumina or stabilized zirconia, are strong but bioinert, while osteoconductive hydroxyapatite is mechanically weak and essentially nonresorbable, and the even weaker osteoconductive tricalcium phosphate is resorbable (see below). Figure 10.4 shows this apparent dichotomy, as well as the trend that is being followed today towards the development of composite biomaterials that consist of a mechanically stable scaffold seeded with cytokine-based growth factors as a second component. The scaffolding materials are either bioinert materials such as a metal (titanium, tantalum) or hard ceramics (alumina, titania, zirconia), or inorganic bioconductive ceramics such as hydroxyapatite and bioglass (Hench, 1991, 2008), as well as calcium phosphates derived from organic templates (bovine spongiosa, corals, sea urchin spines). The third component of the composite biomaterials is the target cells such as osteoblasts, osteocytes, and their progenitor cells (Niedhart and Niethard, 1998).

While this approach appears to provide a promising solution to the many problems that plague conventional biomaterials, there is at least one caveat: namely, that when cytokines such as rhBMP-7 or -2 are released naturally by cells, only tiny quantities (in the nanogram range) are sufficient to trigger the bone repair

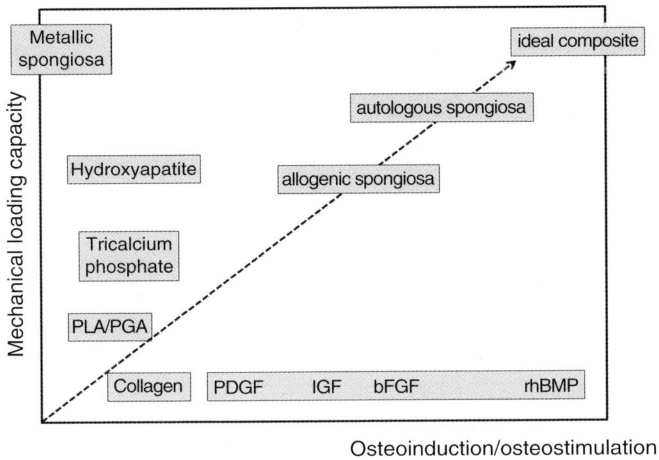

Figure 10.4 Mechanical loading capacity versus osteoinductive capacity of several bone replacement materials. PLA: polylactic acid; PGA: polyglutamic acid; PDGF: platelet-derived growth factor; IGF: insulin-like growth factor; bFGF: basic fibroblast growth factor; rhBMP: recombinant human bone morphogenetic protein. Adapted from Niedhart and Niethard (1998).

cascade. Yet, microgram amounts of BMP must be added intentionally per gram of bone matrix material in order to produce the same effect. Given the cost of the synthesized BMP, such treatment is extremely expensive (Service, 2000); another disadvantage is that these large amounts of BMP (perhaps 1000-fold the natural concentration) may trigger ectopic bone growth in susceptible patients.

10.4 Mechanical Properties of Advanced Bioceramics: Alumina versus Zirconia

As noted in detail above, it is mandatory that any material introduced into the human body with the intent to remain there during the long term must be tolerated by the organism. In particular, biocompatibility must be achieved, as defined above. On the other hand, there are extremely strong quantitative differences of the mechanical properties and responses to external loads between natural bone and bioinert ceramics, as shown in Table 10.2. These differences lead to strong gradients of the modulus of elasticity (Young's modulus) that give rise to so-called "stress shielding"; this means that the load placed on the implant during movement will not be transmitted by the bone but rather through the stiff ceramic femoral ball into the likewise very stiff titanium alloy stem. Since regular tensile loads are required for living bone to stay healthy, an absence of loads will eventually lead to the atrophic loss of cortical bone matter.

Table 10.2 Comparison of mechanical properties of alumina, zirconia (Y-TZP), and bone (Kohn, 1992; Hulbert, 1993; Rieger, 1993; Willmann, 1997).

Property	Alumina	BIOLOX® forte	Y-TZP	ZIOLOX® forte	Bone[a)]
Density ($Mg\,m^{-3}$)	3.98	–	6.08	–	1.7–2.0
Young modulus (GPa)	380–420	380	210	210	3–30
Compressive strength (MPa)	4000–5000	–	2000	–	130–180
Tensile strength (MPa)	350	–	650	–	60–160
Flexural strength (MPa)	400–560	>500	900	>900	100
Fracture toughness ($MN \cdot m^{-3/2}$ or $MPa \cdot m^{1/2}$)	4–6	4	>9	8	2–12
Grain size (μm)	–	<2	–	<0.5	–
Hardness ($HV_{0.1}$)	–	2000		1250	–
Standard	–	ISO 6474	–	ISO/DIS 13356	–

a) The lower value refers to trabecular (spongy) bone, the higher value to cortical (dense) bone.

Whilst alumina is stiffer and has a higher compressive strength than Y-TZP (Cales, 1995), the latter performs better mechanically, in terms of tensile and flexural strengths and, in particular, fracture toughness. This is related to a delay of the well-known martensitic phase transformation from the tetragonal high-temperature to the monoclinic low-temperature modification of zirconia by stabilization with other oxides, most often yttria, but also calcia and magnesia (see Section 7.2.4). The resultant so-called "transformation toughening" accounts for the dissipation of crack energy by a delayed transformation of metastable tetragonal grains to thermodynamically stable monoclinic grains with a lower density. Hence, transformation to a phase with a lower density will exert compressive stresses onto the surrounding ceramic matrix that slows down and eventually arrests any crack movement; for a ceramic material this would lead to exceptionally high fracture toughness values. It should be noted, however, that the fracture toughness of cortical bone exceeds even that of stabilized zirconia, thus confirming Nature's impressive ingenuity to design strong and tough, but lightweight, structures.

Previously, attempts have been made to reduce the stiffness differences between the implant and bone by using "isoelastic" implants. In this case, a sheath of polymer surrounds the metallic shaft of a hip endoprosthetic implant, the aim being to provide a smooth gradient of the modulus of elasticity. Unfortunately, however, degradation of the polymer within the harsh body environment has hampered this approach so far.

10.5 Selected Bioceramic Materials

10.5.1 Bioinert Ceramics

As discussed in Section 10.2, bioinert materials do not release any toxic constituents, but neither do they show any positive interaction with living tissue. As a response of the body to these materials, a nonadherent capsule of connective tissue is usually formed around the bioinert material that, in the case of bone remodeling, manifests itself by a shape-mediated *contact osteogenesis*. Hence, only compressive forces will be transmitted through the bone–material interface ("bony on-growth").

10.5.1.1 Alumina

Extremely pure, fine-grained alumina polycrystals have been used for about 35 years for the femoral heads of hip endoprostheses (Boutin, 1972, 1981). Today, there exists a large variety of clinical options to combine femoral heads and acetabular cups. In Germany, these medical products are marketed under the brandname BIOLOX® and BIOLOX® forte by CeramTec AG (formerly: Feldmühle) (Clarke and Willmann, 1994; Willmann, 2002). In 2000, the German market volume of ceramic femoral heads amounted to about DM 30 million; this corresponded to 90 000 units, 90% of which were made from alumina.

Novel developments in the field of femoral heads for hip endoprostheses rely on high-purity alumina with the addition of 17 vol% tetragonally stabilized zirconia and 1.4 vol% chromia particles (BIOLOX® delta). The former provides mechanical strengthening by transformation toughening (see Section 7.2.4), while the latter acts as a reinforcement, dissipating the crack energy by deflecting the crack paths. The use of these mechanisms leads to an almost threefold increase in the four-point bending strength (to 1400 MPa) when compared to unalloyed alumina (Table 10.3), while the fracture toughness is increased to $6.5\,MN\cdot m^{-3/2}$ (CeramTech, 2006).

This latest trend in load-bearing materials for arthroplastic applications involves the development of highly fracture-resistant alumina/zirconia composites, as an alternative choice to alumina and zirconia monolithic ceramics. Composite materials are designed from both chemical and microstructural viewpoints in order to prevent environmental degradation and fracture events *in vivo*. Based on the experimental determination of an activation energy value for an environmentally driven tetragonal to monoclinic transformation, the long-term *in vivo* environmental resistance of prostheses made from these composite materials can be predicted (Chevalier *et al.*, 2009).

Table 10.3 Properties of clinically utilized alumina ceramics (Müller and Greener, 1970; Hulbert, 1993).

Property	Alumina	Alumina ISO 6474	Alumina ISO 6474/2
Density ($Mg\,m^{-3}$)	3.98	>3.90	>3.94
Al_2O_3 content (%)	>99.7	>99.5	–
$SiO_2 + Na_2O$ (%)	<0.02	<0.1	–
$SiO_2 + Na_2O + CaO$ (%)	–	–	<0.1
Average grain size (μm)	3.6	<7	<4.5
Vickers hardness ($HV_{0.1}$)	2400	>2000	–
Young modulus (GPa)	380–420	–	–
Compressive strength (GPa)	4–5	–	–
Tensile strength (MPa)	350	–	–
Flexural strength (MPa)	400–560	>400	>450
Fracture toughness ($MN \cdot m^{-3/2}$)	4–6	–	–

The properties and required purity of the alumina used in biomedical applications are summarized in Table 10.3, in which empirical values provided by Müller and Greener (1970) are contrasted with existing ISO 6474 (1994)/DIN 58835 (1979) norm and a new ISO 6474/2 norm (Hulbert, 1993). The new ISO 6474/2 norm (established in 1994) deviates from the former in that a much lower average grain size is specified, with a concurrent increase in the flexural strength to beyond 450 MPa. This can be achieved by grain boundary engineering (see, for example, Section 8.4.1) during which the suppression of grain growth at high sintering temperatures is achieved by the addition of small amounts of magnesium oxide (in the region of a few tenths of one percent). The accumulation of magnesium oxide along the grain boundaries of alumina will result in a thin surface layer consisting of spinel ($MgAl_2O_4$), which acts as a barrier towards the grain boundary movement associated with the process of recrystallization. Hence, the formation of large grains by recrystallization will be effectively suppressed. Additional information on the structure and properties of alumina is available in Chapter 7.

Whilst alumina, as the prototype bioinert ceramic, is considered to be extremely stable against corrosion/dissolution even in aggressive body fluids, some concern has been expressed that long-term alumina-bearing implants might lead to elevated levels of aluminum within the body (Lewandowska-Szumiel and Komender, 1990). Such enhanced aluminum levels have been considered etiological agents in dialysis osteomalacia, encephalopathy, and in some forms of anemia (Parkinson *et al.*, 1981; Alfrey, 1985). However, since to date no clinical reports have been made on the upper safety levels of aluminum in human bone, the effect of even minute quantities of aluminum released from alumina femoral heads requires further investigation.

10.5.1.2 Y-Stabilized Zirconia (Y-TZP)

Despite earlier doubts (Willmann, 1993), zirconia materials–and in particular tetragonal zirconia partially stabilized with yttria (Y-PSZ), magnesia (Mg-PSZ) and calcia (Ca-PSZ)–have found various applications in biomedical devices, the most important being as hard and tough structural ceramic materials for femoral balls in hip endoprostheses (Cales and Stefani, 1995; Kokubo *et al.*, 2000) and as materials for restorative dentistry (Cales, 1998). Since stabilized zirconia shows a substantially larger fracture toughness (K_{Ic}) compared to alumina, it might be applied advantageously in prosthetic devices. The routes of synthesis and the general mechanical, chemical, and tribological properties of zirconia, as well as the principles of the toughening mechanism by suppressing the tetragonal-to-monoclinic phase transition, are described in Section 7.2.

Since zirconia is produced from naturally occurring zirconium silicate (zircon, $ZrSiO_4$) or baddeleyite (monoclinic m-ZrO_2), trace amounts of uranium and thorium (replacing the isovalent zirconium ion in the crystal lattice) may remain in the processed material, rendering it slightly radioactive. In fact, this was a major concern that, in the past, has hampered the development of otherwise mechanically superior zirconia ceramics for biomedical applications. However, the use of novel processing routes has led to a reduction in the content of potentially dangerous radioactivity to virtually zero (Cales and Stefani, 1995). Some selected properties of commercially available Y-TZP are listed in Table 10.4 (Hulbert, 1993; Rieger, 1993; Stevens, 1986; ISO/DIS 13356, 1995).

Biocompatibility of Zirconia Today, substantial controversy persists in relation to the cytocompatibility of zirconia when in contact with living tissues. Previous histomorphological and morphometrical studies of the interface of glassy and ceramic biomaterials with a bony implantation bed (Gross and Strunz, 1985) have shown that the presence of ZrO_2 in biomaterials may be generally undesirable. The incomplete transformation of chondroid (Beresford, 1981) and osteoid cells

Table 10.4 Mechanical properties of commercially available zirconia.

Property	Range
Density ($Mg\,m^{-3}$)	6.05–6.09
Zirconia content (%)	95–97
Yttria content (%)	3–5
Average grain size (μm)	0.2–0.4
Vickers hardness ($HV_{0.1}$)	1200–1300
Young modulus (GPa)	150–210
Compressive strength (MPa)	>2000
Tensile strength (MPa)	>650
Flexural strength (MPa)	900–1300
Fracture toughness ($MN \cdot m^{-3/2}$)	7–9

to osteoblasts, as demonstrated in a Sprague-Dawley rat femoral model, suggests that the application of a zirconia bond coat might be counterproductive to bone bonding. However, when Ti6Al4V rods were coated with $CaTiZr_3(PO_4)_6$ and implanted into the femora of sheep, they did not show any adverse reactions, despite earlier reservations expressed by Gross and coworkers (Gross *et al.*, 2003, 2004) that Zr-containing materials have been suspected to inhibit matrix vesicle development and function. On the contrary, the gap-bridging potential of such coatings was excellent, and supported bone apposition without the development of any connective tissue capsule (Heimann *et al.*, 2004; Heimann, 2006). Earlier studies by Hulbert *et al.* (1993) had confirmed that the intramuscular implantation of calcium zirconate ceramics in rabbits had promoted the formation of a 100- to 200-μm-thick pseudomembrane that within six to nine months had gradually densified in the absence of any inflammatory cells, thus suggesting a high degree of biotolerance. Recent studies conducted by Liu *et al.* (2006) have shown that nanostructured ZrO_2 films formed by cathodic arc deposition on silicon surfaces promoted the formation of apatite when incubated in simulated body fluid (SBF). Both, the growth and proliferation of bone marrow mesenchymal stem cells were observed on the apatite film, suggesting a favorable cytocompatibility of zirconia.

Concerns have been raised regarding the mechanical stability of steam-sterilized stabilized zirconia femoral ball heads (see Section 10.6). In 1996, the British Medical Device Agency (MDA) reported that zirconia femoral heads during autoclaving had suffered surface degradation/roughening due to hydrothermal instability, and that consequently the mechanical wear rate had increased. Subsequently, a similar warning was issued by the FDA, although the clinical failure of Y-TZP heads was also reported in the absence of any steam sterilization (Le Mouel, 1997). These findings, in conjunction with thermodynamic calculations that suggested a disastrous decomposition of Y-TZP after an incubation period of 10 years (Pfaff and Willmann, 1998), have placed grave doubt on the long-term performance of femoral ball heads manufactured from stabilized zirconia. Hence, additional studies are required to alleviate these concerns by developing safe manufacturing, sterilization, and implantation protocols. These restrictions notwithstanding, Y-TZP femoral heads presently comprise approximately 25% of the total annual number of hip joint implants in Europe, and 8% in the USA. Between 1985 and 2001, more than 400000 Y-TZP femoral heads were implanted worldwide.

Recently, a novel macroporous (surface area up to $1.5\,m^2\,cm^{-2}$), zirconia-based cell carrier biomaterial (Sponceram®; Figure 10.5), which may be either partially or fully stabilized with Ca, Mg or Y, has been developed that can easily be coated not only with osteoconductive hydroxyapatite but also with the osteoinductive recombinant human bone morphogenetic protein, rhBMP-2. The pore size of this material is about 600μm, and the porosity may be up to 85% (Röker *et al.*, 2009). Subsequent cell proliferation studies performed in a bioreactor with preosteoblastic MC3T3-E1 cells showed that alkaline phosphatase (AP) activity attained a maximum level at day 5 in the presence of rhBMP-2. Likewise, a significant calcification of the extracellular matrix (ECM) occurred in the presence of rhBMP-2,

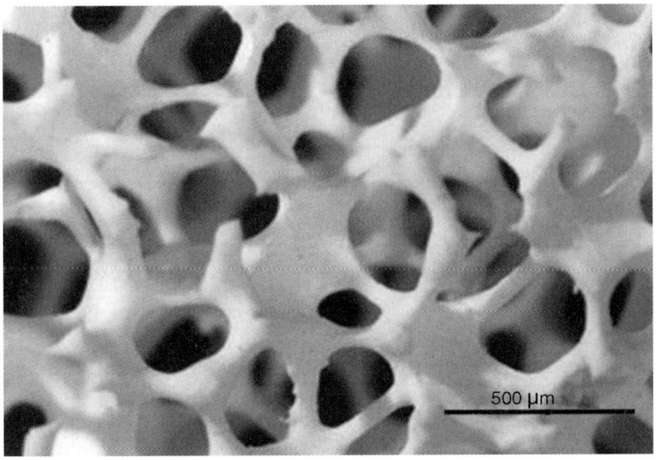

Figure 10.5 Highly porous 3-D scaffold matrix of Sponceram®. Illustration courtesy of Zellwerk GmbH, Eichstädt, Germany.

and some mineralization was observed even in its absence. In addition, the mRNA expression of collagen I, osteocalcin, and bone sialoprotein was detected. Hence, Sponceram® appears to be a suitable cell carrier for the cultivation and differentiation of MC3T3-E1 cells into the osteoblastic phenotype, even in the absence of rhBMP-2 (Suck *et al.*, 2006).

Zirconia Bond Coats and Coating Reinforcement An alternative application of zirconia ceramics has been its use as a bond coat material to improve the performance of bioconductive hydroxyapatite coatings for hip endoprostheses. Since monolithic sintered hydroxyapatite demonstrates poor mechanical properties (in terms of low bending strength, fracture toughness, modulus of elasticity and microhardness) that make it ineffective for load-bearing applications, a number of investigations have been undertaken to improve these properties by the addition of fine zirconia particulates (Tamari *et al.*, 1987; Tamari *et al.*, 1988a; Tamari *et al.*, 1988b). Generally, increases in the bending strength and fracture toughness by a factor of two to three were observed, and were attributed to the formation of reaction phases such as calcium zirconate, and the transition of tetragonal to cubic zirconia.

Hydroxyapatite (HAp) coatings deposited on titanium alloy demonstrate notoriously weak cohesion and adhesion to the substrate surface. Hence, a number of studies were conducted to strengthen these coatings by the addition of reinforcing particles of zirconia. In this way, the composite coatings could be deposited by either radiofrequency suspension plasma spraying (Kumar *et al.*, 2003), atmospheric plasma spraying (Chang *et al.*, 1997a, 1997b; Chou and Chang, 2002), or low-pressure plasma spraying (LPPS) (Heimann *et al.*, 1998). The concept was to apply additional mechanical interlocking between the bond coat layer and substrate, as well as to establish a chemical bond between the bond coat and HAp. However, the scheme was not met with resounding success and, despite claims to the contrary, zirconia-reinforced HAp coatings neither significantly increased the bond strength in as-sprayed coatings, nor slowed down the resorption in SBF (Chang *et al.*, 1997a, 1997b). Moreover $CaZrO_3$, which was thought to be formed as a reaction product at the substrate–coating interface, was suspected of causing a deterioration in the mechanical properties of the coating system (see Wu and Yeh, 1988; Caetano-Zurita *et al.*, 1994). Although, a calcium zirconate bond coat appeared to adhere well to the substrate, it tended to exhibit lateral cracks parallel to the coating interface when subjected to even low tensile forces (Heimann *et al.*, 1998). Likewise, stresses introduced into the HAp coating by thermally induced tetragonal–monoclinic phase transformation within the partially Ca-stabilized zirconia bond coat was found to lead to extensive scaling and concurrent leaching during treatment in a simulated body fluid (Hanks' balanced salt solution; HBSS). These findings were disputed by Chou and Chang (2002), however, who claimed an increase of the adhesive bond strength of a HAp/ZrO_2 composite coating, from 28.6 ± 3.2 MPa for a pure hydroxyapatite coating to 36.2 ± 3.0 MPa, owing presumably to the diffusion of calcium ions from the HAp matrix into the zirconia bond coat. However, the peel adhesion strength[2)] of as-sprayed coatings in the presence of a zirconia bond coat, as measured by Kurzweg *et al.* (1998a) and by Kurzweg *et al.* (1998b), was found to be significantly lower (probability point of a double-sided *t*-test of 2.54 compared to a tabulated value of 2.07 for 22 degrees of freedom, 95% confidence interval) at 18 N m^{-1}, compared to a peel strength of 22 N m^{-1} of a HAp coating without a bond coat. While the reason for this discrepancy is not clear, a different thermal history of the coatings, as well as the use of fully stabilized zirconia by Chou and Chang (2002) as opposed to the partially Ca-stabilized zirconia used by Heimann *et al.* (1998), has been suspected. In particular, the existence of massive residual stresses at the bond coat–HAp coating interface related to a mismatch in the coefficients of thermal expansion (thermal stress, see above) was relieved during leaching for 28 days in SBF (HBSS), resulting in a strong coating delamination (Heimann *et al.*, 1998).

Whilst the HAp/zirconia composite coatings and zirconia bond coat/HAp top coat systems appeared to provide only marginal improvements in coating adhesion strength, the plasma-sprayed gradient coatings behaved much better, increasing the tensile adhesion strength of heat-treated coatings to beyond 50 MPa (Ning *et al.*, 2005). Both, the hardness and modulus were increased gradually from the Ti6Al4V substrate into the coating perpendicular to the interface, while the

2) While the numerical values of a tensile adhesion test according to ASTM C633-01 (2001) and a peel adhesion test according to a modified ASTM D3167-03a (2004) designation (Sexsmith and Troczynski, 1994, 1996) cannot directly be converted into each other, their relative values are proportional. The conventional tensile pull test measures failure stress, expressed as the ratio of applied force to (geometric) coating area in dimension: N m^{-2}, the peel test measures the energy required to separate the coating and the substrate along a line in dimension: N m^{-1} (Kurzweg *et al.*, 1998b).

microstructure and composition varied smoothly without showing any distinct interfaces among adjacent layers.

A metallic Zr coating was deposited on Ti implant surfaces and subsequently oxidized using a continuous-wave Nd:YAG laser to produce a 7 μm-thick, fully dense ZrO_2 layer. Owing to its high surface energy and increased wettability, the wear rate of the laser-oxidized Zr was found to be two orders of magnitude lower than that of as-deposited Zr. However, the oxidized coatings showed comparable *in vitro* biocompatibility to that of pure Ti, and also promoted excellent *in vitro* proliferation and vitality of human osteoblast cells (Krishna Balla *et al.*, 2009).

10.5.2
Bioconductive Ceramics

In contrast to bioinert ceramics, bioactive materials show a positive interaction with living tissues, including chemical bonding to the bone along the interface, which is thought to be triggered by the adsorption of bone growth-mediating proteins at the biomaterials surface. Hence, there will be a biochemically mediated, strong *bonding osteogenesis*. In addition to compressive forces, both tensile and shear forces can to some degree also be transmitted through the interface ("bony in-growth"). Typical bioactive materials are calcium phosphates and bioglasses (see Table 10.1). Although the properties and applications of calcium phosphate ceramics will be described in detail below, only a brief account of the composition and functionality of bioglasses will be provided. Further information on bioglasses and their *in vitro* and *in vivo* reactions is available elsewhere [for example, Hench (1991, 2008); Gross *et al.* (1988); Kokubo (1991); Ducheyne *et al.* (1997); Cerutti (2004); and Ben-Nissan and Ylänen (2006)].

10.5.2.1 **Bioglasses**

Since the discovery during the late 1960s of surface-active bioglasses that bond to living tissues (Bioglass®; see, for example, Hench and Wilson, 1984), various types of bioactive glass and glass-ceramics have been developed with different functions, including high mechanical strength, high machinability, and fast setting ability. The glasses investigated for implantation are based primarily on silica (SiO_2), but containing small amounts of other crystalline phases. The most prominent and successful application of this is Bioglass®, which was developed almost single-handedly by Larry Hench, and has been described in detail in various comprehensive reviews (Hench, 1971; Hench *et al.*, 1972; Hench and Ethridge, 1982). Bioactive glass compositions are positioned in the system $CaO–Na_2O–P_2O_5–SiO_2$ (Figure 10.6). The first development of such a bioglass composition began during the 1970s, when 45S5 Bioglass® was proposed with a composition of 45% SiO_2, 24.5% CaO, 24.5% NaO_2, and 6% P_2O_5 by weight (Hench *et al.*, 1972). Subsequently, Vrouwenvelder *et al.* (1992, 1994) suggested that bioglass® 45S5 had a greater osteoblastic activity than HAp.

One common feature of bioactive glasses is a time-dependent kinetic modification of their surfaces during implantation. While they are generally nonresorbable,

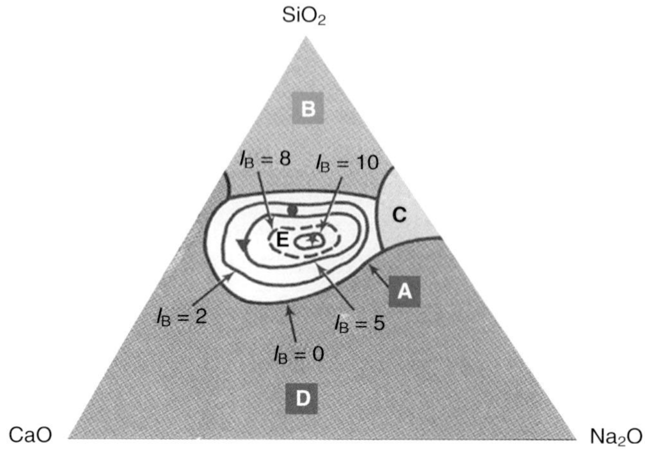

Figure 10.6 Bioglass compositions in the pseudo-ternary system $CaO–Na_2O–SiO_2–(P_2O_5)$ (after Hench, 1991). Region A: Bone-bonding compositions; Region B: Bioinert composition; Region C: Resorbable glasses; Region D: Technically impractical compositions; Region E (inside the dashed contour with $I_B > 8$): Soft tissue-bonding compositions. The isopleths shown inside the region A ($0 < I_B < 10$) are a measure of bone-bonding ability, where $I_B = 100/t_{0.5}$, with $t_{0.5}$ = time required to achieve 50% bone bonding. Reprinted with permission from Wiley-Blackwell.

the release of sodium and calcium ions triggers a cascade of reactions culminating in the nucleation of a thin layer of biological hydroxycarbonate apatite (HCA) that provides a bonding interface with bony tissues (Hench, 1991). This interface is so mechanically strong that, in many cases, the interfacial strength of the adhesion exceeds the cohesive strength of the implant material, or of the tissue to which it is bonded.

Today, many bioactive silica-based formulations exist that have been derived from 45S5 Bioglass® (45 mass% SiO_2, molar ratio Ca/P = 5; Hench *et al.*, 1972). Glasses with a Ca/P ratio substantially below this values do not bond to bone. The classic ternary diagram of the compositional dependence of bone bonding of bioactive glasses, produced originally by Hench (1991), is shown in Figure 10.6. This ternary diagram represents a section through the quaternary diagram $CaO–Na_2O–SiO_2–P_2O_5$, so that all glasses in the bone-bonding range have a constant P_2O_5 concentration of 6 mass%.

However, there are critical differences in composition between bioactive glasses and traditional CNS glasses ($CaO–Na_2O–SiO_2$). Bioglasses, including 45S5 and 30S15B5 (30 mass% SiO_2, 15 mass% B_2O_5, 24.5 mass Na_2O, 24.5 mass% CaO, 6 mass% P_2O_5), have less than 60 mol% SiO_2, high Na_2O and CaO contents, and high CaO/P_2O_5 ratios.

The low hydrolytic stability of these compositions makes the surface of the bioglasses implants highly reactive when exposed to body fluid. Five kinetically different stages have been identified during the interaction of bioglass[3] with liquid, defined as follows (Hench, 1991):

- **Stage 1:** Diffusion-controlled ($t^{-1/2}$ dependence) rapid ion exchange of Na^+ and Ca^{2+} with H_3O^+ from solution according to:

$$\equiv Si{-}O{-}Na + H_3O^+ \rightarrow \equiv Si{-}OH + Na^+ + H_2O \tag{10.1a}$$

- **Stage 2:** Surface-controlled ($t^{1.0}$ dependence) dissolution of silica by breaking bridging oxygen bonds in Si–O–Si, and the formation of Si–OH (silanol) groups at the bioglass–solution interface according to:

$$\equiv Si{-}O{-}Si \equiv + H_2O \rightarrow \equiv Si{-}OH + OH{-}Si \equiv \tag{10.1b}$$

- **Stage 3:** Condensation and repolymerization of the alkali- and alkaline earth-depleted silica-rich surface layer according to:

$$\equiv Si{-}OH + OH{-}Si \equiv \rightarrow \equiv Si{-}O{-}Si \equiv + H_2O \tag{10.1c}$$

- **Stage 4:** Migration of Ca^{2+} and HPO_4^{2-} ions through the silica network and formation of a $CaO{-}P_2O_5$–rich surface film that subsequently transforms into an amorphous layer by incorporation of soluble calcium and phosphate ions from solution (Kim *et al.*, 1989).

- **Stage 5:** The nucleation and growth of nanosized HCA crystals incorporating OH^- and CO_3^{2-} ions from solution.

The adsorption of proteins and other biological moieties occurs concurrently with the first four reaction stages, and is believed to contribute to the biological nature of the HCA layer. Within approximately 3–6 h *in vitro*, the calcium phosphate layer will crystallize into the HCA layer (stage 5). Because this surface is chemically and structurally near-identical to natural bone mineral, the body's tissues are able to attach directly to it. As the reactivity continues, the HCA surface layer grows in thickness to form a bonding zone of 100–150 μm which serves as a mechanically compliant interface that is essential for maintaining the bioactive bonding of the implant to the natural tissue. These surface reactions occur within the first 12–24 h of implantation; thus, by the time the osteogenic cells (such as osteoblasts or mesenchymal stem cells) begin to infiltrate a bony defect – which normally takes 24–72 h – they will encounter a bonelike surface, complete with organic components, and not a foreign material. It is this sequence of events, in which the bioactive glass *participates* in the repair process, that allows for the creation of a direct bond of the material to the tissue. The body's normal healing and regeneration processes begin in concurrence with the formation of these surface layers. Bioactive glasses appear to minimize the duration of the macrophage and inflammatory responses that accompany any trauma, including that imposed by surgery.

3) The sequence is applicable in its simplicity only to single-phase glasses. For multiphase glasses, and in particular glass ceramics, a similar sequence for each individual phase must be established.

Application of Bioactive Glasses As the technology of bioactive glasses for medical use is relatively new, only a relatively few – but highly successful – clinical applications of these materials have been made. Perhaps most important point here is the absence of any reports of adverse responses to these materials in the body, thus confirming the antibacterial and antifungal properties of bioglasses. Today, the most typical applications of bioglasses include:

- Dental implants
- Periodontal pocket obliteration
- Alveolar ridge augmentation
- Maxillofacial reconstruction
- Otolaryngologic applications
- Percutaneous access devices
- Spinal fusion
- Coatings for dialysis catheters made from silicone tubing
- Coatings for surgical screws and wires
- Cochlear implants
- Bone graft substitutes
- Bone tissue engineering scaffolds
- Antibacterial and antifungal applications as wound-healing agents
- Granular filler for jaw defects following tooth extraction

Whilst the results of both laboratory and clinical studies have suggested that bioactivity occurs only within certain compositional limits and very specific ratios of oxides in the Na_2O–K_2O–CaO–MgO–P_2O_5–SiO_2 systems (see Figure 10.6), the extent of these limits and their physical chemical and biochemical reasons are only poorly understood at present. For example, in subcutaneous installations it has been shown that two adjacent samples of S53P4 glasses, when in contact with each other, can establish a bond across their superficial apatite layers. Hence, it would appear that there exists a certain tendency of self-repair if a monolithic bioglass device were to be broken accidentally within the tissue. A similar phenomenon was observed in apatite-wollastonite (A/W) glass-ceramics with a substantially higher phosphorus content but a lower I_b compared to Bioglass®. Bioglasses used in periodontics are marketed as PerioGlas® to dentists and oral surgeons as a bone graft substitute. The market for bone graft materials applied to either enhance or substitute autograft bone is approximately US$ 1 billion per year, and is currently growing at a staggering rate of 20% per annum.

Current and Future Developments The results of recent investigations (Hench, 2008) have indicated that the controlled release of biologically active Ca and Si ions may leads to an upregulation and activation of genes in osteoprogenitor cells. This process results in rapid bone regeneration, and can also be used to induce angiogenesis, thereby offering the tantalizing potential for the design of gene-activating glasses for soft tissue regeneration. On the basis of this discovery, investigations are currently under way to determine the impact of incorporating metabolically important silicon (Carlisle, 1970) into other implantable materials, such as synthetic HAp. Results obtained to date have shown that *in vivo* bone regeneration is enhanced in silicon-doped HAp granules implanted as a bone-filling material, compared to nonsubstituted HAp (Gillespie *et al.*, 2010).

At present, there is also substantial research interest in the use of bioactive glassy materials for tissue scaffolds that mimic the structure of trabecular bone. Tissue scaffolds are three-dimensional (3-D) matrices that can be used as templates to support the growth of tissues *in vitro* (Ducheyne *et al.*, 1997). The challenge here is to prepare resorbable scaffolds of suitable geometry and bioactivity so as to support the growth of artificially seeded tissues which can be tailored to fit specific bone defects. Studies are currently under way to fine-tune the architecture and resorption characteristics of sol–gel-derived bioactive glasses, with foaming agents and surfactants being incorporated into the sol–gel reaction mixture to introduce 3-D interconnected pores which emulate the porous structure of trabecular bone.

These inorganic materials, which can augment the body's own ability to regenerate, are set to become significant in future clinical approaches to restore function in damaged tissues. The possible large-scale manufacture of engineered tissues seeded with the patient's own cells, so as to minimize the risk of rejection, represents an innovative alternative to some of the problems currently associated with prosthetic implants and donor organs. If successful, this approach would lead to dramatic improvements in quality of life for millions of people worldwide.

10.5.2.2 **Hydroxyapatite**

Hydroxyapatite is chemically and structurally very close to naturally occurring biological apatite which forms the inorganic scaffolding materials of bone (see Figure 10.1). Unfortunately, however, as a synthetic ceramic material it is mechanically quite weak, and hence is unable to sustain even moderate tensile, shear or compressive forces. Consequently, HAp cannot be applied as a monolithic material but rather in either granular form to fill larger bone cavities, as coatings for metallic implants (ASTM F1185, 1993), or as a composite material together with biodegradable polymers such as collagen, poly(lactic acid) (PLA), poly(caprolactone) (PCL), poly(etheretherketone) (PEEK), or polyamides (Nylon®6/12).

Structure and Crystal Chemistry of Inorganic and Biological Hydroxyapatite Hydroxyapatite, $Ca_{10}(PO_4)_6(OH)_2$, is a member of a large group of chemically different but structurally identical compounds with the general formula $M_{10}(ZO_4)_6X_2$ (M = Ca, Pb, Cd, Sr, La, Ce, K, Na; Z = P, V, As, Cr, Si, C, Al, S; X = OH, Cl, F, CO_3, H_2O) obeying the hexagonal space group $P6_3/m$[4] (McConnell, 1973; Elliot, 1994). The Ca polyhedra share faces to form chains parallel to the crystallographic c-axis

4) In structures of the apatite group with a space group $P6_3/m$ the cations are usually distributed over two independent sites, I and II. In contrast, in minerals such as belovite and deloneite, large Na^+ cations are distributed over six structural sites, altering the crystallographic hexagonal symmetry to the trigonal space groups $P\bar{3}$ and P 3, respectively (Strunz and Nickel, 2001).

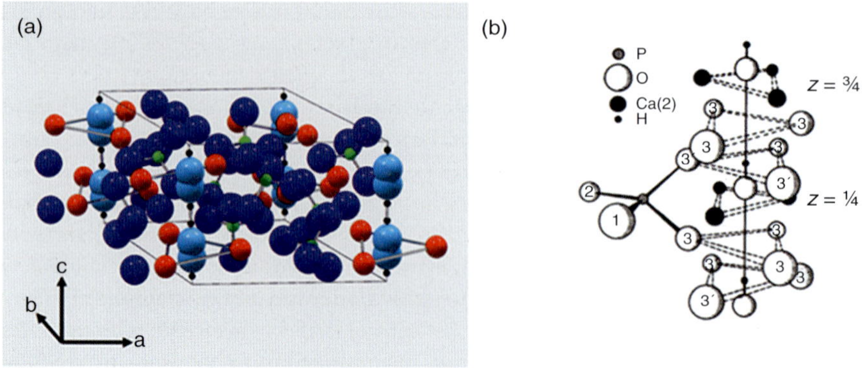

Figure 10.7 (a) Crystallographic structure of hydroxyapatite, showing the orthohexagonal unit cell with a = 943.2 pm, b = a$\sqrt{3}$, c = 688.14 pm (Posner *et al.*, 1958). Color coding of ions: dark blue spheres = O^{2-}; light blue spheres = OH^-; red spheres = Ca^{2+}; green spheres = P^{5+}; black dots = H^+; (b) Detail of the arrangement of ions around the 6_3 screw axis (Elliot, 1994).

[0001]. These chains are linked into a hexagonal array by sharing edges and corners with PO_4 tetrahedra. The OH^- ions are located in wide hexagonal channels // [0001]. Figure 10.7a shows the unit cell of HAp with the orthohexagonal metric; the large dark blue spheres depict oxygen ions, the large light blue spheres aligned parallel to the c-axis are OH^- ions, and the smaller red spheres Ca^{2+} ions, of which six are coordinated with hydroxyl ions to form planar triangles perpendicular to the OH^- columns (Ca_{II}). The small green spheres represent P^{5+} ions, and black dots protons.

Two oxygen atoms of the PO_4 tetrahedra are located on mirror planes through z = ¼ and z = ¾; the other two are symmetrically arranged above and below the mirror plane (Figure 10.7b). The Ca^{2+} ions are situated in two different positions: Ca_I at z = 0 and ½ along the threefold axes a_i, and Ca_{II} at z = ¼ and ¾ along the hexagonal screw axis c parallel [0001]. The Ca_I atoms are coordinated by nine oxygen atoms that belong to six different phosphate tetrahedra, whereas the Ca_{II} atoms have an irregular sevenfold coordination with six oxygen atoms of five phosphate groups in addition to the OH^- ions (Posner *et al.*, 1958). Since each of the 16 OH^- positions in the unit cell are statistically occupied to only 50%, there exist on average eight vacancies/unit cell along the c-axis. Hence, there are direction-dependent differences in the mobility of OH^- ions and also the Ca_{II}^{2+} ions associated with them (Royce, 1973; Takahashi *et al.*, 1978) that are extremely relevant when considering the structural transformation from amorphous calcium phosphate (ACP) to crystalline HAp, as well as the stepwise dehydroxylation of HAp to form oxyapatite (OAp) (see below).

The lattice parameters of hexagonal hydroxyapatite are a = 943.2 pm and c = 688.14 pm (Posner *et al.*, 1958). There appears to be a stoichiometric and ordered form of HAp crystallizing in the monoclinic space group $P2_1/b$, with a = 942.14 pm, b = 2a and c = 688.14 pm, γ = 120° (Elliot *et al.*, 1973). The deviation of the symmetry from the archetypal space group $P6_3/m$ presumably results from local ordering of OH^- ions in [00z] anionic columns. This imposes no restrictions on the mirror symmetry, and leads consequently to a doubling of the b-axis distance and also a doubling of the number of formula units per unit cell to Z = 4[5]. The monoclinic form of HAp transforms to a hexagonal anion-disordered form at approximately 205 °C (Takahashi *et al.*, 2001). However, since this monoclinic form occurs only under such special thermal conditions (Mathew and Takagi, 2001), it can be safely neglected in the present discussion, even though monoclinic HAp has been discovered recently as a natural mineral and named *clinohydroxyapatite* (Chakhmouradian and Medici, 2006). This lack of relevance notwithstanding, there appears to be evidence that the growth of bone-like HAp in an electric field is accelerated by a reorientation of the dipole moments between the O^{2-} and H^+ of the lattice OH^- ions in response to the electric polarization conditions (Hitmi *et al.*, 1986; Yamashita *et al.*, 1996). In this way, the ordered alignment of OH^- columns as present in monoclinic HAp is attained (see Figure 10.26).

While the structural relations are quite clear for inorganically precipitated HAp, they are much less so for biological apatite occurring in the bone substance. Natural bone is a composite material, the water-free substance of which consists of approximately 70 mass% apatite and 30 mass% collagen I. The apatite platelets of 40 × 25 × 3 nm size are orderly arranged along the triple-helical strands of collagen (see Figure 10.1). Hence, the abundant protein collagen I serves as a structural template for the crystallization of nano-HAp, presumably mediated by carboxylate terminal groups (Robey, 1996; Aryal *et al.*, 2006) and osteocalcin (Hoang *et al.*, 2003). Because of its open-channel structure, HAp is able to incorporate other ions by substituting Ca^{2+} cations as well as OH^- and PO_4^{3-} anions, without any major distortion of the lattice. In biological apatite, Ca^{2+} is partially substituted by Na^+, Mg^{2+}, Sr^{2+}, K^+ and some trace elements such as Pb^{2+}, Ba^{2+}, Zn^{2+} and Fe^{2+}. The PO_4^{3-} groups are replaced partially by CO_3^{2-} (carbonate-apatite, dahlite), whereas OH^- can be substituted by CO_3^{2-}, Cl^- and, in particular, F^- in tooth enamel and dentin (LeGeros and LeGeros, 1984). This compositional variability of HAp is responsible for its high biocompatibility and osteoconductivity (e.g., Rejda *et al.*, 1977).

The substitution by other ions reduces the theoretical stoichiometric Ca/P ratio of 1.67 of HAp to values for bone-like apatite of below 1.6 (LeGeros, 1991), or even as low as 1.4 (Weng *et al.*, 1994). The nonstoichiometry of biological apatite can be described by the approximate formula (Young and Holcomb, 1982; Hattori and Iwadate, 1990; Liu *et al.*, 2001):

$$Ca_{10-x}(HPO_4)_x(PO_4)_{6-x}(OH,O,Cl,F,CO_3,\square)_{2-x}\bullet nH_2O;\ 0 < x < 1;\ n = 0-2.5.$$

5) This is akin to the structure of clinomimetite $Pb_{10}(AsO_4)_6Cl_2$ with a space group $P2_1/c$, *i.e.* c = 2a (Dai and Hughes, 1989).

The fact that the OH^- positions can be occupied by mobile O^{2-} ions or by vacancies □ is of vital importance for understanding the kinetics of the dehydroxylation reaction of HAp to oxyhydroxyapatite (OHAp) and OAp, respectively (see below). Rietveld refinements and spectroscopic studies of the structure of Ca-deficient synthetic apatite obtained from suspensions of monetite (dicalcium phosphate anhydrate; DCPA, $CaHPO_4$) showed that crystalline products of the average composition $Ca_{9.303}(PO_4)_6(OH)_{0.606} \cdot 1.97\,H_2O$ were formed with a Ca/P ratio of 1.55 and a = 943.20 pm, c = 687.51 pm (Wilson *et al.*, 2005). The presence of HPO_4^{2-} ions in the otherwise monetite-free product was confirmed by infra-red (IR) and Raman spectroscopy. It should be mentioned that the general formula of Ca-deficient HAp was vigorously disputed during the early 1960s, and in particular the assignment of the IR frequency at 3400 cm^{-1} to either the O–H..O stretching vibration between hydrogen-bonded phosphate groups (Posner and Perloff, 1957; Posner *et al.*, 1960) or to adsorbed water (Winand and Dallemagne, 1962; see also Jones and Smith, 1962). A recent report assigned the broad IR band near 3400 cm^{-1} in plasma-sprayed HAp coatings to "misoriented" hydroxyl ions (Park *et al.*, 2002).

The carbonate substitutional defects are either located in the hydroxyl-occupied channel parallel [001] (type-A defect), or at the position of an orthophosphate group (type-B defect). Computer modeling has shown that the lowest energy configuration exists for type-A defects when two hydroxyl groups are replaced by one carbonate group in such a way that the O–C–O axis is aligned with the c-axis channel of the apatite lattice, and the third oxygen atom of the CO_3^{2-} group lying in the a/b plane (Peroos *et al.*, 2006). The proportions of A and B carbonate ions have been obtained using the out-of-plane bend (ν_2) bands of Fourier transform infrared (FTIR) spectra. Most compositions have ν_2 spectra with one band for type-A carbonate (at 878–880 cm^{-1}) and one for type-B (at 870–872 cm^{-1}). Na-free AB HAp has a third prominent band at 862 cm^{-1}, which has been assigned to the stuffed channel species (A2) (Fleet, 2009).

While the occurrence and behavior of OH^- ions in the HAp lattice has been recognized and widely studied in synthetic material, it came as a complete surprise that natural bone apatite was found to be essentially free of hydroxyl ions when examining not only Raman spectroscopy data (Rey *et al.*, 1995; Pasteris *et al.*, 2001, 2004) but also data acquired by nuclear magnetic resonance (NMR) spectroscopy and inelastic neutron scattering. In fact, with the increasing nanocrystallinity of apatite ranging from tooth enamel to dentin to bone, the intensity of the OH^- Raman band at 3572 cm^{-1} was seen to decrease strongly and essentially to disappear for bone material. Hence, contrary to the general medical nomenclature, bone-like apatite appears to be not hydroxylated. Interestingly there is a strong correlation between the concentration of OH^- ions and the crystallographic degree of atomic order. Whilst it is not obvious how the charge balance is maintained within an OH^--free nanocrystalline apatite, it has been suggested that there is a specific state of atomic order, imposed biochemically by the body, and which is essential for cell metabolism and the ability of the body to carry out tissue-specific functions. It might be speculated that the lack of hydroxyl ions in the lattice of nanocrystalline biological Ca-deficient HAp, causing a high density of vacancies along the 6_3 screw axis (Figure 10.7b). The consequent high mobility of Schottky-type defects in turn influences the solubility, thus providing a mechanism for fast and efficient bone reorganization by dissolution (osteoclastesis) and reprecipitation (osteoblastesis) during bone reconstruction in response to changing stress and load levels, according to Wolff's law (Bassett *et al.*, 1974).

A comprehensive review of HAp and other calcium orthophosphates, including their inorganic and organic occurrences in nature, structure, transformations, and biological and biomedical significances, has been recently produced (Dorozhkin, 2007).

Biomedical Function of Hydroxyapatite Coatings The bond formed between a metallic titanium alloy implant and the bony tissue is mediated by a so-called *contact osteogenesis*. Typically, bone tissue grows one-directionally towards the interface, while the "bony on-growth" that occurs is capable of transmitting strong compressive loads. However, the actual loads to which the interface is subjected during the patient's movement also contains strong tensile and shear components, that must be taken care of. In clinical practice, in many cases a bioconductive HAp layer is provided that will allow a *bonding osteogenesis* which, through "bony in-growth," will be able to transmit these tensile and shear forces. In this case, two ossification fronts will develop, with one front growing from the bone towards the implant, and another from the implant towards the bone (Søballe, 1993). Indeed, strong and convincing clinical evidence is mounting that a 150–200 μm long-term stable bioactive hydroxyapatite coating will elicit a specific biological response at the interface of the implant material by controlling its surface chemistry through the adsorption of noncollagenous proteins such as osteocalcin, osteonectin, silylated glycoproteins, and proteoglycans. This would result in the eventual establishment of a strong and lasting osseoconductive bond between the living tissue and the biomaterial. The advantages of bioactive coatings include:

- Preventing the formation of a fibrous capsule of connective tissue that will surround the implant.
- Fast bone apposition rates through the preferential adsorption of proteins.
- Bonding osteogenesis, to provide a continuous and strong interface between the implant and tissue that is capable of transmitting not only compressive but also tensile and shear loads.
- An accelerated healing compared to implants without a bioactive coating.
- A reduced release of titanium ions to the surrounding tissues, thus minimizing the perceived risk of any cytotoxic response.

Figure 10.8 shows how a plasma-sprayed HAp coating on a porous titanium layer covering the titanium alloy surface (dark bars), depending on the implantation time, will improve the interfacial bond strength compared to uncoated porous titanium (light bars) (Hench, 1991).

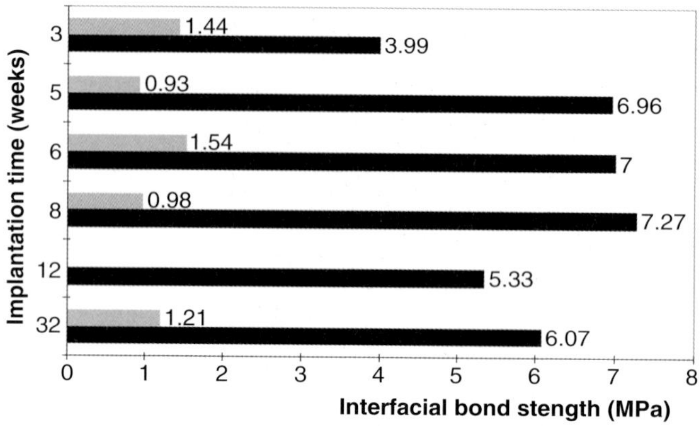

Figure 10.8 Comparison of interfacial bond strength of porous titanium with (dark bars) and without (light bars) a plasma-sprayed hydroxyapatite coating. Modified after Hench (1991).

Formation of Hydroxyapatite Thin Films by Surface Modification In the past, many attempts have been made to deposit, by various surface modification techniques, thin and well-adhering films of HAp onto titanium implant surfaces so as to mediate bone apposition and improve anchoring of the implant within the cavity of the femur. Some of these methods have been reviewed by Heimann *et al.* (1997). By using an electrophoretic technique, Ducheyne *et al.* (1990) deposited HAp from a calcium phosphate solution and subsequently densified the film by sintering. The composition of the resultant film was found to be mixture of HAp and tetracalcium phosphate, the latter being presumably the product of thermal decomposition during sintering. The slow formation of films of <5 μm thickness was achieved by biomimetic control during the reactive immersion of calcium silicate substrates in a simulated, phosphorus-containing body fluid (Kokubo *et al.*, 1991). From a calcium phosphate glass target, Yamashita *et al.* (1994) used radiofrequency sputtering to produce a thin stoichiometric HAp as well as defect apatite films, the composition of which was seen to depend on both the CaO/P_2O_5 ratio and the partial pressure of water. In this way, amorphous calcium phosphate films of <1 μm thickness were deposited (by dynamic ion beam mixing) that, on annealing for 1 h at 600 °C, crystallized to form HAp with an apparent self-repair function through Ca–P–Ti–OH complexes (Ohtsuka *et al.*, 1994).

Formation of Hydroxyapatite Thick Coatings by Plasma Spraying It is well known that thin HAp coatings of less than 50 μm thickness are quickly resorbed *in vivo*. Hence, if long-term stable coatings are desired, then thicker coatings of 150–200 μm will be required that can be produced by a variety of techniques, including the deposition of HAp powder or suspension on the implant and subsequent sintering below 1200 °C, with or without a bond coat (Toriyama *et al.*, 1990, 1991; Kawamoto *et al.*, 1991), or the bonding of HAp granules with an organic glue (e.g.,

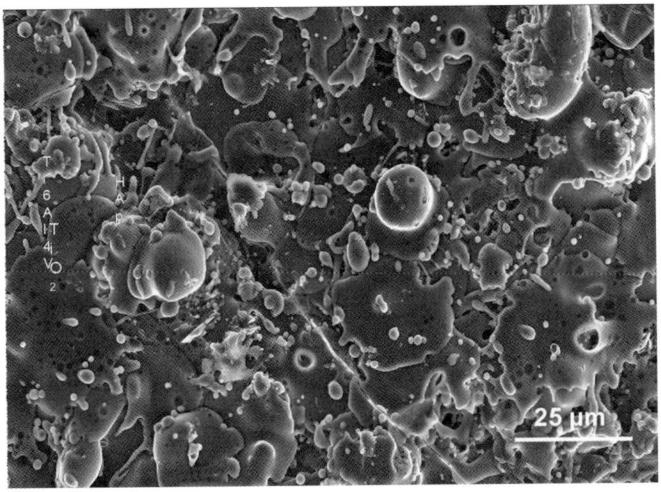

Figure 10.9 Typical plasma-sprayed hydroxyapatite coating for biomedical implants, showing well-developed stacked splats and loosely adhering incompletely melted oversized particles with a spherical shape.

PMMA) to the implant surface. The latter technique is thought to provide an improved load transmission to the bone through a graded structure with quasi-elastic properties. However, the state-of-the-art methods for depositing thick, rather well-adhering and resorption-resistant coatings include atmospheric plasma spraying (APS) (Figure 10.9), high-velocity oxyfuel (HVOF) spraying, or LPPS (Heimann and Vu, 1997). The principles of these techniques have been discussed elsewhere (Heimann, 2008).

Thermal Decomposition of Hydroxyapatite during Plasma Spraying The important performance criteria of plasma-sprayed HAp coatings are: (i) adhesion to the Ti6Al4V substrate, which depends crucially on the microstructural features of the coatings such as porosity, degree of deformation of spray particle splats, and the presence of micro- and macrocracks, as well as residual stresses at the coating–substrate interface; and (ii) a biochemical resistance against aggressive body fluids that depends on the crystallinity of the HAp, and the nature and amount of thermal dehydroxylation (OAp; Trombe and Montel, 1971) and decomposition (tricalcium phosphate, TCP; tetracalcium phosphate, TTCP; CaO) products (McPherson *et al.*, 1995), as well as the presence of amorphous calcium phosphate (ACP). However, a high adhesion strength requires a high plasma enthalpy that, in turn, causes an increased thermal decomposition of HAp and a low biochemical resistance. To solve this conundrum, a compromise is needed. The optimization of coating performance can be achieved principally via three routes: (i) control of the plasma spray parameters; (ii) microstructural engineering of the spray powders prior to deposition; and (iii) the application of bioinert bond coats (Heimann, 1999). Yet,

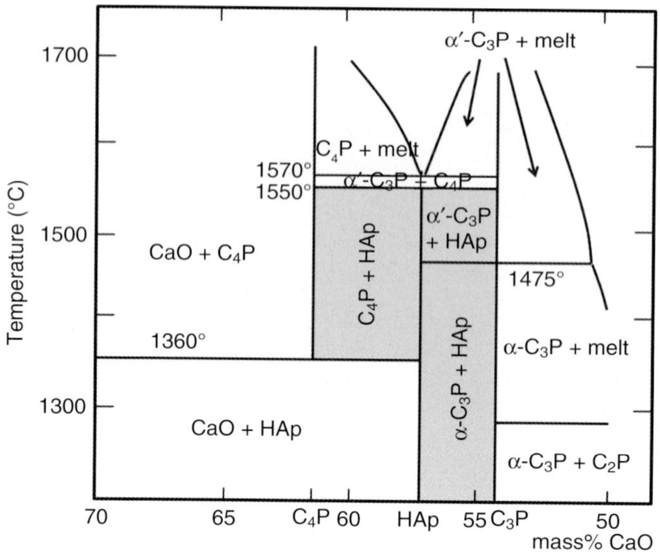

Figure 10.10 Phase diagram of the quasi-binary system $CaO–P_2O_5–H_2O$ at a water partial pressure of 65.5 kPa (after Riboud, 1973; see also Hench, 1991). Note that at 1570 °C, incongruent melting of hydroxyapatite (HAp) occurs under decomposition into α′-C_3P (α′-TCP) and C_4P (TTCP). The shaded areas indicate the regions of relative stability of HAp.

even then the thermal decomposition of HAp in the extremely hot plasma jet is inevitable, owing to its incongruent melting at 1570 °C (Figure 10.10).

The thermal decomposition of hydroxyapatite in the plasma jet occurs in four steps (see Scheme 10.1).

Scheme 10.1 Thermal decomposition scheme of hydroxyapatite subjected to a hot plasma jet.

Step 1:	$Ca_{10}(PO_4)_6(OH)_2 \rightarrow Ca_{10}(PO_4)_6(OH)_{2-2x}O_x\square_x + x\ H_2O$ (hydroxyapatite) (oxyhydroxyapatite)
Step 2:	$Ca_{10}(PO_4)_6(OH)_{2-2x}O_x\square_x \rightarrow Ca_{10}(PO_4)_6O + (1-x)H_2O$ (oxyhydroxyapatite) (oxyapatite)
Step 3:	$Ca_{10}(PO_4)_6O \rightarrow 2Ca_3(PO_4)_2 + Ca_4(PO_4)_2O$ (oxyapatite) (tricalcium phosphate) (tetracalcium phosphate)
Step 4:	$Ca_4(PO_4)_2O \rightarrow 4\ CaO + P_2O_5 \uparrow$ or $Ca_3(PO_4)_2 \rightarrow 3\ CaO + P_2O_5$

Based on this decomposition sequence, a model was developed (Antolotti *et al.*, 1998; Graßmann and Heimann, 2000; Dyshlovenko *et al.*, 2004a) which showed the in-flight evolution of individual calcium phosphate phases. During the short

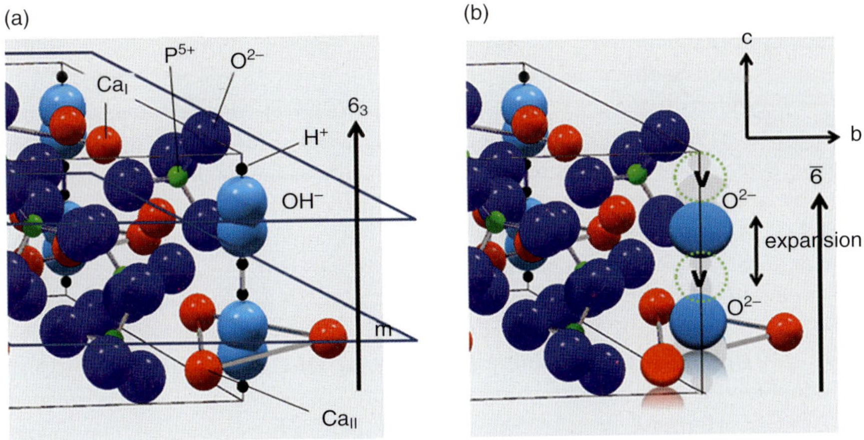

Figure 10.11 (a) Ball-and-spoke model of hydroxyapatite with space group $P6_3/m$. The mirror planes m through z = ¼ and ¾ are shown perpendicular to the screw axis 6_3; (b) Hypothetical structure of oxyapatite with lower symmetry $P\bar{6}$ showing a chain of O^{2-} ions parallel to the polar hexagonal axis $\bar{6}$ separated by vacancies, V. For details of color coding, see Figure 10.7.

residence time of the HAp particle in the plasma jet, the innermost core is still at a temperature below 1823 K owing to its low thermal conductivity, and consists of coexisting HAp, OHAp, and OAp as the stable phases (Steps 1 and 2 in Scheme 10.1; see Figure 10.15, left). The existence of OHAp and OAp had been postulated previously (Trombe and Montel, 1978) and was the subject of controversy for many years, but this has now been well established based on modern analytical techniques. A more recent study (Liao *et al.*, 1999) revealed that: (i) a small amount of OH^- ions is always present in the structure of OAp; and (ii) even a loss of 75% of the chemically bound water maintains the apatite channel structure. The limiting composition is then $Ca_{10}(PO_4)_6(OH)_{0.5}O_{0.75}\square_{0.75}$, corresponding to a water loss of 1.34% (Trombe and Montel, 1971). Hence, partially dehydroxylated HAp (oxyhydroxyapatite) could be described either as a nonstoichiometric defect HAp with a distorted structure (Hartmann *et al.*, 2001; Heimann *et al.*, 2003), or a solid solution of stoichiometric HAp and OAp (Kijima and Tsutsumi, 1979). Vibrational spectroscopy revealed a loss of the intensity of the OH stretching vibration band at 3571 cm^{-1} and the OH vibrational band at 630 cm^{-1} (Park *et al.*, 2002; Tran, 2004).

According to recent investigations into the structure of OAp, there exists a linear chain of O^{2-} ions parallel to the c-axis, with each one followed by a vacancy (Alberius Henning *et al.*, 1999) (Figure 10.11b). Calculations using density-functional theory with local-density approximation (DFT-LDA) and first-principles pseudopotentials (Calderin *et al.*, 2003) postulated the existence of a hexagonal "c-empty" structure $Ca_{10}(PO_4)_6\square_2$ with a stable total-energy minimum. Hence, the defect apatite structure appears to be insensitive to the choice of anion (OH^-, F^-, O^{2-}, $\square$) in the c-axis column; that is, if these are removed completely, the structure is largely unchanged and remains at least metastable, even though a recent computational investigation has suggested that such a pile-up of oxygen defects in the channels would be thermodynamically unfavorable (De Leeuw *et al.*, 2007).

The detection of OAp by X-ray diffraction (XRD) is considered problematic (Gross *et al.*, 1998c), since its somewhat larger c-axis dimension (Trombe, 1973; Montel *et al.*, 1980) accounts for only a small shift of the (00.*l*) interplanar spacings to smaller diffraction angles. Hence, very accurate measurements are required using for example, single-crystal X-ray and neutron diffraction techniques. The lattice expansion during dehydroxylation is presumably caused by the larger Shannon crystal radius of the O^{2-} ion (135 pm) compared to that of the OH^- ion (118 pm). However, deviating results were obtained by Calderin *et al.* (2003), who used *ab initio* calculations by DFT-LDA and first-principles pseudopotentials. Whilst with these calculations, the space group $P\bar{6}$ postulated by Alberius Henning *et al.* (1999) could be faithfully reproduced, the c_0-axis dimension was found to be $c_0 = 0.673$ nm, substantially shorter than the experimental value of $c_0 = 0.6930(8)$ nm inferred from the synchrotron data reported by Gross *et al.* (1998c). The c_0 value of the "c-empty" structure was calculated as 0.682 nm (Calderin *et al.*, 2003).

Since OHAp (and OAp) show only small differences in lattice parameters compared to HAp (Trombe and Montel, 1972), conventional XRD is not the most suitable tool to distinguish among these phases. However, the small shift of the (002) peak position of OAp to lower diffraction angles compared to HAp may be used for this purpose (Gross *et al.*, 1998a). Recent investigations with high-resolution synchrotron radiation (Heimann *et al.*, 2008) have suggested that the atmospheric plasma spraying of stoichiometric HAp results initially in a reversible dehydroxylation to form OAp with $c_0 = 0.6902(3)$ nm (see Scheme 10.1, Step 2). Further heating by the plasma jet causes thermal decomposition under the formation of tri- and tetracalcium phosphate (Step 3), as well as ACP. Figure 10.12 shows the XRD pattern of HAp powder (average grain size 120 μm) when plasma-sprayed (plasma power 11 and 24 kW; total plasma gas flow rate 50 slpm; powder feed rate 18 g min^{-1}; spray distance 100 mm; translation speed 30 m min^{-1}) onto a Ti6Al4V target plate. Here, the only crystalline phases discernible are OAp (PDF 89–649) and β-TCP (PDF 9-016978). However, the strongly elevated background between about 25 and 35 °2Θ suggest the existence of considerable amounts of amorphous calcium phosphate. Indeed, a quantitative estimation of the amorphous content according to the method of Keller and Dollase (2000) revealed ACP contributions of 25% (plasma power 11 kW) and 40% (plasma power 24 kW), respectively (Figure 10.13).

Figure 10.14 shows the high-resolution synchrotron XRD pattern of (002) and (004) of HAp and OAp, respectively. Data for the scattering vectors of the (002) and (004) interplanar spacings obtained from the Powder Diffraction File are given for comparison. These indicate that the scattering vectors measured are between those of HAp (PDF 9–432) and OAp (PDF 89–649). From the measured *Q*-values of 1.821 $Å^{-1}$ and 3.640 $Å^{-1}$, c_0-values of 0.6900(0) nm and 0.6904(6) nm, respectively were calculated, about 0.26% longer than the c_0 of HAp, of 0.6881(4) nm.

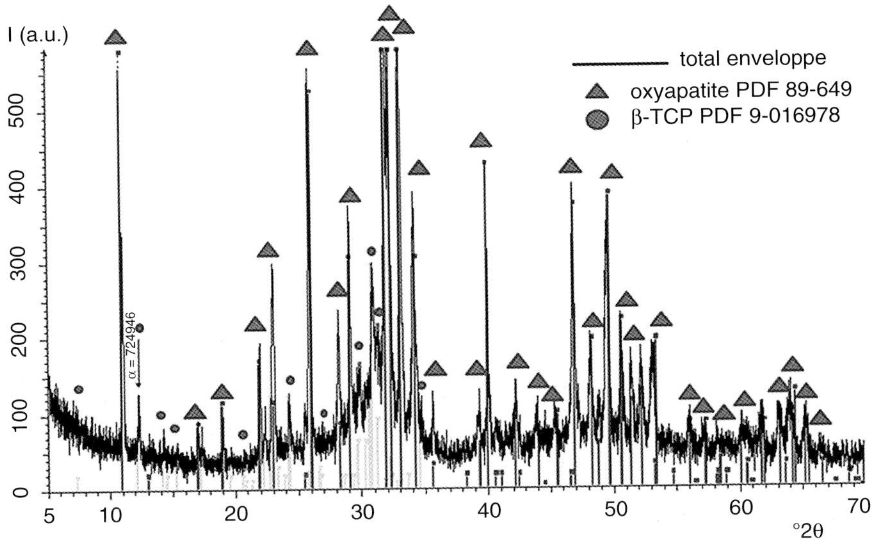

Figure 10.12 Synchrotron X-ray diffraction pattern of as-sprayed hydroxyapatite (Heimann, 2009).

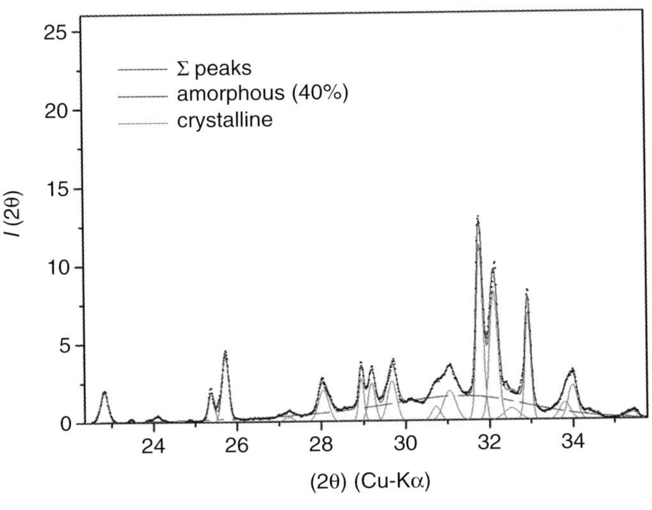

Figure 10.13 Nonlinear least-squares fitting (Keller and Dollase, 2000) to determine the ACP content of a plasma-sprayed hydroxyapatite coating (Heimann *et al.*, 2008).

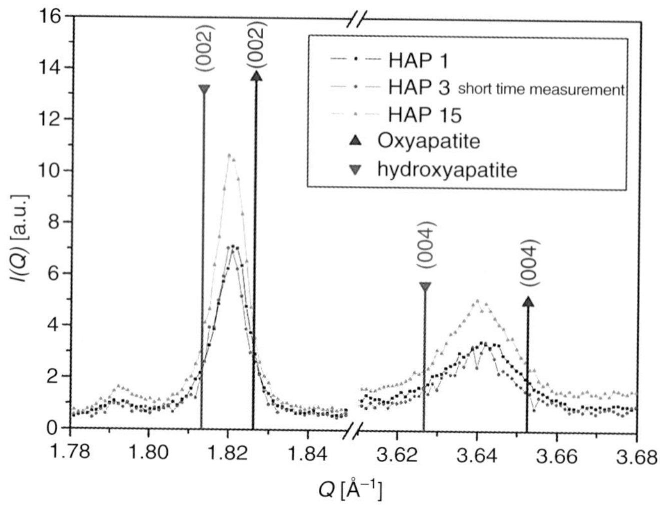

Figure 10.14 Scattering vector $Q = 4\pi\sin\Theta/\lambda$ of synchrotron radiation diffraction profiles of as-sprayed hydroxyapatite coatings showing the vicinities of the (002) and (004) interplanar spacings of hydroxyapatite and oxyapatite.

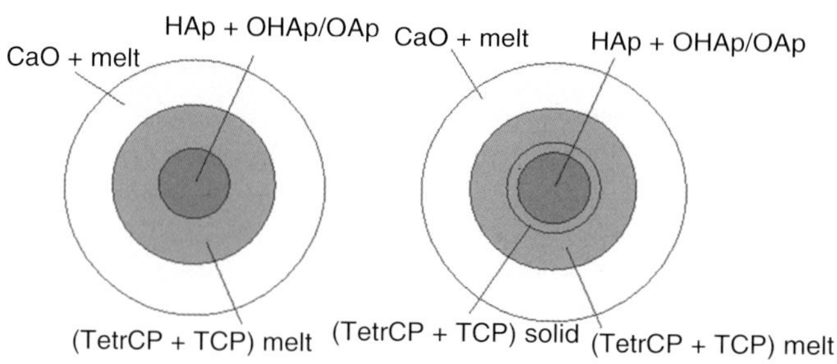

Figure 10.15 Schematic model of the thermal decomposition of a spherical hydroxyapatite particle subjected to high temperature in a plasma jet. Left: at a water partial pressure of 65.5 kPa (Riboud, 1973); Right: at a water partial pressure of 1.3 kPa (Dyshlovenko *et al.*, 2004b).

Figure 10.15 shows a model of thermally induced phase transformation in a spherical HAp powder particle, considering a parabolic temperature gradient according to Fourier's law. During the short residence time of the particle in the plasma jet, the innermost core will remain at a temperature well below 1550 °C owing to the low thermal diffusivity, showing HAp and OAp as stable phases (see Steps 1 and 2 in Scheme 10.1). The second shell has been heated to a temperature above the incongruent melting point of hydroxyapatite (1570 °C), and consists of a mixture of TCP and TTCP (step 3 in Scheme 10.1; see also Figure 10.10). The outermost spherical shells of the particles consist of CaO + melt (L), since the evaporation of P_2O_5 shifts the composition along the liquidus towards CaO-richer phases (Step 4). The temperature increases to well beyond 1800 °C, such that the only nonmelted composition is CaO. When the particle impinges at the metal implant surface to be coated, the clear phase separation will be lost. The result is an inhomogeneous calcium phosphate layer in which HAp, OAp, TCP [α′-TCP, α-TCP, β-TCP (whitlockite)], TTCP (hilgardite) and CaO (oldhamite) are interspersed on a microcrystalline scale (Götze *et al.*, 2001). At the immediate interface to the solid substrate, a very thin layer of ACP may exist, formed by the rapid quenching of the outermost melt layer with heat transfer rates beyond $10^6\,K\,s^{-1}$. This thin layer takes on a special significance, as its high solubility in body fluid may be one of the leading causes of coating delamination.

Recently, this simple model was modified (Dyshlovenko *et al.*, 2004b) to include the product of the solid-state dehydration transformation of HAp into TCP and TTCP between 1633 and 1843 K, that has been shown experimentally (Liao *et al.*, 1999; Santos *et al.*, 1995) to occur at very low water partial pressure (see Figure 10.15, right). Hence, the second shell had been heated to a temperature of $1633 < T < 1843\,K$, just below the incongruent melting point of HAp that underwent solid-state decomposition to a mixture of α′-TCP and TTCP (see Figure 10.10). The third shell was heated to temperatures above 1843 K, and consists of a melt with a Ca/P ratio of 1.67 (Step 3 in Scheme 10.1; Carayon and Lacout, 2003). The outermost shell is composed of CaO and a melt, the Ca/P ratio of which is shifted by the continuous evaporation of P_2O_5 along the liquidus of the binary phase diagram CaO–P_2O_5 towards CaO-richer phases; that is, TTCP (Step 4 in Scheme 10.1). On impact with the target surface, this molten phase solidifies quickly to produce ACP with various Ca/P ratios (Gross *et al.*, 1998b; Carayon and Lacout, 2003; Heimann and Wirth, 2006).

According to the performance requirements of plasma-sprayed coatings discussed above, a list of essential properties has been developed (these are listed in Table 10.5) (Wintermantel and Ha, 1996; Willmann, 1996).

Table 10.5 Performance requirements of bioceramic hydroxyapatite coatings (Wintermantel and Ha, 1996; Willmann, 1996).

Property	Requirement	Reason
Coating thickness	>50 μm	Coating <50 μm will be resorbed
Coating thickness	≤200 μm	Upper limit for sufficient strength
Porosity/roughness	100–100 μm	Minimum porosity for ingrowth
HAp content	>95%	Minimum purity for biocompatibility
HAp content	>98%	Chemical stability
Crystallinity	>90%	Increasing resorption with decreasing crystallinity
Adhesion strength	>35 MPa	Prevention of spalling

10.5.2.3 Calcium–Titanium–Zirconium Phosphates

Calcium–titanium–zirconium phosphate ceramics show solubilities in SBFs that are at least one order of magnitude lower than those of HAp and, in particular, than of TCP. The plasma-spraying of coatings of this composition result in a good adhesion to Ti6Al4V substrates (>40 MPa), even though considerable thermal decomposition has been observed to form zirconium pyrophosphate (ZrP_2O_7), rutile (TiO_2), and baddeleyite (β-ZrO_2). There is some evidence, however, that these bioinert products of incongruent melting of the coating material may lead to a particle-mediated reinforcement of the coating microstructure. *In vitro* biocompatibility tests with primary rat bone marrow cells have demonstrated substantial cell proliferation in the presence of fetal bovine serum. Subsequent studies conducted in sheep indicated that 150 μm-thick coatings based on $CaTiZr_3(PO_4)_6$, when applied to Ti6Al4V rods implanted in the femur, led to a strong neoformation of dense bone at a stable implant–bioceramic interface coating, but without the coating delamination often observed with HAp. The build-up of a Ti6Al4V/TiO_2/NASICON/(HAp) coating system could lead to a "bio"-capacitor which, by correct poling, could be used to store negative electrical charges close to the interface with the growing bone, thus enhancing the bone apposition rate and, presumably, also the bone density (see below).

Structure and Properties of $CaTiZr_3(PO_4)_6$ The $CaTiZr_3(PO_4)_6$ orthophosphate belongs to the NASICON (Na superionic conductor) structural family, and is electrically moderately conductive even at room temperature (Hosono *et al.*, 1992, Szmukler-Moncler *et al.*, 1992, Alamo, 1993; Senbhagaraman *et al.*, 1993). The composition of the NASICONs can be described by the general formula $A_2MX_3O_{12}$, where A = Ti, Zr, Hf, Nb or other transition metals of appropriate size, M = alkali metal, alkaline earth metal, or lattice vacancies, and X = phosphorus or silicon. All members of this family show a low coefficient of linear thermal expansion and a high thermal shock resistance. Figure 10.16 shows the geometry of the $A_2(XO_4)_3$ structural unit, and Figure 10.17 the unit cell of $CaTi_4(PO_4)_6$. The $Ti_2(PO_4)_3$ groups form a 3-D network of two TiO_6 octahedra that are connected through their vertices to three PO_4 tetrahedra. These basic units appear as $-O_3TiO_3-O_3TiO_3-$ bands along the *c*-axis of the hexagonal unit cell. Along the *ab* plane, these bands are connected by PO_4 tetrahedra.

This atomic arrangement results in a hexagonal skeleton of space group *R3c*, with large structural vacancy sites per unit cell of octahedral symmetry (M1) occupied by cations with the coordination number [6], such as titanium or zirconium and trigonal-prismatic symmetry (M2) occupied by vacancies as in $\square Nb_2(PO_4)_3$, by two-valent ions as in $Ca_{0.5}Ti_2(PO_4)_3$ or three-valent ions as in $La_{0.33}Ti_2(PO_4)_3$. These vacancies account for the structural variability of the NASICON family, as well as their ionic conductivity.

Reports of quantitative data of the electric conductivity of NASICONs are few in number, and appear to differ considerably depending on the chemical composition and preparation conditions. Silva *et al.* (2006) measured the d.c. conductivity of $CaTi_4(PO_4)_6$ produced by high-energy dry ball-milling for 15 h of a mixture of

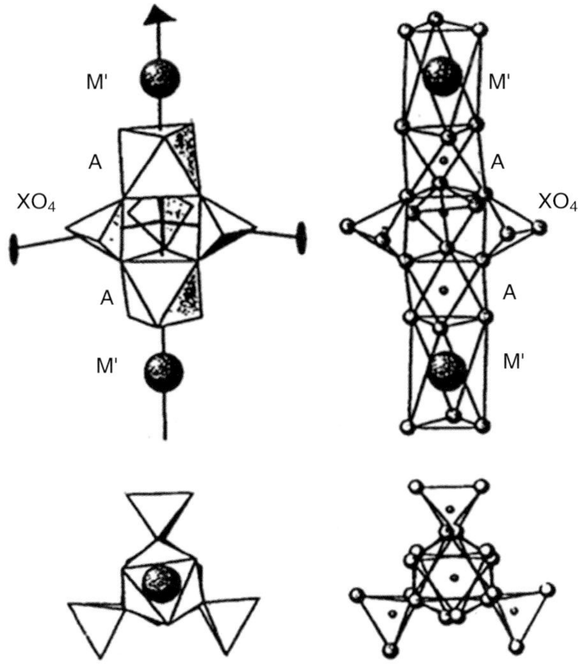

Figure 10.16 Geometry of the $A_2(XO_4)_3$ structural group (Alamo, 1993). Reprinted with permission from Elsevier.

$Ca(H_2PO_4)_2$ and TiO_2 to be about $\sigma = 4.0 \times 10^{-12}\,S\,m^{-1}$, about 30-fold that of HAp. The dielectric permittivity was reported as $\varepsilon = 15.4$, somewhat higher than that of HAp ($\varepsilon = 10$–12). However, the reaction product obtained by milling still contained large amounts of the starting materials, attesting to a very incomplete reaction. Hence, the low conductivity value reported was presumably due to contamination by TiO_2 and calcium dihydrogen phosphate. In contrast, Fuentes *et al.* (2005) prepared submicrometer grains of Y-doped $Na_3Si_2(Zr,Y)_2PO_{12}$ by the sintering of mechanically activated precursor powders, and determined an ionic conductivity of $\sigma = 2.5 \times 10^{-5}\,S\,m^{-1}$, some seven orders of magnitude higher than that found by Silva *et al.* (2006). This high value of electric conductivity may be associated with the high concentration of the charge-carrying sodium ions and their higher electric mobility, compared to the calcium ions in the former structure.

Solubility of As-Synthesized $CaTiZr_3(PO_4)_6$ Figure 10.18 shows solubility data of several bioceramic materials in protein-free simulated body fluid (pf-SBF after Kokubo 1996). The highest solubility was found for α-TCP, followed by HAp. The $CaTiZr_3(PO_4)_6$ exhibited the lowest solubility, and hence was considered to be a biomaterial of the second generation, with extreme biostability (Szmukler-Moncler *et al.*, 1992; Schneider *et al.*, 1998; Schneider, 2002).

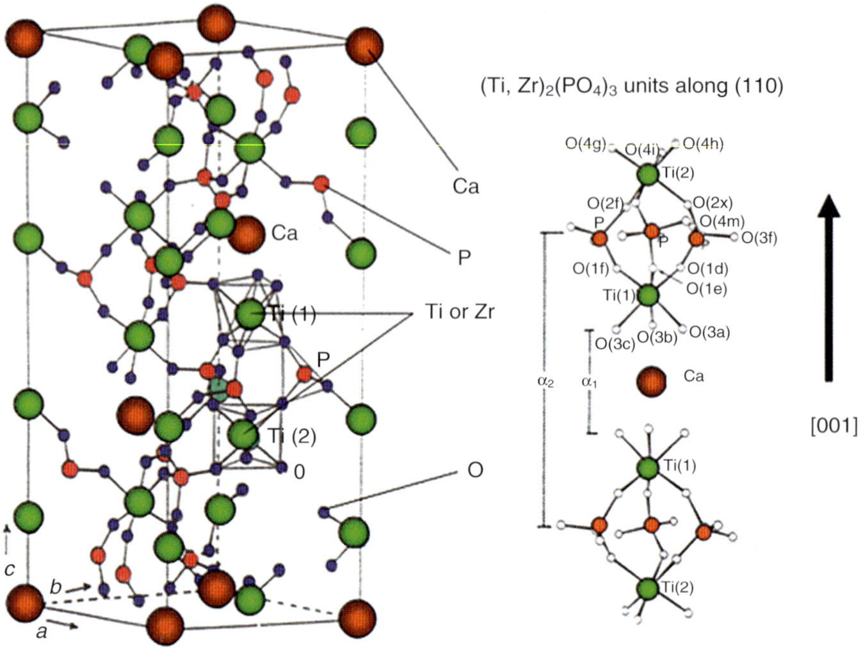

Figure 10.17 Unit cell of $Ca(Ti,Zr)_4(PO_4)_6$ (left) and arrangement of $(Ti,Zr)_2(PO_4)_3$ groups along the crystallographic c-axis (Senbhagaraman *et al.*, 1993). Reprinted with permission from The Royal Society of Chemistry.

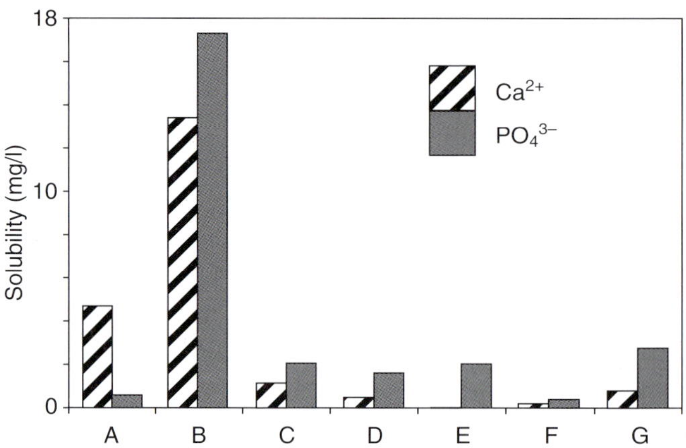

Figure 10.18 Solubilities of several Ca–Ti–Zr orthophosphates (C–G) as well as hydroxyapatite (A) and α-tricalcium phosphate (B) in protein-free simulated body fluid (pf-SBF) at 37 °C and pH 7.4 (Schneider, 2002; Heimann and Schneider, 2000). C: CT_4P_3; D: CT_3ZP_3; E: $CT_2Z_2P_3$; F: CTZ_3P_3; G: CZ_4P_3 (C = CaO, T = TiO_2, Z = ZrO_2, P = P_2O_5).

Table 10.6 Plasma spray parameter levels.

Parameter	High level (+)	Low level (−)
Plasma power (kW)	35	25
Argon flow (slpm)	50	40
Hydrogen flow (slpm)	12	6
Carrier gas flow (slpm)	6	3
Powder feeder settings (%)[a)]	40/80	20/60
Spray distance (mm)	120	80
Grain size (μm)	+45–70	+25–45

a) The first number refers to rotation of the powder feeder disk, the second number to rotation of the stirrer.

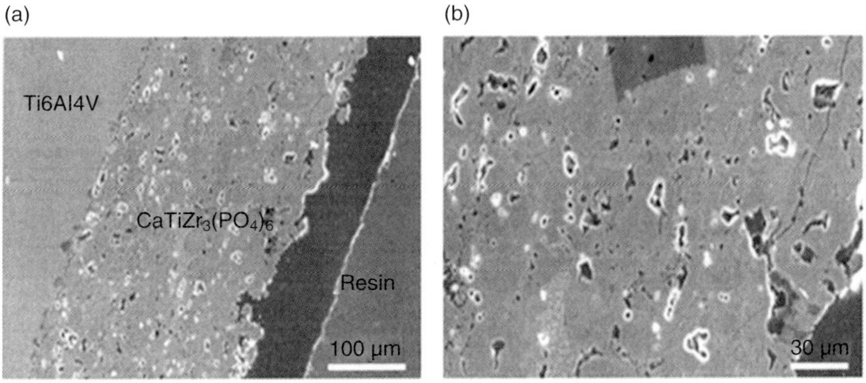

Figure 10.19 (a) Cross-section of a typical plasma-sprayed $CaTiZr_3(PO_4)_6$ coating on a Ti6Al4V substrate; (b) Higher magnification of the coating interior. Note the absence of microcracks and larger interconnected pores.

Atmospheric Plasma Spraying (APS) of $CaTiZr_3(PO_4)_6$ Powder The sol–gel-synthesized and agglomerated material was plasma-sprayed onto Ti6Al4V substrates using a statistical design of experiments methodology (Plackett–Burman design with 12 experiments, seven factors). The spray parameters selected and the factor levels are listed in Table 10.6.

Figure 10.19 shows the cross-sections of a typical coating with an average porosity of 17%, that is thought to promote an easy in-growth of bone cells. One very advantageous feature here involves the low crack concentration of the coatings in relation to the high thermal shock resistance, in conjunction with the low coefficient of thermal expansion ($\alpha_a = 5 \times 10^{-6}\,K^{-1}$; $\alpha_c = 0.9 \times 10^{-6}\,K^{-1}$; Senbhagaraman *et al.*, 1993) of the material that tend to generate compressive residual stress at the substrate–coating interface.

The phase composition of an as-sprayed coating is shown in the XRD pattern of Figure 10.20 (upper trace), compared to the calculated pattern (lower trace). The

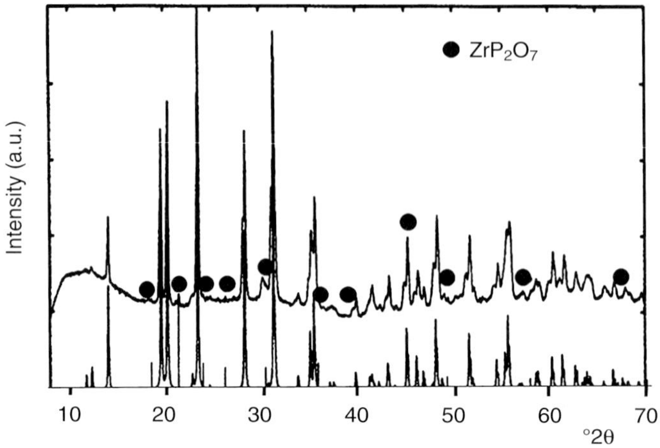

Figure 10.20 X-ray diffraction pattern of a plasma-sprayed $CaTiZr_3(PO_4)_6$ coating on a Ti6Al4V substrate (upper trace) and calculated diffraction pattern (lower trace) (Heimann, 2006).

interplanar spacings marked with dots can be assigned to ZrP_2O_7, which is formed as a thermal decomposition product during plasma spraying.

Problems still exist, however, with regards to thermal stability. In addition to zirconium pyrophosphate (ZrP_2O_7) and very small amounts of rutile (TiO_2) and baddeleyite (m-ZrO_2), amorphous decomposition products appeared during plasma spraying that might be related either to the incongruent melting of the material, or to an inherent structural instability caused by a high lattice vacancy concentration of the NASICON structure. Figure 10.21 shows a highly inhomogeneous APS coating with at least four chemically different phases (labeled 1 to 4). The dark gray phase (1) constitutes about 50% of the coating volume, and consists of rounded particles of the target compound $CaTiZr_3(PO_4)_6$ embedded in a lighter-colored matrix phase (2) of the approximate composition $Ca(Ti,Zr)_{4.5}O_{4.6}(PO_4)_{3.6}$ and a phosphate-poor phase (3), the composition of which was measured [using energy-dispersive X-ray (EDX) analysis] to be $Ca(Ti,Zr)_{4.2}O_{6.85}(PO_4)_{1.7}$. Occasionally, m-$ZrO_2$ (baddeleyite) appears (4). Comparable results were obtained previously with plasma-sprayed $CaTi_4(PO_4)_6$ and $CaTi_2Zr_2(PO_4)_6$ coatings (Reisel, 1996).

Solubility of Plasma-Sprayed $CaTiZr_3(PO_4)_6$ Coatings The *in vitro* solubility in 0.2 M Tris–HCl buffer solution of the phosphorus-depleted phases found in plasma-sprayed coatings (see Figure 10.21) is somewhat higher than that of stoichiometric $CaTiZr_3(PO_4)_6$, and varies according to the spraying conditions (Figure 10.22). Here, samples 2 to 5 show solubilities that are slightly higher than that of the starting powder (sample 1), but much lower than the solubility of APS HAp coatings (sample 6). It is expected that the correct spray parameter engineering would lead to coatings that were optimized in terms of their porosity and phase content, and with minimum solubility under physiological conditions.

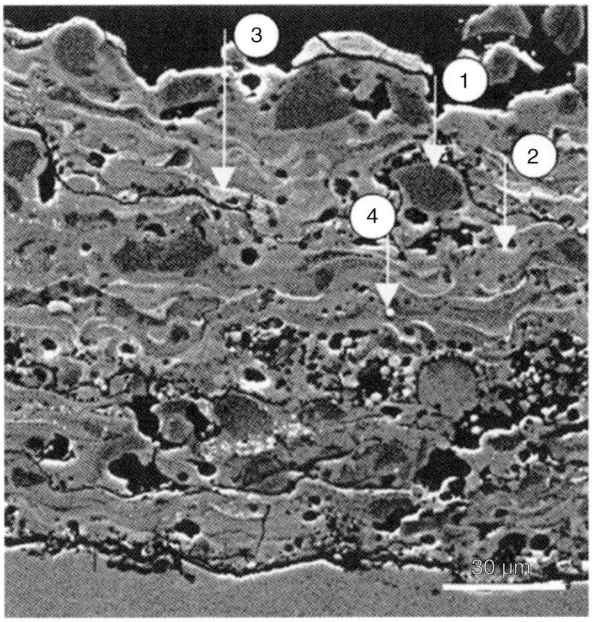

Figure 10.21 Cross-section of a plasma-sprayed $CaO–TiO_2–ZrO_2$ phosphate coating on a Ti6Al4V substrate (lower part of figure) (Heimann, 2006). For an explanation, see the text.

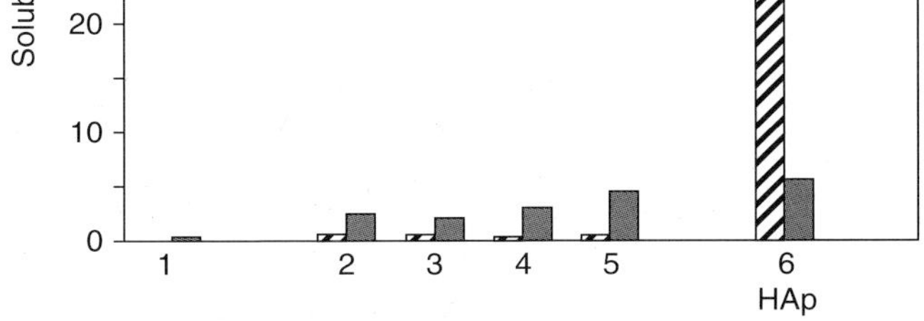

Figure 10.22 Solubility of APS-sprayed $CaTiZr_3(PO_4)_6$ coatings produced with different sets of spray parameters (2–5) compared to the starting powder (1) and APS hydroxyapatite coatings (6) (0.2 M Tris–HCl buffer, pH 7.4, 37 °C) (Schneider, 2002).

***In Vivo* Performance of $CaTiZr_3(PO_4)_6$ Coatings on Ti6Al4V Implants** Earlier *in vivo* investigations with $CaZr_4(PO_4)_6$ disks implanted into the distal epiphyseal region of the dog femura and tibiae showed no deleterious cell responses, but a direct contact between implant and bone was established without any noticeable resorption of the biomaterial. Notably, there appeared to be an earlier bone apposition reaction compared to HAp, with a large bone contact area. After nine months, an extensive remodeling had occurred of the osteons in direct contact with the biomaterial, without any noticeable resorption (Szmukler-Moncler *et al.*, 1992).

Subsequently, animal experiments (Heimann, 2006) were conducted to estimate the biological performance of atmospheric plasma-sprayed $CaTiZr_3(PO_4)_6$ coatings. Figure 10.23a shows a distal section of the coated implant (shown as a black circular disk) with the surrounding trabecular bone tissue as a histological section (Figure 10.23a); the corresponding microradiograph is shown in Figure 10.23b. A larger scale bony integration is apparent in the dense toluidine blue-stained (blue) bridging connections between the implant and the spongy bone. However, the distal part of the implant oriented towards the marrow-filled medulla did not reveal continuous bone apposition but only "spotty" lamellar bony in-growth. Figure 10.24a shows (at higher magnification) the close bony integration of the proximal part of the coated implant. From the proximal section shown in Figure 10.24b, it is apparent that the gap-bridging potential is being reduced at a larger distance between the implanted rod and the surrounding (cortical) bone.

Acceleration of Bone Growth by an Electric Field Electrical currents have been used to heal bone fractures since the mid-1800s (Julius Wolff, 1835–1902; James Prescott Joule, 1818–1889), and the effect of electrical stimulation on bone has long been studied and well documented (Bassett and Becker, 1962; Bassett, 1968; Bassett *et al.*, 1974). It has also long been known that the growth of bone is affected by the presence of materials with different dielectric behavior, as suggested by the so-called "bioelectric phenomenon" in bone (Fukada, 1957; Shamos *et al.*, 1963;

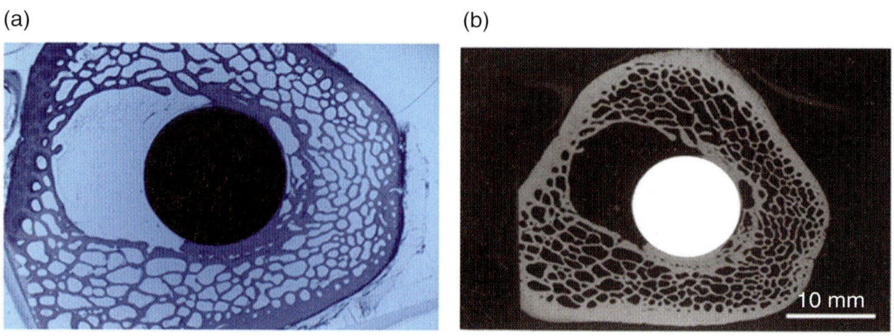

Figure 10.23 Distal section through the femoral medulla of sheep with an implanted Ti6Al4V rod coated with an approximately 150 μm-thick $CaTiZr_3(PO_4)_6$ coating. (a) Toluidine blue-stained histological section; (b) Microradiographic image of the same sample (Heimann, 2006).

(a) (b)

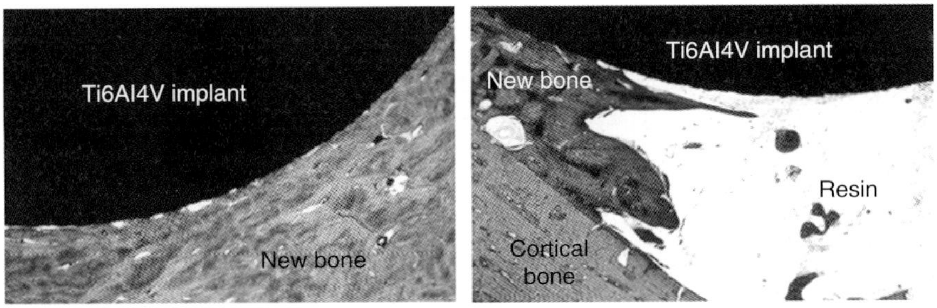

Figure 10.24 Proximal section through the femoral medulla of sheep with an implanted Ti6Al4V rod coated with an approximately 150 μm-thick $CaTiZr_3(PO_4)_6$ coating. (a) Higher magnification showing close integration of the coated implant; (b) The gap-bridging potential of the newly formed bone is reduced when the distance between coated implant and surrounding cortical bone is increased (Heimann, 2006).

Weigert, 1973). In this case, mechanically triggered electrical charge separation acts as the prime mechanism during the remodeling of a broken bone, whereby the structure and function of growing bone cells and extracellular structures are influenced by piezoelectric and stream potentials (Bassett and Becker, 1962; Bassett, 1968). It has been shown in clinical studies that the electrical stimulation of bone growth, with the fields aligned parallel to the axis of a long bone, acts to reduce the time required for endosteal callus remodeling, and hence speeds up the healing process (Weigert, 1973; Weinstein, 1976; Colella *et al.*, 1981; Berry *et al.*, 1986). However, it has also been observed that whilst a pulsed, transverse electric field tended to accelerate the growth of embryonic chick tibiae, a static, non-varying field had no significant effect (Watson *et al.*, 1975).

Since improved bone healing at the negative electrode is accompanied by the resorption of bone (osteolysis) or osteoporosis at the positive electrode, the concept of electrical stimulation of bone healing must be approached judiciously. Although, unfortunately, the origin and function of bioelectric potentials are not yet known with certainty, it is presumed that they establish diffusion gradients; in other words, the ionic currents are able to concentrate polarizable molecules, collagenous proteins, noncollagenous polyanionic proteins, and electrolytes at the wound site. There is some indication that the concentration of Ca^{2+} ions at the (negative) electrode is higher than in a nonstimulated fracture. Typically, the Ca ions originate from the depolymerization of mucopolysaccharides, and P ions from the glycogenolysis of phosphate esters by action of alkaline phosphatase. Ions that initially are stored in the mitochondria are released, by a change in the potential of the mitochondrial membrane, via the cytoplasma and the cell membrane into the ECM (Rubin and Saffir, 1970). The nucleation of calcium phosphate then occurs at the wound site, and acts as a crystallization center for the apatite microcrystals. The important role that the membrane potential plays in the release of Ca and P ions suggests that an external electric field would influence this process.

In these experiments, a substantially higher maximum shear stress was required to push out of its bony bed an implant subjected to electrical stimulation compared to a control (Colella *et al.*, 1981). For this, a commercially pure (cp)-titanium cylinder, acting as a cathode, was inserted into the mid-femoral diaphysis of a dog, and segments of kinked Ti wires (acting as anode) were placed in coils around the bone, above and below the implant, such that the electric field lines were directed along the long axis of the femur. The connection was made using PTFE-coated wires running subcutaneously from the implant to a battery harnessed to the dog's back. While the interfacial shear strength of the stimulated implants was consistently greater than that of controls, no difference could be observed in the quality of bone ingrowth. These data suggest that the rate and quantity of bone ingrowth were enhanced by electrical stimulation.

Similar experiments conducted on the hind legs of rabbits showed conclusively that, after osteotomy of the tibia, healing proceeded much faster in the presence of a d.c. current of 10 μA, as verified by ^{87}Sr-m ($t_{1/2} = 2.8$ h) scintigraphy. In Figure 10.25a, the fractured tibiae are still clearly visible after three weeks, whereas in the presence of a current the ^{87}Sr-m activity was strongly increased (Figure 10.25b), indicating enhanced callus formation and bone healing (Weigert, 1973).

The notion of acceleration of bone growth by a poled electrical field is strongly supported by more recent findings, that the growth of HAp crystals from a SBF is dramatically quickened on negatively polarized dielectric and ferroelectric substrates such as calcium and barium titanate (Yamashita *et al.*, 1996). The results of these studies suggest that a uniform electrical field, rather than localized charges, is the stimulating factor of bone remodeling during the healing process (Calvert and Mann, 1997). There is also evidence that the growth of bone-like HAp in an electric field is accelerated by a reorientation of the dipole moments between

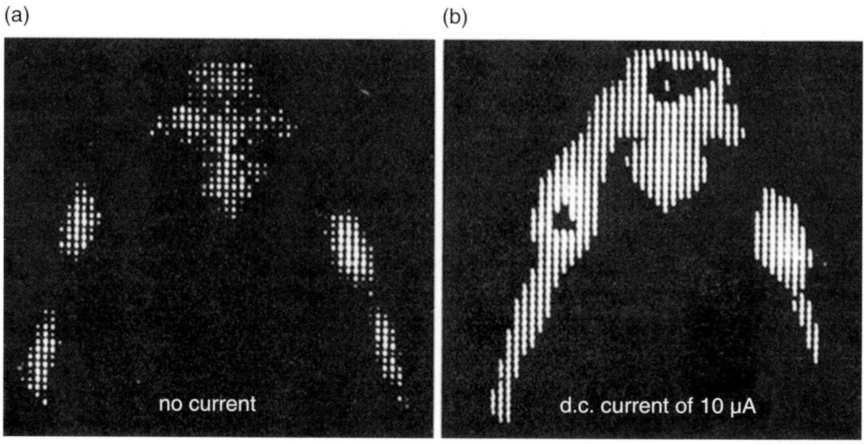

Figure 10.25 ^{87}Sr-m scintigrams of the hind legs of a rabbit subjected to osteotomy of the tibiae. (a) No current; (b) D.c. current of 10 μA and status after 3 weeks (Weigert, 1973).

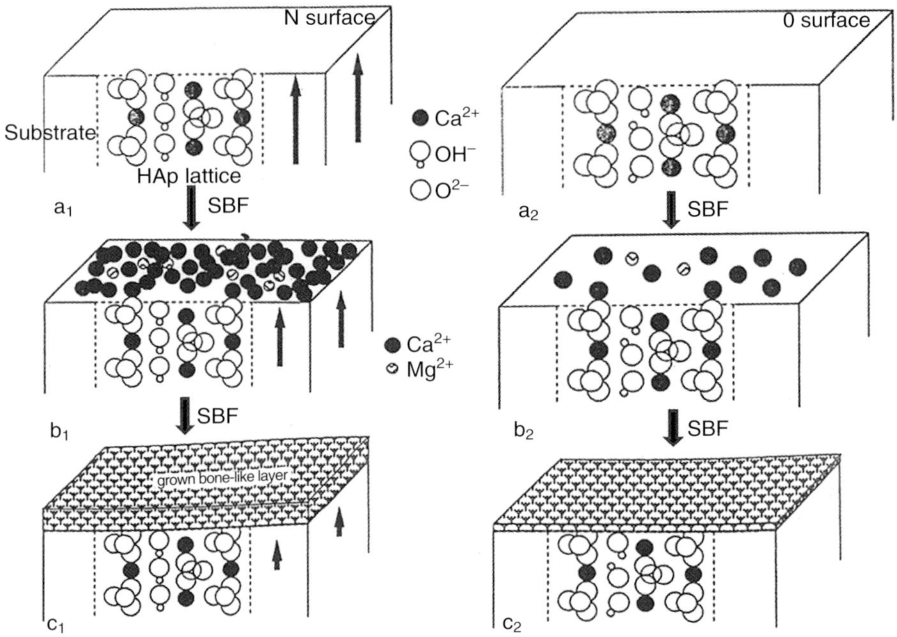

Figure 10.26 Schematics of the accelerated growth of bone-like apatite on a negatively polarized (left column) and nonpolarized (right column) hydroxyapatite surface. Adapted from Yamashita *et al.* (1996). For an explanation, see the text. Reprinted with permission from The American Chemical Society.

O^{2-} and H^+ of the lattice OH^- ions, in response to the electric polarization conditions (Figure 10.26). In this way, an ordered alignment of OH^- columns as are present in monoclinic HAp, is attained (Heimann, 2007). Positively charged Ca^{2+} cations will be adsorbed preferentially onto the negatively charged "N surface," thus attracting a cloud of the counterions HPO_4^{2-}, HCO_3^- and OH^-, and causing a strong nucleation and the eventual precipitation of a thick layer of bone-like apatite (Figure 10.26, left column) that bonds directly to the negatively charged HAp surface (Nakamura *et al.*, 2009). The polarized state would then cause an alignment of the OH^- groups in the HAp structure.

In contrast, the depolarized state of HAp showed randomly oriented OH^- groups and, consequently, a much reduced adsorption of positively charged cations on the "0 surface," in addition to a decelerated growth of bone-like apatite (Figure 10.26, right column). This has important implications: although the nonperiodic fields encountered by Yamashita *et al.* (1996) were weak compared to electrostatic fields, even a small change in the local interfacial energy could be sufficient to show a marked influence on the mechanism of transformation of incipient nuclei into macroscopic mineral phases.

Studies on the biological performance of ferroelectric hydroxyapatite-barium titanate (HABT) ceramics (d_{33} = 58 pC/N) showed that the attachment, proliferation, viability, morphology and metabolic activity of cells cultured on unpoled HABT were comparable to those observed on commercially available HAp, at all time points. However, there were no striking differences between the properties of poled and non-poled ceramics, at least not during the short incubation times employed (Baxter *et al.*, 2009).

Whilst the polarization orientation of the OH^- groups of HAp requires an electric field, it is also conceivable that $CaTiZr_3(PO_4)_6$ behaves differently, in that a spontaneous polarization may occur owing to charge imbalances within the TiO_6 octahedral groups. However, this would impose a ferroelectric behavior which has not yet been reported for the NASICON structure family. In contrast, ferroelectric ceramics such as lead zirconate titanate (PZT; see Section 8.3.2), when implanted into the diaphysis of the metatarsus of roosters in either a polarized or depolarized state, were found to have a negligible effect on the rate and quality of bone formation (Schumacher *et al.*, 1982; Park *et al.*, 1981). This effect was shown to be related to the combined effects of an insufficient pressure load producing a low signal strength, a high internal resistance, and incompatible elastic moduli of ceramics and bone.

On the other hand, calcium titanate ($CaTiO_3$) has been implicated in enhanced osteoblastesis *in vitro* on titanium-coated HAp surfaces. From this, it has been concluded that orthopedic coatings which form $CaTiO_3$ could increase the osseointegration with juxtaposed bone that would be needed for increased implant efficacy (Webster *et al.*, 2003). It is tempting to suggest that the potentially ferroelectric nature of calcium titanate may be a reason for its osseoconductive behavior, but if this were the case then the notion that a titanium oxide surface is an important prerequisite of osteoblastesis would need to be modified. The control of apatite nucleation by calcium titanate surfaces (Kokubo *et al.*, 2003; Webster *et al.*, 2003; Ohtsu *et al.*, 2006) has been explained by Wei *et al.* (2007), on the basis of an epitaxial structural relationship between the (022) lattice plane of calcium titanate and the (0001) lattice plane of hydroxyapatite (Figure 10.27). Here, the (022) plane of the perovskite structure is defined by the position of the oxygen atoms, whereas the (0001) plane of HAp is defined by the position of the hydroxyl ions. A two-dimensional lattice match exists, characterized by a fit of $2X_c \approx X_h$ (mismatch: 0.8%) and $3Y_c \approx Y_h$ (mismatch: 0.09%). Hence, the biomimetic formation of HAp during incubation in SBF, and also during osseoconductive integration in the bone bed *in vivo*, relies on the interfacial molecular recognition determined by the degree of epitaxial crystal lattice matching, as shown schematically in Figure 10.27.

Recently, a hypothetical regulation mechanism for bone modeling and remodeling under the influence of an electromagnetic field has been proposed (Qu *et al.*, 2006). Normally, the damage induced in cortical bone by load changes during daily activities will lead to the generation of microcracks that will be removed by osteoclasts. In turn, growth factors present in latent forms in the osteocytes will be activated and released into the ECM, where they stimulate osteoblasts to refill

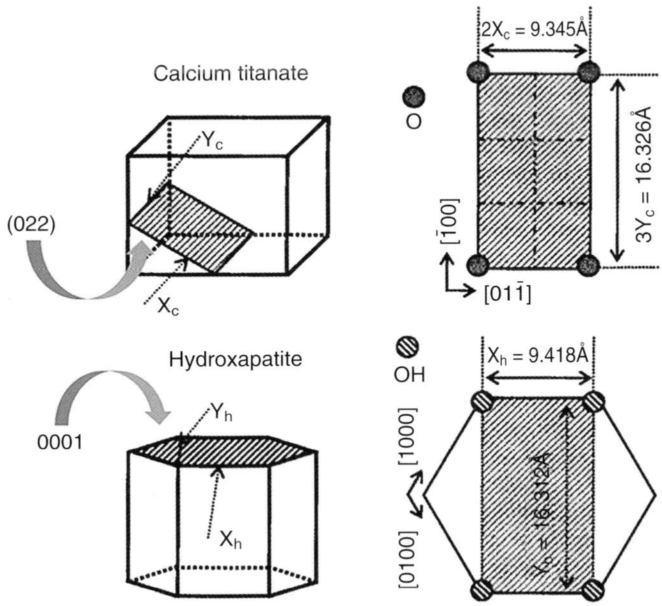

Figure 10.27 Epitaxial relation between (022) of calcium titanate and (0001) of hydroxyapatite (Wei *et al.*, 2007). Reprinted with permission from Elsevier.

the cavities produced during osteolysis. An electromagnetic field is, presumably, capable of stimulating the multiplication of growth factors, and hence accelerating (indirectly) the bone remodeling process. Similar conclusions were drawn by Wang *et al.* (2009), who suggested that an enhanced bone ingrowth by electrical polarization into HAp with a high interpore connectivity would be controlled by the cooperative interaction between the osteoconductivity of HAp and an enhanced osteogenic cell activity induced by large charges stored at the pore surfaces.

10.5.2.4 **Resorbable Calcium Phosphate Ceramics**

The current state-of-the-art materials for producing rapidly resorbable calcium orthophosphate ceramics are TCPs, either as the low-temperature modification (β-TCP) or as one of the high-temperature modifications (α-TCP and α′-TCP) (Bohner, 2010). Whereas, β-TCP is stable below 1125 °C, α-TCP exists at up to and above 1475 °C, whilst the α′-TCP modification is stable up to its melting point at 1756 °C (see Figure 10.10).

α-TCP crystallizes in the monoclinic space group $P2_1/a$ as a large unit cell with parameters a = 1.287 271(9) nm, b = 2.728 034(8) nm, c = 1.521 275(12) nm, β = 126.2078(4)°, and Z = 24 (Yashima *et al.*, 2007). The Ca^{2+} and PO_4^{3-} ions are packed along the c-axis in adjacent columns, with Ca^{2+}–Ca^{2+} and Ca^{2+}–PO_4^{3-} connectivities. Since the space group is identical to that of monoclinic hydroxyapatite ($P2_1/b$), there exist close structural relationships. Consequently, α-TCP can change readily to HAp by replacing the cation–cation columns at the edges of the unit cell with anionic OH^- columns.

β-TCP crystallizes in the rhombohedral space group *R3c*, with a = b = 1.043 52(2) nm, c = 3.740 29(5) nm, α = β = 90.0°, γ = 120°, and Z = 21 (Yashima *et al.*, 2003). Data obtained by Yin *et al.* (2003), using *ab initio* calculation and a DFT approach, yielded slightly shorter lattice constant (a = b = 1.0157 nm, c = 3.6481 nm, α = β = 90.0°, γ = 120°).

The major difference between the structure of α-TCP and β-TCP is the absence of cation–cation columns in the latter case. The results of the Kohn–Sham DFT calculations (Yin *et al.*, 2003) indicates that β-TCP would be much more stable than α-TCP, thus confirming experimental results (see, e.g., Berger *et al.*, 1995a). This situation is thought to be related to the different distributions of Ca atoms, that have a pronounced effect on the stability and the electronic properties of the different modifications of TCP. In particular, the uniformly distributed Ca vacancies stabilize the β-TCP structure, so that its solubility is much less than that of the α-TCP modification.

Tetracalcium phosphate (TTCP), $Ca_4O(PO_4)_2$, is the least stable calcium orthophosphate. It crystallizes in the monoclinic space group $P2_1$ with unit cell parameters a = 0.7023(1) nm, b = 1.1986(4) nm, c = 0.9473(2) nm, β = 90.90(1)° (Dickens *et al.*, 1973; Mathew and Takagi, 2001). The Ca^{2+} and PO_4^{3-} ions are located in four sheets perpendicular to the b-axis, with each sheet containing two Ca^{2+}–PO_4^{3-} columns and one Ca^{2+}–Ca^{2+} column, similar to the structure of glaserite, $K_3Na(SO_4)_2$, but with surplus oxygen ions. Significantly, two adjacent sheets in TTCP form a layer that is closely related to the structure of HAp. In the presence of dicalcium phosphate dihydrate (DCPD, brushite, $CaHPO_4 \cdot 2H_2O$) or DCPA (monetite, $CaHPO_4$), TTCP forms HAp according to:

$$2Ca_4O(PO_4)_2 + 2CaHPO_4 \rightarrow Ca_{10}(PO_4)_6(OH)_2 \qquad (10.2)$$

This reaction describes the formation of the self-setting bone cement that is used ubiquitously in dentistry (Brown and Chow, 1987; Posset *et al.*, 1998).

Figure 10.28 shows the pH dependence of the solubility isotherms of the biologically most important calcium orthophosphates at 25 °C. From this diagram, the following order of solubilities results (De Groot *et al.*, 1990):

$$\text{HAp} < \text{DCPD} < \text{DCPA} < \beta\text{-TCP} < \text{OCP} < \text{TTCP} \qquad (10.3)$$

Although TCP and TTCP are rather easily resorbed within a physiological environment, the quest for products that are even more soluble continues. In the USA, the use of calcium sulfate-containing TCP is promoted (Guo *et al.*, 2006), whereas in Asia there is a trend away from soluble calcium orthophosphates towards calcium diphosphates and calcium polyphosphates. A second development trend is also concerned with alkali-containing calcium orthophosphates in the system $Ca_3(PO_4)_2$–$CaNaPO_4$–$CaKPO_4$ (Berger *et al.*, 1995a; Berger *et al.*, 1995b) that, when used as temporary bone replacement materials, show a high resorbability and biodegradation without imposing any acute or chronic damage to the surrounding tissue. The addition of alkali metal ions (Na, K) to TCP leads to the formation of

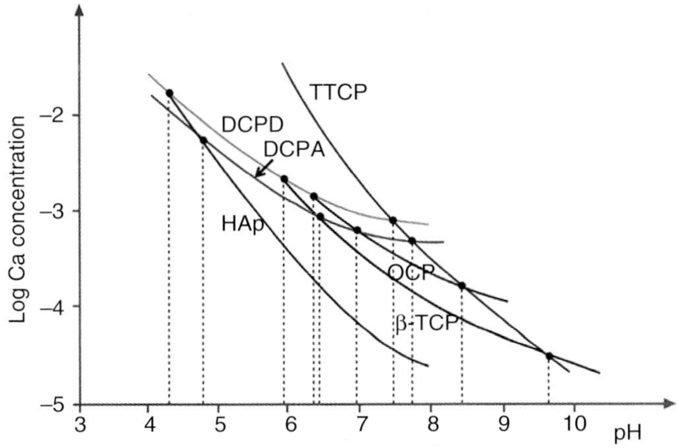

Figure 10.28 Solubility isotherms of calcium orthophosphates at 25 °C. HAp: hydroxyapatite; DCPD: brushite; DCPA: monetite; β-TCP: β-tricalcium phosphate; OCP: octacalcium phosphate; TTCP: tetracalcium phosphate (De Groot *et al.*, 1990).

easily resorbable bone replacement ceramics containing $Ca_{10}Na(PO_4)_7$ and $Ca_{10}K(PO_4)_7$, respectively. Moreover, the structures of these materials seem to very closely resemble that of β-TCP, which makes it difficult to distinguish them from the parent structure. The further addition of alkali ions leads to a crystalline phase of the type $Ca_2M_2(PO_4)_2$, where M = Na, K. In particular, $Ca_2KNa(PO_4)_2$ facilitates enhanced osteoblast growth and ECM elaboration (Berger *et al.*, 1995b; Knabe *et al.*, 1998). Finally, although $CaMPO_4$ (β-rhenanite) will be formed, the solubility of this material is so high that its application as a bone substitute is not feasible. This property notwithstanding, composite HAp/β-rhenanite biomaterials have been developed (Suchanek *et al.*, 1998) in which β-rhenanite acts as a weak interphase with a high bioactivity. Moreover, it significantly enhances the sinterability of HAp at 1000 °C, without forming any undesirable secondary phases.

Currently, phase-pure β-TCP is being used either as sintered ceramic monolithic shapes (e.g., Cerasorb®; Curasan AG, Kleinostheim, Germany) or as coatings for implants (e.g., Chae *et al.*, 1992). In the past, fast-resorbable, alkali-bearing calcium phosphates have been supplied to the medical community predominantly as granular materials, but increasingly they are being produced also as porous spongiosa-like bodies or pastes. Porosity in monolithic ceramic bodies can be introduced by the use of bloating agents and organic spacers, or by using naturally occurring cellular structures such as sponges (the Schwarzwalder–Somers process). Unfortunately, however, owing to the large variability of pore sizes and pore size distributions, it is difficult to obtain products that conform to the stringent quality control requirements imposed by governmental regulations. Consequently, this field is open for the development of process engineering tools to produce these materials with highly reproducible properties.

Recently, a new material was developed that combines the positive properties of bioglasses and calcium phosphates. The biphasic BONIT® (DOT GmbH, Rostock, Germany) consists of 87 mass% crystalline HAp and β-TCP (in the ratio 60 : 40), connected by 13 mass% of a sol–gel-derived silica matrix (Zeggel, 2000; Traykova *et al.*, 2004). The material contains chloride ions at concentrations similar to those in natural bone (39 $mg\,kg^{-1}$), and has a high porosity (60–70%) corresponding to a solid density of 0.9 $Mg\,m^{-3}$. Owing to its high porosity and easy resorbability, BONIT® can be loaded with antibiotics such as gentamicin (Teller *et al.*, 2005) or rhBMPs.

10.6 Performance Requirements and Quality Control of Ceramic Femoral Ball Heads

The femoral ball heads of hip endoprostheses made from bioinert ceramics such as alumina or zirconia, must withstand high mechanical stresses, resorption/corrosion by aggressive body fluids, and abrasive wear over the lifetime of the implant in the human body, for perhaps 15–20 years. Some important properties of ceramic femoral ball heads are listed in Table 10.7 (Willmann, 1995).

The wear performance of the sliding couple ceramic femoral ball/acetabular cup liner is crucially important, as the synovial fluid lubricant that is present in the natural hip joint is absent from the artificial system. Hence, the coefficient of friction must be as low as possible. Data relating to the linear wear of clinically established wear couples, obtained during wear screening tests, are listed in Table 10.8 (Heimann and Willmann, 1998).

Table 10.7 Important mechanical and functional properties of ceramic femoral ball heads (Willmann, 1995).

Materials property	Prerequisite for
High hardness	Wear resistance over many years
No plastic deformation under load	
No elastic deformation under load	
No creep	
Finely grained microstructure	Excellent surface finish, low coefficient of friction
Dense (no porosity)	
High bending strength	Load-bearing capability
High compression strength	High fracture strength
Good fatigue resistance	Reliable product
Extreme corrosion resistance	Biocompatibility
High purity	Bioinertness
No radioactive impurities[a)]	
Electrical insulator	No galvanic reactions (i.e., no fretting) Corrosion

a) In case of zirconia.

Table 10.8 Wear data of combinations used in femoral ball/acetabular cup liner wear pairs (Heimann and Willmann, 1998).

Materials combination	Linear wear ($\mu m\,year^{-1}$)
Metal/PE-UHMW[a)]	200
Alumina/PE-UHMW	<100
Alumina/CFRP[b)]	<4
Y-TZP[c)]/PE-UHMW	<100
Biolox™/Biolox™[d)]	<5
Biolox forte™/Biolox forte™	<1
Y-TZP/Y-TZP	Disastrous

a) PE-UHMW: Polyethylene-ultrahigh-molecular weight.
b) CFRP: carbon fiber-reinforced polymer.
c) Y-TZP: yttria-stabilized tetragonal zirconia polycrystal.
d) Biolox and Biolox forte are trade names of medical alumina ceramics produced by CeramTec AG, Plochingen, Germany.

The reliability of ceramic femoral heads is closely related to their mechanical properties. Previously, measurement of the load to fracture of 107 explanted alumina (Biolox®) femoral ball heads did not reveal any fatigue of the material (Willmann, 1998). This can be related to the very favorable relationship between the crack propagation velocity and the stress intensity factor K_{Ic}. When loaded under standard conditions (i.e., 3 kN, equivalent to normal walking), K_{Ic} was found to be in the range of $0.6\,MPa \cdot m^{1/2}$, a value substantially below the fatigue limit (fracture toughness) of alumina (ca. $1\,MPa \cdot m^{1/2}$).

Figure 10.29 shows a schematic representation of the manufacturing process of ceramic femoral balls (Clarke and Willmann, 1994). Starting from high-purity alumina or Y-stabilized zirconia powders (step a), a cylindrical precursor shape is formed by cold uniaxial pressing (step b), that subsequently is turned on a CNC-lathe to shape the inside taper and outside spherical surface (step c). A laser-etched engraving for identification is then provided (step d), before firing at up to 1600 °C to densify the ceramic green body by sintering (step e). The final finishing is achieved by grinding (using diamond tools) the internal bore and the spherical surface to a very low roughness value, so as to achieve a low coefficient of friction (steps f and g). A final inspection (step h) should assure the extremely high degree of reliability required for the long-term survival of the ball in a hostile and aggressive body environment.

Figure 10.30 shows the three standard diameter sizes (22, 28, and 32 mm) of zirconia femoral balls, complete with laser markings to unequivocally identify the individual parts. Here, the top ball carries the identification code 32-12/14 M 96 @ Z 8179; this refers to a diameter of 32 mm, an inside taper of 12/14 mm, a medium neck length M, the year of manufacture (1996), the type of ceramics (Z = zirconia), and the production number 8179. By using such a complex iden-

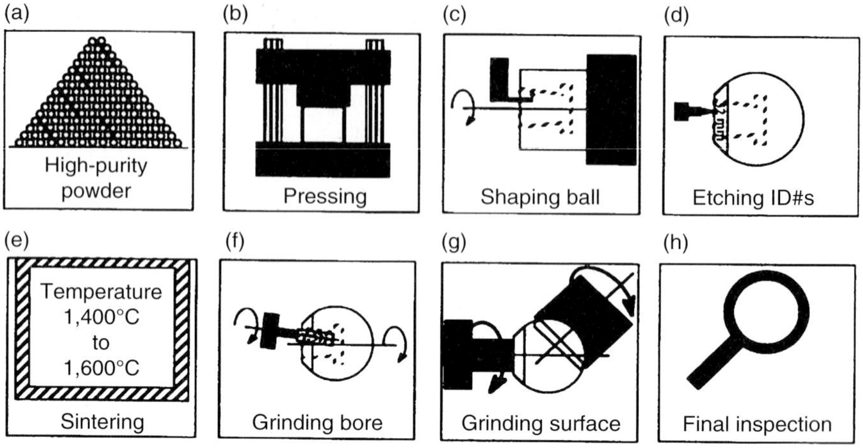

Figure 10.29 Schematic of the manufacturing process of ceramic femoral balls (Clarke and Willmann, 1994). For details, see the text.

Figure 10.30 Laser markings of zirconia femoral ball heads (Heimann and Willmann, 1998). For details, see the text.

tification system, it is possible to establish a quality test protocol and to trace the history of any part. Such an ability may be very important in any subsequent court procedures associated with litigation cases against the producer, the hospital, and the surgeon who has performed the operation. It also provides a unique means of identification in situations where other recognition marks have been obliterated, such as a body recovered from the scene of a fire.

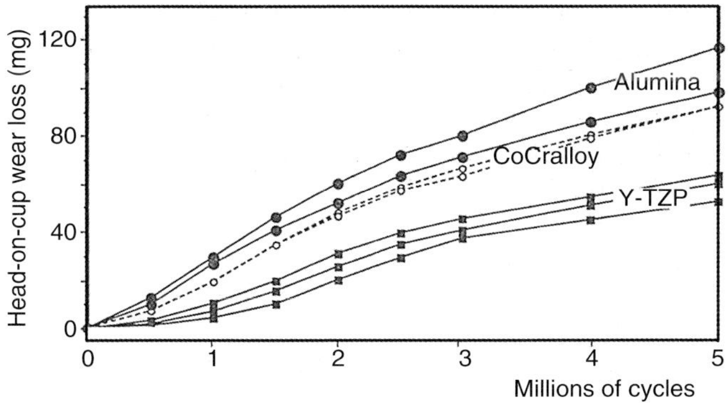

Figure 10.31 Wear loss of PE-UHMW-lined acetabular cups in contact with femoral balls made from alumina, zirconia (Y-TZP) and CoCr alloy in fetal bovine serum +20 mM EDTA solutions (PROZYR™; Ceramiques Techniques Desmarquest, Saint-Gobain Ceramiques Industrielles). The spread of the lines accounts for variations of wear behavior related to differences in the microstructure of the sintered ceramics.

As stringent quality control measures are of the utmost importance during the production of femoral ball heads, wear screening and corrosion tests are generally performed on simplified specimen geometries (cylindrical pins, annulus) in the presence of lubricating SBFs (e.g., Ringer's solution, HBSS, etc.) or protein-containing solutions such as fetal bovine serum and human serum albumin solutions, under physiological conditions. Some of these tests may be continued for up to two million cycles; this corresponds to a service time of two years, taking into account the average number of load changes while walking during that time.

The aging behavior of Y-TZP (Prozyr™; Ceramiques Techniques Desmarquest, Saint-Gobain Ceramiques Industrielles, France) in Ringer's solution confirmed the extraordinary anti-corrosion stability of this material for periods exceeding two years. By using PE-UHMW liners of acetabular cups, articulating against femoral balls made from alumina, zirconia and CoCr alloy, it was possible to perform combined wear and corrosion ("corrabrasion") tests in fetal bovine serum +20 mmol EDTA, corresponding to an exposure time of five years (Figure 10.31). According to these data, zirconia would typically outperform both alumina and metal femoral heads in terms of protection against the synergistic action on PE-UHMW of mechanical wear and chemical corrosion.

This favorable mechanical performance notwithstanding, grave concern has recently been expressed concerning the stability of zirconia femoral ball heads *in vivo*. Despite the initial enthusiasm among orthopedic surgeons, the devices have shown certain shortcomings in terms of wear resistance when articulating against an acetabular lining of the same material (Willmann *et al.*, 1996) as well as an increased risk of fracture (SO.F.C.O.T., 2002). Surface degradation during steam sterilization was also encountered, with an associated roughening leading to increased wear and a substantially elevated risk of osteolysis (MDA SN 9617, 1996). One explanation for the mechanism of degradation of stabilized zirconia assumes that water absorbed at the ceramic surface produces Zr–OH moieties that act as stress concentration sites, while a simultaneous thermally activated dipole reorientation creates localized strain within the crystal lattice. The combination of these stresses, together with residual stresses generated from the thermal expansion anisotropy of the t-ZrO_2 particles, will eventually exceed the critical value for tetragonal phase retention, and the monoclinic zirconia will become nucleated. An autocatalytic reaction chain would then be initiated, such that the tetragonal $\rightarrow$ monoclinic transformation would propagate. If the zirconia grains in Y-ZTP were to become large enough, then micro- and macro-cracking would occur, opening up pathways for the further penetration of water vapor into the bulk of the zirconia ceramic, and the initiation of new, monoclinic nuclei (Birkby and Stevens, 1996).

Currently, the most common reason for the failure of artificial joint implants is aseptic loosening, which manifests itself in reappearing pain. The underlying pathogenesis is a tissue reaction to adverse effects such as mechanical instability, and the release of wear particles from the implant biomaterial. Both conditions result eventually in bone resorption and the development of osteolysis.

- *Mechanical instabilities* allow micro-movements between the anchoring parts of the implant and the adjacent bone that triggers the replacement of bone by fibrous connective tissue.
- *Particle release* into the tissue surrounding an artificial joint is the result of wear of the implant. This initiates foreign body reactions during which macrophages phagocytize the wear particles. Whilst a certain degree of wear can be tolerated as long as the storage and elimination of particles via the lymphatic system balances their production, this equilibrium will decompensate if too many wear particles are being produced. In this case, the macrophages accumulate, form *granulomae*, and invade not only the bone marrow but also the bone–implant interface. Macrophages produce cytokines that trigger a cascade of cellular activities, including bone cell resorption through osteoclasts and widespread osteolysis. In addition, the access of joint fluid to the bone–implant interface and hypersensitivity to implant materials (in particular to metals) will augment the process of aseptic implant loosening. Hence, it is mandatory that wear by particles released from the joint bearings, as well as from the anchoring surfaces, is reduced (Willert, 2000).

10.6.1
Electronic Processes during γ-Ray Sterilization of Zirconia Femoral Heads

The efficient sterilization of metallic and ceramic femoral heads before surgical implantation is an important requirement to guarantee a successful integration of an endoprosthetic implant (Kowalski and Morrisey, 1997). Since the hydrolytic

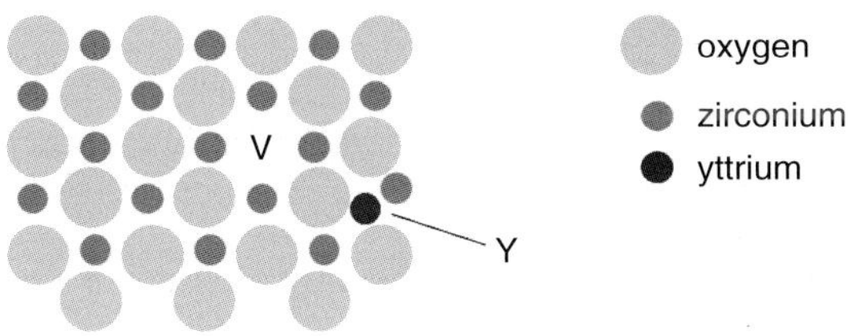

Figure 10.32 Vacancy (V) in the oxygen sublattice of zirconia and interstitial yttrium atom.

stability of zirconia may be compromised during steam sterilization (see above), other approaches must be sought, including sterilization with gases such as ethylene oxide, ozone, chlorine dioxide or vaporized hydrogen peroxide, or with plasmas and ionizing radiation (γ-rays, X-rays, and electron-beam exposure). The effects of the irradiation of Y-ZTP with γ-rays are described in detail below.

Owing to the replacement of Zr^{4+} by Y^{3+}, the Y-TZP ceramic develops vacancies in its oxygen sublattice according to:

$$2Zr_{Zr} + 4O_O + Y_2O_3 \rightarrow 2Y'_{Zr} + V_O^{**} + 2ZrO_2. \tag{10.4}$$

This means that, for 1 mole of Y_2O_3, 2 moles of negative singly charged cation defects Y'_{Zr} and 1 mole of positive doubly-charged oxygen vacancies V_O^{**}, are created and distributed, for example as Frenkel defects (Figure 10.32).

While this Frenkel disorder facilitates ionic conductivity at high temperatures, it is also the source of color centers that will be activated by irradiation with high-energy radiation (such as X-rays or γ-rays), and hence cause the slightly ivory-colored as-sintered ceramics to acquire a grayish-purple shade (Dietrich *et al.*, 1996; Heimann and Willmann, 1998). Since hip endoprostheses are sterilized with γ-irradiation by the manufacturer before shipping to the hospital, such a color change wrongly suggest to the operating surgeon an intolerable degree of contamination with impurities. However, as the actual cause of the reversible discoloration is on an electronic level, it will compromise neither the mechanical nor the biomedical performance of the device. A series of γ-irradiated Y-TZP and alumina femoral heads, and the color changes they undergo during storage, is shown in Figure 10.33.

The color of the Y-TZP samples changed from a slightly grayish hue (designated A2 on the CEC color system) in the as-sintered state, to a purplish-gray color on irradiation. This color is not stable, and transforms during storage in the dark at ambient temperature to a yellow-greenish brown color (C5) after two months. This C5 shade eventually transforms on further storage in daylight for another two months to a light yellowish-brown shade (B3) that is thought to be the stable state. Increasing the storage temperature to 121 °C speeds up the color extinction process, so that after only a 30-min exposure to this typical sterilization temperature, the stable B3 state is obtained.

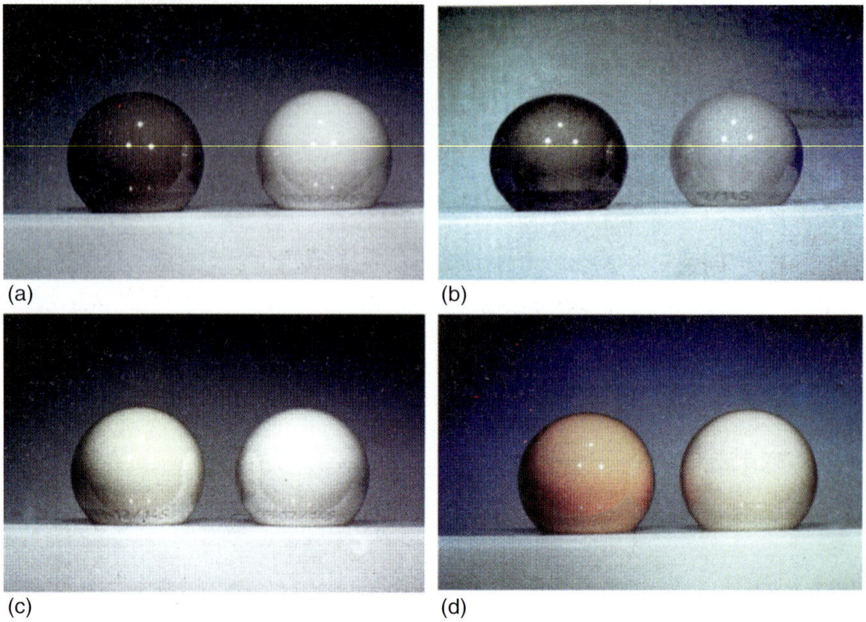

Figure 10.33 Femoral heads produced from medical-grade yttria-stabilized tetragonal zirconia polycrystal (Y-TZP) material (a–c) and alumina (d) irradiated for 10 min with γ-rays obtained from a ^{60}Co source with a minimum energy dose of 25 kGy. The femoral heads shown on the right side in panels a–d are nonirradiated, and serve as comparison. (a) As-irradiated; (b) Stored for 1 week at room temperature (RT) in the dark; (c) Stored for 2 months at RT in the dark and after 30 min steam sterilization at 121 °C, respectively; (d) As-irradiated alumina femoral head.

The residual optical absorption spectrum of irradiated Y-ZTP can be deconvoluted into three individual bands A, B, and C at 2.6 eV (471 nm), 3.2 eV (392 nm), and 3.5 eV (356 nm), respectively (Figure 10.34). Short-term irradiation with either γ-rays (10 min, 25 kGy) or X-rays (10 min, MoK_α, 40 kV, 25 mA) leads to a medium-intensity absorption band B at 3.2 eV and produced, together with the absorption band A of comparable intensity at 2.6 eV, a yellowish-brown coloration. Increasing the irradiation time to 2 and 4 h, respectively, resulted in a more pronounced growth of the intensity of band A. Consequently, the composite color was shifted towards a longer wavelength (471 nm), and the typical purplish color of irradiated zirconia appeared (see Figure 10.33). This increase in intensity of the absorption bands A and B can be described by an exponential function of the type:

$$I(t) = a_1[1-\exp(-t/k_1)] + a_2[1-\exp(-t/k_2)] \quad (10.5a)$$

where $a_1 = a_2 = 15\,\text{eV}\,\text{cm}^{-1}$ and $k_1 = 220\,\text{min}$, $k_2 = 15\,\text{min}$ for band B, and $a_1 = 53\,\text{eV}\,\text{cm}^{-1}$, $a_2 = 15\,\text{eV}\,\text{cm}^{-1}$ and $k_1 = 200\,\text{min}$, $k_2 = 15\,\text{min}$ for band A. Likewise, the fading of the color with storage time (i.e., the kinetics of the reconstruction of the lattice) can be fitted to a similar equation:

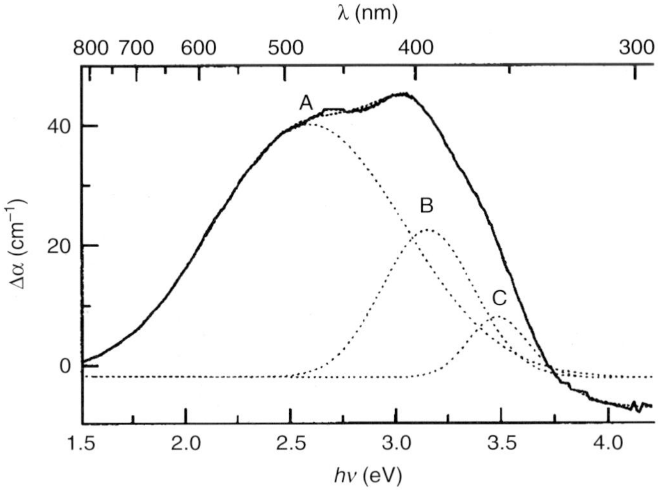

Figure 10.34 Residual optical absorption spectrum of γ-irradiated sintered medical-grade Y-TZP (solid curve) and its Gaussian deconvolutions (dotted curves) (Dietrich *et al.*, 1996). Reprinted with permission from Springer International.

$$I(t) = a_1 \cdot \exp[-(t-t_0)k_1] + a_2 \cdot \exp[-(t-t_0)k_2] \tag{10.5b}$$

where $a_1 = 0.96\,\mathrm{eV\,cm^{-1}}$, $a_2 = 11.83\,\mathrm{eV\,cm^{-1}}$ and $k_1 = 0.83$ days, $k_2 = 257.9$ days and $t_0 = 0.45$ days for band B, and $a_1 = 8.29\,\mathrm{eV\,cm^{-1}}$, $a_2 = 32.35\,\mathrm{eV\,cm^{-1}}$ and $k_1 = 1.55$ days, $k_2 = 257.1$ days and $t_0 = 0.94$ days for band A (Dietrich *et al.*, 1996). Equation (10.5b) suggests that the decay of the color occurs in a two-step process with very different rate constants k_1 (≈1 day) and k_2 (257 days).

The color is thought to be related to a mixture of two bands of the optical absorption spectrum (see Figure 10.34), band C at 3.48 eV (356 nm) and band B at 3.16 eV (392 nm), that may be related to an extrinsic defect of Fe^{3+}_{Zr} type [B center in the electron spin resonance (ESR) spectrum]. Further irradiation for 2–4 h deepens the color to a purple blue-grayish (no shade in the CEC notation) owing to the expression of the 2.63 eV (471 nm) band. This optical absorption band corresponds to an intrinsic $Y'_{Zr}O^*$ defect, and is thermally less stable than the former.

Under the reasonable assumption that the color of the irradiated zirconia is related to electronic defects, attempts have been made to identify these defects by using ESR spectroscopy (Figure 10.35). The recorded ESR signals were correlated with known types of defects X, C, and E reported in the literature (Shinar *et al.*, 1985; Orera *et al.*, 1990).

The high-intensity X center is related to a trigonal T defect (Orera *et al.*, 1990), the C center to the defect type $Zr_{Zr'}O_{8-x}$, (i.e., the transformation of a four-valent to a three-valent zirconium ion), and the E center to an intrinsic defect type $Y'_{Zr}O^*$ (i.e., an electron-hole state caused by an oxygen ion trapped in the neighborhood of a three-valent yttrium ion). This E center is thought to be associated with the

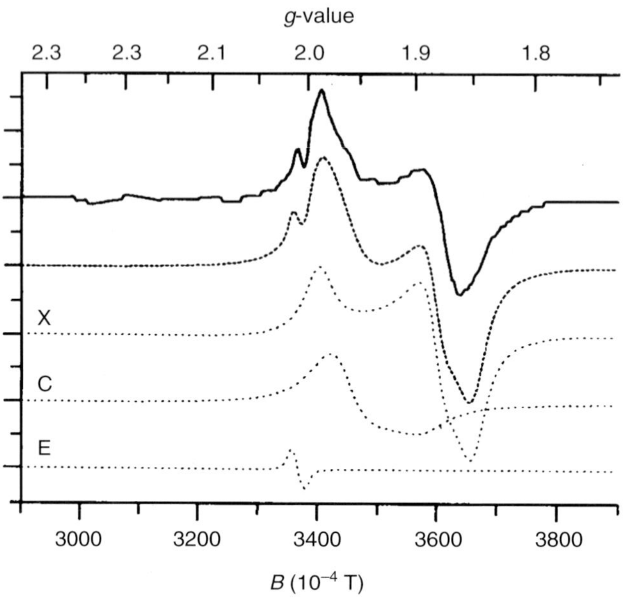

Figure 10.35 Electron spin resonance (ESR) spectrum of γ-irradiated medical-grade Y-TZP (solid curve) and simulated spectrum (dashed curve) obtained by summation of individual ESP spectra of the centers X, C, and E (dotted curves). For details, see the text (Dietrich *et al.*, 1996). Reprinted with permission from Springer International.

optical absorption band A at 2.6 eV (Orera *et al.*, 1990). In addition, there are contributions from trace amounts of iron that can produce extrinsic defects such as the center B (g = 4.3) of defect type Fe^{3+}_{Zr}. A center G could also be identified that may be related to microstructural properties, such as Y_2O_3 clusters constituting defects of F^+ type, coordinated with Y ions (Azzoni *et al.*, 1993). For further detail, see the report by Dietrich *et al.* (1996).

The doubly positively charged V_O^{**} defects created according to Eq. (10.4) are ideal traps for electrons, as new discrete energy states will be created above the valence and below the conduction band.

Figure 10.36a shows the energy term scheme of Y-TZP with electron-sticking terms between 0.7 and 1.3 eV, as well as luminogenic centers caused by electron hole pairs at intrinsic or extrinsic defects between 1.7 and 2.3 eV. During irradiation, the electrons will be trapped at the three sticking terms. Recombination with a luminogenic center will then generate thermoluminescence at 2.2 eV (550 nm, peak A), 2.0 eV (623 nm, peak B), and 1.7 eV (700 nm, peak C), as shown in the complex 3-D glow curve (Figure 10.36b).

In conclusion, color changes of medical-grade Y-TZP induced by irradiation with ionizing radiation such as γ- and X-rays can be explained by two-electron transitions at local lattice defects:

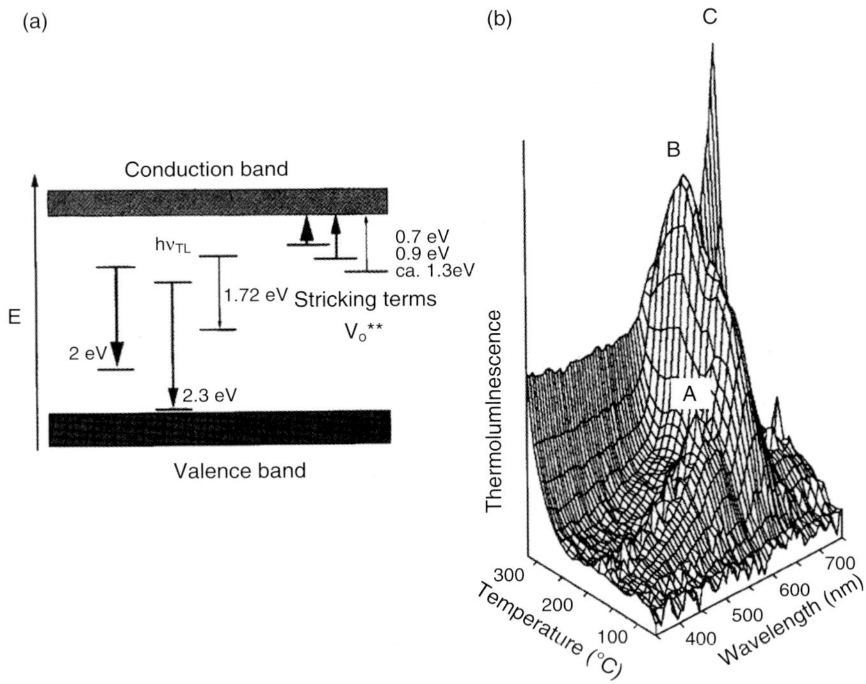

Figure 10.36 Energy term scheme of Y-TZP with (a) electron-sticking terms V_O^{**} (0.7–1.3 eV) and electron-hole (luminogenic) terms hν(TL) (1.7–2.3 eV); and (b) temperature- and wavelength-resolved 3-D thermoluminescence spectrum (Dietrich *et al.*, 1996). Reprinted with permission from Springer International.

- *Intrinsic defects* of type $Y'_{Zr}O^{*}$. This defect is unstable, in particular at elevated temperatures, and can be correlated with an optical absorption band at 2.63 eV (471 nm), creating a purplish-gray hue (see Figure 10.33).
- *Extrinsic defects* at 3.16 eV (392 nm) related to trace amounts of impurity ions are more stable than the former, and hence are still present after steam sterilization at 121 °C. They may be responsible for the yellowish-brown coloration, and hence could be used as a quality control tool characterizing the purity of the medical product.

10.7
Future Developments and Outlook

As discussed in the introduction to this chapter (Section 10.1), the development of novel biomaterials with improved lifetimes, reliability and bioactive functions is high on the R&D agenda of worldwide research. The limited stability of HAp and its thermal decomposition products within the body environment (Fazan and Marquis, 2000) call for bioactive materials with an increased resorption resistance. Such materials could be based, for example, on the quaternary system $CaO–P_2O_5–TiO_2–ZrO_2$ (Schneider *et al.*, 2001; Heimann, 2006; see Section 10.5.2.3), in which stoichiometric compositions exist with much-improved mechanical and chemical stabilities compared to the calcium phosphates.

A different route in the quest for new promising biomaterials is based on ceramic–polymer composites (for an early account, see Bonfield, 1988). These are currently being designed to mimic the mechanical and biological performances of natural bone; examples include composites of PLA and HAp, as well as of collagen and HAP. These novel materials have been designated "intelligent," and are defined as human-friendly materials that can alter their characteristics in response to their surroundings, for example, in varying stress fields. The presence of a polymeric component with a low modulus might serve as an "isoelastic" medium by reducing to a considerable degree the strong gradient of stiffness between natural bone and a metallic implant, and thus also reduce any stress shielding, with its negative consequences (see above). While these composites are at present experimentally applied only under low-loading conditions, their use in loaded implants can be implied in future developments. Ceramic–ceramic composites currently under investigation include HAp with the addition of bioglass (Tancred *et al.*, 2001). Notably, the addition of 5% bioglass, of composition 53% SiO_2, 26% Na_2O, 13% CaO, 5% P_2O_5, 0.5% Al_2O_3, 3% B_2O_3, increases the fracture toughness of the ceramics twofold, but results in a reduced density and hardness. Finally, developments are under way to improve the wear resistance of ceramic femoral heads by the application of ultrahard coatings, for example, diamond-like carbon (DLC) or TiN coatings that not only improve the mechanical abrasion wear resistance of the ceramic when articulating against the PE-UHMW-lined acetabular cup, but also reduce the coefficient of friction (Affatato *et al.*, 2000; Reuter *et al.*, 2006; Stamm *et al.*, 2008).

Despite these advances, it would appear today that a limit has been reached in terms of a current medical practice that proposes the replacement of tissue, using a predominantly materials science-based approach. Some of these aspects have been detailed in this chapter.

First-generation biomaterials have included bioceramics centered on essentially bioinert or biotolerant materials being incorporated into the body to replace missing parts or functions, or to fill bone cavities; these have included dental amalgam, gypsum, ivory, or even wood. This was followed by the development of *second-generation biomaterials* that included bioconductive and bioresorbable ceramics designed to elicit biological responses so as to induce the growth of bone and other tissues. However, in the future a much more biologically based methodology will be applied that will concentrate on the repair and regeneration of tissue through *third-generation biomaterials* rather than simple, more or less mechanical replacement (Planell *et al.*, 2009). This includes the functionalization of bioceramic and biopolymeric surfaces by the adsorption of proteins such as bone morphogenetic proteins (BMPs) or other osseoinductive proteins that provide signals to bone cells to trigger their own growth. These activities are reflected in a recent definition which states that, "... biocompatibility is the property of a material to generate in the host an appropriate response." However, as this response may also be negative, one of the main challenges of biomaterials research will be to minimize and control these negative responses.

Hildebrand *et al.* (2006) reported on surface coatings for biological activation and the functionalization of medical devices, including:

- *Mechanical* functionalization, by the polishing, machining, or irradiation of metallic biomaterials.
- *Physical* functionalization by the application of surface coatings, including APS (atmospheric plasma spraying), CVD (chemical vapor deposition), CGDS (cold gas dynamic spraying), PLD (pulsed laser deposition), MAO (micro-arc oxidation) or plasma electrolytic oxidation (PEO), sol–gel, and polyelectrolyte coatings and films.
- *Chemical and biological* functionalization, including DDS (drug delivery systems); biodegradable collagen and polylactic acid molecules; the release of biomolecules grafted onto cage molecules such as cyclodextrin, biomimetic hydroxyapatite coatings, rhBMPs, and other materials and processes. Increasingly, precursor powder compositions are being tailored to the real composition of biological hydroxyapatite by substituting Ca by Na, Mg, Sr and other metabolic elements (e.g., Xue *et al.*, 2007), as well as PO_4^{3-} and OH^- by CO_3^{2-}.

Consequently, innovation is required in the following areas of tissue engineering (Service, 2000) to solve the socioeconomic challenges of a rapidly aging population. this will include:

- The development of *third-generation biomaterials* that activate genetic repair.
- Tissue engineering by molecular scaffolding.
- Stem cell engineering including marrow stem cell (MSC) therapy.
- Rapid and highly predictive *in vitro*-testing techniques for biomaterial–cell response evaluation.

Among these developments, stem cell engineering in particular remains a highly contentious issue that is hotly debated among various segments of the population and politicians. Rapid and predictive *in vitro*-test methods for biomaterials–cell responses that would alleviate the use of costly (and ethically dubious) animal models, are also high on the agenda of future biomaterials R&D to solve the socioeconomic issues of the aging populations of developed countries worldwide.

The future perspectives of bioceramic materials for dental restoration will be focused on the development of tougher and esthetically more pleasing glass-ceramics, and also on ceramics such as zirconia with optical properties of color and shine that mimic those of natural teeth and provide a high degree of long-term durability (Höland *et al.*, 2008). The development of nanostructured alumina and zirconia ceramics and composites, as well as nanostructured calcium phosphate ceramics and porous bioactive glasses – possibly as composites with organic constituents – will provide the desired properties for bone substitution and tissue engineering for the next 20 years (Chevalier and Gremillard, 2008).

References

Affatato, S., Frigo, M., and Toni, A. (2000) An *in vitro* investigation of diamond-like carbon as a femoral head coating. *J. Biomed. Mater. Res.*, **53** (3), 221–226.

Alamo, J. (1993) Chemistry and properties of solids with the [NZP] skeleton. *Solid State Ionics*, **63/65**, 547–561.

Alberius Henning, P., Landa-Canovas, A., Larsson, A.K., and Lidin, S. (1999) The structure of oxyapatite solved by HREM. *Acta Crystallogr. B*, **55**, 170–176.

Alfrey, A.C. (1985) Aluminum intoxication. *N. Engl. J. Med.*, **310**, 1113–1114.

Antolotti, N., Bertini, S., Fanaro, C., Ranz, X., Rey, C., Rusticchelli, F., and Scrivani, A. (1998) Interface characterization of different apatite coatings, in *Thermal Spray. Meeting the Challenges of the 21st Century , Proceedings 15th International Thermal Spray Conference, 25–29 May, 1998, Nice, France.* (ed. C. Coddet), ASM International, Materials Park, OH, pp. 1121–1126.

Aryal, S., Raj Bhattarai, S., Remant Bahadur, K.C., Khil, M.S., Lee, D.R., and Kim, H.Y. (2006) Carbon nanotubes assisted biomimetic synthesis of hydroxyapatite from simulated body fluid. *Mater. Sci. Eng. A.*, **426**, 202–207.

ASTM 1185 (1993) Standard specification for compositions of ceramic hydroxyapatite for surgical implants, 1988, reapproved 1993.

ASTM C633-01 (2001) Standard Test Method for Adhesion or Cohesive Strength of Thermal Spray Coatings.

ASTM D3167-03a (2004) Standard Test Method for Floating Roller Peel Resistance of Adhesives.

Azzoni, C.B., Paleari, A., Scradina, F., Krajewski, A., Ravaglioni, A., and Meschke, F. (1993) EPR structural study on hydrothermally aged yttria-doped tetragonal zirconia polycrystals. *J. Mater. Sci.*, **28** (14), 3951–3956.

Bassett, C.A.L. (1968) Biologic significance of piezoelectricity. *Calcif. Tissue Res.*, **1** (4), 252–272.

Bassett, C.A.L. and Becker, R.O. (1962) Generation of electric potentials by bone in response to mechanical stress. *Science*, **137**, 1063–1064.

Bassett, C.A.L., Pawluck, R.S., and Pilla, A. (1974) Acceleration of fracture repair by electromagnetic field. *Ann. N. Y. Acad. Sci.*, **238**, 242–262.

Basu, B., Katti, D.S., and Kumar, A. (eds) (2009) *Advanced Biomaterials. Fundamentals, Processing, and Applications*, John Wiley & Sons, Inc., Hoboken, ISBN: 978-0-470-19340-2, 720 pp.

Baxter, F.R., Turner, I.G., Bowen, Ch.R., Gittings, J.P., and Chaudhuri, J.B. (2009) An *in vitro* study of electrically active hydroxyapatite-barium titanate ceramics using Saos-2 cells. *J. Mater. Sci. Mater. Med.*, **20** (8), 1697–1708.

Ben-Nissan, B. and Ylänen, H.O. (2006) Bioactive glasses and glass ceramics, in *Wiley Encycl. Biomed. Eng.* (ed. M. Akay), John Wiley & Sons, Inc., New Jersey, pp. 354–366.

Beresford, W.A. (1981) *Chondroid Bone, Secondary Cartilage and Metaplasia*, Urban & Schwarzenberg, Baltimore.

Berger, G., Gildenhaar, R., and Ploska, U. (1995a) Rapidly resorbable materials based on a new phase: $Ca_2KNa(PO_4)_2$, in *Bioceramics 8* (eds J. Wilson, L.L. Hench, and D.C. Greenspan), Butterworth-Heinemann, Oxford, pp. 453–456.

Berger, G., Gildenhaar, R., and Ploska, U. (1995b) Rapid resorbable, glassy crystalline materials on the basis of calcium alkali orthophosphates. *Biomaterials*, **16**, 1241–1248.

Berry, J.L., Geiger, J.M., Moran, J.M., Skraba, J.S., and Greenwald, A.S. (1986) Use of tricalcium phosphate or electrical stimulation to enhance the bone-porous implant interface. *J. Biomed. Mater. Res.*, **20**, 65–77.

Birkby, I. and Stevens, R. (1996) Application of zirconia ceramics. *Key Eng. Mater.*, **122–124**, 527–552.

Bohner, M. (2010) Resorbable biomaterials as bone graft substitutes. *Materials Today*, **13**, 24–30.

Bonfield, W. (1988) Hydroxyapatite-reinforced polyethylene as an analogous material for bone replacement. *Ann. N. Y. Acad. Sci.*, **253**, 173–177.

Boutin, P.M. (1972) Arthroplastic totale de la hanche par prothese en alumina frittée. *Rev. Chi. Orthop.*, **58**, 229–246.

Boutin, P.M. (1981) THR using alumina-alumina sling and a metal stem: 1330 cases and an 11 year follow up. *Orthop. Ceram. Implants*, **1**, 11–18.

Brown, W.E. and Chow, L.C. (1987) A new calcium phosphate, water setting cement, in *Cement Research Progress*, 1986 (ed. P.W. Brown), American Ceramics Society, Ohio, pp. 152–179.

Caetano-Zurita, J., Bermudez, O., Lopez-Valero, I., Stucchi, E.B., Varella, J.A., Planell, J.A., and Martinez, S. (1994) Mechanical behaviour of a hydroxyapatite-zirconia particle composite, in *Bioceramics, Proceedings 7th International Symposium on Ceramics in Medicine, Turku* (eds O.H. Anderson and A. Yli-Urpo), Butterworth-Heinemann, Oxford, pp. 267–271.

Calderin, L., Stott, M.J., and Rubio, A. (2003) Electronic and crystallographic structure of apatite. *Phys. Rev. B*, **67**, 134106–134112.

Calès, B. (1995) Y-TZP zirconia ceramic hip joint heads: Key issues for a high reability, in *Y-TZP Zirconia Ceramic Hip Joint Heads* (eds A. Ravaglioli and A. Krajewski), Gruppo Editionale Faenza S.p.A., Faenza, pp. 45–52.

Calès, B. (1998) Colored zirconia ceramics for dental applications, in *Bioceramics, 11* (eds T.Z. LeGeros and J.P. LeGeros), World Scientific Publishers, New York, pp. 591–594.

Calès, B. and Stefani, Y. (1995) Yttria-stabilized zirconia for improved orthopedic prostheses, in *Encyclopedic Handbook of Biomaterials and Bioengineering, Part B: Applications*, vol. **1** (eds D.L. Wise, D.J. Trantolo, D.E. Altobelli, M.J. Yaszemski, J.D. Gresser, and E.R. Schwartz), Marcel Dekker, Inc., New York, Basel, Hong Kong, pp. 415–452.

Calvert, P. and Mann, S. (1997) The negative side of crystal growth. *Nature*, **386**, 127–128.

Carayon, M.T. and Lacout, J.L. (2003) Study of the Ca/P atomic ratio of the amorphous phase in plasma-sprayed hydroxyapatite coatings. *J. Solid State Chem.*, **172**, 339–350.

Carlisle, E.M. (1970) Silicon: a possible factor in bone calcification. *Science*, **167**, 279–280.

CeramTech, A.G. (2006) BIOLOX® delta – Nanoverbundwerkstoff für die Endoprothetik. Available at: http://www.ceramtec.com/pdf/biolox_delta_de.pdf.

Cerutti, M.G. (2004) Characterization of bioactive glasses. Effect of the immersion in solutions that simulate body fluids. Ph.D. Dissertation, University of Turin, Italy.

Chae, J.C., Collier, J.P., Mayor, M.B., Suprenant, V.A., and Dauphinais, L.A. (1992) Enhance ingrowth of porous-coated CoCr implants plasma-sprayed with tricalcium phosphate. *J. Biomed. Mater. Res.*, **26**, 93–102.

Chakhmouradian, A.R. and Medici, L. (2006) Clinohydroxylapatite: a new apatite-group mineral from northwestern Ontario (Canada), and new data on the extent of Na-S substitution in natural apatite. *Eur. J. Mineral.*, **18** (1), 105–112.

Chang, E., Chang, W.J., Wang, B.C., and Yang, C.Y. (1997a) Plasma spraying of zirconia-reinforced hydroxyapatite composite coatings on titanium. Part I: phase, microstructure and bonding strength. *J. Mater. Sci. Mater. Med.*, **8**, 193–200.

Chang, E., Chang, W.J., Wang, B.C., and Yang, C.Y. (1997b) Plasma spraying of zirconia-reinforced hydroxyapatite composite coatings on titanium. Part II: dissolution behaviour in simulated body fluid and bonding degradation. *J. Mater. Sci. Mater. Med.*, **8**, 201–211.

Chevalier, J. and Gremillard, L. (2008) Ceramics for medical applications: a picture for the next 20 years. *J. Eur. Ceram. Soc.*, **29** (7), 1245–1255.

Chevalier, J., Grandjean, S., Kuntz, M., and Pezzotti, G. (2009) On the kinetics and impact of tetragonal to monoclinic transformation in an alumina/zirconia composite for arthroplasty application. *Biomaterials*, **30** (29), 5279–5282.

Chou, B.Y. and Chang, E. (2002) Plasma-sprayed zirconia bond coat as an intermediate layer for hydroxyapatite coating on titanium substrate. *J. Mater. Sci. Mater. Med.*, **13**, 589–595.

Clarke, I.C. and Willmann, G. (1994) Structural ceramics in orthopedics, in *Bone Implant Interface* (ed. H.U. Cameron), Mosby, St. Louis, Baltimore, Boston, pp. 203–252.

Colella, S.M., Miller, A.G., Stang, R.G., Stoebe, T.G., and Spengler, D.M. (1981) Fixation of porous titanium implants in cortical bone by electrical stimulation. *J. Biomed. Mater. Res.*, **15**, 37–46.

Currey, J.D. (2004) Tensile yield in compact bone is determined by strain, post-yield behaviour by mineral content. *J. Biomech.*, **37**, 549–556.

Dai, Y. and Hughes, J.M. (1989) Crystal structure refinement of vanadinite and pyromorphite. *Can. Mineral.*, **27**, 189–192.

De Groot, K., Klein, C.P.A.T., Wolke, J.G.C., and de Blieck-Hogervorst, J. (1990) Plasma-spraying of calcium phosphate, in *Handbook of Bioactive Ceramics*, vol. **II** (eds T. Yamamuro, L.L. Hench, and J. Silson), CTC Press, Boca Raton, FL, pp. 3–15.

De Leeuw, N., Bowe, J.R., and Rabone, J.A.L. (2007) A computational investigation of stoichiometric and calcium-deficient oxy- and hydroxyapatite. *Faraday Discuss.*, **134**, 195–214.

Di Silvio, L. (ed.) (2008) *Cellular Response to Biomaterials*, Woodhead Publishing Limited, Cambridge, UK, ISBN: 978-1-84569-358-9, 648 pp.

Dickens, B., Brown, W.E., Kruger, G.J., and Stewart, J.M. (1973) $Ca_4(PO_4)_2O$, tetracalcium diphosphate monoxide. Crystal structure and relationships to $Ca_5(PO_4)_3OH$ and $K_3Na(SO_4)_2$. *Acta Crystallogr. B*, **29**, 2046–2056.

Dietrich, A., Heimann, R.B., and Willmann, G. (1996) The colour of medical-grade zirconia (Y-TZP). *J. Mater. Sci. Mater. Med.*, **7**, 559–565.

DIN 58835 (1979) Chirugische Implantate–Keramische Werkstoffe–Aluminiumoxid.

Dorozhkin, S.V. (2007) Calcium orthophosphates. *J. Mater. Sci.*, **42**, 1061–1095.

Ducheyne, P., Radin, S., Heughebaert, M., and Heughebaert, J.C. (1990) Calcium phosphate coatings on metallic porous surfaces. The effect of structure and composition on the electrophoretic deposition, vacuum sintering and *in vitro* dissolution behavior. *Biomaterials*, **11**, 244–254.

Ducheyne, P., El-Ghannam, A., and Shapiro, I. (1997) Bioactive material template for *in vitro* synthesis of bone. U.S. Patent 5,643, 789 (July 1, 1997).

Dyshlovenko, S., Pateyron, B., Pawlowski, L., and Murano, D. (2004a) Numerical simulation of hydroxyapatite powder behaviour in plasma jet. *Surf. Coat. Technol.*, **179**, 110–117.

Dyshlovenko, S., Pateyron, B., Pawlowski, L., and Murano, D. (2004b) Numerical simulation of hydroxyapatite powder behaviour in plasma jet. Corrigendum. *Surf. Coat. Technol.*, **187**, 408–409.

Elliot, J.C. (1994) *Structure and Chemistry of the Apatites and Other Calcium Orthophosphates*, Elsevier, Amsterdam, London, New York, Tokyo.

Elliot, J.C., Mackie, P.E., and Young, R.A. (1973) Monoclinic hydroxylapatite. *Science*, **180**, 1055–1057.

Fazan, F. and Marquis, P.M. (2000) Dissolution behavior of plasma-sprayed hydroxyapatite coatings. *J. Mater. Sci. Mater. Med.*, **11**, 787–792.

Fleet, M.E. (2009) Infrared spectra of carbonate apatites: ν_2-region bands. *Biomaterials*, **30** (8), 1473–1481.

Fuentes, R.O., Figueiredo, F.M., Soares, M.R., and Marques, F.M.B. (2005) Submicrometric NASICON ceramics with improved electrical conductivity obtained from mechanically activated precursor. *J. Eur. Ceram. Soc.*, **25** (4), 455–462.

Fukada, E. (1957) On the piezoelectric effect in bone. *J. Phys. Soc. Jpn*, **12**, 1159–1162.

Geetha, M., Singh, A.K., Asokamani, R., and Gogia, A.K. (2009) Ti based biomaterials, the ultimate choice for orthopaedic implants–a review. *Prog. Mater. Sci.*, **54** (3), 397–425.

Gillespie, P., Wu, G., Sayer, M., and Stott, M.J. (2010) Si complexes in calcium phosphate biomaterials. *J. Mater. Sci.: Mater. Med.*, **21**, 99–108.

Glimcher, M.J. (2006) Bone: nature of the calcium phosphate crystals and cellular, structural and physical chemical mechanisms in their formation, in *Reviews in Mineralogy & Geochemistry*, vol. **64** (eds N. Sahai and M.A.A. Schoonen), Medical Mineralogy and Geochemistry, pp. 223–282.

Götze, J., Heimann, R.B., Hildebrandt, H., and Gburek, U. (2001) Microstructural investigations into calcium phosphate biomaterials by spatially resolved cathodo-luminescence. *Materialwiss. Werkstofftech.*, **32**, 130–136.

Graßmann, O. and Heimann, R.B. (2000) Compositional and microstructural changes of engineered plasma-sprayed hydroxyapatite coatings on Ti6Al4V substrates during incubation in protein-free simulated body fluid. *J. Biomed. Mater. Res. Appl. Biomater.*, **56**, 685–693.

Gross, K.A., Gross, V., and Berndt, C.C. (1998a) Thermal analysis of amorphous phases in hydroxyapatite coatings. *J. Am. Ceram. Soc.*, **81** (1), 106–112.

Gross, K.A., Berndt, C.C., and Herman, H. (1998b) Amorphous phase formation in plasma-sprayed hydroxyapatite coatings. *J. Biomed. Mater. Res.*, **39** (3), 407–414.

Gross, K.A., Berndt, C.C., Stephens, P., and Dinnebier, R. (1998c) Oxyapatite in hydroxyapatite coatings. *J. Mater. Sci.*, **33**, 3985–3991.

Gross, U. and Strunz, V. (1985) The interface of various glasses and glass ceramics with a bony implantation bed. *J. Biomed. Mater. Res.*, **19**, 251–271.

Gross, U., Kinne, R., Schmitz, H.J., and Strunz, V. (1988) The response of bone to surface active glass/glass-ceramics. *CRC Crit. Rev. Biocompat.*, **4** (2), 25–30.

Gross, U., Müller-Mai, Ch., Berger, G., and Ploska, U. (2003) The tissue response to a novel calcium zirconium phosphate ceramics. *Key Eng. Mater.*, **240–242**, 629–632.

Gross, U., Müller-Mai, Ch., Berger, G., and Ploska, U. (2004) Do calcium zirconium phosphate ceramics inhibit mineralization? *Key Eng. Mater.*, **254–256**, 635–638.

Guo, H., Wei, J., and Liu, C.S. (2006) Development of a degradable cement of calcium phosphate and calcium sulfate composite for bone reconstruction. *Biomed. Mater*, **1**, 193–197.

Hartmann, P., Jäger, C., Barth, St., Vogel, J., and Meyer, K. (2001) Solid state NMR, X-ray diffraction, and infrared characterization of local structure in heat-treated oxyhydroxyapatite microcrystals: an analogy of the thermal deposition of hydroxyapatite during plasma-spray procedure. *J. Solid State Chem.*, **160**, 460–468.

Hattori, T. and Iwadate, Y. (1990) Hydrothermal preparation of calcium hydroxylapatite powders. *J. Am. Ceram. Soc.*, **73**, 1803–1807.

Heimann, R.B. (1999) Design of novel plasma-sprayed hydroxyapatite-bond coat bioceramic systems. *J. Therm. Spray Technol.*, **8**, 597–604.

Heimann, R.B. (2006) *In vitro*- und *in vivo*-Verhalten von osteokonduktiven plasmagespritzten Ca-Ti-Zr-Phosphat-Beschichtungen auf Ti6Al4V-Substraten. *BIOmaterialien*, **7** (1), 29–37.

Heimann, R.B. (2007) Novel approaches towards design and biofunctionality of plasma-sprayed osseoconductive calcium phosphate coatings for biomedical implants: The concept of bond coats, in *Trends in Biomaterials Research* (ed. J.P. Patrick), Nova Science Publishers Inc, USA, pp. 1–80. (See also: Halter, P. and Yamamoto, R. (1988) Markets for biomaterials. *J. Biomater. Appl.*, **2**, 317–327.

Heimann, R.B. (2008) *Plasma-Spray Coating. Principles and Applications*, 2nd edn, Wiley-VCH Verlag GmbH, Weinheim, Ch 9, pp. 277–302.

Heimann, R.B. (2009) Characterisation of as-sprayed and incubated hydroxyapatite coatings with high resolution techniques. *Materialwiss. Werkstofftech.*, **40** (1–2), 23–30.

Heimann, R.B. and Schneider, K. (2000) Entwicklung und Charakterisierung plasma-gespritzter biokeramischer Schichten im quaternären System CaO-TiO_2-ZrO_2-P_2O_5. Final report, DFG-Projekt He 923/9-1, 1 December 2000.

Heimann, R.B. and Vu, T.A. (1997) Low-pressure plasma-sprayed (LPPS) bioceramic coatings with improved adhesion strength and resorption resistance. *J. Therm. Spray Technol.*, **6**, 145–149.

Heimann, R.B. and Willmann, G. (1998) Irradiation-induced colour changes of Y-TZP ceramics. *Br. Ceram. Trans.*, **97**, 185–188.

Heimann, R.B. and Wirth, R. (2006) Formation and transformation of amorphous calcium phosphates on titanium alloy surfaces during atmospheric plasma spraying and their subsequent *in vitro* performance. *Biomaterials*, **27**, 823–831.

Heimann, R.B., Vu, T.A., and Wayman, M.L. (1997) Bioceramic coatings: state-of-the-art and recent development trends. *Eur. J. Mineral.*, **9**, 597–615.

Heimann, R.B., Kurzweg, H., Ivey, D.G., and Wayman, M.L. (1998) Microstructural and *in vitro* chemical investigations into plasma-sprayed bioceramic coatings. *J. Biomed. Mater. Res.*, **43**, pp. 441–450.

Heimann, R.B., Tran, H.V., and Hartmann, P. (2003) Laser-Raman and Nuclear Magnetic Resonance (NMR) studies on plasma-sprayed hydroxylapatite coatings: influence of bioinert bond coats on phase composition and resorption kinetics in simulated body fluid. *Materialwiss. Werkstofftechn.*, **34** (12), 1163–1169.

Heimann, R.B., Schürmann, N., and Müller, R.T. (2004) *In vitro* and *in vivo* performance of Ti6Al4V implants with plasma-sprayed osteoconductive hydroxyapatite-bioinert titania bond coat duplex systems: an experimental study in sheep. *J. Mater. Sci. Mater. Med.*, **15**, 1045–1052.

Heimann, R.B., Precht, U., and Pentinghaus, H.J. (2008) Oxyapatite and amorphous calcium phosphate (ACP) – Formation and transformation in plasma-sprayed calcium phosphate coatings for implants. Abstract 4th Workshop "Biomaterials", AK15, Deutsche Gesellschaft für Kristallographie (DGK), Halle/Saale, 6–7 October 2008.

Hench, L.L. (1971) Mechanisms of interfacial bonding between ceramics and bone. *J. Biomed. Mater. Res.*, **2**, 485–497.

Hench, L.L. (1991) Bioceramics. From concept to clinic. *J. Am. Ceram. Soc.*, **74**, 1487–1510.

Hench, L.L. (2008) Genetic design of bioactive glass. *J. Eur. Ceram. Soc.*, **29** (7), 1257–1265.

Hench, L.L. and Ethridge, E.C. (1982) *Biomaterials. An Interfacial Approach*, Academic Press, New York, London.

Hench, L.L. and Wilson, J.W. (1984) Surface-active biomaterials. *Science*, **226**, 630–636.

Hench, L.L., Splinter, R.J., Allen, W.C., and Greenlee, T.K., Jr (1972) Bonding mechanisms at the interface of ceramic prosthetic materials. *J. Biomed. Mater. Res.*, **2** (1), 116–141.

Hildebrand, H.F., Blanchemain, N., Mayer, G., Chai, F., Lefebvre, M., and Boschin, F. (2006) Surface coatings for biological activation and functionalization of medical devices. *Surf. Coat. Technol.*, **200**, 6318–6324.

Hitmi, N., LaCabanne, C., and Yound, R.A. (1986) On dipole reorientability in hydroxyapatites: effect of tunnel size. *J. Phys. Solids*, **47**, 533–546.

Hoang, Q.Q., Siceri, F., Howard, A.J., and Yang, D.S.C. (2003) Bone recognition mechanism of porcine osteocalcin from crystal structure. *Nature*, **425**, 977–980.

Höland, W., Rheinberger, V., Apel, E., Ritzberger, C., Rothbrust, F., Kappert, H., Krumeich, F., and Nesper, R. (2008) Future perspectives of biomaterials for dental restoration. *J. Eur. Ceram. Soc.*, **29** (7), 1291–1297.

Hosono, H., Sakai, Y., and Abe, Y. (1992) Pore size control in porous glass-ceramics with skeleton of NASICON-type crystal $CaTi_4(PO_4)_6$. *J .Non-Cryst. Solids*, **139**, 90–92.

Hulbert, S.F. (1993) The use of aluminia and zirconia in surgical implants, in *An Introduction to Bioceramics* (eds L.L. Hench and J. Wilson), World Science Publ. Co. Ptc. Ltd., Singapore, pp. 25–40.

ISO 6474 (1994) Implants for surgery - Ceramic materials based on high purity alumina. 2nd ed.

ISO 10993 (2003–2006) Biological evaluation of medical devices. Parts 1–20.

ISO/DIS 13356 (1995) Implants for surgery – Ceramic materials based on yttria-stabilized tetragonal zirconia (Y-TZP).

Jones, D.W. and Smith, J.A.S. (1962) Hydrogen bonding in calcium orthophosphates. *Nature*, **195**, 1090–1091.

Kasemo, B. and Lausmaa, J. (1991) The biomaterials-tissue interface and its analogues in surface science and technology, in *The Bone-Biomaterials Interface* (ed. J.E. Davies), University of Toronto Press, Toronto, pp. 19–32.

Kawamoto, Y., Yokogawa, Y., Toriyama, M., Kawamura, S., and Suzuki, T. (1991) Coating of β-tricalcium phosphate on yttria-partially stabilized zirconia using magnesium metaphosphate as an interlayer. *J. Ceram. Soc. Jpn*, **99**, 19–22.

Keller, L. and Dollase, W.A. (2000) X-ray determination of crystalline hydroxyapatite to amorphous calcium-phosphate ratio in plasma sprayed coatings. *J. Biomed. Mater. Res.*, **49**, 244–249.

Kijima, T. and Tsutsumi, M. (1979) Preparation and thermal properties of dense polycrystalline oxyhydroxyapatite. *J. Am. Ceram. Soc.*, **62** (9/10), 455–460.

Kim, C.Y., Clark, A.E., and Hench, L.L. (1989) Early stages of calcium phosphate layer formation in bioglasses. *J. Non-Cryst. Solids*, **113**, 195–202.

Knabe, C., Ostapowicz, W., Radlanski, R.J., Gildenhaar, R., Berger, G., Fitzner, R., and Gross, U. (1998) *In vitro* investigation of novel calcium phosphates using osteogenic cultures. *J. Mater. Sci. Mater. Med.*, **9**, 337–345.

Kohn, D.H. (1992) Materials for bone and joint replacement, in *Materials Science and Technology*, vol. **14** (ed. D.F. Williams), VCH-Verlag, Weinheim, pp. 31–109.

Kokubo, T. (1991) Bioactive glass ceramics: properties and applications. *Biomaterials*, **12**, 155–163.

Kokubo, T. (1996) Formation of biologically active bone-like apatite on metals and polymers by a biomimetic process. *Thermochim. Acta*, **280/81**, 479–490.

Kokubo, T. (2008) *Bioceramics and Their Clinical Applications*, Woodhead Publ. Ltd., Abington, UK. ISBN: 978-184-569-204-9, 308 pp.

Kokubo, T., Hata, K., Nakamura, T., and Yamamuro, T. (1991) Apatite formation on ceramics, metals and polymers induced by a CaO, SiO_2 based glass in a simulated body fluid. *Bioceramics*, **4**, 113–120.

Kokubo, T., Kim, H.-M., Kawashita, M., and Nakamura, T. (2000) Novel ceramics for biomedical applications. *J. Aust. Ceram. Soc.*, **36** (1), 37–46.

Kokubo, T., Kim, H.M., and Kawashita, M. (2003) Novel bioactive materials with different mechanical properties. *Biomaterials*, **24**, 2161–2175.

Kowalski, J.B. and Morrisey, R.F. (1997) Sterilization of implants, in *Biomaterials Science: An Introduction to Materials in Medicine* (eds B.D. Ratner, *et al.*), Academic Press, Section 9.2, p. 415.

Krishna Balla, V., Xue, W., Bose, S., and Bandyopadhyay, A. (2009) Laser assisted Zr/ZrO_2 coating on Ti for load bearing implants. *Acta Biomater.*, **5** (7), 2800–2809.

Kumar, R., Cheang, P., and Khor, K.A. (2003) Radio frequency (RF) suspension plasma sprayed ultra-fine hydroxyapatite (HA)/zirconia composite powders. *Biomaterials*, **24**, 2611–2621.

Kurzweg, H., Heimann, R.B., and Troczynski, T. (1998a) Adhesion of thermally sprayed hydroxyapatite-bond coat systems measured by a novel peel test. *J. Mater. Sci. Mater. Med.*, **9**, 9–16.

Kurzweg, H., Heimann, R.B., Troczynski, T., and Wayman, M.L. (1998b) Development of plasma-sprayed bioceramic coatings with bond coats based on titania and zirconia. *Biomaterials*, **19**, 1507–1511.

Le Mouel, S. (1997) Premiers résultats alarmants du couple zircon/polyéthylène dans les prothèses totales de hanche. *Rev. Chir. Orthop.*, **83** (Suppl. II), 44–46.

Lee, K.Y., Park, M., Kim, H.M., Lim, Y.J., Chun, H.J., Kim, H., and Moon, S.H. (2006) Ceramic bioactivity: progress, challenges and perspectives. *Biomed. Mater.*, **1**, R31–R37.

LeGeros, R.Z. (1991) *Calcium Phosphates in Oral Biology and Medicine*, Monographs in Oral Sciences, vol. **15** (ed. H. Myers), Karger, Basel, Switzerland, pp. 109–127.

LeGeros, R.Z. and LeGeros, J.P. (1984) Phosphate minerals in human tissues, in *Phosphate Minerals* (eds J.O. Nriagu and

P.B. Moore), Springer, New York, pp. 351–385.

Lewandowska-Szumiel, M. and Komender, J. (1990) Aluminium release as a new factor in the estimation of alumina bioceramic implants. *Clin. Mater.*, **5**, 167–175.

Liao, C.J., Lin, F.H., Chen, K.S., and Sun, J.S. (1999) Thermal decomposition and reconstitution of hydroxyapatite in air atmosphere. *Biomaterials*, **20**, 1807–1813.

Lindgren, M., Åstrand, M., Wiklund, U., and Engquist, H. (2009) Investigation of boundary conditions for biomimetic HA deposition on titanium oxide surfaces. *J. Mater. Sci. Mater. Med.*, **20** (7), 1401–1408.

Liu, C., Huang, Y., Shen, W., and Cui, J. (2001) Kinetics of hydroxyapatite precipitation at pH 10 and 11. *Biomaterials*, **22**, 301–306.

Liu, X., Huang, A., Ding, C., and Chu, P.K. (2006) Bioactivity and cytocompatibility of zirconia (ZrO_2) films fabricated by cathodic arc deposition. *Biomaterials*, **27** (21), 3904–3911.

McConnell, D. (1973) *Apatite. Its Crystal Chemistry, Mineralogy, Utilization and Geologic and Biologic Occurrences*, Springer, New York.

McPherson, R., Gane, N., and Bastow, T.J. (1995) Structural characterization of plasma-sprayed hydroxyapatite. *J. Mater. Sci. Mater. Med.*, **6**, pp. 327–334.

Marks, S.C. and Popoff, S.N. (1988) Bone cell biology: the regulation of development, structure, and function in the skeleton. *Am. J. Anat.*, **183**, 1–44.

Mathew, M. and Takagi, S. (2001) Structures of biological minerals in dental research. *J. Res. Natl. Inst. Stand. Technol.*, **106** (6), 1035–1044.

MDA (1996) SN9617. Zirconia ceramic heads for modular total hip femoral components: advice to users on resterilization. *Medical Device* Agency Adverse Incident Center, Department of Health, London, UK. Safety notice, June 1996.

Montel, G., Bonel, G., Trombe, J.C., Heughebaert, J.C., and Rey, C. (1980) Progress dans le domaine de la chimie des composes phosphore solides a structure d'apatite. *Pure Appl. Chem.*, **52**, 973–987.

Müller, H.I. and Greener, E.H. (1970) Polarization studies of surgical materials in Ringer's solution. *J. Biomed. Mater. Res.*, **4**, 29–41.

Nakamura, S., Kobayashi, T., Nakamura, M., and Yamashita, K. (2009) Enhanced *in vivo* response of osteoblasts in electrostatically activated zones by hydroxyapatite electrets. *J. Mater. Sci. Mater. Med.*, **20** (1), 99–103.

Niedhart, C. and Niethard, F.U. (1998) Klinische anforderungen an knochenersatzstoffe, in *Proceedings, 3rd BIOLOX® Symposium* (ed. G. Willmann), Georg Thieme Verlag, Stuttgart, New York, vol. **2-1**, pp. 46–50.

Ning, C.Y., Wang, Y.J., Chen, X.F., Zhao, N.R., Ye, J.D., and Wu, G. (2005) Mechanical performance and microstructural characteristics of plasma-sprayed biofunctionally gradient HA-ZrO_2-Ti coatings. *Surf. Coat. Technol.*, **200** (7), 2403–2408.

Ohtsu, N., Saito, K., Asami, K., and Hanawa, T. (2006) Characterization of $CaTiO_3$ thin films prepared by ion-beam assisted deposition. *Surf. Coat. Technol.*, **200**, 5455–5461.

Ohtsuka, Y., Matsuura, M., Chida, N., Yoshinori, M., Sumii, T., and Derand, T. (1994) Formation of hydroxyapatite coating on pure titanium substrates by ion beam dynamic mixing. *Surf. Coat. Technol.*, **65**, 224–230.

Orera, V.M., Merino, R.I., Chen, Y., Cases, R., and Alonso, P.J. (1990) Intrinsic electron and hole defects in stabilized zirconia single crystals. *Phys. Rev. B: Condens. Matter*, **42** (16), 9782–9789.

Park, E., Condrate, R.A., Lee, D., Kociba, K., and Gallagher, P.K. (2002) Characterization of hydroxyapatite: before and after plasma spraying. *J. Mater. Sci. Mater. Med*, **13**, 211–218.

Park, J.B., Kelly, B.J., Kenner, G.H., von Recum, A.F., Grether, M.F., and Coffeen, W.W. (1981) Piezoelectric ceramic implants: *in vivo* results. *J. Biomed. Mater. Res.*, **15**, 103–110.

Parkinson, I.S., Ward, M.K., and Kerr, D.N.S. (1981) Dialysis encephalopathy, bone disease and anaemia: the aluminum intoxication syndrome during regular haemodialysis. *J. Clin. Pathol.*, **34**, 1285–1294.

Pasteris, J.D., Wopenka, B., Freeman, J.J., Rogers, K., Valsami-Jones, E., and van der Houwen, J.A.M. (2001) Apatite in bone is not hydroxylapatite: there must be a reason. *Geol. Soc. Am. Abstr.*, **33**, A-384.

Pasteris, J.D., Wopenka, B., Freeman, J.J., Rogers, K., Valsami-Jones, E., van der Houwen, J.A.M., and Silva, M.J. (2004) Lack of OH in nanocrystalline apatite as a function of degree of atomic order: implications for bone and biomaterials. *Biomaterials*, **25** (2), 229–238.

Pasteris, J.D., Wopenka, B., and Valsami-Jones, E. (2008) Bone and tooth mineralization: why apatite? *Elements*, **4**, 97–104.

Peroos, S., Du, Z., and de Leeuw, N.H. (2006) A computer modelling study of the uptake, structure and distribution of carbonate defects in hydroxyapatite. *Biomaterials*, **27** (9), 2150–2161.

Pfaff, H.G. and Willmann, G. (1998) Stability of Y-TZP zirconia, in *Proceedings 3rd BIOLOX® Symposium* (ed. G. Willmann), Georg Thieme Verlag, Stuttgart, New York, vol. **1–6**, pp. 29–31.

Piconi, C. (2000) Materials for medical devices, in *Applied Mineralogy in Research, Economy, Technology, Ecology and Culture*, vol. **1** (eds D. Rammlmair, J. Mederer, Th. Oberthür, R.B. Heimann, and H. Pentinghaus), Proceedings 6th International Congress of Applied Mineralogy, ICAM 2000, Göttingen, Germany, 17–19 July, 2000, Balkema, Rotterdam, pp. 63–65.

Planell, J.A., Lacroix, D., Best, S., and Merolli, A. (2009) *Bone Repair Biomaterials*, Woodhead Publishing Ltd, 420 pp.

Posner, A.S. and Perloff, A. (1957) Apatites deficient in divalent cations. *J. Res. Natl Bur. Stand.*, **58**, 279–286.

Posner, A.S., Perloff, A., and Diorio, A.F. (1958) Refinement of the hydroxyapatite structure. *Acta Crystallogr.*, **11**, 308–309.

Posner, A.S., Stutman, J.M., and Lippincott, E.R. (1960) Hydrogen bonding in calcium-deficient hydroxyapatite. *Nature*, **188**, 486–487.

Posset, U., Löcklin, E., Thull, R., and Kiefer, W. (1998) Vibration spectroscopic study of tetracalcium phosphate in pure polycrystalline form and as a constituent of a self-setting bone cement. *J. Biomed. Mater. Res.*, **40**, 640–645.

Qu, C., Qin, Q.H., and Kang, Y. (2006) A hypothetical mechanism of bone remodeling and modeling under electromagnetic loads. *Biomaterials*, **27** (21), 4050–4057.

Reisel, G. (1996) Entwicklung von HVOF- und APS-gespritzten biokeramischen Schichten für die Endoprothetik. Unpublished Master's Thesis, RWTH Aachen.

Rejda, B.V., Peelen, J.G.J., and de Groot, K. (1977) Tricalcium phosphate as a bone substitute. *J. Bioeng.*, **1**, 93–97.

Reuter, S., Weßkamp, B., Büscher, R., Fischer, A., Barden, B., Löer, F., and Buck, V. (2006) Correlation of structural properties of commercial DLC coatings to their tribological performance in biomedical applications. *Wear*, **261**, 419–425.

Rey, C., Hina, A., Tofighi, A., and Glimcher, M.J. (1995) Maturation of poorly crystalline apatites: chemical and structural aspects *in vivo* and *in vitro*. *Cells Mater.*, **5** (4), 345–365.

Riboud, P.V. (1973) Composition et stabilité des phases À structure d'apatite dans le système CaO-P_2O_5- de fer-H_2O à haute température. *Ann. Chim.*, **8**, 381–390.

Rieger, W. (1993) Biocompatibility studies on zirconia and aluminia in orthopaedic joint applications. Proceedings, 9th Monte Veritá Conference on Biocompatible Materials Systems, Ascona, Switzerland.

Roba, M., Naka, M., Gautier, E., Spencer, N.D., and Crockett, R. (2009) The adsorption and lubrication behavior of synovial fluid proteins and glycoproteins on the bearing-surface materials of hip replacements. *Biomaterials*, **30** (11), 2072–2078.

Robey, P.G. (1996) Vertebrate mineralized matrix proteins: structure and function. *Connect. Tissue Res.*, **35**, 185–190.

Röker, S., Diederichs, S., Stark, Y., Böhm, S., Ochoa, I., Sanz, J.A., García-Aznar, J.M., Doblaré, M., van Griensven, M., Scheper, T., and Kasper, C. (2009) Novel 3D biomaterials for tissue engineering based on collagen and macroporous ceramics. *Materialwiss. Werkstofftech.*, **40** (1–2), 54–60.

Royce, B.S.H. (1973) The effect of structure and ionic transport properties of calcium apatite. *J. Phys. Suppl.*, **34**, 327–332.

Rubin, B. and Saffir, A. (1970) Calcification and the ground substance: Precipitation of calcium phosphate crystals from a nutrient gel. *Nature*, **225**, 78–80.

Santos, C., Pazo, A., and Guitian, F. (1995) Water vapour pressure influence on $CaO\text{-}P_2O_5$ system phase diagram, in *Materials in Clinical Applications* (ed. P. Vincenzini), Techna, Faenza, Italy, pp. 11–18.

Schneider, K. (2002) Entwicklung und Charakterisierung plasmagespritzter biokeramischer Schichten im quaternären System $CaO\text{-}TiO_2\text{-}ZrO_2\text{-}P_2O_5$. Unpublished Master Thesis, Technische Universität Bergakademie, Freiberg.

Schneider, K., Heimann, R.B., and Berger, G. (1998) Untersuchungen im quaternären System $CaO\text{-}TiO_2\text{-}ZrO_2\text{-}P_2O_5$ im Hinblick auf die Verwendung als langzeitstabiler Knochenersatz am Beispiel des $CaTiZr_3(PO_4)_6$. *Ber. Dtsch. Mineral. Ges.*, **1**, 259.

Schneider, K., Heimann, R.B., and Berger, G. (2001) Plasma-sprayed coatings in the system $CaO\text{-}TiO_2\text{-}ZrO_2\text{-}P_2O_5$ for long-term stable endoprostheses. *Materialwiss. Werkstofftech.*, **32**, 166–171.

Schumacher, D., Strunz, V., and Gross, U. (1982) Beeinflusst Piezokeramik die Knochenbildung? (Does piezoceramics influence bone formation?), *Osteogenese und Knochenwachstum* (eds M.H. Hackenbroch, H.J. Refior, and M. Jäger), Georg Thieme Verlag, Stuttgart, New York, pp. 231–235.

Senbhagaraman, S., Guru Row, T.N., and Umarji, A.M. (1993) Structural refinement using high-resolution powder x-ray diffraction data of $Ca_{0.5}Ti_2P_3O_{12}$, a low thermal expansion material. *J. Mater. Chem.*, **3**, 309–314.

Service, R.F. (2000) Tissue engineers build new bone. *Science*, **289**, 1498–1500.

Sexsmith, M. and Troczynski, T. (1994) Peel adhesion test for thermal spray coatings. *J. Therm. Spray Technol.*, **3** (4), 404–414.

Sexsmith, M. and Troczynski, T. (1996) Peel strength of thermal sprayed coatings. *J. Therm. Spray Technol.*, **5**, 196–206.

Shamos, M.H., Lavine, L.S., and Shamos, M.I. (1963) Piezoelectric effect in bone. *Nature*, **197**, 81.

Shinar, J., Tannhauser, D.S., and Silver, B.L. (1985) ESR study of color centers in yttria-stabilized zirconia. *Solid State Commun.*, **56** (2), 221–223.

Silva, C.C., Graça, M.P.F., Valente, M.A., and Sombra, A.S.B. (2006) AC and DC conductivity analysis of hydroxyapatite and titanium calcium phosphate formed by dry ball milling. *J. Non-Cryst. Solids*, **352** (9–20), 1490–1491.

SO.F.C.O.T. (2002) Arbre decisionnel en presence d'un malade porteur d'une tete zircone faisant partie des lots a risque de rupture. Société Francaise de Chirurgie Orthopédique et Traumatologique.

Søballe, K. (1993) Hydroxyapatite ceramics for bone implant fixation. Mechanical and histological studies in dogs. *Acta Orthop. Scand.*, **64**, 58.

Spinelli, M., Affatato, S., Corvi, A., and Viceconti, M. (2009) Ceramic-on-ceramic vs. metal-on-metal in total hip arthroplasty (THA): do 36-mm diameters exhibit comparable wear performance? *Materialwiss. Werkstofftech.*, **40** (1–2), 94–97.

Stamm, B., Filipov, O., Reuter, S., Erdmann, I., Deuerler, F., and Buck, V. (2008) Adjustment of wear particle size distribution of DLC coatings for tribological metal-to-metal pairing in artificial hip joints, Abstract 4th Workshop 'Biomaterials', AK 15 Deutsche Gesellschaft für Kristallographie (DGK), Halle/Saale, 6–7 October 2008.

Stevens, R. (1986) *Zirconia and Zirconia Ceramics*, Magnesium Elektron Ltd.

Strunz, H. and Nickel, E.H. (2001) *Mineralogical Tables. Chemical-Structural Mineral Classification System*, E. Schweizerbart, Stuttgart, Germany, pp. 465–468.

Suchanek, W., Yashima, M., Kakihana, M., and Yoshimura, M. (1998) β-Rhenanite (β-$NaCaPO_4$) as weak interphase for hydroxyapatite ceramics. *J. Eur. Ceram. Soc.*, **18**, 1923–1929.

Suck, K., Behr, L., Fischer, M., Hoffmeister, H., van Griensven, M., Stahl, F., Scheper, Th., and Kasper, C. (2006) Cultivation of MC3T3-E1 cells on a newly developed material (Sponceram®) using a rotating bed system bioreactor. *J. Biomed. Mater. Res.*, **80** (2), 268–275.

Szmukler-Moncler, S., Daculsi, G., Delécrin, J., Passuti, N., and Deudon, C. (1992) Calcium-metallic phosphates: a new coating biomaterial? *Adv. Biomater.*, **10**, 377–383.

Takahashi, T., Tanase, S., and Yamamoto, O. (1978) Electrical conductivity of some hydroxyapatites. *Electrochim. Acta*, **23**, 369–373.

Takahashi, H., Yashima, M., Kakihana, M., and Yoshimura, M. (2001) A differential scanning calorimeter study of the monoclinic ($P2_1/b$) → hexagonal ($P6_3/m$) reversible phase transition in hydroxyapatite. *Thermochim. Acta*, **371**, 53–56.

Tamari, N., Mouri, M., and Kondo, I. (1987) Mechanical properties and existing phases of composite ceramics obtained by sintering of a mixture of hydroxyapatite and zirconia. *J. Ceram. Soc. Jpn, Int. Ed.*, **95**, 758–761.

Tamari, N., Kondo, I., and Mouri, M. (1988a) Mechanical properties of tricalcium phosphate-zirconia composite ceramics. *J. Ceram. Soc. Jpn, Int. Ed.*, **96**, 108–110.

Tamari, N., Kondoh, I., Mouri, I., and Kinoshita, M. (1988b) Effect of calcium fluoride addition on densification and mechanical properties of hydroxyapatite-zirconia composite ceramics. *J. Ceram. Soc. Jpn, Int. Ed.*, **96**, 1170–1172.

Tancred, D.C., Carr, A.J., and McCormick, B.A.O. (2001) The sintering and mechanical behavior of hydroxyapatite with bioglass additions. *J. Mater. Sci. Mater. Med.*, **12**, 81–93.

Teller, M., Becker, P., and Neumann, H.G. (2005) Antibiotic loaded sol-gel derived calcium phosphate/silica composite for bone regeneration. *Key Eng. Mater.*, **284–286**, 415–418.

Toriyama, M., Kawamura, S., Kawamoto, Y., Suzuki, T., and Yokokawa, Y. (1990) Beta-tricalcium phosphate coating on aluminia ceramic. *J. Ceram. Soc. Jpn, Int. Ed.*, **99**, 1061–1063.

Toriyama, M., Kawamoto, Y., Yokokawa, Y., Nishiziwa, Y., and Nagae, H. (1991) Tricalcium phosphate coating on titanium. *J. Ceram. Soc. Jpn, Int. Ed.*, **99**, 1231–1233.

Tran, H.V. (2004) Investigation into the thermal dehydroxylation and decomposition of hydroxyapatite during atmospheric plasma spraying: NMR and Raman spectroscopic study of as-sprayed coatings and coatings incubated in simulated body fluid. Unpublished Ph.D. Thesis, Department of Mineralogy, Technische Universität Bergakademie Freiberg, D-09596 Freiberg, Germany.

Traykova, T., Böttcher, R., Neumann, H.G., Henkel, V., Bienengraeber, V., and Gerber, Th. (2004) Silica/calcium phosphate sol-gel derived bone grafting material – From animal tests to first clinical experience. *Key Eng. Mater.*, **254–256**, 679–682.

Trombe, J.C. (1973) Contribution à l'étude de la decomposition et de la réactivité de certaines apatites hydroxylées et carbonates. *Ann. Chim. (Paris)*, 14th ser., **8**, 335–347.

Trombe, J.C. and Montel, G. (1971) Sur la préparation de l'oxyapatite phospho-calcique. *Compt. Rend. Acad. Sci. Paris*, **273**, 452–465.

Trombe, J.C. and Montel, G. (1972) Sur l'oxyapatite phospho-calcique. *Compt. Rend. Acad. Sci. Paris*, **274**, 1169–1172.

Trombe, J.C. and Montel, G. (1978) Some features of the incorporation of oxygen in different oxidation states in the apatite lattice. *J. Inorg. Nucl. Chem.*, **40**, 15–21.

Urist, M.R. (1965) Bone: formation by autoinduction. *Science*, **150** (698), 893–899.

Vrouwenvelder, W.C.A., Groot, C.G., and de Groot, K. (1992) Behaviour of fetal rat osteoblasts cultured *in vitro* on bioactive glass and nonreactive glasses. *Biomaterials*, **13**, 382–392.

Vrouwenvelder, W.C.A., Groot, C.G., and de Groot, K. (1994) Better histology and biochemistry for osteoblasts cultured on titanium-doped bioactive glass: Bioglass 45S5 compared with iron-, titanium-, fluorine- and boron-containing bioactive glasses. *Biomaterials*, **15**, 97–110.

Wang, W., Itoh, S., Tanaka, Y., Nagai, A., and Yamashita, K. (2009) Comparison of enhancement of bone ingrowth into hydroxyapatite ceramics with highly and lowly interconnected pores by electrical polarization. *Acta Biomater.*, **5** (8), 3132–3140.

Watson, J., De Haas, W.G., and Hauser, S.S. (1975) Effect of electric fields on growth rate of embryonic chick tibiae *in vitro*. *Nature*, **254**, 331–332.

Webster, T.J., Ergun, C., Doremus, R.H., and Lanford, W.A. (2003) Increased osteoblast adhesion on titanium-coated hydroxyapatite that form $CaTiO_3$. *J. Biomed. Mater. Res. Part A*, **67A** (3), 976–9809.

Wei, D., Zhou, Y., Jia, D., and Wang, Y. (2007) Structure of calcium titanate/titania bioceramic composite coatings on titanium alloy and apatite deposition on their surfaces in a simulated body fluid. *Surf. Coat. Technol.*, **201**, 8715–8722.

Weigert, M. (1973) *Anregung der Knochenbildung durch Elektrischen Strom*, Hefte zur Unfallheilkunde, vol. **115**, Springer, Berlin, Heidelberg, New York. 101 pp.

Weinstein, A.M. (1976) Electrical stimulation of bone growth in porous Al_2O_3. *J. Biomed. Mater. Res.*, **10**, 231–247.

Weng, J., Liu, X., Zhang, X., and Ji, X. (1994) Thermal decomposition of hydroxyapatite structure induced by titanium and its dioxide. *J. Mater. Sci. Lett.*, **13**, 159–161.

Willert, H.G. (2000) Aseptic implant loosening. Abstract, Leopoldina Symposium, "Implant Materials. Infection, tissue integration, advances in new materials", Deutsche Akademie der Naturforscher Leopoldina, Cologne, 22–23 September 2000.

Williams, D.F. (1985) *Biocompatibility of Tissue Analogs*, vol. **11**, CRC Press, Boca Raton, FL.

Williams, D.F. (1990) Biocompatibility: an overview, in *Concise Encyclopedia of Medical & Dental Materials* (ed. D.F. Williams), Pergamon Press, Oxford etc., p. 52.

Willmann, G. (1993) Zirconia – a medical-grade material? *Bioceramics*, **6**, 271–276.

Willmann, G. (1995) *Bioceramics. Materials and Application* (eds G. Fishman, A. Clare, and L. Hench), Ceramic Transactions, vol. **48**, The American Ceramic Society, Westerville, Ohio, p. 83.

Willmann, G. (1996) Hydroxyapatite coating. State-of-the-art. *Br. Ceram. Trans.*, **95**, 212–216.

Willmann, G. (1997) Wie darf man keramische Komponenten für Hüftendoprothesen sterilisieren? 2nd BIOLOX® Symposium, CeramTec AG, Plochingen, Germany, pp. 139–141.

Willmann, G. (1998) Überlebensrate und Sicherheit von keramischen Kugelköpfen für Hüftendoprothesen. *Materialwiss. Werkstofftech.*, **29**, 595–604.

Willmann, G. (2002) Ceramic inserts for acetabular components of THR, in *Bioceramics in Joint Arthroplasty* (eds P. Garino and G. Willmann), Proceedings 7th International BIOLOX® Symposium, 15–16 March 2002. Georg Thieme Verlag, Stuttgart, New York, pp. 59–64.

Willmann, G., Früh, H.J., and Pfaff, H.G. (1996) Wear characteristics of sliding pairs of zirconia (Y-TZP) for hip endoprostheses. *Biomaterials*, **17**, 2157–2162.

Wilson, R.M., Elliot, J.C., Dowker, S.E.P., and Rodriguez-Lorenzo, L.M. (2005) Rietveld refinements and spectroscopic studies of the structure of Ca-deficient apatite. *Biomaterials*, **26** (11), 1317–1327.

Winand, L. and Dallemagne, M.J. (1962) Hydrogen bonding in the calcium phosphates. *Nature*, **193**, 369–370.

Wintermantel, E. and Ha, S.W. (1996) *Biokompatible Werkstoffe und Bauweisen. Implantate für Medizin und Umwelt*, Springer-Verlag, Berlin, Heidelberg, Tokio.

Wu, J.M. and Yeh, T.S. (1988) Sintering of hydroxyapatite-zirconia composite materials. *J. Mater. Sci.*, **23** (10), 3771–3777.

Xue, W., Hosick, H.L., Bandyopadhyay, A., Bose, S., Ding, C., Luk, K.D.K., Cheung, K.M.C., and Lu, W.W. (2007) Preparation and cell-materials interaction of plasma sprayed strontium-containing hydroxyapatite coating. *Surf. Coat. Technol.*, **201**, 4685–4693.

Yamashita, K., Arashi, T., Kitagaki, K., Yamada, S., and Ogawa, K. (1994) Preparation of apatite thin films through rf-sputtering from calcium phosphate glasses. *J. Am. Ceram. Soc.*, **77**, 2401–2407.

Yamashita, K., Oikawa, N., and Umegaki, T. (1996) Acceleration and deceleration of bone-like crystal growth on ceramic hydroxyapatite by electric poling. *Chem. Mater.*, **8**, 2697–2700.

Yashima, M., Sakai, A., Kamiyama, T., and Hoshikawa, A. (2003) Crystal structure analysis of β-tricalcium phosphate $Ca_3(PO_4)_2$ by neutron diffraction. *J. Solid State Chem.*, **175** (2), 272–277.

Yashima, M., Kawaike, Y., and Tanaka, M. (2007) Determination of precise unit cell parameters of the α-tricalcium phosphate $Ca_3(PO_4)_2$ through high-resolution synchrotron powder diffraction. *J. Am. Ceram. Soc.*, **90** (1), 272–274.

Yin, X., Stott, M.J., and Rubio, A. (2003) α- and β-tricalcium phosphate: a density functional study. *Phys. Rev. B*, **68**, 205205–205213.

Young, R.A. and Holcomb, D.W. (1982) Variability of hydroxylapatite preparations. *Calcif. Tissue Int.*, **34** (Suppl. 2), S17–S32.

Zeggel, P. (2000) Bioaktive Calcium-Phosphat-Beschichtungen auf Implantaten. Eine zusammenfassende Charakterisierung von BONIT-FBR (Oraltronics). *J. Implant.*, **4**, 90–94.

11
Non-Oxide Ceramics: Structure, Technology, and Applications

11.1
Introduction

Advanced non-oxide ceramics can be divided into four categories that are distinguished by their bonding types and the properties resulting thereof, such as high melting points, hardness, strengths, and stiffness as follows:

i) hard elements such as carbon (diamond) and boron;
ii) hard ceramics formed as compounds of transition metals of groups IVa (Ti, Zr, Hf), Va (V, Nb, Ta) and VIa (Cr, Mo, W) of the PSE with B, C, N and Si;
iii) binary and ternary compounds of elements of Groups IIIb to Vb of the Periodic System of the Elements (PSE), that is, B, C, N and Si; and
iv) some salt-like halogenides and chalcogenides such as refractory fluorides (CaF_2) and sulfides (BaS, CeS).

Figure 11.1 shows a graphical representation of major types of non-oxide ceramics, together with their predominant bonding relations. Metallic bonding prevails in carbides, nitrides, borides and silicides of category (ii). Mostly covalently bonded are compounds of category (iii), that is, B_4C, SiC, BN, AlN, Si_3N_4, and diamond, whereas ionic bonding is found predominantly in oxides of Be, Al, and Zr. However, all of these compounds are characterized by hybrid bonding types with varying contributions from covalent and ionic bonds (Figure 11.2).

Although non-oxide ceramics do not form a materials group with qualitatively and quantitatively unified properties, they show comparable technical performance parameters based on their type of bonding, as shown in Figure 11.2. This triangular plot reveals that hard ceramics of category (iii) and hard oxides straddle the join between pure covalent and ionic bonding types, without any significant contribution from metallic bonding. Also, several ceramics of category (ii), such as TiC, VC and WC, are characterized by a mix of metallic and covalent bonding with only a weak contribution of ionicity. There are, however, some ceramics, such as TiN and VN, that possess a significant portion of ionic bonding. This singular property has been exploited to improve the adhesion of ultrahard surface coatings of TiN to a chemical Al_2O_3 barrier layer on WC cutting tools (see Section 7.1.4.3).

Classic and Advanced Ceramics: From Fundamentals to Applications. Robert B. Heimann

ISBN: 978-3-527-32517-7

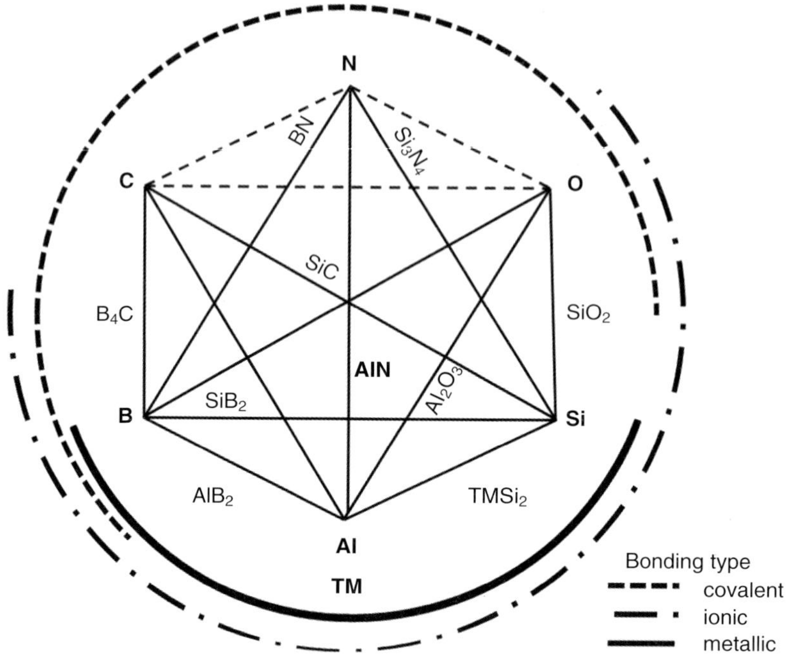

Figure 11.1 Graphic representation of the pseudo-hexanary system C–N–O–B–Si–(Al,TM). TM, transition metal.

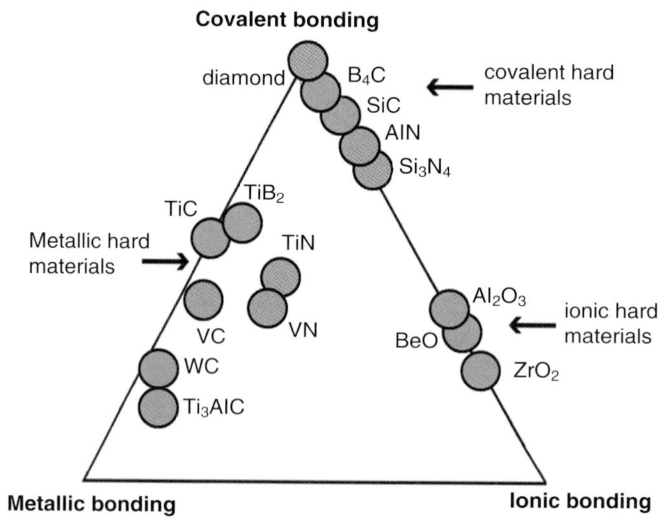

Figure 11.2 Chemical bonding types of different hard ceramic materials. Adapted from Holleck (1991).

Metallic hard materials, such as WC, TiC or TiN, are characterized by their high proportion of metal–metal bonds in the hexagonal or cubic close-packed arrangement of metal atoms, whereas the small carbon or nitrogen atoms occupy the octahedral interstitial sites. On the other hand, covalent ceramic materials have a high proportion of highly directional covalent bonds, and heteropolar (ionic) ceramic materials possess simple sublattices of large close-packed oxygen anions with the small metal atoms at interstitial sites, opposite to the metal bonding structure.

In general, all materials used for wear- and erosion-resistant coating layers (Heimann, 2008) are characterized by mixed bonding types. Their properties change with the position of the material in the bonding triangle (Figure 11.2). TiN, which is close to the center of the triangle, has a particularly favorable combination of metallic, covalent and ionic bonding types. Carbide, nitrides and borides of the transition metals (Ti, V, W) crystallize in simple, densely packed lattices of large metal atoms, with the non-metallic atoms joined to the metal lattice by covalent or mixed covalent/ionic bonds. If metallic and ionic structures are combined (e.g., TiC–Al_2O_3) there are no corresponding metal planes at the interface, but the Ti-planes of TiC will match the oxygen planes of Al_2O_3 in positioning and size. Because the covalent structures have highly directional, saturated bonds, there is only a small tendency to interact with other materials at the interface.

Table 11.1 shows a compilation of important properties of non-oxide ceramic materials of categories (i) to (iii). Many properties depend critically on the processing conditions, the purity of the materials, porosity, grain size distributions, and measurement techniques, and also reflect the anisotropy of the crystal lattice in non-cubic materials. Hence, rather wide ranges of properties have been listed in Table 11.1.

Table 11.1 Ranges of properties of some non-oxide ceramic materials (Matweb, 2008).

Material	Melting point (°C)	Density ($Mg\,m^{-3}$)	Modulus (GPa)	Hardness (GPa)	CTE ($\times 10^7\,K^{-1}$)	Thermal conductivity ($W\,m^{-1}\,K^{-1}$)
Diamond	>4000	3.5	900–1200	>150	10	2000
Boron	2300	2.6	(0.4)	>20	80	27
TiC	3065	4.9	300–450	32	75	20–35
ZrC	3400	6.5	385–405	26	67	20
VC	2730	5.7	270–430	30	72	–
WC	2870	15.8	700	22	40–70	120
TiN	2950	5.4	600	23	95	20–40
TiB_2	3225	4.5	370–540	35	75	27–80
SiC	(2700)	3.2	400–500	40	40–50	70–100
B_4C	2450	2.5	450	70	50	20–90
c-BN	3027	3.5	–	90	5–15	20
h-BN	3027	1.8–2.3	30–90	55	5–40	10–30
Si_3N_4	(1900)	3.2	220–350	17–30	36	20–40

Recently, an amorphous silicon boronitride ceramic of composition a-$Si_3B_3N_7$ was discovered, the near-stoichiometric ideality of which suggests a structure akin to a ternary nitride analogon of vitreous silica, a-SiO_2. Structural investigations, including nuclear magnetic resonance (NMR) studies (Jansen *et al.*, 2006) revealed the existence of local network polyhedra of types SiN_4, BN_3, and $NB_{3-x}Si_x$. A central B atom is thought to be statistically surrounded by four to five B and one or two Si atoms in the second coordination sphere. In addition, a Si atom is connected–through four nitrogen atoms in the first coordination sphere–to six Si and two B atoms as next-nearest neighbors.

11.2
Carbides

The crystal structures of transition metal carbides of category (ii) are determined by the atomic radius ratio $r = r_X/r_{TM}$, where X = carbon and TM = transition metal. For $r < 0.59$, the metal atoms form very simple structures with close-packed cubic or hexagonal stacking order of lattice planes. The carbon atoms are positioned at interstitial sites which must be somewhat smaller than the carbon atom, because otherwise there would be insufficient bonding resulting in an essentially unstable structure (Baumgart, 1984). The following monocarbides crystallize in the cubic B1 (NaCl) structure: TiC, ZrC, HfC, VC, NbC, and TaC. Although molybdenum carbide has an *r*-value of 0.556, its cubic phase is non-stoichiometric MoC_{1-x}, with a low-temperature α-form and a high-temperature β-form. The more stable structure with higher hardness and wear resistance is Mo_2C; this crystallizes in the hexagonal TM_2C structure in which only half of the interstitial sites are occupied by carbon. The same structure type is found for β-W_2C.

Although *titanium carbide* obeys only the cubic B1 structure, it has an extraordinarily wide compositional range, and is thus stable between $TiC_{0.97}$ and $TiC_{0.50}$. The melting point is 3065 °C near a composition $TiC_{0.80}$. *Tungsten carbide* is stable at room temperature as hexagonal α-WC that melts at 2870 °C. The hexagonal β-W_2C-phase melts at a slightly lower temperature of 2750 °C and transforms to the cubic B1 structure just below its melting point. WC loses carbon at appreciable rate above 2200 °C and will form a surface layer of W_2C.

For $r > 0.59$, more complicated structures arise in which the transition metal atoms no longer form a close-packed arrangement. An example is chromium carbide, the complicated phase relationships of which include a peritectically melting cubic $Cr_{23}C_6$ high-temperature phase, a hexagonal Cr_7C_3 phase, and a peritectically melting orthorhombic low-temperature Cr_3C_2 phase (T_P = 1810 °C). All three phases show a narrow range of homogeneity. Precipitation-strengthening of Cr_3C_2 by the formation of Cr_7C_3 due to decarburization may be the reason for the exceptionally high solid-particle erosion resistance of those ceramics (Heimann, 2008).

11.2.1 Boron Carbide

11.2.1.1 General Properties and Applications

Boron carbide is an extremely hard ceramic material second only to diamond. This property, coupled with a high erosion resistance, gives rise to many applications, including nozzles for other hard abrasive materials such as alumina or carborundum, grinding and lapping materials, thread guides, bearings, and tooling dies. The light weight, high modulus and high sound velocity of boron carbide (Table 11.1) are utilized in a large variety of military and personnel armor applications (see Section 7.1.4.3). The high thermal neutron capture cross-section of 600 barns ($6 \times 10^{-22}\,m^2$) provides neutron absorption capability for wet and dry spent nuclear fuel storage and transport applications, and possibly for the first wall protection of nuclear fusion reactors (Van der Laan *et al.*, 1994; Buzhinskij *et al.*, 2009).

11.2.1.2 Processing of Boron Carbide

Boron carbide is generally produced by the carbothermal reduction of boron trioxide with coke (see for example, Alizadeh *et al.*, 2006) in an electric arc furnace. After cleaning and comminution, dense bodies can be obtained by hot pressing, hot isostatic pressing, or pressureless sintering at 2200 °C under the addition of sintering aids such as monocarbides of Ti, Zr, Hf, V, Nb or Ta (ESK, 1996), fine-grained carbon or beryllium carbide (Figure 11.3), or alumina + TiC (Sun *et al.*, 2009).

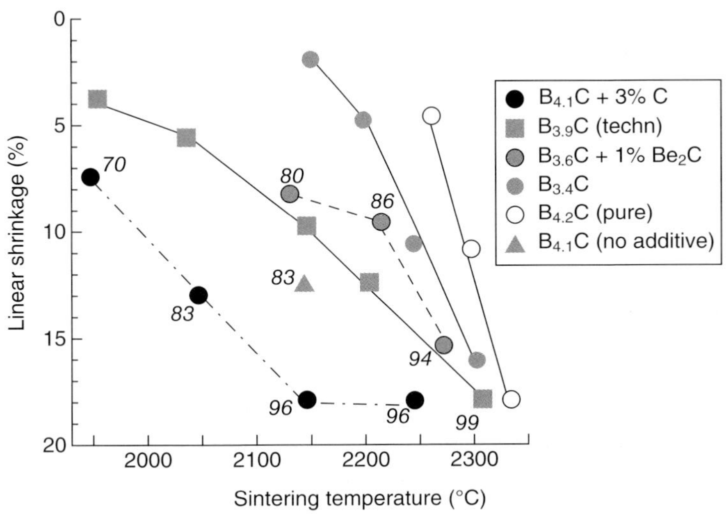

Figure 11.3 Results of pressureless sintering of boron carbide of different stoichiometric compositions (green density 60%, 1 h soaking time at final temperature, 1 bar argon atmosphere). The numbers in italics refer to percentage theoretical density.

The mechanical properties of self-bonded boron carbides produced by different densification techniques are summarized in Table 11.2.

11.2.1.3 Structure and Bonding of Boron Carbide

The crystallographic structure of boron carbide is still not known in all its complex details. Figure 11.4 shows that boron atoms occupy the vertices of icosahedra ($B6h_1$; $B6h_2$) that are positioned at the corners of a rhombohedral unit cell. Another boron atom occupies the center of the rhombohedral cell (B1b). Along the space diagonal of the cell a C–B–C chain is situated. The carbon atoms (C2c) are coordinated with four boron atoms forming three B_4C clusters per unit cell ($B_{12}C_3$). Recent structure determination by non-resonant X-ray scattering revealed a rhom-

Table 11.2 Mechanical properties of self-bonded boron carbide ceramics.

Property	HPBC[a)]	SBC[b)]	HIPSBC[c)]
Carbon content (mass%)	21.7	22.5	22.5
Porosity (%)	<0.5	<2	<0.5
Density ($Mg\,m^{-3}$)	2.51	2.44	2.51
Average grain size (μm)	5	8	8
Flexural strength (25 °C, 4-point) (MPa)	480	351	401
Modulus of elasticity (GPa)	441	390	433
Shear modulus (GPa)	188	166	183
Poisson number	0.17	0.17	0.18
Fracture toughness ($MPa \cdot m^{1/2}$)	3.6 ± 0.3	3.3 ± 0.2	3.4 ± 0.3

a) Hot-pressed (axial).
b) Pressureless sintered.
c) Sintered and HIPed.

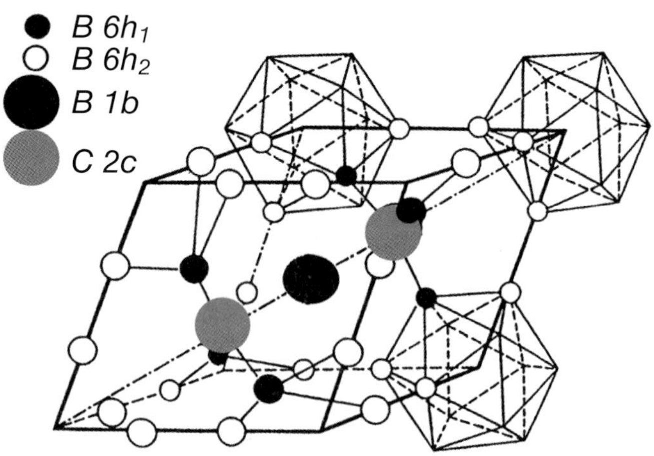

Figure 11.4 Structure of boron carbide with $B_{13}C_2$ composition.

bohedral unit cell with crystal symmetry group $R\bar{3}m$. Hence, boron carbide contains a maximum of three carbon atoms located at the two ends of the chains and in one of two nonequivalent sites on the icosahedra (Marx *et al.*, 2008). Carbon deficiencies are permissible, but at least one carbon atom must be present in the structure in order for it to be stable.

The general formula of boron carbide is $B_{12-x}C_x$ ($0.06 \leq x \leq 1.7$). Surprisingly, the structure accommodates a large variation in carbon composition without undergoing any basic structural change. The carbon deficiency originates from the need of the material to attain a higher stability by strengthening of the π-bonds at the expense of σ-bonds so as to meet the special constraints enforced by the C–B–C chains (Balakrishnarajan *et al.*, 2007). Hence, the framework of boron carbide must expel some electrons to maximize the contribution of the π-bonding, thus introducing structural disorder. This reduces the repulsive steric interactions between the bulky B_{12} icosahedral units that surround the carbon atom in the center of the rhombohedron. The localization of electronic states arising from this disorder leads to semiconducting properties. This is in contrast to the existence of a forbidden energy gap of around 3 eV that points to insulating properties. However, measurements have revealed a noticeable and measurable electrical conductivity of about 140 S at 25 °C. It is thought that defect formation is responsible for this conductivity, based on the hopping conduction of bipolarons. The formation of such bipolarons may be associated with a defective C–B–C chain connecting the icosahedral B_{12} (or $B_{11}C$) units (Wang and Marshall, 2001). The potential p-type semiconductor properties with a band gap of 0.8 eV, coupled with its high melting point and chemical stability, make boron carbide an interesting candidate semiconductor material for high-temperature applications.

11.2.1.4 **Selected Applications of Boron Carbide**

As indicated above, boron carbide is light in weight, very hard, erosion-resistant, stiff, and an excellent neutron absorber. Selected *structural applications* include:

- Nozzles for distributing hard materials (dental, spray, blasting nozzles)
- Nozzles for ceramic injection molding
- Grinding and lapping materials
- Lapping stones for grinding wheels
- Tooling dies
- Precision tool parts
- Thread guides
- Ceramic bearings
- Mortar and pestle materials.

Functional applications include:

- Antioxidants for carbon-based refractories
- Evaporation boats for materials testing and physical vapor deposition (PVD) applications
- Rocket propellants

- Tank, aircraft, and personnel armor
- Neutron absorbers for handling spent nuclear fuel, and radiation shields and control rods for nuclear reactors, often as Al/B_4C metal matrix composites (MMCs)
- Thermocouples for temperatures up to 2200 °C (U at 2000 °C about 500 mV)
- p-type semiconductors for high-temperature (HT) applications with a band gap of 0.8 eV (potential).

Ballistic Armor Today, strong and light ceramics such as silicon carbide and boron carbide, coupled with improved polymeric backing materials such as Kevlar™ replacing conventional E-glass as reinforcing fibers, are considered state of the art (Matchen, 1996). Depending on the type of projectile and the expected performance, several classes of ballistic armor (Ceradyne Inc., 2008) have been developed that include:

- *Aircraft armor* systems for wing and seating panels, as well as cockpit and floor armor systems for MH-47 Chinooks, MH-60 Blackhawks, and UH1H Super Huey aircraft.
- *Vehicle armor*, of several types.
- *Special operations forces* protection, such as low-visibility side and shoulder armor.
- *Advanced special operations force* protection devices such as hip, leg, and arm protection.
- *Personnel protection* systems with customized ergonomic shapes and curvature, including multi-hit performance technology.

The functional principle of composite armor materials and the processes occurring during impact of a projectile are shown in Figure 11.5.

During impact, the projectile tip is assumed to be destroyed within 9 μs (phase 1), and the backup plate begins to yield at the fracture conoid interface. Between 9 and 15 μs (phase 2), the projectile is eroded by the ceramic fragments, and approximately 40% of the impact energy is carried off by the eroded projectile material. Finally, after 15 μs (phase 3), the backup plate begins to absorb, by dynamic deflection, the remaining 60% of the impact energy contained in the fractured projectile and ceramic plate (Matchen, 1996). The stopping power increases with increasing areal weight of both the ceramic and backing materials. Since the extension of the duration of phase 2 by a mere 2 μs increases the ballistic limit by about 10%, the integrity of the ceramic must be maintained for as long as possible (Wilkins *et al.*, 1969). Possible ways to achieve this might include the selection of a stiffer backing material to delay the onset of ceramic tensile failure, and/or to utilize a new ceramic with improved tensile properties. The critical properties of several ceramic armor materials are listed in Table 7.3.

Boron Carbide Coatings Several attempts have been reported whereby thick boron carbide coatings are deposited by plasma spray technology (e.g., Van der Laan

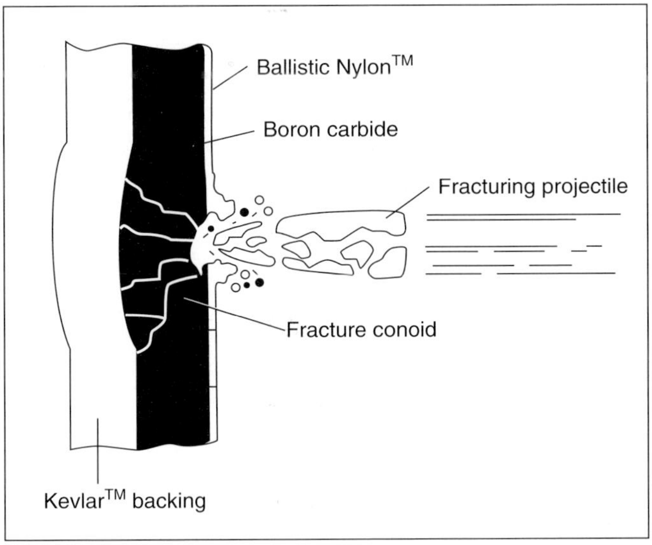

Figure 11.5 Schematic rendering of the impact of a projectile on a ceramic composite armor.

et al., 1994; Bolt *et al.*, 1996). However, the high porosity and rather weak adhesion to the substrate metals of such coatings has led to approaches to increase the particle velocity using a pulsed high-current arc discharge. Hence, electromagnetically accelerated plasma spraying (EMAPS) with a jet pressure of 1 MPa and velocities of up to 2.5 km s^{-1} yields less-porous (4%) crystalline coatings with a rhombohedral β-B_4C structure that adhere well to mirror-polished stainless steel surfaces (Kitamura *et al.*, 2003).

11.2.2 Silicon Carbide

11.2.2.1 General Properties and Applications

> "Silicon carbide (SiC) is no stranger to our universe, having been generated in the atmosphere of stars for billions of years, and hexagonal shaped microcrystals of SiC found in meteorites have their origin outside of our solar system." (Bergman *et al.*, 1998).

Indeed, SiC in the form of the naturally occurring mineral moissanite has been found in rocks of the Canyon Diablo meteorite crater in Arizona, as well as in the Murchinson carbonaceous chondrite, with uncommon isotope ratios of both silicon and carbon that point to an extrasolar origin, possibly from carbon-rich stars of the asymptotic giant branch (AGB) of the Hertzsprung–Russel diagram. Hence, we are dealing with a type of ceramics that, despite its conceived elusiveness, belongs to the basic building materials of the universe.

Its ubiquity notwithstanding, silicon carbide ("carborundum") has long been used exclusively as a hard and durable material in abrasive metal-machining operations such as grinding, honing, sandblasting, and water-jet cutting. Much later, owing to its positive-temperature coefficient of electrical conductivity at elevated temperatures, it was used for heating elements in high-temperature furnaces to melt and cast metals, as well as for heat exchanger tubes in the chemical process industry. Today, many more applications exist that exploit the exceptionally high hardness, thermal conductivity, dielectric strength, chemical stability and, most importantly, the semiconducting properties of silicon carbide. Several selected applications are described in greater detail below.

Increasingly high-purity and largely defect-free silicon carbide is utilized as an intrinsic semiconductor material with a wide (indirect) bandgap of 3.28 eV of the 4H polytype and 3.03 eV of the 6 H polytype. This property, together with the high thermal conductivity, high electric field strength, and tolerance against strong radiation fields, make silicon carbide an ideal choice for device fabrication for applications in different environmentally challenging areas. The latter include light-emitting diodes (LEDs) operating in the blue range of the visible spectrum, high-temperature and high-power electronics, high-power microwave devices, micro-electromechanical system (MEMS) technology, and high-λ substrates for gallium nitride. However, until recently the poor material quality has not allowed the fabrication of high-quality devices, and consequently extensive research has been undertaken to grow high-quality (i.e., virtually dislocation-free) single crystals of SiC (Nakamura *et al.*, 2004) with increasingly sophisticated technologies, and to adapt their properties for use over a wide range of electronic devices (Willander *et al.*, 2006). This new generation of near-perfect single crystals promises to reduce the energy losses of electrical systems operating with high-power SiC devices (Madar, 2004).

11.2.2.2 Processing of Silicon Carbide

The classic method of manufacturing industrial-grade silicon carbide is the *Acheson process*, as patented in 1893 (Acheson, 1893). Today, in excess of 800 000 tons of SiC are manufactured worldwide via this process, which is based on the carbothermal reduction of SiO_2 whereby quartz sand and petroleum coke are mixed and heated electrically to a temperature of about 2000 °C. Consequently, the intermediate product Si combines with carbon to form essentially hexagonal α-SiC in one or more of its polytypic structures according to:

$$SiO_2 + 2C \rightarrow Si + 2CO \tag{11.1a}$$

$$Si + C \rightarrow SiC \tag{11.1b}$$

The reaction begins at about 1600 °C, but leads only to a crystalline product at temperatures beyond 1900 °C.

The main disadvantages of the Acheson process include: (i) the need for very high temperatures to achieve a complete reaction; (ii) the large blocks of SiC product must be broken up, ground and comminuted to yield the classified granulated end-product, thereby expending a large amount of energy; (iii) the discontinuous nature of the process; and (iv) the rather wide grain size distribution of the ground product (this is advantageous in that it provides great strength of the green body, but disadvantageous with regard to the Weibull modulus of the sintered monolithic SiC bodies). To this list must be added the environmentally contentious issue of CO generated by the process [see Eq. (11.1a)].

The problem of CO evolution has been addressed by the modern, environmentally compatible ESK process, which relies on the basic precursor materials of the classic Acheson process but captures the CO within a canopy covering the reaction mixture and funnels it into a gas collection duct. In addition, any unwanted liquid byproducts are collected into a porous (sand) bed beneath the reacting mass.

Several other similar or alternative production processes have been developed and patented. These include a two-step process that generates β-SiC first by reacting SiO_2 with a carbon source at 1500–1800 °C, and then continues with a heat treatment at 1800–2300 °C to convert the β-SiC stable at low temperature into α-SiC (Slashed *et al.*, 2000). The General Electric Co. has developed a synthesis route starting from a silica gel formed within a sugar solution, thus yielding a very homogeneous distribution of Si and C. Heat treatment at 1800 °C in an inert atmosphere results in the formation of fine-grained β-SiC. As rice hulls contain up to 20% SiO_2, and are available in huge amounts as waste products in rice-growing countries, their pyrolysis at between 1300 and 1600 °C produces extremely fine (ca. 0.1 μm) powders of predominantly β-SiC (Martinez *et al.*, 2005). Alternatively, SiC whiskers that are useful for the reinforcement of metals can be manufactured by ashing rice hulls in air at approximately 620 °C; this produces amorphous SiO_2 which is then blended with fine carbon particles. When the mixture is rapidly heated under a CO atmosphere to 1500 °C, SiC whiskers form after flushing the CO atmosphere with argon (Horne and Ramsey, 1981).

The processing and densification technologies of SiC powder to form ceramic monolithic shapes include: pressureless sintering in the solid state (SSiC); reaction-bonded silicon carbide (RBSiC) (Figure 11.6); the infiltration of porous silicon carbide (SiSiC) by molten silicon; the reaction of carbon preforms by infiltration with chemical vapor deposition (CVD)-methyltrichlorosilane (Amirthan *et al.*, 2009); and hot pressing (HPSiC) and hot isostatic pressing (HIPSiC) (Table 11.3). Yet, each of these methods has definite drawbacks. For example, the SSiC materials that are usually densified with the aid of carbon and boron have fine-grained equiaxed microstructures with grain sizes of between 1 and 4 μm. However, although they show good high-temperature creep and oxidation resistance and exhibit moderate fracture toughness, they display a highly flaw-sensitive strength at room temperature. Moreover, due to the high temperatures applied during sintering, and to the relatively long soaking times at these temperatures, an abnormal grain growth can often occur, resulting in anisotropic α-SiC phase formation. The SiSiC materials are poor candidates for high-temperature applications as they are limited by the rather low melting point of silicon. The limitations of hot pressing (HPSiC) and hot isostatic pressing (HIPSiC) as a densification technique are well known.

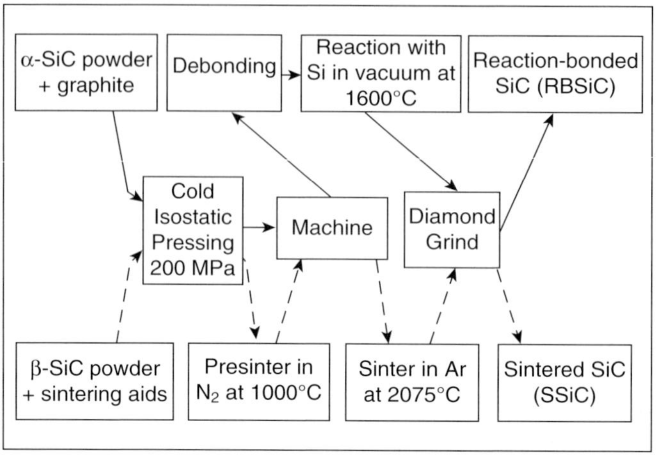

Figure 11.6 Flow chart of silicon carbide processing (T & N Technology Ltd., Rugby, UK).

Further research into the possibilities of SiC densification have shown that a pressureless sintering of SiC can be achieved with liquid-phase-forming additives, such as Al_2O_3 and Al_2O_3 + Y_2O_3 (or rare-earth oxides). Such an approach enables a sufficient lowering of the sintering temperature to 1850 °C, but still achieves complete densification. The majority of materials processed with sintering aids have equiaxed fine-grained microstructures, with secondary phases located at the triple junctions of SiC grains. This densification technique (as shown in Figure 11.6) is comparable to that used for silicon nitride (Si_3N_4) (see Figure 11.15). The resultant microstructure, which consists of crystalline and intergranular noncrystalline phases, is also similar to that formed by Si_3N_4 during liquid-phase sintering (LPS). Graphite added to α-SiC powder reacts with the subsequently added Si to form β-Sic, which then acts as a binder for the α-SiC grains, thus improving the microstructure of RBSiC (Figure 11.6).

A compilation of the properties of SiC monolithic shapes produced by a variety of densification methods is provided in Table 11.3.

Crystalline high-end silicon carbide of exceptional purity can be produced at comparatively moderate temperatures of about 1000 °C by the chemical reaction of high-purity silicon-bearing gaseous compounds such as silane (SiH_4) or silicon tetrachloride ($SiCl_4$) with carbon-bearing gases such as methane (CH_4), ethylene (C_2H_4), or carbon tetrachloride (CCl_4). Alternatively, it can be produced through the pyrolysis of carbon–silicon compounds such as methyl trichlorosilane (CH_3SiCl_3) or methyl silane (CH_3SiH_3). Arc plasma irradiation of bulk silicon in a methane atmosphere has also been utilized to produce ultrafine SiC particles (Inoue *et al.*, 1989).

The growth of single crystals of SiC with desired size, homogeneous doping level, and with minimum defect density, represents a major challenge. Since SiC does not melt congruently, but rather decomposes at high temperature, techniques to grow high-quality Si single crystals from a melt (Czochralski crystal pulling or Bridgman techniques) are not applicable. Instead, alternative technolo-

Table 11.3 Properties of SiC sintering products.

Property	Temperature (°C)	DSSiC[a)]	SSiC	HPSiC	HIPSiC	HIPSSiC	RBSiC	SiSiC
Density ($Mg\,m^{-3}$)	–	3.15	3.12	3.21	3.21	3.21	2.65	3.04
CTE ($\times 10^6\,K^{-1}$)	20–1000	4.4	4.9	5.8	4.9	4.9	4.9	4.3
Thermal conductivity ($W\,m^{-1}\,K^{-1}$)	20	150	110	130	120	120	30	160
	1000		50	45	50	45	23	50
Flexural strength (MPa)	20	480	410	650	650	540	120	400
	1000		410	650	530	460	140	
Weibull modulus	20		7–10	7–10	11–14	11–14	7–10	10
Fracture strength ($MN \cdot m^{-3/2}$)	20	4–5	4.6	4.0	4.7	4.7		4.7
Elastic modulus (GPa)	20	410	410	450	450	450	230	410
Poisson number	–	–	0.14	0.16	0.16	0.16	0.16	0.24

a) Direct-sintered (Coors).

SSiC: pressureless sintered SiC; HPSiC: hot-pressed SiC; HIPSiC: hot isostatically pressed SiC; HIPSSiC: hot isostatically pressed sintered SiC; RBSiC: reaction-bonded SiC; SiSiC: silicon-infiltrated SiC) (Matweb, partly).

gies have been developed based on the so-called *Rayleigh process*, which involves the sublimation of SiC to produce a vapor that is then condensed onto a target seed usually held at a somewhat lower temperature (Tairov and Tsvetkov, 1981). Although such single crystals display a large degree of lattice misalignment, and hence defect densities up to 10^7 cm^{-2}, this situation can be remedied by the heteroepitaxial growth of a layer of SiC on a high-quality Si substrate by CVD or liquid-phase epitaxy (LPE). By using this technology, layers of single-crystalline n-type 6H-SiC doped with Al were grown with impurity levels <1 ppm and macrodefect densities $<10^2$ cm^{-2} (Furukawa *et al.*, 1995). A review of the progress in the industrial production of SiC substrates up to the year 2001 was produced by Müller *et al.* (2001).

In order to minimize intrinsic defects such as micropipes – that is, small holes in the <00.1> direction and screw dislocations – the orientation of the seed crystal is slightly tilted relative to a (00.1) or $(00.\bar{1})$ lattice plane (Oyanagi, 2003). With a modified Rayleigh process, large SiC ingots are obtained in the form of a single-crystal polymorph. However, these ingots usually contain micropipes with a density of about 1 to 50 cm^{-2}, and screw dislocations with a Burgers vector in the c-direction with a density of about 10^3 to 10^4 cm^{-2} (Shiomi *et al.*, 2004). Consequently, reducing the number of macrodefects and increasing the diameter of the wafers is high on the agenda of the main industrial players in power semiconductors, such as the Siemens/Infineon joint venture SICED GmbH, Cree Inc., Northrop Grumman, Rohm, SemiSouth, Toshiba, Hitachi, Sumitomo Electric Industries, Dow Corning, SiCrystal AG, and many others.

The "gold standard" of SiC wafers at the time of writing this chapter appeared to be Cree Inc.'s standard n-type 4" (100 mm) 4H-SiC wafer, with a lowest micropipe density of a stunning 0.31 MP cm^{-2} and a monthly median of 5 MP cm^{-2} in production (Anonymous, 2006a). However, whilst it was predicted that it should be possible to grow high-quality 15 cm (6-inch) wafers and make them commercially available by the year 2010, the recent economic downturn has halted this development (SiC 10, 2009).

Some selected magnetic and electric properties of semiconductor α-SiC are listed in Table 11.4.

Table 11.4 Properties of semiconductor α-SiC.

Property	Value
Lattice constant a_0 (pm)	308.17
Lattice constant c_0 (pm)	1511.83
Magnetic susceptibility	-6.4×10^{-6}
Dielectric constant	10.2
Band gap (eV)	2.86
Electron mobility (cm^2 V^{-1}s^{-1})	400
Melting point (°C)	2797

11.2.2.3 Structure of Silicon Carbide

Silicon carbide exists in two crystallographically distinct modifications: (i) hexagonal α-SiC, with a wurtzite-type lattice and a huge number of polytypic variants, depending on the stacking order of the Si–C lattice planes; and (ii) cubic β-SiC, with a zincblende-type lattice (a = 0.437 nm).

The crystal lattice of hexagonal α-SiC can be derived from a diamond lattice by replacing half of the carbon atoms with silicon, thus forming the SiC_3 and CSi_3 tetrahedral units that are arranged in layers, the various stacking sequences of which give rise to an enormous number of polytypes that are structurally different but maintain an identical chemical composition. The distance C–Si is 0.190 nm, which is about the arithmetic mean of the C–C distance in diamond (0.154 nm) and the Si–Si distance in silicon (0.234 nm). The simplest polytype of α-SiC is of 2H symmetry; that is, a stacking order ABAB... exists. High-temperature polytypes are 4H, 6H, or 15R (Figure 11.7).

Impurity elements that were shown to have the most pronounced influence on polytype formation are those in Groups III and V of the PSE–that is, electron acceptors (Al, B, etc.) and electron donors (N, P). Aluminum was first noted to stabilize the 4H structure and, since this primary observation, other investigations have noted the same effect. The same stabilizing effect is also attributed to boron. In contrast, electron donors such as nitrogen and phosphorus appear to stabilize cubic β-SiC [for details and references, see Izhevskyi *et al.* (2000)].

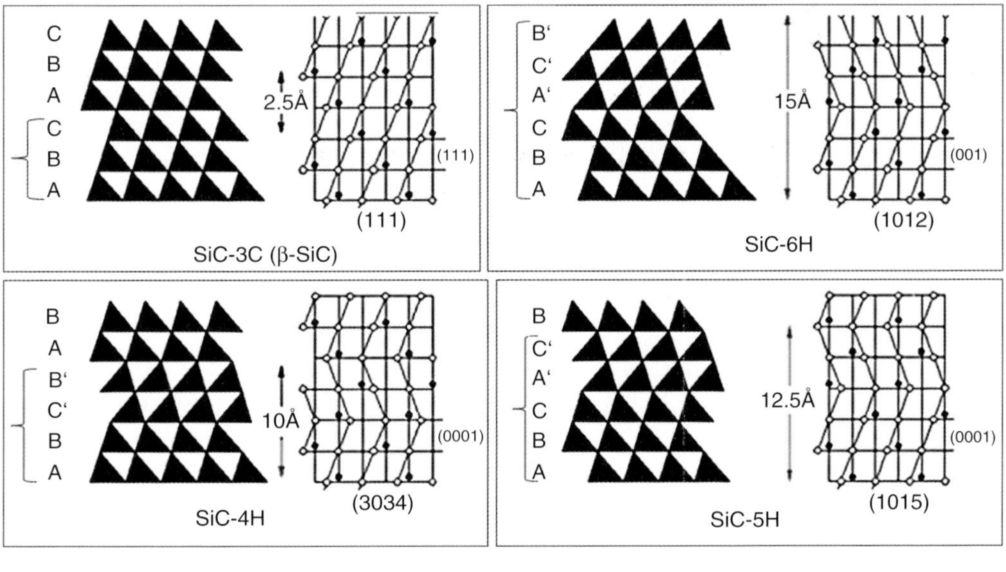

Figure 11.7 Stacking sequences of SiC_3 and CSi_3 tetrahedra, respectively in the most common SiC polytypes. Modified after Page (1990).

11.2.2.4 Selected Applications of Silicon Carbide

Owing to its high proportion of covalent bonding (see Figure 11.2), silicon carbide shows very high hardness and mechanical resilience, and hence has been traditionally used in applications such as:

- Grinding media for abrasive machining processes
- Cutting tools either as monolithic SiC or as composite Al_2O_3 + SiC whisker ceramics
- Wire-drawing tools
- Diesel particulate filters
- Bearings and rotary seals
- Cross-flow heat exchangers
- Recuperators for ceramic kilns
- Wear- and corrosion-resistant machine parts
- Ballistic armor
- Automotive applications, for example, as silicon-infused carbon/carbon composite brake discs and clutch linings/plates.

Increasingly, other applications have emerged that exploit the numerous advantageous functional properties of silicon carbides, such as the high-temperature stability, high thermal conductivity, low coefficient of thermal expansion, positive coefficient of electrical conductivity, high dielectric strength, varistor properties, chemical stability, radiation hardness, electroluminescence, and semiconducting properties including wide band gap, moderate electron mobility, and high maximum current density. Applications include, but are not limited to:

- Heating elements and heating pipes
- Heat exchanger tubes
- Dielectric coatings
- Varistors, such as lightning arresters in electrical power systems
- Heterogeneous catalyst supports
- High-temperature (HT) integrated electronics and sensors
- Thermal packaging of integrated circuits
- Blue LEDs
- Ultrafast Schottky barrier diodes and metal oxide semiconductor field-effect transistors (MOSFETs)
- HT thyristors for high-power switches
- Photovoltaics based on amorphous SiC
- Substrates for GaN-based semiconductors
- Nuclear fuel cladding for advanced light water and HT gas-cooled nuclear fission reactors
- Nuclear fuel encapsulation of pebble bed reactors
- Potential first wall-protecting ceramics and coolant channels of blankets for nuclear fusion reactors (TOKAMAKs)
- Desoxidation of steel

Lightning Arrester One of the earliest applications of SiC was that of a lightning arrester in electrical power systems. The principle of this is based on the varistor (= *var*iable res*istor* or voltage-dependent resistor, VDR) properties of SiC; that is, the dependence of the resistance of the semiconducting SiC on the voltage applied. A varistor consists of bonded grains of SiC or metal oxides such as ZnO doped with oxides of Bi, Co or Mn, sandwiched between two metal electrodes (Figure 11.8). The boundary between each grain and its neighbor forms a diode junction that allows current to flow in only one direction. The mass of randomly oriented grains is electrically equivalent to a network of back-to-back diode pairs, each pair in parallel with many other pairs. During the application of a small or moderate voltage across the electrodes, only a tiny current flows that originates from reverse leakage through the diode junctions. However, when a large voltage is applied – as during a voltage surge caused by a lightning strike – the diode junctions break down due to a combination of thermionic emission and electron tunneling, and a large current flows. The result of this behavior is a highly nonlinear (i.e., non-Ohmic current–voltage) characteristic, in which the device has a high resistance at low voltages, but a low resistance at high voltages.

A high follow-through current as a result of a lightning strike may generate an excessive current that permanently damages the varistor by localized heating as an effect of thermal runaway. As a varistor can absorb only a part of a power surge, its protective effect depends on the equipment and details of the selected varistor. Varistors do not absorb a significant percentage of a lightning strike, as the energy that must be conducted elsewhere is many orders of magnitude greater than what is absorbed by the small device.

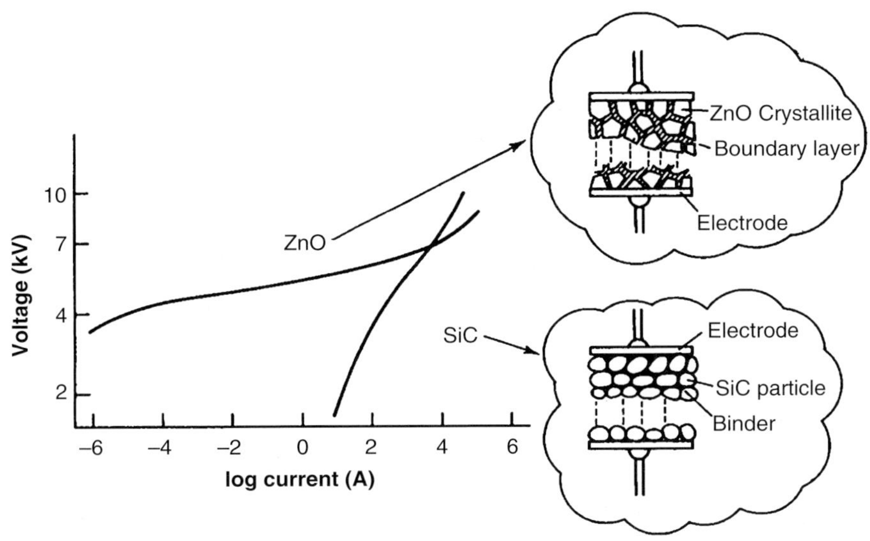

Figure 11.8 Voltage–current characteristics of ZnO and SiC junctions of a lightning arrester (after Ichinose, 1987). Reprinted with permission from Wiley-Blackwell, Oxford, UK.

As shown in Figure 11.8, a metal oxide varistor (MOV) based on ZnO is superior to the classic SiC varistor, since its *V–I* characteristics has a much lower slope, thus producing a much lower voltage rise per unit current increase. The ZnO crystallites (10 μm) constitute the semiconductor section of a boundary layer capacitor with a resistance of between 1 and 1000 Ω·cm, whereas the bismuth oxide-containing grain boundaries, with their higher resistance, act as the insulator section. The SiC device, with its low-resistance binder holding together the SiC particles of around 200 μm in size, does not allow the transmission of direct current that would be a requirement for long-distance transmission.

Thermal Packaging of Integrated Circuits Electronic packaging is a critical enabling technology for a variety of electronic applications that exploit the effective use and combination of ceramic and metallic parts. Since ceramic packaging materials will generally impose constraints on an electronic chip by limiting its electrical performance, increasing its physical size and weight, reducing its reliability and adding costs, an enhanced electronics performance will require: (i) ceramic materials with improved electrical properties so as to increase the speed of signal transmission; (ii) a high thermal conductivity for rapid heat dissipation; and (iii) a coefficient of thermal expansion (CTE) that matches that of silicon, so as to minimize the thermal stresses and thus increase the reliability of the device (Roy, 1996). Whilst alumina, beryllia, and aluminum nitride command a dominant role in ceramic packages for high-temperature cofire and multichip modules (MCMs), as well as high-power/heat-dissipating packages, SiC doped with 3% BeO is currently under investigation on the basis of its high thermal conductivity of $270\,W\,m^{-1}\,K^{-1}$ and low CTE of $3.7 \times 10^{-6}\,K^{-1}$, thus making SiC fit for high-power, high-temperature device applications (Savrun, 2002). A high dielectric constant of about 40 and a rather high dielectric loss of 0.05 at 1 MHz render SiC problematic, however, as this will reduce both the speed and quality of signal transmission. An extensive review of existing and potential applications of SiC for power electronics has been produced by Marckx (2006).

Blue Light-Emitting Diodes (LEDs), UV Photodiodes, and HT Rectifiers Single-junction devices composed of SiC have been developed for use as blue LEDs, as ultra-violet (UV) photodiodes, and as HT rectifiers. As a light emitter, the 6H-SiC junction can be tailored to emit light across the visible spectrum. The most widely commercialized device is the blue LED, which emits light with a peak wavelength of 470 nm and a spectral half-width of ~70 nm. The optical power output is typically between 12 and 18 μW for a forward current of 20 mA at 3 V (Edmond *et al.*, 1993). Super-bright blue LEDs are now produced from a combination of materials, notably a gallium nitride (GaN) layer grown epitaxially on top of a SiC wafer. These LEDs are over 20-fold brighter than the previously available SiC-only LEDs, with the intensity of their emitted light exceeding 800 mcd. The power emitted by these novel blue LEDs (several mW at 435 nm) can be used effectively for outdoor signs and other applications that must be brighter and more power-efficient.

In addition to blue emission, the wide energy bandgap of SiC of ~3.0 eV allows for inherently low dark currents and high quantum efficiencies for UV photodiode detectors made from 6H-SiC, even at high temperatures. These devices typically exhibit a quantum efficiency of 80–100%, and a peak response of ~250–280 nm that can be maintained to at least 623 K.

Rectifiers with blocking voltages as high as ~1400 V and a forward current rating of 400 mA at ~3.0 V, have also been fabricated from SiC. For a 710 V rectifier, the reverse bias leakage current density at 200 V has been shown to increase from ~10^{-9} to ~10^{-7} A cm^{-2} between 300 and 673 K, respectively. However, for a ~1400 V rectifier, the reverse bias leakage current at 1375 V is less than 1 μA at room temperature (Edmond *et al.*, 1993).

Power Semiconductors The rise in popularity of SiC-based semiconductor power devices began during the early 1990s, when the technology was developed to grow the first wafer-like SiC single crystals, using the Rayleigh process (see Section 11.2.2.2). In 1993, the first high-voltage Schottky diode was demonstrated, followed by thyristor-based switching devices in the mid-1990s. This was followed by the first MOS interface and the first commercial SiC Schottky barrier diodes in 2001, as developed and supplied by Infineon and Cree Research Inc. (Durham, NC, USA). A major breakthrough was achieved in 2003 when 8 cm (3-inch) SiC wafers with a macro defect density <10 cm^{-2} were first grown. The initial problems of a limited surge mode and failure due to overheating of the SiC Schottky diodes, were successfully overcome such that, today, 1700 V devices are now in production while 3300 V units are at the prototype stage. The production of 15 cm (6-inch) SiC wafers was forecast for 2009 or 2010; however, as the price of silicon has more than doubled in the past few years, the reduced price differential for SiC has led to it becoming economically more viable and attracting long-overdue interest from many mainstream silicon producers (Anonymous, 2006a).

Today, SiC-based electronic devices continue to advance from the research stage to production, although many commercial power devices are now approaching their performance limits in terms of their ability to block high voltages, to provide low on-state voltage drops, and to switch at high frequency (Singh and Pecht, 2008). Currently, the most suitable and well-established SiC polytype for high-temperature power electronics is the hexagonal 4H polytype (see below), which is used in applications such as pn-diodes, junction field effect transistors, thyristors/inverters (Anonymous, 2006b), and MOSFETs (the latter application is still under development, having encountered problems of low channel mobility). For microwave applications, 4H-SiC is today competing with Si and GaAs for frequencies below 10 GHz, and also for systems that require efficient cooling, such as power amplifiers.

Photovoltaic Applications For future use in the photovoltaic industry, solar modules will need to be produced on the Terawatt scale. Since, until now, polycrystalline silicon wafers have been the most widely investigated solar cell concept, and continue to demonstrate substantial cost advantages over their competitors, the use of alternative materials is at present economically unfavorable.

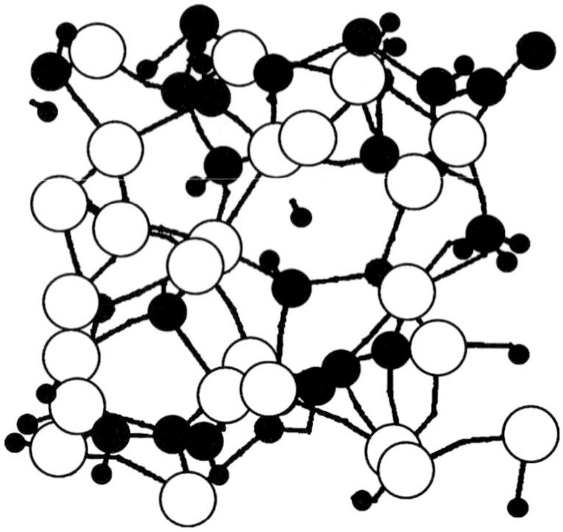

Figure 11.9 One of several possible structures of a-SiC:H. The open circles indicate silicon; filled circles represent carbon; black dots represent hydrogen (after McCulloch, 2003). Reprinted with permission from Elsevier.

Nonetheless, owing to its temperature stability, dopability and widely tunable optical behavior, amorphous SiC doped with hydrogen (a-SiC:H) has been investigated for its application in photovoltaic devices. One of several structural renderings of amorphous SiC is shown in Figure 11.9. In this case, the refractive index can be varied from 2.3 to 3.7, and the band gap from 2.4 to 2.0 eV, simply by altering a few parameters of the plasma-enhanced CVD technique that is used to deposit an amorphous silicon carbide layer with low defect density onto, for example, a substrate of RSiC. Attention has also been focused on the surface passivation performance of Si_xC_{1-x} layers deposited on crystalline silicon. Indeed, investigations into the temperature stability of Si_xC_{1-x} on nontextured solar cell structures have demonstrated an excellent photovoltaic performance, with a V_{oc} of 672 mV (Janz, 2006), such that alternative solar cell processing becomes possible.

Recently, monolithic electrochromic (EC) window-coating devices powered by integral semitransparent amorphous silicon carbide (a-SiC:H) photovoltaics (PV) have been developed by Gao *et al.* (2000). These self-powered PV-EC windows might be used to modulate solar energy gains in buildings, so as to reduce both heating and cooling requirements. One major requirement for this would be the prevention of pinhole shunting in the photovoltaic portion of the coating, as well as a resistive power loss in the transparent conductor contact layers.

Nuclear Fission and Fusion Reactors Despite a mostly politically motivated reluctance in several countries to continue generating electric energy by *nuclear fission* reactors, a combination of worldwide rising energy costs, environmental risks posed by the burning of fossil fuels and, not least, the development of inherently

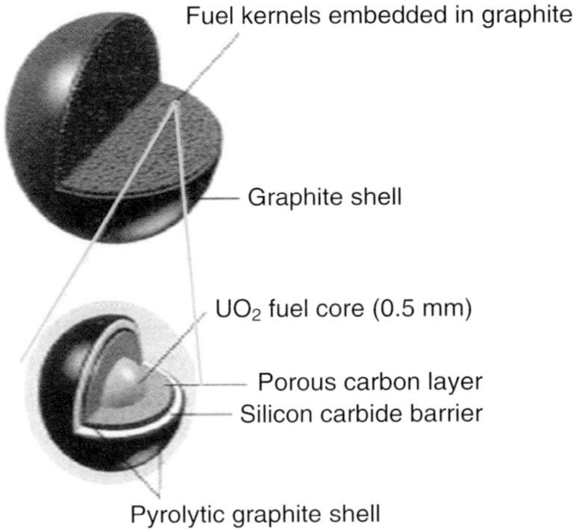

Figure 11.10 Design of a spherical fuel element ("pebble") of a very high-temperature (VHTR) gas-cooled pebble bed reactor (PBR).

safe nuclear power reactors such as the pebble bed reactor (PBR), will very likely lead to a renaissance of this technology in the years to come.

However, the situation is quite different in the case of *nuclear fusion*, as such technology is still far removed from a positive energy balance, and large sums of development money are still being spent to achieve this goal. Despite the fact that several experimental fusion reactors have been built, none of these has yet produced more thermal energy than the electrical energy consumed. Consequently, whilst it is very unlikely that a fully functioning fusion reactor will become available before the year 2050, the ITER project is currently leading the race to produce fusion power on a commercial basis.

Silicon carbide is an important ceramic material for a variety of potential applications in the nuclear field. Notably, it has been utilized as a gas-tight, high-temperature-resistant barrier surrounding the uranium oxide (or carbide) fuel kernels in a pebble bed reactor (Figure 11.10) of the Generation IV initiative. In this case, the graphite acts as the neutron moderator, while the SiC coating surrounding the tristructural-isotropic (TRISO) fuel microparticles will retain the fission products (including helium) at elevated temperatures, and hence must be gastight. In addition, the SiC should provide structural integrity to the pebbles.

The need for the annihilation of military plutonium, as triggered by the recent dismantling of nuclear weapons in the USA and Russia (Dastur *et al.*, 1995), coupled with the associated risks of proliferation due to the potential diversion of plutonium, has led to research being conducted into the use of SiC as a radiation-resistant carrier material. In this case, plutonium and other toxic actinides, such as ^{237}Np, ^{241}Am, and ^{244}Cm, could be incorporated into an inert-matrix (SiC) carrier fuel and transmuted in thermal or fast-breeder reactors. The main requirements for such a container material would include a low neutron absorption, a high melting point, chemical compatibility with water and Li ceramics (Nogami *et al.*, 2008), phase stability, a high thermal conductivity, and radiation hardness (Krstic *et al.*, 1996).

Likewise, SiC has been considered a suitable material for the coolant channels of the blankets of fusion reactors created from a SiC composite (Ward and Dudaev, 2008), and also as a low-activation structural material to protect against the excessive heat loads of the metal first wall of a potential fusion reactor (Hopkins, 1974). The key figure of merit for the latter application is a high thermal shock resistance, R'', which is necessary to withstand the stresses introduced by startups and plasma disruptions, together with the thermal cycling associated with normal pulsed mode operation. In this case, $R'' = \sigma_b \cdot k \cdot (1 - \nu)/E \cdot \alpha$, where σ_b is the flexural strength, k the thermal conductivity, ν the Poisson ratio, E the modulus of elasticity, and α the coefficient of thermal expansion.

11.3 Nitrides

11.3.1 Boron Nitride

11.3.1.1 General Properties and Applications

Boron nitride shows many similarities to carbon owing to its isoelectronic valence electron structure; that is, both compounds possess an equal sum of eight valence electrons (4 + 4 in carbon; 3 + 5 in boron nitride) and an equal sum of (two) atoms each (C_2; BN). Consequently, boron nitride shares with carbon a multiple of polymorphic modifications with very similar properties (Table 11.5).

Table 11.5 Space groups, lattice dimensions, and densities of boron nitride compared to carbon modifications.

	Boron nitride				Carbon			
	h-BN	r-BN	w-BN	c-BN	Graphite	r-C	Lonsdaleite	Diamond
Crystal class	6/mmm	$\bar{3}$m	6mm	$\bar{4}$3m	6/mmm	$\bar{3}$m	6/mmm	m$\bar{3}$m
Space group	$P6_3/mmc$	R$\bar{3}$m	$P6_3mc$	F$\bar{4}$3m	$P6_3/mmc$	R$\bar{3}$m	$P6_3/mmc$	Fd$\bar{3}$m
a_0 (pm)	250.4	250.4	255.3	361.6	246.0	245.6	251.0	356.7
c_0 (pm)	665.6	1001.0	422.8	–	671.0	1006.0	412.0	–
Density ($Mg\,m^{-3}$)	2.34	2.29	3.454	3.487	2.284	2.280	3.51	3.514

The graphite-like (g-BN, h-BN) polymorph with sp^2-hydridized bonds has many applications that rely on a sheet structure that imparts superior tribological properties for solid lubricants of high-temperature sliding bearings, as a separating medium for the poring and pressing of glasses and metals, as an additive for lubricating oils, a pressure-transmitting medium for hot isostatic pressing, or as an isolation for heating coils in high-temperature furnaces. On the other hand, the high-pressure cubic polymorph (c-BN) with sp^3-bonding is the extreme III–V semiconductor, with the largest band gap known to date of about 6.4 eV, this being larger than that of diamond (5.5 eV) and comparable to that of AlN (6.3 eV). Hence, although interesting electronic applications abound, the use of c-BN on a large scale is hampered by its very involved and expensive high-pressure synthesis (e.g., see Heimann and Kleiman, 1988). The combination of c-BN's mechanical strength, hardness, high thermal conductivity, good chemical stability, and large cross-section for slow neutrons, may bring about many novel applications in the future (Mishima and Era, 2000).

11.3.1.2 Synthesis and Processing of Boron Nitride

Hexagonal Boron Nitride (h-BN) The industrial-scale synthesis of boron nitride is accomplished by reacting $B(OH)_3$, either chemically synthesized or as the mineral sassolite, with gaseous ammonia at elevated temperatures. The two-stage process proceeds according to:

$$B(OH)_3(s) + 3NH_3(g) \rightarrow B(NH_2)_3(s) + 3H_2O(g). \tag{11.2a}$$

On further heating at 900 °C, the boron triamide reacts to:

$$B(NH_2)_3(s) \rightarrow BN(s) + 2NH_3(g). \tag{11.2b}$$

Boron trioxide can also be reacted at $T > 1000\,°C$ with other nitrogen carriers such as urea, guanidine, dicyandiamide or melamine, in a nitrogen atmosphere, to yield:

$$B_2O_3 + CO(NH_2)_2 \rightarrow 2BN + CO_2 + 2H_2O \tag{11.3}$$

A third synthesis route which is used today begins from calcium hexaboride, according to:

$$B_2O_3 + 3CaB_6 + 10N_2 \rightarrow 20BN + 3CaO\ (T > 1500\,°C). \tag{11.4}$$

The resultant h-BN is well-crystallized, and shows platelets with a dimension of 1–10 μm and a thickness of 0.1–0.5 μm. Purification may be achieved by reheating the product to temperatures in excess of 1500 °C, so as to evaporate off any remaining boric oxide.

Starting from as-synthesized powder, dense sintered h-BN can be obtained by axial hot pressing in graphite molds at a temperature of about 2200 °C, to yield monolithic ceramic bodies of 88–97% theoretical density. Any residual oxygen in the BN at the grain boundaries will form B_2O_3 that, subsequently, will act as an efficient sintering aid. The typical anisotropic properties of uniaxially hot pressed

Table 11.6 Mechanical and thermal properties of axially hot pressed BN and hot isostatically pressed BN.

Property	Temperature (°C)	HPBN (6 mass% B_2O_3)[a]	HPBN (1.7 mass% B_2O_3)[a]	HIPBN (0.1 mass% B_2O_3)
Density ($Mg\,m^{-3}$)	–	2.1	2.0	2.2
Flexural strength (MPa)	25 1000	a:115; b: 50 a: 20; b: 10	a: 95; b: 70 a: 35; b: 20	60 45
Modulus of elasticity (GPa)	25	a: 68; b: 44	a: 71; b: 35	32
Thermal conductivity ($W\,m^{-1}\,K^{-1}$)	25 1000	a: 55; b: 50 a: 28; b: 24	a: 48; b: 43 a: 28; b: 24	50 20
CTE ($\times 10^{-6}\,K^{-1}$)	20–1000	a: 1.1; b: 8.6	a: 1.0; b: 8.4	4.4
Electrical resistivity ($\Omega\,cm$)	25 1000	9.3×10^{10}	4.9×10^{12} 10^7	6.3×10^{13} 10^8
Dielectric permittivity	25	4.6	4.0	5.4
Dielectric loss factor (100 kHz)	25	3×10^{-4}	3×10^{-4}	3×10^{-4}

a) a: perpendicular; b: parallel to loading (pressing) direction.

boron nitride, as opposed to hot isostatically pressed boron nitride, are listed in Table 11.6.

Cubic Boron Nitride (c-BN) Similar to diamond, c-BN can be manufactured by high-pressure/high-temperature synthesis under static pressure conditions, using either a belt-type, multianvil, or toroid-type apparatus (for details, see Mishima and Era, 2000) to yield single crystals by the catalyst (solvent) method. Fine c-BN powders can be produced under dynamic pressure conditions using cylindrical containers to confine any single or multiple shock waves created by an explosive charge (e.g., Adadurov *et al.*, 1982) or a flying plate-planar shock wave generator (Akashi *et al.*, 1977; Sato *et al.*, 1982). For additional details on the application of dynamic pressure methods to manufacture c-BN powders, see Heimann and Kleiman (1988).

Cubic BN can also be obtained at low pressure in the field of stability of h-BN under nonequilibrium conditions, in analogy to the formation of diamond or diamond-like carbon films. As in the case of diamond, CVD, plasma-assisted chemical vapor deposition (PACVD), and PVD methods such as ion beam bombardment, radiofrequency sputter deposition, laser ablation, and magnetron sputtering can each be applied.

11.3.1.3 Structure of Boron Nitride

The structure of all four crystalline BN modifications consists of hexagonal rings composed of three boron atoms and three nitrogen atoms. The h- and r-BN sheet structures are built up by planar rings with sp^2-hybridization stacked in ABA...and ABCA... fashion, respectively. In the w-BN and c-BN structures, the sp^3-hybridized rings are "buckled" with boat- (w-BN) and "puckered" with chair-form (c-BN) configurations, respectively, analogous to lonsdalite and diamond. The distance B–N in sp^2 configuration is 144 pm, and consequently shorter by about 8% compared to the sp^3 bond length of 157 pm. This implies that the bonding within the 6-ring planes of h-BN and r-BN is considerably stronger than that in w-BN and c-BN. On the other hand, the interlayer distance between the planar 6-ring sheets is 333 pm, indicating very weak binding forces between the layers. This accounts for the fact that the layers are easily separated when subjected to shear forces, and that the high-pressure transformation of h-BN to w-BN requires a comparatively low activation energy (see below).

Graphite-Like Boron Nitride (h-BN) As in graphite, h-BN forms a structure of hexagonal sheets stacked along the c-axis with ABA... sequence (Figure 11.11a).

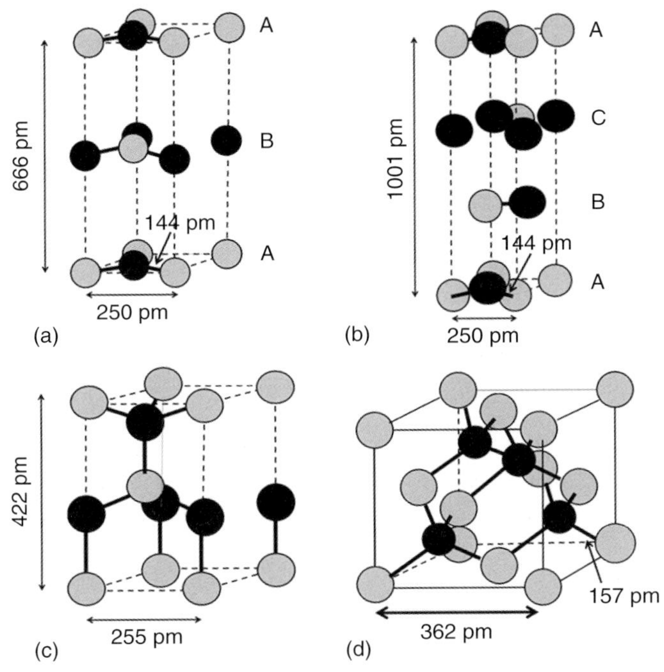

Figure 11.11 Units cells of: (a) Hexagonal sheet-type (h-BN); (b) Rhombohedral sheet-type (r-BN); (c) Wurtzite-type (w-BN); (d) Zincblende-type (z-BN) polymorphic modifications of boron nitride. The lengths of the sp^2-hybridized B–N bonds are 144 pm, those of the sp^3-hybridized bonds are 157 pm. Boron atoms are shown as light circles, nitrogen atoms as dark circles.

There is, however an important difference, as the smaller boron atoms lay over and above the larger nitrogen atoms, so that there are sublayers composed of either boron or nitrogen atoms. This introduces polarity and reduces the covalency of the B–N bonds with an associated reduction in electrical conductivity, as opposed to the well-conducting graphite with an extended network of π-bonds. The reduced electron delocalization in h-BN results in a large band gap; consequently, h-BN is white, in contrast to the black graphite.

The thermal conductivity of h-BN is highly anisotropic, showing in single crystals very high values of about $400\,W\,m^{-1}\,K^{-1}$ in the basal plane, but low values of about $2\,W\,m^{-1}\,K^{-1}$ in the direction perpendicular to the stacked 6-ring planes.

Boron Nitride with Wurtzite (w-BN)- and Zincblende (c-BN) Structures As shown in Figure 11.11c and d, the sp^3-hybridized boron nitride structures consist of B_4N tetrahedra. The B–N bond length is 157 pm. The transfer of one electron from N^{5+} to B^{3+} results in a strong covalent B^-N^+ bond, analogous to diamond. Alternatively, the transfer of all three valence electrons of B to the N atom results in an ionic $B^{3+}N^{3-}$ bond. A study of the electron density distribution of c-BN revealed an electric polarity of $B^{\delta+}N^{-\delta}$, with $\delta \sim 0.4$ indicating a charge transfer of 0.4e from the boron to the nitrogen atoms (Will *et al.*, 1986). The shift of the valence electrons is characteristic of Group III–V compound materials, and provides properties that differ from those of diamond, which shows a symmetric electron distribution between equivalent carbon atoms. Thus, c-BN is considered to have a mixed bonding of about 75% covalency and 25% ionicity.

Owing to the crystallographic polarity along <111> of the zincblende structure with crystal class $\bar{4}3m$, there exist (111)B and $(\bar{1}\bar{1}\bar{1})$N surfaces. Likewise, the non-centrosymmetrical wurtzite structure with crystal class 6mm exhibits polarity along the <00.1> direction that results in (00.1)B and $(00.\bar{1})$N surfaces. This polar nature accounts for the direction-dependent properties of c-BN and w-BN that affect crystal growth, dislocation density, lattice strain, mechanical and tribological properties, chemical reactivity including etching behavior (Heimann, 1975), impurity concentration, lattice dynamics, electronic band structure, and many other properties [for details, see Mishima and Era (2000)].

The transformation of h-BN into w-BN can be achieved in a diffusionless manner via a martensitic mechanism that involves splitting of the planar B–N 6-rings, thus generating lattice planes occupied solely by boron and nitrogen atoms, respectively (Figure 11.12a). In the ABAB... stacking of h-BN, layer A splits into two sublayers a′ and b, while layer B splits into sublayers b′ and a. Hence, in w-BN there exists a stacking order ab′ba′ab′...As is evident from Figure 11.12b, the orientation relation between parent h-BN and daughter w-BN is $(00.1)_h \| (00.1)_w$, and $\langle 2\bar{1}.0\rangle_h \| \langle 2\bar{1}.0\rangle_w$. For details on the orientation relations and the transformation mechanism, see Heimann and Kleiman (1988).

The pressure-induced transformation of h-BN to c-BN is more complex. As noted previously (Corrigan and Bundy, 1975), w-BN is not a thermodynamically stable phase in the boron nitride system. Metastable w-BN can form only from (metastable) h-BN at low temperature (Figure 11.13) by an athermal martensitic

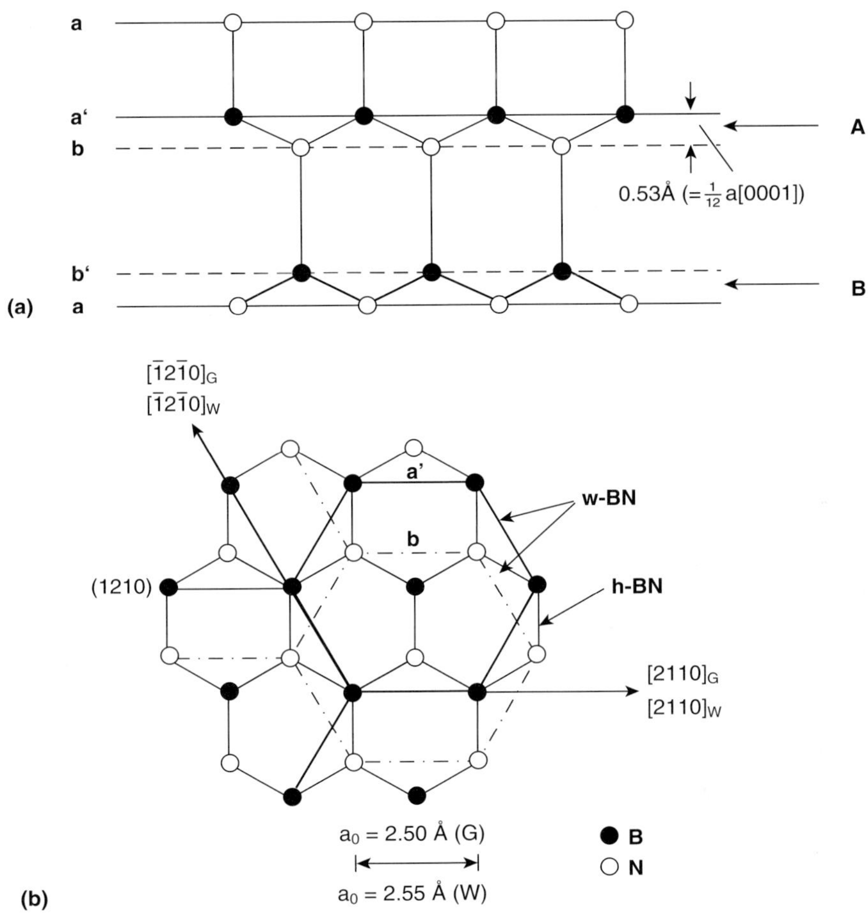

Figure 11.12 Pressure-induced transformation of h-BN into w-BN. (a) Projection onto (10.0) of w-BN; (b) Projection onto (00.1) of h-BN (layer A), (thin lines), and projection onto (00.1) of w-BN (layers a′ and b), (heavy lines) (Heimann and Kleiman, 1988).

transformation mechanism, whereas the formation of c-BN from h-BN (or w-BN) requires either a high temperature to provide the high activation energy necessary to disrupt the lattice in order to rearrange it in a puckered c-BN configuration, or the use of a catalyst to lower the energy threshold for the disruption/rearrangement sequence.

11.3.1.4 **Selected Applications of Boron Nitride**

Hexagonal boron nitride (h-BN) is used in its own right, and also as an additive to other ceramics to improve their properties.

Applications of h-BN Powder Some typical applications of h-BN powders, and the relevant properties exploited in particular applications, are listed in Table 11.7.

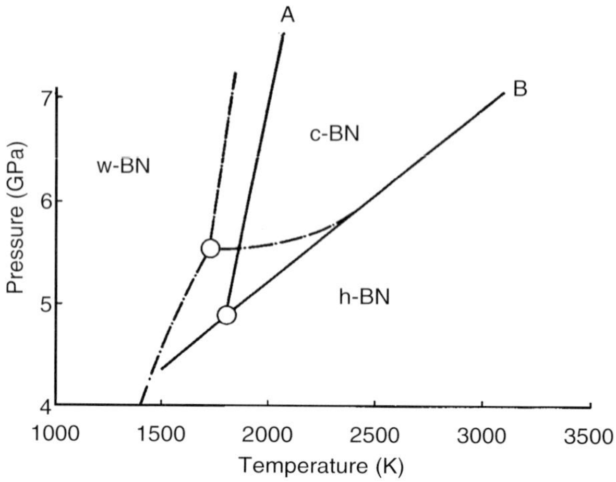

Figure 11.13 Phase relations of boron nitride polymorphs. Solid lines A and B according to Corrigan and Bundy (1975); chain lines according to Tani *et al.* (1975). The triple point w-BN/h-BN/c-BN is approximately at 1780 K/4.85 GPa and 1725 K/5.5 GPa, respectively. The activation energy for the transformation h-BN → c-BN was estimated to be 600–1050 kJ mol^{-1} at 6.5 GPa, and for w-BN → c-BN about 850 kJ mol^{-1} (Corrigan and Bundy, 1975). Reprinted with permission from The American Institute of Physics.

Applications of Monolithic h-BN Some applications of hot-pressed BN monolithic bodies are listed in Table 11.8.

Application in Composite Ceramics The addition of BN to other advanced ceramics, such as SiC, ZrO_2, Si_3N_4 or sialons, results in sintered composite ceramic materials with improved thermal conductivity, thermal shock resistance, dielectric strength, and tribological properties such as self-lubrication for ceramics bearings. On the other hand, a hot-pressed BN matrix acquires flexural strength and modulus by being reinforced with ceramics fibers (Kim *et al.*, 2005). Some properties of BN-reinforced, hot-pressed and hot isostatically-pressed composite ceramics are summarized in Table 11.9. Polymeric materials to which BN has been added demonstrate decreases in thermal expansion, increases in electrical insulation and thermal conductivity, but a reduction in mechanical wear caused by sliding abrasion.

Applications of Cubic Boron Nitride Owing to the similarity of the synthesis routes and properties of c-BN and diamond, the applications of c-BN listed in Table 11.10 resemble those of diamond.

Amorphous and Nanostructured Boron Nitride Amorphous boron nitride (a-BN) can be produced by the chemical decomposition of B-trichloroborazine with metallic Cs, or by CVD via the abstraction of chlorine and hydrogen, and subsequent

Table 11.7 Applications of h-BN powder.

Application	Property utilized				
	Thermal shock resistance ΔT (°C)	Thermal conductivity k ($W m^{-1} K^{-1}$)	Electrical insulation	Corrosion resistance	Friction properties
Solid lubricant for HT glide bearings					√
Mold separation materials for shaping of metals and glasses	√			√	√
Additive for lubricant oils and greases				√	√
Filler for rubbers, resins and polymers	√	√	√	√	√
Pressure transmitter for ultrahigh pressures	√			√	√
Coating for PVD systems	√			√	
Coating for graphite molds using in hot pressing	√			√	√
Insulators for heating elements in furnaces	√		√	√	
Preparation of other boron compounds					
Cosmetic uses for skin care					√

Table 11.8 Applications of hot-pressed sintered boron nitride.

Application	Property utilized			
	Thermal shock resistance ΔT (°C)	Thermal conductivity, k ($W\,m^{-1}\,K^{-1}$)	Electrical insulation	Corrosion resistance
Separation rings for horizontal pouring of steel bars	√			√
Crucibles for melting metals and glasses	√	√		√
Parts for electrical HT furnaces	√	√	√	√
Construction elements for magnetohydrodynamic generators	√			√
Dielectrics for radar antennas			√	
Insulators for HT capacitors	√		√	
Insulators for plasma jet furnaces and electric arc impulse generators	√		√	
Insulating parts for low- and high-frequency furnaces	√		√	
Mounting fixtures and heat sinks for power transistors		√	√	
Pump parts and delivery tubes for molten metals	√			√
Protective tubes for thermocouples	√	√		√
Protective tubes for weld metal sticks in automatic welding	√		√	
Doping material for semiconductors				
Matrices for hot pressing of special advanced ceramics	√			√
Neutron absorber and shielding in nuclear reactors	√	√		√

Table 11.9 Properties of selected BN-reinforced mixed ceramics.

Property	$SiC–ZrO_2–BN$		SiC–BN	Si_3N4-BN	Sialon-BN
	HP[a)]	HIP	HP[a)]	S	RS
BN content (mass%)	60	60	80	20	10
Density ($Mg\,m^{-3}$)	2.95	3.05	2.25	2.20	2.35
Flexural strength (MPa) @ 25 °C	137/78	115	123/58	196	127
Modulus of elasticity (GPa)	84/34	78	95/31	–	–
Knoop hardness (HK 0.1)	57	98	33	–	–
Thermal conductivity ($W\,m^{-1}\,K^{-1}$) @ 25 °C	54/27	31	61/30	18	6
CTE (20–1000 °C) ($\times 10^{-6}\,K^{-1}$)	3.5/8.5	5.7	1.5/6.5	2.4	2.0

a) The first value refers to the direction perpendicular, the second to the direction parallel to the pressing direction.

Table 11.10 Potential applications of cubic boron nitride.

Mechanical applications	Electronic applications	Optical applications
Abrasive powders	Electrical insulators	Light emitters and detectors
Cutting tools for cast iron	IC substrates und heat sinks	Optical windows
Drilling bits for hardened steel	pn-junction diodes and transistors	Optical protectors
Mirror-finishing of hardened steel	Be-doped varistors	UV-LEDs
General metal machining	Sensors for harsh conditions	X-ray lithography
Wear resistant machine parts	Neutron detectors	Windows for nuclear reactors
Ultrahigh pressure anvils	Photovoltaic devices	Windows for spacecraft
Tribological thin CVD films	Photochemical devices	
Speakers and oscillators	Piezoelectric devices	
	C/BN heterojunctions	

ring opening that involves breaking the strong B–N bonds (Stolle and Wahl, 1995; Auwärter *et al.*, 2004). Thin films of a-BN are utilized in some metal-insulator-semiconductor field-effect transistor (MISFET) semiconductor devices.

Like carbon, boron nitride forms structures such as nanotubes, nanomeshes, and BN-based fullerenes.

- *Boron nitride nanotubes* (BNNTs) are potential space materials with radiation-shielding properties owing to the ability of ^{10}B to absorb neutrons. For this property, and given the lightweight nature of BN, BNNTs may find applications as composite structural materials in future manned interplanetary spacecraft, their role being to absorb neutrons arising from cosmic ray spallation (Yu *et al.*, 2006). In contrast to carbon nanotubes (CNTs), BNNTs of different chiralities all have semiconducting properties with an identical band gap.
- *Boron nitride nanomeshes* consisting of a single layer of h-BN on a Rh or Ru substrate are stable up to 700 °C. The distance between two hole centers is 3.2 nm, and the pore depth is 0.05 nm. These meshes are capable of trapping molecules and metallic clusters in a variety of applications.
- *Boron nitride fullerenes* can be doped with transition metal atoms. Interestingly, the doped BN fullerenes have the same magnetic moment as the isolated dopants (Batista *et al.*, 2007).

11.3.2 Aluminum Nitride

11.3.2.1 General Properties and Applications

Aluminum nitride shows high thermal conductivity in excess of $200\,W\,m^{-1}\,K^{-1}$, high electric resistance ($>10^{14}\,\Omega\cdot cm$), high rigidity ($E > 320\,GPa$), and a coefficient of thermal expansion comparable to that of silicon ($\geq 4.2\,\mu m\,m^{-1}\,K^{-1}$). Although it is inert against melts of Group III–V compounds and transparent towards IR radiation, aluminum nitride will be corroded at $T > 1000\,°C$ by oxygen and atmospheric environments containing water. Owing to its positive thermal properties, the main application of AlN is for electronic substrates, although a rather high dielectric constant ($\varepsilon \geq 9$) prevents its use in integrated circuits that require a high signal transmission speed (see below).

11.3.2.2 Synthesis and Processing of Aluminum Nitride

Powders Aluminum nitride is customarily synthesized by the carbothermal reduction of a mixture of aluminum oxide and carbon black, under a nitrogen pressure of 0.2–0.3 MPa at 1800–1900 °C, and with a variable isothermal soaking time. The best results are obtained at 1800 °C and with a nitrogen pressure of 0.3 MPa, from a charge of stoichiometric composition. The resultant powders, with an AlN content up to 99 mass%, consist of isometric grains with a narrow size range of between 2 and 6 µm (Makarenko and Mahajan, 2002).

Aluminum nitride powder has been synthesized by combustion and thermal explosion modes of the self-propagating high-temperature synthesis (SHS). The addition of small amounts of ammonium chloride appears to activate the reaction (Rosenband and Gany, 2004).

Alternatively, aluminum nitride can be obtained by the direct nitridation of aluminum powder compacts between 500 and 700 °C, under a nitrogen atmosphere, between 0.5 and 7 MPa pressure. Complete nitridation was achieved at a temperature as low as 540 °C by controlling the nitrogen pressure. The nitridation process and the structures of the aluminum nitride obtained were drastically influenced by the nitrogen pressure (Okada *et al.*, 2000). As it is extremely difficult to obtain AlN without some unreacted metallic aluminum in the core of the particles, milling and acid washing is required to remove the metallic contaminant (Baik and Drew, 1996).

The ammonolysis of $AlCl_3$ results in a crystalline AlN which remains contaminated with chlorine that, even under thermal treatment at 900 °C in an ammonia atmosphere, cannot be completely removed (Pätzold, 2001).

Chemical Vapor Deposition In this process, semiconductor wafers are heated to between 300 °C and about 550 °C in a diffusion furnace. Ammonia and/or hydrazine are introduced into the chamber, followed by the introduction of an aluminum carrier such as triethylaluminum. With time, a layer of AlN is simultaneously formed at the surface of each wafer. A subsequent annealing of the AlN layer can be carried out *in situ*, thus improving its crystallinity (Kraus *et al.*, 2002).

Powder Compaction The densification of AlN powder can be achieved at high temperature by LPS, or at a somewhat lower temperature in the presence of sintering aids such as yttria (see, e.g., Watari *et al.*, 1992a; Baik and Drew, 1996) or lanthania (Li *et al.*, 2008), multicomponent oxides (Troczynski and Nicholson, 1989), or alumina, calcia, and boria (Harris *et al.*, 1996). The mechanical and thermal properties of pressureless sintered and hot-pressed AlN, with and without yttria sintering aids, are listed in Table 11.11.

Green Processing The most commonly applied technique to produce substrates is *tape-casting*, using organic solvents in conjunction with plasticizers and binders. *Dry-pressing* as well as *slip-casting* have also been attempted. The low-temperature hydrolysis of aluminum nitride causes problems, with AlN being converted to γ-AlOOH which, during subsequent sintering, forms corundum that in turn causes a deterioration in thermal conductivity by promoting phonon scattering (see below). Although technologies are currently available to prevent this hydrolysis, the oxygen pickup during sintering must still be controlled (see Section 11.3.2.4).

11.3.2.3 Structure of Aluminum Nitride

Aluminum nitride crystallizes in the hexagonal wurtzite-type structure with a space group $P6_3mc$ (186) and the lattice parameters a = 311 pm and c = 498 pm.

Table 11.11 Mechanical and thermal properties of pressureless and hot-pressed AlN, with and without an yttria sintering aid.

Property	Pressureless sintered AlN		Hot-pressed AlN	
	AlN	AlN + Y_2O_3	AlN	AlN + Y_2O_3
Density ($Mg\,m^{-3}$)	2.61–2.93	3.26–3.50	3.20	3.26–3.50
Porosity (%)	10–20	0	2	0
Fracture strength (MPa)	10–30	45–65	30–40	50–90
Hardness (HV)	–	1200–1600	1200	1200–1600
Bulk modulus (GPa)	–	310	351	279
Thermal conductivity ($W\,m^{-1}\,K^{-1}$)	–	–	150	270
Thermal expansion ($\mu m\,m^{-1}\,K^{-1}$)	5.7	–	5.6	4.9

It forms two interpenetrating hexagonal close-packed (h.c.p.) arrangements of aluminum and nitrogen atoms in tetrahedral coordination. Viewed along the [00.1] direction, the layers are stacked ABAB...

11.3.2.4 Selected Applications of Aluminum Nitride

The applications of AlN lie almost exclusively in the realm of electronics, and include:

- Electronic substrates and chip carriers in power electronics
- Microelectronics [large-scale integrated (LSI) circuits, sensor carriers, high-frequency modules]
- Dielectric layers in optical storage media
- Naval radio systems
- Radiofrequency (RF) filters in mobile phones (microelectro-mechanical systems; MEMS)
- LEDs for deep UV emissions for use in photolithography, medicine, and the cleaning of contaminated water
- Inverters for railway drive systems
- Telecommunications for aeronautical systems and research satellites
- Environmental emission control of hazardous nanosized particles through surface acoustic wave (SAW) sensors, based on epitaxially grown AlN films on Si wafers.

Market Position Although, initially, the world market for AlN products was predicted to reach US$ 550 million in 2000 (Baik and Drew, 1996), contrary to expectations the use of AlN ceramics has been restricted to only a few products. AlN ceramics have always been compared to alumina, the typical ceramic package material, but today the consumption of AlN ceramic packages has been affected by the ratio of performance and reliability, and cost. Although the price of AlN ceramic packages has decreased over the past few years, the cost reduction is still considered inadequate, and the advantages of the material's characteristics do not serve as compensation in today's cost-oriented climate. Today, few companies are able to produce AlN powder in large volumes, and many changes have occurred in the list of suppliers over the past 10 years. Currently, the main vendors of AlN powders include: Advanced Refractory Technologies Inc. (USA); H.C. Starck and Elf Atochem (Europe); and Toyo Aluminum and Tokuyama Soda (Japan). Today, approximately 2000 tons of AlN powder are produced worldwide each year, with 75% of this being utilized by Japanese companies. Currently, the end value of AlN devices is approximately US$ 250 million, about half of the sales originally forecast for the year 2000 (Anceram, 2008).

Electronic Substrates and IC Carriers Thermal conductivity represents one of the most important properties of advanced AlN ceramics in the application as electronic substrates, as it is vital for the efficient dissipation of heat generated by electronic switching processes in modern very large-scale integrated (VLSI) or ultra-large-scale integrated (ULSI) circuit technology. The typical energy density of an 8 × 8 mm VLSI chip is approximately 10 W cm^{-2}, which corresponds to that of a heating plate.

Although the theoretical thermal conductivity of AlN is estimated as 320 W m^{-1} K^{-1} (Slack and McNelly, 1976), it is difficult to achieve this value as the presence of oxygen within the lattice leads to secondary phases with lower thermal conductivities. As the thermal conduction of electrically insulating ceramic materials depends on phonon-mediated processes, any phonon scattering at lattice defects, impurities, and pores and grain boundaries will lead to a substantial reduction in thermal conductivity. In addition, whilst at higher temperatures the photon mean free path length is decreased, the frequency of phonon scattering events will be increased due to phonon–phonon interactions. As determined by Watari *et al.* (1992b), the photon mean free path length of sintered AlN at room temperature is about 10–30 nm, and much less than the typical grain size (which ranges from 1 to 40 μm). Consequently, at room temperature the thermal conductivity of AlN appears to be independent of the grain size and the presence of grain boundaries, but to be controlled only by the presence of oxygen substituting for nitrogen in the lattice. Hence, the efficient gettering of oxygen during powder production and sintering is essential if the theoretical limits of thermal conductivity are to be achieved for AlN devices.

Of equal importance for this particular application, however, is the dielectric permittivity, ε, which determines the velocity of propagation (V_p) of electric signals. As $V_p \propto 1/\sqrt{\varepsilon}$, the dielectric permittivity should be minimized to maximize the speed of signal transmission. Figure 11.14 shows different ceramic substrate materials in a λ–ε_r plot. Clearly, a compromise is always required, and target materials suitable for electronic substrate and chip carrier application should be

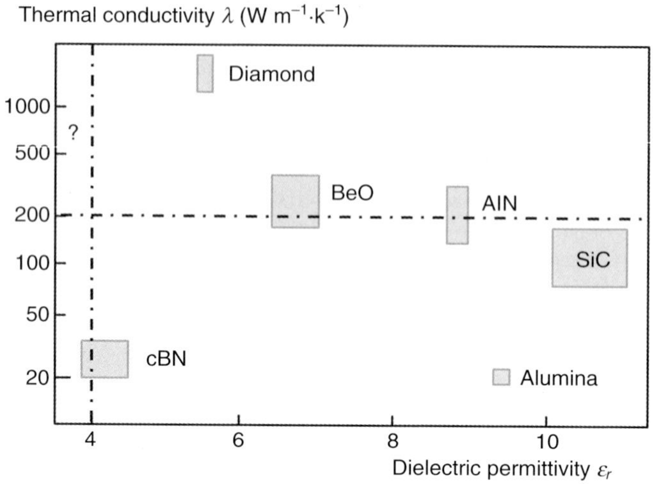

Figure 11.14 Thermal conductivities and dielectric permittivities of potential ceramic electronic substrate materials.

in the range $\lambda > 200\,\text{W}\,\text{m}^{-1}\,\text{K}^{-1}$, $\varepsilon_r < 4$, and CTE $\sim 3.5 \times 10^{-6}\,\text{K}^{-1}$, close to the CTE of silicon. The inclusion of a question mark at the lower range in Figure 11.14 indicates that research and development is currently ongoing to achieve this technological goal.

Although, AlN fares well in terms of thermal conductivity, its dielectric permittivity is rather high. As polymeric materials such as polyimide possess ε-values as low as 2.5, and are photodefinable, AlN/polyimide substrates with high thermal conductivity and low dielectric permittivity for combined thermal and signal delay management are now available.

Light-Emitting Diodes Today, compact, high-efficiency UV solid-state light sources such as LEDs and laser diodes are of considerable technological interest as alternatives to large, toxic, low-efficiency gas lasers and mercury lamps. Both, microelectronic fabrication processes and environmental sciences require light sources with shorter emission wavelengths; the former for improved resolution in photolithography, and the latter for sensors that, for example, can detect nanosized environmentally hazardous particles. In addition, UV solid-state light sources continue to attract attention for their potential application as high-density optical data storage media, and also in biomedical research, water and air purification, and sterilization systems. Although wide-bandgap materials, such as diamond and Group III–V nitride semiconductors (GaN, AlGaN, AlN) represent potential materials for UV LEDs and laser diodes, they suffer from problems in controlling electrical conduction. For example, an undoped AlN layer sandwiched between n- and p-type AlN layers will produce only a very small output power of about 0.02 μW, as compared to 1–10 mW for conventional LEDs operated at a higher wavelength. In addition, the operating voltage is very high at 25 V. One way of overcoming such a problem and improving this performance is to use appropriate doping, and the successful control of both n-type and p-type doping in AlN with a very wide direct bandgap of 6.2 eV has recently been achieved (Taniyasu *et al.*, 2006). This doping strategy has led to the development of an AlN PIN (p-type/intrinsic/n-type) homojunction LED with an emission wavelength of 210 nm, the shortest reported to date for any type of LED. Such emission has been attributed to an exciton transition mechanism.

11.3.3
Silicon Nitride

11.3.3.1 General Properties and Applications

Silicon nitride ceramics show excellent *mechanical properties* (high bending strength, elastic modulus, and fracture toughness up to 1400 °C, high abrasion wear and solid particle erosion resistance, low density); *thermal properties* (low coefficient of thermal expansion, high thermoshock resistance; see Figure 7.1), and *chemical properties* (stability against most acids and bases, corrosive gases, and liquid metals) that today are being exploited in diverse, technologically challenging fields (Michalowski, 1994; Riley, 1996). These include – but are not limited to – applications as roller and ball bearings, as coil and disc spring materials, as cutting tools for the high-speed machining of hard steels and superalloys, as heat exchangers and heat pumps, for inert gas welding and brazing fixtures and pins, as stationary blades and burner nozzles of gas turbines, as dies for wire drawing, as molds for the pressure casting of light metals, and as ladles and tundishes for the horizontal casting of steel tubes. Particular promising areas of application exist in gasoline and diesel engines, including exhaust valves, valve spring retainers, bucket tappets, rocker arm pads, pistons of internal combustion engines, and turbocharger rotors (Mörgenthaler and Bühl, 1994). Unfortunately, there are major barriers against the incorporation of these parts into the standard power trains and engines of passenger cars, most of which relate to the cost of processing, as well as to the huge technological problem of mass-producing complex ceramic components with a very high degree of reproducibility and long-term reliability.

In many cases, the use of monolithic silicon nitride parts poses problems during both manufacture and service. For example, experimental turbocharger rotors and the pistons of internal combustion engines are prone to failure in service, owing to their inherently brittle nature. However, if these parts were to be produced from tough, high-temperature-resistant metals that would provide superior mechanical stability, and could be protected against attack from highly corrosive combustion gases by the deposition of a mechanically, thermally, and chemically compatible silicon nitride coating, then a longer service life, an increased combustion temperature, and hence an environmentally beneficial engine performance, could be expected. Consequently, much effort is currently being expended to develop silicon nitride-base coatings that will function under extreme environmental conditions (Heimann *et al.*, 1997; Usuba and Heimann, 2006, Heimann 2008; see also below).

Amorphous silicon nitride thin films produced by CVD or PACVD are widely used in semiconductor processing, as masking layers for growing silicon oxide layers on silicon wafers, as barriers to prevent the diffusion of doping atoms (such as boron, phosphorus, arsenic or gallium), to protect doped wafers against sodium ion contamination and mechanical damage, and as a dielectric to manufacture thin film capacitors, as gate insulators for insulated-gate field-effect transistors (IGFETs), and as components for gate dielectrics in metal insulator-semiconductor (MIS) memory devices (Riley, 1996).

During the past fifty years, an extraordinary amount of research has been conducted on silicon nitride (and the sialons; see Section 11.3.4), with much of the acquired data being recorded in monographs (e.g., Belyi *et al.*, 1988; Hoffmann *et al.*, 1994; Sangster, 2005) and handbooks (e.g., Gmelin, 1994; Gmelin, 1995; Petzow and Herrmann, 2002).

11.3.3.2 **Synthesis and Processing of Silicon Nitride**

In this chapter, only a very cursory assessment of the complexity of producing silicon nitride powders, and the processing of such powders to obtain monolithic shapes, will be provided. Further details of recent developments in this area may be found elsewhere (Petzow and Herrmann, 2002; Lin *et al.*, 2006).

Powders The synthesis of silicon nitride powders can be achieved in several ways that involve a variety of silicon carriers such as elemental silicon, silicon dioxide, and gaseous silicon tetrachloride and monosilane. Routes to silicon nitride powders include the following technologies (Petzow and Herrmann, 2002):

- Direct nitridation of silicon
- Carbothermal nitridation of silica, including rice hulls
- Diimide synthesis
- Vapor-phase synthesis
- Plasmachemical synthesis
- Laser-induced reactions
- Pyrolysis of organosilicon compounds

For example, the diimide process involves the reaction of silicon tetrachloride or organyl aminosilane with ammonia and the subsequent thermal decomposition of the silicon diimide according to:

$$3SiCl_4 + 18NH_3 \rightarrow 3Si(NH)_2 + 12NH_4Cl \qquad (11.5a)$$

$$3Si(NH)_2 \rightarrow Si_3N_4 + 2NH_3 \qquad (11.5b)$$

In a different process, the high-temperature ammonolysis of $SiCl_4$ in the gaseous phase between 800 and 1100 °C leads to amorphous silicon nitride that, on subsequent treatment at 1500 °C, can be crystallized to form α-Si_3N_4 (Pätzold, 2001).

Silicon Nitride Compacts A flow chart of the production routes for porous reaction-bonded silicon nitride (RBSN), dense sintered silicon nitride (SSN), and sintered reaction-bonded silicon nitride (SRBSN) is shown in Figure 11.15.

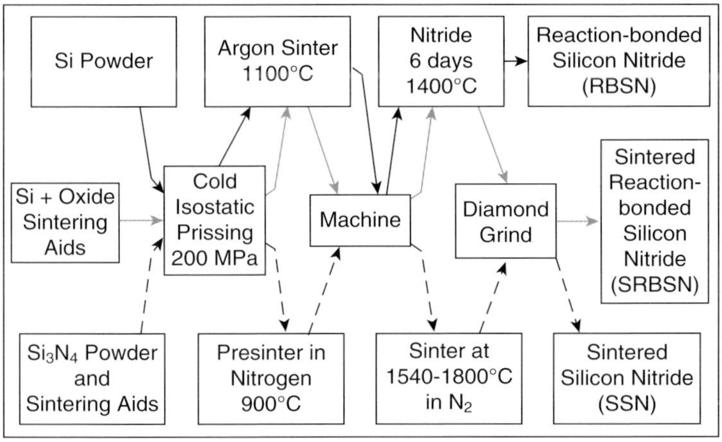

Figure 11.15 Flow chart of silicon nitride processing.

The direct nitriding of fine compacted silicon powder leads to the formation of a mixture of α- and β-Si_3N_4 by prolonged heating in either nitrogen or ammonia between 1000 and 1450 °C. Detailed investigations of this reaction have indicated that the nitridation of silicon proceeds mainly through vapor-phase reactions that involve silicon monoxide, generated from contaminant oxygen, according to:

$$2Si(s) + O_2(g) \rightarrow 2SiO(g) \tag{11.6a}$$

$$3SiO(g) + 2N_2(g) \rightarrow Si_3N_4 + 3/2O_2(g) \tag{11.6b}$$

The volume increase of approximately 22% is completely accommodated by the interparticle void space of the compacted silicon powder. Hence, no shrinkage occurs during nitriding, and the original dimensions of the green component will be retained, and this is the underlying reason for using RBSN for complex-shaped components. Moreover, as expensive diamond grinding and machining can be avoided, those parts manufactured from RBSN are economically highly competitive. Although the porosity of RBSN is between 20% and 30%, the mechanical strength is remarkable, with flexural strengths of 200–300 MPa being readily attained. An additional benefit is that this strength can be retained to about 1400 °C (Figure 11.16), so that designing with RBSN will generally lead to high Weibull factors.

In order to achieve fully dense silicon nitride monolithic bodies, it is necessary to employ either the hot pressing (HP) or hot isostatic pressing (HIP) of silicon nitride powder, with the addition of metal oxides (for a previous review, see Maccagno, 1989). At temperatures above 1550 °C, these additives form with contaminant silicon dioxide films present around individual silicon nitride grains a liquid silicon oxynitride phase in which the silicon nitride may be readily dissolved.

As these HP and HIP techniques are highly involved and are restricted to simple shapes, today the most frequently used method is that of pressureless sintering.

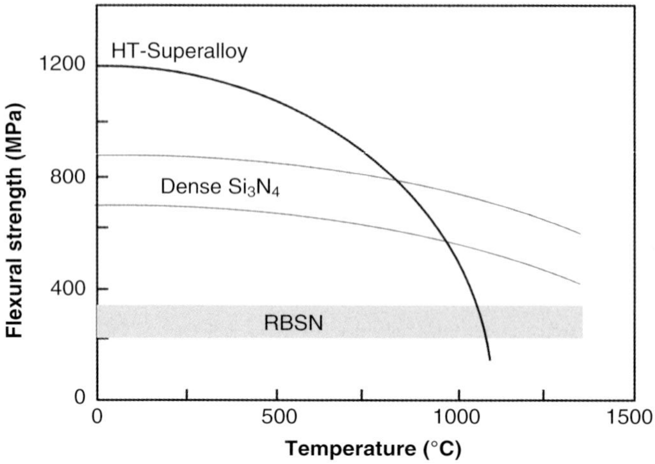

Figure 11.16 Flexural strengths of reaction-bonded silicon nitride (RBSN) and dense sintered silicon nitride as a function of temperature. The strong deterioration of the strength of HT superalloys with temperature is given for comparison.

In this case, silicon nitride powder can be sintered economically to full density by the addition of sintering aids such as magnesia, alumina and yttria, or a combination thereof. Here, the critical parameters are the use of nanosized powders (in the range of 100–200 nm), the ability of the added oxides to form a film of liquid phase surrounding the silicon nitride grains, an atmosphere of high-pressure nitrogen to counteract any dissociation, and a sintering temperature of between 1700 and 1850 °C. During processing in the presence of alumina + yttria, a significant proportion of Al and O will enter the Si_3N_4 lattice to form the family of SiAlONs as α'- and β'-phases (see Section 11.3.4). On cooling, the intergranular liquid will solidify as a glassy or crystalline (YAG) phase. Notably, the quinary Si–Al–N–O–Y glass is generally the continuous phase that controls the high-temperature properties of the silicon nitride ceramics, such as strength, fracture toughness, creep, and oxidation behavior. The degree of *in-situ* reinforcement of the structure of pressureless sintered silicon nitride by the glass-forming intergranular phase depends on the ionic radius of the rare-earth element (REE) cation. There exists a (negative) linear dependence of the flexural strength on the radius of the glass-forming REE ion, such as $\sigma = -18.2 \cdot r + 2238\,\text{MPa}$ at 25 °C, and $\sigma = -20.9 \cdot r + 2261\,\text{MPa}$ at 1200 °C (where r is measured in picometers).

Some of the mechanical and thermophysical properties of RBSN and SSN are listed in Table 11.12.

Textured Silicon Nitride A modern technique of manufacturing strong and mechanically resilient silicon nitride monolithic shapes relies on the concept of texturing. As the α-Si_3N_4 phase that is stable at low temperature is converted to β-Si_3N_4 at the high temperatures applied during densification, the powder com-

Table 11.12 Selected properties of RBSN and SSN.

Property	RBSN	SSN
Density ($Mg\,m^{-3}$) (% of theoretical density)	70–88	95–100
Flexural strength (4-point, 25 °C) (MPa)	150–350	500–1 000
Fracture toughness (25 °C) ($MPa \cdot m^{1/2}$)	1.5–3	5–8
Fracture energy ($J\,m^{-2}$)	4–10	~60
Modulus of elasticity (25 °C) (GPa)	120–220	300–330
Thermal conductivity (25 °C) ($W\,m^{-1}\,K^{-1}$)	4–30	15–50
Thermal shock resistance R (K)	220–580	300–780
Thermal shock fracture toughness R′ ($W\,m^{-1}$)	500–10 000	7 000–32 000
Coefficient of thermal expansion ($10^{-6}\,K^{-1}$)	3.2	3.2

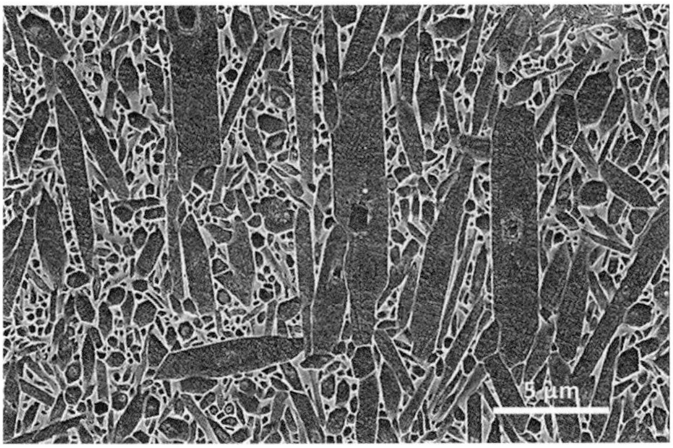

Figure 11.17 Scanning electron microscopy image of a polished and plasma-etched surface of a textured silicon nitride sample with rodlike β-Si_3N_4 grains aligned by a rotating magnetic field of 12 T, subsequently sintered at 1800 °C for 3 h (Zhu and Sakka, 2008). Reprinted with permission from IOP Publishing Ltd., Bristol, UK.

pacts develop a rodlike morphology generated by "texturing" and attain a self-reinforcing microstructure that greatly enhances the mechanical and thermal properties of the product. Consequently, seeds of β-Si_3N_4 will be added to the raw powder that consists predominantly of α-Si_3N_4. These rodlike crystals will be aligned by either hot working (such as hot pressing, hot- and sinter-forging), or by template grain growth (TGG) that involves alignment by cold pressing alignment (CPA), the extrusion alignment (EA) of a slurry, tape casting alignment (TCA), and either static magnetic field alignment (SMFA) or rotating magnetic field alignment (RMFA). These aligned seeds grow during subsequent high-temperature treatment into a reinforcing microstructure, as shown in Figure 11.17. A comprehensive review of the processing technologies and properties achieved has recently been provided by Zhu and Sakka (2008).

By using this novel technique, ceramic materials with typical flexural strengths of 1.4 GPa, stress intensity factors exceeding 14 $MPa \cdot m^{1/2}$, and elastic moduli of 280 GPa (perpendicular to the c-axis) and 540 GPa (parallel to the c-axis of the crystals) were achieved (Zhu and Sakka, 2008; see also Imamura *et al.*, 2000).

Coatings The deposition of pure silicon nitride coatings by conventional thermal spraying has been considered impossible, since Si_3N_4 does not melt but rather dissociates and, in turn, sublimates above 1800 °C. It is also subject to oxidation in the presence of an oxygen-containing atmosphere at elevated temperature. Thin amorphous silicon nitride films are frequently deposited by CVD (see below), and applied as masking layers for semiconductor integrated circuits during profile etching, diffusion barriers in VLSI production lines, and for damage protection of optical fibers. However, attempts at the mechanical deposition of stable, thick silicon nitride coatings for nonelectronic applications by thermal spraying using metallic (Lugscheider and Limbach, 1990; Lugscheider *et al.*, 1990; Limbach, 1992) or silicate glass binders (Kucuk *et al.*, 2000), conversion in the as-deposited state through a reactive spray process (Eckardt *et al.*, 1994), or *in situ* nitridation in flight (Eckardt *et al.*, 1996) were not met with resounding success, as the coatings produced contained very small amounts of silicon nitride but, instead, substantial amounts of embrittling metal silicides. More successful were endeavors to prepare coatings with higher Si_3N_4 contents, starting from β'-$Si_{6-z}Al_zO_zN_{8-z}$ powders with different degrees of substitution (z) (Sodeoka *et al.*, 1992), or clad-type powder consolidation using alloy bond coats (Tomota *et al.*, 1988). Recently, it was recognized that high particle velocities generated by detonation spraying (DS), Top Gun™ technology, high-frequency pulse detonation (HFPD), and APS with axial powder injection were conducive to depositing dense and well-adhering silicon nitride coatings (Thiele *et al.*, 1996; Berger *et al.*, 1998; Heimann *et al.*, 1998, 1999; Thiele *et al.*, 2002; Heimann, 2010). In addition, the optimization of heat transfer into the powder particles to control the critical viscosity of the oxide binder phase was found to be one of the most decisive factors (Bao *et al.*, 2008). This necessitates special powder preparation procedures (Berger *et al.*, 1997, 2000). Extremely high powder particle velocities of up to 3 $km\,s^{-1}$, generated by using EMAPS, were exploited to deposit dense, homogeneous silicon nitride coatings with desirable mechanical properties that adhered well to polished steel surfaces (Usuba and Heimann, 2006). Examples of these dense and well-adhering, mechanically stable silicon nitride-based coatings on steel substrates, as deposited by detonation spraying and ultrahigh-velocity pulsed plasma spraying (EMAPS) are shown in Figure 11.18 a and b (respectively).

Amorphous Thin Films Currently, thin amorphous films of silicon nitride for applications as masking layers and as diffusion barriers during semiconductor processing are produced by gas-phase reactions of silicon tetrachloride or silane with ammonia, in the presence of hydrogen as carrier gas. Today, the standard CVD process is augmented by complex molecular excitation methods that include PACVD, laser-excited CVD (LECVD) and photosensitized CVD (PHCVD) enhance-

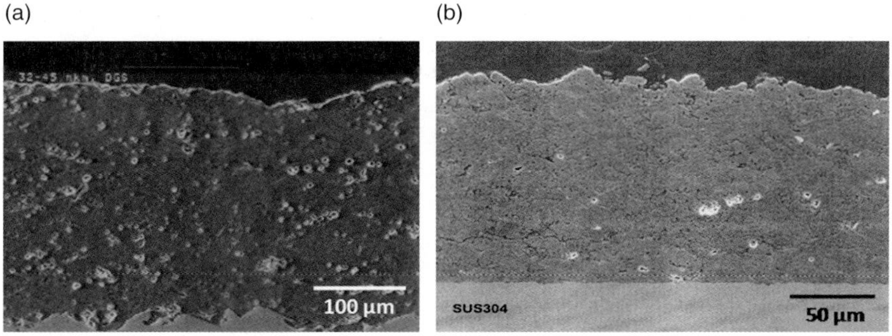

Figure 11.18 Silicon nitride-based coatings on steel produced by (a) detonation spraying (Berger *et al.*, 1998) and (b) ultrahigh-velocity pulsed plasma spraying (Usuba and Heimann, 2006).

ment, and microwave and electron cyclotron resonance (Riley, 1996). Details on these applications are provided later in the chapter.

11.3.3.3 Structure of Silicon Nitride

Silicon nitride crystallizes in two hexagonal forms (α- and β-Si_3N_4) and also a cubic γ-modification with spinel structure that is stable at high pressure and temperature (>15 GPa, >2000 K; Zerr *et al.*, 1999; Leitch *et al.*, 2004). The α-phase occurs in nature as a rare mineral (nierite) in ordinary and enstatite chondritic meteorites (Lee *et al.*, 1995).

The crystal structures were determined by Hardie and Jack (1957) and Turkdogan *et al.* (1958). The space group of α-Si_3N_4 is $P31c$ (154) with a = 774.8(1) pm and c = 561.7(1) pm. The space group of β-Si_3N_4 is $P6_3/m$ (173), with a = 760.8(1) pm and c = 291.1(1) pm. The symmetry of γ-Si_3N_4 is $Fd\overline{3}m$ (227), with a = 773.8(1) pm.

As shown in Figure 11.19, the structure consists of slightly distorted corner-sharing SiN_4 tetrahedra that form distorted hexagonal rings arranged in layers with ABCDABCD... stacking (α- Si_3N_4) and ABAB... stacking (β-Si_3N_4). In β-Si_3N_4, the bond lengths between Si and N1 are 173.0 and 173.9 pm, respectively, and between Si and N2 174.5 pm. In α-Si_3N_4, the bond lengths vary between 156.6 pm (N1-Si1) and 189.6 pm (N1-Si2).

While the AB layer is identical in both the α- and β-phases, the CD layer in the α-phase is related to AB by a c-glide plane. Hence, a double layer in α-Si_3N_4 can be thought of as a superposition of a β-Si_3N_4 layer and its counterpart inverted by 180°. Consequently, there are twice as many atoms per unit cell in α-Si_3N_4 (Z = 4) than in β-Si_3N_4 (Z = 2). The Si_3N_4 tetrahedra in β-Si_3N_4 are interconnected in such a way that channels are formed extending parallel to the c-axis of the unit cell. Owing to the c-glide plane, the α-structure contains two (isolated) interstitial sites per unit cell instead of channels (Figure 11.20a). The channel structure of the

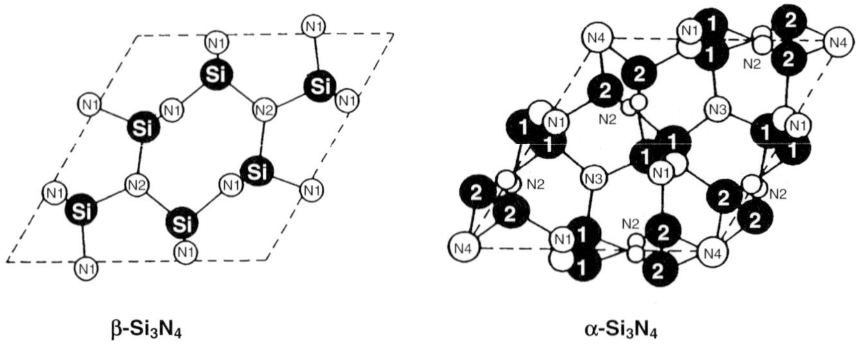

Figure 11.19 Structures of hexagonal β- and α-silicon nitride. Black circles represent silicon atoms; white circles represent nitrogen atoms.

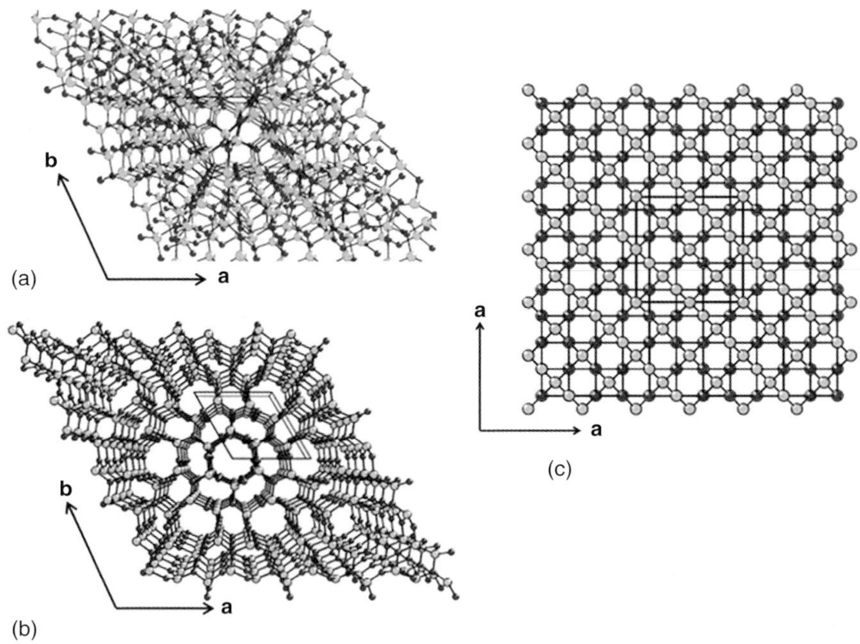

Figure 11.20 Crystal structures of (a) α-Si_3N_4, (b) β-Si_3N_4, and (c) spinel-type high-pressure polymorph of γ-Si_3N_4. The structure of α-Si_3N_4 shows the interstitial sites, the structure of β-Si_3N_4 the through-channel. Light balls represent silicon atoms; dark balls represent nitrogen atoms (Peng, 2004). Reprinted with kind permission from Dr Hong Peng.

β-phase (Figure 11.20b) supports an easy diffusion of ions through the lattice, thus greatly promoting sinterability at high temperatures.

The structure refinement of α-Si_3N_4, using synchrotron radiation powder diffraction, was more recently accomplished by Toraya (2000), yielding $a = 775.193\,(3)$ pm, $c = 561.949\,(4)$ pm, and $Z = 4$.

The cubic γ-Si_3N_4 has a spinel-type structure in which two silicon atoms each coordinate six nitrogen atoms octahedrally, and one silicon atom which coordinates four nitrogen atoms tetrahedrally (Figure 11.20c).

11.3.3.4 **Selected Applications of Silicon Nitride**

Owing to its excellent mechanical, thermal, and thermophysical properties, silicon nitride finds applications in aerospace gas turbines, internal combustion engines, heat engines, and a variety of dies, molds and tools for metal manufacturing and forming (Von Hocking *et al.*, 1994; Riley, 1996; Petzow and Herrmann, 2002; Lin *et al.*, 2006).

The advantageous properties of silicon nitride that lead to such uses include: a low density; sufficient HT stability, up to 1800 °C; a high degree of hardness; a low frictional coefficient; high corrosion resistance; high thermoshock resistance compared to most other ceramics; and various tribological properties, including high contact fatigue resistance.

Silicon nitride does, however, have a number of drawbacks relating to its use, which include: (i) an inability to cope with high service-imposed stresses (brittleness); (ii) difficulties when machining monolithic parts, with the exception of RBSN; and (iii) difficulties in joining silicon nitride to metallic construction elements, which leads in turn to a need for complex technologies such as brazing, gas-pressure sintering, or HIPing.

Typical applications of porous RBSN include crucibles and stirrers for handling molten aluminum, tin, zinc and lead alloys, thermoelement sheaths, nozzles, spouts, arc-welding nozzles, weld location pins, and specialized kiln furniture with low thermal masses and high thermal shock resistances for use in firing components such as dental porcelain, when repeated thermal cycling is required. Other applications include burner nozzles, heat-exchanger panels, and soldering jigs for the assembly of electronic devices such as diodes, thyristors, and transistors.

The quest for more efficient *internal combustion engines*, combined with increasingly stringent emissions standards, has resulted in increased cylinder operating and fuel injection pressures in diesel engines. These conditions produce higher contact stresses in engine key components that must be accommodated by novel materials options that also include gas-pressure-sintered reaction-bonded silicon nitride (GPS-RBSN; Ceralloy® 147-31N) parts for improved wear resistance of the valve train and fuel delivery systems for both light- and heavy-duty diesel engines (Mikijelj *et al.*, 2004). Consequently, the current largest market for silicon nitride components is in reciprocating internal combustion engines, for combustion components and wear parts such as glow plugs, precombustion (swirl) chambers, piston caps, water pump seals, cam rollers; as turbocharger rotors in diesel engines; and as rocker armpads, turbocharger rotors, and exhaust gas control

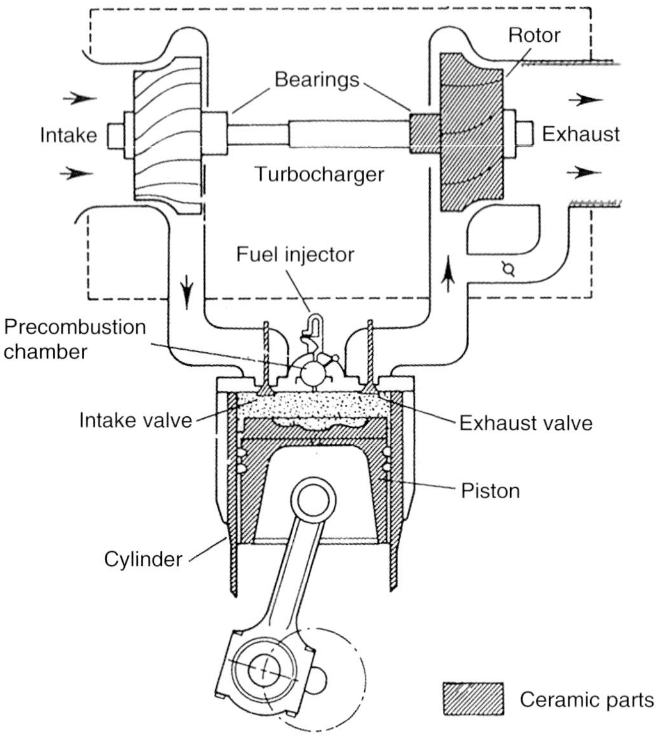

Figure 11.21 Cross-sectional view of the combustion section of an internal combustion engine. The shaded parts are manufactured from silicon nitride (piston head, valve seats, turbocharger rotor), alumina (cylinder liner), or aluminum titanate (exhaust pipe lining).

valves and valve springs for spark-ignition engines. Other applications include heat pumps and pipe linings in future nuclear fusion reactors.

In pressureless sintered or hot-pressed silicon nitride and sialons, the hard ceramic grains are held together by a quinary glass that is formed during reaction with sintering aids, such as alumina and yttria. Consequently, the high-temperature stability of these ceramic bodies is limited by the thermophysical properties of the grain boundary glass to about 1300 °C, even after long-term annealing processes during which the glass is largely converted to crystalline yttrium aluminum garnet (YAG). This limitation – apart from other issues of long-term reliability and cost considerations – impacts severely on the full incorporation of silicon nitride-based ceramics as moving hot-zone engine components. A cross-section of the combustion area of an internal combustion engine, highlighting those parts considered for replacement by silicon nitride or other advanced ceramic compositions, is shown in Figure 11.21.

Gas turbine engine technology is an area likely to benefit most from future growth, as reduced fuel consumption and reduced emissions become ever stronger drivers.

However, the introduction of silicon nitride for stationary blades of gas turbines, burner cans, and nose cone shrouds has been slower than expected, mainly because small to medium-sized turbine engines have yet to find wide application.

Silicon nitride is used for *roller and ball bearings* in applications where corrosion and/or the presence of strong electric and magnetic fields prohibit the use of metals. The wear resistance, low friction and high stiffness of fully dense silicon nitride each leads to improvements in the performance of high-temperature, non-lubricated roller and ball bearings, which preferentially are operated in aggressive environments. For example, hot-pressed silicon nitride (HPSN) bearings have demonstrated an increased service life, a better speed capability and a greater corrosion resistance compared to conventional higher-density steel and hard metal bearings. The vast majority of silicon nitride bearings are used in hybrid ceramic/steel ball bearings; typical applications include machine tool spindles, vacuum pumps, and dental drills.

In *metal working*, dense sintered silicon nitride cutting tool inserts with high hardness, reasonable fracture toughness and high thermoshock resistance, can be used to cut cast-iron, hard steel and nickel-based alloys, with surface speeds that outperform those achieved with conventional materials such as tungsten carbide. The current market value for silicon nitride cutting tools is about $50 million per year; however, since Si_3N_4 is not suited to machining high-silicon aluminum alloys, its future growth will presumably be slow as the automotive industry is poised to move from cast-iron engine blocks to aluminum blocks. Other applications are related to drawing dies and punching tools for wire and sheet metal.

Amorphous thin films of silicon nitride produced by either CVD or PACVD are widely used as efficient barriers against the diffusion of dopant elements such as boron, phosphorus, arsenic and gallium, during the production of silicon-based microelectronic chips. In particular, relatively oxidation-resistant silicon nitride is used as a masking layer controlling the stepwise local oxidation of semiconductor silicon to grow an insulating layer of silicon oxide (Riley, 1996; Figure 11.22). After depositing the silicon nitride masking layer onto doped silicon (step a), the required microstructures are etched (see Heimann, 1982) into the silicon (step b). During thermal oxidation, the exposed silicon is covered by a thin silicon oxide layer (step c). After selective removal of the mask by hot orthophosphoric acid, and of surplus oxide by dilute hydrofluoric acid, a fully recessed field or gate silicon oxide microstructure is obtained (step d).

As silicon nitride has a higher density than amorphous silicon oxide, it will serve as a better barrier to metal atom diffusion, and consequently it is used in VLSI production to prevent the cross-penetration of dopant atoms from one region of the device to another. During the manufacture of devices based on gallium arsenide, silicon nitride coatings will prevent the evaporation of (toxic) arsenic. Further information on the use of silicon nitride in semiconductor processing is available elsewhere (Belyi *et al.*, 1988).

Silicon nitride thin films have also been applied as gate dielectrics for specific MIS memory devices, and as a moisture barrier for organic light-emitting diode (OLED) displays.

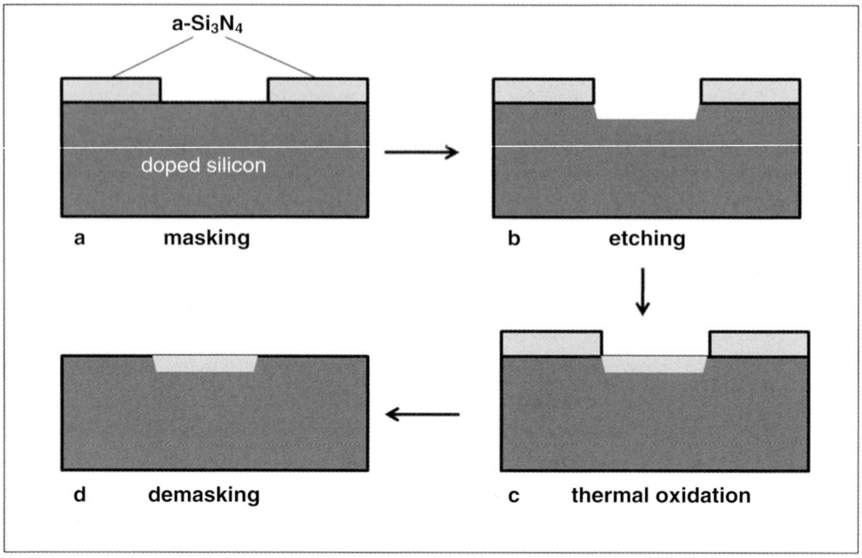

Figure 11.22 Production stages of a fully recessed silicon field or gate oxide island in doped semiconductor silicon.

11.3.4 SiAlONs

11.3.4.1 General Properties and Applications of Sialons

Sialons, as ceramic alloys of silicon nitride and aluminum oxide, were developed as an economically and functionally superior alternative to HPSN (see above). This alloying imposes increased high-temperature mechanical (flexural and tensile strengths, fracture toughness, hardness, wear resistance), thermal (thermal shock resistance) and chemical properties (corrosion and oxidation resistance) compared to unalloyed silicon nitride (see Table 11.13).

Hence, sialons are used for applications where high mechanical and chemical stability and strengths need to be retained over a wide range of temperatures, including roles as cutting tool inserts for machining cast-iron and hard nickel-based superalloys (see below), as heater tubes, as welding and brazing fixtures, and as a variety of tools and bearings in contact with corrosive, nonferrous metal melts; typical examples are metal feeder tubes for the die casting of aluminum, thermocouple sheaths, and burner nozzles.

11.3.4.2 Synthesis and Processing of Sialons

The synthesis of sialons is a complex process, many details of which still have not been elucidated. The reaction of silicon nitride with alumina and/or aluminum nitride involves the formation of intermediate products, and their complete and

Table 11.13 Selected properties of sialon specifications. Values for Si_3N_4 are given for comparison.

Property	Selected SiAlON specifications[a)]					Si_3N_4
	HCN-10	HCH-40	HSN-65	Syalon 101	Syalon 050	
Density ($Mg\,m^{-3}$)	3.26	4.00	3.20	3.23	3.23	3.2
Flexural strength (MPa)	880	830	1050	–	–	600
Fracture toughness ($MPa \cdot m^{1/2}$)	7.5	6	–	7.7	6.5	5.9
Hardness (GPa)	15.5	15.2	14	15	20	14.7
Modulus of elasticity (GPa)	290	330	–	288	300	300
Thermal shock resistance (ΔT)	710	400	–	900	600	>800
CTE ($\times 10^6\,K^{-1}$)	3	3.1	3	3	3.2	3
Thermal conductivity ($W\,m^{-1}\,K^{-1}$)	17	25	65	21	20	>80
Electrical resistivity ($\Omega \cdot m$)	$>10^9$	$\sim 10^7$	$>10^{13}$	10^{10}	10^{10}	$>10^{13}$

a) HCN: Hitachi Metals America, Ltd., Syalon: International Syalons Ltd., Newcastle, UK.

stoichiometric reaction is dominated by time factors and basic materials processing factors. Hence, the synthesis of single phases and complete control of the external conditions are central to the development of specific properties of the various grades and specification of commercially manufactured sialons.

One of the oldest industrial methods used to synthesize sialons (see Section 11.3.4.3) involves the carbothermal nitriding of alumosilicates (e.g., kaolinite) according to:

$$3Al_2[Si_2O_5(OH)_4] + 15C + 5N_2 \rightarrow 2Si_3Al_3O_3N_5 + 6H_2O + 15CO, \tag{11.7}$$

a process resulting in β-sialon $Si_{6-z}Al_zO_zN_{8-z}$ with $z = 3$. Similarly, the reaction of SiO_2-rich puzzolanic ash with metallic aluminum in a stream of flowing nitrogen leads to β-sialon with $z = 4$, according to:

$$2SiO_2 + 4Al + 2N_2 \rightarrow Si_2Al_4O_4N_4 \tag{11.8}$$

Rice hulls can also be processed with comparable results.

During the pressureless sintering of silicon nitride between 1800 and 1900 °C (see Section 11.3.3.2), in the presence of sintering aids such as alumina and yttria, β-sialon with $z = 1$ is formed [Eq. (11.9a)] that, on annealing at 1400 °C, continues to react with the embedding quinary grain boundary Y–Si–Al–O–N glass [Eq. (11.9b)] according to:

$$2Si_3N_4 + Al_2O_3 + Y_2O_5 \rightarrow Si_5AlON_7 + Y{-}Si{-}Al{-}O{-}N \text{ glass} \tag{11.9a}$$

$$Si_5AlON_7 + Y{-}Si{-}Al{-}O{-}N \text{ glass} \rightarrow Si_{5+x}Al_{1-x}O_{1-x}N_{7+x} + Y_3Al_5O_{12} \tag{11.9b}$$

The crystalline YAG phase collects at grain boundaries and acts as a structurally reinforcing constituent.

The second type of sialon, α-sialon (see Section 11.3.4.3), can be synthesized by reacting silicon nitride with aluminum nitride and a stabilizing alkali, alkali earth, or REE ions according to:

$$0.5Ca_3N_2 + 3Si_3N_4 + 3AlN \rightarrow Ca_{1.5}(Si_9Al_3N_{16}) \text{ or} \tag{11.10a}$$

$$CaO + 3Si_3N_4 + 3AlN \rightarrow Ca(Si_9Al_3ON_{15}) \tag{11.10b}$$

The general formula of α-sialon is $M_x^{p+}Si_{12-(m+n)}Al_{(m+n)}O_nN_{16-n}$. Hence, Eq. (11.10a) corresponds to $x = 1.5$, $n = 0$, $m = 3$, and Eq. (11.10b) corresponds to $x = 1$, $n = 1$, $m = 2$, where p relates to the valency of M and $m = p \cdot x$.

Sialon powders produced by one of the techniques discussed above can be shaped by slip casting and subsequently consolidated to a fully dense state by HIPing, or by pressureless sintering in the presence of sintering aids such as yttria, sometimes under addition of seed crystals of β- or α-sialon (Zenotchkine *et al.*, 2004).

11.3.4.3 **Structure of Sialons**

The α- and β-sialons are isostructural with α-Si_3N_4 and β-Si_3N_4, respectively. Substituting Al–O for Si–N in β-Si_3N_4 yields the β-sialons with the general formula $Si_{6-z}Al_zO_zN_{8-z}$ $(0 < z < 4.2)$. The structure is built up by Si and Al tetrahedra coordinated with oxygen and nitrogen (Figure 11.23). The unit cell contains two Si_3N_4 units. The α-sialons are solid solution based on the α-Si_3N_4 structure, with the general formula $M_x^{p+}Si_{12-(m+n)}Al_{(m+n)}O_nN_{16-n}$, where M is a metal ion such as Li, Ca, Ba, Y, REE with a valence of p^+ and $m = p \cdot x$.

The unit cell of α-sialon comprises four Si_3N_4 units that are formed by the replacement of $(m + n)$(Si–N) by m(Al–N) and n(Al–O). As in α-Si_3N_4 (Figure 11.20a), there are two interstitial sites per unit cell in α-sialon, thus constraining x to be ≤2 in the formula given above. The added M^{p+} cations compensate for the valency mismatch created by the replacement of Si by Al. Thus, the range of the stability of α-sialon solid solution has a 2-D extension in the Si_3N–Al_2O_3–AlN-MN plane located in the Jänecke prism (Figure 11.24), which is used to visualize the phase relationships among α-sialon, β-sialon and other phases in the M-Si–Al–O–N system. In the special case M = Y, the Jänecke prism contains a series of stable binary crystalline phases such as YAG ($Y_3Al_5O_{12}$), thortveitite ($Y_2Si_2O_7$), and mullite solid solution ($Al_6Si_2O_{13}$) (Figure 11.25).

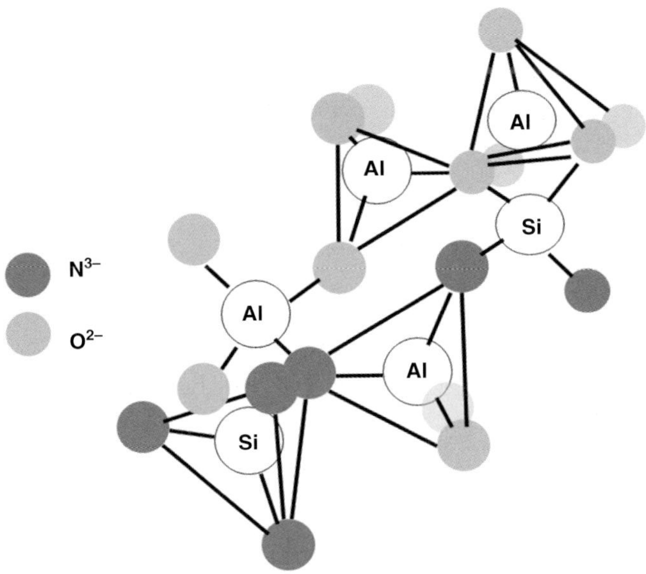

Figure 11.23 Ceramic alloying: the atomic arrangement of tetrahedral units in sialons.

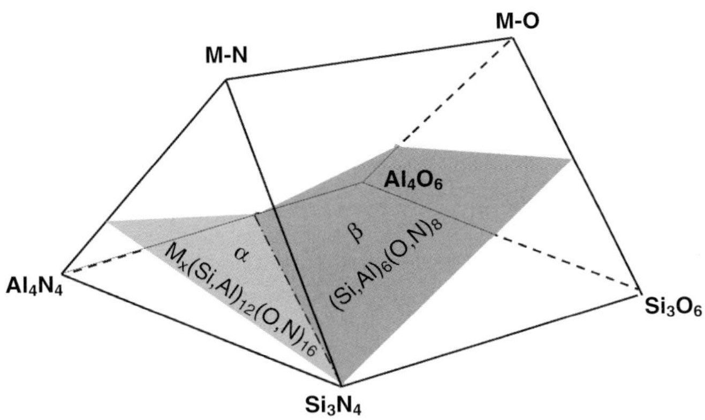

Figure 11.24 Location of α-sialon and β-sialon stability planes in the hexanary M-Si–Al–O–N system.

In the case of β-sialon, the replacement of Si–N by Al–O results only in a very small change of the unit cell, because the Si–N and Al–O bond lengths are very similar. In α-sialon, the replacement of $(m + n)$(Si–N) by m(Al–N) and n(Al–O) results in a considerable increase of the unit cell size, since the Al–N bond (187 pm) is much longer than Si–N (172–174 pm).

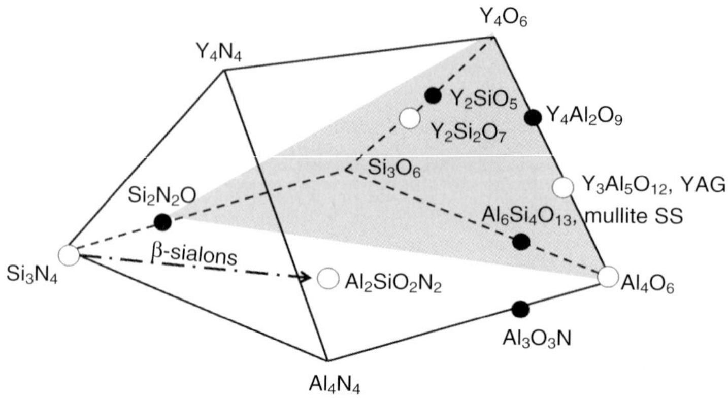

Figure 11.25 Jänecke prism (hexanary system Si_3N_4–Al_2O_3–SiO_2–AlN–YN–Y_2O_3) to visualize major crystalline phases (empty circles) and minor accessory phases (full circles). On the shaded plane Si_2N_2O–Y_4O_6–Al_4O_6 several crystalline solid solutions are situated in which oxygen is partially replaced by nitrogen such as N-apatite, N-wollastonite, and N-melilite (not shown).

The β-sialon phase often has a whisker-like morphology, akin to that of β-Si_3N_4 (see Figure 11.19), that promotes the formation of a tough interlocking microstructure yielding improved mechanical properties, such as increased fracture toughness. Although, during the past two decades both α- and β-sialon ceramics have been intensively studied, until now only β-sialons have been widely used on a commercial basis. Increasingly, α/β-sialons with improved hardness are utilized for modern cutting tools for the high-speed machining of cast-iron or hard superalloys (Figure 11.26). Physically, β-sialon resembles β-Si_3N_4, especially in terms of its coefficient of thermal expansion (2.7 versus 3.5 ppm), but chemically it is similar to α-Al_2O_3.

As in the Si_3N_4 system, there exists a high-pressure γ-sialon corresponding to the spinel-type γ-Si_3N_4 (Sekine *et al.*, 2001; Gross *et al.*, 2005).

11.3.4.4 **Selected Applications of Sialons**

Owing to their mechanically strong, hard, and tough properties, the sialons find application in fields where high-temperature strength and corrosion resistance are required, including:

- Cutting tools
- Grinding media
- Metal wire and tube drawing tools
- Wear components
- Location pins for resistance welding of automotive components
- Seals and bearings
- Metal melt feeder tubes for die casting of non-ferrous metals
- Heater tubes
- Burner nozzles

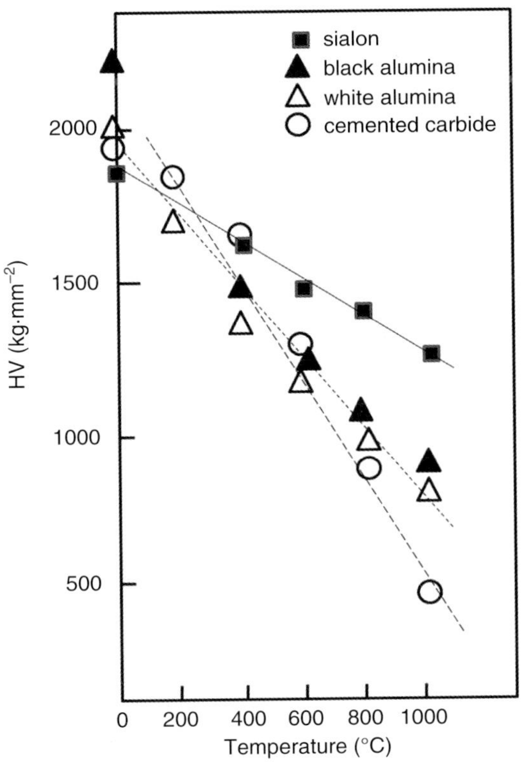

Figure 11.26 Vickers hardness versus temperature of some important hard cutting tool materials.

- Thermocouples sheaths
- Roller bearings in continuous galvanizing

Cutting Tools The hot hardness, fracture toughness and thermal shock resistance of fully dense sialon makes it well suited for use in cutting tools. The material is an attractive low-cost alternative to HPSN for machining gray cast iron for automotive applications, as it provides both an increased machining speed and a longer tool service life compared to conventional cutting tools. Sialons have also replaced cemented carbide tools for the machining of the nickel- and titanium-based superalloys that are used on the basis of their heat resistance in aerospace applications (including gas turbines) and are notoriously difficult to machine. When machining these alloys, pressureless sintered sialon can also increase tool life up to tenfold compared to silicon nitride tools.

Whilst, in the past only β-sialon was utilized, today α/β-sialons with improved hardness are increasingly applied for modern cutting tools in the high-speed machining of cast-iron or hard superalloys. Sialons retain strengths to substantially higher temperatures compared to either cemented carbide or white and black

(Al_2O_3/TiC) alumina, as shown in Figure 11.26. The hot hardness of α/β-sialon increases with the increasing proportion of α-sialon, and at 1000 °C is substantially higher than that of alumina.

References

Acheson, E.G. (1893) Production of artificial crystalline carbonaceous materials. U.S. Patent 492,767.

Adadurov, G.A., Ananin, A.V., Bavina, T.V., and Breusov, O.N. (1982) Method of obtaining diamond and/or diamond-like modification of boron nitride. G.B. Patent 2,090,239.

Akashi, T., Sawaoka, A., and Saito, S. (1977) Effect of multiple shock compression on the structural changes of boron nitride. *Report Res. Lab. Eng. Mater.*, vol. **2**, Tokyo Institute of Technology, Tokyo, Japan, pp. 133.

Alizadeh, A., Taheri-Nassaj, E., Ehsani, N., and Baharvandi, H.R. (2006) Production of boron carbide powder by carbothermic reduction from boron oxide and petroleum coke or carbon active. *Adv. Appl. Ceram.*, **105** (6), 291–296.

Amirthan, G., Udayakumar, A., Bhanu Prasad, V.V., and Balasubramanian, M. (2009) Properties of Si/SiC ceramic composite subjected to chemical vapour infiltration. *Ceram. Int.*, **35**, 2601–2607.

Anceram (2008) Available at: http://www.anceram.de (Accessed 18 December 2009).

Anonymous (2006a) SiC driving interest for power semiconductors. *Semiconductor Today*, **1** (5), 36–39.

Anonymous (2006b) Silicon carbide inverter demonstrates higher power output. *Power Electronics Technology*, 1 February 2006. Penton Media, Inc., 2007. Available at: http://powerelectronics.com/news/silicon-carbide-inverter/. Accessed 19 December 2009.

Auwärter, W., Suter, H.U., Sachdev, H., and Greber, T. (2004) Synthesis of one monolayer of hexagonal boron nitride on Ni(111) from B-trichloroborazine $(ClBNH)_3$. *Chem. Mater.*, **16**, 343–345.

Baik, Y. and Drew, R.A.L. (1996) Aluminum nitride: processing and applications, in *Advanced Ceramic Materials*, (ed. H. Mostaghaci), Key Engineering Materials, Trans Tech Publ. Inc., Zurich, Switzerland, vol. **122–124**, pp. 553–570.

Balakrishnarajan, M.M., Pancharatna, P.R., and Hoffmann, R. (2007) Structure and bonding in boron carbide: the invincibility of imperfections. *New J. Chem.*, **31**, 473–485.

Bao, Y., Zhang, T., Gawne, D.T., and Mason, P. (2008) Quantitative model for the viscous flow and composition of two-phase silicon nitride-based particles in plasma-spray deposition. *J. Eur. Ceram. Soc.*, **29**, 1521–1528.

Batista, R.J.C., Mazzoni, M.S.C., and Chacham, H. (2007) Boron nitride fullerene $B_{36}N_{36}$ doped with transition metal atoms: first principle calculations. *Phys. Rev. B*, **75** (3), 035417.

Baumgart, W. (1984) Hard materials, in *Process Mineralogy of Ceramic Materials* (eds W. Baumgart, A.C. Dunham, and G.C. Amstutz), Ferdinand Enke, Stuttgart, p. 177.

Belyi, V.I., Vasilyeva, L.L., Ginovker, A.S., Gritsenko, V.A., Repinsky, S.M., Sinitsa, S.P., Smirnova, T.P., and Edelman, F.L. (1988) *Silicon Nitride in Electronics, Materials Science Monographs*, vol. **34**, Elsevier, Amsterdam.

Berger, L.-M., Herrmann, M., Nebelung, M., Heimann, R.B., and Wielage, B. (1997) Modifizierte Siliciumnitrid-Kompositpulver für thermische Beschichtungstechnologien und Verfahren zu ihrer Herstellung, Offenlegungsschrift DE 196 12 926 A 1, Deutsches Patentamt, 2 Oktober 1997 (in German).

Berger, L.-M., Herrmann, M., Nebelung, M., Thiele, S., Heimann, R.B., Schnick, T., Wielage, B., and Vuoristo, P. (1998) Investigations on thermal spraying of silicon nitride-based powders, in *Thermal Spray: Meeting the Challenges of the 21st*

Century (ed. C. Coddet), ASM International, Materials Park, OH, pp. 1149–1154.

Berger, L.-M., Herrmann, M., Nebelung, M., Heimann, R.B., and Wielage, B. (2000) Modified composite silicon nitride powders for thermal coatings and process for their production, US Patent 6,110,853 (August 23, 2000).

Bergman, J.P., Janzén, E., and Choyke, W.J. (1998) Multiple bound exciton associated with the nitrogen donor in 3C silicon carbide. *Phys. Status Solidi B*, **210** (2), 407–413.

Bolt, H., Araki, M., Linke, J., Malléner, W., Nakamura, K., Steinbrech, R.W., and Suzuki, S. (1996) Heat flux experiments of first wall mock-ups coated by plasma sprayed B_4C. *J. Nucl. Mater.*, **233** (237), 809–813.

Buzhinskij, O.I., Barsuk, V.A., and Otroshchenko, V.G. (2009) Renewable boron carbide coating in plasma shots of tokamak T11-M. *J. Nucl. Mater.*, **391**, 996–999.

Ceradyne Inc. (2008) Available at: http://www.ceradyne.com/products/armor (Accessed 18 December 2009).

Corrigan, F.R. and Bundy, F.P. (1975) Direct transitions among the allotropic forms of boron nitride at high pressure and temperature. *J. Chem. Phys.*, **6** (9), 3812–3820.

Dastur, A.R., Meneley, D.A., and Verrall, R.A. (1995) Annihilation of plutonium in CANDU reactors. *Proceedings, IAEA Technical Communications Meeting Unconventional Options for Plutonium Deposition, IAEA-TECDOC-840*, p. 103.

Eckardt, T., Malléner, W., and Stöver, D. (1994) Reactive plasma spraying of silicon in controlled nitrogen atmosphere, in *Thermal Spray Industrial Applications* (eds C.C. Berndt and S. Sampath), ASM International, Materials Park, OH, pp. 515–519.

Eckardt, T., Malléner, W., and Stöver, D. (1996) Development of plasma sprayed silicon/silicon nitride coatings by in-situ nitridation, in *Thermische Spritzkonferenz: TS96*, vol. **175** (ed. E. Lugscheider), DVS Berichte, Deutscher Verlag für Schweisstechnik, Düsseldorf, pp. 309–312 (in German).

Edmond, J.A., Kong, H.S., and Carter, C.H. (1993) Blue LEDs, UV photodiodes and high-temperature rectifiers in 6H-SiC. *Physica B. Phys. Condens. Matter*, **185** (1-4), 453–460.

ESK (Elektroschmelzwerk Kempten GmbH) (1996) Method of making polycrystalline dense shaped bodies based on boron carbide by pressureless sintering. European Patent EP 19940109275.

Furukawa, K., Tajima, Y., and Suzuki, A. (1995) Method of producing silicon carbide single crystals by sublimation recrystallization process using a seed crystal. US Patent 5,433,167.

Gao, W., Liu, P., Crandall, R.S., Lee, S.-H., Benson, D., and Branz, H.M. (2000) Approaches for large-area a-SiC:H photovoltaic-powered electrochromic window coatings. *J. Non-Cryst. Solids*, **266–69(b)**, 1140–1144.

Gmelin Handbook of Inorganic and Organometallic Chemistry (1994) Silicon Supplement Volume **B5e**, *Non-Electronic Applications of Silicon Nitride*, (eds W. Kurtz and F. Schröder), Springer, Berlin, pp. 429–441. ISBN 3-540-93651-3.

Gmelin Handbook of Inorganic and Organometallic Chemistry (1991) Silicon Supplement Volume **B5c**, *Silicon Nitride in Microelectronics and Solar Cells*, (eds A. Pebler and F. Schröder), Springer, Berlin, ISBN 3-540-93630-0.

Gross, T., Schwarz, M., Knapp, M., Kroke, E., and Fuess, H. (2005) Thermal expansion of gamma-sialon. Annual Report Hasylab, Hamburg, pp. 551–553.

Hardie, D. and Jack, K.H. (1957) Crystal structure of silicon nitride. *Nature*, **180**, 332–333.

Harris, J.H., Shinde, S.L., Takamori, T., Youngman, R.A., Herron, L.W., and Fasano, B.V. (1996) Low temperature sintering route for aluminum nitride ceramics. US Patent 5,541,145 (July 30, 1996).

Heimann, R.B. (1975) *Auflösung von Kristallen. Theorie und Technische Anwendung*. Springer, Wien, New York. 270 pp.

Heimann, R.B. (1982) Principles of Chemical Etching–the art and science of etching crystals, in *Crystals. Growth, Properties, and Applications*, vol. **8** (ed.

J. Grabmaier), Springer, Berlin, Heidelberg, New York, pp. 173–224.

Heimann, R.B. (2008) *Plasma Spray Coating. Principles and Applications*, 2nd edn, Wiley-VCH Verlag GmbH, Weinheim.

Heimann, R.B. (2010) Thermal spraying of silicon nitride coatings using highly accelerated presursor powder particles. *Surf. Coat. Technol.*, doi: 10.1016/j.surfcoat.2010.02.049.

Heimann, R.B. and Kleiman, J. (1988) Shock-induced growth of superhard materials, in *Crystals. Growth, Properties, and Applications*, vol. **11** (ed. H.C. Freyhardt), Springer, Berlin, Heidelberg, New York, pp. 1–73.

Heimann, R.B., Thiele, S., Berger, L.-M., Herrmann, M., Nebelung, M., Wielage, B., Schnick, T.M., and Vuoristo, P. (1998) Thermally sprayed silicon nitride-based coatings on steel for application in severe operation environments: preliminary results, in *Microstructural Science: Analysis of In-Service Failures and Advances in Microstructural Characterization*, vol. **26** (eds E. Abramovici, D.O. Northwood, M.T. Shehata, and J. Wylie), ASM International, Materials Park, OH, pp. 389–394.

Heimann, R.B., Thiele, S., Wielage, B., Zschunke, M., Herrmann, M., and Berger, L.M. (1997) Entwicklung von Schichten auf Siliziumnitrid-Basis als HT-erosionsbeständiger Werkstoff für thermische Anlagen. *Freiberger Forschungshefte (Maschinenbau)*, **A848**, 166–182.

Heimann, R.B., Thiele, S., Berger, L.-M., Herrmann, M., and Vuoristo, P. (1999) Detonation spraying of silicon nitride for high temperature erosion-resistant ceramic coatings. *Proceedings. Coatings for Aerospace and Automotive Industries (CAAI-6), Toronto, October*, pp. 14–15.

Hoffmann, M.J., Becher, P.F., and Petzow, G. (eds) (1994) *Silicon Nitride 93*, vol. **89–91**, Key Engineering Materials, Trans Tech Publ. Inc., Zurich, Switzerland.

Holleck, H. (1991) Designing advanced coatings for wear protection. *Surf. Eng.*, **7** (2), 137–144.

Hopkins, G.R. (1974) Silicon carbide and graphite materials for fusion reactors. *Proceedings, IAEA Plasma Physics and Controlled Nuclear Fusion, IAEA-CN-33/s*, vol. 2, pp. 275–281.

Horne, O.J., Jr and Ramsey, D.E., Jr (1981) Production of SiC whiskers. US Patent 4,283,375.

Ichinose, N. (ed.) (1987) *Introduction to Fine Ceramics. Applications in Engineering*, John Wiley & Sons, Ltd, Chichester, ISBN: 0-471-91445-2.

Imamura, H., Hirao, K., Brito, E., Toriyama, M., and Kanzaki, S. (2000) Further improvement in mechanical properties of high anisotropic silicon nitride ceramics. *J. Am. Ceram. Soc.*, **83** (3), 495–500.

Inoue, Y., Nariki, Y., and Tanaka, K. (1989) Mechanism of production of ultra-fine silicon carbide powder by arc plasma irradiation of silicon bulk in methane-based atmospheres. *J. Mater. Sci.*, **24** (11), 3819–3823.

Izhevskyi, V.A., Genova, L.A., Bressiani, C., and Bressiani, A.H.A. (2000) Review article: Silicon Carbide. Structure, properties and processing. *Cerâmica (São Paulo, Brazil)*, **46** (297), 4–13.

Jansen, M., Schön, J.C., and van Wüllen, L. (2006) Der Weg zur Struktur amorpher Festkörper–eine Studie am Beispiel der Keramik $Si_3B_3N_7$. *Angew. Chem.*, **118**, 4350–4370.

Janz, S. (2006) Amorphous silicon carbide for photovoltaic applications. Doctoral dissertation, University Konstanz, Germany. Fraunhofer Institute für Solare Energiesysteme, Freiburg, Germany.

Kim, D.P., Gofer, C.G., and Economy, J. (2005) Fabrication and properties of ceramic composites with a boron nitride matrix. *J. Am. Ceram. Soc.*, **78** (6), 1546–1552.

Kitamura, J., Usuba, S., Kakudate, Y., Yokoi, H., Yamamoto, K., Tanaka, A., and Fujiwara, S. (2003) Formation of boron carbide coating by electromagnetically accelerated plasma spraying. *Surf. Coat. Technol.*, **169–170**, 324–327.

Kraus, B.D., Moore, J.T., and DeBoer, S.J. (2002) Batch processing for semiconductor wafers to form aluminum nitride and titanium aluminum nitride. US Patent 6,265,519 (April 2, 2002).

Krstic, V.D., Vlajic, M.D., and Verrall, R.A. (1996) Silicon carbide ceramics for nuclear application, in *Advanced Ceramic Materials*, (ed. H. Mostaghaci), Key

Engineering Materials, Trans. Tech. Publ. Inc., Zurich, Switzerland, vol. **122–124**, pp. 387–396.

Kucuk, A., Lima, R.S., and Berndt, C.C. (2000) Composite coatings of Si_3N_4-soda lime silicate produced by the thermal spray process. *J. Mater. Eng. Perform.*, **9**, 603–608.

Lee, M.R., Russell, S.S., Arden, J.W., and Pillinger, C.T. (1995) Nierite (Si_3N_4): a new mineral from ordinary and enstatite chondrites. *Meteoritics*, **30**, 387–398.

Leitch, S., Moewes, A., Ouyang, L., Ching, W.Y., and Sekine, T. (2004) Properties of non-equivalent sites and band gap of spinel-phase silicon nitride. *J. Phys. Condens. Matter*, **16** (26), 6469–6476.

Li, X.L., Ma, H.A., Zheng, Y.J., Liu, Y., Zuo, G.H., Liu, W.Q., Li, J.G., and Jia, X. (2008) AlN ceramics prepared by high-pressure sintering with La_2O_3 as a sintering aid. *J. Alloys Compd.*, **463** (1–2), 412–416.

Limbach, R. (1992) Development of thermally sprayed wear resistant coatings with favorable tribological behavior, in *Techn.-Wiss. Berichte Lehr- und Forschungsgebiet Werkstoff-Wissenschaften der RWTH Aachen*, Nr. 37.03.12.92 (ed. E. Lugscheider), RWTH, Aachen (in German).

Lin, H.T., Plucknett, K., and Brito, M.E. (2006) *Silicon Based Structural Ceramics for the New Millenium*, vol. **142**, Ceramic Transactions, John Wiley & Sons, Inc.

Lugscheider, E., and Limbach, R. (1990) Plasma spraying of agglomerated powders on the basis of Si_3N_4, *Dtsch. Verlag Schweisstechnik. Ber.*, **130**, 224–225.

Lugscheider, E., Limbach, R., Liden, A., and Lodin, J. (1990) Plasma spraying of silicon nitride (Si_3N_4), in *Proceedings, Conference on High-Temperature Materials and Powder Engineering.*, Liege, Belgium, vol. **1** (ed. F. Bachelet), Kluwer Academic Publishers, Dordrecht, The Netherlands, pp. 877–880.

Maccagno, T.M. (1989) Processing of advanced ceramics which have potential for use in gas turbine aero engines. Aeronautical Note, NAE-AN-58, NRC No. 30057, National Research Council Canada: Ottawa, February 1989.

Madar, R. (2004) Material science: silicon carbide in contention. *Nature*, **430**, 974–975.

Makarenko, G.N. and Mahajan, Y.R. (2002) Features of carbothermal synthesis of aluminum nitride under nitrogen pressure. *Powder Metall. Met. Ceram.*, **41** (5–6), 1068–1302.

Marckx, D.A. (2006) Breakthrough in power electronics from SiC. Report NREL/SR-500-38515, March 2006. National Renewable Energy Laboratory, Golden CO, USA.

Martinez, V., Valencia, M.F., Cruz, J., Mejia, J.M., and Chejne, F. (2005) Production of β-SiC by pyrolysis of rice husk in gas furnace. *Ceram. Int.*, **32** (8), 891–897.

Marx, D., Seidler, G., Fister, T., Nagle, K., and Segre, C. (2008) Structure of boron carbide: where's the carbon?, in *Proceedings, APS Meeting, Session Q40: Metals: Nanoparticles, Compounds and Thermodynamics, March 10–14, New Orleans, Louisiana*, Abstract Q40.00010.

Matchen, B. (1996) Applications of ceramics in armor products, in *Advanced Ceramic Materials*, vol. **122–124** (ed. H. Mostaghaci), pp. 333–342.

Matweb (2008) Available at: www.matweb.com/ (Accessed 19 December 2009).

McCulloch, D.G. (2003) Wannier function analysis of tetrahedral amorphous networks. *Diamond Relat. Mater.*, **12**, 2026–2031.

Michalowski, L. (1994) *New Ceramic Materials*, Deutscher Verlag für Grundstoffindustrie, Leipzig, Stuttgart (in German).

Mikijelj, B., Mangels, J., Belfield, E., and MacQueen, A. (2004) Silicon nitride applications in modern diesel engines. Report Ceradyne 2004-01-1448.

Mishima, O. and Era, K. (2000) Science and technology of boron nitride, in *Electric Refractory Materials* (ed. Y. Kumashori), Marcel Dekker, Inc., Ch. 21, pp. 495–556.

Mörgenthaler, K.D., and Bühl, H. (1994) Ceramics for engines, in *Tailoring of Mechanical Properties of Ceramics* (eds M.J. Hoffmann and G. Petzow), NATO ASI Series E, Applied Sciences, Kluwer, Dordrecht, The Netherlands, pp. 429–441.

Müller, St-G., Glass, R.C., Hobgood, H.M., Tsvetkov, V.F., Brady, M., Henshall, D., Malta, D., Singh, R., Palmour, J., and Carter, C.H., Jr (2001) Progress in the industrial production of SiC substrates for semiconductor devices. *Mater. Sci. Eng. B*, **80** (1–3), 327–331.

Nakamura, D., Gunjishima, I., Yamaguchi, S., Ito, T., Okamoto, A., Kondo, H., Ondo, S., and Takatori, K. (2004) Ultrahigh-quality silicon carbide single crystals. *Nature*, **430**, 1009–1012.

Nogami, S., Hasegawa, A., Murayama, T., Otake, N., Satou, M., and Abe, K. (2008) Compatibility between SiC and Li ceramics for solid breeding blanket system. *J. Nucl. Mater.*, **386–388**, 628–630.

Okada, T., Toriyama, M., and Kanzaki, S. (2000) Direct nitridation of aluminum compacts at low temperature. *J. Mater. Sci.*, **35** (12), 3105–3111.

Oyanagi, N. (2003) Method for growth of silicon carbide single crystal, silicon carbide seed crystal, and silicon carbide single crystal (sic!). WO/2004(111318).

Page, T.F. (1990) *The Physics and Chemistry of Carbides, Nitrides and Borides*, NATO ASI Series E, no. 185, (ed. R. Freed), Kluwer Academic Publishers, Dordrecht, pp. 197–214.

Pätzold, C. (2001) Herstellung, Charakterisierung und neue Einsatzgebiete von keramischen Nitrid- und Oxinitridpulvern. Dissertation, Technische Universität Bergakademie, Freiberg.

Peng, H. (2004) Spark plasma sintering of Si_3N_4-based ceramics. Ph.D. Thesis, Department of Inorganic Chemistry, Stockholm University, Sweden (urn_nbn_se_su_diva-129.1_fulltext.pdf).

Petzow, G. and Herrmann, M. (2002) Silicon nitride ceramics, in *High Performance Non-Oxide Ceramics II. Structure and Bonding*, vol. **102** (eds M. Jansen, D.M.P. Mingos, and R. Haubner), Springer, Berlin, pp. 47–168.

Riley, F.L. (1996) Application of silicon nitride ceramics, in *Advanced Ceramic Materials* (ed. H. Mostaghaci), vol. **122/124**, pp. 479–488.

Rosenband, V., and Gany, A. (2004) Activation of combustion synthesis of aluminum nitride powder. *J. Mater. Proc. Technol.*, **147** (2), 197–203.

Roy, R. (1996) Ceramic packaging for electronics, in *Advanced Ceramic Materials*, vol. **122–124** (ed. H. Mostaghaci), pp. 17–34.

Sangster, R.C. (2005) *Formation of Silicon Nitride, From the 19th to the 21st Century.* Materials Science Foundations (Monograph Series), Trans. Tech. Publications, Zurich, Switzerland, pp. 22–24 and 948–952. ISBN 0-878-49492-8.

Sato, T., Ishii, T., and Setaka, N. (1982) Formation of cubic boron nitride from rhombohedral boron nitride by explosive shock compression. *J. Am. Ceram. Soc.*, **65**, 162–165.

Savrun, E. (2002) Packaging considerations for very high temperature microsystems. *Proceedings, IEEE Sensors 2002, Orlando, FL, 12–14 June 2002*, vol. 2, pp. 1139–1143.

Sekine, T., He, H., Kobayashi, T., Tansho, M., and Kimoto, K. (2001) Cubic $Si_{6-z}Al_zO_zN_{8-z}$ (z = 1.8 and 2.8) spinels formed by shock compression. *Chem. Phys. Lett.*, **344** (3-4), 395–399.

SiC 10 (2009) SiC 10 – How SiC will impact electronics: A 10 year projection. Bharat Book Bureau, Executive summary, p. 4. Available at: http://www.bharatbook.com/Market-Research-Reports/SiC-10-How-SiC-will-impact-electronics-A-10-year-project.

Shiomi, H., Kimoto, T., and Matsunami, H. (2004) SiC wafer, SiC semiconductor device, and production method of SiC wafer. US Patent 6,734,461.

Singh, R. and Pecht, M. (2008) Commercial impact of silicon carbide. *Ind. Electron. Mag. IEEE*, **2** (3), 19–31.

Slack, G.A. and McNelly, T.F. (1976) Growth of high purity AlN crystals. *J. Cryst. Growth*, **34**, 263–279.

Slashed, S., Le Kjell, A.O., and Velken, S.V. (2000) Process for producing silicon carbide. US Patent 6,022,515.

Sodeoka, S., Ueno, K., Hagiwara, Y., and Kose, S. (1992) Structure and properties of plasma-sprayed SiAlON coatings. *J. Therm. Spray Technol.*, **1**, 153–159.

Stolle, R. and Wahl, G. (1995) Deposition of boron nitride films from BB′B″-trichloroborazine. European Conference on CVD, 10, Venice, Italy. *J. Phys. IV*, **5** (2), C5.761–C5.768.

Sun, J., Liu, C., and Wang, R. (2009) Low pressure hot pressing of B_4C matrix ceramic composites improved by Al_2O_3 and TiC additives. *Mater. Sci. Eng. A.*, **519** (1–2), 27–31.

Tairov, Y.M. and Tsvetkov, V.F. (1981) General principles of growing large-size single crystals of various silicon carbide polytypes. *J. Cryst. Growth*, **52**, 146–150.

Tani, E., Soma, T., Sawaoka, A., and Saito, S. (1975) Stability of wurtzite-type boron nitride under high pressure. *Jpn. J. Appl. Phys.*, **14** (10), 1605–1606.

Taniyasu, Y., Kasu, M., and Makimoto, T. (2006) An aluminum nitride light-emitting diode with a wavelength of 210 nanometers. *Nature*, **441**, 325–328.

Thiele, S., Heimann, R.B., Herrmann, M., Berger, L.-M., Nebelung, M., Zschunke, M., and Wielage, B. (1996) Thermal spraying of silicon nitride-based powders, in *Thermal Spray: Practical Solutions for Engineering Problems* (ed. C.C. Berndt), ASM International, Materials Park, OH, pp. 325–331.

Thiele, S., Heimann, R.B., Berger, L.-M., Herrmann, M., Nebelung, M., Schnick, T., Wielage, B., and Vuoristo, P. (2002) Microstructure and properties of thermally sprayed silicon nitride-based coatings. *J. Therm. Spray Technol.*, **11** (2), 218–225.

Tomota, T., Miyamoto, N., and Koyama, H. (1988) Formation of Thermal Spraying Ceramic Layer, Patent JP 63 169371, A. Int. Cl.[4]: C23C 4/10, Filing date 29.12.1986.

Toraya, H. (2000) Crystal structure refinement of α-Si_3N_4 using synchrotron radiation powder diffraction data: unbiased refinement strategy. *J. Appl. Crystallogr.*, **33**, 95–102.

Troczynski, T.B. and Nicholson, P.S. (1989) Effect of additives on the pressureless sintering of aluminum nitride between 1500 °C and 1800 °C. *J. Am. Ceram. Soc.*, **72** (8), 1488–1491.

Turkdogan, E.T., Bills, P.M., and Tippett, V.A. (1958) Silicon nitrides: some physico-chemical properties. *J. Appl. Chem.*, **8**, 296–302.

Usuba, S. and Heimann, R.B. (2006) Dense Si_3N_4 coatings with high friction coefficient deposited by high-velocity pulsed plasma spraying. *J. Therm. Spray Technol.*, **15** (3), 356–363.

Van der Laan, J.G., Schnedecker, G., van Osch, E.V., Duwe, R., and Linke, J. (1994) Plasma sprayed boron carbide coatings for first wall protection in ITER. *J. Nucl. Mater.*, **211**, 135–140.

Von Hocking, E.F., Krimmel, E.F., Kurtz, W., Popper, P., and Schröder, F. (eds) (1994) *Non-Electronic Applications of Silicon Nitride*, 8th edn, Springer, Berlin.

Wang, J. and Marshall, D.S. (2001) Defect formation in boron carbide–an ab-initio electronic structure study. *Proceedings, Fall Meeting MRS, Boston, 26–30 November, Symposium G, paper G8.6*. Materials Research Society. Available at: http://www.mrs.org/s_mrs/

Ward, D.J. and Dudaev, S.L. (2008) Economically competitive fusion. *Mater. Today*, **11** (12), 46–53.

Watari, K., Tsuzuki, A., and Torii, Y. (1992a) Effect of rare-earth oxide addition on the thermal conductivity of sintered aluminum nitride. *J. Mater. Sci. Lett.*, **11** (22), 1508–1510.

Watari, K., Ishizaki, K., and Fujikawa, T. (1992b) Thermal conduction mechanism of aluminum nitride ceramics. *J. Mater. Sci.*, **27**, 2627–2630.

Wilkins, M.L., Cline, C.F., and Honodel, C.A. (1969) Fourth progress report of light armor program. Lawrence Livermore Laboratory, Livermore, CA, Report No. UCRL 50694.

Will, G., Kirfel, A., and Josten, B. (1986) Charge density and chemical bonding in cubic boron nitride. *J. Less-Common Met.*, **117**, 61–71.

Willander, M., Friesel, M., Wahab, Q., and Straumal, B. (2006) Silicon carbide and diamond for high temperature devise applications. *J. Mater. Sci.: Mater. Electron.*, **17** (1), 1–25.

Yu, J., Chen, Y., Elliman, R.G., and Petrovic, M. (2006) Isotopically enriched ^{10}BN nanotubes. *Adv. Mater.*, **18**, 2157–2160.

Zenotchkine, M., Shuba, R., and Chen, I.W. (2004) Liquid-phase growth of small crystals for seeding α-SiAlON ceramics. *J. Am. Ceram. Soc.*, **86** (6), 1040–1046.

Zerr, A., Miehe, G., Serghiou, G., Schwarz, M., Kroke, E., Riedel, R., Fueß, H., Kroll, P., and Boehler, R. (1999) Synthesis of cubic silicon nitride. *Nature (London)*, **400** (6742), 340–342.

Zhu, X. and Sakka, Y. (2008) Textured silicon nitride: processing and anisotropic properties. *Sci. Technol. Adv. Mater.*, **9**, 033001. (47 pp).

12
Advanced Ceramic Processing and Future Development Trends

12.1
Design of Monodisperse Ceramic Powders

Many of the current limitations to the performance of ceramics in high-technology areas are a direct consequence of non-optimum physico-chemical properties of the powders used in their fabrication. The average particle size, particle size distribution, degree of agglomeration, and dispersion characteristics of the precursor powders combine to limit the macroscopic homogeneity of the ceramics in the green and final sintered states. The microstructural homogeneity of the ceramics can be optimized by processing powders with tightly controlled characteristics. Monodisperse metal oxide powders derived from sol–gel processes are prime examples of these materials. Such powders offer many processing advantages over conventional ceramic powders, since their spherical shape and narrow particle size distribution allow for a tight control over the packing density of the powder particles in the green ceramic. Since there are very few agglomerates, the particles pack very uniformly, with residual pore sizes in the green ceramic often being on the order of just two particle diameters – that is, less than 1 μm. While dispersions of those powders settle only slowly, very high green densities are attainable due to statistical ordering of the particles with residual pore size that often are on the order of one particle diameter. This is a crucial advantage, since residual porosity or agglomerates in a fired structural ceramic will act as crack initiation sites or create defects in the material, hence lowering its overall performance and strength. In processing conventional ceramic powders, it is very difficult to eliminate any residual porosity and flaws that might result from poor powder packing and the presence of agglomerates. Densification is typically accomplished by applying pressure and/or adding sintering aids to control grain growth (see Chapter 2). Processing by hot isostatic pressing ("HIPing") is extremely costly and therefore undesirable, and is also limited to simple geometric shapes. The addition of sintering aids can adversely affect the high-temperature properties of the material by lowering the melting point, and therefore is also undesirable. With monosized powders, such processing steps are not necessary to achieve good densities, microstructural uniformity, and superior sinterability.

Classic and Advanced Ceramics: From Fundamentals to Applications. Robert B. Heimann

ISBN: 978-3-527-32517-7

Many methods of synthesizing monodisperse powders via sol–gel technologies have been described in the past. To quote only one typical example, Olsen and Liss (1988) outlined a route for the preparation of monodisperse submicron powders of titania, via a sol–gel route, that possessed a spherical shape, uniform dimensions, and a low state of aggregation. The powder was prepared by mixing equal volumes of ethanolic solutions of titanium tetraethoxide and water, and stirring briefly. Powder isolation procedures by centrifuging must be initiated within 30 minutes of the onset of precipitation; otherwise, undesirable hard-necked aggregates of individual powder particles would be formed. The powder was then washed with ethanol, followed by washing with alkaline water to impart a negative charge to the surface of the titania particles, thereby providing a net repulsive interparticle potential that effectively inhibits flocculation.

12.2
Ceramic Processing in a Microgravity Environment

The previous statements notwithstanding, ceramic processing science has progressed little beyond the needs of traditional ceramics. Yet, the emergence of new and improved technologies hinge critically on rational improvements in the wet (or suspension) processing of ceramic powders (Lind, 1978). As indicated above, previous investigations have focused on model systems consisting of single-phase, monosized ceramic particles. However, relevant novel ceramic systems will either be single or multiphase with engineered particle size distributions, for example, bimodal or even trimodal. Hence, a fundamental understanding of these systems is required. The development of this science base could benefit significantly from research conducted in a microgravity environment, as density-driven phase separation and/or particle size-induced segregation effects will be dramatically reduced relative to ambient–that is, $g = 1$ – conditions.

Microgravity conditions offer the possibility of producing composite structures that are either impossible or not easily obtainable on Earth. Of particular interest are ceramic–metal or other composites consisting of phases of widely differing densities, and also composites which consist of discrete areas of a second phase contained within a major phase. In the first case, approaches to fabricating metal–ceramic composites on Earth are restricted by density differences, and thus the segregation of phases during fabrication. For example, efforts to impregnate ceramic preforms or to mix ceramic and metal powders or other precursors are impeded by problems associated with ceramic preforms floating to the surface and the metal phases being nonuniformly distributed. Processing approaches which allow uniform mixtures to be prepared on Earth and heat-treated under microgravity conditions, should allow the formation of layered ceramic–metal structures, uniform interconnecting two-phase structures, or homogeneous particulate-reinforced structures. The removal of gravity-driven phase segregation will allow wetting interactions between two or more phases to be more clearly understood.

In the second case, mixing of whiskers or other second-phase particles with ceramic powders to produce composites is difficult, and can result in nonuniform structures. However, mixing powders and whiskers in an aqueous medium in space would allow the preparation of more uniform, novel, or controlled structures, because settling due to gravitational forces could be completely avoided, or at least controlled. For example, mixing can be achieved by the application of mechanical force only without the effects of convection. Although an aggregation of particles would occur, the clusters would not settle as they would in the Earth's gravity; consequently, it should be possible to grow clusters of certain sizes and to engineer layered structures by artificially applying a gravitational force of specified duration (e.g., in a centrifuge), or by filtration. A uniform composite containing dispersed clusters or individual particles in a major phase could also be prepared by filter-pressing.

The combustion synthesis of ceramics generates high temperatures at and ahead of the reaction front (Merzhanov, 2004). In turn, these high temperatures generate liquids (and possibly also gases) which are subject to gravity-driven flow. Heat loss occurs exclusively through radiation, since convection is lost under zero-gravity conditions. The removal of such gravitational effects is likely to provide an increased control of the reaction front, with a consequent improvement in control of the microstructure of the synthesized ceramic products. Currently, experiments are being conducted to determine the role that gravity and a lack of gravitational forces, respectively, play in the synthesis process such as the self-propagating high-temperature synthesis (SHS; see Section 12.4) of tricalcium phosphate under low-gravity conditions for biomedical applications (Castillo *et al.*, 2003).

12.3
Powder Preparation under Reduced Gravity

The preparation of ceramic powders by sol–gel technology under ambient gravity conditions is plagued by many problems that relate to rapid flocculation and particle sedimentation, as well as inhomogeneous particle size distribution. Hence, it is desirable to prepare powders with a spherical shape and a uniformly small size that will possess superior space-filling properties. Such properties would allow the production of ceramic green bodies, and also of thin films with a high packing density. The quest for very small, ideally spherical powders with extremely narrow grain size distributions would best be met in an environment where gravity-driven convection can be minimized. Hence, the reduction of convective flows in a microgravity environment would benefit the preparation of high-quality superconducting and ferroelectric ceramics film, particularly when epitaxy is desired. Such microgravity conditions can be provided, with increasing durations of reduced gravity conditions, by drop towers, aircraft in parabolic trajectories, sounding rockets, and Earth-orbiting laboratories such as the NASA-operated Spacelab, and the International Space Station (ISS). However, the microgravity research studies conducted to date by space-faring nations have been limited by the complexity of the experiments, the time constraints, and the huge expenditures that are not in the least caused by the need for extreme redundancy in safe operability in space. Early accounts of research opportunities for materials processing in space have been provided by the National Research Council (1978) of the USA (see also Elwell, 1978; Biswas, 1989).

In the field of ceramics, a low-gravity environment will reduce buoyancy-driven flow in liquids. Most ceramic syntheses and processing are carried out at high temperature, either by true solid-state processes or by processes such as sintering, where only a small amount of viscous liquid phase is present. The preparation of ceramic powders from aqueous solutions or colloidal suspensions requires the stringent control of supersaturation, so that powders with desirable properties can be synthesized. Research and development studies of ceramic processing under reduced gravity conditions may yield potential advantages, in particular for high-quality bulk ceramics and thin films for sensors, microelectronic memory devices, and antireflection coatings. These areas could include (Sirignano, 1995):

- The processing of ceramics at high temperatures requires refractory containers that may react with the ceramic through the formation of eutectic melts with low melting temperatures. A general capability for containerless, high-temperature processing to at least 1800 °C would allow the contamination-free synthesis of ceramics and glasses (e.g., for optoelectronic applications) and for studying homogeneous nucleation and crystal growth unaffected by a container wall that would introduce heterogeneous nucleation.
- Fundamental studies on the kinetics of crystal nucleation and growth in glassy melts under the influence of surface tension-driven (Marangoni) convection that, under 1 g conditions, would be superseded by gravity-driven (Bénard or Rayleigh) convection.
- The determination of accurate data on diffusion in ceramics melts unavailable under terrestrial conditions, where mass transport is affected by convective forces.
- The suppression of evaporation from the free surface of a melt during single crystal growth, using Czochralski and Bridgman techniques that would allow the use of higher temperatures than would be achievable under terrestrial gravity conditions.
- The epitaxial growth of single crystal films from aqueous solution or fluxes on oriented crystal surfaces, including the biomimetic synthesis of self-assembling monolayers of functional advanced ceramics.
- The synthesis of microsized to nanosized ceramic powders with a spherical geometry and a narrow grain size distribution, from aqueous solutions or colloidal suspensions.

At the peak of the high-temperature superconductor (HTSC) ceramic hype during the late 1980s (see Chapter 9), the quest for YBCO (yttrium barium copper oxide)-based materials with improved critical temperature T_c and critical current density j_c was also extended to processing in space. In particular, films of HTSCs require a low proportion of nonconducting pathways – that is, a reduction of the grain boundaries and their interaction – so as to yield improved current-carrying properties and an increased resistance to high magnetic fields. Owing to the absence of gravity-driven convection, mass transport in the bulk of the growth medium would be slow, and consequently the particle size would be small. Furthermore, the growth of particles formed by homogeneous nucleation that are suspended in the growth medium would result in an ideal sphericity. The preparation of such superconducting powders from solution requires the rapid introduction and mixing of reagents into a vigorously stirred mixing cell. The reaction would then occur in microseconds to milliseconds, so that the convection and particle settling would be minimized.

In order to obtain information on the general trends of the kinetics of particle precipitation from aqueous solutions in the absence of gravity, a simple model system was adopted by Heimann in 1989 based on the hydrolytic polycondensation of titanium tetraisopropoxide, $Ti(iPrO)_4$, dissolved in water-free toluene with 5% isopropanol added, and injected into an aqueous solution. A short period of microgravity (20–30 s) was achieved during a parabolic flight trajectory of an aircraft. The experiment was statistically designed according to a Box–Behnken response surface methodology with three parameters, and flown by NASA on a KC 135 plane in October 1989 under the direction of the Canadian Space Agency (CSA). The reactions occurring during hydrolysis are rather complex, but can be summarized (Yoldas, 1986) as:

$$\equiv \text{Ti-OR} + H_2O \rightarrow \equiv \text{Ti–OH} + \text{ROH} \tag{12.1}$$

$$\equiv \text{Ti–OH} + \text{RO-Ti} \equiv \rightarrow \equiv \text{Ti–O–Ti} \equiv + \text{ROH}, \tag{12.1a}$$

where $R = CH_3(CH_2)_2–$.

Precipitation in a microgravity environment was monitored by the attenuation of a laser beam directed via optical windows through the magnetically stirred solution in the mixing vessel. Figure 12.1, which is based on unpublished research conducted by Heimann (1990), shows the light intensity (in μW) of the recorded laser beam for $g \approx 0$ and $g = 1$. In the absence of gravity, and immediately after injection of the $Ti(iPrO)_4$ solution ($t = 0$), very strong fluctuations of the laser light intensity were recorded that reflected both the microgravity-induced levitation of the precipitated $TiO(OH)_2$ particles and their vigorous swirling around throughout the volume of the reaction vessel. In contrast, under normal gravity conditions ($g = 1$) the particles settled very quickly, within about 5 s.

In this experiment, the total duration of zero gravity[1] lasted for about 20 s. Scanning electron microscopy (SEM) images of the precipitated particles are shown in Figure 12.2. The size of the particles, and their deviation from sphericity, were

1) In fact, the reduced gravity achieved during the flight of the aircraft along a parabolic trajectory was not exactly zero, but about 10^{-4} g.

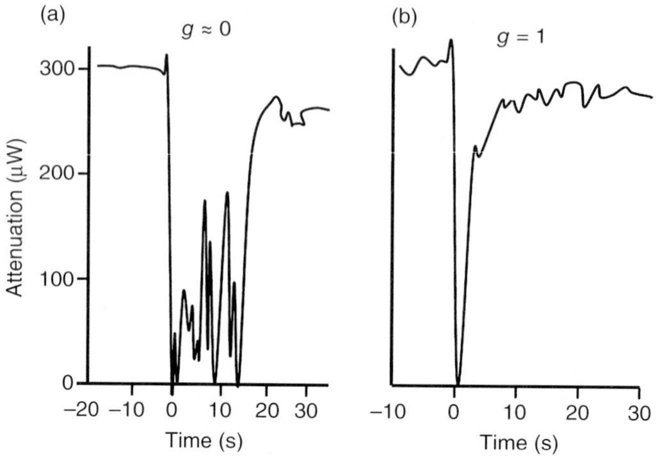

Figure 12.1 Laser light attenuation during precipitation of titanium oxyhydroxide particles in the absence (a) and presence (b) of gravity. The $Ti(iPro)_4$ concentration was 2 mmol, and the stirring rate 500 r.p.m (Heimann, 1990).

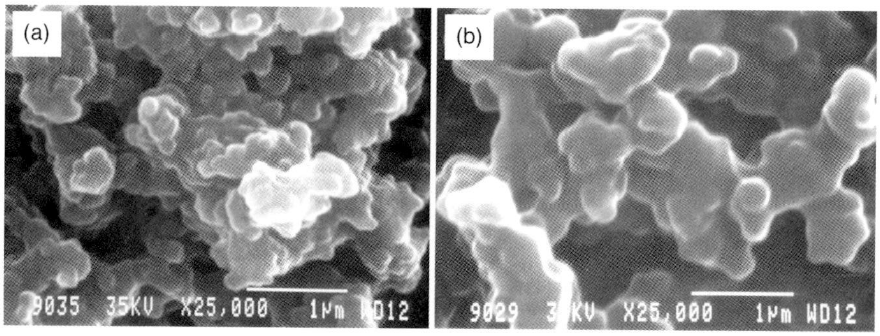

Figure 12.2 Titanium oxyhydroxide particles precipitated through hydrolysis and polycondensation of titanium tetraisopropoxide at (a) $g \approx 0$ and (b) $g = 1$. Stirring rate $\omega = 250$ r.p.m., concentration $c = 4$ mmol (Heimann, 1990).

much larger when formed at $g = 1$ (Figure 12.2b) compared to those formed at $g \approx 0$ (Figure 12.2a), the average size of which was about 200 nm.

However, larger agglomerates were formed after a return to normal gravity at the end of the flight parabola, because the time of 20 s was not sufficiently long to complete the precipitation process (Figure 12.2a). Consequently the small, spherical particles formed at zero gravity were cemented together by material deposited *post factum*. This underscores the need for efficient particle separation, for example, by centrifuging (as noted above) or by shock freezing of the aqueous solution within a few milliseconds. To account for this effect, the recovered samples were subjected to a very gentle milling procedure and their grain size distributions investigated by laser diffraction, with statistical grain population parameters such as sorting σ^*, skewness α^* (2nd momentum), and kurtosis β^* (4th momentum) being determined. The influence of the three experimental parameters g (level of gravity; range: 0, 1/6–lunar gravity, 1), ω (rotation speed of the magnetic stirrer; range: 250, 500, 1000 r.p.m), and c (concentration of $Ti(iPrO)_4$; range: 2, 4, 8 mmol) on the three response parameters σ^*, α^* and β^* was tested in a polynomial model. Sorting σ^* was found to increase only moderately with increasing two-factor interaction ($\omega \cdot g$), but very strongly with decreasing g^2. Skewness increased moderately with decreasing ($\omega \cdot g$) and decreasing g^2. Kurtosis increased only weakly with decreasing ($c \cdot \omega$), and thus was independent of the level of gravity. In conclusion, the gravity level g was shown to strongly control both sorting and skewness in a parabolic relationship: low levels of gravity led to high sorting and an only very slightly positively skewed ($\alpha^* = 0.05$) distribution. These findings confirmed that powders precipitated under the regime of microgravity attain an (almost) ideal log-normal distribution, with very narrow grain size distribution around a median size of 200 nm.

Despite much effort having already been expended on ceramic powder production and crystal growth in space, it is highly speculative whether the technical and economic conditions can ever be attained in future so as to allow routine ceramic powder preparation or even high-temperature processing under reduced gravity conditions. Nonetheless, experiments performed in the extreme environment of zero gravity have already led to important insights into the nucleation and crystal growth processes otherwise concealed by gravitational perturbation under terrestrial conditions. For example, ideal, diffusion-controlled, steady-state segregation, which has never been accomplished on Earth, was achieved during the growth of Te-doped InSb crystals in Skylab experiments (Witt *et al.*, 1975; see also Elwell, 1978; Kroes *et al.*, 1992; Ge *et al.*, 2001; Minakuchi *et al.*, 2004). Likewise, the synthesis of high-purity Group II–VI semiconductors, such as HgCdTe with a band gap of 1.5 eV, a very high electron mobility (in the range of several hundred thousand $cm^2 V^{-1} s^{-1}$) and a mean free path length of several micrometers, represent target areas of interest for liquid-phase epitaxy (LPE) or molecular beam epitaxy (MBE) under reduced gravity.

12.4
Ceramic Powder Synthesis by SHS

Self-propagating high-temperature synthesis (SHS) represents the primary process for manufacturing ceramics and intermetallic compounds by highly exothermic reactions in which the heat of reaction is large enough to sustain the propagation of a combustion wave through the reactants. These may be mixed powders, such as Al + C to react with Cr_2O_3 to a mixture of Cr_3C_2 + $(Al,Cr)_2O_3$, or a powder with a gas, such as Ti + C + N_2 to yield Ti_2CN (Merzhanov, 1995, 2004). The process, the details of which were first reported by Merzhanov *et al.* (1967), is based on the metallothermic Goldschmidt process that was discovered previously in 1893 (Wang *et al.*, 1993) and widely applied for the welding of railroad tracks. Two variants of the process have been recognized: (i) a plane wave propagation mode, which involves triggering the reaction by rapid heating at one end of the specimen; and (ii) a thermal explosion mode, which involves the uniform heating of a compacted powder sample at a constant rate (Roy *et al.*, 1993).

SHS is particularly suited to the synthesis of refractory ceramic powders and compacts such as carbides of Ti, Si, Cr, Ta, and B, borides of Ti, V and Cr, nitrides of B, Ti, Al, Si, silicides of Mo, Ti and V, or even more complex compounds such as $YBa_2Cu_3O_{7-x}$. The thermodynamic basis of the process, the individual types of SHS processing techniques, and the equipment and post-synthesis processing to obtain powder compacts have been reviewed by Yukhvid (1992).

Intermetallic compound coatings on carbon steel, such as NiAl with excellent bond strength, were synthesized by solar energy concentrated by a Fresnel lens (Rogriguez *et al.*, 1999). The SHS technique can also be used to immobilize radioactive reactor graphite waste by the self-sustaining reaction:

$$4Al + 3TiO_2 + 3C = 3TiC + 2Al_2O_3$$

whereby ^{14}C is bound in the TiC (Karlina *et al.*, 2007), and also to synthesize porous mullite ceramics from kaolinitic clays by adding Mg powder and combusting the mixture in oxygen (Esharghawi *et al.*, 2010).

12.5
Nanosized Ceramic Powders

Today, the research into, and manufacture and applications of, nanosized ceramics powders are emerging as one of the fastest growing segments of the ceramics industry. Nanoparticles are defined as particles having at least one dimension less than 100 nm, whereby in the case of platelet (nanoplatelets) or rod-like (nanowires) morphologies one dimension may be as small as 1 nm, corresponding to roughly four or five bonded atoms (quantum dots). As nanoscale particles have a high proportion of atoms near their surfaces, strong deviations from the behavior of the bulk structure (including chemical composition) may occur. Such size may also affect a number of important bulk properties, owing to restrictions on wave function radius, the separation of lattice defects, and interacting strain fields (Waychunas and Zhang, 2008).

12.5.1
Unique Properties of Nanoparticles

In the nanosize range, many ceramic compounds begin to show commercially useful, novel and, in some cases, quite unexpected properties and unusual performances, partly as a result of *size confinement quantum effects* (Gammon, 2000;

Gleiter, 2000; Valiev, 2002). In this case, the delocalized wave function associated with an exitonic state (i.e., a band gap transition to the lowest states in the conduction band) may be restricted in its extent at small diameters, thus raising its energy relative to the bulk state (Kayanuma, 1988). However, the increase in reactivity which is frequently associated with nanosized particles cannot generally be related to quantum confinement (QC) (Waychunas and Zhang, 2008) but is, in many cases (if not in most cases) due to surface-state modifications and structural disorders. For example, if the surface energies of polymorphs differ significantly at small enough sizes, the order of phase stability can be altered (Banfield and Zhang, 2001); moreover, the crystallite habit can be changed if the order of equilibrium form faces is reversed (Wang and Feng, 2003). An example of the first option is the observation that, for TiO_2, the stable phase of bulk crystals is rutile but becomes anatase for nanosized particles. An example of the second option is the change in crystal morphology of ceria (CeO_2), from {111} for larger crystals to {111} + {100} for crystals <10 nm. It should also be noted that QC is most important for semiconducting materials. In insulators such as quartz, corundum, and also in many other ceramic materials with a wide band gap, an overlap of wave functions will not occur since, even at higher dopant concentrations, the localized dopant wave functions will be too far apart.

As discussed above, the role of QC versus the alteration of surface states in explaining the size-related effects is not entirely clear. However, both *surface charge* and *surface bond relaxation effects* in nanoscale materials can affect the electronic energy states of surface atoms and adsorbed species, and hence alter the reactivity due to modified electronic transfer properties. It is well known that paramagnetic materials show size-dependent effects. For example, in nanosized goethite and hematite particles, the bulk antiferromagnetic coupling is reduced and superparamagnetic activity enhanced (Carbone *et al.*, 2005). An example of this is superparamagnetic magnetite nanoparticles of average size of 12 nm which have been incorporated into polymer microbubbles and used as contrast agents in magnetic resonance and ultrasound medical imaging (Yang *et al.*, 2009).

Particle size also plays a decisive role in *solubility*. According to classic nucleation theory (CNT), larger particles are more stable than smaller ones; indeed, the size of the critical nucleus and its curvature, respectively, determine the chemical potential and hence its solubility as expressed by the Gibbs–Thompson equation in CNT. This fact is at the heart of the effect of Ostwald ripening, which presumes the existence of surface energy that is independent of particle size. However, this assumption will break down for particles near the critical nucleus size, and consequently Ostwald ripening may not proceed classically in the nanoscale region (Waychunas and Zhang, 2008). Likewise, while the melting point of bulk materials is independent of size, decreasing the size towards the atomic scale results in a strong depression of the melting point $T_M(r)$ that is also governed by the Gibbs–Thompson equation which, in this case for a spherical particle in isotropic approximation, reads

$$\Delta T_M = k/2r, \tag{12.2}$$

where the melting point depression factor $k = 2\sigma^* \cdot T_M/\Delta H_f \cdot \rho_s$, where σ^* is the surface energy at the liquid–solid interface, T_M is the bulk melting temperature, r the particle radius, ΔH_f the bulk enthalpy of fusion, and ρ_s the density of the solid.

12.5.2
Application of Ceramic Nanopowders

While nanomaterials are in existence for quite some time in products such as catalysts, carbon black, pigments and other simple oxide nanopowders, in today's developing nanoceramic materials industry the emergence of low-cost, high-volume, novel manufacturing processes offer the chance of complex compositions with a level of precision and a range of properties that were, in the past, either too difficult to achieve or economically not feasible.

Currently, both the *electronic* and *information industries* require nanoclusters, nanowires, inks and materials for high-density interconnects and layers. In particular, information technology is faced with the need for continuously increasing the data storage densities of integrated circuits (ICs). However, the progress of such advances is governed by the famous "Moore's law" that, according to Kurzweil (2005), will hold true long after 2020 as exponential growth in storage capacity is thought to continue beyond the age of ICs into the future advent of novel computing technologies, that is, optical and/or quantum computers. Kurzweil proposes that, eventually, this staggering growth will lead to the technological singularity – that is, a period where progress in technology occurs (almost) instantaneously. To aid the increasing speed of data storage and processing, not only will novel magnetic compositions be needed but also energy storage and battery systems that require innovation in ceramic materials at the nanometer scale.

One decisive advantage of ceramic nanopowders is that they sinter faster and at lower temperatures than their more coarse-scale counterparts [see Eq. (12.2)] – properties which enable not only their low-cost manufacture but also a much improved flexibility of processing routes. For example, one innovative method of reducing the firing temperature of porcelain tableware involves comminution of the raw materials down to submicron and nanoscale sizes, leading to greater reactivity. For example, a customary porcelain slurry was ground in an agitator ball mill from $d_{50} = 5.0\,\mu m$ to $0.9\,\mu m$, green bodies were prepared, and glost firing simulated in a dilatometer. Then the sintering temperature could be reduced by approximately 180°C (Kivitz *et al.*, 2009). Ceramic coatings, catalysts, biomedical antimicrobial materials and coatings, drug-delivery agents, dental and bone implants (Gonsalves *et al.*, 2008), optical storage devices and systems, gas sensors and SOFCs (Lamas *et al.*, 2009) and nanometer-scale precision polishing powders for silicon wafer, hard disk, and micro-electromechanical system (MEMS) surfaces all require supplies of high-quality, reliable, complex nanoscale powders in large quantities, and at an economically viable cost. Several present and future applications of nanoscale ceramic materials are listed in Table 12.1.

However, in order to achieve commercial use, nanoscale materials must abide by certain characteristic properties, the wide range of which is listed in Table 12.2.

Table 12.1 Applications of nanoscaled ceramic powders. Adapted from Yadav (2006).

Ceramic material	Application
Barium titanate, strontium titanate and other perovskite ceramics	Capacitors, sensors, embedded electronics, security inks
Alumina, aluminum nitride, silicon carbide	Electronic substrates, packaging
Lead zirconate titanate, lithium titanate	Piezoelectric transducers
Ferrites, REE borides	Magnets
Alumina	High-performance insulators
Zinc oxide, bismuth oxide, titanium oxide	Varistors
Titanates, manganates	Thermistors
Silicon carbide, zirconia, titanium carbide, titanium nitride, boron carbide, boron nitride	Mechanical components, sealants, gaskets, structural components
Aluminum silicate, hydroxyapatite, alumina, zirconia, titania	Implants, biomedical applications
Indium tin oxide, titania, zirconia, tungsten oxide, doped oxides	Ceramic coatings, transparent conductive films
Oxynitrides, zinc oxide, zirconium silicate, zirconia, doped oxides, transition metal oxides, REE oxides	Pigments
Aluminum silicate, alumina, mixed metal oxides, zirconia, metal doped oxides, zeolites	Catalysts
Aluminum silicate, zirconium silicate, alumina, zirconia, ceria, magnesia, tin oxide, silicon carbide, boron carbide, tungsten carbide	Hard coatings, abrasives, polishing media

REE: rare earth element.

In addition to these emerging applications of nanoparticles in engineering, there is an increasing recognition of the enormous roles that nanominerals and mineral nanoparticles (Hochella, 2008) play in the environment. Research into the formation and properties of nanoparticles based on silica and silicates have important implications for the transport and storage of heavy-metal and toxic organic contaminants in the environment. In fact, a new branch of the geosciences–"nanogeoscience"–has been defined which caters to this novel issue (Hochella, 2008). In natural waters, silica clusters have a critical nucleus size of approximately 3 nm

Table 12.2 Desired characteristics in nanoscale materials enabled by precision manufacturing (after Yadav, 2006).

Product features	Desired characteristics
Mean particle size	1 to 250 nm
Grain size distribution	Narrow or broad
Particle shape	Spherical, needles, platelets, wires
Degree of agglomeration	None or controlled
Composition	Complex oxides and nonoxides
Purity level	High
Dispersability	High, in water and solvents
Production scale	Tons per year
Reproducibility	Very high (6 σ)
Safety	Inherent, contained
Affordability	€5 to €250 kg^{-1}
Uniqueness, patents	Well established

under near-neutral pH conditions, and exhibit variable water contents. The structure of the silica clusters reorganizes itself during dewatering, from opal-A (amorphous) to opal-CT (cristobalite-tridymite), to opal-C (cristobalite), moganite, cristobalite, to quartz (Icopini *et al.*, 2005). Natural nanominerals such as ferrihydrite appear to be important geochemical nanovectors for metal transport in soil–river systems (Hassellöv and von der Kammer, 2008). For example, Al-(hydr)oxides and nanosilicates, such as allophane and imogolite, play important roles in the adsorption of humic substances and viruses, in the sequestering of arsenic and phosphorus in effluents, and in the remediation of As-contaminated soils (Theng and Yuan, 2008).

12.5.3
The Manufacture of Nanopowders

Current production methods of ceramic nanoscale powders include attrition, solution, and vapor technologies:

- During *attrition milling*, coarse micron-sized powders are comminuted by the application of directed mechanical (milling) (see, for example, Santos *et al.*, 2009) or ultrasonic energies within carefully controlled thermal, chemical, and shear environments. Unfortunately, in the case of milling there is a major risk of the product becoming contaminated with particles that have been released from the milling media by abrasion.
- *Solution methods*, such as sol–gel processing, involve the precipitation of nanoparticles from liquid precursors. Whilst this technique has a low capital cost, is normally carried out at low temperatures, and also demonstrates

advantages such as an easy control of purity, problems relating to crystal size and shape may arise from the fact that the thermodynamic constraints frequently limit the attainment of complex material compositions. The need to collect by filtration, and to calcine the precipitates at elevated temperatures, will not only add to the cost but also impose difficulties in controlling the particle size, purity, and yield.

- *Vapor technologies* require the creation of a vapor of atoms or atomic clusters that must be condensed and quenched to yield the desired size range of nanoscale particles. Whilst few restrictions exist in terms of compositional flexibility, the high temperatures needed to vaporize the precursor materials, the need to add coolants, and the difficult separation procedures involved render the process expensive, and may also impact on the agglomeration and purity of the product.

Recently, a novel route to nanoscale ceramic powders has been developed that utilizes the Joule–Thompson expansion, and which avoids some of the negative features of vapor-based powder generation (Vigliotti *et al.*, 2003; Yadav and Pfaffenbach, 2003). This process uses solid, liquid, or gaseous precursor mixtures of the desired composition, sometimes with added fuel. The mix is first combusted into a product stream at temperatures of 1600 K or higher, after which the combustion product stream is further heated in a plasma jet to temperatures as high as 3000 K, at supersonic velocities. A subsequent cooling of the vapor results in the nucleation of nanoscale particles, after which the gas stream containing the powder nuclei is quenched at near-sonic to supersonic velocities, using a Joule–Thompson expansion step. The quenching reduces the collisions between particles, thus preventing agglomeration and leading to a free-flowing nanoscale powder that can be harvested on a continuous basis. A further advantage is that, if the desired powder composition is already available as a micron-sized ceramic powder, it can be fed directly into a d.c. plasma jet for evaporation and then processed further, as described above. Additional details relating to Joule-Quench™ technology are available elsewhere (Yadav, 2006).

12.5.4 Structural Characterization of Ceramic Nanoparticles

The exceptionally small size of nanoparticles renders most commonly applied characterization techniques unsuitable. This pertains, in particular to X-ray diffraction (XRD) in the range of small wave vectors, Q. In this case, information is acquired by observing how the aggregated (nano)particles cause diffraction of the X-rays, whilst observations of their size and shape, and the bonding relationships of segregated particles, may be impeded. In contrast to these limitations, average structural parameters can be acquired by using techniques such as wide-angle X-ray scattering (WAXS; see Appendix C), neutron scattering, extended X-ray absorption fine structure (EXAFS) analysis, and electron diffraction-transmission electron microscopy (ED-TEM), sometimes in conjunction with molecular dynamics simulations (MDS) (Huang *et al.*, 2008). Improvements in synchrotron-source, X-ray intensity and brightness has led to a powerful pair distribution function (PDF) analysis that is, essentially, a powder diffraction carried out to large scattering vectors Q beyond 4 nm^{-1}. This is equivalent to using extremely short X-ray wavelengths, and the Fourier transform of the weighted structure factors obtained yields high-precision interatomic distances as large as 4 nm, or more (Gateshki *et al.*, 2007). Details on the technique and results of neutron-scattering experiments on TiO_2, alumina, and ZnS nanoparticles have been reported by Waychunas and Zhang (2008).

Some of these analytical techniques allow the definition of particle size, either directly from the PDF or from X-ray scattering peak broadening (XSPB), small-angle X-ray scattering (SAXS), and dynamic light scattering (DLS) (Waychunas, 2001).

12.6 Future Development Trends

In order to chart the course for future research and development in the area of advanced ceramics, a German consortium of stakeholders from industry and academia have recently developed three "roadmaps" for the period from 2010 to 2025 on fields of applications, and two "roadmaps" within the realm of research. The former roadmaps are centered on: electronics, information and communication; energy and environment; and mechanical engineering, and the latter roadmaps on: research into structural and functional properties; and process technology. In the related reports (Rödel *et al.*, 2008, 2009), the growth opportunities for advanced ceramics as an enabling technology were highlighted, and a plethora of suggestions provided for future developments.

The electronics, information and communication roadmap includes piezoelectrics, ceramic MEMS, ceramic packaging materials for ICs, ceramic thermoelectric converters, chemosensors and biosensors, lasers and adaptive optics, and ultrahigh integration-level semiconductor chips (Figure 12.3, left).

The energy and environment roadmap predicts high-temperature thermal barrier coatings (HT-TBCs), self-healing ceramics, solid oxide fuel cells (SOFCs), substrates for thin-film solar cells, novel HT-superconductor ceramics, and ceramic membranes for a variety of applications to play important roles in the international advanced ceramic market (Figure 12.3, right). The mechanical engineering market will likely benefit from the development of novel tribological coatings, extremely hard tools based on c-BN and diamond, and from ceramic-forming dies and casting tools, cavitation-resistant heat exchangers, ceramic filters, sensors and actuators, as well as adaptronic ceramics for ultra-precision machining.

Research needs will occur in the study of structural and functional properties, including very small structures, the modeling of tribological systems, scale-dependent properties, the tailoring of defect structures, the thermodynamics and kinetics of ceramic interfaces, and an atomistic simulation of crystal structures

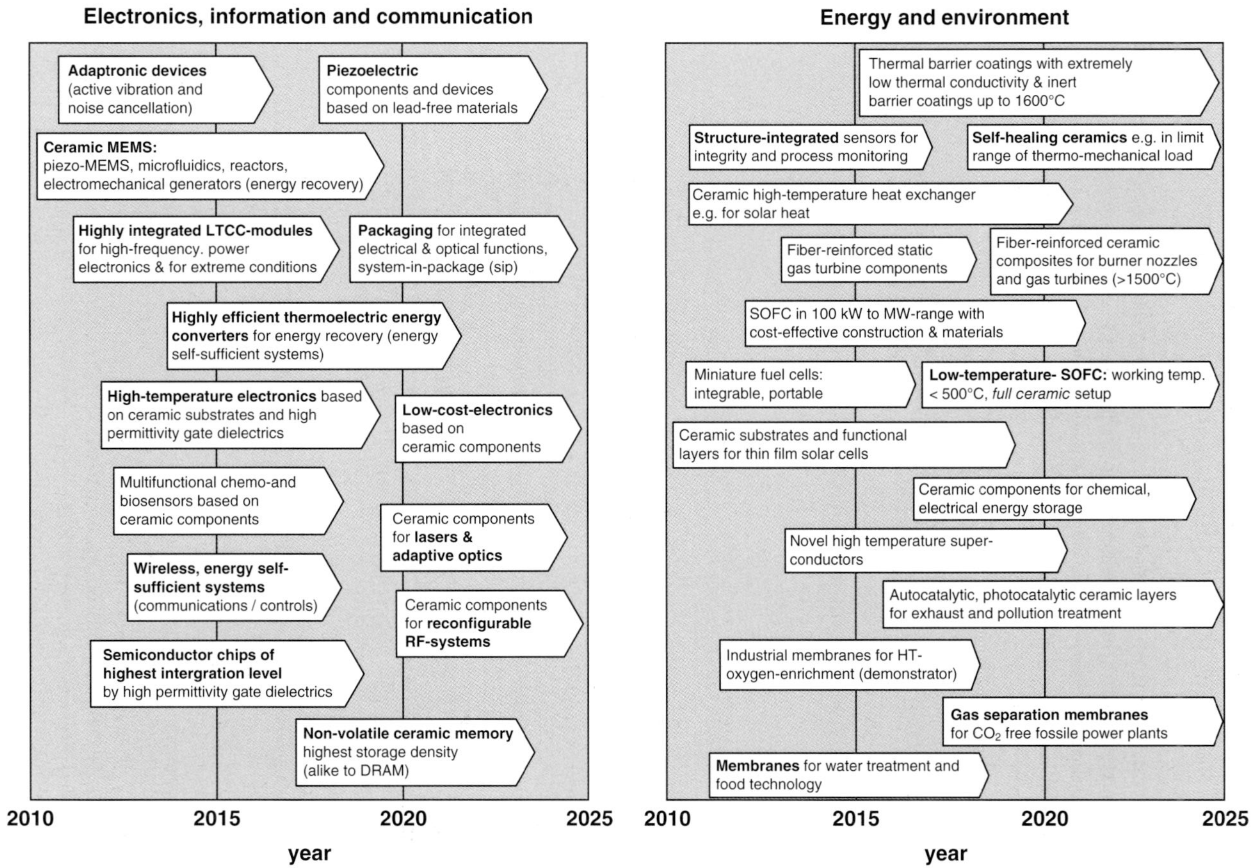

Figure 12.3 Timelines of the development of advanced ceramics for electronics, information and communication (left), and energy and environment (right) (Rödel *et al.*, 2009). Reprinted with permission from Elsevier.

and properties. In the area of process technology, research needs will arise in robust, simple and cost-effective processing, the role of process additives, pore structure tailoring, miniaturization, biomimetic processing, neat-net shape processing, freeform manufacturing, the development of neuro-interface structures, and modeling on all size-scales, from nano to macro.

According to Rödel *et al.* (2009), these perceived research and development requirements can be grouped into four major research topics that should be addressed in future to secure a sustainable development and healthy growth of the advanced ceramics market:

- Novel ceramics with enhanced and new properties
- High-performance key components for systems applications
- Highly efficient processing technology
- Holistic modeling and simulation techniques.

References

Banfield, J.F. and Zhang, H. (2001) Nanoparticles in the environment, in *Nanoparticles and the Environment*, (eds J.F. Banfield and A. Navrotsky), Reviews of Mineralogy and Geochemistry, vol. **44**, Mineralogical Society of America, Chantilly, VA, pp. 1–58.

Biswas, D.R. (1989) Development of novel ceramic processing. *J. Mater. Sci.*, **24** (10), 3791–3798.

Carbone, C., Di Benedetto, F., Marescotti, P., Sangregorio, C., Sorace, L., Lima, N., Romanelli, M., Lucchetti, G., and Cipriani, C. (2005) Natural Fe-oxide and –oxyhydroxide nanoparticles: an EPR and SQUID investigation. *Mineral. Petrol.*, **85**, 19–32.

Castillo, M., Moore, J.J., Schowengerdt, F.D., and Ayers, R.A. (2003) Effect of gravity on porous tricalcium phosphate and nonstoichiometric titanium carbide produced via combustion synthesis. NASA/CP-2003-21276/REV1, pp. 61–64.

Elwell, D. (1978) Crystal growth. Priorities in the initial use of Spacelab for crystal growth. *Proc. R. Soc. London*, **A361**, 151–156.

Esharghawi, A., Penot, C., and Nardou, F. (2010) Elaboration of porous mullite based materials via SHS reaction. *Ceram. Int.*, **36**, 231–239.

Gammon, D. (2000) Semiconductor physics: electrons in artificial atoms. *Nature*, **405**, 899–900.

Gateshki, M., Chen, Q., Peng, L.M., Chupas, P., and Petkov, V. (2007) Structure of nanosized materials by high-energy x-ray diffraction. *Z. Kristallogr.*, **222**, 612–616.

Ge, P., Nishinaga, T., Li, C., Huo, C., Nakamura, T., Huang, W., Voloshin, A.E., and Lomov, A.A. (2001) Growth of GaSb single crystal in space. *Sci. China Ser. A.*, **44** (6), 762–769.

Gleiter, H. (2000) Nanostructure materials: basic concepts and microstructure. *Acta Mater.*, **48**, 1–29.

Gonsalves, K.E., Halberstadt, C.R., Laurencin, C.T., and Nair, L.S. (eds) (2008) *Biomedical Nanostructures*, John Wiley & Sons, Ltd, 507 pp.

Hassellöv, M. and von der Kammer, F. (2008) Iron oxides as geochemical nanovectors for metal transport in soil-river systems. *Elements*, **4**, 401–406.

Hochella, M.F. (2008) Nanogeoscience: from origins to cutting-edge applications. *Elements*, **4**, 373–379.

Huang, W.J., Sun, R., Tao, J., Menard, L.D., Nutto, R.G., and Zuo, J.M. (2008) Coordination dependent surface atomic contraction in nanocrystals revealed by coherent diffraction. *Nat. Mater.*, **7**, 308–313.

Icopini, G.A., Brantley, S.L., and Heaney, P.J. (2005) Kinetics of silica oligomerization and nanocolloid formation as a function of pH and ionic strength at 25 °C. *Geochim. Cosmochim. Acta*, **69**, 293–303.

Karlina, O.K., Klimov, V.L., Pavlova, G.Y., and Ojovan, M.I. (2007) Thermodynamic simulation and experimental study of irradiated reactor graphite waste processing with REE oxides, in *Symposium on the Scientific Basis for Nuclear Waste Management XXXI, 16–21 September, Sheffield, UK*, Materials Research Society, vol. **1107**, pp. 109–116.

Kayanuma, Y. (1988) Quantum site effects of interacting electrons and holes in semiconductor microcrystals with spherical shape. *Phys. Rev. B*, **38**, 797–805.

Kivitz, E., Palm, B., Heinrich, J.G., Blumm, J., and Kolb, G. (2009) Reduction of the porcelain firing temperature by preparation of the raw material. *J. Eur. Ceram. Soc.*, **29** (13), 2691–2696.

Kroes, R.L., Reiss, D.A., and Lehoczyk, S.L. (1992) Crystal growth in a microgravity environment. US Patent 5,173, 087.

Kurzweil, R. (2005) *The Singularity is Near*, Penguin Books, ISBN: 0-670-03384-7.

Lamas, D.G., Bianchetti, M.F., Cabezas, M.D., and Walsöe de Reca, N.E. (2009) Nanostructured ceramic materials: applications in gas sensors and solid-oxide fuel cells. *J. Alloy Compds.*, doi: 10.1016/j.jallocom.2009.10.014.

Lind, M.D. (1978) Crystal growth from solution in low gravity. *AIAA J.*, **16** (5), 458–462.

Merzhanov, A.G. (1995) History and recent developments in SHS. *Ceram. Int.*, **21** (5), 371–379.

Merzhanov, A.G. (2004) The chemistry of self-propagating high-temperature synthesis. *J. Mater. Chem.*, **14**, 1779–1786.

Merzhanov, A.G., Shkiro, V.M., and Borovinskaya, I.P. (1967) Synthesis of refractory inorganic compounds. Certif. No. 255221, Invent. Bull., 10 (1972). US Patent 3,726,643 (1973).

Minakuchi, H., Okano, Y., and Dost, S. (2004) A three-dimensional numerical simulation study of the Marangoni convection occurring in the crystal growth of Si_xGe_{1-x} by the float-zone technique in zero gravity. *J. Cryst. Growth*, **266** (1–3), 140–144.

National Research Council (1978) *Materials Processing in Space*. Space Science Board, National Academy of Science, Washington, DC.

Olsen, W.L. and Liss, W.E. (1988) Preparation of monodisperse titania by titanium alkoxide hydrolysis. US Patent 4,732,750 (March 22).

Rödel, J., Weissenberger-Eibl, M., Kounga, A., Koch, D., Bierwisch, A., Rossner, W., Hoffmann, M.J., and Schneider, H. (eds) (2008) *Hochleistungskeramik 2025. Strategieinitiative für die Keramikforschung in Deutschland.* Koordinierungsausschuss Hochleistungskeramik der DKG und DGM, Werkstoffinformationsgesellschaft mbH, Frankfurt, ISBN: 978-3-88355-364-1.

Rödel, J., Kounga, A.B.N., Weissenberger-Eibl, M., Koch, D., Bierwisch, A., Rossner, W., Hoffmann, M.J., Danzer, R., and Schneider, H. (2009) Development of a roadmap for advanced ceramics: 2010–2025. *J. Eur. Ceram. Soc.*, **29** (9), 1549–1560.

Rogriguez, G.P., Garcia, I., and Vázques, A.J. (1999) Coating processing by self-propagating high-temperature synthesis (SHS) using a Fresnel lens. *J. Phys. IV Fr.*, **09** (Pr3), 411–416.

Roy, S.K., Biswas, A., and Banerjee, S. (1993) Self-propagating high-temperature synthesis of titanium borides. *Bull. Mater. Sci.*, **16** (5), 347–356.

Santos, C., Koizumi, M.H., Daguano, J.K.M.F., Santos, F.A., Elias, C.N., and Ramos, A.S. (2009) Properties of Y-TZP/Al_2O_3 ceramic nanocomposites obtained by high-energy ball milling. *Mater. Sci. Eng. A*, **502** (1-2), 6–12.

Sirignano, W.A. (1995) *Microgravity Research Opportunities for the 1990s*, National Academy of Science, 148 pp.

Theng, B.K.G. and Yuan, G. (2008) Nanoparticles in the soil environment. *Elements*, **4**, 395–399.

Valiev, R.Z. (2002) Nanomaterials advantage. *Nature*, **419**, 887–889.

Vigliotti, A., Yadav, T., Kostelecki, C., and Wyse, C. (2003) Reducing manufacturing and raw material costs for devise manufacture with nanostructured powders. US Patent 6,641,775.

Wang, Z.L. and Feng, X. (2003) Polyhedral shapes of CeO_2 n anoparticles. *J. Phys. Chem.*, **B107**, 13563–13566.

Wang, L.L., Munir, Z.A., and Maximov, Y.M. (1993) Thermite reactions: their utilization in the synthesis and processing of materials. *J. Mater. Sci.*, **28** (14), 3693–3708.

Waychunas, G.A. (2001) Structure, aggregation and characterization of nanoparticles, in *Nanoparticles and the Environment*, Reviews of Mineralogy and Geochemistry, vol. **44** (eds J.F. Banfield and A. Navrotsky), Mineralogical Society of America, Chantilly, VA, pp. 105–166.

Waychunas, G.A. and Zhang, H.Z. (2008) Structure, chemistry, and properties of mineral nanoparticles. *Elements*, **4**, 381–387.

Witt, A.F., Gatos, H.C., Lichtensteiger, M., Lavine, M.C., and Herma, C.J. (1975) Crystal growth and steady-state segregation under zero gravity–InSb. *J. Electrochem. Soc.*, **122**, 276–283.

Yadav, T. (2006) Nanoscale ceramics: the building blocks of nanotechnology. Available at: www.ceramicindustry.com/ (Accessed 19 December 2009).

Yadav, T. and Pfaffenbach, K. (2003) Nano-dispersed powders and methods for their manufacture. US Patent 6,652,967.

Yang, F., Li, Y., Chen, Z., Zhang, Y., Wu, J., and Gu, N. (2009) Superparamagnetic iron oxide nanoparticle-embedded encapsulated microbubbles as dual contrast agents of magnetic resonance and ultrasound imaging. *Biomaterials*, **30** (23–24), 3882–3890.

Yoldas, B.E. (1986) Hydrolysis of titanium alkoxide and effects of hydrolytic polycondensation parameters. *J. Mater. Sci.*, **21** (3), 1087–1092.

Yukhvid, V.I. (1992) Modifications of SHS processes. *Pure Appl. Chem.*, **64** (7), 977–988.

Appendices

Classic and Advanced Ceramics: From Fundamentals to Applications. Robert B. Heimann

ISBN: 978-3-527-32517-7

Appendix A
Construction of the Phase Diagram of a Binary System A–B with Ideal Solid Solution

The variation of the Gibbs function G of phase α of a binary system A–B with the intensive variables temperature T, pressure P, and mole numbers n_i can be expressed by:

$$dG^\alpha = -s^\alpha dT + v^\alpha dP + \sum \mu_i^\alpha dn_i^\alpha, \tag{A.1}$$

where s and v are the molar entropy and volume, respectively, and μ is the chemical potential. Hence, according to Gibbs' phase rule, there are four variables in a binary ($C = 2$) system: the temperature T, the pressure P, and the concentrations of A and B, expressed by their mole numbers n_A and n_B.

For a single-phase situation ($\alpha = 1$), the degrees of freedom F are 3 (F = $C - P + 2$); that is, there are three independent variables, T, P, and either n_A or n_B. The molar fractions of the components A and B in phase α are defined as:

$$X_A^\alpha = n_A^\alpha/(n_A^\alpha + n_B^\alpha);\ X_B^\alpha = n_B^\alpha/(n_A^\alpha + n_B^\alpha);\ X_A^\alpha + X_B^\alpha \equiv 1 \tag{A.2}$$

From this it follows

$$g^\alpha = X_A^\alpha \cdot \mu_A^\alpha(P, T, X_A) + (1 - X_A^\alpha) \cdot \mu_B^\alpha(P, T, X_A). \tag{A.3}$$

The lower-case letter g refers to the *molar* Gibbs function; that is, the Gibbs free energy per mol.

For the two-phase situation ($\alpha = 2$), the degrees of freedom F reduce to 2, since

$$\mu_A^1(P, T, X_A^1) = \mu_A^2(P, T, X_A^2), \text{ and } \mu_B^1(P, T, X_B^1) = \mu_B^2(P, T, X_B^2), \tag{A.4}$$

where the superscripts (1) and (2) refer to phase 1 and phase 2, respectively. Consequently, there are only two independent variables P and T, or P and X_A, or T and X_A, since the molar fraction X_B is fixed according to $X_B = 1 - X_A$.

The molar Gibbs function of a phase of a *binary mechanical mixture* (A + B), g^m can be derived from the thermodynamic potential equation $g = h - Ts = \mu \cdot n$, and can be expressed by:

$$g^m = X_A \cdot h_A + X_B \cdot h_B - T(X_A \cdot s_A + X_B \cdot s_B) = X_A \cdot \mu_A^* + X_B \cdot \mu_B^*, \tag{A.5}$$

where the enthalpy h and the entropy s are molar entities independent of the presence of the other component. This can be safely assumed, since the particles are

Classic and Advanced Ceramics: From Fundamentals to Applications. Robert B. Heimann

ISBN: 978-3-527-32517-7

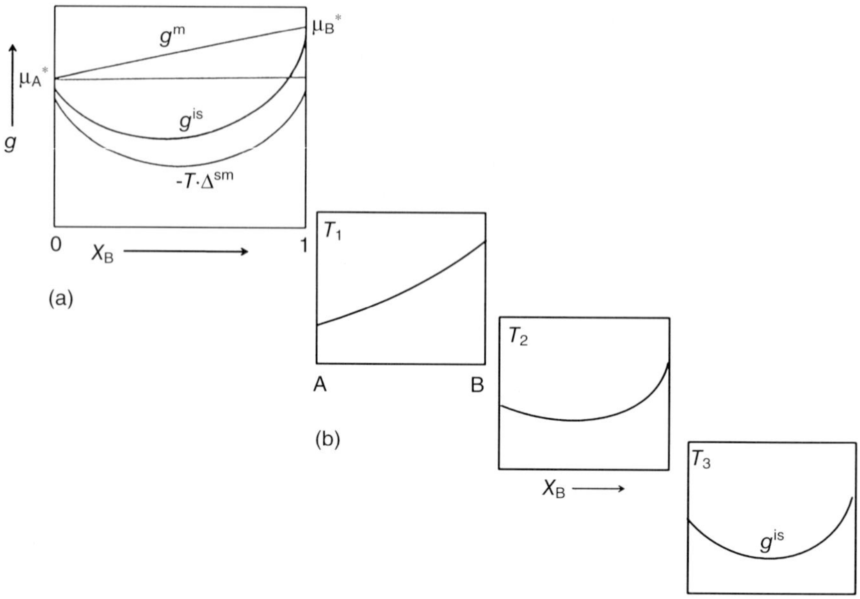

Figure A.1 (a) Gibbs free energy of a mechanical mixture (g^m), the configurational or mixing entropy term $-T\cdot\Delta s_m$ and the dependence of the Gibbs free energy of an ideal solution (g^{is}) on the composition for a binary system A–B; (b) Temperature dependence of the Gibbs free energy g^{is} at temperatures $T_1 < T_2 < T_3$.

so small that the number of atoms at the phase boundary can be neglected compared to those in the volume of the phases. This means that the phases are completely spatially separated, and consequently the Gibbs function is additive in terms of the pure components (see line g^m in Figure A.1a). The term μ^* refers to the standard chemical potential of a condensed phase.

This additivity is lost in true *solutions*, since the particles of the components A and B interact with each other. The potential and kinetic energies of A and B in solution are different compared to their individual pure states. Both energy changes influence not only the internal energies u_i of the components, but also their entropies s_i owing to purely kinetic effects, for example, variations of oscillation and rotation frequencies of the atoms of the two components. This is the reason why the chemical potentials μ_A and μ_B in solutions depend on the composition, and on T and P. The rules of equilibrium thermodynamics require replacement of the molar parameters h and s by the partial molar parameters $\bar{h}$ and $\bar{s}$. However, there is still another aspect to consider, namely the configurational or mixing entropy Δs_m. Even if the atoms of A and B were to be identical in every aspect, so that their entropy in the pure substance and the solution would be (almost) equal, their mixing would cause a state of lower order, and hence higher entropy. This increase of configurational entropy must be added to the molar Gibbs function of a solution, g^s, yielding

$$g^s = X_A \cdot \bar{h}_A + X_B . \bar{h}_B - T(X_A \cdot \bar{s}_A + X_B \cdot \bar{s}_B) - T \cdot \Delta s_m. \tag{A.6}$$

In a thermodynamic sense, solutions can be subdivided into *ideal* and *real* solutions, but in this context it suffices to consider only *ideal solutions* with a Gibbs function g^{is}. Ideal solutions are characterized by the identities $\bar{h}_A = h_A$, $\bar{h}_B = h_B$ and $\bar{s}_A = s_A$, $\bar{s}_B = s_B$. This does not mean that no interactions take place among atoms of A and B; rather, it means that the interaction A–B is (almost) identical to that of the interactions A–A and B–B in the pure components A and B. Hence, mixtures of isotopes and isomorphic components approximate ideal solutions (see below). The molar Gibbs function of an ideal solution is then

$$g^{is} = g^m - T \cdot \Delta s_m, \text{ with } \Delta s_m = -R[X_A \cdot \ln(X_A) + X_B \cdot \ln(X_B)]. \tag{A.7}$$

Using the relationship $\mu_i = \mu_i^* + RT \ln(X_i)$, this equation can be transformed to

$$g^{is} = X_A \cdot \mu_A + X_B \cdot \mu_B. \tag{A.7a}$$

The configurational entropy Δs_m is always positive for all compositions and finite temperatures. Consequently, it holds that $g^{is} < g^m$; that is, under ideal conditions a solution is always more stable than a mechanical mixture. Likewise, a homogeneous solution is more stable than a mechanical mixture of solutions.

To construct the phase diagram of an *ideal solid solution* (A,B) of the two components A and B, the concentration dependence of the configurational entropy will be considered, as shown in Figure A.1a. The convex curvature of the configurational entropy term, and hence g, increase with increasing temperature (Figure A.1b). This is also shown in Figure A.2. At high temperature, T_6, the free energy g^L of the melt is lower than that of the solid g^s over the entire range of compositions ($g^L < g^s$); hence, the only phase stable is that of the liquid melt. At decreasing temperature, the free energy of the liquid increases faster than that of the solid phase, so that the former will eventually overtake the latter. Consequently, at T_5 the two curves intersect at the composition of pure A. This is the melting point, T_A. At this point liquid A and solid A are in equilibrium, that is, their chemical potentials and hence their free energies are equal. A further decrease in temperature to T_4 leads to an intersection of the free energy curves g^L and g^s at X_1. At this point, the state with the lowest possible free energy is a mixture of a solid with composition *a* on the g^s curve and a liquid of compositions *b* on the g^L curve, given by the common tangent. Between the limits X_a and X_b the liquid has always the composition X_a and the solid always the composition X_b, even though zonation is possible. The relative amounts of liquid and solid can be determined by the lever rule (see below). At T_3, the point of intersection of the two free energy curves has shifted to a composition richer in B, and at T_2 the melting point of the pure component B, T_B, has been reached. At this point the last trace of melt disappears, and at temperatures lower than T_2 the system contains only a single solid solution ($g^s < g^L$).

Based on the dependence of free energy on temperature and composition shown in Figure A.2, the phase diagram can now be constructed as shown in Figure A.3. The temperatures T_1–T_6 define unequivocally the compositions of the liquid and solid phases in equilibrium with each other. The curve connecting all freezing

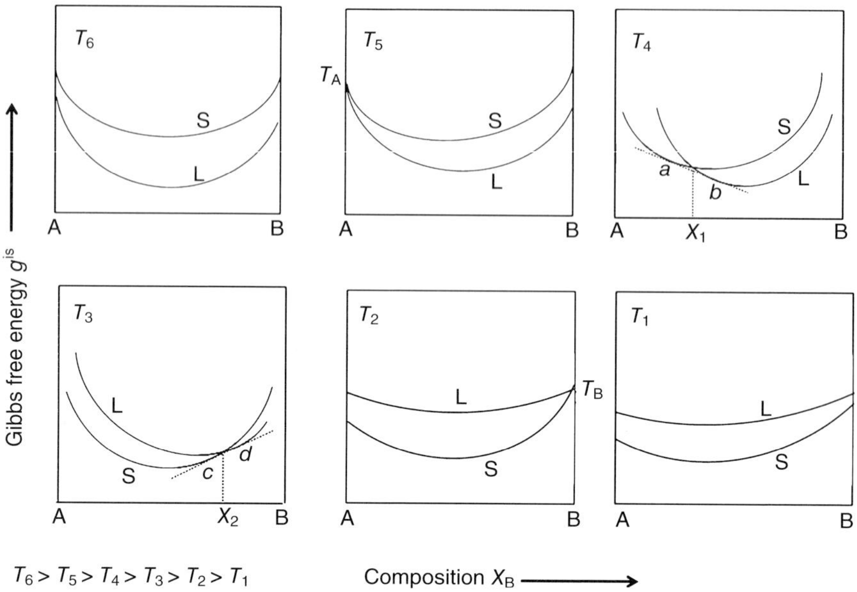

Figure A.2 Gibbs free energy curves of the liquid (L) and solid (s) phases of an ideal solid solution system between T_1 (low temperature) and T_6 (high temperature).

points of the liquid is the *liquidus*, while the curve connecting all melting points of the solid solutions is the *solidus*. Lines of constant temperature between those two curves are called *conodes*. Hence, the liquidus and the solidus divide the phase diagram into three regions: (i) the single-phase field of the liquid; (ii) the single-phase field of the solid solutions; and (iii) the two-phase liquid-solid field between the curves.

The amounts of coexisting melt and solid can be determined by the "lever rule," using the simple geometric relation that the amounts of the coexisting phases are inversely proportional to their distance from the point of the global composition measured along a conode. For example, at T_4 the relationship holds

$$(n_A^s + n_B^s)/(n_A^L + n_B^L) = m^s/m^L = \text{Ob}/\text{Oa}, \tag{A.8}$$

where $m^s = (n_A^s + n_B^s)$ is the total mass of solid and $m^L = n_A^L + n_B^L$ is the total mass of liquid.

There are several real solid solution systems that approximate this ideal system, for example, the binary systems Ge–Si and Cu–Ni owing to the energetic similarities of the two limiting components that crystallize in identical space groups (Si and Ge: S.G. $Fd\bar{3}m$ (No. 227); Cu and Ni: S.G. $Fm\bar{3}m$ (No. 225)).

It should be briefly mentioned that the construction of binary phase diagrams with invariant phase transformations requires three Gibbs free energy curves – two for the solid phases α and β, and one for the liquid phase, L. At equilibrium temperature a common tangent exists for g^L, g^α and g^β. The position of the

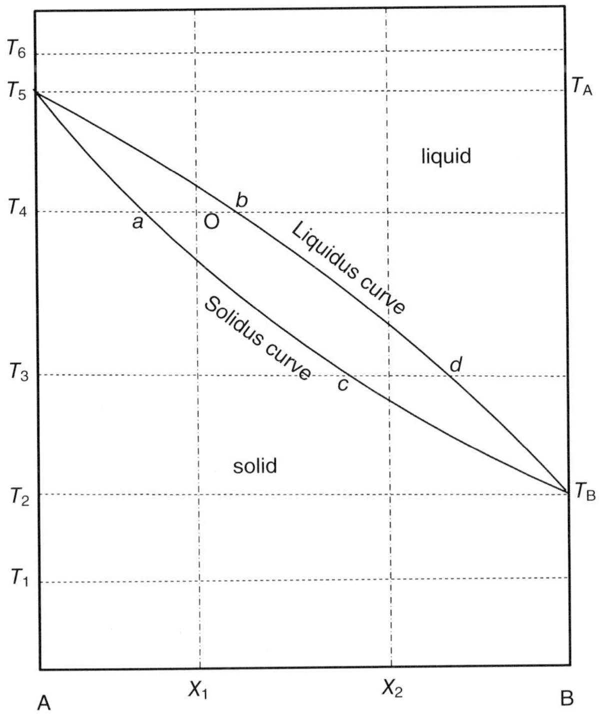

Figure A.3 Equilibrium phase diagram of an ideal solid solution, constructed from the Gibbs free energy curves shown in Figure A.2.

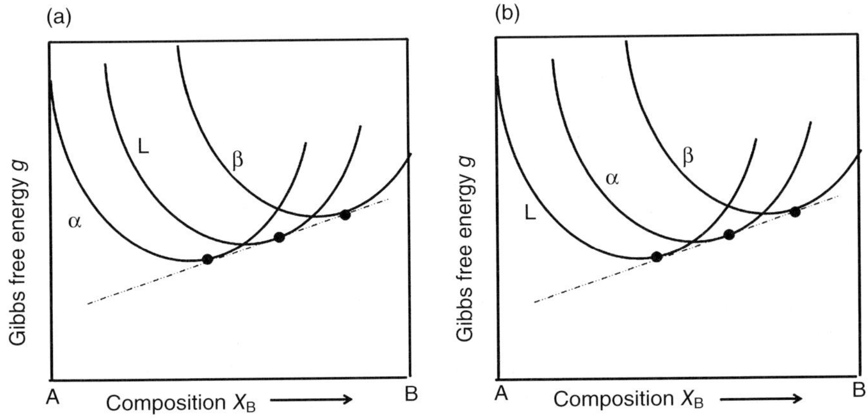

Figure A.4 Gibbs free energy curves for invariant phase transitions, showing an eutectic (a) and a peritectic (b) reaction between a liquid (L) and two solid phases (α and β).

Table A1 Types of three-phase invariant reactions in binary phase diagrams. (Smith, W.F. (1996) *Principles of Materials Science and Engineering*, 3rd edition, McGraw-Hill, Inc., New York, p. 464).

Type of reaction	Reaction equation
Eutectic	$L \rightarrow \alpha + \beta$
Eutectoid	$\alpha \rightarrow \beta + \gamma$
Peritectic	$\alpha + L \rightarrow \beta$
Peritectoid	$\alpha + \beta \rightarrow \gamma$
Monotectic	$L_1 \rightarrow \alpha + L_2$

minimum of g^L between those of the solid phases g^α and g^β characterizes a *eutectic reaction* (Figure A.4a). If the minimum g^L is outside the minima of g^α and g^β, a *peritectic reaction* takes place during which the liquid reacts with one of the solid phases to form a second solid phase of different composition (Figure A.4b). If the decomposing phase and product phases are all solid, then the reaction type is called *eutectoid*. Likewise, if two solid phases react to form a new solid phase, one speaks of a *peritectoid reaction*. Hence, peritectic and peritectoid reactions are the inverse of the corresponding eutectic and eutectoid reactions. The relations are depicted schematically in Table A1.

Appendix B
Thermodynamics of Displacive Phase Transitions

Displacive Transition in Crystals with Perovskite Structure

Although, experimentally, $SrTiO_3$ is not ferroelectric even at low temperatures, it is very close to the ferroelectric threshold. The isotopic replacement of oxygen or partial cation substitution reduces quantum fluctuations and makes it ferroelectric. Hence, $SrTiO_3$ will serve as a model material to apply the Landau theory to a quantitative description of the displacive phase transition. Similar descriptions of structural phase transitions have been provided for many minerals, including feldspars (e.g., Carpenter, 1988), garnets (Carpenter and Boffa Ballaran, 2001), quartz (Carpenter *et al.*, 1998), or cristobalite (Schmahl *et al.*, 1992).

The soft optical phonon mode at the center of the Brillouin zone is a rigid unit mode (RUM), whereby the BO_6 octahedra of the perovskite structure or the SiO_4 tetrahedra in silicates are taken as rigid units that rotate by the angle Φ around their vertices to attain phase transition (Figure B.1; Dove *et al.*, 2000). The softening of the transverse optical (TO) phonon mode causes the frequency of the associated transverse acoustic (TA) branch to be depressed with increasing wave vector **k** to zero frequency close to $\mathbf{k} = 0$ (see Figure 8.6). As a result, an incommensurate displacive phase transition occurs. The angle of rotation Φ of the RUM octahedra is the order parameter Q in the Landau–Ginzburg theory of phase transition (Landau and Lifshitz, 1980), that converges towards zero when the Curie temperature has been reached (Carpenter, 1992).

As indicated above, the transition from a low-temperature tetragonal antiferroelectric phase to a high-temperature cubic paraelectric phase in strontium titanate can be described by the rotation of rigid $[TiO_6]$ octahedra in opposite directions around the [001] direction (Figure B.1a) The angle of rotation, Θ, characterizes the average displacement of oxygen atoms in the crystal lattice, and hence may be taken as an order parameter Q in the Landau–Ginzburg thermodynamic theory of phase transition (as explained in the following paragraph). As evidenced from experimental data obtained by Müller and von Waldkirch (1973), the angle Θ decreases with increasing temperature, and reaches zero once the transition temperature T_C–and hence the cubic high-temperature phase–have been attained (Figure B.1b).

Classic and Advanced Ceramics: From Fundamentals to Applications. Robert B. Heimann

ISBN: 978-3-527-32517-7

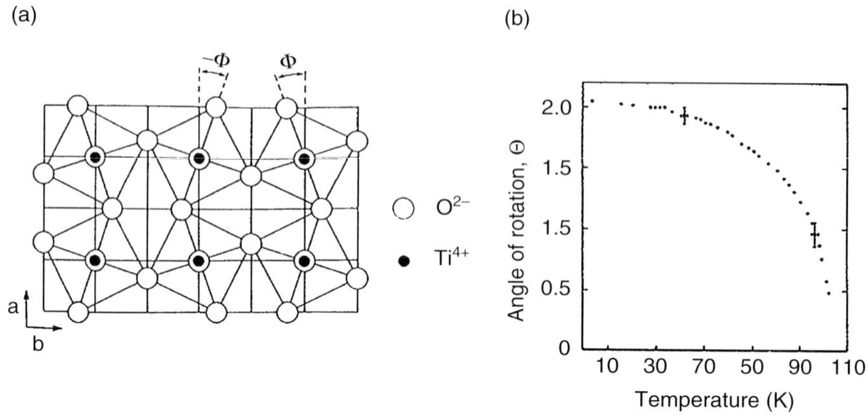

Figure B.1 Displacive phase transition in $SrTiO_3$. (a) Neighboring $[TiO_6]$ octahedra rotate in opposite directions by an angle Θ around [001]; (b) Decrease of Θ with increasing temperature until T_C of about 106 K has been attained (after Müller and von Waldkirch, 1973; Carpenter, 1992). Reprint courtesy of International Business Machines Corp., © 1973 International Business Machines Corporation.

The rotation of the $[TiO_6]$ octahedra causes a reduction of the unit cell parameter that is proportional to cos Θ; for small values of Θ it may be expanded as $[1-(\Theta^2/2)]$, so that the lattice distortion varies with Q^2.

Landau Theory

A new look at the thermodynamics of structural phase transitions has been provided by the Landau theory (Landau and Lifshitz, 1980; Tolédano and Tolédano, 1987). Its principal element is a Gibbs free energy expansion in the form of a Taylor series in an order parameter Q. As the value of Q cannot be directly assessed, the coefficients of the power series

$$G = AQ + BQ^2 + CQ^3 + DQ^4 + \ldots \tag{B.1}$$

are calibrated using measurable properties such as excess enthalpy, heat capacity, lattice strain, and so on. Consequently, values of Q may be quantified only by reference to some measurable properties that reflect the degree of transformation, such as magnetization in a ferromagnetic transition, polarization in a ferroelectric transition (see below), or lattice strain or birefringence in a ferroelastic transition (Carpenter, 1992), as well as Si–O–Si bonding lengths and angles, as in the displacive phase transitions of silica polymorphs (Schmahl *et al.*, 1992).

The order parameter Q is scaled to vary between 0 in the high-symmetry form and 1 in the low-symmetry form at 0 K. For a thermodynamically continuous phase transition, there are three regimes between 0 K and the transition (critical) tem-

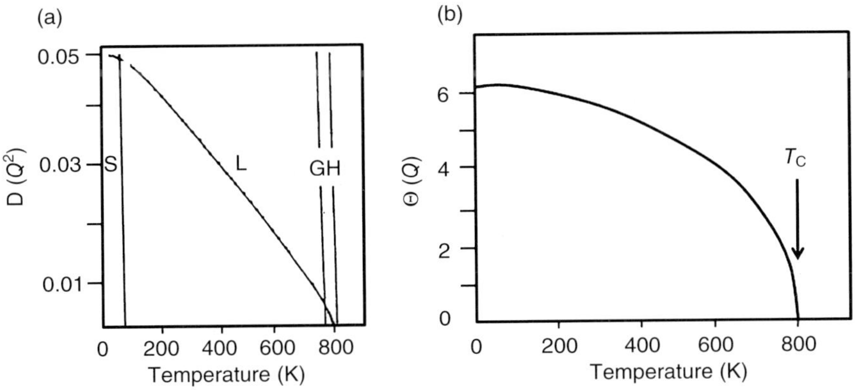

Figure B.2 Temperature dependence of the order parameter Q for the displacive transition in $LaAlO_3$. The quasilinearity of the Landau regime L, $Q^2 \propto (T_C - T)$ is in accordance with the predicted second-order transition (Salje, 1989; Carpenter, 1992). With kind permission of Springer Science and Business Media.

perature T_C (see Figure B.2): (i) the temperature-invariant "saturation" regime S, which is close to 0 K and related possibly to quantum effects; (ii) the "Landau" regime L, where Q or Q^2 vary linearly with T; and (iii) the "Ginzburg" interval G, which is near the transition temperature, owing to statistical fluctuations of the order parameter Q (Salje, 1989). Figure B.2 shows, schematically, the temperature dependence of the order parameters $D \propto Q^2$ (Figure B.2a) and $\Theta \propto Q$, respectively (Figure B.2b) of the displacive phase transition of $LaAlO_3$.

The perovskite $LaAlO_3$, although not ferroelectric, possesses excellent microwave dielectric properties. Moreover, its lattice constants closely match those of layered perovskite copper oxide superconductors (see Chapter 9), so that it can be applied as a low-strain substrate material for superconducting devices (Simon *et al.*, 1999). Its rhombohedral low-temperature phase (S.G. $R\bar{3}c$, $a_0 = 0.5357\,\text{nm}$) transforms to the (pseudo)cubic high-temperature phase (S.G. $Fm\bar{3}c$) at approximately 800 K. During transformation, the $[AlO_6]$ octahedra rotate by the angle Θ around the [111] axis (see Figure B2a). According to Salje (1989), the variation of a measurable spectroscopic parameter D, obtained by paramagnetic resonance spectroscopy on Fe^{3+}-doped samples, is proportional to Θ^2 and hence Q^2.

The equilibrium state of a crystal is characterized by the conditions

$$(\delta G/\delta Q) = 0 \text{ and} \tag{B.2}$$

$$\delta^2 G/\delta Q^2 > 0. \tag{B.3}$$

This means that the stability of the high-symmetry form ($Q = 0$) can only be satisfied if, in Eq. (B.1), the linear term is absent (A = 0) and B > 0. For B < 0, the low-symmetry form is stable, and the equilibrium values of Q is between 0 and 1. Equations (B.2) and (B.3) can further be satisfied by setting all odd-order terms in Eq. (B.1) to zero. The transition is then continuous between the states character-

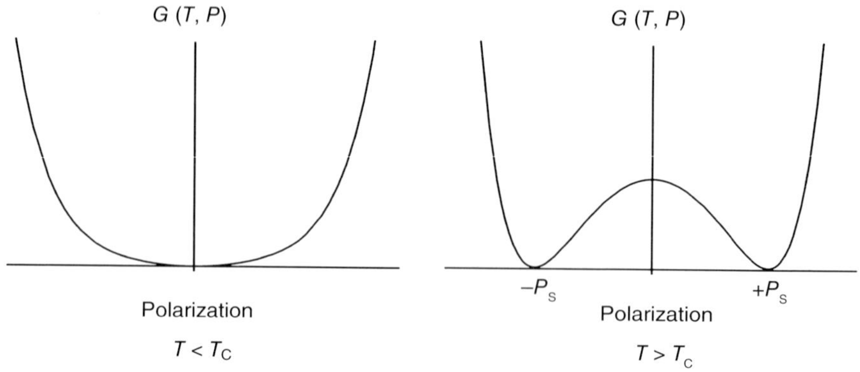

Figure B.3 Gibbs free energy of a ferroelectric material above (left) and below (right) the Curie temperature, T_C.

ized by $Q = 0$ (high symmetry) and $Q > 0$ (low symmetry). According to Landau, with changing temperature there must be a crossover in the signs of B at $T = T_C$, and this temperature dependence can be expressed as a linear function of T to yield:

$$G = \frac{1}{2}a(T - T_C)Q^2 + \frac{1}{4}bQ^4 + \frac{1}{6}cQ^6 + \ldots \tag{B.4}$$

If $b > 0$ (positive) and $c \approx 0$, the Landau expansion describes a *second-order transition* in which the equilibrium value of Q varies as

$$Q = [(T_C - T)/T_C]^{\beta}\,(T < T_C) \tag{B.5}$$

with $\beta = ½$ (see Figure B.3). If $b = 0$, $c > 0$, and all higher-order terms are negligible, the expansion relates to a *tricritical transition* with $\beta = 1/4$. Finally, if $b < 0$ (negative) and $c > 0$ (positive), the transition is of *first-order* with a discontinuity of Q at $T = T_C$.

From the Landau expansion of G the extensive thermodynamic variable entropy (S) can be obtained, such as the Gibbs–Helmholtz equation:

$$-S = (\delta G/\delta T) = 1/2aQ^2, \tag{B.6}$$

as well as the thermodynamic potential function H (excess enthalpy) such as

$$H = G - T \cdot S = -1/2aT_CQ^2 + 1/4bQ^4 \tag{B.7}$$

and the heat capacity

$$C_p = (a^2/2b)T. \tag{B.8}$$

Application of the Landau theory to order–disorder (OD) transitions, the effect of pressure, and the coupling of composition with strain and Q, respectively, as well as symmetry constraints, are dealt with in detail by Carpenter (1992).

In conclusion, the Landau theory provides a welcome quantitative description of the thermodynamics and kinetics of phase transitions in minerals, including ferroelectric ceramics, by using macroscopic order parameters and their relationship to physical properties and symmetry, as shown below. For an excellent review of the application of Landau theory to displacive phase transitions in minerals, see Dove (1997).

Application of the Landau Theory to Ferroelectric Transitions

The critical quantity investigated to describe ferroelectric transitions is the macroscopic order parameters P (polarization).

A general approach is given by the equation (Devonshire, 1949)

$$G(T, P, E) = \alpha(T) + 1/2\beta P^2 + 1/4\gamma P^4 + 1/6\delta P^6 - EP. \tag{B.9}$$

The temperature-dependent functions $\alpha(T)$ and $\beta(T)$, and the coefficients γ and δ are selected in such a way that in a $G(T,P)$ versus P plot below the Curie temperature T_C, the Gibbs free energy shows two minima (Figure B.3).

In a somewhat reduced approach, the elastic Gibbs free energy of a ferroelectric crystal is given (Uchino 1995) by:

$$G = 1/2\alpha P^2 + 1/4\beta P^4 - 1/2sX^2 - QP^2X \tag{B.10}$$

where $\alpha = (T - T_C)/\varepsilon_0 \cdot C$ (Curie–Weiss law), P is polarization, s is elastic compliance, X is applied stress, and Q is the electrostrictive coefficient.[1] The dielectric permittivity ε, the field-induced strain x, and the spontaneous polarization P_s, as well as the temperature dependence of the piezoelectric constant d, can be calculated starting from the classic thermodynamic equation of the Gibbs free energy:

$$\mathrm{d}G = E\mathrm{d}P - x\mathrm{d}X - S\mathrm{d}T. \tag{B.11}$$

The derivatives $E = (\delta G/\delta P)$, $x = -(\delta G/\delta X)$ and $S = -(\delta G/\delta T)$ are the electric field strength, electric-field induced strain and entropy, respectively. Connecting Eqs (B.10) and (B.11) yields:

$$E = (\delta G/\delta P) = aP + \beta P^3 + 2QPX, \tag{B.12a}$$

$$x = -(\delta G/\delta X) = sX + QP^2. \tag{B.12b}$$

Without external stress ($X = 0$), three characteristic equation can be deduced such as:

$$E = \alpha P + \beta P^3 \tag{B.13a}$$

1) The electrostrictive coefficient Q links the field-induced strain x and the polarization P by a parabolic relationship, and should not be confused with the general order parameter Q in Landau theory.

Table B.1 Dielectric permittivity and strain in paraelectric and ferroelectric states.

Paraelectric phase ($T > T_c$; $P_s = 0$, $P = \varepsilon_o\varepsilon E$)
Dielectric permittivity: $\varepsilon = C/(T - T_c)$ (Curie–Weiss law)
Field-induced strain: $x = QP^2 = Q\varepsilon_0^2\varepsilon^2E^2$ (electrostriction)
Ferroelectric phase $T < T_C$; $P_s^2 = -\alpha/\beta$; $P = P_s + \varepsilon_o\varepsilon E$
Dielectric permittivity: $\varepsilon = C/2(T_C - T)$
Total strain: $x = Q(P_s + \varepsilon_o\varepsilon E)^2 \approx QP_s^2 + 2\varepsilon_o\varepsilon QP_sE = QP_s^2 + d\cdot E$
Spontaneous strain: $x_s = QP_s^2$
Piezoelectric constant: $d = 2\,\varepsilon_o\varepsilon QP_s$

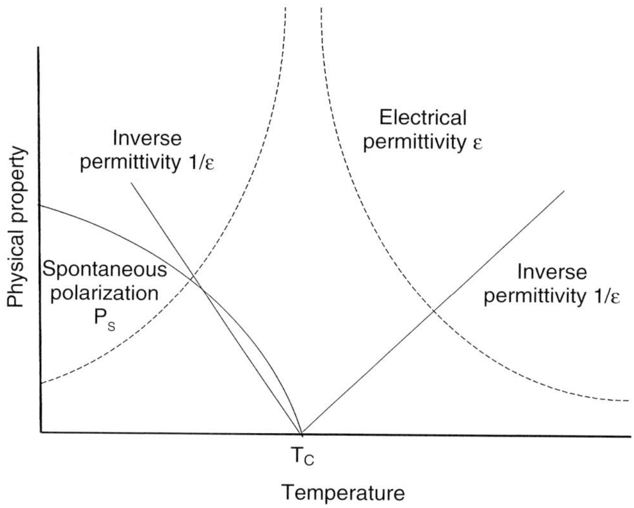

Figure B.4 Second-order transition in a ferroelectric material.

$$x = QP^2, \text{ and} \tag{B.13b}$$

$$1/\varepsilon_0\varepsilon = \alpha + 3\beta P^2. \tag{B.13c}$$

If the external electric field is also zero ($E = 0$), the following two states $P_s^2 = 0$ (paraelectric) and $P_s^2 = -\alpha/\beta$ (ferroelectric) are obtained (Table B.1).

Figure B.4 shows the theoretical temperature dependence of the spontaneous polarization P_s and the dielectric permittivity ε. Considering that, for the ferroelectric state, $\varepsilon = C/2(T_C - T)$ and $P_s = \sqrt{((T_C - T)/\varepsilon_0 C\beta)}$, the temperature dependence of the piezoelectric strain constant d is obtained as

$$d = Q\sqrt{\varepsilon_0 C/\beta(T_C - T)} \tag{B.14}$$

References

Carpenter, M.A. (1988) Thermochemistry of aluminium/silicon ordering in feldspar minerals, in *Physical Properties and Thermodynamic Behaviour of Minerals* (ed. E.K.H. Salje), D. Reidel Publishing Company, Dordrecht, pp. 265–323.

Carpenter, M.A. (1992) Thermodynamics of phase transitions in minerals: a macroscopic approach, in *The Stability of Minerals* (eds G.D. Price and N.L. Ross), Chapman & Hall, London, ch. 5, pp. 172–215.

Carpenter, M.A. and Boffa Ballaran, T. (2001) The influence of elastic strain heterogeneities in silicate solid solutions. *EMU Notes Mineral.*, **3**, 155–178.

Carpenter, M.A., Salje, E.K.H., Graeme-Barber, A., Wruck, B., Dove, M.T., and Knight, K.S. (1998) Calibration of excess thermodynamic properties and elastic constant variations due to the $\alpha \leftrightarrow ß$ phase transition in quartz. *Am. Mineral.*, **83**, 2–22.

Devonshire, A.F. (1949) Theory of ferroelectrics. *Philos. Mag. Suppl.*, **40**, 1040–1063.

Dove, M.T. (1997) Theory of displacive phase transitions in minerals. *Am. Mineral.*, **82**, 213–244.

Dove, M.T., Trachenko, K.O., and Tucker, M.G. (2000) Rigid unit modes in framework structures: theory, experiment and application, in *Transformation Processes in Minerals*, vol. **39** (eds S.A.T. Redfern and M.A. Carpenter), Reviews in Mineralogy and Geochemistry, Mineralogical Society of America, Washington, DC, ch. 1, pp. 1–33.

Landau, L.D. and Lifshitz, E.M. (1980) *Statistical Physics*, 3rd edn, Part 1, Pergamon Press, Oxford, New York, Toronto, Sydney, Paris, Frankfurt.

Müller, K.A. and von Waldkirch, T. (1973) Structural phase transition studies in $SrTiO_3$ and $LaAlO_3$ by EPR. IBM Research report RZ620, # 20576 Solid State Physics.

Salje, E. (1989) Characteristics of perovskite-related materials. *Philos. Trans. R. Soc. London*, **A328**, 409–416.

Schmahl, W.W., Swainson, I.P., Dove, M.T., and Graeme-Barber, A. (1992) Landau free energy and order parameter behavior of the α-ß phase transition in cristobalite. *Z. Kristallgr.*, **201**, 124–145.

Simon, R.W., Platt, C.E., Lee, A.E., and Lee, G.S. (1999) Superconductor Josephson junction comprising lanthanum aluminate. U.S. Patent 5,906,963 (May 25, 1999).

Tolédano, J.-C. and Tolédano, P. (1987) *The Landau Theory of Phase Transition*, World Scientific, Singapore, New Jersey, Hong Kong.

Uchino, K. (1995) New piezoelectric devices for smart actuator/sensor systems, in *Smart Materials Symposium* (ed. S. Sirisoonthorn), National Metal and Materials Technology Center (MTEC), Bangkok, Thailand, pp. 1–110.

Appendix C
Radial Electron Density Distribution (RED) Function

Well-ordered crystalline lattices with long-range ordering of atoms interact with X-rays or neutrons to generate sharp Bragg peaks when the positive interference conditions expressed by the Laue equations are met. However, the diffuse elastic scattering leads to a flat and featureless background. In contrast, in disordered solids such as glasses and amorphous materials, the background features between the Bragg peaks can become meaningful as they provide information on both the long-range and short-range ordering of atoms. Hence, data are collected that cover a large range of scattering vectors $Q = 4\pi \cdot (\sin \Theta)/\lambda$. Figure C.1 shows a plot of Q for β-cristobalite, silica gel, and silica glass. The sharp Bragg peaks of cristobalite attest to its crystalline nature, whereas in the amorphous silica structures only a very broad scatter peak exists at the position of the (111) reflection of cristobalite. Historically, this finding has led to the so-called "crystallite model of glasses" (Lebedev, 1921), in which the structure of glass was viewed as being composed of small ordered crystalline domains of compositions corresponding to compounds present in the system and, in the case of silica glass, consisting of cristobalite. The size of the proposed crystallites ranges between 0.7 and 0.8 nm (Warren, 1937) and 1.5 and 2.0 nm (Randall *et al.*, 1930). This intuitive model, however, is no longer supported in its original form (Kingery *et al.*, 1976).

Inelastic neutron scattering experiments have shown that strong structural similarities exist between β-cristobalite and amorphous silica, but also HP-tridymite over a length scale up to 1.0 nm. This has been interpreted in terms of the internal flexibility of the SiO_4 tetrahedra, as deduced from the rigid unit mode (RUM) model (see Appendix B). This internal flexibility will distort the local structure of the crystalline phase such as β-cristobalite to achieve a disorder similar to that found in the amorphous phases, silica gel, or silica glass.

A harmonic Fourier integral analysis of experimentally obtained X-ray or neutron scatter curves leads to the average radial distribution of atoms around each other atom in the (disordered) structure. The fundamental functional relationship between the scattering angle 2Θ and the intensity of the coherently scattered X-ray or neutron radiation can be expressed by

$$I(Q) = I_0\left(\frac{e^4}{m^2c^2R^2}\right)(1+\cos^2 2\Theta)\left(\frac{N}{2}\right)\cdot \sum_m \sum_n f_m f_n \cdot \left[\frac{\sin Q\cdot r_{mn}}{Q\cdot r_{mn}}\right] \qquad \text{(C.1)}$$

Classic and Advanced Ceramics: From Fundamentals to Applications. Robert B. Heimann

ISBN: 978-3-527-32517-7

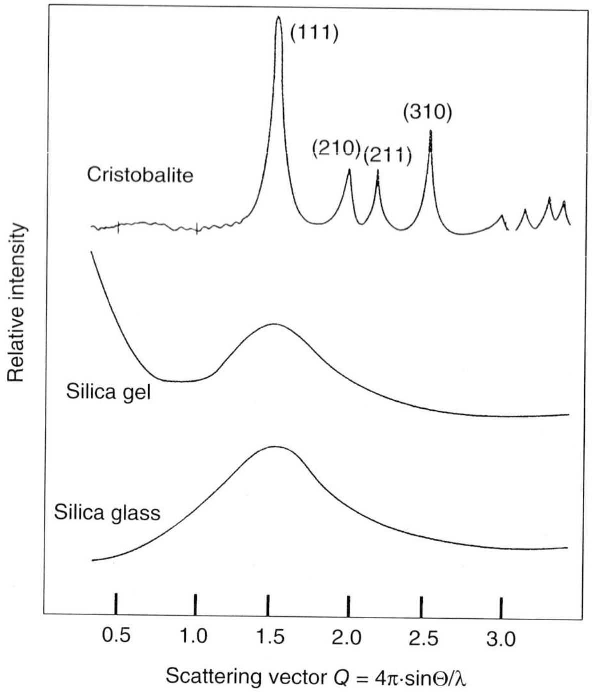

Figure C.1 X-ray diffraction patterns of crystalline β-cristobalite, and amorphous silica gel and glass.

The intensity of X-rays scattered at free electrons is

$$I_e = I_0\left(\frac{e^4}{m^2c^2R^2}\right)(1+\cos^2 2\theta)/2, \tag{C.2}$$

and the ratio is

$$I(Q)/I_e = I(Q) = N\cdot\frac{\sum_m \sum_n f_m f_n(\sin Q\cdot r_{mn})}{Q\cdot r_{mn}}, \tag{C.3}$$

where I_0 is the intensity of the incident beam, e the charge of electron, m the mass of the electrons, c the velocity of light, R the distance between the scatter center and the observation point, f_m and f_n are scattering factors of atoms m and n, N is the number of scatter centers, and r_{mn} the distance of atoms m and n.

Fourier transformation of experimentally obtained intensities yields (Warren *et al.*, 1936; Warren 1990) the weighted radial electron density (RED) distribution

$$\sum_m \overline{K_m} 4\pi r^2 \rho_m = \sum_m \overline{K_m} 4\pi r^2 \rho_0 + \left(\frac{2r}{\pi}\right)\int_0^\infty Q[i(Q)]\sin Q\cdot r dQ, \tag{C.4}$$

where K_m is the effective number of electrons in atom m, r is the distance of atoms, ρ_m is the (unknown) electron density, ρ_0 the average electron density, and $i(Q)$ the intensity of scattered radiation corrected for air scattering and polarization [see Eq. (C.5)]. The effective number of electrons in atom m, K_m is defined as:

$$\overline{K_m} = Q_{\max}^{-1} \int_0^{Q_{\max}} K_m dQ, \tag{C.4a}$$

that is, the averaged K_m value within the reciprocal sphere of radius $Q_{\max}$. The average electron density ρ_0 is

$$\rho_0 = \frac{[(N \cdot d \cdot 10^{-24}) \sum_m K_m]}{M}, \tag{C.4b}$$

where N is Avogadro's number, d is the density of the material, and M the molecular mass.

From experimental data, the following relationship can be obtained:

$$i(Q) = \frac{[I_{\exp,corr} - (I_{mod} + \sum_m f_m^2)]}{f_e^2}, \tag{C.5}$$

where $I_{exp,corr}$ is the experimentally determined intensity corrected for air scattering and polarization, I_{mod} is the contribution of Compton scattering (inelastic scattering at electrons), f_m is the atomic scattering factor of atom m, and f_e the average molecular electron scattering factor (= $\Sigma f_m / \Sigma Z_m$, where Z_m is the atomic number of atom m).

For a homoatomic material ($f_m = f_n$), Eq. (C.3) simplifies to:

$$I(Q) = N \cdot f^2 \sum_m [(\sin Q \cdot r_{mn})/Q \cdot r_{mn}] \tag{C.6}$$

The intensity $I(Q)$ can be thought of as being composed of the sum of two intensities, I_{int} and I_{ext}. I_{int} applies for $m = n$, and describes the intensity scattered by different regions of the electron cloud around an atom m. I_{ext} describes the situation when waves scattered by different atoms will interfere. Hence, $I(Q)$ is a function of the distance of the atoms:

$$I(Q) = N \cdot f^2 \sum_m^{m=n} [(\sin Q \cdot r_{mn})/Q \cdot r_{mn}] + N \cdot f^2 \sum_m^{m \neq n} [(\sin Q \cdot r_{mn})/Q \cdot r_{mn}] \tag{C.7}$$

Since $r_{mn} = 0$ when $m = n$, and since $(\sin Q \cdot r_{mn})/Q \cdot r_{mn} \rightarrow 1$ when $r_{mn} \rightarrow 0$ it follows

$$I(Q) = N \cdot f^2 \{1 + \sum_m^{m \neq n} [(\sin Q \cdot r_{mn}/Q \cdot r_{mn})]\} \tag{C.8}$$

The continuous distribution of atoms around each reference atom requires

$$I(Q) = N \cdot f^2 \left\{1 + \int_0^{\infty} 4\pi r^2 \rho(r) [(\sin Q \cdot r)/Q \cdot r] dr\right\} \tag{C.9}$$

where $\rho(r)$ is the density of atoms at a distance r from the reference atom, also called *pair distribution function*. Hence, $4\pi r^2 \rho(r) dr$ is the density of atoms in a spherical shell of radius r and thickness dr.

Equation (C.9) can be transformed to:

$$Q\left\{\frac{I(Q)}{N} \cdot f^2 - 1\right\} = 4\pi \int_0^{\infty} r \cdot \rho(r) \cdot \sin Q \cdot r dr \tag{C.9a}$$

and because of Eq. (C.5) it follows

$$Q \cdot i(Q) = 4\pi \int_0^{\infty} r \cdot \rho(r) \cdot \sin Q \cdot r dr. \tag{C.9b}$$

According to Fourier's integral theorem, it further follows that

$$r \cdot \rho(r) = \frac{\pi^2}{2} \int_0^{\infty} Q \cdot i(Q) \sin Q \cdot r dQ \tag{C.10}$$

If ρ_0 is the average density of atoms at a distance r from the reference atom, then

$$2\pi^2 \cdot r[\rho(r) - \rho] = \int_0^{\infty} Q \cdot i(Q) \sin Q \cdot r\, dQ \tag{C.11}$$

When multiplied by $2r/\pi$ and ordered, this finally yields the radial electron density distribution in a homoatomic material:

$$4\pi r^2 \rho(r) = 4\pi r^2 \rho_0 + \left(\frac{2r}{\pi}\right) \int_0^{\infty} Q \cdot i(Q) \cdot \sin Q \cdot r\, dQ \tag{C.12}$$

For heteroatomic materials ($f_m \neq f_n$)n the electron density distribution function has been determined by Warren and Gingrich (1934) whereby

$$f_m = K_m \cdot fe; \text{ with } fe = \frac{\sum_m f_m}{\sum_m z_m} \tag{C.13}$$

with Z_m = atomic number.

As an example, Figure C.2 displays the radial electron density (RED) curve $\rho(r)$ of silica glass as a function of the distance r.

The peak at 1.62 Å corresponds to the Si–O distance in a SiO_4 tetrahedron – that is, the distance between the Si atom located at the center of the tetrahedron and an O atom at the vertex. The intensity – that is, the area under the Si–O peak – has been determined to be 970 barns [see Eq. (C.4b)]. From f_{Si}/f_e and f_O/f_e, it follows that the effective number of electrons per Si atoms, K_{Si} = 16 and K_O = 7. The

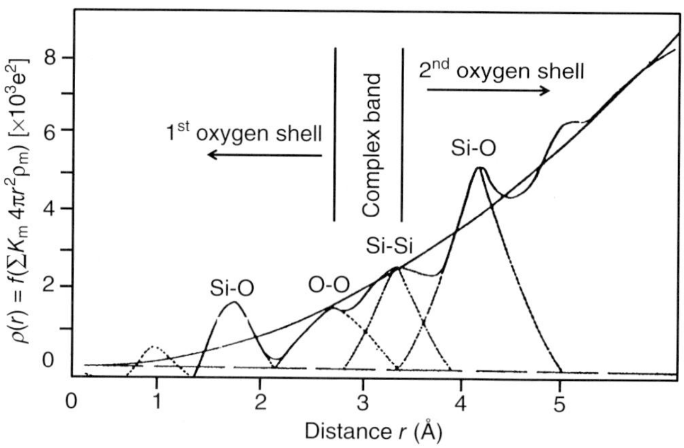

Figure C.2 Electron density distribution as function of the distance r (in Å) for amorphous silica (modified after Warren *et al.*, 1936).

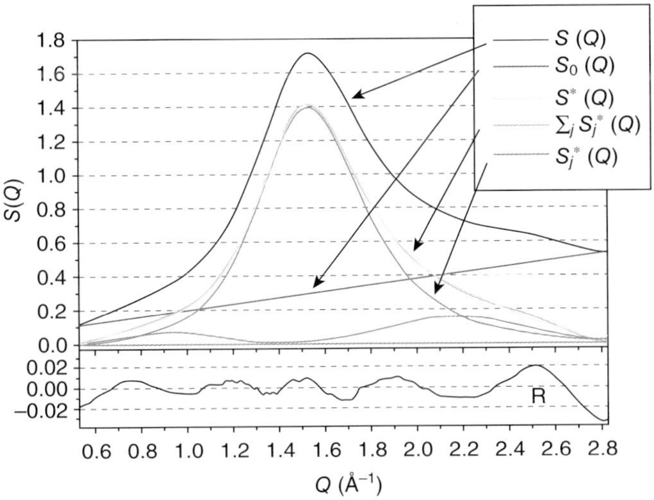

Figure C.3 Scattering curve $S(Q)$ for SiO_2 glass at 20 °C, and its deconvolution (Precht and Pentinghaus, 2000). R denotes the residual of the fitting procedure.

number of oxygen atoms surrounding each silicon atom in SiO_2 can be (roughly) calculated from the peak area as $n_O = I \cdot [\nu_{Si} \cdot (f_{Si}/f_e) \cdot \nu_O \cdot (f_O/f_e)]^{-1} = 970/(1 \times 16 \times 7 \times 2) = 4.3 \approx 4$, where ν refers to the stoichiometric factors of Si and O, respectively. From the geometry of the tetrahedron the O–O distance from vertex to vertex (i.e., the edge length of the tetrahedron) is $\sqrt{8/3} \times 1.62$ Å=2.65 Å, corresponding to the third peak in the curve of Figure C.2. The intensity of this peak is 1240 barns. Hence, the average number of Si atoms surrounding each O atom is $n_{Si} = 1240/(2 \times 16 \times 7) = 5.5$ (Warren *et al.*, 1936).

A higher resolution can be obtained today by applying synchrotron radiation. For this, the general procedure involves determining the structure factor $S(Q)$ in the medium-range order (MRO), that is, $Q = 4\pi(\sin\Theta)/\lambda \sim 0.5 \ldots 3\,Å^{-1}$, where glasses are characterized by the so-called first sharp diffraction peak (FSDP). This peak is often of composite nature, and can be deconvoluted into up to three partial profiles. Figure C.3 shows the experimentally obtained $S(Q) = i(Q)$ curve [Eq. (C.5)] of silica glass analogous to Figure C.1, and its deconvolution into partial profiles from which the fitting parameters Q_j, ΔQ_j and A_j (position of maximum of the scatter curve, width at half height, and area of peak, respectively) can be calculated. With these three parameters, the composition and amount of amorphous compounds can be characterized, for example, in binary Na_2O–SiO_2 glasses (Pentinghaus *et al.*, 2004).

After correction for background, distance and polarization, the structure factors are calculated as:

$$S(Q)-1=i(Q)=I_{cor}(Q)-N\sum_{i}^{uk} f_i(Q)^2/Nf_e(Q)^2. \tag{C.14}$$

This equation is equivalent to Eq. (C.5) above. Separating the structure factor $S(Q)$ from the base line $S_0(Q)$ by linear interpolation between adjacent local minima, the resultant function $S^*(Q)$ can be divided by a nonlinear least-squares fit into a minimum number ($j \leq 3$) of partial profiles $S_j^*(Q)$:

$$S(Q) = S_0(Q) + S^*(Q), \text{ with } S^*(Q) = \sum_j S_j^*(Q), \qquad \text{(C.15)}$$

where $S_j^*(Q) = f(Q_j, \Delta Q_j, A_j)$ and $a_j = A_j/\Sigma A_j$ (Pentinghaus *et al.*, 2004; Heimann *et al.*, 2007).

Note that in Figure C.3 there is only one partial profile $S_j^*(Q)$ since, in pure silica glass, there is only one type of topological order of the constituting SiO_4 building blocks.

References

Heimann, R.B., Pentinghaus, H.J., and Wirth, R. (2007) Plasma-sprayed 2:1-mullite coatings deposited on aluminium substrates. *Eur. J. Mineral.*, **19**, 281–291.

Kingery, W.D., Bowen, H.K., and Uhlmann, D.R. (1976) *Introduction to Ceramics*, 2nd edn, John Wiley & Sons, Inc., New York, Chichester, Brisbane, Toronto, Singapore, p. 95.

Lebedev, A.A. (1921) On the polymorphism and annealing of glass. *Trud. Gos. Opt. Inst.*, **2**, 1–20. (in Russian).

Pentinghaus, H.J., Precht, U., and Göttlicher, J. (2004) Fundamental material constants of silicate glasses and their melts from high-resolution WAXS applying synchrotron radiation. *International Congress on Glass (ICG) Kyoto*, Ceramic Society of Japan; Glass Section. ISBN 4-931-298-43-5, C3858.

Precht, U. and Pentinghaus, H.J. (2000) Metric characterization of medium range order in silicate glasses and their melts by WAXS. *Proceedings, International Conference on Applied Mineralogy (ICAM)*, (eds D. Rammlmair *et al.*), 17–19 July, Göttingen. Germany, Balkema, Rotterdam, pp. 209–212.

Randall, J.T., Rooksby, H.P., and Cooper, B.S. (1930) The structure of glasses: the evidence of X-ray diffraction. *J. Soc. Glass Technol.*, **14**, 219.

Warren, B.E. (1937) X-ray determination of the structure of liquids and glass. *J. Appl. Phys.*, **8** (10), 645–654.

Warren, B.E. (1990) *X-Ray Diffraction*, Dover Publications, New York.

Warren, B.E. and Gingrich, N.S. (1934) Fourier integral analysis of X-ray powder patterns. *Phys. Rev.*, **46** (5), 368–372.

Warren, B.E., Krutter, H., and Morningstar, O. (1936) Fourier analysis of X-ray patterns of vitreous SiO_2 and B_2O_3. *J. Am. Ceram. Soc.*, **19**, 202–206.

Appendix D
Thermodynamics of Spinodal Decomposition of Crystals

There are two types of phase transition, characterized by: (i) changes of concentration that are initially small but extend over a wide spatial range; and (ii) changes of concentration that are initially large but affect only a narrow spatial range. The former are termed *spinodal decomposition*, and the latter *nucleation and growth* (Kingery *et al.*, 1976).

Discussion of the theory of spinodal decomposition usually starts with the Gibbs free energy function, ΔG. The free energy of a hypothetical binary system A–B exhibits below the critical point of the miscibility gap two points of equal slope on the tangent, x_e^α and x_e^β, that is, the equilibrium compositions of the coexisting phases (Figure D.1). In a T–x diagram, these points mark the width of the region of immiscibility, the strain-free solvus, also called a binodal (Figure D.2). The points of inflection of the free energy curve ($d^2G/dx^2 = 0$) are the so-called *spinodes*, *s*. Their position in the T–x plot defines several curves: the strain-free solvus (a); the coherent solvus (b); and the chemical (c) and coherent spinodales (d).

Even in homogeneous solid solutions, such as marked by the points P or Q in Figure D.1, many statistical oscillations of the compositions exist on an atomic scale. These oscillations can accumulate if the composition is located between the spinodes. From a thermodynamic point of view, the composition P decreases its free energy; that is, it moves towards an energetically more favorable configuration by enhancing the fluctuations during decomposition into the adjacent compositions p and p'. This process does not require nucleation of the phases p and p' since the compositional fluctuations are of statistic and continuous nature. This is equivalent to the variation of the amplitude (= composition) of a sine wave, the wavelength of which is a distance parameter in the crystal structure (Yund, 1975b). Composition p is richer in A than composition p', composition p' is richer in B than composition p. This decomposition reaction often results in formation of coherent lamellae composed of p and p' that can be observed, for example, during the spinodal decomposition of alkali feldspars of the system albite-potassium feldspar. At low pressure, there exists a coherent spinodal at T_{max} = 580 °C that causes the exsolution of two coexisting sanidine compositions with different Na/K ratios. Such alkali feldspars with coherent lattices formed by spinodal decomposition are termed *cryptoperthites* (e.g., Smith, 1974).

Classic and Advanced Ceramics: From Fundamentals to Applications. Robert B. Heimann

ISBN: 978-3-527-32517-7

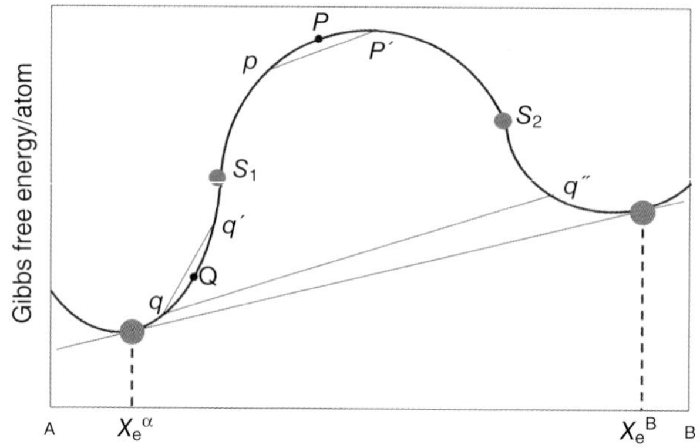

Figure D.1 Free energy curve of a binary system A–B in the region of a miscibility gap between the equilibrium compositions x_e^α and x_e^β. Spinodes s_1 . s_2. For an explanation, see the text. Reprinted with permission from The Mineralogical Society of America.

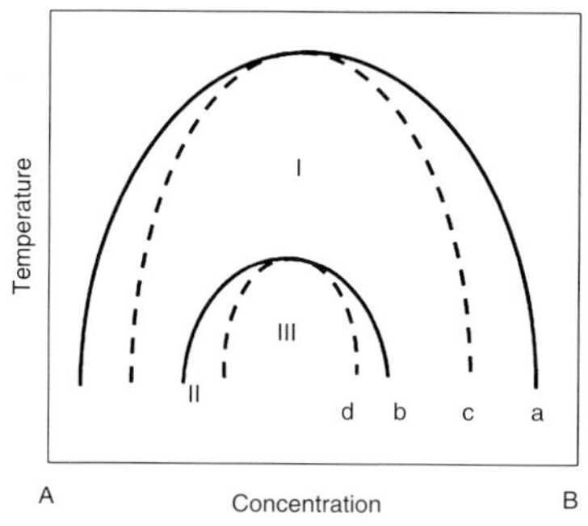

Figure D.2 *T–x* section of a binary phase diagram A–B with strain-free solvus (binodal) *a*, chemical spinodal *c*, coherent solvus *b*, and coherent spinodal *d*. Reprinted with permission from The Mineralogical Society of America.

Between the strain-free solvus and the spinode $(x_e^\alpha - s_1)$, the free energy of the system increases as the compositional difference between the initial statistical fluctuation and the average composition of the surrounding increases. This is shown in Figure D.1 (as line *q*–*q*′). In contrast to *p*–*p*′, the demixing of a phase located between the chemical spinodal and the strain-free solvus requires the nucleation of phase *q*″. In this case the system consists of phases *q* and *q*″, with a much decreased free energy.

It should be emphasized that Figure D.1 does not take into account the energy required to nucleate *q*″. By considering the additional interfacial and elastic strain energies, a curve emerges that is positioned higher on the *G*–*x* diagram. Such a curve is the Cahn free energy curve (Cahn, 1962; Hilliard, 1970) that, in addition

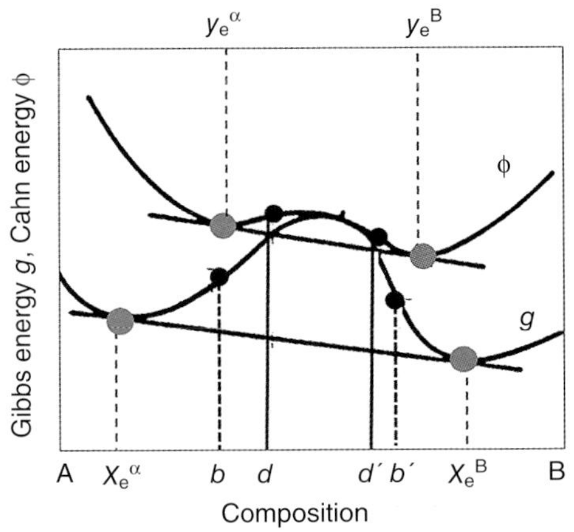

Figure D.3 Gibbs (chemical) free energy g and Cahn (strain) free energy Φ as a function of the composition (after Yund, 1975a). Reprinted with permission from The Mineralogical Society of America.

to the Gibbs (chemical) free energy curve, contains energy increments related to the coherent strain energy (Figure D.3).

In Figure D.3, the points y_e^α and y_e^β on the Cahn free energy curve Φ mark the coherent solvus, while the points x_e^α and x_e^β on the Gibbs (chemical) free energy *g* curve mark the strain-free solvus. The points of inflection of the Φ curve, *d* and *d′*, are located inside those on the *g* curve, *b* and *b′*. These so-called *coherent spinodes* define the limit of the spinodal decomposition by considering the elastic strain energy due to nucleation introduced by the Cahn–Hilliard approach. As shown in Figure D.2, in the *T*–*x* diagram the position of the coherent spinodes defines the coherent spinodale *d*–*d′* and the coherent solvus *b*–*b′*, respectively. In region (I) between the strain-free solvus (*a*) and the coherent solvus *b*, demixing can occur only by classical nucleation and subsequent crystal growth. This is characteristic for the metastable part of a phase diagram. Region (III) below the coherent spinodal *d* allows demixing without any energetic barrier – that is, without nucleation. In this area the kinetics of exsolution is limited only by the diffusivity of ions in the crystal lattice. Finally, region (II) between the coherent solvus *b* and the coherent spinodal *d* is part of the metastable part of the phase diagram, but the formation of coherent nuclei is required by homogeneous nucleation (Yund, 1975b). As expected, there is a symmetry between the *g* and Φ curves such that the strain-free solvus (*a*) and the chemical spinodal (*c*) on the *g* curve bear the same relation to the coherent solvus (*b*) and the coherent spinodal (*d*) on the Φ curve.

This very basic discussion of spinodal decomposition is entirely phenomenological, as it does not involve any mathematical treatment of thermodynamics and kinetics of the reaction. Such details can be found in Kingery *et al.* (1976) and Langer (1971, 1973). More recently, the theory of spinodal decomposition was extended using dynamic density functional theory (DFT). For the early stages of spinodal decomposition, this nonlinear theory is equivalent to the (generalized) linear Cahn–Hilliard theory, but for later stages it incorporates coupling between different Fourier components of the density fluctuations (modes) and therefore goes beyond the Cahn–Hilliard theory discussed above (Archer and Evans, 2004).

References

Archer, A.J. and Evans, R. (2004) Dynamical density functional theory and its application to spinodal decomposition. *J. Chem. Phys.*, **121** (9), 4246–4254.

Cahn, J.W. (1962) Coherent fluctuations and nucleation in isotropic solids. *Acta Metall.*, **10**, 907–913.

Hilliard, J.E. (1970) Spinodal decomposition, in *Phase Transformations* (ed. H.I. Aaronson), Am. Soc. Metals, Ohio, pp. 497–560.

Kingery, W.D., Bowen, H.K., and Uhlmann, D.R. (1976) *Introduction to Ceramics*, John Wiley & Sons, Inc., New York, Chichester, Brisbane, Toronto, Singapore, pp. 320–328.

Langer, J.S. (1971) Theory of spinodal decomposition in alloys. *Ann. Phys. (N.Y.)*, **65**, 53–86.

Langer, J.S. (1973) Statistical methods in the theory of spinodal decomposition. *Acta Metall.*, **21**, 1649–1659.

Smith, J.V. (1974) *Feldspar Minerals*, vol. **2**, Springer, Berlin, Heidelberg, New York, ch. 19.1.4, pp. 493–504.

Yund, R.A. (1975a) Subsolidus phase relations in the alkali feldspars with emphasis on coherent phases, in *Feldspar Mineralogy. Short Course Notes*, Mineral Society of America, vol. **2**, pp. Y1–Y28.

Yund, R.A. (1975b) Microstructure, kinetics, and mechanisms of alkali feldspar exsolution, in *Feldspar Mineralogy. Short Course Notes*, Mineral Society of America, vol. **2**, pp. Y29–Y57.

Appendix E[1]
Theory of Superconductivity

This appendix summarizes some basic aspects of the theory of superconductivity, covering the classical London equations, the BCS theory of weakly coupled low-T_c materials, the unconventional superconductivity of strongly coupled high-T_c compounds with its electronic and magnetic origin, and contributes some aspects to the understanding of the behavior of type II superconductors.

London Equations

In the classical model of superconductivity, the London equations (London and London 1935) are equivalent to Ohm's law $\mathbf{j} = \sigma \cdot \mathbf{E}$ for a normal electric conductor. The first of the London equations [Eq. (E.1)] represents a conductor with $R = 0$, while the second [Eq. (E.2)] is equivalent to the Meissner–Ochsenfeld effect (Figure E.1), and describes the decay of a magnetic field within a thin surface layer characterized by the penetration depth, λ_L.

$$\mathbf{E} = \frac{\partial}{\partial t}(\Lambda \mathbf{J}) \quad \text{where } \Lambda = \frac{m_s}{n_s q_s^2} \quad \text{First London equation} \tag{E.1}$$

$$\mathbf{B} = -\nabla \times (\Lambda \mathbf{J}) \quad \text{Second London equation} \tag{E.2}$$

where $\mathbf{J}$ is the current density, m_s and q_s are twice the electron mass and electron charge, respectively, and n_s consequently is half of the electron number density, $\mathbf{E}$ is the electric field vector, and $\mathbf{B}$ the magnetic field vector. The London penetration depth, λ_L, is given as

$$\lambda_L = \sqrt{\frac{\Lambda}{\mu_0}} \tag{E.3}$$

where μ_0 is the permeability of the vacuum.

It is important to note that these London equations are not an explanation of superconductivity at all. They were introduced only as a restriction on Maxwell's equations, so that the behavior of superconductors deduced from these equations

1) Contributed by H.H. Otto.

Classic and Advanced Ceramics: From Fundamentals to Applications. Robert B. Heimann

ISBN: 978-3-527-32517-7

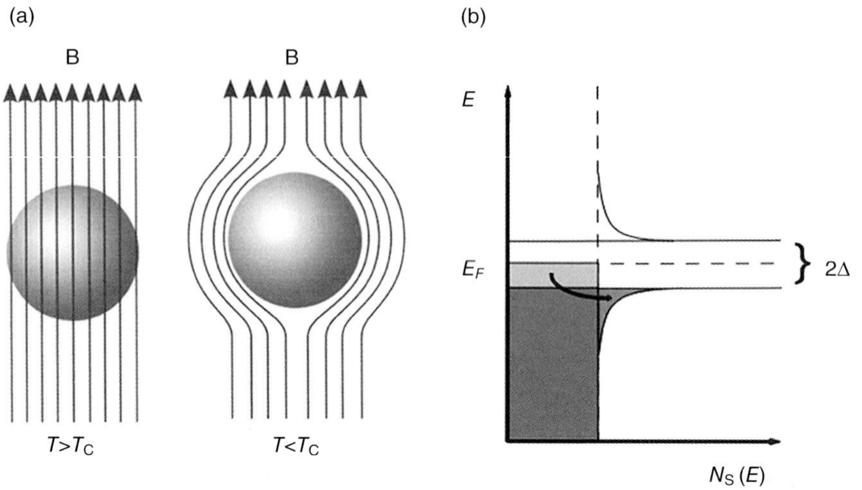

Figure E.1 (a) Schematic rendering of the Meissner–Ochsenfeld effect of magnetic flux expulsion below T_c (created by Piotr Jaworski, Poznan, Poland); (b) The energy gap 2Δ of a superconductor. The electronic states of a normal electrically conducting metal are filled up to the Fermi energy E_F, including the region outlined in yellow. In a superconductor an energy gap 2Δ opens up and the yellow states are redistributed to the blue region. $N_s(E)$ indicates the number of states.

was consistent with experimental observations and, in particular, with the Meissner–Ochsenfeld effect. Consequently, their status is somewhat similar to Ohm's law which, useful as it is to describe the behavior of many normal metals, does not provide any explanation of the conduction process at a microscopic level.

Bardeen–Cooper–Schrieffer (BCS) Theory

The Bardeen–Cooper–Schrieffer theory (Bardeen *et al.*, 1957) has been extremely successful for explaining the weak coupling of electrons underlying the mechanism of superconductivity of low-T_c metals and alloys. The electrons interact with lattice vibration excitations (phonons) to pair in s-wave state with zero total orbital angular momentum. A traveling electron can locally polarize the lattice, attract positively charged ions, and expel negatively charged ions. The polarization is still sustained when attracting a second approaching electron, and thus both electrons couple to pairs. The Coulombic repulsion is weak because the first electron is already far away when the second approaches. The mean distance between such electron couples is large compared to the mean distance between the free conducting electrons. The most probable distance ξ_{co} (the coherence length) expands in low-T_c materials from 100 to 1000 nm. In this theory the single electron properties are renormalized through interaction with other electrons to form quasiparticles, hence tracing back the material properties to remaining weak interaction between them. The BCS theory has been used to successfully explain the following experimental results:

- The second-order phase transition at T_c (Ginzburg and Landau, 1950).
- The existence of an energy gap 2Δ in the electronic excitation spectrum (Figure E.1b). This energy gap 2Δ is highest at low temperature and decreases asymptotically to zero at T_c as $2\Delta = 3.2T_c\{1 - (T/T_c)\}^{1/2}$. Since for $T = 0$ the energy gap does not depend on the material, $2\Delta(0) = 3.5\ k_B T_c$.
- The exponential decay of the specific heat near $T = 0$.
- The Meissner–Ochsenfeld effect (Figure E.1a).
- the infinite d.c. conductivity at $T = 0$.
- The dependence of T_c on the mass M of isotopes, $T_c \propto M^{-1/2}$.

The superconducting transition temperature can be expressed in terms of the electron–phonon coupling potential U and the Debye cutoff energy $E_D = h \cdot \omega_D / 2\pi$ as:

$$k_B T_c = 1.14 E_D \cdot \exp\{-1/(N(E_F)U)\}, \tag{E.4}$$

where $N(E_F)$ represents the density of states at the Fermi energy that is the highest occupied quantum state of the system, and k_B is the Boltzmann constant.

Allowing for stronger coupling in a conventional superconductor corresponding to a higher T_c, the Allen–Dynes–McMillan semi-empirical formula (McMillan, 1968; Allen and Dynes, 1975) is widely used for qualitative predictions,

$$k_B T_C = \frac{\hbar\omega_C}{1.2} \exp\left\{-\frac{1.04(1+\lambda)}{\lambda - \mu^*(1-0.62\lambda)}\right\} \tag{E.5}$$

where λ is the electron–phonon coupling parameter and μ^* the renormalized pseudopotential of the Coulombic repulsion,

$$\mu^* = \frac{\mu}{1 + \ln(E_F/\omega_D)} \tag{E.6}$$

that may be considered a vertex correction of the screened Coulombic interaction.

The recently discovered MgB_2 with a very high transition temperature of $T_c = 40\,K$, is still a BCS superconductor, dubbed by Pickett (2008) the "queen of phonon coupling," in recognition of its unusually high T_c. For MgB_2, the first-principle calculations explain indeed the most unusual behavior (Cho *et al.*, 2002). Two different types of electron can couple to pairs, creating multiple energy gaps in the superconducting regime that converge towards the transition temperature of $T_c = 40\,K$. Partly occupied σ-bonds in the boron layer of the crystal structure (see Figure 9.2b) may strengthen the quantized lattice vibrations driving the superconductivity. The high T_c is reproduced for the pseudopotential value $\mu^* = 0.12$.

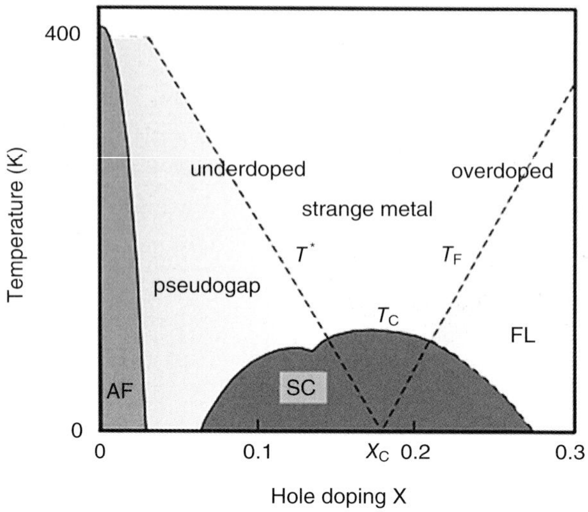

Figure E.2 Temperature as a function of hole concentration in high-T_c cuprates.

High-Temperature Superconductor (HTS) Cuprates (Unconventional Superconductors)

It is widely accepted that the unconventional superconductivity of the cuprates is not mediated by phonon interaction, but rather has its origins in the electric and magnetic peculiarities of the CuO_2 layers, and the doped holes therein. However, the indirect nature of experimental evidence for the rich variety of states that can coexist, cooperate or compete, may be one of the reasons why the theory of high-T_c superconductivity is still incomplete, and hence unsatisfactory. Appropriately, Jan Zaanen of Leiden University speaks of a "... graveyard of theories" in this research field (Zaanen, 2004). This outlook notwithstanding, a short survey of promising theoretical approaches will be given below.

Using Varma's first proposal (1997), and based on new experimental evidence, a universal temperature-doping phase diagram for high-T_c superconductors has been drawn up (Figure E.2). This maps the phase boundaries between the antiferromagnetic insulating Mott phase (AF), the superconducting dome (SC), the metallic Fermi-liquid phase (FL), and, far above T_c, the weakly metallic pseudogap (spin-gap) phase that develops on stronger doping into a peculiar "strange" metal phase. Considerable similarities can be recognized between this approach and those published for organic superconductors (Powell and McKenzie, 2005) and for heavy fermion superconductors, respectively, that show even lower T_c. The crossover temperature lines T^* and T_F are thought to meet in the center of the superconducting dome at a quantum critical point with hole doping concentration x_c (Varma, 1997). The opening of a pseudogap far above the superconducting gap in the underdoped region is clearly connected with the confinement of mobile charge

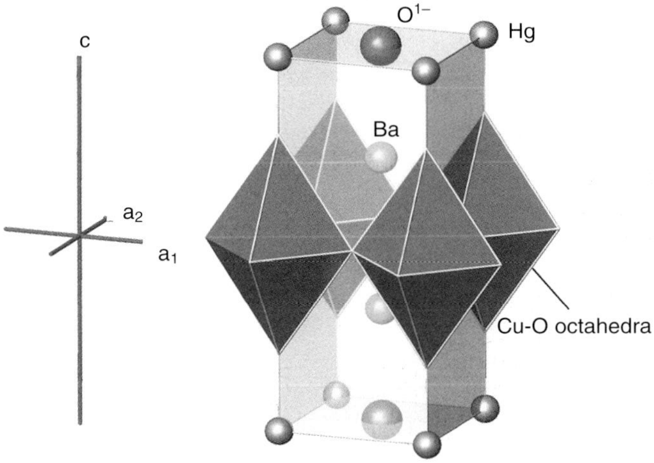

Figure E.3 Crystal structure of tetragonal $HgBa_2CuO_{4+\delta}$. The symmetry may be reduced compared to space group *I4/mmm*; a = 0.3880 nm, c = 0.951 nm. The extra O_δ^{1-} oxygen ions may actually occupy an eightfold split position.

carriers in the cuprate layers within one-dimensional (1-D) stripes that are separated by regions of antiferromagnetic spin order. Previously predicted by mean-field theoretical results, these were first observed by Tranquada *et al.* (1995) in neutron-scattering experiments, and the existence of pre-formed electron pairs in this region of the phase diagram was recently confirmed (Yang *et al.*, 2008). Since not only the stripe–spin-charge ordered states but also these preformed electron pairs disappear slowly at higher doping levels, they compete with rather than assist superconductivity. By crossing the T^* borderline in the direction of the "strange" metal phase, the static flux lattice observed in the pseudogap phase is assumed to melt.

Li *et al.* (2008) proved, by using polarized neutron diffraction measurements on differently doped samples of the very simple Hg1201 compound (Figure E.3), the existence of an extended charge current loop orders below T^* that involves the apical oxygen atom. The magnetic signal has a large spatial extent and disappears when the temperature is decreasing towards T_c; thus, the effect seems to compete with superconductivity. Nevertheless, the possibility exists that the appearance of superconductivity may be associated with fluctuations and quantum criticality. Previously, Aji and Varma (2007) proposed the appearance of current loops and quantum critical behavior.

Finally, some interesting scaling laws will be reported that exist for unconventional superconductors. Yu *et al.* (2009) demonstrated for such unconventional superconductors (including cuprates, pnictides and heavy-fermions) the existence of a linear relationship between the magnetic resonance energy of a collective spinon excitation mode and the superconducting energy gap, $E_\gamma \propto 2\Delta$. This relationship extends over two orders of magnitude in energy, and is valid for distinct doping levels. The magnetic resonance mode is a soft spin mode related to the stripe–spin-charge ordered state in the cuprates, and the antiferromagnetic fluctuations causing this mode may be therefore connected to unconventional superconductivity.

Some time ago, Uemura *et al.* (1991) proposed a universal linear relationship $T_c \propto \rho_s$ between the superconducting transition temperature and the superfluid density ρ_s of charge carriers for a group of superconducting compounds. Although the Uemura law describes underdoped cuprates well, it does not work for optimal or overdoped cuprates.

Recently, the superfluid density ρ_s was related to the normal state d.c. conductivity at T_c and to T_c itself by the simple scaling rule $\rho_s \propto \sigma(T_c) \cdot T_c$ (Homes *et al.*, 2004). This linear relation holds for all high-T_c cuprates, regardless of the nature of the dopants (electrons or holes) and the doping level, as well as the crystallographic direction of the CuO_2 layers. With Homes' and Uemura's law scaling laws in mind, Zaanen (2004) demonstrated by dimensional analysis that Planck's timescale of dissipation can be expressed by $\tau(T_c) = h/(2\pi k_B T_c)$, converting energy into time through the uncertainty principle. This means that the normal electron fluid in the high-T_c cuprates may be considered to be at the quantum limit of dissipation, pointing to the suspected quantum critical nature of the normal state dominating the physical properties.

The high transition temperature of the cuprate superconductors and the short coherence length ξ_{co} (on the order of 1 nm) indicate that strongly correlated electron systems exist in which the interaction between nearest neighbors leads to real space pairing, in contrast to the BCS superconductors with pairing in momentum space. Also, in contrast to low-T_c superconductors, recent experiments on high-T_c materials have confirmed that the electron pairs in the latter show the d-wave state, a superposition of states with a nonzero angular momentum. The small coherence length in unconventional superconductors is responsible for the occurrence of spatial extended fluctuations in the order parameter. The energy barrier against the creation of fluctuations is proportional to the volume, ξ_{co}^3.

In the following section, two mainstream theoretical concepts will be briefly described, namely the resonating valence bond (RVB) theory of Anderson, and Varma's pairing model of quantum critical fluctuation (QCF), as the origin of high-T_c superconductivity.

For strongly correlated systems, the so-called *t–J* model (Spałec *et al.*, 1978) was successfully applied. A doped Mott insulator can be described by the *t–J* Hamiltonian, the simplest operator dealing with holes moving about an antiferromagnetic background:

$$\hat{H} = -t\sum_{ij\sigma} \hat{a}^{\neq}_{i\sigma}\hat{a}_{j\sigma} + J\sum_{ij}\left(\vec{S}_I \cdot \vec{S}_J - n_i n_j/4\right) \quad \text{(E.7)}$$

where $J = 4t^2/U$ represents the coupling constant with the Coulombic repulsion, U. The $\hat{a}$ terms are fermion creation and annihilation operators for spin polarization σ, and the $S_{i,j}$ are spins on i and j nodes.

In the *t–J* model, t_{pd} electron hopping from local ε_p, ε_d energy levels (Figure E.4) to states with higher energy U is leveled out in favor of an antiferromagnetic

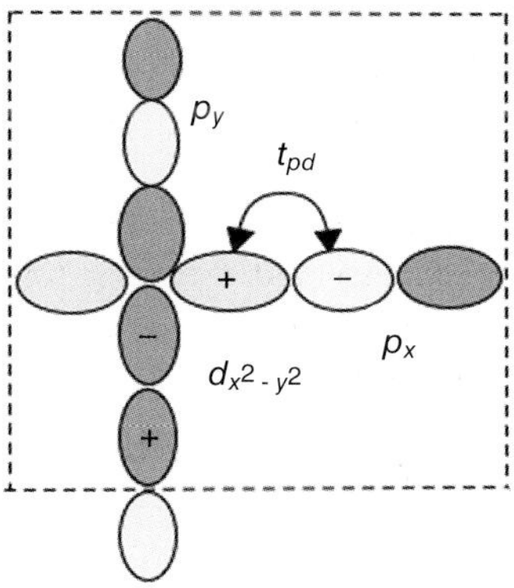

Figure E.4 CuO_2 orbital model represented in the planar (physical) unit cell (dashed outline) with three atoms. The kinetic t_{pd} hopping term is indicated.

exchange interaction, J. The kinetic exchange between nearest neighbors is regarded a source of real space pairing.

One of the theories which uses the t–J model and is favored by low structural dimensionality, low spin and magnetic frustration, is the RVB theory of Anderson (1987), which was actualized to explain new experimental evidence. By first removing any double occupancies, only singly occupied sites with spin will remain in the half-filled band of the Mott insulator that can pair up to generate a bond pair liquid, commonly termed the RVB configuration (Figure E.5). It is assumed that, in the Mott insulator, after sufficiently strong doping with holes, the resonating singlet pairs will be retained in the resulting metal and will be capable of carrying the charge and supporting the super-currents.

By using the d-wave concept of Kotliar and Liu (1988), a "conventional" (BCS-like) single-band singlet-pair wave function (BCS-like) will be achieved when electron double occupancies are removed by applying the Gutzwiller method (Gutzwiller, 1963).

In contrast to the Anderson theory, Varma (Aji and Varma, 2007) calculated the momentum dependence of fermion scattering with quantum critical fluctuations (QCFs) in the pseudogap (underdoped) regime. Coupling of the QCF to the local angular fermion momentum clearly promotes attractive d-wave pairing and rejects repulsive s-wave coupling. Fluctuations are considered between four domains, with the resultant circulating currents flowing from the O–Cu–O loop with negative flux to the loop with positive flux in the plane of the CuO_2 plaquette (Figure E.6). The phase diagram (see Figure E.2) shows that the circulating current phase terminates at the point x_c at $T = 0$. Varma's result, leading to a high T_c and explaining the important properties, is quite astonishing since the experiment of Li *et al.* (2008) indicates for such circulating currents no contribution of the copper ion but rather a canted momentum, owing to the contribution of the apical oxygen ion (Figure E.6).

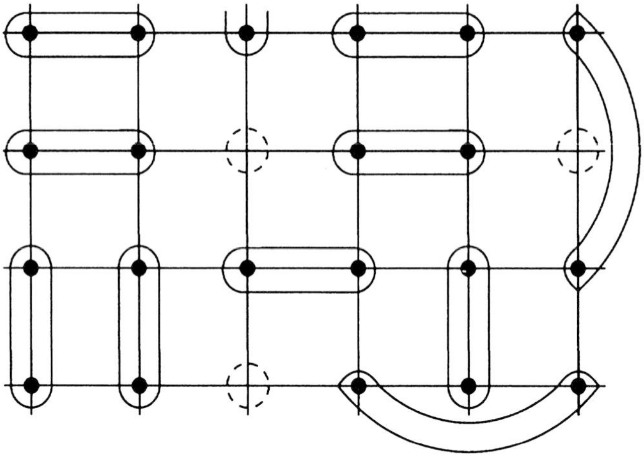

$$r \; (\bullet \quad \bullet) \; r' = \frac{|\uparrow_r \downarrow_{r'}\rangle - |\downarrow_r \uparrow_{r'}\rangle}{\sqrt{2}}$$

Figure E.5 Singlet pair liquid configuration and some doped holes (dashed circles) according to Anderson's resonating valence bond (RVB) model (as depicted by Anderson *et al.*, 2004). The asymmetric spin part of the spin-singlets (total spin $S = 0$) is also shown.

To conclude, some thoughts of the present author (H.H. Otto) may be expressed. The current loops reported extend in the direction of the doped O_δ^{1-} ion in the Hg1201 model compound. When the temperature decreases towards T_c, increasingly the currents smear out and may finally overlap diagonally from the CuO_2 layer below to the layer above via the oxygen dopant. This will allow for a cooperative interlayer effect that results in pairing. In contrast, the Varma concept does not consider (or need) any interlayer coupling at all.

The apical oxygen-to-copper bond length in superconducting cuprates dominantly contributes to the bond valence sum, showing that the critical temperature is the higher the more the formal charge distribution favors Cu^{2+}–O^{1-} instead of Cu^{3+}–O^{2-} allocation (De Leeuw *et al.*, 1990). This relation may be also inverse exponential, as was demonstrated for the mean cationic charge $\langle q_c \rangle$ indicating maximum T_c near $\langle q_c \rangle = 2$ (Otto, 2008). Any superconductivity theory that does not consider the effect of the apical oxygen may at best be approximating the whole scientific truth.

A purely electronic transition to a non-BCS superconducting state under pressure with $T_c = 39\,K$ has been described recently by Takabayashi *et al.* (2009) for the disorder-free Cs_3C_{60} cesium fulleride with A15 crystal structure. At ambient pressure, this compound behaves as an antiferromagnetic insulator.

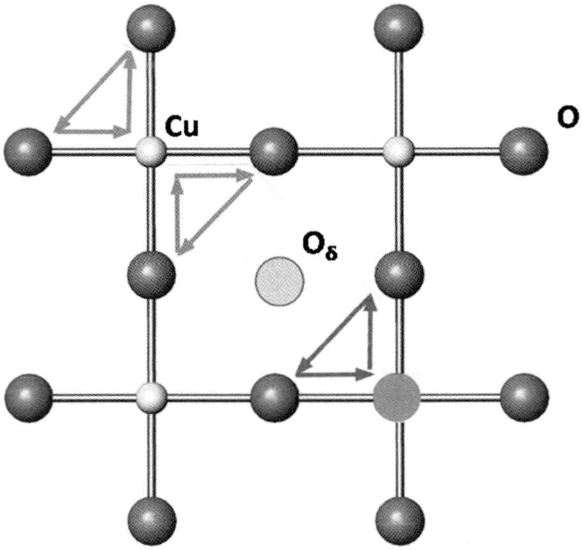

Figure E.6 Current loops between oxygen (large spheres) and copper ions emerging from the Varma model (arrows at top left), and extending only in the CuO_2 plane, violate time reversal symmetry but conserve translational symmetry. The experimental result of Li *et al.* (2008) shows, on the right-hand side, canted circulating current loops between the planar and apical oxygen atoms (arrows at bottom right). The magnetic scattering extends in the direction of the extra oxygen O_δ in the center of the unit cell of Hg1201.

Heavy Fermion Superconductors

Intermetallic compounds with localized or itinerant f-electrons, based upon rare earth elements such as cerium or actinide elements such as uranium, show a large effective electronic mass that exceeds the free electron mass more than a hundredfold; these masses are therefore termed "heavy-electron metals" or "heavy fermions." Despite the strong antiferromagnetic interaction of local magnetic moments in a sublattice with conducting electrons at low temperature (Kondo effect), low-T_c superconductivity was observed in these compounds, initially by Steglich *et al.* (1979) in $CeCu_2Si_2$. The pairing mechanism leading to superconductivity may be distinct for different compounds. For example, in the case of $CeCu_2Si_2$ two distinct superconducting phases have been observed under ambient and high pressure (Yuan *et al.*, 2003). In this case, superconductivity is likely to be mediated by coupling of the heavy quasiparticles to spin degrees of freedom in the low-pressure phase, but to charge degrees of freedom in the high-pressure phase. A mechanism mediated by magnetic interaction is suggested for the pressure-induced superconductivity of $CeIn_3$ and $CePd_2Si_2$.

Type II Superconductors

Besides the critical temperature, the technical applicability of a superconducting material depends on important properties, such as the critical current-carrying capability and the critical magnetic field that does not destroy the superconductivity. If a type II superconductor is placed in a magnetic field, the magnetic flux lines are able to penetrate the material above a critical field H_{C1}, thus forming a triangular magnetic vortex lattice (the Abrikosov lattice; Abrikosov, 1957) with a lattice spacing that depends on the strength of the field (see Figure 9.1). Vortices can move and dissipate energy under the influence of the Lorentz force acting according to the Lenz rule at a right-angle to an applied current flow. It is of great importance for applications to hinder such vortex movement (flux creeping) by pinning the flux lattice at impurity centers. This may be introduced either by intrinsic defects and phase inhomogeneities, or by pointedly tailored (extrinsic) defects such as radiation-generated lattice defects or substituted Zr in REBCO (rare-earth barium copper oxide). However, above a critical field, H_{C2} superconductivity will be destroyed such that a normal metallic behavior results. Figure E.7a shows the magnetization behavior of a type II superconductor under the influence of the operating magnetic field with force, H. The superconducting phase below $\mathbf{B}_{C1}$, with a completely expelled magnetic flux, is referred to as the Meissner phase, and the mixed phase below $\mathbf{B}_{C2}$ with a partly penetrated magnetic flux as the Shubnikov phase (Figure E.7b).

The critical field strength, H_c–that is, the magnetic field capable of destroying the superconductivity–depends on the temperature according to:

$$H_C(T) = H_C(0)\left[1-(T/T_C)^2\right] \qquad \text{(E.8)}$$

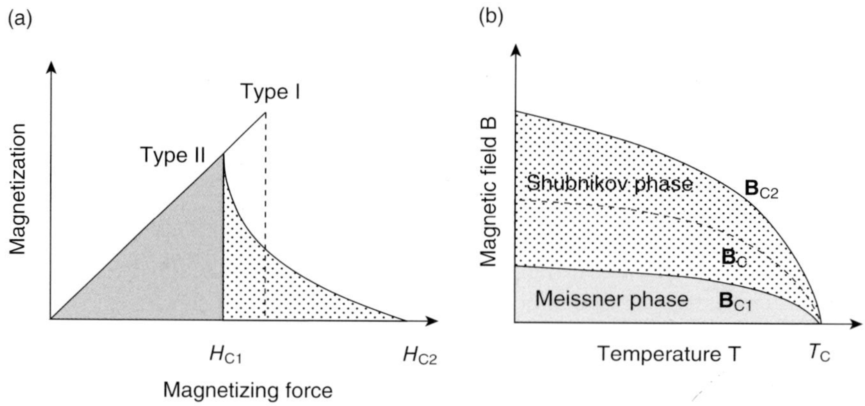

Figure E.7 (a) Magnetic phase diagram of type II superconductors; (b) B-T phase diagram of type II superconductors indicating the regions of the Meissner and the Shubnikov phases with magnetic flux penetration into the material.

The critical magnetic field density $\mathbf{B}_{C2} = \mu_0 \cdot H_{C2}$ can be estimated in relation to the onset of overlap of magnetic flux lines. A single flux line (fluxoid), representing the region of a circulating supercurrent of extension λ_L (penetration depth) with a normal conducting core of radius ξ_{co} (correlation length), carries one flux quantum $\Phi_0 = 2.1 \times 10^{-15}$ Tm^{-2}. The normal conducting state results if fluxoids touch each other at a distance of $2\xi_{co}$. Then, the critical magnetic field density (per unit area) evolves as:

$$B_{C2} = \frac{\Phi_0}{2\pi\xi^2} \tag{E.9}$$

Type II superconductors are normally limited by the Ginzburg–Landau ratio $k = \lambda/\xi > 1/\sqrt{2}$. One special attribute of a high-temperature superconductor (HTS) materials is its small coherence length in comparison with a low-temperature superconductor (LTS), and its anisotropy due to the two-dimensional (2-D) character of the layered crystal structures. The coherence length is on the order of unit cell dimensions, where $\xi_{ab} > \xi_c$. For example, yttrium barium copper oxide (YBCO) shows values of $\xi_{ab} \approx 2\,\text{nm}$ and $\xi_c \approx 0.4\,\text{nm}$ as well as $\lambda_{ab} \approx 150\,\text{nm}$ and $\lambda_c \approx 800\,\text{nm}$, with an anisotropy parameter of $\xi_{ab}/\xi_c = \lambda_c/\lambda_{ab} \approx 5$ corresponding [according to Eq. (E.9)], to the anisotropy ratio of the critical fields in the ab-plane and perpendicular to it, $\mathbf{B}_{C2}^{ab}/\mathbf{B}_{C2}^{c}$.

The critical current density j_c, the second important property, attains a maximum at $T = 0$ and $H = 0$, and decreases in a characteristic but complicated manner on increasing T and/or H. It can be determined experimentally from magnetization loops. The maximum current density of a superconductor is the depairing density $j_c^{dep} = H_c/\lambda_L$, but in reality the current density is several orders lower than this value. Further details may be obtained from textbooks on applied superconductivity.

References

Abrikosov, A.A. (1957) On the magnetic properties of superconductors of the second group. *Zh. Eksp. Teor. Fiz.*, **32**, 1442–1452 (in Russian).

Aji, V. and Varma, C.M. (2007) Theory of the quantum critical fluctuations in cuprate superconductors. *Phys. Rev. Lett.*, **99**, 067003 (4 pages).

Allen, P.B. and Dynes, R.C. (1975) Transition temperatures of strong-coupled superconductors reanalyzed. *Phys. Rev.*, **B12** (2), 905–922.

Anderson, P.W. (1987) The resonating valence bond state in La_2CuO_4 and superconductivity. *Science*, **235** (4793), 1196–1198.

Anderson, P.W., Lee, P.A., Randeria, M., Rice, T.M., Trivedi, N., and Zhang, F.C. (2004) The physics behind high-temperature superconducting cuprates: The "Plain Vanilla" version of RVB. Available at: http://arxiv.org/pdf/0311.467v2.

Bardeen, J., Cooper, L.N., and Schrieffer, J.R. (1957) Microscopic theory of superconductivity. *Phys. Rev.*, **108**, 1175–1204.

Cho, H.J., Roundy, D., Sun, H., Cohen, M.L., and Louie, S. (2002) The origin of the anomalous superconducting properties of MgB_2. *Nature (London)*, **418** (6899), 758–760.

De Leeuw, D.M., Groen, W.A., Feiner, L.F., and Hawinga, E.E. (1990) Correlation between the superconducting transition temperature and crystal structure of high-T_c cuprate compounds. *Physica C (Superconductivity)*, **166** (1-2), 133–139.

Ginzburg, V.L. and Landau, L.D. (1950) On the theory of superconductivity. *Zh. Eksp. Teor. Fiz.*, **20**, 1064–1082.

Gutzwiller, M.C. (1963) Effect of correlation on the ferromagnetism of transition metals. *Phys. Rev. Lett.*, **10**, 159–161.

Homes, C.C., Dordevic, S.V., Strongin, M., Bonn, D.A., Liang, R., Hardy, W.H., Komiya, S., Ando, Y., Yu, G., Kameko, N., Zhao, X., Greven, M., Bason, D.N., and Timusk, T. (2004) A universal scaling relation in high-temperature superconductors. *Nature (London)*, **430** (6999), 539–541.

Kotliar, G. and Liu, J. (1988) Superexchange mechanism and d-wave superconductivity. *Phys. Rev.*, **B38** (7), 5142–5149.

Li, Y., Balédent, V., Barisic, N., Cho, Y., Fauqué, B., Sidis, Y., Yu, G., Zhao, X., Bourges, P., and Greven, M. (2008) Unusual magnetic order in the pseudogap region of the superconductor $HgBa_2CuO_{4+\delta}$. *Nature (London)*, **455** (7211), 372–375.

London, F. and London, H. (1935) The electromagnetic equations of the supraconductor. *Proc. R. Soc. London A*, **149**, 71–88.

McMillan, W.L. (1968) Transition temperatures of strong-coupled superconductors. *Phys. Rev.*, **167** (2), 331–344.

Otto, H.H. (2008) Family tree of perovskite-related superconductors. Available at: http://arxiv.org/abs/0810.350.

Pickett, W.E. (2008) The next breakthrough in phonon-mediated superconductivity. *Physica C (Superconductivity)*, **468**, 126–135.

Powell, B.J. and McKenzie, R.H. (2005) Half-filled layered organic superconductors and the resonating-valence-bond theory of the Hubbard-Heisenberg model. *Phys. Rev. Lett.*, **94**, 047004 (4 pages).

Spałec, J., Oleś, A.M., and Chao, K.A. (1978) Kinetic exchange interaction for magnetic impurities in metals: non-degenerate Wolff model. *Phys. Rev.*, **B18** (7), 3748–3752.

Steglich, F., Aarts, J., Bredl, C.D., Lieke, W., Meschede, D., Franz, W., and Schäfer, H. (1979) Superconductivity in the presence of strong Pauli paramagnetism: $CeCu_2Si_2$. *Phys. Rev. Lett.*, **43**, 1892–1895.

Takabayashi, A.Y., Ganin, P., Jeglic, D., Arcon, T., Takano, Y., Iwasa, Y., Ohishi, M., Takata, N., Takeshita, K., Prassides, K., and Posseinsky, M.J. (2009) The disorder-free non-BSC superconductor Cs_3C_{60} emerges from an antiferromagnetic insulator parent state. *Science*, **323** (5921), 1585–1590.

Tranquada, M., Sternlieb, B.J., Axe, J.D., Nakamura, Y., and Uchida, S. (1995) Evidence for stripe correlations of spins and holes in copper oxide superconductors. *Nature (London)*, **375** (6532), 561–564.

Uemura, Y.J., Lee, L.P., Luke, G.M., Sternlieb, B.J., Wu, W.D., Brewer, J.H., Riseman, T.M., Seaman, C.L., Maple, M.B., Ishikawa, M., Hinks, D.G., Jorgensen, J.D., Saito, G., and Yamoshi, H. (1991) Basic similarities among cuprate, bismuthate, organic, Chevrel-phase, and heavy-fermion superconductors shown by penetration-depth measurements. *Phys. Rev. Lett.*, **66** (20), 2665–2668.

Varma, C.M. (1997) Non-Fermi liquid states and pairing instability of a general model of copper oxide metals. *Phys. Rev.*, **B55**, 14554–14580.

Yang, H.-B., Rameau, J.D., Johnson, P.D., Valla, T., Tsvelik, A., and Gu, G.D. (2008) Emergence of preformed Cooper pairs from the doped Mortt insulating state in $Bi_2Sr_2CaCu_2O_{8-\delta}$. *Nature*, **456**, 77–80.

Yu, G., Li, Y., Motoyama, E.M., and Greven, M. (2009) Universal relation between magnetic resonance and superconducting gap in the unconventional superconductors. Available at: http://arxiv.org/abs/0903.2291.

Yuan, H.Q., Grosche, F.M., Deppe, M., Geibel, C., Sparn, G., and Steglich, F. (2003) Observation of two distinct superconducting phases in $CeCu_2Si_2$. *Science*, **302**, 2104–2107.

Zaanen, J. (2004) Superconductivity: why the temperature is high. *Nature (London)*, **430** (6999), 512–513.

Index

Classic and Advanced Ceramics: From Fundamentals to Applications. Robert B. Heimann

ISBN: 978-3-527-32517-7

c

d

t